Proceedings in Life Sciences

The Blue Light Syndrome

Edited by H. Senger

With 432 Figures

Springer-Verlag
Berlin Heidelberg New York 1980

Professor Dr. Horst Senger
Fachbereich Biologie-Botanik
der Philipps-Universität Marburg
Lahnberge, 3550 Marburg/Lahn, FRG

ISBN 3-540-10075-X Springer-Verlag Berlin Heidelberg New York
ISBN 0-387-10075-X Springer-Verlag New York Heidelberg Berlin

Library of Congress Cataloging in Publication Data. Main entry under title: The blue light syndrome. (Proceedings in life sciences) Bibliography: p. Includes index. 1. Light – Physiological effect. 2. Photobiology. I. Senger, Horst. [DNLM: 1. Light – Congresses. 2. Plants – Radiation effects – Congresses. 3. Plants – Physiology – Congresses. 4. Photochemistry – Congresses. 5. Photoreceptors – Congresses. QK757 B658] QH515.B57   574.19'153   80-16523

Offsetprinting and bookbinding: Brühlsche Universitätsdruckerei, Giessen.
2131/3130-543210

Preface

Investigations on the specific effects of blue light on plants began some fifty years ago. In recent years the growing awareness of blue-light-induced phenomena in plants, microorganisms, and animals has accelerated and expanded this research into an ever-increasing variety of blue-light effects in biological systems.

In 1977, J.A. Schiff and W.R. Briggs proposed a specific meeting to present and summarize the various blue-light effects and to discuss their mechanisms and possible photoreceptors. In view of the variety of responses and the range of organisms affected by blue light the term *Blue Light Syndrome* seemed to be the only appropriate one for the meeting.

With the help of the International Advisory Committee (W.R. Briggs, Stanford; J. Gressel, Rehovot; W. Kowallik, Bielefeld; S. Miyachi, Tokyo; W. Rau, Munich, and J.A. Schiff, Waltham), and the very generous financial support provided by the Deutsche Forschungsgemeinschaft as well as by the Bundesministerium für Forschung und Technologie, the Kultusminister des Landes Hessen, and the Philipps-Universität Marburg, the "International Conference on the Effect of Blue Light in Plants and Microorganisms" was held in July 1979 in the Philipps-Universität Marburg.

In the name of all participants I would like to thank my colleagues, the above-mentioned organizations, and last but not least my secretary Mrs. Ilse Krieger and my unnamed co-workers whose indefatigable efforts made this conference and the publication of this book possible. It is also a pleasure to thank Springer-Verlag and its editing staff for their helpful expertise and their patience.

The conference was attended by 140 scientists from 18 different countries. Necessarily, much time was devoted to the various reviews giving an up-to-date assessment of the Blue Light Syndrome. It was particularly stimulating to cross the limits of "Plants and Microorganisms" set by the conference's title to include a review of blue-light effects in vision. This again demonstrated the universality of the Blue Light Syndrome. The largest amount of time was dedicated to the photoreceptor problem, but the attempt to decide between flavins or carotenoids failed and, consequently, the list of possible candidates for the primary photoreceptor of blue light increased. Five additional sections were (more or less wisely) devised to categorize the various contributions, but some of these papers could appear as well under a different or even an additional heading.

This book contains all contributions presented at this conference, supplemented by a few articles by those colleagues who were prevented at the last minute from attending. I wish to thank all contributors for their quick, friendly, and effective cooperation. May this book be of future help in advancing our mutual interest in the fascinating field of research on the Blue Light Syndrome.

Marburg, August 1980   HORST SENGER

Contents

Photoreceptors and Primary Reactions

Contents

Carotenogenesis

Carbon Metabolism and Respiration

Contributors

You will find the addresses at the beginning of the respective contribution

AKOYUNOGLOU, G. 473
ALLEGRINI, P.R. 30
ANNI, H. 473
APARICIO, P.J. 411
BECKMAN, J. 465
BJÖRN, L.O. 455
BLATT, M.R. 261
BRIGGS, W.R. 261
BRINKMANN, G. 526
BUSCHMANN, C. 485
BUTLER, W.L. 205
CALERO, F. 411
CARLILE, M.J. 3
CASTILLO, F. 422
CODD, G.A. 392
CONRADT, W. 368
DALEY, L.S. 381
DE FABO, E. 187
DÖRNEMANN, D. 541
EDMUNDS, L.N., Jr. 584
EGER-HUMMEL, G. 555
FUJITA, Y. 597
FURUYA, M. 119
GALLAND, P. 563
GNANAM, A. 435
GREEN, M.S. 269
GRESSEL, J. 133
HABIB MOHAMED, A. 435
HARTMANN, E. 221
HARTMANN, K.M. 422
HASE, E. 512
HOLZAPFEL, A. 444
HUBER, A. 299
ISONO, T. 597
KADOTA, A. 119
KALOSAKAS, K. 473

KAMIYA, A. 321, 605
KLEIN, O. 541
KLEMM-WOLFGRAMM, E. 238
KOWALLIK, W. 344
KUHN SILK, W. 643
KULANDAIVELU, G. 372
KUMAGAI, T. 251
LICHTENTHALER, H.K. 485
LIPSON, E.D. 110
LÖSER, G. 244
MIYACHI, SHIGETOH 321, 329, 332, 429, 605
MIYACHI, SHIZUKO 321, 329, 429
MOHR, H. 97
NINNEMANN, H. 238
NULTSCH, W. 38
OGAWA, T. 622
OH-HAMA, T. 332
OHKI, K. 597
OSAFUNE, T. 269
PORRA, R.J. 541
RAHMSDORF, U. 485
RAKOCZY, L. 570
RAU, W. 283
REIHL, W. 465
REINHARDT, B. 401
RICHTER, G. 465
RUSSO, V.E.A. 563
RUYTERS, G. 361, 368
SAROJINI, G. 372
SCHÄFER, E. 244
SCHÄTZLE, S. 344
SCHIFF, J.A. 269, 495
SCHMID, G.H. 198
SCHMID, K. 221

Reviews on Various Fields of Blue Light Action

The Biological Significance and Evolution of Photosensory Systems

M.J. CARLILE[1]

The earliest organisms are thought to have originated in a "primaeval soup" of photo-chemically generated organic molecules, and to have been extreme heterotrophs, requiring complex organic nutrients. Later, intense competition for diminishing supplies of nutrients would have led to the evolution of motility and then to chemosensory systems [1—4] permitting movement towards nutrients (positive chemotaxis). The ability to utilise less complex nutrients, culminating in carbon dioxide fixation and in the utilization of light energy to fix carbon dioxide (photosynthesis) would also have evolved. At this stage one can envisage, in a motile photosynthetic organism, the modification of the chemotactic system to respond to light. This first photosensory system would have had photosynthetic pigments as the photoreceptor and the organism would have moved towards the light by a biassed random walk, as with the positive phototaxis of present-day photosynthetic bacteria [5, 6].

Positive phototaxis [7—9] is of obvious advantage to photosynthetic micro-organisms, but may also be of value to non-photosynthetic organisms, leading them to sites where fruit-body production can result in effective spore dispersal. Negative phototactic responses can guide organisms away from damagingly high light intensities. It is suggested that phototaxis can be acquired readily by organisms that have light-absorbing molecules in their plasma membrane and are already chemotactic [10—12] and that phototaxis has arisen independently many times in evolution. A variety of photoreceptors are involved, many responding to blue light, and the biassed random walk has often been supplemented or replaced by a more direct path to favourable conditions of illumination.

In non-motile organisms, ranging from fungi [13—17] and sea-weeds to higher plants [18, 19], phototaxis is replaced by phototropism, light-oriented growth. The roles of positive phototropism in positioning for photosynthesis and in achieving spore dispersal are often obvious; negative phototropism may guide rootlets and rhizoids into the substratum and the germ-tubes of some pathogenic fungi into their host plants.

The appropriateness of various forms of behaviour depends on the time of day, hence light interacts with endogenous circadian rhythms [20—23], "setting" the "biological clock". It depends also on time of year, and there day-length is a better guide than temperature, so in many plants there is a photoperiodic control of reproduction; this, however, is a red light rather than a blue-light effect. Light is also needed to initiate sporula-

---

1 Department of Biochemistry, Imperial College of Science and Technology, London SW7 2AZ, England

tion in many fungi [14, 24, 25], and in some micro-organisms stimulates the production of pigments [26] – especially carotenoids and melanin – that protect the organisms from damage by intense light [27].

References

1. Adler J (1975) Annu Rev Biochem 44: 341–356
2. Berg HC (1975) Annu Rev Biophys Bioeng 4: 119–136
3. Koshland DE (1977) Science 196: 1055–1063
4. Macnab RM (1978) CRC Crit Revs Biochem 5: 291–341
5. Clayton RK (1964) In: Giese AC (ed) Photophysiology, vol II, pp 51–77. Academic Press, London New York
6. Hildebrand E (1978) In: Hazelbauer GL (ed) Taxis and behaviour, pp 37–73. Chapman and Hall, London
7. Bendix S (1960) Bot Rev 26: 145–208
8. Lenci F, Colombetti G (1978) Annu Rev Biophys Bioeng 7: 341–361
9. Nultsch W (1975) In: Carlile MJ (ed) Primitive sensory and communication systems, pp 29–90. Academic Press, London New York
10. Macnab R, Koshland DE (1974) J Mol Biol 84: 399–406
11. Taylor BL, Koshland DE (1974) J Bacteriol 119: 640–642
12. Taylor BL, Koshland DE (1975) J Bacteriol 123: 577–569
13. Buller AHR (1934) Researches on fungi, vol VI. Longmans, London
14. Carlile MJ (1970) In: Halldal P (ed) Photobiology of micro-organisms, pp 309–344. Wiley, New York
15. Cerda-Olmedo E (1977) Annu Rev Microbiol 31: 535–547
16. Foster KW (1977) Annu Rev Biophys Bioeng 6 419–443
17. Page RM (1968) In: Giese AC (ed) Photophysiology, vol III, pp 65–90. Academic Press, London New York
18. Briggs WR (1963) Annu Rev Plant Physiol 14 311–352
19. Curry GM (1969) In: Wilkins MB (ed) Physiology of growth and development, pp 245–273. McGraw Hill, London
20. Brady JB (1979) Biological clocks. Arnold, London
21. Hastings JW, Schweiger HG (eds) (1976) The molecular basis of circadian rhythms. Dahlem Konferenzen, Berlin
22. Njus D, Gooch VD, Mergenhagen D, Sulzman F, Hastings JW (1976) Fed Proc 35: 2353–2357
23. Schweiger HG, Schweiger M (1977) Int Rev Cytol 51: 315–342
24. Leach CM (1971) In: Booth C (ed) Methods in microbiology, vol IV, pp 609–664. Academic Press, London New York
25. Tan KK (1978) In: Smith JE, Berry DR (ed) The filamentous fungi, vol III, pp 334–357. Arnold, London
26. Batra PP (1971) In: Giese AC (ed) Photophysiology, vol VI, pp 47–76. Academic Press, London New York
27. Krinsky NI (1968) In: Giese AC (ed) Photophysiology, vol III, pp 123–195. Academic Press, London New York

Short Wavelength Light in Invertebrate Visual Sense Cells – Pigments, Potentials and Problems

D.G. STAVENGA[1]

1 Introduction

A treatise of current questions in vision of invertebrates may seem an odd issue in a symposium devoted to phenomena induced by blue light in lower organisms and plants. Yet, the title chosen by the organizer for this conference is well applicable to certain aspects of invertebrate vision which are presently hotly debated. I refer to the role of the bi-stable visual pigments in the process of phototransduction (Hamdorf 1979; Hochstein 1979) and the functions of photostable pigments in the photoreceptor cells (Kirschfeld et al. 1977, 1978; Stark et al. 1979). These topics will be surveyed and illustrated by a few exemplary cases; the eye of flies will recurrently be highlighted.

At the end of my contribution I will indicate possible parallels between invertebrate vision with other photobiological areas, notably that of phytochrome-governed systems.

2 Visual Pigments

2.1 Rhodopsin

Visual excitation starts when light impinges upon a photoreceptor cell and is absorbed by the visual pigment molecules, which are embedded in the membrane of the cell. From this it is evident that the spectral absorption characteristics of the visual pigment determine the spectral sensitivity of the visual sense cell. Spectral sensitivities of several photoreceptor cell types have been measured by electrophysiological methods and were found to span a range from the UV up to the red, in agreement with visual pigment studies performed with optical techniques either on extractions or by microspectrophotometry of tissue (Dartnall 1972; Goldsmith 1972, 1975).

Such a wide wavelength range is surprising when one notices the unique feature of visual pigments; all are proteins and have the same chromophore: retinal(dehyde), the oxidized derivate of the alcohol vitamin A.

Actually, there are two classes of visual pigments. The vast majority of visual pigments, the rhodopsins, contain retinal (see Dartnall 1972). There is a minor group of pigments, called prophyropsins, which is only encountered in the vertebrate kingdom, having 3 dehydroretinal as the chromophore.

1 Biophysical Department, Rijksuniversiteit Groningen, Groningen, The Netherlands

Table 1. Wavelengths of maximal absorption of invertebrate visual pigment states

	Rhodopsin	Metarhodopsin
Owlfly *Ascalaphus* (1)	345	475
Sphingid moth *Deilephila* (2)	345	480
Horseshoe crab *Limulus* (1)	360	480
Fruitfly *Drosophila* (3)	370	470
Sphingid moth *Deilephila* (2)	440	480
Honeybee drone *Apis* (4)	446	505
Drone fly *Eristalis* (5)	460	550
Octopus *Eledone* (6)	470	520
Fruitfly *Drosophila* (3, 6)	480	580
Housefly *Musca* (7)	490	580
Spider crab *Libinia* (8)	493	498
Squid *Loligo* (8)	493	498
Mud crab *Hemigrapsus* (9)	495	495
Blowfly *Calliphora* (1, 5)	495	580
Scallop *Pecten* (10)	495	580
Blue crab *Callinectes* (11)	500	500
Lobster *Panuliris* (9)	504	495
Moth *Galleria* (8)	510	484
Rock crab *Leptograpsus* (9)	513	495
Lobster *Homarus* (9)	515	490
Mosquito *Aedes* (16)	515	480
Shrimp *Penaeus* (9)	516	475
Sphingid moth *Deilephila* (2)	520	480
Lacewing *Chrysopa* (12)	530	485
Crayfish *Orconectes* (13)	530	515
Barnacle *Balanus* (1, 14)	532	492
Prawn *Palaemonetes* (9)	539	497
Metalmark butterfly *Apodemia* (15)	610	510

(1) For references see Hamdorf 1979; (2) Schwemer and Paulsen 1973; (3) Harris et al. 1976; (4) Muri 1978; (5) Stavenga 1976; (6) Ostroy et al. 1974; (7) Kirschfeld et al. 1977; (8) For references see Goldsmith 1975; (9) For references see Goldsmith 1972; (10) Cornwall and Gorman 1979; (11) Bruno and Goldsmith 1974; (12) Seitz and Haeckel 1979; (13) Goldsmith 1978a; (14) Minke and Kirschfeld 1978; (15) Bernard 1979; (16) Brown and White 1972

Generally, rhodopsins have a two-peaked absorption spectrum. The long wavelength α-band has approximately a Gaussian shape, with a halfwidth of roughly 100 nm. The versatility to set the peak wavelength λ_{max} from 345 nm up to 620 nm (see the tables of Bridges 1972; Lythgoe 1972; Goldsmith 1972; and Table 1), is attributed to the intimate interaction of the chromophore and the protein environment. Retinal, absorbing maximally at 380 nm in its 11-cis configuration, is connected to the ε-amino group of a lysine residue in the protein moiety by a Schiff base linkage. Protonation of this linkage and secondary noncovalent bonds enabled by additional conformational changes will cause the shifted absorption spectra toward longer wavelengths (reviews Dartnall 1972; Honig and Ebrey 1974; Ebrey and Honig 1975).

2.2 Metarhodopsin

Light absorption by a rhodopsin molecule induces isomerization of the retinal chromophore principally from an 11-cis to an all-trans conformation (reviews Dartnall 1972; Kropf 1969, 1972; Honig and Ebrey 1974). In vertebrates this step leads to a chain of conformational changes of the protein, which proceeds via intermediate states, until retinal splits from the protein part by hydrolysis (Dartnall 1972).

The rhodopsins of invertebrates, however, photoconvert into a thermostable metarhodopsin state (Goldsmith 1972, 1975). This characteristic may well be due to the different properties of the cell membrane: the diffusional mobility of the molecules is low in the photomembrane of invertebrates, and thus it may be hypothesized that the lipid environment stiffens the visual pigment molecules in this case (Goldsmith 1975; Hamdorf 1979).

Protonation appears to have a distinct effect on the absorption spectrum of the metarhodopsin state: metarhodopsins can exist in two forms, dependent on the acidity of the tissue. The basic form, occurring at high pH, invariably absorbs maximally in the UV. In the living tissue the so-called acid form is encountered (Hamdorf 1979). Table 1 lists the wide variation in the location of the absorption spectrum of the acid metarhodopsin form.

The respective location of the absorption maxima of the rhodopsin and its metarhodopsin does not seem to be completely independent. When the rhodopsin absorbs in the short wavelength range the metarhodopsin appears bathochromically shifted, whereas the long wavelength absorbing rhodopsins have hypsochromically shifted metarhodopsins. Inversion appears to occur when the peak absorption of the rhodopsin is at $\lambda_{max} \cong 500$ nm (Table 1).

Rather independent from the relative positions of the absorption spectra is the relative absorption of the two states: metarhodopsin always has a higher molecular absorption coefficient, pointing to a more stretched, all-trans shape of the chromophore. The explanation of the absorption spectra of the various visual pigments as well as that of the dynamics of the photochemical processes is a busy task fulfilled by several, both experimentally and theoretically active, groups. Considerable thrust into this field was recently given from the side of bacteriorhodopsin (Callender and Honig 1977; Sperling et al. 1977).

2.3 An Exemplary Case: the Eye of Flies

A typical and well-studied case of invertebrate visual pigment is that of the eye of flies. Before going into an account of the present state of knowledge on this matter, however, I must turn to a short anatomical description. The unit of the faceted eye of flies (as that of other insects as well as that of Crustacea) is the ommatidium; two ommatidia are drawn in Fig. 1. Each fly ommatidium contains eight photoreceptor cells (also called visual sense cells or retinular cells), screening pigment cells, the Semper cells, a pseudocone, and a facet lens. The latter components serve to focus incident light on those structures of the photoreceptor cells, which contain the visual pigment. These rhabdomeres are excessive invaginations of the cell membrane forming a cylindrical pile of close-packed microvilli. Optically a rhabdomere acts like a waveguide which is quite useful

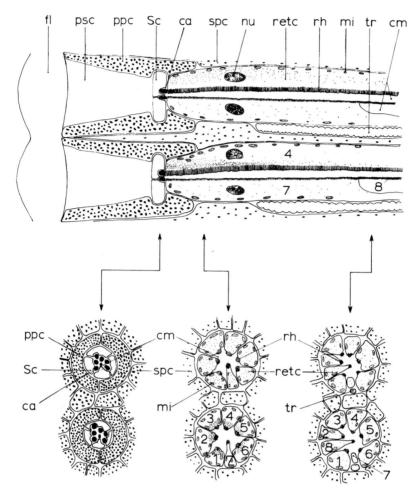

Fig. 1. Two ommatidia from a fly retina. A fly ommatidium consists of a facet lens (*fl*), a pseudo-cone (*psc*), four Semper cells (*Sc*), eight retinular or sense cells (*retc*), numbered R1 to R8, two primary pigment cells (*ppc*), six secondary pigment cells (*spc*) and a trachea (*tr*). The rhabdomeres (*rh*) have a tubular structure. Distally from the rhabdomeres the caps (*ca*) are located. The rhabdomeres are separated by the ommatidial cavity or central matrix (*cm*). Inside the photoreceptor cell nucleus (*nu*) and mitochondria (*mi*) are indicated. The upper ommatidium is dark-adapted, the lower ommatidium shows the light-adapted state (Stavenga 1975)

since the enormous increase in the probability of absorption of axially directed light facilitates good spatial acuity for the eye.

The retinular cells can be distinguished in two classes: six large peripheral cells R1–6 and two smaller central cells R7,8. The rhabdomeres of the cells are separate except for those of the central cells which are stacked on top of one another.

Due to the spherical shape of fly eyes, and due to the perfect repetitive order of the retinular cells in the retinal lattice an extremely nice optical phenomenon occurs when one observes the eye with a microscope focused on its center of curvature. With

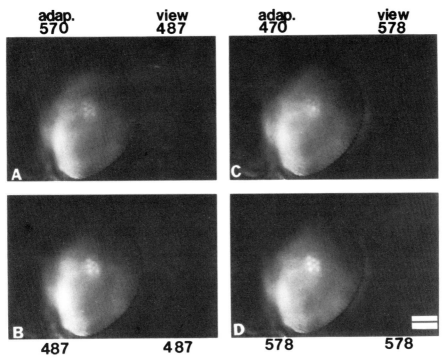

Fig. 2A–D. Deep pseudopupil of a *Drosophila* white-eyed mutant (cnbw) in two different stages of photoequilibrium of the visual pigments. (The deep pseudopupil is the superposition of magnified virtual images of photoreceptor endings located in several adjacent ommatidia, which is observed at the level of the center of curvature of the eye; the number of the participating ommatidia depends on the aperture of the observing microscope and the local interommatidial angle and is here 20–30.) The photoequilibrium stages were established respectively by short wavelength (487 nm in **B**, 470 nm in **C**) and by long wavelength adapting light (570 nm in **A**, 578 nm in **D**). These stages are photographed with antidromic illumination of 487 nm (**A, B**) and 578 nm (**C, D**) respectively. Since the peripheral retinular cells R1–6 contain a visual pigment R480-M580 (Ostroy et al. 1974) long wavelength adapting light results in a high rhodopsin content, thus giving a low transmittance for the peripheral rhabdomeres in the blue (**A**) and a high transmittance in the yellow (**D**). Conversely, short wavelength adapting light yields a high metarhodopsin content, resulting in a high transmittance in the blue (**B**) and a lower transmittance in the yellow (**C**). The transmittance of the central rhabdomeres is high in the yellow and lower in the blue, probably due to the existence of pigments absorbing in the shorter wavelength range only (Kirschfeld and Franceschini 1977; Hardie et al. 1979; Smola and Meffert 1979). Calibration mark 100 μm (figure by courtesy of Dr. William Stark)

antidromic illumination, i.e., light coming from the back of the eye, one then observes a superimposed image of the rhabdomere pattern of numerous ommatidia due to light radiating from each individual rhabdomere. This phenomenon is called the deep pseudopupil and is extensively described by Franceschini (e.g., 1975; for references see also Franceschini and Kirschfeld 1976).

The deep pseudopupil facilitates visual pigment studies in an intact fly as demonstrated in Fig. 2. A white-eyed mutant *Drosophila* was investigated with antidromic light of wavelength 487 nm and 583 nm respectively. The transmission of the rhabdomeres at these wavelengths distinctly depends on the wavelength of the previous (long-

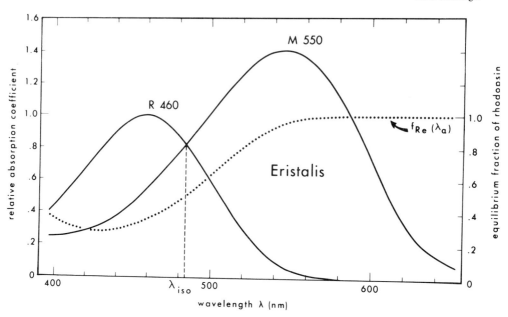

Fig. 3. The absorption curves for rhodopsin (R460) and metarhodopsin (M550) from photometric measurements on retinular cells 1–6 of the drone fly *Eristalis* (Stavenga 1976). The *dotted line* shows the equilibrium fractional concentration of rhodopsin $f_{Re}(\lambda_a)$ which is reached after illumination at each wavelength, with ordinate on the right hand side (Tsukahara and Horridge 1977)

lasting) illumination, which affected the visual pigment localized in the rhabdomeres. From measurements of the transmission of (single) rhabdomeres difference spectra can be calculated and subsequently the absorption spectra of both the rhodopsin and metarhodopsin state. Figure 3 presents the curves deduced for the dominant visual pigment contained in the peripheral retinular cells R1–6 of the dronefly *Eristalis*. Very similar spectra were obtained for other flies (for further details see Hamdorf 1979, and Table 1). (The central cells R7 and R8 of flies contain, among others, a UV visual pigment, but the picture for these cells is certainly not perfectly clear; see Hardie et al. 1979, and Smola and Meffert 1979).

The bistability of the visual pigments of flies (and other invertebrates) is a nice property for the experimenter because the photoconversion of rhodopsin into metarhodopsin is reversible at will: blue light converts predominantly rhodopsin into metarhodopsin until a photoequilibrium is reached with a high metarhodopsin fraction (Fig. 2A,C), and yellow or red light does the opposite (Fig. 2B, D).

Since rhodopsin conversion is the important step for vision, depletion of rhodopsin is counteracted by adaptations for good screening pigment properties. The red leaky pigment screen in the eyes of flies is discussed in some detail below.

3 Photostable Pigments

3.1 Photostable Pigment in the Pigment Cells

Photostable pigments occur abundantly in the eyes of invertebrates in the form of granules, notably in the pigment cells. The general function of the pigment cells is optical; namely, the (photostable) pigment creates a selective screen so that paraxial light enters the photoreceptors whereas off-axial, stray light is absorbed. The effectiveness of the screening depends on the optical density which may vary with wavelength. (On several occasions the effectiveness of the screening greatly depends on dark-light adaptation and/or circadian rhythm systems: reviews Waterman 1961; Goldsmith and Bernard 1974; Stavenga 1979a).

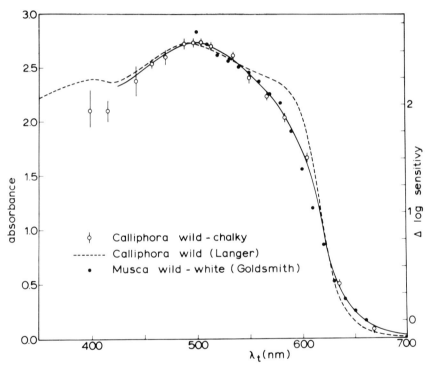

Fig. 4. Absorbance spectrum of fly screening pigment calculated from transmission measurements on the eyes of wild-type *Calliphora* and the pigmentless mutant Chalky. The *dotted line* represents measurements on pigment granules in squash preparations of *Calliphora* eyes (performed by Langer 1967). Also included are data inferred for housefly screening pigment from electrophysiological measurements of sensitivity on wild-type and white-eyed mutants respectively (*right hand ordinate;* data taken from Goldsmith 1965)

Figure 4 presents the absorbance spectrum of the pigment screen determined for the blowfly *Calliphora* by measuring both the transmission of the eye of a wild-type fly and that of a pigmentless mutant (Chalky). The spectrum obtained is compared with a spectrum determined by microspectrophotometry on histological sections (Langer 1967) and with a spectrum determined from the difference in the electroretinogram (ERG) in a wild-type housefly and a white-eyed mutant respectively (Goldsmith 1965). Clearly, the pigment absorbs substantially at wavelengths up to 600 nm, but at longer wavelengths absorption rapidly diminishes. This results in enhanced red sensitivity of the receptors especially at off-axis illumination.

The pigment screen thus was thought not to fulfill its task properly (see also Burkhardt 1962; Goldsmith 1965) until the photochemistry of fly visual pigments was understood (Hamdorf et al. 1973; Stavenga et al. 1973). Now we know that the leakage in the red serves to photoconvert created metarhodopsin M580 back into the rhodopsin state R495 (e.g., Stavenga et al. 1973; Cosens 1979).

A red leaky screen would be of no value for the commonly found green rhodopsin-blue metarhodopsin visual pigment (Table 1). Generally, therefore, invertebrate eyes possess an abundant amount of black screening pigments (see Stavenga 1979a, for discussion).

Yet, although the visual pigment system of crayfish is R530-M515, a long wavelength leaky pigment screen exists in crayfish eyes (Goldsmith 1978a, b). Hence, stray light, when sufficiently intense, would deplete the rhodopsin fraction. This danger appears to be negligible since crayfish are active at low light levels and, moreover, most probably the rhodopsin is regenerated from its photoproduct in the dark (demonstrated in the lobster *Homarus,* which has a crayfish-like visual system; see Goldsmith 1975, for references). The actual reason for the long wavelength leakage seems here to be a desired longer wavelength sensitivity: the sensitivity spectra of the receptors peak at 562 nm, shifted well apart from the λ_{max} = 530 nm of the rhodopsin (Goldsmith 1978a, b).

3.2 Photostable Pigments in the Photoreceptor Cell Soma

Numerous pigment granules usually exist within the soma of the visual sense cells of invertebrates. The main function exhibited by the assembly of intracellular pigment granules is that of a light controlling pupil (see Figs. 1 and 5; for references see Mazokhin-Porshnyakov 1969; Franceschini and Kirschfeld 1976; Stavenga 1979a). The granules exist displaced from the rhabdomere in the dark-adapted state of the cell, whereas they gather against the rhabdomere upon light adaptation (Fig. 1); then light is absorbed and scattered away from the light-sensitive structures. Whereas generally the absorbance spectrum of pupils is rather flat (e.g., hymenopterans: Stavenga and Kuiper 1977), in flies the spectrum peaks in the UV or blue and is minor in the yellow (*Drosophila:* Franceschini 1972; Franceschini and Kirschfeld 1976; *Calliphora:* Stavenga et al. 1973; *Musca:* Hardie 1979). Obviously, the blue band filter function has a similar protection function as is fulfilled by the pigments of the screening pigment cells: at high light intensities the closed pupil selectively suppresses the rhodopsin-converting blue wavelength range (Fig. 5). This function in the photochemical cycle of the visual pigment probably is of vital importance for those flies which are active under bright light conditions, because the receptors can be severely desensitized by blue light (see Sect. 4.3).

Again, a pigment filter absorbing predominantly in the short wavelength range only is beneficial for a visual pigment system where the metarhodopsin is bathochromically

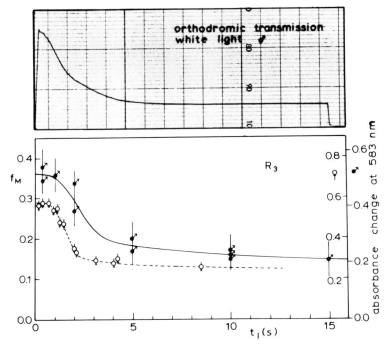

Fig. 5. Influence of the intracellular pupil on photoequilibrium of the visual pigment in a peripheral retinular cell R3 of the blowfly *Calliphora.* White light from a 150 W Xe-lamp was delivered in the normal, so-called orthodromic way. The intensity was sufficient to establish a photoequilibrium within 0.2 s. Initially a pure rhodopsin concentration was created by red light. Subsequently, after 1 min darkness, the white illumination was applied. The pupillary granules cause a decrease in transmission with a time constant in the order of seconds. The pupil absorbs predominantly in the blue and little in the yellow and red (Stavenga et al. 1973). At onset of the illumination, when the pupil is still open, a photoequilibrium with a metarhodopsin fraction $f_M \cong 0.3$ is rapidly created by the white light. When the pupil closes the remaining long wavelength light shifts the equilibrium so that a smaller metarhodopsin fraction results ($f_M \cong 0.15$). The metarhodopsin fraction was calculated from absorbance measurements at 583 nm after 1 min dark adaptation time (in order to let the pupil relax again) following the variable illumination time t_ϱ (for methodology, see further Stavanga 1976). Data for a male and a female fly are shown

shifted in respect of the rhodopsin. Such a short wavelength rhodopsin does not seem to exist in the proximal receptors of pierid butterflies, but still in those cells a short wavelength absorbing, red-colored photostable pigment is located in granules against the rhabdomeres even in the dark-adapted state (Ribi 1978, 1979). The spectral sensitivity of those receptors must be shifted toward longer wavelengths in respect of the rhodopsin spectrum. The red granules are absent in the dorsal ommatidia which are usually directed toward the (bright) skies. Whether this is due to a specific difference in color detection in the various eye regions, or rather was necessary in order to avoid severe depletion of the rhodopsin, must presently remain a matter of speculation (see further Ribi 1978, 1979).

3.3 Photostable Pigments in the Photosensitive Membrane of Visual Sense Cells

3.3.1 Filtering Pigment

Photostable pigments within the membranes of visual sense cells are now known for a few animal species (see Kirschfeld 1979). By observing eye-slices of houseflies Kirschfeld and Franceschini (1977) discovered that the rhabdomeres of a set of R7 receptors appeared to be yellow. The extinction spectrum of these 7y rhabdomeres (which is very reminiscent of β-carotene) facilitated the explanation of anomalous narrow spectral sensitivity curves which are occasionally encountered (see Kirschfeld et al. 1978; Hardie et al. 1979; Smola and Meffert 1979). The severe blue absorption by the β-carotene-like substance in R7 together with a normally shaped rhodopsin peaking at 520 nm in R8 results in a narrow sensitivity spectrum with a peak around 540 nm (Fig. 6; from Hardie et al. 1979). The absorption spectrum of the metarhodopsin of the R8 visual pigment is unknown but can be assumed to coincide approximately with that of the rhodopsin, from the impossibility to adapt the R8 cells selectively (e.g., Harris et al. 1976).

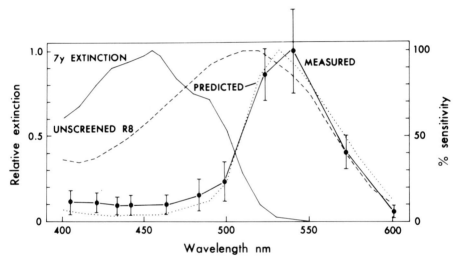

Fig. 6. Spectral sensitivity of a *Calliphora* R8 cell ($\cdot - \cdot$) compared with the theoretical prediction which assumes a 520 nm rhodopsin screened by the photostable pigment in 7y rhabdomeres. The 7y extinction curve was determined from microspectrophotometry by Kirschfeld et al. (1978). The unscreened R8 curve represents a typical rhodopsin curve adjusted for self-screening (Hardie et al. 1979)

3.3.2 Sensitizing Pigments

Also the peripheral retinular cells R1–6 of flies probably have a photostable pigment localized in the visual membrane. Direct in vivo evidence of this pigment can be obtained by incident fluorescence microscopy. Figure 7a presents the fluorescing deep pseudopupil of a white-eyed mutant fruitfly *Drosophila* under UV excitation. The six

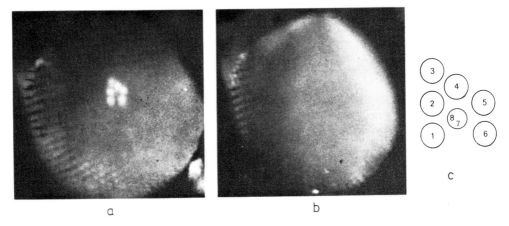

Fig. 7a–c. UV-induced fluorescence photographed in the deep pseudopupil of the compound eye of completely intact and living white-eyed *Drosophila* [mutants cinnabar (cn) and brown (bw)], reared on a high (**a**) and a low (**b**) vitamin A diet respectively. The characteristic fly rhabdomere pattern is drawn in **c**. Only the rhabdomeres of cells R1–6 of high vitamin A flies (**a**) give an observable fluorescence to UV excitation in the deep pseudopupil. The vitamin A-dependent rhabdomere emission is pink (>550 nm). However, when superimposed on the blue autofluorescence of the eye, which is not vitamin A-dependent, the fluorescent deep pseudopupil in vitamin A-enriched flies has a whitish appearance. No distinct fluorescence is observed from the rhabdomeres of the cells R7, 8 (see Fig. 1c) or from any photoreceptor in vitamin A-deprived flies (**b**) (Stark et al. 1979)

bright spots show that a fluorescing substance is located in the rhabdomeres of cells R1–6. This fluorescing substance is clearly vitamin A-dependent, since the fluorescence is absent in vitamin A-deprived flies (Fig. 7b). Since retinal is a vitamin A derivate, in the latter case the visual pigment content is extremely lowered (Razmjoo and Hamdorf 1976; Harris et al. 1977). Hence, it might be assumed that the visual pigment causes the fluorescence. However, the magnitude of the emission appears to be virtually identical whether the visual pigment molecules are either predominantly in the rhodopsin or in the metarhodopsin state (Stark et al. 1979). Furthermore, the excitation spectrum has a distinct peak in the UV with a minor tail in the blue. The underlying vitamin A-dependent pigment thus is concluded to be a UV-absorbing photostable pigment (Stark et al. 1979). Such a pigment was recently hypothesized by Kirschfeld and Franceschini (1977; see also Kirschfeld et al. 1977, and Kirschfeld 1979) in order to explain the high sensitivity of fly photoreceptors in the UV. The UV-absorbing pigment will act as a sensitizing pigment when visual pigment conversion is induced by energy transfer. Data suggesting that the metarhodopsin of R1–6 also has a functional high UV maximum have been explained in terms of this same photostable pigment also transferring energy to metarhodopsin (Stark et al. 1977; Kirschfeld 1979).

For a discussion on other functions of photostable pigments in the photoreceptor membrane see Kirschfeld and Franceschini (1977) and Kirschfeld (1979).

4 Potentials

4.1 Early Receptor Potential

Upon light absorption by the visual pigment molecules a conformational change of initially the chromophore and subsequently the protein complex occurs. Hence, charges in the molecule are displaced and, since the visual molecule is a constituent of the photoreceptor cell membrane, a local change in the polarization of the membrane is equilibrated by a current flowing along the membrane of the cell. The potential caused by the current over a section of extracellular space is called the (extracellular) early receptor potential (ERP). The potential across the receptor cell membrane which is measured by an intracellular electrode, the (intracellular) ERP, is the external ERP integrated by the time constant of the cell membrane (Cone 1969; Cone and Pak 1971; Hodgkin and O'Bryan 1977; Minke and Kirschfeld 1979; see also Hamdorf 1979; Järvilehto 1979).

Calculations using quantitative measurements of the intracellular ERP and estimates of membrane capacitance and density of the visual molecules in the membrane yielded that the charge displacements are in the order of 0.1 electronic charge per converted molecule (e.g., *Limulus:* Lisman and Bering 1977; fruitfly: Stephenson and Pak 1979). Generally, rhodopsin conversion induces a hyperpolarization (outward displacement current) of the cell membrane and metarhodopsin a depolarization (e.g., barnacle: Hillman et al. 1972; *Limulus:* Lisman and Sheline 1976; scallop: Cornwall and Gorman 1976, 1979).

The case of the scallop *Pecten* is particularly interesting because the visual pigment is (virtually) indistinguishable from that of the blowfly *Calliphora* (Table 1). Whereas in *Pecten* both a hyperpolarizing and a depolarizing ERP can be obtained (dependent on the converting pigment state), in the blowfly only a depolarizing ERP exists. Thus, only metarhodopsin seems to be active; the reason why fly rhodopsin conversion remains silent in the ERP is presently unclear (see Minke and Kirschfeld 1979; Stephenson and Pak 1979). All the same, the virtue of the fly ERP, as that of the ERP in general, is the possibility to study the photochemistry of a visual pigment by electrical methods, because the ERP signal is linearly proportional to the number of visual pigment molecules converted (Cone and Pak 1971; Hillman et al. 1972; Minke et al. 1973).

ERP's are only detectable when large amounts ($> 10^6$) of visual pigment molecules are converted within the cell membrane's integration time (a few ms). At the necessary extreme light intensities a much larger potential, the late receptor potential (LRP), immediately follows the early receptor potential. Whether the charge displacements expressed in the ERP are directly or rather indirectly involved in LRP induction is a still unsolved question of extreme interest.

4.2 Late Receptor Potential

The principal step in the visual process is the generation of the late receptor potential. Even a single photon or light quantum is capable of eliciting an appreciable change (mV) in the membrane potential of the receptor cell (Fig. 8). Such an event, called a quantum bump, lives in the order of 100 ms (see Järvilehto 1979). The spectral sensitivity of bump induction appears to be completely determined by the spectral sensitivity for

R1-6

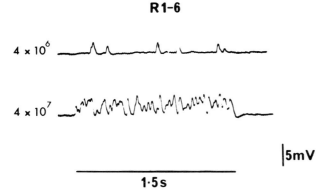

Fig. 8. Quantum bumps measured in a retinular cell of the class R1−6 of the housefly *Musca*. In the upper curve 4×10^6 blue quanta are incident per cm^2 per second axial on the photoreceptor. In the lower curve the intensity is ten times higher, resulting in fusing of the bumps (Hardie 1979)

converting visual molecules from their rhodopsin state, confirming the general rule for all visual sense cells, namely that rhodopsin conversion triggers the principal electric effects in the membrane. Single bumps can only be noticed in dark-adapted cells at low levels of illumination. At higher light levels the bumps fuse into the graded late receptor potential (Figs. 8 and 9).

Illumination of a visual sense cell at intermediate and higher intensities induces an LRP with a transient peak at onset which levels off into a plateau. The resting potential

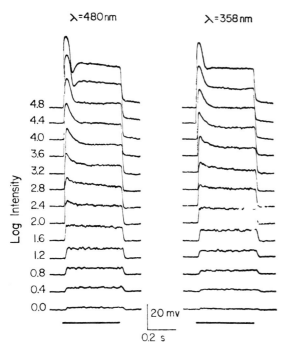

Fig. 9. Retinular cell responses from a white-eyed blowfly *Calliphora* to a sequence of increasing intensity flashes having wavelengths of 480 and 358 nm. The line segment below each column of responses represents the stimulus presentation. The two families of responses can be made nearly congruous by shifting the 480 nm family up one intensity level (McCann and Arnett 1971)

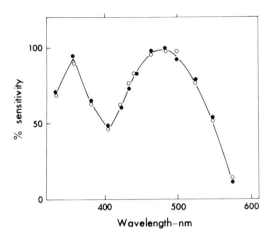

Fig. 10. Spectral sensitivity measured in a dark-adapted (o–o–o) and a light-adapted cell (●–●–●) of a white-eyed mutant housefly; light adaptation was achieved by intense rhodopsin generating (591 nm) light. A high UV sensitivity is noticed in addition to a typical blue-absorbing rhodopsin peak. The UV peak is attributed to the action of a sensitizing pigment (Hardie 1979)

being usually in the order of −50 mV, is approached again at illumination off (Fig. 9). The magnitude of the LRP is a sigmoidal function of log intensity. The shape of this curve is independent of wavelength (e.g., Hardie 1979). A spectral sensitivity curve like that of Fig. 10 is obtained by determining the number of quanta at a variety of wavelengths necessary to induce a certain criterion LRP.

Generally in flies the UV peak to blue peak ratio of the photoreceptor's sensitivity spectrum is much higher than that of normal rhodopsins. The high UV peak of the photoreceptor was explained by assuming the existence of a sensitizing, UV-absorbing photostable pigment (Kirschfeld et al. 1977; see section 3.3.2).

Virtually all invertebrate visual sense cells are depolarizing upon illumination. This is due to a light-induced increase in membrane permeability, principally to sodium ions (see Fuortes and O'Bryan 1972). A secondary increase in internal calcium has a regulatory influence on both sodium and potassium permeability, causing the fall from peak to plateau in the LRP (see Goldsmith 1975; Muijser 1979).

Some invertebrates possess hyperpolarizing photoreceptor cells, however. A most curious case is created by the eyes of the scallop *Pecten* which have a bilayered retina. The proximal receptors are depolarizing, but the distal receptors hyperpolarize upon illumination. For the latter type it was shown that the initial phototransduction step is a light-induced increase of the conductance to potassium ions (McReynolds 1976; Cornwall and Gorman 1976; Cornwall and Gorman 1979). Curiously enough, both distal and proximal receptors appear to have the same type of visual pigment (McReynolds and Gorman 1970).

4.3 Prolonged De- (or Hyper-) Polarizing Afterpotential

When the illumination is selectively converting rhodopsin into metarhodopsin so that a high metarhodopsin content in the photoreceptor results (this occurs with intense blue light in a previously red illuminated fly photoreceptor) one observes a remarkable phenomenon, when recording the membrane potential. Upon light off the membrane

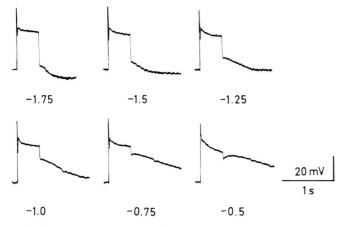

Fig. 11. The development of the PDA as an effect of intensity above that required to saturate the receptor potential measured in the dronefly *Eristalis*. The stimulating flash (wavelength 451 nm, duration 1 s) causes a depolarizing receptor potential. With increasing intensities a PDA develops. The stimulus intensity is indicated below each trace by the value of log I; log I = 0.0 corresponds to 4.45×10^{14} quanta $cm^{-2} s^{-1}$ at the eye surface (Tsukahara et al. 1977)

potential, instead of rapidly returning to its dark, resting value (see Fig. 9) stays as latched on to its light value (Fig. 11). In fly photoreceptors thus a prolonged depolarizing afterpotential (PDA) occurs, while in the distal receptors of *Pecten* a prolonged hyperpolarizing afterpotential (PHA) results.

Only intense light of blue wavelengths can induce a PDA in flies (Fig. 11), whereas intense long wavelength light elicits an afterhyperpolarization which also occurs at moderate intense blue light (Figs. 11 and 12). Yet, a long wavelength flash following a PDA-inducing blue flash suppresses the PDA or annihilates it completely (Fig. 12). The underlying cause is that the long wavelength light reconverts the metarhodopsin into the rhodopsin state (review Hamdorf 1979). Similarly, long wavelength illumination suppresses the PHA of *Pecten* distal photoreceptors (Cornwall and Gorman 1979).

The occurrence of the PDA in flies makes the red leaky screening pigments of both the pigment cells and the visual sense cells intelligible. Those flies which are active under bright day conditions have a considerable part of their receptors looking toward the blue skies. Those receptors may become severely desensitized when there is no means to keep the metarhodopsin fraction low. The red light leaking through the screening pigment and the intracellular pupillary pigment photoreconverts the metarhodopsin into the rhodopsin (Stavenga et al. 1973, 1975; Cosens 1979; Stavenga 1979a, b).

Interestingly, the pupil mechanism of fly photoreceptor cells on the other hand offers a unique means to study phototransduction and the PDA properties by optical measurements (see Stavenga et al. 1975; Stavenga 1979b).

The prolonged afterpotentials have initiated various models for the process of phototransduction. Extensive discussions and criticisms are expounded by Hillman et al. (1977), Tsukahara et al. (1977), Hamdorf (1979), Hamdorf and Razmjoo (1979), Hochstein (1979) and Minke (1979).

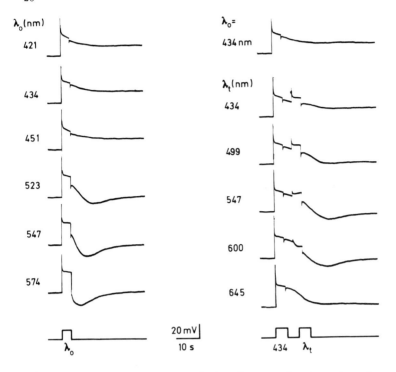

Fig. 12. Afterpotentials in response to flashes of constant duration and equal quantum numbers over a range of wavelengths (dronefly *Eristalis*). The intensity applied (log I = −0.15) saturates the receptor potential at all wavelengths shown. The afterpotential changes gradually from a prolonged depolarizing afterpotential (PDA) into an afterhyperpolarization when going to longer wavelengths (*left column*). In the *right column* the effect of a second pulse of selected wavelength λ_t on the PDA created by a 434 nm flash is shown. Duration of flashes and interval is 4 s; intensity of first flash is log I = −0.15 and of second flash log I = −0.4. The second flash causes an additional depolarization in all cases except at red wavelengths. The longer wavelengths cancel the created PDA or even give rise to an afterhyperpolarization; log I = 0.0 corresponds to 4.45×10^{14} quanta $cm^{-2} s^{-1}$ at the eye surface (Tsukahara et al. 1977)

5 Conclusion

Short wavelength illumination induces a number of phenomena in invertebrate photo-receptors which still require extensive further elucidation. Specifically, the mechanisms of the sensitizing pigments and the prolonged afterpotentials deserve considerable interest. The unraveling of these systems will help a better understanding of the details of the phototransduction process. Hopefully this research will have a fructifying effect on other photobiological investigations. The reverse has already proved true, as sensitizing pigments are well recognized in the work on photosynthesis. A future parallel may become practicable in the phenomenon of electrochromism. It has been shown in chloroplasts and bacterial chromatophores that the absorption spectrum of carotenoids shifts under a membrane potential created by illumination (e.g., Witt 1971). Possibly, the carotenoids now shown to be located in visual photoreceptors (Sect. 3.3.1) can serve as an optical tool for measuring membrane potential changes.

Furthermore, I should like to point to the correspondence between the pigment screen above the visual pigments in fly eyes and the screening effects exhibited by the "short" wavelength absorbing leaves upon the photoequilibrium of the bistable phytochrome system (Holmes and Smith 1975). The research on phytochrome is particularly interesting, since in a number of organisms phytochrome was evidenced to be a membrane pigment with dichroic properties. A specific orientation of rhodopsin molecules causing polarization sensitivity is a well-known characteristic of invertebrate eyes. Despite these suggestive parallels, however, it may turn out that the processes of phototransduction in animals and in plants have more differences than similarities.

A Final, Personal Remark. During the conference I was impressed by the tremendous complexity of the field of blue light-induced processes in plants and algae. I severely felt the need for a filling in of the gaps of knowledge by detailed biophysical and physiological studies. Except for a wider application of the microstimulation techniques performed by Haupt and Furuya and their co-workers, in my opinion, the field calls for a number of approaches well-established and shown to be highly effective in other areas of photobiological and membrane research; e.g., microspectrophotometry of photopigments in situ (transmission, fluorescence, polarization), electrophysiology (ERP — fast photovoltages, membrane potentials and resistances), ionic and pharmacological techniques. A rapidly developing tool with large potential, furthermore, might be that of (electrochromic) dyes revealing processes in biomembranes. Yet, these remarks must be considered as put forward by a benevolent outsider highly appreciating the progress already achieved on a difficult frontier.

Acknowledgment. Fruitful discussions with Dr. C.J.P. Spruit and several participants to the conference, especially Drs. E. De Fabo and W.K. Silk, are acknowledged.

References

Bernard GD (1979) Red-absorbing visual pigment of butterflies. Science 203: 1125–1127
Bridges CDB (1972) The rhodopsin-porphyropsin visual system. In: Dartnall HJA (ed) Handbook of sensory physiology, vol VII/1, pp 417–480. Springer, Berlin Heidelberg New York
Brown PK, White RH (1972) Rhodopsin of the larval mosquito. J Gen Physiol 59: 401–414
Bruno MS, Goldsmith TH (1974) Rhodopsin of the blue crab *Callinectes:* Evidence for absorption differences *in vitro* and *in vivo.* Vision Res 14: 653–658
Burkhardt D (1962) Spectral sensitivity and other response characteristics of single visual cells in the arthropod eye. Symp Soc Exp Biol 16: 86–109
Callender RH, Honig B (1977) Resonance Raman studies of visual pigments. Annu Rev Biophys Bioeng 6: 33–55
Cone RA (1969) The early receptor potential. In: Reichardt W (ed) Processing of optical data by organisms and by machines, pp 187–200. Academic Press, London New York
Cone RA, Pak WL (1971) The early receptor potential. In: Loewenstein WR (ed) Handbook of sensory physiology, vol I, pp 345–365. Springer, Berlin Heidelberg New York
Cornwall MC, Gorman ALF (1976) Color dependent potential changes of opposite polarity in single visual receptors of *Pecten irradians.* Biophys J 16: 146a
Cornwall MC, Gorman ALF (1979) Wavelength dependent changes in sensitivity related to rhodopsin metarhodopsin photo-interconversion in scallop photoreceptors. Biophys J 25: 317a

Cosens D (1979) Blue adaptation: An experimental tool for the study of visual receptor mechanisms and behaviour of *Drosophila.* Biophys Struct Mech 5: 211–222

Dartnall HJA (1972) Photosensitivity. In: Dartnall HJA (ed) Handbook of sensory physiology, vol VII/1, pp 122–145. Springer, Berlin Heidelberg New York

Ebrey TG, Honig B (1975) Molecular aspects of photoreceptor function. Q Rev Biophys 8: 129–184

Franceschini N (1972) Sur le traitement optique de l'information visuelle dans l'oeil à facettes de la drosophile. Thesis, Grenoble

Franceschini N (1975) Sampling of the visual environment by the compound eye of the fly: Fundamentals and applications. In: Snyder AW, Menzel R (eds) Photoreceptor optics, pp 98–125. Springer, Berlin Heidelberg New York

Franceschini N, Kirschfeld K (1976) Le contrôle automatique du flux lumineux dans l'oeil composé des Diptères. Propriétés spectrales, statiques et dynamiques du mécanisme. Biol Cyberné 21: 181–203

Fuortes MGF, O'Bryan PM (1972) Generator potentials in invertebrate photoreceptors. In: Fuortes MGF (ed) Handbook of sensory physiology, vol VII/2, pp 279–320. Springer, Berlin Heidelberg New York

Goldsmith TH (1965) Do flies have a red receptor? J Gen Physiol 49: 265–287

Goldsmith TH (1972) The natural history of invertebrate visual pigments. In: Dartnall HJA (ed) Handbook of sensory physiology, vol VII/1, pp 685–719. Springer, Berlin Heidelberg New York

Goldsmith TH (1975) Photoreceptor processes: some problems and perspectives. J Exp Zool 194: 89–102

Goldsmith TH (1978a) The spectral absorption of crayfish rhabdoms: Pigment, photoproduct and pH sensitivity. Vision Res 18: 463–473

Goldsmith TH (1978b) The effects of screening pigments on the spectral sensitivity of some Crustacea with scotopic (superposition) eyes. Vision Res 18: 475–482

Goldsmith TH, Bernard GD (1974) The visual system of insects. In: Rockstein M (ed) The physiology of insecta, vol II, pp 165–272. Academic Press, London New York

Hamdorf K (1979) The physiology of invertebrate visual pigments. In: Autrum H (ed) Handbook of sensory physiology, vol VII/6A, pp 145–224. Springer, Berlin Heidelberg New York

Hamdorf K, Razmjoo S (1979) Photoconvertible states and excitation in *Calliphora;* the induction and properties of the prolonged depolarizing afterpotential. Biophys Struct Mech 5: 137–161

Hamdorf K, Paulsen R, Schwemer J (1973) Photoregeneration and sensitivity control of photoreceptors. In: Langer H (ed) Biochemistry and physiology of visual pigments, pp 155–166. Springer, Berlin Heidelberg New York

Hardie RC (1979) Electrophysiological analysis of fly retina. I. Comparative properties of R1–6 and R7 and R8. J Comp Physiol 129: 19–33

Hardie RC, Franceschini N, McIntyre PD (1979) Electrophysiological analysis of fly retina. II. Spectral mechanisms in R7 and R8. J Comp Physiol 133: 23–39

Harris WA, Stark WS, Walker JA (1976) Genetic dissection of the photoreceptor system in the compound eye of *Drosophila melanogaster.* J Physiol 256: 415–439

Harris WA, Ready DF, Lipson ED, Hudspeth AJ, Stark WS (1977) Vitamin A deprivation and *Drosophila* photopigments. Nature (London) 266: 648–650

Hillman P, Hochstein S, Minke B (1972) A visual pigment with two physiologically active stable states. Science 175: 1486–1488

Hillman P, Keen ME, Winterhager J (1977) Discussion of selected topics about the transduction mechanism in photoreceptors. Biophys Struct Mech 3: 183–190

Hochstein S (1979) On the implications of bistability of visual pigment systems. Biophys Struct Mech 5: 129–136

Hodgkin AL, O'Bryan PM (1977) Internal recording of the early receptor potential in turtle cones. J Physiol 267: 737–766

Holmes MG, Smith H (1975) The function of phytochrome in plants growing in the natural environment. Nature (London) 254: 512–514

Honig B, Ebrey TG (1974) The structure and spectra of the chromophore of the visual pigments. Annu Rev Biophys Bioeng 3: 151–177

Järvilehto M (1979) Receptor potentials in invertebrate visual cells. In: Autrum H (ed) Handbook of sensory physiology, vol VII/6A, pp 315–356. Springer, Berlin Heidelberg New York

Kirschfeld K (1979) The function of photostable pigments in fly photoreceptors. Biophys Struct Mech 5: 117–128

Kirschfeld K, Franceschini N (1977) Photostable pigments within the membrane of photoreceptors and their possible role. Biophys Struct Mech 3: 191–194

Kirschfeld K, Franceschini N, Minke B (1977) Evidence for a sensitising pigment in fly photoreceptors. Nature (London) 269: 386–390

Kirschfeld K, Feiler R, Franceschini N (1978) A photostable pigment within the rhabdomere of fly photoreceptor no. R7. J Comp Physiol 125: 275–284

Kropf A (1969) Photochemistry of visual pigments. In: Reichardt W (ed) Processing of optical data by organisms and by machines, pp 28–43. Academic Press, London New York

Kropf A (1972) The structure and reactions of visual pigments. In: Fuortes MGF (ed) Handbook of sensory physiology, vol VII/2, pp 239–278. Springer, Berlin Heidelberg New York

Langer H (1967) Ueber die Pigmentgranula im Facettenauge von *Calliphora erythrocephala*. Z vergl Physiol 55: 354–377

Lisman JE, Bering H (1977) Electrophysiological measurement of the number of rhodopsin molecules in single *Limulus* photoreceptors. J Gen Physiol 70: 621–633

Lisman JE, Sheline Y (1976) Analysis of the rhodopsin cycle in *Limulus* ventral photoreceptors using the early receptor potential. J Gen Physiol 68: 487–501

Lythgoe JN (1972) List of vertebrate visual pigments. In: Dartnall HJA (ed) Handbook of sensory physiology, vol VII/1, pp 604–624. Springer, Berlin Heidelberg New York

Mazokhin-Porshnyakov GA (1969) Insect vision. Plenum Press, New York

McCann GD, Arnett DW (1972) Spectral and polarization sensitivity of the dipteran visual system. J Gen Physiol 59: 534–558

McReynolds JS (1976) Hyperpolarizing photoreceptors in invertebrates. In: Zettler F, Weiler R (eds) Neural principles in vision, pp 394–409. Springer, Berlin Heidelberg New York

McReynolds JS, Gorman ALF (1970) Membrane conductances and spectral sensitivities of *Pecten* photoreceptors. J Gen Physiol 56: 392–406

Minke B (1979) Transduction in photoreceptors with bistable pigments: Intermediate processes. Biophys Struct Mech 5: 163–174

Minke B, Kirschfeld K (1978) Microspectrophotometric evidence for two photo-interconvertible states of visual pigment in the barnacle lateral eye. J Gen Physiol 71: 37–45

Minke B, Kirschfeld K (1980) Fast electrical potentials arising from activation of metarhodopsin in the fly. J Gen Physiol in press

Minke B, Hochstein S, Hillman P (1973) Early receptor potential evidence for the existence of two thermally stable states in the barnacle resual pigment. J Gen Physiol 62: 87–104

Muijser H (1979) The receptor potential of retinular cells of the blowfly *Calliphora:* The role of sodium, potassium and calcium ions. J Comp Physiol 132: 87–95

Muri RB (1978) Microspectrophotometry of rhabdomes in the honeybee drone. Neurosci Lett Suppl 1: S410

Ostroy SE, Wilson M, Pak WL (1974) *Drosophila* rhodopsin: Photochemistry, extraction and differences in the *norp* A^{P12} phototransduction mutant. Biochem Biophys Res Commun 59: 960–966

Razmjoo S, Hamdorf K (1976) Visual sensitivity and the variation of total pigment content in the blowfly photoreceptor membrane. J Comp Physiol 105: 279–286

Ribi WA (1978) Ultrastructure and migration of screening pigments in the retina of *Pieris rapae* L. (Lepidoptera, Pieridae). Cell Tissue Res 191: 57–73

Ribi WA (1979) Coloured screening pigments cause red eye glow hue in pierid butterflies. J Comp Physiol 132: 1–9

Schwemer J, Paulsen R (1973) Three visual pigments in *Deilephila elpenor* (Lepidoptera, Sphingidae). J Comp Physiol 86: 215–229

Scitz G, Haeckel U (1980) Reflexionsspectrokopische Karakterisierung eines Sehpigmentsystems im Superpositionsauge der Florfliege *Chrysopa carnea* STEPH. J Comp Physiol in press

Smola U, Meffert P (1979) The spectral sensitivity of the visual cells R7 and R8 in the eye of the blowfly *Calliphora erythrocephala.* J Comp Physiol 133: 41–52

Sperling W, Carl P, Rafferty ChN, Dencher NA (1977) Photochemistry and dark equilibrium of retinal isomers and bacteriorhodopsin isomers. Biophys Struct Mech 3: 79–94

Stark WS, Ivanyshyn AM, Greenberg RM (1977) Sensitivity and photopigments of R1–6, a two peaked photoreceptor, in *Drosophila, Calliphora,* and *Musca.* J Comp Physiol 121: 289–305

Stark WS, Stavenga DG, Kruizinga B (1979) Fly photoreceptor fluorescence is related to UV sensitivity. Nature (London) 280: 581–583

Stavenga DG (1975) Optical qualities of the fly eye. An approach from the side of geometrical, physical and waveguide optics. In: Snyder AW, Menzel R (eds) Photoreceptor optics, pp 126–144. Springer, Berlin Heidelberg New York

Stavenga DG (1976) Fly visual pigments. Difference in visual pigments of blowfly and dronefly peripheral retinula cells. J Comp Physiol 111: 137–152

Stavenga DG (1979a) Pseudopupils of compound eyes. In: Autrum H (ed) Handbook of sensory physiology, vol VII/6A, pp 357–439. Springer, Berlin Heidelberg New York

Stavenga DG (1979b) Visual pigment processes and prolonged pupillary responses in insect photoreceptor cells. Biophys Struct Mech 5: 175–185

Stavenga DG, Kuiper JW (1977) Insect pupil mechanisms. I. On the pigment migration in the retinula cells of Hymenoptera (suborder Apocrita). J comp Physiol 113: 55–72

Stavenga DG, Zantema A, Kuiper JW (1973) Rhodopsin processes and the function of the pupil mechanism in flies. In: Langer H (ed) Biochemistry and physiology of visual pigments, pp 175–180. Springer, Berlin Heidelberg New York

Stavenga DG, Flokstra JH, Kuiper JW (1975) Photopigment conversions expressed in pupil mechanism of blowfly visual sense cells. Nature (London) 253: 740–742

Stephenson RS, Pak WL (1980) Heterogenic components of a fast electrical potential in *Drosophila* compound eye and their relation to visual pigment photoconversion. J Gen Physiol in press

Tsukahara Y, Horridge GA (1977) Visual pigment spectra from sensitivity measurements after chromatic adaptation of single fly retinula cells. J Comp Physiol 114: 233–251

Tsukahara Y, Horridge GA, Stavenga DG (1977) Afterpotentials in dronefly retinula cells. J Comp Physiol 114: 253–266

Waterman TH (1961) Light sensitivity and vision. In: Waterman TH (ed) The physiology of Crustacea, vol II, pp 1–64. Academic Press, London New York

Witt HF (1971) Coupling of quanta, electrons, fields, ions and phosphorylation in the functional membrane of photosynthesis. Q Rev Biophys 4: 365–477

Bacteriorhodopsin and its Position in the Blue Light Syndrome

R.J. STRASSER[1]

1 Introduction

The blue light syndrome represents a large heterogenous group of biological phenomena which are triggered by blue light. Different classifications of all the known blue light phenomena are possible: e.g.,

1. Functional classification between phenomena where the absorbed blue light energy is stored chemically in the organism and phenomena where the absorbed blue light is used as a source of information which influences the regulation of metabolic reactions.
2. Molecular classification between the different types of photoreceptor molecule e.g., flavins, carotenoids, rhodopsins, chlorophylls etc.
3. Topological classification indicating in which organelle the photoreception takes place.
4. Energetic classification of the primary photoreactions as indicated below.

2 Three Types of Primary Photoreaction

All the following primary reactions depend on the excitation of the photoreceptor pigment.

$P + Energy \rightarrow P^*$.

1. The excited pigment-protein-complex provokes a chain of conformational modifications in the protein in such a way that ions (mostly protons) are taken up at one side and released at the other side. That means the pigment protein complex undergoes a photocycle, pumping protons from one location to another. In this case a *primary proton flux* is established.

$P^* + H^+_{here} \longrightarrow P + H^+_{there}$.

2. The excited pigment-protein-complex becomes oxidized while reducing a primary electron acceptor. A suitable electron donor can now rereduce the oxidized pigment to its original form. In this case a *primary electron flux* is established.

1 Institut of Plant Physiology, University of Geneva, 1211 Geneve, Switzerland

$P* + \ominus \downarrow \rightarrow P + \ominus \uparrow.$

3. The excited chromophor transfers its excitation energy into the protein or to a neighbor pigment-protein-complex. In this case an *exciton flux* (energy transfer) is established.

$P_1^* + P_2 \longrightarrow P_1 + P_2^*.$

3 The Transformation of Energy Fluxes

The three types of primary photoreaction as described above plus the utilization of energy in an electron transport chain (e.g., respiratory chain) can be presented like building blocks of energy transformation units:

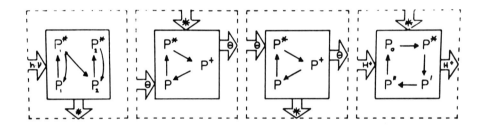

Each single unit or the combination of several units can be found in nature:

1. Excitation of rhodopsin provoking a proton flux.

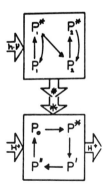

2. Excitation of flavoproteins provoking an electron flux.

3. Electron flux driving a proton flux in the respiratory chain.

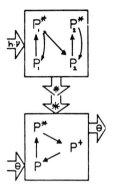

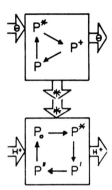

4. Photosynthetic energy transformations where excitons are driving electrontransport and electrontransport is pumping protons.

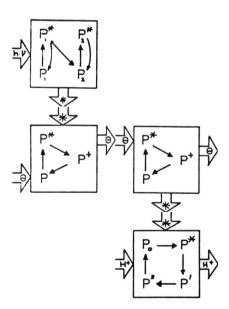

All four examples show in essence the transformation of one type of energy flux into another.

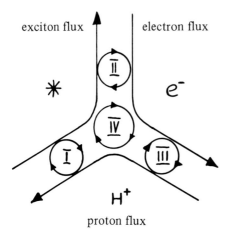

In a very simplified way each blue light phenomena can energetically be classified as described above.

4 Bacteriorhodopsin

Bacteriorhodopsin is a typical example of a blue light-driven proton pump. It is therefore an excellent object to investigate the mechanisms of proton pumping. The indicated review articles refer to all the details about bacteriorhodopsin known today [1].

The behavior of bacteriorhodopsin may perhaps be taken as a model type of some blue light phenomena. The techniques used for the studies of bacteriorhodopsin and retinarhodopsin [2] systems may also help to investigate and to understand other blue light phenomena. An important step is to distinguish between the photocycle of bacteriorhodopsin and its consequence, the proton pumping activity. For this reason artificial membrane labels are introduced covalently into the bacteriorhodopsin. The influence on the photocycle and the proton-pumping activity due to the chemical modification of such a membrane label can be followed by different spectroscopic techniques. Such an example is given in the following paper. The same experiments reported with bacteriorhodopsin could be done on other systems to find more information about the blue light syndrome.

References

1. Henderson R (1977) The purple membrane from halobacterium halobium. Annu Rev Biophys Bioeng 6: 87–109

   Jost PC, McMillen DA, Morgan WD (1978) Lipid-protein interaction in the purple membrane. In: Dearner DW (ed) Light transducting membranes, pp 141–155. Academic Press, London New York

   Lanyi JK (1978) Light energy conversion in halobacterium halobium. Microbiol Rev 42: 682–706

   Ovchinnikov YA, Abdulaev NG, Feigina MY, Kiselev AV, Lobanov NA (1979) The structural basis of functioning of bacteriorhodopsin: an overview. FEBS Lett 100: 219–224

   Stoeckenius W, Lozier RH, Bogomolni RA (1979) Bacteriorhodopsin and the purple membrane of halobacteria. Biochim Biophys Acta 505: 215–278

2. Darszon A, Strasser RJ, Montal M (1979) Rhodopsin-phospholipid complexes in apolar environments: photochemical characterization. Biochemistry 18: No 23, 5205–5213

   Montal M, Darszon A, Strasser RJ (1978) Rhodopsin and bacteriorhodopsin in model membranes. In: Frontiers of biological energetics, vol II, pp 1109–1117. Academic Press, London New York

Chemical Modification of Bacteriorhodopsin by Phenylisothiocyanate: Effect on the Photocycle

H. SIGRIST[1], P.R. ALLEGRINI[1], R.J. STRASSER[2], and P. ZAHLER[1]

Bacteriorhodopsin, the only polypeptide in the purple membrane of *Halobacterium halobium,* is a light-driven proton pump [1, 2]. Based on kinetic resonance Raman spectroscopy [3, 4] and protein modification studies [5, 6], a gating mechanism for proton translocation across the purple membrane has been proposed. In addition to the Schiff-base linkage between the retinal chromophore and the lysine residue 41 [7, 8], the amino acids lysine, arginine, aspartic acid, and tyrosine were suggested to participate in the light-induced transfer of protons [4, 5].

One approach to elucidate the molecular mechanism of proton translocation is chemical modification of the membrane-integrated bacteriorhodopsin. The labeling reagents used to date were of a predominantly hydrophilic character and are therefore expected to react with aqueous-accessible functional groups [5, 6, 9–12]. Recently phenylisothiocyanate was applied for covalent modification of integral membrane proteins [13]. The probe has been used for covalent modification of erythrocyte band 3 protein [13], cytochrome C oxidase [14] and sarcoplasmic Ca^{2+}-stimulated ATPase [15]. The interaction of phenylisothiocyanate with bacteriorhodopsin has been presented in a preliminary form [16].

Phenylisothiocyanate is a small hydrophobic molecule, virtually insoluble in aqueous solutions [13]. The reagent is reactive with nucleophiles in their nonprotonated form $(RS^- \gg RO^- > RNH_2)$ [17]. Modification of α- and ϵ-amino groups, cysteine SH and tyrosine OH, if present in the reactive form, is therefore feasible.

It is conceivable that chemical modification of bacteriorhodopsin might influence proton pumping activity. Due to the apparent link between proton translocation and the photochemical cycle [18, 19] optical methods have been applied to monitor label-induced effects on function. The extensively studied photocycle of bacteriorhodopsin is initiated by the absorption of a photon by the light-adapted ground state (bR_{570}). Via a series of dark reactions the excited state decays in microseconds through intermediates $K_{590}, L_{550}, M_{412}$, and O_{640} back to the ground state [18]. The transition of the M_{412} intermediate to the bR_{570} ground state is slowed by a factor of 1000 when diethylether is present in the system [20].

In this study the covalent modification of bacteriorhodopsin by phenylisothiocyanate is reported. The subsequent effect of labeling on the turnover rate of the photocycle is investigated by analysis of the formation and decay of the M_{412} intermediate.

1 Institut für Biochemie and Medizinisch-Chemisches Institut, University of Bern, Freiestrasse 3, 3012 Bern, Switzerland
2 Present address: Institut de Physiologie Végétale, Université de Genève, 1211 Genève, Switzerland

1 Experimental Procedures

1.1 Phenylisothiocyanate Labeling and Optical Measurement of the M_{412} Intermediate

A detailed description for phenyl[14C]isothiocyanate incorporation into bacteriorhodopsin is given in the legend to Fig. 1; prior to optical measurements purple membranes (0.374 mM bacteriorhodopsin) were treated with up to 8.35 mM phenylisothiocyanate for defined periods of time under conditions as described for phenyl[14C]-
isothiocyanate labeling. The labeled membranes were then transferred into the ether-
saturated salt medium as described by Oesterhelt and Hess [20].

Absorption changes were detected by a multifiber optic system connected to a
fluorimeter (Perkin Elmer MPF 2A) on line with a microcomputer system ARETE 100
(Arete Instruments, San Diego, USA) as previously described [21, 22]. The samples
were analyzed at ambient temperature in stoppered round quartz cells (Hellma 121-QS,
path length 1 mm or 5 mm). For excitation a tungsten light source was used with a
Corning glass filter combination, resulting in an excitation beam of about 1.2 mW at
565 ± 18 nm. Changes in the 412 transmission of the samples were detected with a
photomultiplyer which was shielded from actinic illumination by two Corning glass
filters and a Balzers B-40-567 interference filter giving a bandpass of 412 ± 4 nm. A
second photomultiplyer simultaneously detected the intensity and fluctuations of a
measuring beam not passed through the sample. Both signals were recorded simultaneously, digitized and stored in the microcomputer.

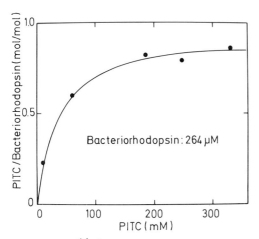

Fig. 1. Phenyl[14C]isothiocyanate labeling of bacteriorhodopsin in purple membranes. For phenyl-
[14C]isothiocyanate incorporation studies suspensions of purple membranes (0.265 mM bacteriorhodopsin in 25 mM sodium phosphate buffer pH 7.0) were combined with various amounts of
phenyl[14C]isothiocyanate. For light adaptation the membranes were exposed to unfiltered white
light for 1 min and then incubated under continuous agitation in covered glass tubes at $37°C$ for
90 min. The reaction was stopped by the addition of 5 vol precooled ($-20°C$) acetone. The precipitated protein was sedimented at $4°C$ for 10 min at 1500 g. The pellet formed was suspended
in water by sonication in a bath-type sonicator. The acetone extraction procedure was repeated
twice. The final pellet was solubilized in 2% sodium dodecylsulfate and analyzed for protein and
radioactivity

2 Results and Discussion

2.1 Covalent Chemical Modification of Bacteriorhodopsin

The extent of bacteriorhodopsin modification by phenylisothiocyanate is dependent upon the amount of label present during incubation (Fig. 1). At low label to protein ratios identical incorporation data have been obtained using either purple membrane suspensions in water or phosphate buffer. However, buffered suspensions are recommended when excess label is used ($\geqslant 50$ mol phenylisothiocyanate per mol bacteriorhodopsin) in the incubation mixture. pH changes, caused by side-reaction byproducts ($HCO_3{}^-$), can be therewith prevented. Phenylisothiocyanate incorporation, determined in a repeatedly sonicated and acetone-precipitated fraction, tends to saturate at a stoichiometry of approximately one mol label bound per mol bacteriorhodopsin. Equimolar modification was only obtained using a large excess of label. This implies a rather specific interaction even under extreme conditions, since random modification of reactive nucleophiles is not expected to saturate at a 1 : 1 ratio.

Evidence for covalent protein modification is given by the fact that the radioactivity co-migrates in sodium dodecylsulfate gels with the monomeric protein band (Fig. 2). Furthermore, label and protein can be repeatedly precipitated with acetone from sonicated suspensions. Upon proteolytic digest of bacteriorhodopsin with various proteases the label remains peptide-associated.

Functional studies reported in the following section have been performed with phenylisothiocyanate-modified purple membranes equivalent to and not exceeding an incorporation of 0.125 mol of label bound per mol bacteriorhodopsin.

2.2 Effect of Phenylisothiocyanate on the Photocycle

The interpretation of the presented data is based on the cyclic reaction described by Lozier et al. [18]. In a simplified scheme (Scheme 1) k_{on} represents the rate constant for M_{412} intermediate formation, k_{off} is the decay rate constant of the transition M_{412} into an undefined cyclic intermediate X.

$$bR_{570} \;+\; h\nu \;\xrightarrow{\;k_{on}\;}\; M_{412} \;\xrightarrow{\;k_{off}\;}\; X \qquad\qquad \text{(Scheme 1)}$$

The rate equation for the "light on" reaction is therefore

$$dM_{412}/dt = k_{on} \cdot bR_{570} - k_{off} \cdot M_{412} \tag{1}$$

and for the "light off" reaction

$$dM_{412}/dt = - k_{off} \cdot M_{412}. \tag{2}$$

The reaction constants k_{on} and k_{off} are given as relative values with the dimension time^{-1}.

In Fig. 3 the observed 412 nm absorbance changes of the "light on" (Fig. 3A) and the "light off" reaction (Fig. 3B) are recorded. Equal protein concentrations and light intensities were used in the compared samples. Phenylisothiocyanate-modified purple

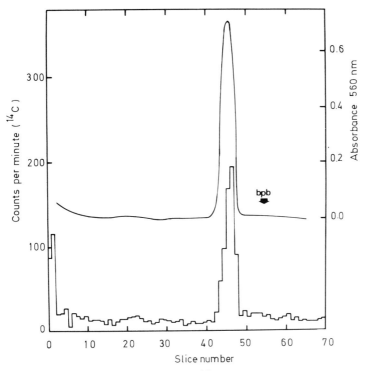

Fig. 2. Electrophoretic analysis of phenyl[^{14}C]isothiocyanate labeled bacteriorhodopsin. Sodium dodecylsulfate gel electrophoresis was carried out according to Weber and Osborn [23] with 10% acrylamide tube gels. Before electrophoresis the samples were treated for 15 min at 100°C in 10 mM sodium phosphate buffer pH 7.3 containing 1% sodium dodecylsulfate. The electrophoresis was performed in 0.1% sodium dodecylsulfate, 50 mM sodium phosphate buffer pH 7.3 with 80 to 100 μg protein per gel. Gels were stained with Coomassie brilliant blue and destained with 7.5% acetic acid. For radioactivity measurements the gels were cut into 1 mm slices immediately after the electrophoresis. The individual slices were extracted with 0.2 ml 0.66% Triton X 100 at 37°C overnight and analyzed for radioactivity. Radioactive content is displayed below the densitometric trace. *bpb* bromophenol blue

membranes (+) showed a significantly different absorption pattern than the control sample (−). If bR_{570}^{tot} represents the total amount of bacteriorhodopsin initially present, Eq. (1) can be transformed in Eq. (3):

$$dM_{412}/dt = k_{on}(bR_{570}^{tot} - X) - (k_{on} + k_{off}) M_{412}. \qquad (3)$$

At initial time points, however, X and M_{412} are negligible, resulting in

$$dM_{412}/dt = k_{on} \cdot bR_{570}^{tot} \quad \text{at} \quad t = 0. \qquad (4)$$

Based on rate Eq. (4), it was inferred from Fig. 3G that changes of k_{on} are not detectable upon phenylisothiocyanate modification, since the two functions intercept in the zero-time ("on") position.

For the "light off" reaction a relative measurement for the M_{412} decay rate constant was obtained from Eq. (5), representing the integrated form of Eq. (2).

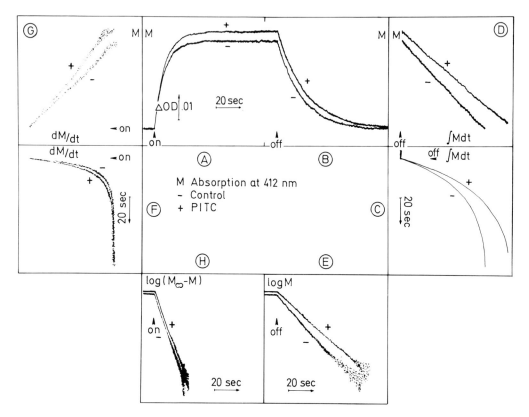

Fig. 3. Effect of phenylisothiocyanate on the kinetics of the M_{412} intermediate. Purple membranes (0.374 mM bacteriorhodopsin) were treated with (+) or without (−) 2.8 mM phenylisothiocyanate for 1 h at 37°C. M represents the absorbance at 412 nm in the "light on" (*A*) and the "light off" (*B*) reaction. Integration (*C*) or differentiation (*F*) of the recorded signals (*A, B*) are correlated in plots *G* and *D* with the respective absorbance changes. Kinetic order parameters are obtained from diagram *E* and *H*

$$M_{412} = - k_{off} \int_0^t M_{412} \, dt. \tag{5}$$

Since phenylisothiocyanate modification affected the disappearence rate of the M_{412} intermediate, the time and label concentration dependence of bacteriorhodopsin modification was analyzed with respect to changes in the relative M_{412} decay rate constant (k_{off}). In Fig. 4A the normalized 412 nm absorbance changes of phenylisothiocyanate-labeled purple membranes are recorded. The k_{off} values obtained from Fig. 4C were found to be related with the labeling time and tend to saturate after 1 h at about 60% of the control. Additional control experiments showed that the kinetic behavior of phenylisothiocyanate-modified purple membranes in ether-saturated salt solution remained unchanged during several hours. Secondary modifications in the analyzing medium could therefore be excluded.

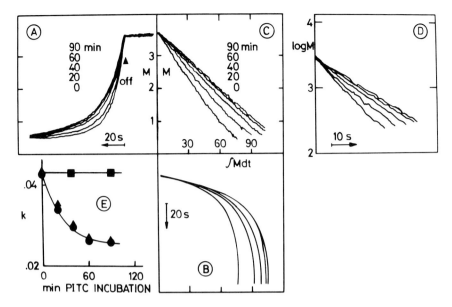

Fig. 4. Time-dependent inhibition of the relative M_{412} decay rate constant in the "light off" reaction. Purple membranes containing 0.374 mM bacteriorhodopsin were incubated with 8.35 mM phenylisothiocyanate for various periods of time. *A* Normalized 412 absorbance changes. *B* The integrated signal. The correlation of the signal *A* with its integrated form *B* is shown in *C*. *D* Logarithm of the M_{412} decay rate. *E* Time dependent inhibition. k_{off} is the relative M_{412} decay rate constants as obtained from plot *C* (●) and plot *D* (▲) respectively. (■) Control without phenylisothiocyanate

The effect of phenylisothiocyanate concentration on the M_{412} decay rate constant (Fig. 5) was analyzed in an analogous fashion. Maximal reduction of k_{off} to about 55% of the control sample was found when bacteriorhodopsin (0.374 mM) in purple membranes was modified for 1 h with 8.3 mM phenylisothiocyanate.

2.3 Discussion of Functional and Chemical Interaction Sites

The reported results suggest that phenylisothiocyanate acts on the protein moiety of bacteriorhodopsin. The inhibitory site appears to be located between the M_{412} intermediate and the regenerated ground state bR_{570}, the site where proton exchange is reported to occur [18, 19]. Chemical modification of one out of eight molecules bacteriorhodopsin induced 40% to 45% reduction of the k_{off} rate constant. This may suggest that monomer modification effects functional properties of the trimer.

The interaction of phenylisothiocyanate with amino acid side chain nucleophilic groups of bacteriorhodopsin is limited. The N-terminal amino acid is blocked (pyroglutamic acid [7, 24]), cysteine and histidine are not present [8, 24, 25]. Therefore, the only amino acid side chains available for phenylisothiocyanate modification are amino groups of lysine or tyrosine OH, characterized by a unique pK value. Modification of membrane lipids by phenylisothiocyanate is improbable in that nitrogen-containing

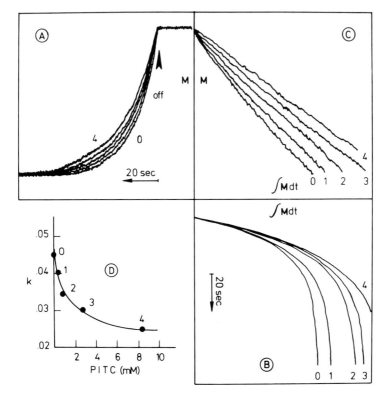

Fig. 5. Phenylisothiocyanate concentration dependence of the M_{412} decay rate constant on the "light off" reaction. Purple membranes (0.374 mM bacteriorhodopsin) were incubated with various amounts of phenylisothiocyanate. The normalized 412 nm absorbance changes (*A*) were transformed as described in Fig. 4. *D* Phenylisothiocyanate concentration dependence of the relative M_{412} decay rate constant (k_{off}). For further explanation (*B* and *C*) see Fig. 4

lipids are reported not to be present in the purple membrane [26]. Furthermore, phenylisothiocyanate-binding stoichiometry was determined in a repeatedly acetone-precipitated fraction.

Investigations are now centering on identification of the phenylisothiocyanate binding site(s). It is hoped that forthcoming information may give more insight into the structural and functional relationship of the light driven proton pump.

Acknowledgments. We wish to thank E. Lighthart, Arete Instruments, San Diego, for programming the computer system. The advise of Dr. A. Boschetti for setting up the *Halobacterium halobium* culture is greatefully acknowledged. This work was supported by the Swiss National Science Foundation (grant Nro. 3.133.77) and by the Central Laboratories of the Swiss Blood Transfusion Service SRK, Bern, Switzerland.

References

1. Oesterhelt D, Stoeckenius W (1971) Nature New Biol 233: 149–152
2. Henderson R (1977) Annu Rev Biophys Bioenerg 6: 87–109
3. Ehrenberg E, Lewis A (1978) Biochem Biophys Res Commun 82: 1154–1159
4. Lewis A, Marcus MA, Ehrenberg B, Crespi H (1978) Proc Natl Acad Sci USA 75: 4642–4646
5. Konishi T, Packer L (1978) FEBS Lett 92: 1–4
6. Konishi T, Packer L (1978) FEBS Lett 89: 333–340
7. Ovchinnikov YuA, Abdulaev NG, Feigina MYu, Kiselev AV, Lobanov NA (1977) FEBS Lett 84: 1–4
8. Bridgen J, Walker ID (1976) Biochemistry 15: 792–798
9. Konishi T, Packer L (1976) Biochem Biophys Res Commun 72: 1437–1442
10. Konishi T, Packer L (1977) FEBS Lett 79: 369–373
11. Dellweg HG, Sumper M (1978) FEBS Lett 90: 123–126
12. Henderson R, Jubb JS, Whytock S (1978) J Mol Biol 123: 259–274
13. Sigrist H, Zahler P (1978) FEBS Lett 95: 116–120
14. Sigrist-Nelson K, Sigrist H, Ariano B, Azzi A (1978) FEBS Lett 95: 140–142
15. Sigrist H, Schnippering Ch, Azzi A, Zahler P (1979) Experientia in press
16. Allegrini PR, Sigrist H, Strasser RJ, Zahler P (1979) Experientia in press
17. Drobnika L, Kristian P, Augustin J (1977) In: Patai S (ed) The chemistry of cyanates and their thio derivatives , part 2, pp 1002–1222. John Wiley and Sons, New York
18. Lozier RH, Bogomolni RA, Stoeckenius W (1975) Biophys J 15: 955–962
19. Hellingwerf KJ, Schuurmans JJ, Westerhoff HV (1978) FEBS Lett 92: 181–186
20. Oesterhelt D, Hess B (1973) Eur J Biochem 37: 316–326
21. Strasser RJ (1973) Arch Int Physiol Biochim 81: 935–955
22. Strasser RJ, Butler WL (1976) Biochim Biophys Acta 449: 412–419
23. Weber K, Osborn M (1969) J Biol Chem 244: 4406–4412
24. Ovchinnikov YuA, Abdulaev NG, Feigina MYu, Kiselev AU, Lobanov NA (1979) FEBS Lett 100: 219–224
25. Kaplan H, Cheng DCH, ODA G, Kates M (1978) Can J Biochem 56: 517–520
26. Kushawa SC, Kates M, Martin WG (1975) Can J Biochem 53: 284–292

Effects of Blue Light on Movement of Microorganisms

W. NULTSCH[1]

Since the classical investigations on phototaxis and photoaccumulation of *Euglena, Phacus,* and *Trachelomonas* by Engelmann (1892), Oltmanns (1917) and Mast (1917), who observed maximal activity at wavelengths between 470 and 490 nm and, in addition, at 405 nm (Oltmanns 1917), phototactic reactions of flagellates are regarded as typical blue light responses, comparable to the phototropism. This has been confirmed by studies on the spectral phototactic sensitivity of *Eudorina, Volvox* and *Chlamydomonas* carried out by Luntz (1931). He observed phototactic reactions in the range between 360 and 578 nm with a maximum at 494 nm (Fig. 1). According to Oltmanns and Mast, however, the maximum of *Chlamydomonas* lies between 500 and 510 nm.

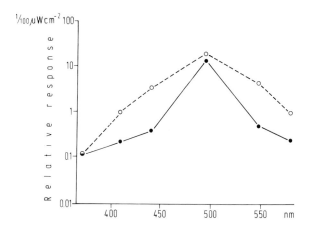

Fig. 1. Spectral phototactic sensitivity of *Eudorina (dashed line)* and *Volvox (solid line). Abscissa* wavelength in nm; *ordinate* phototactic threshold values. (After Luntz 1931; modified from Haupt 1959)

First action spectra of positive phototaxis and the step-up photophobic response have been measured by Bünning and Schneiderhöhn (1956) and Gössel (1957) with green and colorless strains of *Euglena gracilis.* They revealed that in the green form with stigma the whole range between 400 and 540 nm was phototactically active with a main maximum at 495 nm and a second maximum around 420 nm, whereas in the colorless form without a stigma a sharp maximum was found at 410 nm and a second smaller one at 425 nm. The action spectra of the step-up photophobic response of the green and the colorless forms with and without a stigma agreed in that maximal activity was

1 Fachbereich Biologie der Universität, Marburg, FRG

found around 410 nm, while the spectral range up to 540 nm was less though significantly effective. Since the presence of the stigma seems not to be necessary, these findings have been interpreted to indicate that the phototactic response of *Euglena* is the
result of a periodic shading of an unknown photoreceptor, the absorption spectrum of
which should be similar to the action spectrum of the step-up photophobic response,
by the carotenoids of the stigma.

The photophobic responses of *Euglena gracilis* were measured by Diehn (1969)
with the aid of a population method (phototaxigraph). The action spectrum of the step-
up photophobic response (Fig. 2) shows peaks at 365, 412, 450, and 480 nm and supposedly points to a flavin pigment as photoreceptor. It must be mentioned, however,
that this action spectrum is composed of the action spectra obtained by polarized light
vibrating parallel and perpendicular to the long axis of the cell. The action spectrum of
the step-down photophobic response shows peaks at 375 and 480 nm and resembles
that of positive phototaxis. Diehn (1969) suggested that in this spectrum the spectral
sensitivity of the flavin photoreceptor is modified by the spectral absorption characteristics of screening pigments. The inhibition of the step-down photophobic response
by high concentrations of potassium iodide (150 mM), a quencher of flavin excited
states, has also been interpreted to indicate flavins as photoreceptors (Diehn and Kint
1970; Mikolajczyk and Diehn 1975), although potassium iodide does not inhibit the
step-up photophobic response and negative phototaxis.

Contrary to *Euglena,* the phototactic action spectra of other flagellates do not display stronger activity of the UV region absorbed by flavins, as shown by Halldal (1958)
for *Platymonas, Dunaliella,* and *Stephanopetra,* which have a maximum at 493 nm
(Fig. 3), and the Dinophyceae *Peridinium* and *Goniaulax,* which have a maximum at
475 nm (Fig. 4). The boundary to longer wavelengths ranges between 530 and 570 nm,
depending on the species.

With *Platymonas subcordiformis* Halldal (1961) measured action spectra of positive
and negative phototaxis as well, even in the UV. As shown (Fig. 5), both action spectra
are essentially identical. Besides the main maximum at 495 nm, a shoulder at 450 nm
and a small maximum at 405 nm, two maxima in the UV at 335 nm and 275 nm were
found. The latter one corresponds with the absorption band of aromatic amino acids.
Radiation of 220 nm is also active. UV around 370 nm, however, indicative for flavins,

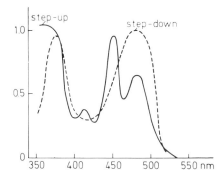

Fig. 2. Action spectra of step-down (*dashed line*)
and step-up (*solid line*) photophobic responses of
Euglena gracilis. (After Diehn 1969; from Häder
1979)

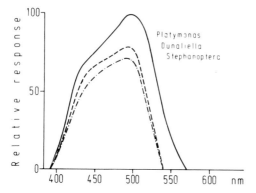

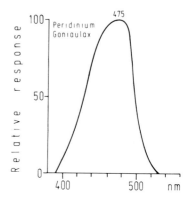

Fig. 3. Phototactic action spectra of *Platymonas sub-cordiformis, Dunaliella viridis,* and *Stephanoptera gracilis. Abscissa* wavelength in nm; *ordinate* relative response. (After Halldal 1968)

Fig. 4. Phototactic action spectrum of *Peridinium trochoideum* and *Goniaulax catenella. Abscissa* wavelength in nm, *ordinate* relative response. (After Halldal 1958)

is only slightly effective, and no distinct peak exists in this range of wavelengths. Since these action spectra in the near UV and in the visible region are similar to the absorption spectra of certain carotenoids, Halldal (1961) concluded that a carotenoprotein might be the photoreceptive pigment in this organism. A similar action spectrum, indicating a carotenoprotein as photoreceptor, has been measured by Forward (1974) with *Gymnodium* (Fig. 6).

More recently, Nultsch et al. (1971) measured the phototactic action spectrum of *Chlamydomonas reinhardtii,* which also lacks a peak in the near UV, and which shows maximal activity at 503 nm (Fig. 7). This action spectrum also suggests a carotene rather than a flavin as photoreceptor pigment.

Besides the flagellates, there are other algae such as diatoms and the red alga *Porphyridium,* in which violet, blue and blue-green light, and in some of them near UV, is active in triggering photomotive responses. Among the diatoms, the phototactic action

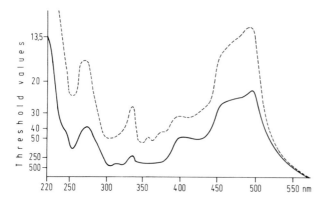

Fig. 5. Action spectrum of positive (*dashed line*) and negative (*solid line*) phototaxis of *Platymonas subcordiformis. Abscissa* wavelength in nm; *ordinate* threshold value of response in quanta cm^{-2} s^{-1} X 10^8. (Modified after Halldal 1961)

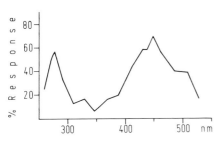

Fig. 6. Phototactic action spectrum of *Gymnodinium splendens. Abscissa* wavelength in nm; *ordinate* percentage of cells showing positive phototaxis upon stimulation with a fixed number of quanta. (After Forward 1974)

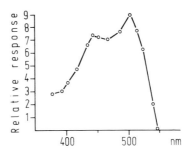

Fig. 7. Phototactic spectrum of *Chlamydomonas reinhardtii. Abscissa* wavelength in nm; *ordinate* relative quantum efficiency. (After Nultsch et al. 1971)

spectrum of *Nitzschia communis* was measured by Nultsch (1971). It shows maxima at 430 and 490 nm in the visible region, but strongest activity in the near UV with a maximum around 380 nm (Fig. 8). This action spectrum favors flavins as photoreceptors. The phototactic action spectrum of *Porphyridium cruentum,* measured by Nultsch and Schuchart (unpublished) most recently, shows a maximum at 443 and shoulders at 416 and around 470 nm, but no peak in the UV at 370 nm (Fig. 9). Consequently, it favors carotenoids rather than flavins as photoreceptors.

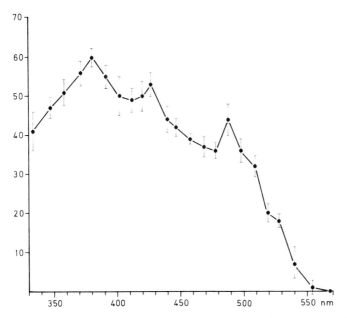

Fig. 8. Phototactic action spectrum of the diatom *Nitzschia communis. Abscissa* wavelength in nm; *ordinate* relative response. (After Nultsch 1971)

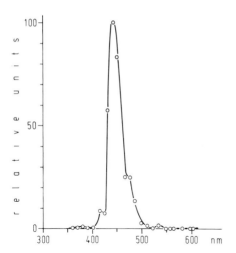

Fig. 9. Phototactic action spectrum of the red alga *Porphyridium cruentum. Abscissa* wavelength in nm; *ordinate* relative response. (After Nultsch and Schuchart, unpublished)

In conclusion, the aforementioned action spectra of phototaxis, photoaccumulation and photophobic responses point to yellow pigments as photoreceptors. Some of them are indicative for flavins, others for carotenoids, both probably bound to proteins within the living cell.

Arguments favoring carotenoids and carotenoproteins, respectively, are:

1. Maximal activity is found between 450 and 520 nm, very often between 500 and 510 nm.
2. Many action spectra lack a UV peak around 370 nm.
3. Light has no effect on the movement of the carotenoid-free diatom *Nitzschia alba* (Nultsch and Wenderoth 1973).

On the other hand, the following arguments favor flavins and flavoproteins, respectively, as photoreceptors:

1. Occurrence of a UV peak around 370 nm in some action spectra, such as those of *Euglena gracilis* and *Nitzschia communis.*
2. Extremely short lifetime ($< 10^{-13}$ s) of the singlet excited state (S_1) of carotenoids.
3. The triplet energy levels of carotenoids and the singlet to triplet quantum efficiency are very low (Bensasson et al. 1976; Bensasson 1977).
4. High concentrations of potassium iodide, a quencher of flavin singlet excited states, inhibit phototaxis of *Euglena.*

Numerous attempts have been made to identify the blue light-absorbing photoreceptor organelle. Most of them deal with *Euglena.* In this organism two organelles come into question as photoreceptor sites: the paraflagellar body (PFB) and the stigma (Kivic and Vesk 1972). Recently, most of the investigators agree that the PFB is the photoreceptor site, whereas the stigma functions as a screen which periodically shades the photoreceptor during rotation (cf. Feinleib and Curry 1971). The most convincing experimental result that the PFB and not the stigma is the photoreceptor site has been reported by Gössel (1957), who found that *Euglena* cells without stigma react phototactically, while cells without PFB do not. Checcucci et al. (1976) found that the action spectra for photoaccumulation and photodispersal of green and dark bleached cells, containing a colored stigma, and a streptomycin bleached mutant with a colorless stigma were essentially identical. Moreover, Benedetti and Checcucci (1975) and Benedetti and Lenci (1977) have demonstrated the presence of flavins in the PFB, using fluorescence microscopic and scanning microphotometric methods. Photomicrographs and microfluorescence measurements gave no hint that the pigment content of the PFB was changed by the streptomycin treatment (Ferrara and Banchetti 1976).

Investigations by Batra and Tollin (1964) and by Bartlett et al. (1972) have shown that carotenoids are the main pigments of the stigma. This has been confirmed by Benedetti et al. (1976), who measured microspectrophotometrically the absorption of the *Euglena* stigma within green and dark grown cells and of isolated stigma granules (Fig. 10). They found absorption maxima around 410, 450, and 490 nm.

Supposing that flavins were the photoreceptor pigments, these findings would give further evidence that the PFB is the photoreceptor site, whereas the stigma acts as a screen, supporting the periodical shading hypothesis. It must be mentioned, however, that according to Sperling Pagni et al. (1976) flavins are also present in the stigma of *Euglena*. With isolated stigmata they measured a fluorescence emission spectrum with a maximum around 540 nm, while the excitation spectrum has two maxima, one between 360 and 370 nm and a second one at 450 nm. Estimation of the flavin content yielded an approximate value of 5×10^{-4} µg flavin/ml stigma suspension. Thus, the photoreceptor site in *Euglena* seems to be still an open question.

In other algae the stigmata vary widely in structure and complexity, so that details cannot be reported here (cf. Nultsch and Häder 1979). In Chlorophyceae and Prasinophyceae, the stigma constitutes a part of the chloroplast and is placed at some distance from the flagella. Therefore, it seems to be unlikely that it could function as a shading device. In Chrysophyceae, Xanthophyceae, and Phaeophyceae, the stigma is also a part of the chromatophore, but is situated more or less adjacent to the flagellar base. The

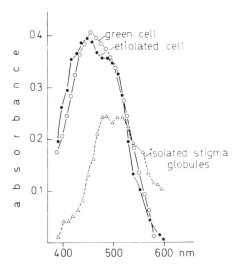

Fig. 10. Absorption of the stigma in *Euglena gracilis. Circles* stigma inside a green cell; *dots* stigma inside an etiolated cell; *triangles* isolated stigma globules. (After Benedetti et al. 1976; from Häder 1979)

structure of all these stigmata is relatively simple, so that they are at least not predestinated to act as photoreceptor organelles. Some more complex organelles are described for Wanowiaceae, a family of naked Dinoflagellates, and for the zoospores of the estuarine fungus *Phlyctochytrium*. However, these organisms are unique types, so that they may be disregarded here.

With *Volvox* Schletz (1976) has shown that the action spectrum of phototaxis resembles the combined action of the shading devices (stigma and chloroplast) and the photoreceptor. He has interpreted these results as supporting the shading function of stigma and chloroplasts. As photoreceptor pigment he assumes a carotenoprotein, as photoreceptor site the plasmalemma covering the concave side of the stigma.

Summarizing the results reported here we must concede that we have no certain knowledge of the photoreceptor site, not even in *Euglena*. Moreover, all findings indicate that several different photoperceptive structures are developed in microorganisms.

Besides the aforementioned phototactic reactions in which UV, violet, and blue light are active, several organisms display photomotive responses which cannot, or at least not exclusively, be explained on the basis of the carotenoid/flavin concept, because their action spectra are extended to longer wavelengths. Since in most of these organisms UV and visible light of shorter wavelengths is also active, flavins and carotenoids are not excluded, but additional photoreceptor pigments must be active. Edmondson and Tollin (1971) have found that the absorption of some flavoprotein complexes can extend to 600 nm, and Delbrück et al. (1976) have shown that the weak peak at 595 nm in the phototropic action spectrum of *Phycomyces* corresponds to the singlet-triplet absorption band of flavins. The action spectra in question, however, cannot be interpreted this way since characteristic maxima point to pigments other than flavins.

Phototactic action spectra of the blue-green algae *Phormidium autumnale* and *Ph. uncinatum* show three main peaks (Fig. 11). The broad one between 350 and 400 nm can be regarded as indicative for flavins, the second one at about 490 nm with a shoulder at 460 nm points to carotenoids rather than flavins, and the third one around 560 nm is obviously due to C-phycoerythrin (Nultsch 1961, 1962). In addition, a small but significant peak at 615 nm indicates C-phycocyanin. The phycobiliproteins which belong to the photosynthetic apparatus are also active as photoreceptors of the photophobic response of these species (cf. Nultsch and Häder 1979). However, contrary to

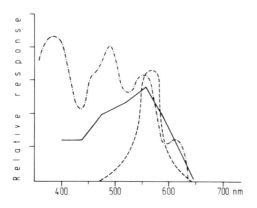

Fig. 11. Phototactic action spectra of the blue-green alga *Phormidium uncinatum* (*dashed-dotted line*) and the flagellates *Prorocentrum micans (dashed line)* and *Cryptomonas* sp. (*solid line*). *Abscissa* wavelength in nm; *ordinate* relative response. (After Nultsch 1962; Halldal 1958; Watanabe and Furuya 1974)

the photophobic reaction which is triggered by the photosynthetic electron transport, no correlations exist between phototaxis and photosynthesis, since red light above 640 nm is completely ineffective in phototaxis. The mechanism of the phototactic response in these forms is not yet understood.

The phototactic action spectrum of the flagellate *Prorocentrum micans* was measured by Halldal (1958). In this organism only radiation between 480 and 640 nm with a maximum around 580 nm is active (Fig. 11). *Prorocentrum* does not contain biliproteins. The photoreceptor is unknown.

In *Cryptomonas,* a biliprotein containing flagellate, biliproteins are also active as photoreceptors of phototaxis, as indicated by a maximum near 560 nm in the action spectrum (Fig. 11), coinciding with the absorption maximum of phycoerythrin (Watanabe and Furuya 1974). Blue light is less effective, while red light is again completely ineffective as in the blue-green algae of the genus *Phormidium*. In addition, 10^{-5} mol DCMU which completely inhibits photosynthetic O_2-production has only a slight effect on phototaxis. Consequently, also in this organism no correlations between phototaxis and photosynthesis exist (Watanabe et al. 1976).

The phototactic action spectrum of the pseudoplasmodia of *Dictyostelium discoideum* was measured by Poff et al. (1973). They found maxima around 430 and 560 nm. Between 520 and 600 nm the spectrum agreed fairly well with the action spectrum of absorbance changes at 411 nm. The isolated and purified photoreceptor pigment shows a strong absorption band at 430 nm and two broad bands between 530 and 590 nm (Poff et al. 1974). Low-temperature spectroscopy revealed the heme protein nature of this photoresponsive pigment (Poff and Butler 1974), which is supposed to control the phototactic responses. The results of an unpublished investigation by Hong et al. (personal communication by D.-P. Häder) have shown that the phototactic action spectrum of *Dictyostelium* amoebae is different from that of pseudoplasmodia, indicating that in the single amoebae another photoreceptor might be active.

Most recently, other photoresponses of *Dictyostelium,* such as photoaccumulation and photodispersal, have been studied by Häder and Poff (1979a, b). The action spectrum of photoaccumulation of amoebae in a light trap measured at low photon fluence rates (Fig. 12) shows a main maximum at 405 nm, a broad band in the blue, blue-green and green, and two sharp peaks at 570 and 640 nm (Häder and Poff 1979a). A microvideographic analysis of movement of single amoebae revealed that the amoebae react phototactically towards the light scattered from the cells within the trap.

Higher photon fluence rates cause a dispersal from the light trap as a result of negative phototaxis from the light scattered by the cells within the trap (Häder and Poff 1979b). The action spectrum of photodispersal (Fig. 12) resembles that of photoaccumulation. It shows a main maximum at 405 nm and a broad band extending to the red with a secondary maximum at about 635–640 nm. Both action spectra are different from that of phototaxis of pseudoplasmodia, and do not coincide with any pigment known to be present in *Dictyostelium*. Thus the chemical nature of the photoreceptor is an open question.

Most recently, Häder and Poff (personal communication by d.-P. Häder) have shown that ionophores, such as gramicidin, and the protonophore TPMP$^+$ do not inhibit phototaxis of pseudoplasmodiae, but impair their movement. Ionophores specifically inhibit photoaccumulation of amoebae, but do not affect their photodispersal, whereas

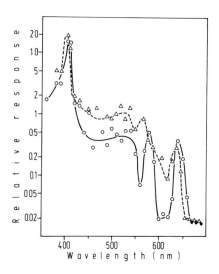

Fig. 12. Action spectra of photoaccumulation (*circles* and *solid line*) and photodispersal (*triangles* and *dashed line*) of *Dictyostelium* amoebae. *Abscissa* wavelength in nm; *ordinate* relative response. (After Häder and Poff 1979a, b)

TPMP⁺ has the opposite effect. These results give further evidence for the existence of different transduction chains in pseudoplasmodia and amoebae, respectively.

The effects of cations, especially calcium, on phototaxis of *Chlamydomonas* and *Euglena* have been studied by several authors. Hyams and Borisy (1975, 1978) observed that a change from forward to reverse motion of the isolated flagellar apparatus of *Chlamydomonas* could be induced by manipulation of exogenous calcium ions. Flagel-

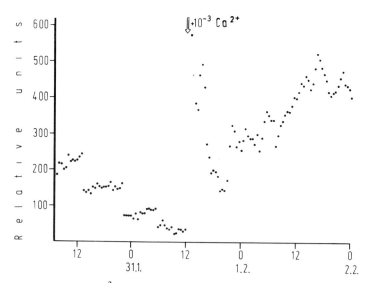

Fig. 13. Effect of 10^{-3} mol calcium ions on phototaxis of a *Chlamydomonas* culture grown in a Ca^{2+} deficient medium. *Abscissa* days and hours of the single experiments; *ordinate* positive phototactic reaction after 10 min in relative units. Each *point* represents the result of one experiment. (After Nultsch 1979)

lar beating of a ciliary type occurs between 10^{-9} and 10^{-6} mol Ca^{2+}, a flagellar type of motion above 10^{-5} mol calcium ions. Schmidt and Eckert (1976) reported that changes in the calcium concentration markedly modify the light-induced response. They proposed the hypothesis that photostimulation normally results in an influx of calcium ions through the cell membrane, and that calcium couples the flagellar responses to photostimulation. Nichols and Rikmenspoel (1978) controlled the flagellar motion of *Chlamydomonas* and *Euglena* by microinjection of magnesium and calcium ions. Long-term experiments with *Chlamydomonas* carried out in our laboratory have shown that any decrease of the intracellular calcium concentration, either by removal of calcium from the medium, by culturing the cells in a calcium-free medium (Fig. 13), or by blocking the calcium uptake with the aid of the calcium antagonist langthanum (Fig. 14), decreases the phototactic activity to very low reaction values or to zero. This inhibition can be reversed by the addition of calcium. Thus it seems to be well established that calcium ions are a link of the sensory transduction chain, coupling photostimulation and flagellar motion in *Chlamydomonas*.

Finally, it should be mentioned that there are other photomotive responses which are caused by blue and red light as well. However, since in most of these reactions chlorophyll a is one of the photoreceptors, and many of them are directly coupled with photosynthesis, they are beyond the scope of this review.

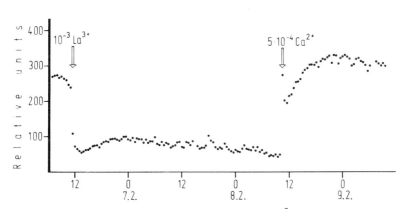

Fig. 14. Inhibition of phototaxis of *Chlamydomonas* by 10^{-3} mol lanthanum ions. The inhibition is reversed by addition of 5 $\cdot$ 10^{-4} mol calcium ions. For further explanations see Fig. 13. (After Nultsch 1979)

References

Bartlett CJ, Walne PL, Schwarz OL, Brown DH (1972) Large scale isolation and purification of eyespot granules from *Euglena gracilis* var. *bacillaris*. Plant Physiol 49: 881–885

Batra PP, Tollin G (1964) Phototaxis in *Euglena:* I. Isolation of the eye-spot granules and identification of the eye-spot pigments. Biochim Biophys Acta 79: 371–378

Benedetti PA, Checcucci A (1975) Paraflagellar body (PFB) pigments studied by fluorescence microscopy in *Euglena gracilis*. Plant Sci Lett 4: 47–51

Benedetti PA, Lenci F (1977) In vivo microspectrofluorometry of photoreceptor pigments in
    Euglena gracilis. Photochem Photobiol 26: 315–318

Benedetti PA, Bianchini G, Checcucci A, Ferrara R, Grassi S (1976) Spectroscopic properties and
    related functions of the stigma measured in living cells of *Euglena gracilis*. Arch Microbiol 111,
    73–76

Bensasson R (1977) Pigments involved in photomotion. In: Castallani A (ed) Research in photo-
    biology, pp 85–94. Plenum Press, New York London

Bensasson R, Land EJ, Maudinas B (1976) Triplet states of carotenoids from photosynthetic bac-
    teria studied by nanosecond ultraviolet and electron pulse irradiation. Photochem Photobiol 23:
    189–193

Bünning E, Schneiderhöhn G (1956) Über das Aktionsspektrum der phototaktischen Reaktionen
    von *Euglena*. Arch Mikrobiol 24: 80–90

Checcucci A, Colombetti G, Ferrara R, Lenci F (1976) Further analysis of the mass photoresponses
    of *Euglena gracilis* Klebs (flagellata euglenoidina). Monit Zool Ital 10: 271–277

Delbrück M, Katzir A, Presti D (1976) Responses of *Phycomyces* indicating optical excitation of
    the lowest triplet state of riboflavin. Proc Natl Acad Sci USA 73: 1969–1973

Diehn B (1969) Action spectra of the phototactic responses in *Euglena*. Biochim Biophys Acta
    177: 136–143

Diehn B, Kint B (1970) The flavin nature of the photoreceptor molecule for phototaxis in *Euglena*.
    Physiol Chem Phys 2: 483–488

Edmondson DE, Tollin G (1971) Circular dichroism studies of the flavin chromophore and of the
    relation between redox properties and flavin environment in oxidase and dehydrogenase. Bio-
    chemistry 10: 113–124

Engelmann ThW (1882) Über Licht- und Farbenperception niederster Organismen. Pflügers Arch
    Ges Physiol 29: 387–400

Feinleib ME, Curry GM (1971) The nature of the photoreceptor in phototaxis. In: Loewenstein
    WR (ed) Receptor mechanisms, pp 366–395. Springer, Berlin Heidelberg New York (Handbook
    of sensory physiology, vol II)

Ferrara R, Banchetti R (1976) Effect of streptomycin on the structure and function of the photo-
    receptor apparatus of *Euglena gracilis*. J Exp Zool 198: 393–402

Forward RB (1974) Phototaxis by the dinoflagellate *Gymnodinium splendens* Lebour. J Protozool
    21: 312–315

Gössel I (1957) Über das Aktionsspectrum der Phototaxis chlorophyllfreier Euglenen und über die
    Absorption des Augenfleckes. Arch Mikrobiol 27: 288–305

Häder D-P (1979) Photomovement. In: Haupt W, Feinleib ME (eds) Encyclopedia Plant Physiology
    NS, vol VII, pp 268–309. Springer, Berlin Heidelberg New York

Häder D-P, Poff KL (1979a) Light-induced accumulations of *Dictyostelium discoideum* amoebae.
    Photochem Photobiol 29: 1157–1162

Häder D-P, Poff KL (1979b) Photodispersal from light traps by amoebae of *Dictyostelium discoi-
    deum*. Exp Mycol 3: 121–131

Halldal P (1958) Action spectra of phototaxis and related problems in Volvocales, *Ulva*-gametes
    and Dinophyceae. Physiol Plant 11: 118–153

Halldal P (1961) Ultraviolet action spectra of positive and negative phototaxis in *Platymonas sub-
    cordiformis*. Physiol Plant 14: 133–139

Haupt W (1959) Die Phototaxis der Algen. In: Bünning E (ed) Encyclopedia of plant physiology,
    vol 17/1, pp 318–370. Springer, Berlin Heidelberg New York

Hyams JS, Borisy GG (1975) The dependence of the waveform and direction of beat of *Chlamy-
    domonas* flagella on calcium ions. J Cell Biol 67: 186a

Hyams JS, Borisy GG (1978) Isolated flagellar apparatus of *Chlamydomonas*. Characterization of
    forward swimming and alteration of waveform and reversal of motion by calcium ions in vitro.
    J Cell Sci 33: 235–253

Kivic PA, Vesk M (1972) Structure and function of the eyespot. J Exp Bot 23: 1070–1075

Luntz A (1931) Untersuchungen über die Phototaxis. I. Die absoluten Schwellenwerte und die re-
    lative Wirksamkeit von Spektralfarben bei grünen und farblosen Einzelligen. Z Vergl Physiol
    14: 68–92

Mast SO (1917) The relation between spectral color and stimulation in the lower organisms. J Exp Zool 22: 472–528

Mikolajczyk E, Diehn B (1975) The effect of potassium iodide on photophobic responses in *Euglena:* Evidence for two photoreceptor pigments. Photochem Photobiol 22: 269–271

Nichols KM, Rikmenspoel R (1978) Control of flagellar motion in *Chlamydomonas* and *Euglena* by mechanical microinjection of Mg^{2+} and Ca^{2+} and by electric current injection. J Cell Sci 29: 233–247

Nultsch W (1961) Der Einfluß des Lichtes auf die Bewegung der Cyanophyceen. I. Phototopotaxis von *Phormidium autumnale.* Planta 56: 632–647

Nultsch W (1962) Phototaktische Aktionsspektren von Cyanonphyceen. Ber Dtsch Bot Ges 75: 443–453

Nultsch W (1971) Phototactic and photokinetic action spectra of the diatom *Nitzschia communis.* Photochem Photobiol 14: 705–712

Nultsch W (1979) Effect of external factors on phototaxis of *Chlamydomonas reinhardtii* III. Cations. Arch Microbiol 123: 93–99

Nultsch W, Häder D-P (1979) Photomovement of motile microorganisms. Photochem Photobiol 29: 423–437

Nultsch W, Wenderoth K (1973) Phototaktische Untersuchungen an einzelnen Zellen von *Navicula peregrina* (Ehrenberg) Kützing. Arch Microbiol 90: 47–58

Nultsch W, Throm G, Rimscha v I (1971) Phototaktische Untersuchungen an *Chlamydomonas reinhardtii* Dangeard in homokontinuierlicher Kultur. Arch Mikrobiol 80: 351–369

Oltmanns F (1917) Über Phototaxis. Z Bot 9: 257–338

Poff KL, Butler WL (1974) Spectral characteristics of the photoreceptor pigment of phototaxis in *Dictyostelium discoideum.* Photochem Photobiol 20: 241–244

Poff KL, Butler WL, Loomis Jr WF (1973) Light-induced absorbance changes associated with phototaxis in *Dictyostelium.* Proc Natl Acad Sci 70: 813–816

Poff KL, Loomis Jr WF, Butler WL (1974) Isolation and purification of the photoreceptor pigment associated with phototaxis in *Dictyostelium discoideum.* J Biol Chem 249: 2164–2167

Schletz K (1976) Phototaxis bei *Volvox* – Pigmentsysteme der Lichtperzeption. Z Pflanzenphysiol 77: 189–211

Schmidt JA, Eckert R (1976) Calcium couples flagellar reversal to photostimulation in *Chlamydomonas reinhardtii.* Nature (London) 262: 713–715

Sperling Pagni PG, Walne PL, Whery EL (1976) Fluorometric evidence for flavins in isolated eyespots of *Euglena gracilis* var. *bacillaris.* Photochem Photobiol 24: 373–375

Watanabe M, Furuya M (1974) Action spectrum of phototaxis in a cryptomonad alga, *Cryptomonas* sp. Plant Cell Physiol 15: 413–420

Watanabe M, Miyoshi Y, Furuya M (1976) Phototaxis in *Cryptomonas* sp. under condition suppressing photosynthesis. Plant Cell Physiol 17: 683–690

Blue Light-Induced Intracellular Movements

J. ZURZYCKI[1]

1 Displacement of Chloroplasts

In the group of intracellular movement phenomena induced by light the rearrangements of chloroplasts are among the most spectacular. Depending on the intensity and direction of light, the arrangement of chloroplasts – cell organelles having very distinct light absorption – can be modified in the space of time from 20 min to 2 h, which evokes marked changes in the microscopic appearance of the cell and measurable differences in light absorption by the tissues.

1.1 Occurrence

The ability of chloroplasts to change their position in response to light conditions was discovered by Böhm (1856) in Crassulaceaen leaves. A very common occurrence of this ability among plants was already known at the beginning of the 20th century and was described in the classical monograph by Senn (1908).

Light-dependent rearrangements of chloroplasts take place in many algae among Chlorophyceae: *Vaucheria, Bryopsis, Eremosphaera, Hormidium;* Conjugatophyta: *Mougeotia, Mesotaenium;* Phacophyta: *Dictyota, Padina* (Senn 1908; Haupt 1959a; Haupt and Scholz 1966; Fischer-Arnold 1963; Pfau et al. 1974). More detailed description of chloroplast movements in algae, with special attention to *Mougeotia* type in which the position of chloroplast is controlled not only by the blue absorbing system but also by phytochrome, is reviewed in this volume in the chapter by Schönbohm.

Very convenient material for demonstration and study of chloroplast rearrangements are mosses where this phenomenon was found in protonema cells (*Leptobryum*) as well as in leaves (*Funaria, Mnium, Fissidens, Cirriphyllum*) and in the thalli of liverworts (*Lophocolea, Rodula, Marchantia, Pellia*) (Senn 1908; Voerkel 1934; Zurzycki 1967a; Zurzycki and Lelatko 1969).

Among Pteridophyta the chloroplast movements were found in prothalia (*Ceratopteris*) and leaves (*Aspidium, Allosurus, Osmunda, Selaginella*) (Senn 1908; Mayer 1964).

In angiosperms the phenomenon of light-controlled translocation of chloroplasts is very common. Senn (1908) presented a wide list of sun- and shade-plants, in which on the leaf cross-sections he found displacements of chloroplasts related to light condi-

1 Laboratory of Plant Physiology, Institut of Molecular Biology, Jagellonian University, Grodzka str. 53, 31-001 Kraków, Poland

tions. Very useful material is supplied by water plants with thin leaves easy for microscopic observations, i.e., *Lemna trisulca, Elodea,* and *Vallisneria* (Senn 1908; Zurzycki 1962; Seitz 1964, 1967). For the leaves of land plants, rearrangements of chloroplasts are usually detected by measurements of changes in light absorption by the leaves. Recently Lelątko (1970) and Inoue and Shibata (1973) investigated many species and in all cases found the dependence of light transmission on the illumination conditions. In some cases the effect of light pretreatment was rather strong (*Begonia, Setaria, Ajuga, Galeopsis*), in others very small but detectable (*Vicia faba, Rhus*). Only in *Oryza* leaves did Inoue and Shibata note some increase in light absorption on weak light irradiation, but there was practically no response to strong light.

Changes in light transmission due to irradiation were also found in *Zea* leaves belonging to the group of C_4 plants (Inoue and Shibata 1973). Anatomical studies on C_4 leaves showed the rearrangements as taking place only in the mesophyll cells, whereas the chloroplasts in the bundle sheath cells had always a stable position either centripetal (*Panicum miliaceum, Portulaca oleracea*) or centrifugal (*Zea, Chloris gayana*), independently of illumination (Lechowski 1980).

The very common occurrence of chloroplast rearrangements arouses the question of where this phenomenon does not occur. In numerous algae no chloroplast rearrangement was observed. In some cases this seems to be comprehensible, for instance, where the chloroplasts are imbedded in the thick cortical layer (Charophyta), where the shape and dimensions of chromatophore make the displacements impossible (many Conjugatophyta), or where the free-swimming microalgae have no stable position in respect to light.

1.2 Morphology and Quantitative Evaluation

The morphology of the chloroplast rearrangements controlled by light depends to a great extent on the number and dimensions of chloroplasts, as well as on the dimensions and special relations of the cell. Among many types of morphological diversity three sorts of cell will be presented as examples:

Selaginella. In the upper epidermis of the leaves of this fern there is only one chromatophore per cell. This chromatophore possesses the ability to change not only its position but also its shape in a nearly ameboid manner. At the low intensity of the light falling perpendicularly on the leaf, the cup-shaped chloroplast is situated at the bottom of the cell. Looking from the direction of incident light, nearly the whole optical cross-section of the cell is covered by the chloroplast. At high light intensity the chloroplast moves to one of the side walls, and the area of the cell inspected from the direction of light is partly free of chloroplast (Fig. 1). The mean value of the chloroplast-covered part of the cell may be treated as a quantitative criterion of the chloroplast position. This value expressed in percent is nearly 100% for the low light intensity position and about 50% for the high light position (Zurzycki and Zurzycka 1951). Principally a similar type of measurement but based on visual, statistical evaluation was used by Mayer (1964). In darkness the chloroplast takes no special position characteristic of this condition. At least for some hours the position of the chloroplast remains the same as in the previous light period.

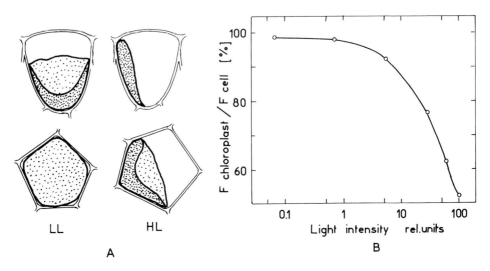

Fig. 1A, B. *Selaginella martensii.* **A** Cross-sections *upper row* and view from above *lower row* of the cells with low light intensity (*LL*) and high light intensity (*HL*) positions of chloroplast. **B** Dependence of the position of chloroplasts on the intensity of white light (Zurzycki and Zurzycka 1951)

Funaria. The cells of the one-layer leaf lamina of this moss contain many disc-shaped chloroplasts. At low light intensity the chloroplasts take their position on the walls perpendicular to the light rays, i.e., on the upper — directed towards light — and on the lower cell wall. At high light intensity the chloroplasts move to the side walls and occupy all of them without any discrimination. In darkness the position of the chloroplast of this species is similar to its position at high light intensity. Most quantitative experiments with *Funaria* were based on the counting of chloroplasts in flat position, i.e., situated on the upper and lower cell walls, and subsequent calculation of this number in percent of all chloroplasts in chosen cells (Zurzycki 1967a). This percentage may vary from 100% for low light intensity to 0% for high intensity. The intensity curve shows (Fig. 2) that in some range of intensity practically all chloroplasts are in flat position, but that they discriminate between the cell wall turning toward light and that turned away from light. At higher light intensities the chloroplasts move away at first from the cell wall facing the light. When only a part of the cell is illuminated, the chloroplasts react to light in this illuminated part alone.

For quantitative measurements of the chloroplast position in thin and transparent leaves other methods can be used as well, such as, for instance, measurements of changes in light transmission (Pfau et al. 1964) or microscopic scanning with the aid of a microphotometer equipped with an electrodynamic moving condenser (Nultsch and Benedetti 1978).

Tradescantia. The chloroplast-containing mesophyll cells form only a thin layer in the middle part of the leaf cross-section and are covered on both sides by the giant transparent cells of the water tissue. At low light intensity the chloroplasts tend to take position on the walls perpendicular to the light direction, whereas at high light intensity

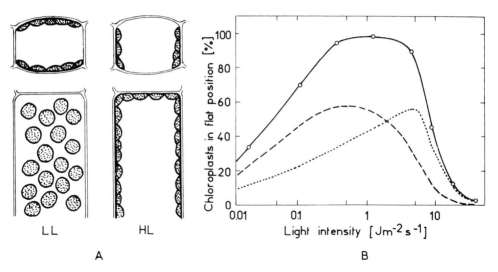

Fig. 2A, B. *Funaria hygrometrica.* **A** Cross-sections and a view of the cells from above. **B** Light intensity curve of the position of chloroplasts for 454 nm; *broken line* percentage of chloroplasts on the walls directed towards light, *dotted line* on the walls facing away from the source of light (Zurzycki 1967b)

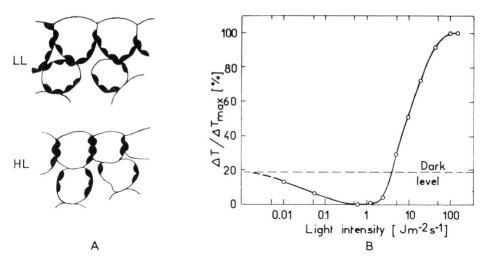

Fig. 3A, B. *Tradescantia viridis.* **A** Cross-sections of the mesophyll cells with chloroplasts in low light (*LL*) and high light (*HL*) intensity position. **B** Changes in the light transmission of the leaf as measured at 660 nm in dependence on the white light intensity (Walczak 1980)

they move to the side walls situated parallel to the light rays (Fig. 3). In the dark the chloroplasts become evenly distributed on all cell walls. Rearrangement of the chloroplasts influences the light absorption of the leaf. Recently a new method has been described for continuous recording of light transmission, using very weak red light impulses with the frequency of 800 Hz, which permits the use of illumination with one or two beams of actinic light in a very wide range of intensities and wavelengths (Walczak and Gabryś 1980).

For establishing changes in the light absorption of leaves many other methods have been proposed. For instance, Iwamura and Shibata (1973) used spectrophotometric measurements of light transmission at two wave-length 673 and 750 nm, adopting the value $\Delta A = A_{678} - A_{750}$ as a criterion of chloroplast translocation.

The above presented examples allowed some general conclusions which are also valid for other rearrangement processes, at least in mosses and higher plants:

a) In each case where chloroplast rearrangements were detected, at least two specific positions of chloroplasts can be distinguished: low light intensity position and high light intensity position.

b) In the low intensity position the chloroplasts tend to become situated at the best illuminated cell walls — they show a positive response to light.

c) Quite inversely, in the high light intensity position the chloroplasts are situated at the cell walls or cell parts where the intensity of illumination is the smallest, this being connected with the angle at which these walls are turned to the light rays, with local shading, etc. In this case the chloroplasts show light-avoiding reaction or negative response.

d) The behavior in darkness may be different. In most cases the chloroplasts are distributed uniformly on all cell walls (*Tradescantia, Lemna*), but in others they may remain in the position previously induced by illumination (*Selaginella*) or take a position specific for darkness (*Funaria*).

1.3 Action Spectra

Using any method of quantitative evaluation of chloroplast position, it is possible to investigate the activity of different wavelengths of radiation and to calculate the action spectra of light. Such measurements were made for the moss *Funaria* (Zurzycki 1967a), the fern *Selaginella* (Mayer 1964), water plants *Lemna* and *Vallisneria* (Zurzycki 1962; Seitz 1967) and some land plants: *Setaria* (Inoue and Shibata 1973) and *Ajuga* (Lechowski 1973). They are combined (Fig. 4) with one example of action spectrum for alga *Vaucheria* (Fischer-Arnold 1963).

From the comparision of the presented curves for different species some general conclusions may be drawn:

a) In the cases where positive and negative responses of chloroplasts were studied (*Vaucheria, Selaginella, Lemna, Setaria*), the shape of action spectra for both reactions is principally the same, this suggesting one and the same photoreceptor controlling both types of rearrangement.

b) Spectral activity of radiation is restricted to blue light and ultraviolet. The wavelengths above 500 nm are unable to evoke a typical light-dependent arrangement. Only for *Vallisneria* did Seitz (1967) report a very low activity of red in the range of 680 nm.

c) In all cases the highest activity of the visible part occurs in the region 450–480 nm, usually at 450 nm. It was only in *Selaginella* and *Ajuga* that the highest activity was noted at 472 nm and at 480 nm respectively (Mayer 1964; Lechowski 1973). On the other hand, in other species either an accessery maximum at 480 nm (Zurzycki 1962,

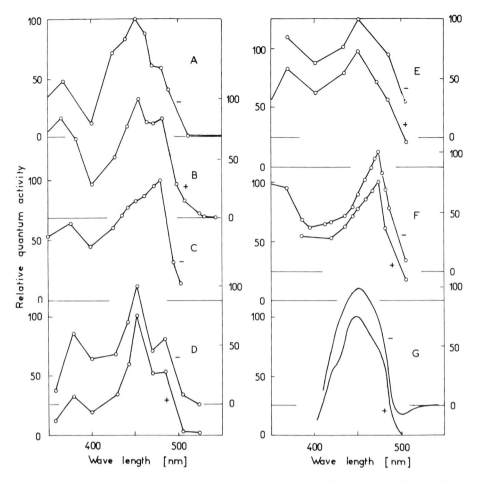

Fig. 4A–G. Action spectra of the chloroplast rearrangements. + positive response, – light-avoiding response. **A** *Vallisneria spiralis* (Seitz 1967), **B** *Funaria hygrometrica* (Zurzycki 1967a), **C** *Ajuga reptans* (Lechowski 1973), **D** *Lemna trisulca* (Zurzycki 1962), **E** *Vaucheria sessilis* (Fischer-Arnold 1963), **F** *Selaginella martensii* (Mayer 1964), **G** *Setaria viridis* (Inoue and Shibata 1972)

1967a) or at least a shoulder in this range was found (Seitz 1967; Fischer-Arnold 1963; Inoue and Shibata 1973).

d) In the near UV region there always appears an accessory maximum at about 360 nm (Zurzycki 1962, 1967a; Fischer-Arnold 1963; Mayer 1964; Lechowski 1973). In the short UV the maximal activity, higher than in the visible range, was shown for *Selaginella* and *Funaria* (Mayer 1964; Zurzycki 1967a), its peak being situated near 266 nm (Fig. 7).

All the facts presented are not in contradiction with the assumption that flavin pigment acts as a photoreceptor controlling the position of chloroplasts.

1.4 Localization of the Photoreceptor

Several data suggest that the blue-absorbing photoreceptor, active in controlling the chloroplast position, is situated not in chloroplasts themselves but in the outer layer of cytoplasm, most probably in the plasmalemma complex.

1.4.1 Local Irradiation with the Light Spot

As was shown by Fischer-Arnold (1963), when the light spot is directed on the chloro-plast of *Vaucheria,* which streams with the cytoplasm and is moved together with the chloroplast, no special effect of irradiation is detectable. But when the light spot is applied to one place of the cell, after some time of irradiation the illuminated part ac-quires the ability to fix the chloroplasts which migrate into this area (Fig. 5). Local ir-radiation of the *Selaginella* cells, even in the place free of chloroplasts, shows that the irradiated part becomes the place of attraction for a chromatophore (Fig. 5).

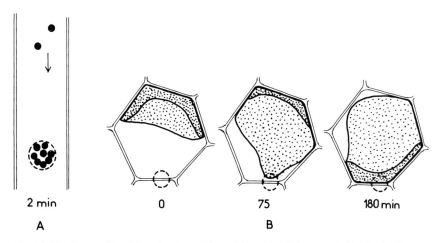

Fig. 5A, B. Effects of local illumination of the cell by a blue light spot. **A** *Vaucheria* (Fischer-Ar-nold 1963), **B** *Selaginella* (Mayer 1964)

1.4.2 Experiments with Linear Polarized Light

A response of the chloroplast to irradiation with polarized light differs from that of the chloroplast exposed to normal unpolarized light (Mayer 1964; Zurzycki 1967b; Zurzycki and Lelatko 1969). The side walls situated parallel to the E-vector of polariza-tion become at low light intensity as attractive for chloroplasts as are the upper and lower cell walls, and at high light intensity, forbidden to chloroplasts (Fig. 6). The wide occurrence of this response, encountered in all cases where light reaches the cell in the polarized form, means that this is a general phenomenon for chloroplast rearrangements (Zurzycki and Lelatko 1969). The explanation of such behavior is possible on assump-

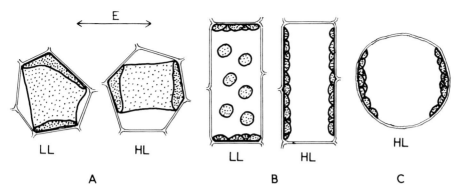

Fig. 6A–C. Response of chloroplasts to linear polarized light of low (*LL*) and high (*HL*) intensity. *E* the plane of electrical vector. **A** *Selaginella* (Mayer 1964), **B** *Funaria* (Zurzycki 1967b), **C** *Sedum* (Zurzycki and Lelatko 1969)

tion that the photoreceptor molecules are dichroic and oriented in the cell in such a way that the vector of electronic transition is always situated parallel to the cell surface. In the plane of the cell surface the orientation of transition vectors is random, showing no preferred direction (Zurzycki 1967b). This specific response of chloroplasts to polarized light occurs only in the visible range above 400 nm. In the UV region the chloroplasts cannot distinguish between the cell walls situated parallel and those lying

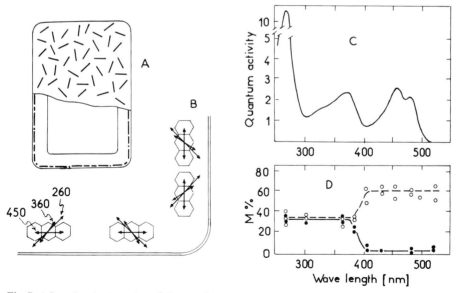

Fig. 7. **A** Postulated orientation of the transition vectors responsible for blue absorption in the *Funaria* cell (Zurzycki 1967b). **B** Situation of the transition vectors for visible and ultraviolet peaks in the oriented flavin molecules (Zurzycki 1972). **C** Action spectrum for *Funaria* cells (Zurzycki 1967a). **D** Action dichroism in *Funaria* cells. *M* measure of the occupation of the unit of length of the side walls by chloroplasts. *Continuous line* side walls parallel to the E-vector; *broken line* side walls perpendicular to the E-vector of polarization (Zurzycki 1967b)

perpendicular to the E-vector and react to polarized and to unpolarized radiation alike (Fig. 7). To explain this phenomenon it was assumed that the riboflavin molecules have the transition vector responsible for the blue peak of absorption situated almost parallel to the longer axis of the isoalloxazine ring, whereas the vectors responsible for UV absorption are oriented at the angle of about $45°$ to that direction (Kurtin and Song 1968). If such molecules are situated with their plane invariably always perpendicular to the cell surface and their longer axis always parallel to that surface, the absorption of polarized radiation on the parallel and perpendicular situated side walls will vary in the visible but not in the UV range (Fig. 7).

1.4.3 Calculation of Local Absorption

The results of such calculations are the third argument for localization of the photoreceptor at the cell surface. Recently Gabryś-Mizera (1976) has elaborated the method of computer analysis of a geometric model of *Funaria* and *Lemna* cells, which permits the local distribution of light intensity and absorption to be calculated. The best agreement of the theoretical and experiental data is obtained by assuming that the transition vector at 450 nm is oriented parallel to the surface of the cell.

1.5 Kinetics of the Chloroplast Movements

The speed and paths of individual chloroplasts were measured by the time lapse film analysis mainly for *Lemna* cells (Zurzycka and Zurzycki 1957; Zurzycki 1962, 1967b). At the steady state low light intensity the arrangement of chloroplasts does not change, but they are not motionless. They move out of the main position in various directions by small irregular jerks. The amplitude of such oscillations is $1-2.5$ μm and the average speed 0.1 μm min^{-1}. By application of low intensity ultrasound both the speed and the amplitude range increase, without any change of the general arrangement which remains typical for low light intensity (Jelesnianska et al. 1980).

A typical light-induced rearrangement can proceed according to two morphological types. In type I the chloroplasts migrate along undulated or almost stright paths to their place of destination at an average speed of $0.3-0.5$ μm min^{-1}. In type II the chloroplasts trace complicated meanderings with many loops and changes of direction. Their average speed is much higher, ranging from 3 to 5 μm min^{-1} (Fig. 8). Type I occurs in the high light-low light rearrangement and in the reverse one, provided the incident light intensity is not too high. Type II (observed also for *Funaria* and *Elodea*) occurs in the low light–high light rearrangement, when the light intensity is higher. Type I can be found only in the blue region, whereas type II is encountered also in longer wavelengths and in darkness exhibiting, however, a much lower speed of translocation.

2 Centrifugation of Chloroplasts

The movement of chloroplasts induced by the application of centrifugal force is not a typical "blue light-induced movement" but will be shortly reviewed because it is essen-

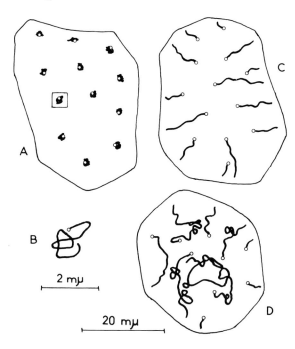

Fig. 8A–D. Paths traced by chloroplasts as studied by time lapse film analysis in *Lemna* cells. **A** Steady state low light intensity position. **B** Path of one chloroplast in higher scale. **C, D** Low light–high light intensity rearrangements at 480 nm, induced by the intensity of 3.5 and 5.9 J m^{-2} s^{-1} respectively (Zurzycki 1962)

tial for the discussion of the light-induced chloroplast rearrangements. It has been known since the experiments of Northen (1938) that for the displacement of chloroplasts by centrifugal force a given threshold of acceleration must be exceeded. Application of lower acceleration even for a longer time does not move the chloroplasts which counteract the centrifugal force. The ability of chloroplasts to centrifuge depends on various factors, among which light is a very important one. The light effects have been studied extensively by Virgin (1951), who found the chloroplasts to be most resistant to centrifugation when the cells were kept in light of low intensity, and the easiest to move after illumination with the high intensity of light.

The action spectra of the light-controlled ability to centrifuge are shown in Fig. 9 (Virgin 1952, 1954; Seitz 1967). This spectral activity is very similar to the light activity which influences chloroplast translocations (Fig. 4). In *Elodea* no effect of long wave-length part of the spectrum was found, whereas in *Vallisneria* some effect was observed but it was much weaker than in blue.

In connection with the light-controlled chloroplast translocations some other facts resulting from the centrifugal experiments are important:

a) Light induces changes in the chloroplasts' ability to centrifuge, even in such cells where no rearrangement takes place. Ohiwa (1977) demonstrated that local illumination of a *Spirogyra* cell by blue light of low intensity evokes greater resistance of the illuminated part of the chromatophore ribbon to dislocation in the centrifugation.

b) After illumination with linear polarized light of high intensity the greatest ability to dislocate is shown by the chloroplasts situated at the cell walls parallel to the E-vector of polarization (Fig. 10) (Seitz 1967, 1971).

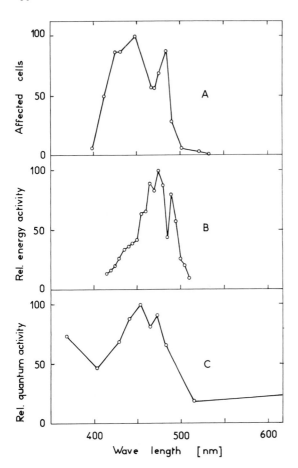

Fig. 9A–C. Action spectra for the light-controlled ability of chloroplasts to dislocation by centrifugation. **A** *Elodea* (Virgin 1952) **B** *Elodea* (Virgin 1954) **C** *Vallisneria* (Seitz 1967)

c) The frequency curves characterizing the ability of individual chloroplasts to centrifuge are shown in Fig. 11. In the steady state condition specific for low and high light intensities position of chloroplasts the curves exhibit slight differences. But when centrifugation was performed in the course of rearrangements (5–8 min after application of high light intensity), a considerable proportion of chloroplasts was very easy to centrifuge even at very low acceleration (Zurzycki 1960) (Fig. 11).

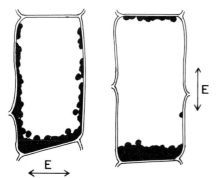

Fig. 10. Displacements of chloroplasts in *Vallisneria* after illumination with high intensity polarized blue light. *E* plane of the electrical vector of polarization (Seitz 1967)

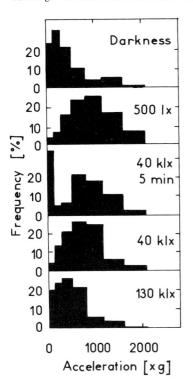

Fig. 11. Frequency curves of *Lemna* chloroplasts displaced in dependence on the acceleration used. The cells adapted to constant intensities of white light except in the middle diagram where the cells adapted to low light intensity were illuminated during 5 min by light of high intensity (Zurzycki 1960)

3 Protoplasmic Streaming

In presenting light effects on protoplasmic streaming, let us consider two cases: light effects on the cells without chloroplasts and on the cells containing chloroplasts. In the second case the response to light must be more complex because it includes effects of photosynthesis and relations between chloroplasts and protoplasm.

As an example of chloroplast-free tissue *Avena* coleoptile may be considered. Bottelier (1934) found that the constant rate of protoplasmic streaming may decrease or increase under illumination. Such effects may be found at constant illumination as well as after exposure to short-lasting light impulses; in the latter case the effect is related to the dose of light. At low intensity (or low dose), the streaming rate was found to decrease, whereas at high light intensity this rate increased (Fig. 12). Spectral activity was measured at some wavelengths only (Fig. 13), showing the activity of the blue and near UV regions. More recently Keul (1976) found in barley root hairs that the increase of the streaming rate to a constant final value depended on the wavelength and intensity of illumination used. The action spectrum shows the main activity in blue and near UV, but also an accessory peak located between 547—554 nm.

Cells containing chloroplasts (*Elodea, Vallisneria*) have been studied by many researchers. Earlier results (review see Haupt 1959b) must be treated rather critically, because in most cases no proper light measurements were performed. Recent studies (Seitz 1964, 1967) have shown the speed of protoplasm (as measured by observation of protoplasmic granules) to be accelerated after illumination in the blue as well as in

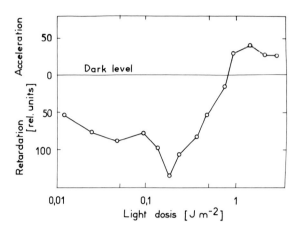

Fig. 12. Effect of blue light (436 nm) on the protoplasmic streaming in *Avena* coleoptile cells (Bottelier 1934)

the red region, however, red being about 10 times less effective than blue (Seitz 1964). The action spectra were determined only for the light-induced chloroplast cyclosis and they show (Fig. 13) no activity in red for *Elodea* and low activity for *Vallisneria*. These action spectra reflect rather the ability of chloroplasts to move with the stream of protoplasm than the effect of light on the speed of protoplasm

4 The Mechanism of Blue Light Effects

The similarity of light effects on chloroplast rearrangements, ability to centrifuge, and protoplasmic streaming (blue range activity, positive and negative response according to intensity, and specific action of polarized light) suggests that there is one photoreceptor and probably the same primary events for all these phenomena. The chain of reactions leading from the excited photoreceptor molecules to the observed end effects on movement is certainly complex and at many points unknown. Nevertheless one can propose some speculative and hypothetical explanation.

It is clear that the cell does not react to illumination as a unit, but that the local conditions of illumination within the cell are important for some locomotive phenomena. One must assume that, depending on the local absorption in the photoreceptor molecules, some physicochemical differences are created between some regions in one and the same cell and that a gradient of these differences is generated.

4.1 Primary Processes

Recent studies on the flavin-mediated photoreactions in microorganisms (Munoz and Butler 1975) and in artificial systems (Schmidt and Butler 1976) have shown that the excitation of flavin leads to its reduction, and reoxidation of the pigment induces transport of the electron through cytochrome to oxygen. If we assume that this model works for the flavin photoreceptor, localized in the plasmalemma also, it is possible to postulate such an orientation of the intermediates in the membrane that the reaction pathway will act as a vectorial, transmembrane electron or proton transport (Fig. 14). In such a case one can expect:

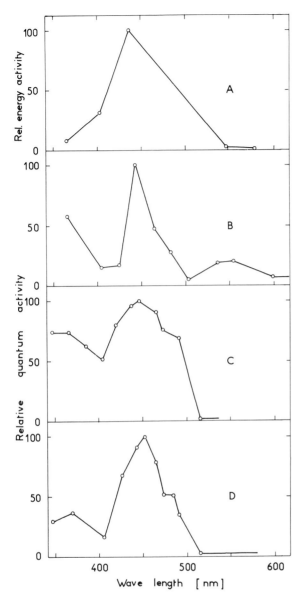

Fig. 13A–D. Action spectra of photoplasmic streaming and cyclosis of chloroplasts. **A** *Avena* coleoptile (Bottelier 1934), **B** root hair cells of *Hordeum* (Keul 1976), **C** *Elodea* (Seitz 1964), **D** *Vallisneria* (Seitz 1967)

a) the proton accumulation at the cytoplasmic side of the membrane, which in turn can modify the permeability of plasmalemma,
b) oxidation of an unknown hydrogen donor in the cytoplasm,
c) extra oxygen uptake.

Some of these expectations are in agreement with the experimental data.

A decrease in the intracellular pH after illumination, as measured with intracellular antimony-covered microelectrodes, was reported for *Phaeoceros* cells (Davis 1974). Recently Polański (1980), working on the plasmalemma-enriched subcellular fraction,

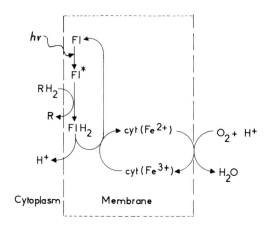

Fig. 14. Scheme of the postulated primary reactions

was able to find the acidification of the medium after illumination. In this fraction fragments of the plasmalemma membrane form reversed vesicles, so that a decrease in pH of the medium corresponds to H^+ transport inside the intact cell. Changes in pH can be induced only by blue light, whereas red is not effective (Fig. 15).

Extra O_2 uptake induced by illumination was discovered by Kowallik (1966–1969) in *Chlorella*. The measurements performed on *Lemna* showed that the extra O_2 uptake is closely correlated with the dependence of light intensity on the chloroplast translocation (Zurzycki 1970). The leaves of some mosses (*Climatium, Brachythecium*) contain elongated spindle-shaped cells oriented parallel to the leaf axis. If the gas exchange in such leaves is measured by the microrespirometric technique which permits the leaves to be oriented in one direction, the extra oxygen uptake shows the dependence on the direction of polarization of blue light used. This uptake begins at lower intensity, when

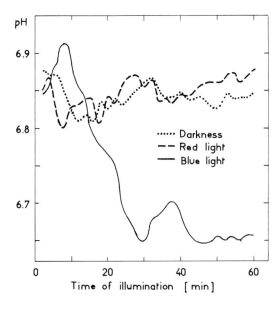

Fig. 15. Light-induced pH changes in the plasmalemma fraction of *Phaseolus* leaves (Polański 1980)

the E-vector of light is directed parallel to the cell axis (Zurzycki 1971). This suggests that the extra O_2 uptake is controlled by the photoreceptor of the same character as the photoreceptor for rearrangements of chloroplasts.

The scheme presented above of the primary events concerning light activity is only one of several possible explanations. Some other propositions have been postulated (Seitz 1971, 1972, 1974; Zurzycki 1972).

4.2 Secondary Processes

Many recent data show that in the plant cell there exists an actomyosin system responsible for various locomotive phenomena. In Characeae, the actin fibrils are identified by decoration with heavy meromyosin (Palevitz and Hepler 1974; Williamson 1974). They are situated on the border layer between gel ectoplasm and endoplasm. Actin was also identified in some higher plant cells (Condeelis 1974; Forer and Jackson 1975) . The actin filaments may have the dimensions below the limit of microscopic resolution, but they are often visible as very thin threads in light microscopy (O'Brien and McCully 1970). The activity of actomyosin system is responsible for generation of the motive force for protoplasmic streaming (Allen 1974; Williamson 1976). Actin filaments may be bound to the cortical layer of protoplasm, but they may also be fixed to the chloroplasts (Kamiya and Kuroda 1964; Palevitz and Hepler 1975; Kersey et al. 1976).

Basing on these and similar data, the following scheme of chloroplast movements can be postulated. In the cells where chloroplasts can change their position, the chloroplasts are situated in the endoplasm but touch the cortical layer and may be fixed at that layer by the actomyosin system. The anchoring connections are labile, the binding filaments can be broken and restored, their turnover being regulated by the local milieu developed by spatial distribution of incident light. The strongest anchoring forces are developed in places where the local conditions are optimal for chloroplasts. When the conditions are changed the binding forces decrease, and the chloroplasts become free and are transported to other planes where the local conditions are optimal. There they get fixed again.

The successive steps of this hypothesis can be supported by experimental data. Binding forces between chloroplasts and the cortical layer are the greatest at low light intensity (greatest resistance to centrifugation). But even under these conditions the irregular type of protoplasmic streaming (agitation) induces small oscillations of chloroplasts (Fig. 8). An increase in light intensity induces a rapid break of the binding forces (Fig. 11). After transportation to the new position the binding forces increase again. On the other hand, at extremely high light intensity the centrifugability of chloroplasts becomes very easy (Fig. 11) (Zurzycki 1960), probably because the local light intensity is supraoptimal even on the side walls. In such conditions the chloroplasts do not keep the high light intensity position, but form irregular clumps (Zurzycki 1957, 1972). In some objects, in the light conditions optimal for the "trapping" of chloroplasts, the formation of a filament network of probably actin filaments was described (Schönbohm 1972, 1973; Blatt and Briggs 1978).

Transportation of chloroplasts can proceed in various ways but the chloroplasts are always passively moved by the protoplasm. The two morphological types of rearrangement described (Fig. 8) reflect either transport by the irregular streaming of cytoplasm varying in speed and direction (agitation – type II), or transport with the mass flow of cytoplasm (Type I). In some objects where the streaming of protoplasm is detectable owing to the motion of cytoplasmic granules (*Vaucheria, Elodea*) the movement of the chloroplast is seen to proceed always in the same direction as the surrounding proto-plasm, but usually at a much lower speed. In other objects the analysis of chloroplast paths suggests a close connection with the turbulent movement of protoplasm (Zurzycka and Zurzycki 1957).

The mass flow of protoplasm and its accumulation in some cell regions can explain type I rearrangements. The occurrence of this type in the blue region alone (Zurzycki 1962) is in agreement with the observation of Fischer-Arnold (1973), who detected the accumulation of protoplasm in the blue light spots in *Vaucheria*. Also Mouravieff (1960) described the dislocation of the mass of protoplasm in the cell effected by blue irradia-tion. The question arises why sometimes type I and sometimes type II appears in the blue region. The possible explanation lies in the intensity of radiation. When the inten-sity is not too high there exists a gradient of absorption along the walls, which can de-termine the direction of mass dislocation of protoplasm. At too high intensity the ab-sorption at these walls is supraoptimal, the gradient is not perceptible, and light evokes only the acceleration of turbulent streaming.

One of the most important questions is to explain the connection between primary and secondary translocation processes. What is the molecular link between the postu-lated electron transport in plasmalemma and the activity of the actomyosin system? Several possibilities such as transformation of actin from a soluble form to filamentous F-actin (Isenberg and Wolfarth-Bottermann 1976) or regulation of the free calcium level can be taken into consideration.

Red light is as a rule not active in inducing or maintaining typical light-controlled positions of chloroplasts, but seems to accelerate the protoplasmic streaming (Seitz 1964). It can also accelerate the translocation of chloroplasts to the dark position (Zur-zycki 1962) or to the position induced by simultaneously applied blue light (Zurzycki 1964). These accelerating effects, found only in photosynthetic cells, may be explained by photosynthetic ATP formation and its transport to cytoplasm (Heber and Santarius 1970), where ATP may serve as an accessory source of energy for the actomyosin sys-tem.

References

Allen NS (1974) Endoplasmic filaments generate the motion force for rotationel streaming in *Nitel-la.* J Cell Biol 63: 270–287

Blatt MR, Briggs WR (1978) Blue-light-induced chloroplast aggregation and cortical fiber reticula-tion in the alga *Vaucheria.* Carnegie Inst Wash Yearb 77: 333–336

Böhm JA (1856) Beiträge zur näheren Kenntnis des Chlorophylls. Sitzungsber der math-naturw Klasse der kais Akad der Wissensch Wien 22: 479ff

Bottelier HP (1934) Über den Einfluß äußerer Faktoren auf die Protoplasmaströmung in der *Avena*-Koleoptile. Rec Trav Bot Neerl 31: 474–582

Condeelis JP (1974) The identification of F-actin in the pollen tube and protoplast of *Amaryllis belladona.* Exp Cell Res 88: 435–439

Davis RP (1974) Photoinduced changes in electrical potentials and H^+ activities of the chloroplast, cytoplasm and vacuole of *Phaeoceros laevis.* In: Zimmermann U, Dainty J (eds) Membrane transport in plants, pp 197–201. Springer, Berlin Heidelberg New York

Fischer-Arnold G (1963) Untersuchungen über die Chloroplastenbewegung bei *Vaucheria aessilis*. Protoplasma 56: 495–520

Forer A, Jackson WT (1975) Actin in the higher plant *Haemanthus Katherinae* Baker. Cytobiologie 10: 217–226

Gabryś-Mizera H (1976) Model consideration of the light conditions in noncylindrical plant cells. Photochem Photobiol 25: 453–461

Haupt W (1959a) Die Chloroplastendrehung bei *Mougeotia*. I. Über den quantitativen und qualitativen Lichtbedarf der Schwachlichtbewegung. Planta 53 484–501

Haupt W (1959b) Photodinese. In: Bünning E (ed) Handbuch der Pflanzenphysiologie, vol XVII/1, pp 388–402. Springer, Berlin Heidelberg New York

Haupt W, Scholz A (1966) Nachweis des Linseneffektes bei der Chloroplastenorientierung von *Hormidium flaccidum*. Naturwissenschaften 53: 388

Heber U, Santarius KA (1970) Direct and indirect transfer of ATP and ADP across the chloroplast envelope. Z Naturforsch 25b: 718–728

Inoue Y, Shibata K (1973) Light-induced chloroplast rearrangements and their action spectra as measured by absorption spectrophotometry. Planta 114: 341–358

Isenberg G, Wohlfarth-Bottermann KE (1976) Transformation of cytoplasmic actin. Cell Tissue Res 173: 495–528

Jelesnianska M, Zurzycki J, Walczak T (1980) The effect of low power ultrasounds on the motility of chloroplasts in *Funaria* leaves. Acta Phys Plant in press

Kamiya N, Kuroda K (1964) Mechanical impact as a means of attacking structural organization in living cells. Annu Rep Sc Works Fac Sci Osaka Univ 12: 83–97

Kersey YM, Hepler PK, Palevitz BA, Wessells NK (1976) Polarity of actin filaments in *Characean* algae. Proc Natl Acad Sci USA 73 165–172

Keul M (1976) Das Wirkungsspektrum der Photodinese in den Wurzelhaaren der Gerste (*Hordeum vulgare* L.). Z Pflanzenphysiol 79: 40–52

Kowallik W (1966) Chlorophyll-independent photochemistry in algae. In: Energy conversion by the photosynthetic apparatus. Brockhaven Symp Biol 19: 467–477

Kowallik W (1969) Der Einfluß von Licht auf die Atmung von *Chlorella* bei gehemmter Photosynthese. Planta (Berl) 86 372–384

Kurtin WE, Song PS (1968) Photochemistry of the model phototropic system involving flavins and indoles. I. Fluorescence polarization and MO calculation on the direction of electronic transition moments in flavin. Photochem Photobiol 7: 263–273

Lechowski Z (1973) The action spectrum in chloroplast translocation in multilayer leaf cells. Acta Soc Bot Pol 42: 461–572

Lechowski Z (1980) Chloroplasts translocation in the leaves of C-4 plants. Acta Soc Bot Pol in press

Lelatko Z (1970) Some aspects of chloroplast movement in leaves of terrestial plants. Acta Soc Bot Pol 39: 453–468

Mayer F (1964) Lichtorientierte Chloroplasten-Verlagerung bei *Selaginalla martensii*. Z Bot 52: 346–381

Mouravieff I (1960) Polarisation phototactique du protoplasma dans cellules epidermique d'*Aponogeton distachyus* L. CR Acad Sci 250: 1104–1105

Munoz V, Butler WL (1975) Photoreceptor pigment for blue licht in *Neurospora crassa*. Plant Physiol 55 421–426

Northen HT (1938) Studies of protoplasmic structure in *Spirogyra*. I. Elasticity. Protoplasma 31: 1–8

Nultsch W, Benedetti PA (1978) Microspectrophotometric measurements of the light induced chromatophore movements in a single cell of the brown alga *Dictyota dichotoma*. Z Pflanzenphysiol 87: 173–180

O'Brien TP, McCully MG (1970) Cytoplasmic fibres associated with streaming and saltatory-particle movement in *Heracleum mantegazzianum*. Planta 94: 91–94

Ohiwa T (1977) Response of *Spirogyra* chloroplast to local illumination. Planta 136: 7–11

Palevitz BA, Hepler PK (1975) Identification of actin in situ at the ectoplasm-endoplasm interface of *Nitella*. J Cell Biol 65: 29–38

Palevitz BA, Ash JF, Hepler PK (1974) Actin in the green alga *Nitella*. Proc Natl Acad Sci USA 71: 363–366

Pfau J, Throm G, Nultsch W (1974) Recording microphotometer for determination of light idnuced chromatophore movements in brown algae. Z Pflanzenphysiol 71: 242–260

Polański M (1980) Dr Thesis, Kraków

Schmidt W, Butler WL (1976) Flavin-mediated photoreactions in the artificial system: a possible model for the blue-light photoreceptor pigment in living system. Photochem Photobiol 24: 71–75

Schönbohm E (1972) Experiments on the mechanism of chloroplast movement in light oriented chloroplast arrangement. Acta Protozool 11: 211–223

Schönbohm E (1973) Kontraktile Fibrillen als aktive Elemente bei der Mechanik der Chloroplasten-verlagerung. Ber Dtsch Bot Ges 86: 407–422

Seitz K (1964) Das Wirkungsspektrum der Photodinese bei *Elodea canadensis*. Protoplasma 58: 621–640

Seitz K (1967) Wirkungsspektren für die Starklichtbewegung der Chloroplasten, die Photodinese und die lichtabhängige Viskositätsänderung bei *Vallisneria spiralis* ssp. *torta*. Z Pflanzenphysiol 65: 246–261

Seitz K (1971) Die Ursache der Phototaxis der Chloroplasten: ein ATP Gradient? Z Pflanzenphysiol 64: 241–256

Seitz K (1972) Primary processes controling the light induces movement of chloroplasts. Acta Protozool 11: 227–235

Seitz K (1974) Lichtabhängige Orientierungsbewegungen und ihre Regelung. Ber Dtsch Bot Ges 87: 195–206

Senn G (1908) Die Gestalts- und Lageveränderung der Pflanzen-Chromatophoren, p 397. Engelmann, Leipzig

Virgin HI (1951) The effect of light on the protoplasmic viscosity. Physiol Plant 4: 255–237

Virgin HI (1952) An action spectrum for the light induced changes in the viscosity of plant protoplasm. Physiol Plant 5: 575–582

Virgin HI (1954) Further studies of the action spectrum for light-induced changes in the protoplasmic viscosity of *Helodea densa*. Physiol Plant 7: 343–353

Voerkel SH (1934) Untersuchungen über die Phototaxis der Chloroplasten. Planta 21: 156–205

Walczak T (1980) Dr Thesis, Kraków

Walczak T, Gabryś H (1980) New type of photometer for measurements of transmission changes corresponding to chloroplast movements in leaves. Photosynthetica in press

Williamson RE (1974) Actin in the alga *Chara corralina*. Nature (London) 241: 801–802

Williamson RE (1976) Cytoplasmic streaming in *Characean* algae. In: Transport and transfer processes in plants, pp 51–58. Academic Press, London New York

Zurzycka A, Zurzycki J (1951) Cinematographic studies on phototactic movements of chloroplasts. Acta Soc Bot Pol 26: 177–206

Zurzycki J (1957) The destructive effect of intense light on the photosynthetic apparatus. Acta Soc Bot Pol 26: 157–175

Zurzycki J (1960) Studies on the centrifugation of chloroplasts in *Lemna trisulca*. Acta Soc Bot Pol 29: 385–393

Zurzycki J (1962) The action spectrum for the light dependent movements of chloroplasts in *Lemna trisulca*. Acta Soc Bot Pol 31: 489–528

Zurzycki J (1964) The effects of simultaneous action of the short and long wave part of spectrum on the movements of chloroplasts. Acta Soc Bot Pol 33: 133–139

Zurzycki J (1967a) Properties and localization of the photoreceptor active in displacements of chloroplasts in *Funaria hygrometrica*. I. Action spectrum. Acta Soc Bot Pol 36: 133–142

Zurzycki J (1967b) Properties and localization of the photoreceptor active in displacements of chloroplasts in *Funaria hygrometrica*. II. Studies with polarized light. Acta Soc Bot Pol 36: 143–152

Zurzycki J (1970) Light respiration in *Lemna triculsa*. Acta Soc Bot Pol 39: 485–495

Zurzycki J (1971) Effect of linear polarized light on the O_2 uptake in leaves. Biochem Physiol Pflanz 162: 319–327

Zurzycki J (1972) Primary reactions in the chloroplast rearrangements. Acta Protozool 11: 189–199

Zurzycki J, Lelatko Z (1969) Action dichroism in the chloroplasts rearrangements in various plant species. Acta Soc Bot Pol 38: 493–506

Zurzycki J, Zurzycka A (1956) Investigation on the phototactic movements of chloroplasts in *Selaginella martensii*. Spring Bull Acad Sci Cracovie B 1: 235–251

Phytochrome and Non-Phytochrome Dependent Blue Light Effects on Intracellular Movements in Fresh-Water Algae

E. SCHÖNBOHM[1]

This review will only report on phytochrome- and nonphytochrome-controlled blue light effects on light-oriented chloroplast movements in fresh-water algae, not on light-induced cytoplasmic streaming or on migration of nuclei.

Results obtained from investigations on some algae which are very suitable for analyzing problems of the "blue light syndrom" in chloroplast movement will be presented here. Figure 1 shows that each single cell of *Hormidium*, *Mesotaenium*, and *Mougeotia* contains only one large flat chloroplast (Fig. 1a); these chloroplasts are oriented to face or to profile depending on the intensity (small arrows: weak light; broad arrows: strong

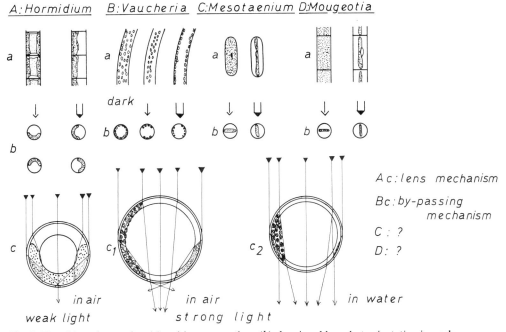

Fig. 1. The objects in top-view (*a*) and in cross-sections (*b*) showing chloroplast orientation in weak (↓) and strong light (⋃). (After Senn 1908)

1 Botanisches Institut, Universität Marburg, Fachbereich Biologie, Auf den Lahnbergen, 3550 Marburg, FRG

light) and quality of unilateral light (Fig. 1b). The small lens-like chloroplasts of the coenocytic algae *Vaucheria* (Fig. 1B) gather at the side walls in strong light or at front and rear in weak light. Additionally during darkness the chloroplasts gather at all walls equally (Fig. 1B,b).

The response of the small chloroplasts in *Vaucheria* is graded; this fact favors *Mougeotia, Mesotaenium,* and *Hormidium* much more for exact quantitative analyses in chloroplast movements than *Vaucheria.*

Analyzing blue light effects on light-oriented chloroplast movements many questions have to be solved.

1. What can we report on the most likely photoreceptors active in these movements?
2. Do we know anything about the mechanisms which are active in establishing absorption gradients in cells exposed to unilateral given light?
3. Where in the cells of the four algae are the blue light-sensitive photoreceptors localized?
4. What can we report on synergistic or antagonistic effects of blue light in light-oriented chloroplast movements?
5. Do we know anything about processes in the mechanism of light-induced chloroplast movement, which are controlled by blue light?
6. Do we know anything about the chance to realize a blue light induced chloroplast movement in lowered temperatures?

In the following we will try to give answers on these central questions. Most of the detailed results in the past were obtained from experiments on chloroplast movements in *Mougeotia;* consequently in the middle of this review blue light effects in light-oriented chloroplast movement in *Mougeotia* will be reported preferably. But first we will start with a report on *Vaucheria,* followed by some problems solved in *Hormidium.*

Figure 2 shows five action spectra: The action spectra for weak- and strong-light movement of *Vaucheria* chloroplasts are nearly identical and they coincide with the absorption curve of flavin. Thus it is postulated that the photoreceptor for strong- and weak-light movement in *Vaucheria* should be a flavin. Not only in *Mougeotia* but also in *Mesotaenium* do we have different action spectra for strong- and weak-light move-

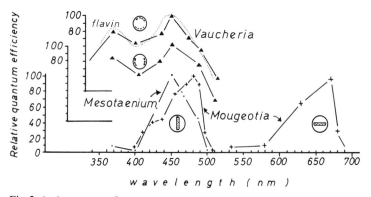

Fig. 2. Action spectra of strong- and weak-light chloroplast movements. (After Haupt and Schönbohm 1970)

Mesotaenium : 1. phase   2.phase

Fig. 3. Action spectra of two phases in weak-light chloroplast movement of _Mesotaenium._ (After Dorscheid and Wartenberg 1966)

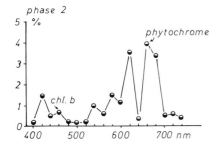

phase 1

chlorophyll b

chlorophyll a

(dark control:~37%)

400   500   600   700 nm

phase 2

%

phytochrome

chl. b

400   500   600   700 nm

ment of chloroplasts. Strong-light movement seems to be mediated by a yellow pigment, probably by a flavin, whereas weak-light movement is controlled by red light which is absorbed by phytochrome. Figure 3 shows two further action spectra for the weak-light movement of _Mesotaenium_ chloroplasts. Here we distinguish between two phases in this movement: The starting phase should be dependent on the chlorophylls (a,b), whereas the second phase, the retardation phase of the movement, in which the chloroplast reaches face position, should be controlled by phytochrome and probably also by chlorophyll b. I think we need further data and we cannot discuss these action spectra here.

1 Chloroplast Movement in _Vaucheria_

Let us look for the mechanisms active in establishing light-absorption gradients in _Vaucheria._ As shown in Fig. 1 there should be two active principles: The lens mechanism and the by-passing mechanism. In weak-light movement the chloroplast should be oriented by the lens principle; but the chloroplast moves not only to the rear but also to the front of a _Vaucheria_ filament. In strong-light movements the chloroplast should be oriented by the by-passing principle; but as shown in Fig. 1, the chloroplast not only moves to the "dark zones" (pointed zones in c_1 and c_2), but also to such parts of the side walls where some light could be absorbed. As we know, chloroplast movement in _Vaucheria_ is mediated both in strong light and weak light by one and the same photoreceptor (Fig. 2). Now we will answer the question about the localization of the blue

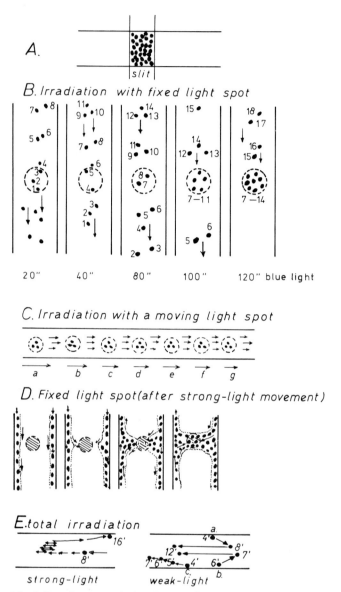

Fig. 4. Experiments on the localization of the photoreceptor in weak-light movement of *Vaucheria*. (After Haupt and Schönbohm 1970; Fischer-Arnold 1963)

light photoreceptor in the filament. By experiments, the results of which are given in Fig. 4, it could be proved that the photoreceptor is localized in the ectoplasm, but not confined to the chloroplasts. If only a small light spot becomes fixed at a small area of a *Vaucheria* filament, no reaction could be observed during the first 40–70 s: The chloroplasts were transported by the cytoplasmic streaming to and through this light spot (Fig. 4B). But after about 80 s suddenly all chloroplasts which happened to come by cytoplasmic streaming into the light spot were caught in this light trap. In contrast to these experiments with a fixed light spot the chloroplasts never stopped if the light spot is moved continuously together with the distinct group of chloroplasts as is shown in Fig. 4C.

If you irradiate a chloroplast-free area of the cytoplasm with weak blue light, the streaming becomes retarded or is stopped in the irradiated zone; together with the cytoplasm also movement of the chloroplast becomes inhibited in this zone (Fig. 4D).

The way of chloroplast movement during strong light differs from that during weak light, as is shown by the arrows and numbers (the numbers mean minutes blue-light irradiation) in Fig. 4E. The chloroplasts perform oscillations (double arrows) only in strong light but not in weak light. So we can conclude that strong-light response is not only an inversion movement of weak-light response.

2 The Chloroplast Movement in *Hormidium*

There are no action spectra for strong- and weak-light chloroplast movement in *Hormidium*. But it is known that blue light is the most effective area of the spectrum for inducing strong- and weak-light movement (Scholz 1976b).

The orientation of chloroplast movement in *Hormidium* is controlled by an absorption gradient which becomes established by the lens principle (Figs. 1 and 5). If the refractive index becomes increased, *Hormidium* changes from a collecting to a diverging lens (Fig. 5). The consequence is that with increasing refractive indexes of the medium, the tetrapolar gradients become inverted and the chloroplasts change their orientation from rear to front. This principle is only active within a distinct intensity range: In very low and in high weak-light intensities the lens mechanism with inversion in paraffin oil is inactive (Fig. 6). So we have to differentiate between two types (Fig. 7): In system "A" the lens mechanism is dominant in establishing a light-absorption gradient, whereas in system "B" the activity of the lens mechanism becomes suppressed and the "chloroplast" mechanism seems to dominate over the lens mechanism. In system "B", I think,

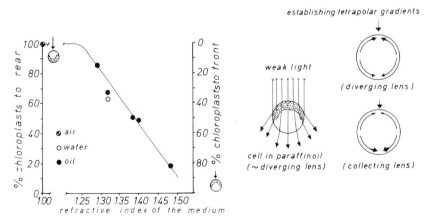

Fig. 5. The lens-principle in weak-light movement of *Hormidium*. (After Scholz 1976a; Haupt 1977)

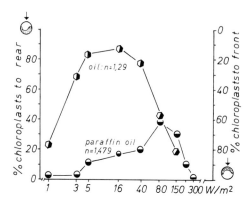

Fig. 6. The dependence of the validity of the lens principle on light intensity. (After Scholz 1976b)

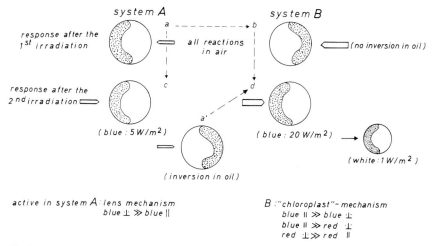

Fig. 7. The lens principle and the chloroplast mechanism as two principles active in establishing absorption gradients. (After Scholz 1976b)

the active principle could be the shadowing mechanism. The dichroic principle is also involved in both systems. The different principles seem to compete with each other. If *Hormidium* is darkened for some days, the content of FMN in the cells is decreased; Parallel with this decrease of FMN, system "A" loses its activity whereas system "B" becomes active (Fig. 7A). It is postulated that there is some correlation between reaction type "A" and flavin that should be demonstrated by the fact that system "A" regains its activity if the material becomes incubated during a single short phase in a solution of FMN (cf. in Fig. 8, B with C). We can, however, not conclude from these positive results that the added dye must be identical with the responsible endogenous photoreceptor.

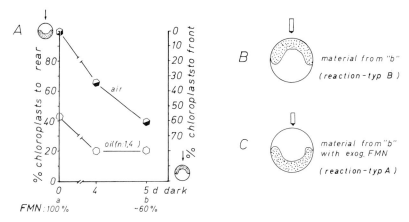

Fig. 8. The effect of prodarkening (*abscissa*) on the content of FMN and on the activity of the two orientation principles. (After Scholz 1976b)

3 The Chloroplast Movement in *Mougeotia* and *Mesotaenium*

In the center of this review there stands a report on the "blue light syndrom" in *Mougeotia;* it will be of interest to compare some of the results obtained from *Mougeotia* with those obtained from *Mesotaenium*.

3.1 Strong-Light Movement in *Mougeotia* and *Mesotaenium*

Not only in *Mougeotia* but also in *Mesotaenium* strong-light movement is controlled by blue light, whereas weak-light movement becomes mediated by red light which is absorbed by phytochrome (Fig. 2). Comparing the effects of polarized strong blue light in its dependence on the vibrating plane of the E-vector and from the intensity of blue, we can see (Fig. 9) that in *Mougeotia* parallel vibrating blue is much more effective than

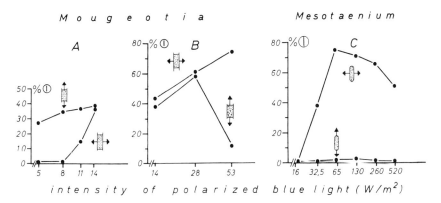

Fig. 9. The action dichroism in strong-light movement of *Mougeotia* and *Mesotaenium*. (After Schönbohm 1963; Gärtner 1970b)

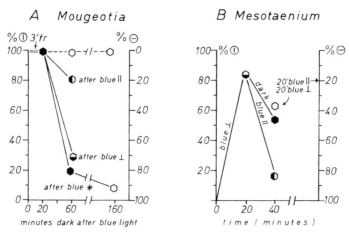

Fig. 10. "Profile-to-face" movement in darkness induced by strong blue light given during the strong blue light phase. (After Schönbohm 1963; Gärtner 1970b)

perpendicularly vibrating blue; but this is valid only if relative low strong-light intensities were given (Fig. 9A). With increasing intensities this action dichroism decreases and then converts (Fig. 9B). In contrast to *Mougeotia,* parallel vibrating blue given alone cannot induce strong-light movement in *Mesotaenium* at all (Fig. 9C).

If *Mougeotia* is kept in darkness immediately after strong blue-light position (profile) has been reached, the chloroplasts perform a *second* oriented movement, namely from "profile-to-face" position (Fig. 10A). This "weak-light movement" shows an action dichroism: The response was much higher after an irradiation with polarized strong blue light perpendicularly vibrating than after parallel vibrating blue (Fig. 10A). This dark response could be completely inhibited by a single short far red phase (arrow in Fig. 10A), which was given immediately after the strong blue light phase. In *Mesotaenium* also a "profile-to-face" response could be observed in darkness after a strong-light phase with perpendicularly vibrating blue; this response becomes increased if parallel vibrating strong blue light was given as a second strong-light phase instead of a dark phase (Fig. 10B).

These experiments with *Mougeotia* indicate that this "profile-to-face" movement in darkness is controlled by phytochrome which must be activated by strong blue light during the strong-light phase. In the "blue light syndrom" in *Mougeotia* phytochrome seems to be involved. Thus it is necessary to give some information about phytochrome and the weak-light chloroplast movement in *Mougeotia.* As shown schematically in Fig. 11, phytochrome is localized with its transition moments parallel to the cell surface, oriented screw-like in the cortical plasm (Fig. 11A). In the cross-sections (Fig. 11B) the vectors of the transition moments are given as dashes and points. Activated by red light P_r is transformed into P_{fr}. The transition moments of this active phytochrome are oriented quite normally to the cell surface. Establishing light absorption gradients in *Mougeotia* the dichroic principle is active (Fig. 11C): A P_{fr}-gradient becomes established by absorption of unpolarized (a) or perpendicularly vibrating polarized red light (b), but

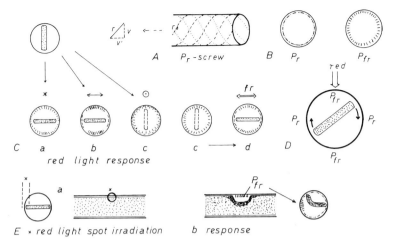

Fig. 11. Data on the localization of the transition moments of P_r and P_{fr} in *Mougeotia* and on the orientation of chloroplasts in P_{fr}-gradients. (After Haupt 1972, 1977)

not by parallel vibrating red light, which induces P_r to P_{fr}-transformation not only at front and rear, but also at the flanks and other zones of the cortical plasm (c). If such a cell (c) is irradiated with perpendicularly vibrating far red light, preferably P_{fr} on the flanks becomes transformed to P_r; so a P_{fr}-gradient can also be established (d). In weak light movement the *Mougeotia* chloroplast always *avoids* such zones of a cell in which P_r was transformed to the active phytochrome (Fig. 11D). This could also be demonstrated by experiments with microbeams (Fig. 11E).

Now let us come back to the strong-light effect in chloroplast movement of *Mougeotia*.

As shown in Fig. 12 I strong blue light response (A) becomes increased by additionally given, perpendicularly vibrating red (Fig. 12B), whereas it becomes decreased by parallel red or by far red light (Fig. 12D, E). The same effects were obtained from experiments with intermittantly given irradiation programs (Fig. 12 II, III). In such experiments the increasing effect of red could be canceled by a following far red phase (Fig. 12 II D).

In *Mesotaenium* the strong blue light effect can be increased by polarized strong blue light simultaneously given with polarized red light, whenever the vibration planes of the E-vectors of both irradiations were crossed to each other. The increasing red light effect could be canceled by simultaneously given far red (Figs. 13 and 14A, B). The results obtained from blue light double-irradiation experiments in *Mesotaenium* are unexpected: It could be demonstrated (Fig. 14) that strong blue light responses not only can be increased by polarized red light (Fig. 14A, B) but also by polarized weak blue light, provided the vibrating planes of polarized strong and polarized weak blue light are oriented perpendicularly to each other (Fig. 14C, D). It seems probable that some of the blue light effects are controlled by phytochrome, which is the most likely photoreceptor for establishing an orientation gradient in the strong-light movement, not only in *Mougeotia* but also in *Mesotaenium*. This assumption is confirmed by ex-

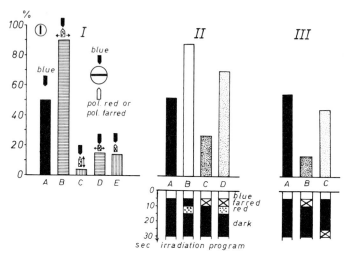

Fig. 12. The effect on strong light response of red or far red light given simultaneously or intermittently with strong blue light. (After Schönbohm 1963, 1965)

periments with polarized far red light in *Mougeotia* (Fig. 15): In *Mougeotia* strong blue light responses can be enhanced by simultaneously given polarized far red light with perpendicularly vibrating E-vector, whereas parallel vibrating far red has an inhibiting effect (Fig. 15 I). These results are interpreted in II: A weak, but effective P_{fr}-gradient, established by unpolarized blue light (Fig. 15A) could be increased by transforming P_{fr} to P_r at the flanks; for this transformation perpendicularly vibrating low-intensity far red is effective (Fig. 15C). Parallel vibrating far red becomes absorbed by P_{fr} at front and rear, so the weak P_{fr}-gradient decreases and chloroplast movement is inhibited (Fig. 15B, D). Similar effects are supposed in some of the double-irradiation experiments in *Mesotaenium*.

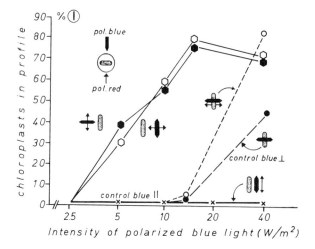

Fig. 13. Strong blue light responses as influenced by simultaneously given polarized red light. (After Gärtner 1970b)

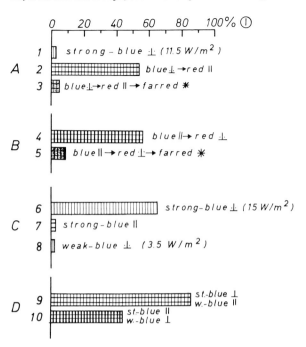

Fig. 14. Strong blue light responses as influenced by simultaneously given polarized red or polarized weak blue light and the inhibiting effect of simultaneously given far red. (After Gärtner 1970b)

We suppose that in strong blue light movement the orientation gradient is established by blue light absorption in phytochrome. In this function red light can be substituted for blue light.

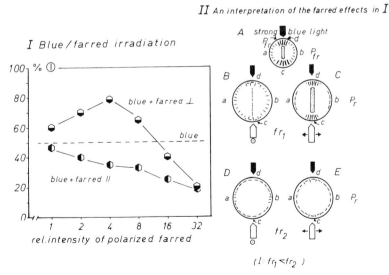

Fig. 15. Strong blue light responses and the action dichroism of simultaneously given polarized far red. (After Schönbohm, unpublished)

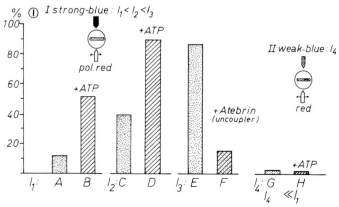

Fig. 16. The increasing effect of exogenous ATP on strong-light movement induced by strong light of different intensities. (After Schönbohm and Schönbohm, unpublished)

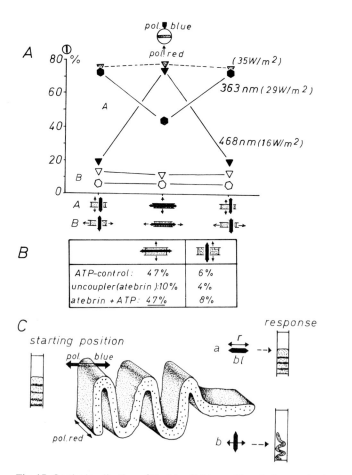

Fig. 17. On the localization of the blue light-sensitive photoreceptor in strong-light movement. (After Schönbohm 1971a, b)

At this point of the review we have to answer the question about the essential function of blue light in strong-light movement in *Mougeotia*. Where in the cell is the blue light-sensitive photoreceptor localized?

As demonstrated in Fig. 16 blue light has an important function in the energetics of strong-light movement. If ATP is added to *Mougeotia* only some minutes before strong blue light is given, the intensity of blue light can be reduced drastically without reducing the response. But as shown in Fig. 16G and H strong blue light is an essential factor in strong light movement and cannot be replaced by exogenous ATP.

We have heard that perpendicularly vibrating blue is much more effective than parallel vibrating. We have now to reexamine the effect of polarized blue in experiments in which an orientation gradient becomes established by simultaneously given red light. The results obtained are shown in Fig. 17: Now parallel blue is much more effective in inducing strong-light movement than perpendicularly vibrating. So we can postulate that a blue light-sensitive photoreceptor is oriented about parallel to the cell axis, and parallel to the cell surface or to the surface of the chloroplast broad side. Coming from blue to UV the action dichroism is changed; this can be a hint at flavin as the photoreceptor for blue light in *Mougeotia*.

In further experiments we demonstrated that the action dichroism in blue does not depend on a supposed action dichroism of photophosphorylation; the strong inhibiting effect of the uncoupler "atebrin" could be canceled completely by additionally given ATP. The response of the algae treated shows the same action dichroism as in the untreated ones. If the photoreceptor is localized in the chloroplast it must have an orientation preferably parallel to the long axis of the chloroplasts and parallel to their surface. In the experiment which is demonstrated in Fig. 17C, the chloroplasts in face position are folded by a short, weak centrifugation; by this process only a very reduced part of the chloroplast remains oriented with its surface parallel to the cell surface. These folded chloroplasts are able to perform exact light-oriented chloroplast movements. The results obtained show the same action dichroism for the strong blue-light response as in non-centrifugated algae (cf. responses in Fig. 17a, b). So we conclude that the responsible blue-light photoreceptor is localized dichroic in the cortical plasm but not in the chloroplast itself.

Iodide quenches the triplet excited state of flavins. High concentrations of iodide ($\geqslant 0.1$ m) inhibit many blue light-induced reactions. In *Mougeotia* not only does the blue light-induced strong-light movement become inhibited by iodide, but also the phytochrome-mediated weak-light movement (Fig. 18 I). Furthermore we could demonstrate that iodide also inhibits weak-light movement if it was given only after the light treatment, but during the dark reaction phase (Fig. 18 II). No inhibiting effect was obtained in such experiments in which iodide was given only during the light phase, but not during the following reaction phase. We think that iodide experiments must be interpreted very cautiously.

We know that the orientation of chloroplasts in strong light is usually exactly opposite the orientation in weak light. In *Mougeotia* the orientation gradient seems to be the same in strong- and in weak-light movement, namely a P_{fr}-gradient. We suppose that a tonic photoreceptor system may change the response of the movement in one and the same gradient or may change the nature of a secondary orientation gradient.

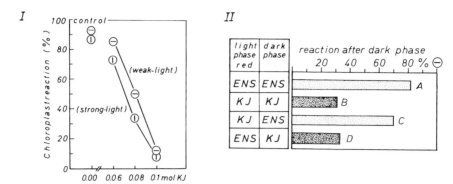

Fig. 18. The inhibiting effect of iodide on weak- and strong-light movement (*I*) and the demonstration of dark responses as the only iodide-sensitive phase in light-oriented movement (*II*). (After Schönbohm 1967)

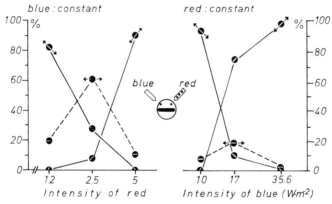

Fig. 19. The regulation of chloroplast orientation in strong-light movement by both P_{fr}-gradients and the tonic blue light. (After Schönbohm 1966)

The results shown in Fig. 19 demonstrate that in *Mougeotia* strong-light response utilizes phytochrome to sense light direction, while a blue light-sensitive photoreceptor system senses the intensity of blue light: Simultaneous irradiation with separate beams of blue and red light situated at various angles to one another show that chloroplasts are always oriented with respect to the P_{fr}-gradient (chloroplast orientation is given in the figure by the orientation of the double arrows at the curves). The sign of response (face or profile to red) was determined by the intensity of blue light. Both processes, the formation of a P_{fr}-gradient and the blue light-controlled tonic effect, are in very tight correlations, as could be shown by the results given in Fig. 20: The response "profile-to-red" only becomes realized in such double-irradiation experiments, in which red and blue light are given simultaneously during the whole reaction-phase (program Fig. 20C). If red light was given only during the first quarter (Fig. 20A) or during the first half of the blue light irradiation time (Fig. 20B), the chloroplasts change their orientation sense.

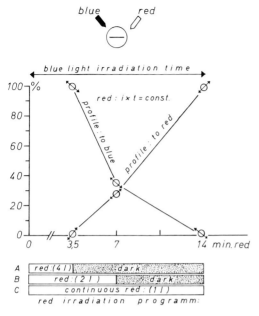

Fig. 20. Dependence of chloroplast orientation from the duration of the red phase (A−C) during blue light irradiation. (After Schönbohm 1966)

This model of the double function of strong blue light in strong-light chloroplast movement becomes confirmed by "blue-red-far red" irradiation experiments. The results obtained are shown in Fig. 21: The transition moments of phytochrome are given schematically by dashes normal (P_{fr}) and parallel (P_r) to the cell surface. Different irradiation programs are given together with the responses by cross-sections drawn in Fig. 21A−D. The postulated P_{fr}-gradients after the irradiation programs A and B are given in Fig. 21, 1 and 2, respectively. Figure 21, 3 and 4 demonstrate two irradiation programs of simultaneously given strong blue light, red light and far red light. Only in program No. 4 (Fig. 21) can far red be absorbed optimally by P_{fr}, which was established

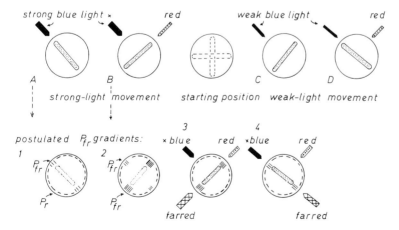

Fig. 21. Double-irradiation experiments and the effect of far red on the sense of orientation in dependence on the angles of the far red beams (3, 4). (After Schönbohm et al. 1979)

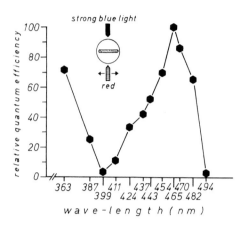

Fig. 22. An action spectrum for the tonic blue light effect. (After Schönbohm 1971b)

by red light, but not by P_{fr}, which was established by blue light (see: Fig. 21, 2). The consequences of this fact are that in program No. 4 (Fig. 21) the chloroplast becomes oriented with profile to blue light and no longer to red light. By these experiments the tonic effect of strong blue light has been proved.

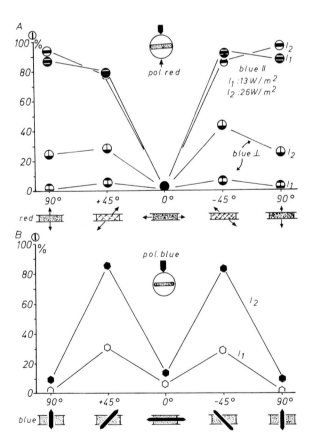

Fig. 23. The action dichroism of polarized strong blue in dependence on the orientation of the E-vector of simultaneously given polarized red (**A**) and in experiments without red (**B**). (After Schönbohm 1968)

A detailed action spectrum for this tonic blue-light effect is given in Fig. 22. We can see a second high peak near 363 nm, which coincides with the UV-peak of flavin, whereas the main peak of the action spectrum is shifted more to longer-waved regions of the spectrum.

If blue light is not given simultaneously with red but alone it is most effective if polarized with its E-vector at an angle of about 45° to the long cell axis, as shown in Fig. 23. This angle represents a compromise between establishing a P_{fr}-gradient via light absorption in the "phytochrome screw" and exciting the tonic blue light system via absorption in an about parallel oriented yellow pigment, both by polarized blue light (cf. both diagrams in Fig. 23).

If the beams of polarized red and/or polarized blue light were situated at various angles to one another not oriented radially (Fig. 24A), but sagitally as shown in Fig.

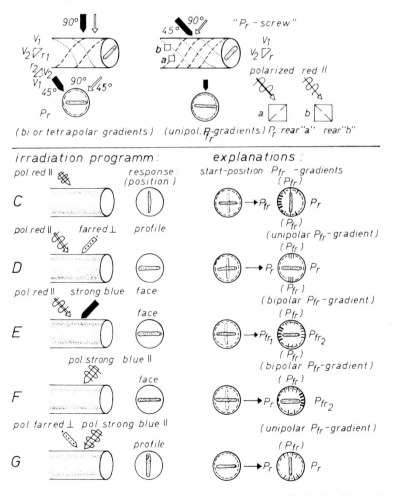

Fig. 24. Weak-(C, D) and strong-light chloroplast orientation (E, F, G) in P_{fr}-gradients with only one (C, E, F) or with two symmetry-axis (D, G). (After Schönbohm et al. 1979)

24B, some very unexpected results would be obtained. The consequences for the absorption conditions in phytochrome for parallel red are now extremely favored at one flank (Fig. 24Ba) in contrast to the opposite flank (Fig. 24Bb). At front and rear red can also be absorbed by phytochrome, but not as well as at the favored flank. So we receive P_{fr}-gradients as shown on the right hand of Fig. 24 in C–G. Only in parallel vibrating red light do the chloroplasts show profile position (C). By additionally given unpolarized strong blue light, the tonic effect forces the chloroplasts to perform strong-light movement (E): that is "profile" to P_{fr} (flanks). Parallel vibrating strong blue light has a double function: Additionally to the tonic effect it has to establish a P_{fr}-gradient, so the chloroplast orients "profile" to P_{fr} (Fig. 24F). Additionally given polarized far red light perpendicularly vibrating cancels the P_{fr} at the flank but not at front and rear: so a tetrapolar-P_{fr}-gradient (front-to-flank; rear-to-flank gradient) becomes established and the chloroplast orients "profile" to P_{fr} again (Fig. 24B). By these experiments the double function of blue light could be demonstrated once more.

Let us sum up some functions of blue light in strong-light movement:

1. Establishing a P_{fr}-orientation gradient (blue is replaceable by red light).
2. Sensing the intensity by the blue light-sensitive tonic system.
3. Producing ATP via photophosphorylation (blue can be replaced by exogenous ATP).

3.2 The "Blue Light Syndrome" in Weak-Light Movement in *Mougeotia*

As we know, weak-light movement in *Mougeotia* and *Mesotaenium* can not only be induced by red light, but also by short-waved light.

In blue, but not in red light you can change the sign of response via simple increase or decrease of the temperature; the only condition is that weak blue light is given with relative "high" intensities (Fig. 25A). If we compare the responses in continuously given weak blue light of low intensities and in red light during lowered temperatures (2°C), we see (Fig. 25B), that movement is completely inhibited in red light, whereas it is only retarded in blue light. Weisenseel (1968) supposed that movement in blue light should be controlled by a flavin and not by phytochrome. Only the phytochrome-mediated response should be inhibited completely at 2°C.

Later on we will try to look in more detail at this problem. First let us have a look at weak blue light responses at standard temperatures of about 20°C.

As shown by Haupt and Schönfeld (1963) weak-light response is inducable by red but also by UV (Fig. 26): In contrast to a red light induction, an induction by UV cannot be canceled by a following far red phase (II). "Cross"-red can be canceled by parallel vibrating red, but not by parallel vibrating UV (III). Moreover: The red/far red antagonism can be inhibited by a short UV-phase, which is given between red and far red (IV); that is the so-called stabilizing UV effect. For these UV effects a separate photoreceptor has been postulated.

Quite similar results were obtained by Haupt (1971) in a detailed analysis of weak-light movement induced by blue light and not by UV: As shown in Fig. 27A, parallel blue is much more effective than crossed blue. In contrast to similar experiments with strong blue light (Schönbohm 1963), far red light has no sensitizing effect if given before parallel vibrating weak blue light; rather, it has a decreasing one. Far red given after

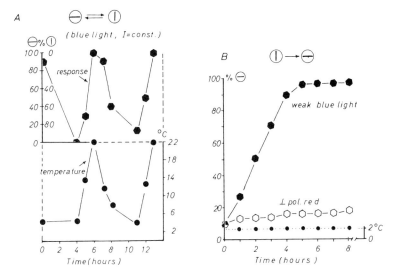

Fig. 25. Regulation of the sense of blue light reactions by changing temperatures (**A**). **B** demonstrates the retarding or inhibiting effect on weak-light movement in blue or red respectively. (After Weisenseel 1968; Schönbohm and Hellwig 1979)

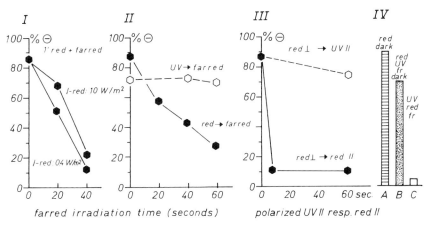

Fig. 26. Weak-light movement and its dependence on the inductive irradiations (red or UV) and the effect of far red. (After Haupt and Schönfeld 1963)

the blue light phase has no effect. An induction by perpendicularly vibrating red light becomes decreased by a following short blue-light phase (Fig. 27B), whereas parallel vibrating red becomes increased in its effects by a following given blue light irradiation (Fig. 27C). All these results contradict the opinion that phytochrome should be also the responsible photoreceptor for the blue light-mediated weak-light movement in *Mougeotia*.

The same conclusions were drawn from experiments with microbeams which are presented in Fig. 27 II, A and B: The chloroplasts do not avoid such regions of a cell

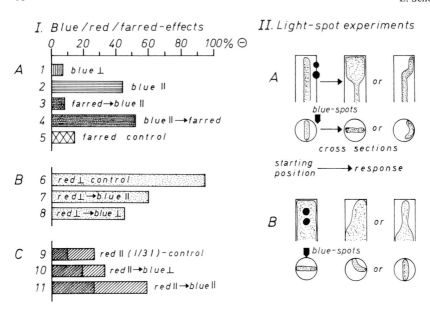

Fig. 27. Weak-light movement and its dependence on the inductive irradiations (red or blue) and the effect of far red. (After Haupt 1971)

which were irradiated by a small blue light spot, but rather seek these regions. These results also demonstrate that blue light does not have a supposed effect like far red. Thus it was postulated by Haupt (1971) that not phytochrome, but a flavin should be the more likely photoreceptor in the blue light-induced weak-light movement in *Mougeotia*. Hartmann and Cohnen-Unser (1973) reexamined the data of Haupt in a theoretical paper and concluded that the data of Haupt are not consistent with flavin, but with phytochrome as a photoreceptor for the blue light-induced weak-light movement. It is suggested that P_{fr} is also an effector molecule for the blue-UV-mediated photoresponses, if one considers that in phytochrome actions the stationary P_{fr} concentration interacts with the formation rate of P_{fr}.

Haupt and Wachter (1980) report on some new results on the blue light problem in weak-light movement: After nonsaturating unilaterally given weak blue light, the edge of a chloroplast in profile facing the light starts with movement, that means that the chloroplast avoids the front, because more blue light is absorbed at the front than at the rear (front-to-rear gradient). After saturating inductions, however, both edges of a chloroplast simultaneously start not only in red light but also in blue light-induced weak-light movements. These results demonstrate, according to Haupt, that in blue light the chloroplast not only orients in a "front-to-rear" gradient (bipolar) but also in a "front/rear to flanks" gradient (= tetrapolar gradient), as is known from the red light-induced chloroplast movement, but also from strong blue light movement. But the question remains how parallel vibrating blue light can establish a "front-rear-to-flanks" gradient by light absorption.

At this point, I think, the results obtained by Brühl (unpublished) with polarized blue light are very interesting: The curves given in Fig. 28 demonstrate that only in low

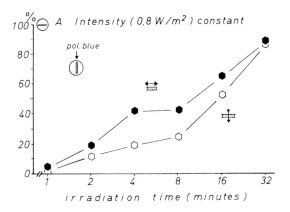

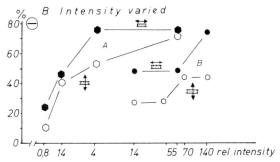

Fig. 28. Weak blue light movement in dependence on varied irradiation times (**A**) or varied intensities (**B**). (Brühl, unpublished)

energy range is parallel vibrating blue light much more effective than perpendicularly vibrating blue. At the curves which give the responses for parallel blue there are three parts discernible. Between the first and the second inclination we see a plateau reaching over a wide range. The angles of inclination of the first and second increasing part of these curves seem to be identical. It is concluded that one and the same process is regulated by two different mechanisms which have different thresholds. Brühl supposes that parallel vibrating polarized blue light of low intensity induces red chlorophyll fluorescence. This red fluorescence light is an intracellular red light source and transforms P_r to P_{fr} only at the front; so a sharp "front-to-rear" orientation P_{fr}-gradient is established, as shown schematically in Fig. 29A, B.

This hypothesis could be proved by experiments with DCMU which is a strong inhibitor of photosystem II. DCMU-incubation causes a visible high increase of red chlorophyll fluorescence in weak blue light. Parallel with this increase of fluorescence the blue light-dependent chloroplast movement could be enhanced (Fig. 29D) up to about 400% in relation to the controls without DCMU. The action dichroism in weak polarized blue light seems to he caused by a red-difluorescence of a *Mougeotia* chloroplast being preoriented to profile position.

The question is open, whether the response shown by the second increase of the reaction curves is controlled by phytochrome or by a specific blue light photoreceptor. As indicated in Fig. 30, I, not only parallel (Fig. 30A) but also perpendicularly vibrating polarized weak blue light (Fig. 30B) induces saturating weak-light responses. Far red

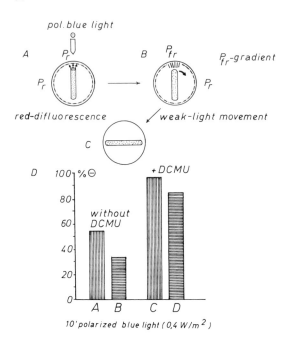

Fig. 29. Experiments on the function of red chlorophyll fluorescence as an intracellular red light source establishing a P_{fr}-gradient. (After Brühl, unpublished)

cancels a blue light induction only if polarized blue is given perpendicularly vibrating (Fig. 30E) but not if irradiated parallel vibrating (Fig. 30B). Polarized far red light is only effective in inhibiting this blue light induction, if polarized far red is vibrating parallel (Fig. 30E), but not if vibrating perpendicularly to the cell axis (Fig. 30F). An interpretation for this action dichroism of far red is given in Fig. 30 II. Perpendicularly vibrating blue establishes a "front/rear to flanks"-gradient (Fig. 30d). Parallel far red can be absorbed preferably at the front and the rear (Fig. 30e) and so P_{fr} is transformed to P_r. Thus the P_{fr}-gradient becomes weakened. In contrast to parallel vibrating far red, perpendicularly far red increases this gradient (Fig. 30f).

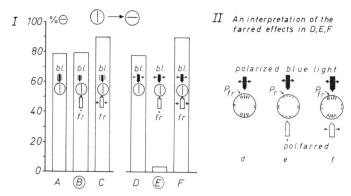

Fig. 30. The effect of polarized weak blue light and of simultaneously given polarized far red light on weak-light response. (After Brühl, unpublished)

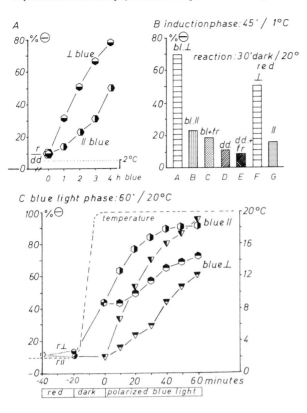

Fig. 31. The dependence of the sense of action dichroism in blue light-induced weak-light movement on the temperature given during the inductive phase. (After Schönbohm and Hellwig 1979)

At this state of the experiments we do not know whether parallel vibrating blue is absorbed by phytochrome or by a specific photoreceptor for blue.

I think the existence of the "blue-light syndrome" in weak blue-light movement of *Mougeotia* can be demonstrated typically by experiments, the results of which are given in Fig. 31: As you already have seen (Fig. 25B) in this review, weak-light movement is inhibited completely by temperatures of about 2°C in red light but not in blue light. In continuous blue the movement is only retarded by lowered temperatures (Fig. 25).

If you compare the action dichroism for the blue light-mediated weak light movement obtained by Haupt (1971) and by Brühl (unpublished) with the action dichroism for the blue light movement during a cold phase (2°C), we recognize that the action dichroism is converted during the cold phase (curves in Fig. 31A). This inverse action dichroism remains unchanged also in such experiments in which only the blue-light induction was given during the cold phase (Fig. 31 column A, B) whereas the chloroplasts move only at 20°C during a following dark phase. In these experiments a blue light induction could be canceled by far red given after blue during the cold phase (Fig. 31 column C). Further we found that polarized blue light given at 20°C after a red-cold or dark-cold period shows the well-known dichroism: Parallel vibrating polarized blue is again much more effective than perpendicularly vibrating (diagram C). For a detailed interpretation of these results further data are necessary.

3.3 The Effects of Blue Light on the Mechanism of Chloroplast Movement in *Mougeotia*

In former experiments we could show that in red light-induced chloroplast movement there are many phytochrome-dependent processes which show very similar kinetics: Only a few seconds after a short red light induction was given:

1. the chloroplasts begin to move,
2. the chloroplasts become fastened to a high degree in the cytoplasmic coating of the cell wall (Fig. 32A, 2),
3. in the zone between the edges of the chloroplasts and the cortical plasm numerous plasmic filaments become established (Schönbohm 1972; 1973a, b; Fig. 33 I),

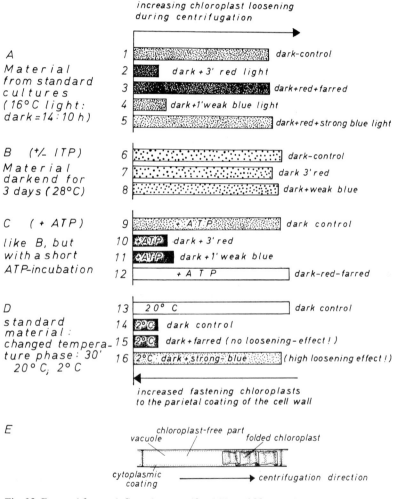

Fig. 32. External factors influencing centrifugability of *Mougeotia* chloroplasts in vivo. (After Schönbohm 1971c, 1972, and unpublished)

4. globular tannin-vesicles which also contain Ca-ions (Wagner, unpublished) are transported with a high speed if they contact these filaments (Schönbohm 1975).

Most of these processes depend on ATP (Fig. 32B, C) and can be reversed or stopped by a following far red phase; thus far red light increases the centrifugability of the chloroplasts (Schönbohm 1971c, 1972, 1973a; see also Fig. 32A, 3).

Quite similar effects can also be observed during the weak blue-light phase. In contrast to weak blue light which decreases centrifugability, strong blue light increases centrifugability (Schönbohm 1971c; Fig. 32A). As we know, centrifugability of chloroplasts also becomes decreased by lowering temperature. In contrast to decrease of centrifugability which is induced by red or by weak blue light, the decrease of centrifugability caused by low temperature cannot be canceled by far red light but by strong blue light (Fig. 32D).

Thus we see that one has to differentiate whether the chloroplasts are enchored within the cortical plasm of a cell by phytochrome-activated plasmic filaments, a process in which the transport of globular Ca-containing tannin-vesicles seems to be involved (see also Fig. 33), or by changed viscosity of the cytoplasm. It is known that blue light of high and of low intensities is very effective in changing plasm viscosity. A flavin is postulated as a responsible photoreceptor in these processes. We think that it might be probable that viscosity can also be changed by blue light at distinct small areas of a cell. Then the consequences might be an antagonistic change dependent on a possible dichroism. Thus it can be supposed that oriented chloroplast movement in *Mougeotia* becomes regulated not only by phytochrome- or flavin-activated contractile plasmic filaments with actomyosin character, but also by local changes of plasm viscosity.

I think it could be demonstrated by this review that the "blue light syndrome" exists indeed in blue light-induced intracellular movements of the four investigated fresh-water algae.

References

Dorscheid T, Wartenberg A (1966) Chlorophyll als Photoreceptor bei der Schwachlichtbewegung des *Mesotaenium*-Chloroplasten. Planta 70: 187–192

Fischer-Arnold G (1963) Untersuchungen über die Chloroplastenbewegung bei *Vaucheria sessilis*. Protoplasma 56: 495–520

Gärtner R (1970a) Die Bewegung des *Mesotaenium*-Chloroplasten im Starklichtbereich. I. Zeit- und Temperaturabhängige Sekundär-Prozesse. Z Pflanzenphysiol 63: 147–161

Gärtner R (1970b) II. Aktionsdichroismus und Wechselwirkungen des Photoreceptors mit Phytochrom. Ibid 63: 428–443

Hartmann KM, Cohnen-Unser I (1973) Carotenoids and flavins versus phytochrome as the controlling pigment for blue-UV-mediated photoresponses. Ibid 69: 109–124

Haupt W (1971) Schwachlichtbewegung des *Mougeotia* Chloroplasten im Blaulicht. Ibid 65: 248–265

Haupt W (1972) Localization of phytochrome within the cell. In: Mitrakos K, Shropshire W jr. (eds) Phytochrome, pp 553–569. Academic Press, London New York

Haupt W (1977) Bewegungsphysiologie der Pflanzen. Thieme, Stuttgart

Haupt W, Schönbohm E (1970) Light-oriented chloroplast movements. In: Halldal P (ed) Photobiology of microorganisms, pp 283–307. Wiley Interscience, London New York Sydney Toronto

Haupt W, Schönfeld I (1962) Über das Wirkungsspektrum der "Negativen Phototaxis" der *Vaucheria* Chloroplasten. Ber Dtsch Bot Ges 75: 14–23

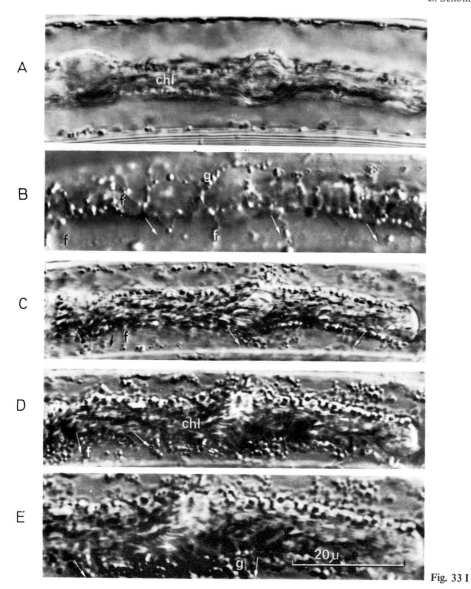

Fig. 33 I

Fig. 33. Series of micrographs of some details of *Mougeotia*-cells taken with interference-phase-contrast techniques during the first phases of weak-light movement (**I, B—E**) or during some phases of strong-light movement in blue (**II, B—E**). **I A** or **II A** demonstrate the chloroplasts (*chl*) in starting position, whereas **II F** shows the end of strong-light movement (profile). Plasmic fibrills (*f*) and rows of moving globular vesicles (*g*) are established only while movement takes place (cf. **II E** with **F**). (After Schönbohm, unpublished)

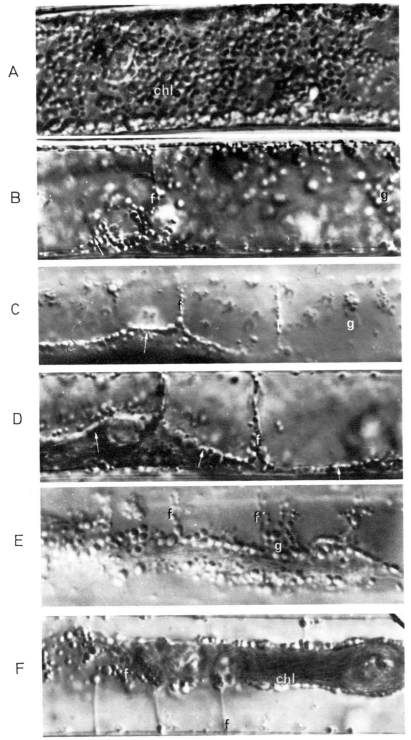

Fig. 33 II

Haupt W, Schönfeld I (1963) Die Wirkung von kurzwelliger Strahlung auf die Schwachlichtbewegung des *Mougeotia* Chloroplasten. Z Bot 51: 17–31

Haupt W, Thiele R (1961) Chloroplastenbewegung bei *Mesotaenium*. Planta 56: 388–401

Haupt W, Wachter W (1980) Steuerung der Chloroplastenbewegung von Mougeotia durch Absorp-
tionsgradienten in Rot- und Blaulicht. Z Pflanzenphysiol 96: 211–216

Schönbohm E (1963) Untersuchungen über die Starklichtbewegung des *Mougeotia* Chloroplasten.
Z Bot 51: 233–276

Schönbohm E (1965) Die Beeinflussung der negativen Photo-Taxis des *Mougeotia* Chloroplasten
durch linear polarisierte langwellige Strahlung. Z Pflanzenphysiol 53: 344–355

Schönbohm E (1966) Die Bedeutung des Phytochromsystems für die negative Phototaxis des *Mougeotia* Chloroplasten. Ber Dtsch Bot Ges 79: 131–138

Schönbohm E (1967) Die Hemmung der positiven und negativen Phototaxis des *Mougeotia* Chloroplasten durch Jodid-Ionen. Z Pflanzenphysiol 56: 366–374

Schönbohm E (1968) Aktionsdichroismus bei der Starklichtbewegung des Chloroplasten von *Mougeotia* spec. Ber Dtsch Bot Ges 81: 203–209

Schönbohm E (1971a) Über die Lokalisierung des Photorezeptors für den tonischen Blaulicht-Effekt
bei der Verlagerung des *Mougeotia*-Chloroplasten im Starklicht. Z Pflanzenphysiol 65: 453–457

Schönbohm E (1971b) Untersuchungen zum Photorezeptorproblem beim tonischen Blaulicht-Effekt
der Starklichtbewegung des *Mougeotia* Chloroplasten. Ibid 66: 20–33

Schönbohm E (1971c) Die Wirkung von SH-Blockern sowie von Licht und Dunkel auf die Verankerung des *Mougeotia* Chloroplasten im cytoplasmatischen Wandbelag. Ibid 66: 113–132

Schönbohm E (1972) Experiments on the mechanism of chloroplast movement in light-oriented
chloroplast arrangement. Acta Protozool 11: 211–223

Schönbohm E (1973a) Kontraktile Fibrillen als aktive Elemente bei der Mechanik der Chloroplastenverlagerung. Ber Dtsch Bot Ges 86: 407–422

Schönbohm E (1973b) Die lichtinduzierte Verankerung der Plastiden im cytoplasmatischen Wandbelag: Eine phytochromgesteuerte Kurzzeitreaktion. Ibid 86: 423–430

Schönbohm E (1975) Der Einfluß von Colchicin sowie von Cytochalasin B auf fädige Plasmastruk-
turen, auf die Verankerung der Chloroplasten sowie auf die orientierte Chloroplastenbewegung.
Ibid 88: 211–224

Schönbohm E, Hellwig H (1979) Zum Photorezeptor-Problem der Schwachlichtbewegung des *Mougeotia* Chloroplasten im Blau, bzw. Hellrot bei niederen Temperaturen. Ber Dtsch Bot Ges
92: 749–762

Schönbohm E, Schönbohm E, Lücke G (1979) Die Entstehung symmetrischer und asymmetrischer
Phytochrom-Gradienten bei *Mougeotia* und deren Bedeutung für die Chloroplastenorientierung.
Ibid 92: 297–304

Scholz A (1976a) Lichtorientierte Chloroplastenbewegung bei *Hormidium* flaccidum: Perzeption
der Lichtrichtung mittels Sammellinsen-Effekt. Z Pflanzenphysiol 77: 406–421

Scholz A (1976b) Lichtorientierte Chloroplastenbewegung bei *Hormidium* flaccidum: Verschiedene
Methoden der Lichtrichtungsperzeption und die wirksamen Pigmente. Ibid 77: 422–436

Senn G (1908) Die Gestalts- und Lageveränderungen der Pflanzenchromatophoren. Wilhelm Engelmann, Leipzig

Weisenseel M (1968) Vergleichende Untersuchungen zum Einfluß der Temperatur auf die lichtinduzierte Chloroplastenverlagerung. Z Pflanzenphysiol 59: 56–69

Interaction Between Blue Light and Phytochrome in Photomorphogenesis

H. MOHR[1]

1 Introduction

As far as we know at present photomorphogenesis (including light-induced carotenoid synthesis) in fungi is exclusively mediated by a photoreceptor pigment which absorbs only in the blue and near UV part of the electromagnetic spectrum ("blue-UV photoreceptor") (Rau 1975; Presti and Delbrück 1978). This blue-UV photoreceptor is probably identical with the ubiquitous blue-UV photoreceptor involved in the phototropic response (Shropshire 1974). Isolated reports about the involvement of phytochrome in light control of fungal development, e.g., Schneider and Murray (1979), require confirmation. Trials to demonstrate an interaction between blue light effects and phytochrome in carotenogenesis in *Fusarium aquaeductuum* were without success (W. Rau, pers. comm.).

On the other hand, photomorphogenesis (including chlorophyll and carotenoid biogenesis) in the mustard seedling (*Sinapsis alba*) seems to be exclusively under the control of phytochrome (Steinitz et al. 1976; Wildermann et al. 1978). So far no specific blue light effect was detected in young seedlings of this species, except phototropism. ("Specific" means that the blue light effect cannot be attributed to phytochrome.)

These are the extremes. In many plants phytochrome as well as the blue light photoreceptor operate, whereby the mode of cooperation (or, interaction) differs conspicuously among species. The following categories have been found useful in describing the phenomena: Obligatory sequential interaction between blue light and light operating through phytochrome; facultative sequential interaction between blue light and light operating through phytochrome; summative behavior of blue light effect and phytochrome-mediated effect; predominant operation of phytochrome with some blue light effect; predominant operation of the blue light photoreceptor with some phytochrome effect.

2 Obligatory Sequential Interaction Between Blue-UV Light and Light Operating Through Phytochrome

Example: anthocyanin synthesis in the mesocotyl = first internode of milo *Sorghum vulgare.*

---

1 Biologisches Institut II der Universität, Lehrstuhl für Botanik, Schänzlestraße 1, 78 Freiburg, FRG

Accumulation of anthocyanin (predominantly red anthocyanin previously identified as an acylated cyanidin-3-glucoside, see Stafford 1965, 1966) was measured in the
first internode (mesocotyl) of the milo seedling (Drumm and Mohr 1978). Without light
no formation of the red anthocyanin was observed in this organ. The principal results
we obtained are summarized in Table 1: If only light above 500 nm is applied, no anthocyanin synthesis takes place. Phytochrome (P_{fr}) can only act once a blue-UV light
effect has occurred. On the other hand, the expression of the blue-UV light effect (action peaks in the blue and in the near UV around 370 nm) is controlled by P_{fr}. It was
concluded that there is an obligatory dependency (or sequential interaction) between
the blue-UV light and the light operating through phytochrome. In dichromatic experiments, i.e., simultaneous irradiation with two kinds of light to establish very different
levels of P_{fr}, it was shown that the blue-UV mediated photoreaction per se is not affected
by the presence or almost absence of P_{fr}. The data obtained this way substantiate the
conclusion that the initial action of the blue-UV light is totally independent of phytochrome even though the expression of the blue-UV light effect is controlled by the
level of P_{fr}.

Table 1. Induction (or lack of induction) of anthocyanin synthesis in the mesocotyl of milo seedlings
(*Sorghum vulgare*) by light of different qualities (white light: Xenon light, similar to sunlight, 250
Wm^{-2}; white light ($>$305 nm): Short wavelenght UV below 305 nm eliminated by cut-off filters;
white light ($>$ 395): Long wavelength UV below 395 nm eliminated by cut-off filters; blue-UV light:
Fluorescent tubes, Osram L 40/73, λ_{max} at 366 nm, half bandwidth 40 nm, fluence rate 3.6 Wm^{-2}).
In the case of a 3-h light rreatment the seedlings were kept in the dark for 24 h before extraction
of anthocyanin. (After Drumm and Mohr 1978)

Light treatment (onset, 60 h after sowing)	Amount of anthocyanin (measurement, 87 h after sowing) (relative units)
27 h dark	0
27 h white light	185
27 h red light	0
27 h far-red light	0
3 h white light	19
3 h white light ($>$305 nm)	16
3 h white light ($>$395 nm)	11
3 blue-UV	19
3 h white light + 5 min red light[a]	19
3 h white light + 5 min 756 nm-light[a]	6
3 h white light + 5 min 756 nm-light + 5 min red light	20
3 h blue-UV + 5 min red light	19
3 h blue-UV + 5 min 756 nm-light	5
3 h blue-UV + 5 min 756 nm-light + 5 min red light	19

[a] Photoequilibria of the phytochrome are of the order of $\phi_{red} \approx 0.8$ and $\phi_{756} < 0.01$ (see Schäfer
et al. 1975). A 5 min light pulse suffices under the present circumstances to virtually establish the
photoequilibrium $\phi_\lambda = \dfrac{[P_{fr}]\lambda}{[P_{tot}]}$. For details of light sources see Table 2

3 Facultative Sequential Interaction Between Blue Light and Light Operating Through Phytochrome

Example: longitudinal growth of the mesocotyl of milo, *Sorghum vulgare.*

Table 2. Effect of red and blue light pretreatments on the effectiveness of phytochrome (P_{fr}) in controlling longitudinal growth of the mesocotyl of milo, *Sorghum vulgare.* (Data obtained by H. Drumm)

Pretreatment	Increase of mesocotyl length within 24 h (54–78 h after sowing) in darkness	
	[mm]	[%]
54 h dark	13.8	100
54 h dark + 5 min red$_{10}^{a}$	9.3	67
54 h dark + 5 min 756 nm light^{b}	12.8	93
54 h dark + 5 min red$_{10}$ + 5 min 756 nm light	13.2	96
48 h dark + 6 h red$_{1}^{a}$ + 5 min red$_{10}$ + 5 min 756 nm light	10.5	96
48 h dark + 6 h red$_{1}$ + 5 min red$_{10}$	5.3	48
48 h dark + 6 h blue^{a} + 5 min red$_{10}$ + 5 min 756 nm light	8.1	96
48 h dark + 6 h blue + 5 min red$_{10}$	1.5	18
48 h dark + 6 h blue + 5 min 756 nm light	8.0	93
48 h dark + 6 h far-red^{a} light + 5 min red$_{10}$ + 5 min 756 nm light	13.4	96
48 h dark + 6 h far-red light + 5 min red$_{10}$	7.4	53
48 h dark + 6 h far-red light + 5 min 756 nm light	13.9	99

[a] Light fields: Blue, λ_m at 450 nm, half bandwidth $\approx$45 nm, fluence rate 7 Wm^{-2}; red$_{10}$, λ_m at 658 nm, half bandwidth $\approx$25 nm, fluence rate 6.8 Wm^{-2}; red$_1$, 0.68 Wm^{-2}; far-red, λ_m at 765 nm, half bandwidth 120 nm, 3.5 Wm^{-2} (for details regarding light fields see Schäfer 1977)
[b] Obtained with an AL interference filter, fluence rate 7 Wm^{-2}

Growth rate of the milo mesocotyl responds slightly to a saturating red light pulse, i.e., to phytochrome (P_{fr}) without pretreatment (Table 2). To this extent the light growth response of the mesocotyl differs conspicuously from the light-mediated anthocyanin synthesis of the same organ (see Sect. 2). However, a blue light pretreatment strongly increases the responsiveness towards P_{fr} whereas red and far-red light pretreatments are much less effective. Without light pretreatment a saturating red light pulse reduces the growth rate by approximately 33%, following a 6 h blue light pretreatment the inhibitory effect of the red light pulse is of the order of 83%.

This kind of interaction between blue and red light treatments seems to be quite common. As an example, in dark-grown buckwheat seedlings (*Fagopyrum esculentum*) induction of anthocyanin synthesis can hardly be achieved by red light pulses only; there is an inductive effect via phytochrome but it is very small (Mohr and van Nes 1963). However, if a 3 h white light treatment is given prior to red or far-red light pulses, a very much higher responsiveness towards changes of the P_{fr}-level can be demonstrated

(Scherf and Zenk 1967). A rough action spectrum of the potentiating light effect shows its peak in the blue range even though there is considerable action in the far-red range as well (Mohr and van Nes 1963).

4 Summative Behavior

Example: Betalain synthesis in seedlings of *Amaranthus caudatus*, var. *viridis*.

Betalain synthesis in the genus *Amaranthus* is controlled by phytochrome and in some way by a blue light-dependent photoreaction (Köhler 1972; Colomas and Bulard 1975). The mode of interaction, if any, between the two photoreceptors is not clear so far. The present account is a preliminary report about some pertinent results obtained in our laboratory by V.K. Kochhar (unpublished). It was found (Table 3) that a 3-h induction period with blue light exerts a strong induction effect. The considerable effect of an additional 5 min red light pulse (given immediately after the blue light induction period) was completely reversible by a 5 min 756 nm light pulse, and the amount of pigment synthesized after a treatment with 3 h blue light plus 5 min red light was about equal to the added amounts induced by the blue and the red lights given separately (about 150). This result suggested that phytochrome and the blue light photoreceptor operate independently of each other under these experimental conditions. However, an interaction was indicated by the fact that a 5 or 20 min 756 nm light pulse (7 Wm^{-2}) given at the end of 3 h of high fluence rate blue light is hardly effective even though one expects that the level of P_{fr} is strongly decreased by the 756 nm light pulse.

If high fluence rate white light (17,000 lx) is used instead of blue light the extent of the reversible response $[(3 \text{ h wl} + 5 \text{ min red}_{10}) - (3 \text{ h wl} + 5 \text{ min 756 nm light})]$ is indeed increased, but only slightly (Table 4). On the other hand, if the fluence rate of

Table 3. Effect of blue, red and 756 nm light on betalain synthesis in seedlings of *Amaranthus caudatus*, var. *viridis*. For details about the light fields, see legend to Table 2. (Data obtained by V.K. Kochhar)

Treatment (onset, 24 h after sowing)	Amount of betalain (measurement, 51 h after sowing) (A at 545 nm)
3 h dark[a]	97
3 h dark + 5 min red$_{10}$	151
3 h dark + 5 min red$_{10}$ + 5 min 756 nm light	113
3 h dark + 5 min 756 nm light	113
3 h blue	203
3 h blue + 5 min red$_{10}$	251
3 h blue + 5 min red$_{10}$ + 5 min 756 nm light	199
3 h blue + 5 min 756 nm light	195
3 h red	165
3 h red + 5 min 756 nm light	136

[a] The dark level remains constant between 24 and 51 h after sowing.

Table 4. Effect of white light (wl, 17,000 lx), red light and 756 nm light on betalain synthesis in seedlings of *Amaranthus candatus,* var. *viridis.* The white light was obtained from a xenon arc (Osram XQO, 10 KW). After filtering through heat absorbing glass (KG 1, 3 mm) and 6 mm of Thermopane glass the spectral distribution of the light is very similar to sunlight. (Data obtained by V.K. Kochhar)

Treatment (onset 24 h after sowing	Amount of betalain (measurement, 51 h after sowing) (A at 545 nm)
3 h dark[a]	97
3 h wl	244
3 h wl + 5 min red_{10}	263
3 h wl + 5 min red_{10} + 5 min 756 nm light	188
3 h wl + 5 min 756 nm light	188
3 h (blue[b] + red, given simultaneously)	209
3 h (blue + red) + 5 min red_{10}	220
3 h (blue + red) + 5 min red_{10} + 5 min 756 nm light	174
3 h (blue + red) + 5 min 756 nm light	178
3 h blue	198
3 h blue + 5 min 756 nm light	189

[a] The dark level remains constant between 24 and 51 h after sowing

[b] For technical reasons the blue light field could only be used at a fluence rate of 3.1 Wm^{-2}

blue light is decreased in an experiment with simultaneously applied red and blue light, the extent of the reversible response is slightly decreased (Table 4). Thus, it seems that the extent of the blue light effect determines the extent of the potential phytochrome response, but the interaction is not strong.

A second example for independent action of two separate photoreceptors in a growth response is control of hypocotyl lengthening in partly de-etiolated seedlings of *Cucumis sativus* (Gaba and Black 1979). It was found that blue light inhibits virtually only during irradiation, whereas phytochrome-mediated red light inhibition has a delayed onset and recovery. That blue light does not operate through phytochrome was shown by kinetic studies and further substantiated by the inability of simultaneously applied far-red light to alter either the extent or the timing of blue light inhibition even though the ratio P_{fr}/P_{tot} ($= \phi_\lambda$, see Table 1) was modulated from 0.44 down to 0.14. In blue light there was no correlation between ϕ and the kinetics of photoinhibition. It seems that the P_{fr} established by blue light does not operate to any detectable extent in the presence of high fluence rate blue light. In white light, however, which contains a considerable amount of red light and establishes a relatively high ϕ-value (approximately 0.70) the simultaneous operation of both phytochrome and the blue light photoreceptor is clearly indicated by the time course of the inhibition.

In the preceding two examples blue light and red light were of the same order of effectiveness with regard to a particular photoresponse. We now consider two examples where the operation of either blue light photoreceptor or phytochrome is predominant, while the simultaneous involvement of the second photoreceptor is only detectable if particular experimental precautions are taken.

5 Predominant Operation of Phytochrome with Some Blue Light Effect

Example: Chlorophyll synthesis in the shoot of milo, *Sorghum vulgare*.

As shown above (Sect. 2) two nonphotosynthetic photoreceptors (phytochrome and the blue-UV photoreceptor) are involved in light-mediated anthocyanin synthesis in the mesocotyl of the milo seedling. The report in this paragraph deals with the control by light of chlorophyll a and b synthesis in the shoot of the milo seedling. So far, no clear picture exists with regard to the details of the light control of chlorophyll synthesis in monocotyledonous plants (Virgin 1961; Raven and Spruit 1972). In particular, we posed the question of whether the dominating effect of blue-UV light (as observed in anthocyanin synthesis) is also detectable in light-mediated chlorophyll (Chl) synthesis of the milo seedling.

Background information: If milo seedlings grown for 60 h in the dark (25°C) are placed in continuous fluorescent white light, a light flux as low as 100 lx saturates the protochlorophyll → chlorophyllide-a photoconversion. Chl-b becomes only detectable approximately 2 h after the onset of white light (lag-phase) provided that the seedling has not received any light treatment before 60 h after sowing. Pretreatment of the milo seedlings with light pulses (e.g., at 48 h after sowing) eliminates the lag-phase of Chl-b synthesis in continuous white light (saturating the PChl → Chl-a and Chl-b photoconversion) and strongly increases the rates of Chl-a and Chl-b accumulation. The relative effect of a light pretreatment on Chl-b formation is considerably stronger than on Chl-a formation.

Results directly relevant for our particular problem:

1. The potentiating effect of red or blue light pulses, applied at 48 h, on chlorophyll accumulation in saturating white light between 60 and 64 h after sowing is fully reversible by a saturating 756 nm light pulse (7 Wm^{-2}) indicating that the effect of a

Table 5. The potentiating effect of light pulses, applied at 48 h after sowing, on chlorophyll accumulation in saturating white light (100 lx) between 60 and 64 h after sowing. (Data obtained by S. Sawhney)

Light treatment (at 48 h after sowing)	Chlorophyll accumulated between 60 and 64 h after sowing [pmol · $shoot^{-1}$]	
	Chl-a	Chl-b
Dark control	150	37
4 min red[a] light	329	65
4 min red light + 7.5 min 756 nm[a] light	208	39
7.5 min 756 nm light	209	39
4 min blue [a] light	322	65
4 min blue light + 7.5 min 756 nm light	209	38
30 min red light	344	84
30 min red light + 7.5 min 756 nm light	209	40
30 min blue light	349	85
30 min blue light + 7.5 min 756 nm light	209	40

[a] Light sources were described in Table 2

pulse pretreatment is due to phytochrome exclusively (Table 5). There is no detectable escape from reversibility up to approximately 40 min after the onset of the potentiating light.

2. The potentiating effect of dichromatic light pulses (30 min of red and blue lights applied together, yielding "white light") is not fully reversible by a saturating 756 nm light pulse (7 Wm^{-2}) (Table 6), in contrast to the full reversibility of the red or blue pulse treatments applied singly (see Table 5). This fact indicates the operation of a blue light photoreceptor, although only in collaboration with phytochrome.

Table 6. "Superinduction" by dichromatic irradiation, i.e., red and blue light beams applied simultaneously, and lack of full reversibility of the induction effect of dichromatic irradiation by 756 nm light. (Data obtained by S. Sawhney)

Licht treatment (onset 48 h after sowing)	Chlorophyll accumulated between 60 and 64 h after sowing [pmol · shoot^{-1}]	
	Chl-a	Chl-b
Dark control	150	37
4 min red light	329	65
4 min blue light	322	65
4 min (red + blue) lights	335	70
30 min red light	344	84
30 min blue light	349	85
30 min (red + blue) lights	418	90
12 h red light	630	141
12 h blue light	621	135
12 h (red + blue) lights	726	167
4 min (red + blue) lights	335	69
4 min (red + blue) lights + 7.5 min 756 nm light	216	40
30 min (red + blue) lights	418	90
30 min (red + blue) lights + 7.5 min 756 nm light	249	54
7.5 min 756 nm light	209	39

3. Dichromatic irradiation (red and blue lights applied together, yielding "white light") applied at 48 h as a pulse (30 min) or given as long-term exposure (for 12 h) promotes chlorophyll accumulation in saturating white light between 60 and 64 h after sowing considerably more than one would expect from the effects exerted by single saturating pulses via phytochrome (Table 6), in particular since the level of P_{fr} established by simultaneously applied red and blue light must be expected to be somewhat lower than the P_{fr} level in red light. The ϕ_{blue} value (see Table 1) is of the order of 30% only.

The results (Tables 5 and 6) have so far led us to conclude that some specific blue light effect exists which becomes apparent only in the presence of considerable red light fluxes which establish a high level of P_{fr} in the photoequilibrium (or photo steady state, respectively) of the phytochrome system. An alternative explanation which is not ex-

cluded at present is that "superinduction" and "lack of full reversibility" in dichromatic light are due to the operation of a "High Irradiance Reaction" of the phytochrome
system (see Schäfer 1975), which comes into play at higher fluence rates. Irrespective
of the actual existence of a specific blue light effect, it is clear that Chl-a and Chl-b
synthesis in *Sorghum vulgare* are predominantly under the control of phytochrome.
Thus, light control of anthocyanin synthesis (see Sect. 2) and light control of chlorophyll synthesis (this Section) in the same system are totally different.

6 Predominant Operation of the Blue-UV Photoreceptor with Some Phytochrome Effect

Example: longitudinal growth of the protonema of the fern *Dryopteris filix-mas.*

6.1 Germination of Spores

Like most fern spores the gonospores of *Dryopteris filix-mas* are not able to germinate
in the dark. They need some light to be able to germinate. This obligatory light requirement can be satisfied by a relatively brief period of illumination of the fully imbibed,
turgid spores. That is, the germination can be induced by light. If the spores are placed
in darkness after the light treatment, complete germination occurs without any further
supply of light. It was shown previously (Mohr 1956a; Sugai and Furuya 1967) that induction by light of fern spore germination (*Dryopteris filix-mas, Pteris vittata*) is mediated through phytochrome, i.e., is due to the formation of P_{fr}. Besides the operation
of the phytochrome system, however, a specific inhibitory effect of blue light was noticed with peaks of action around 480, 440, and 380 nm (Mohr 1956a; Sugai 1971)
strongly suggesting the involvement of the blue-UV photoreceptor.

6.2 Growth of the Protonema

It was noticed more than a century ago (Borodin 1867) that light controls the development of fern gametophytes which originate from the germinating gonospores. The sporelings can in fact develop normally from the transient filamentous stage, called the
"protonema", to the two- or three-dimensional stage, called the "prothallus" only if
they receive enough short-wavelengths visible light below 500 nm. Under long-wavelength visible light, e.g., red or far-red light of low or higher intensities, the sporelings
continue to grow as cellular filaments (protonemata) which are similar to the filaments
of the dark controls (Fig. 1). A characteristic of protonematal growth is that synthesis
of the new cell wall takes place only at the very tip ("apical growth") and that the actual growing point at the apex can be shifted by light leading to a phototropic (or polar-
tropic) response (Mohr 1956b; Etzold 1965). Both photoreceptors, phytochrome as
well as the blue-UV photoreceptor, can mediate the phototropic (or polarotropic) response (Steiner 1967). The pertinent photoreceptor molecules are very probably located
in a dichroic layer within or close to the plasma membrane (Etzold 1965).

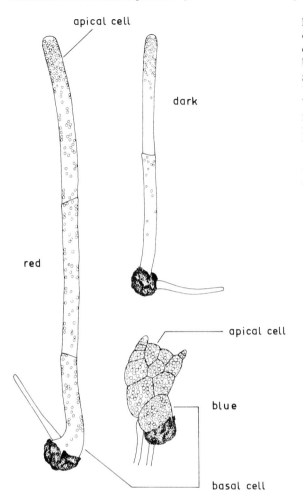

apical cell

dark

red

apical cell

blue

basal cell

Fig. 1. Representative sporelings of *Dryopteris filix-mas* grown in darkness, continuous red light, or blue light. The sporelings were grown for 6 days after germination on inorganic nutrient solution. The lights were adjusted so as to permit about the same rate of photosynthesis. Correspondingly, the blue-grown and the red-grown sporelings have about the same dry mass. (After Mohr and Ohlenroth 1962)

A convenient gauge of the photomorphogenetic effect on the fern gametophytes has been the "morphogenetic index", i.e., the length of the protonema (or prothallus) divided by its maximum width. This index can be applied in quantitative studies about the photoreceptors involved because the morphogenetic effect of the blue light, i.e., the causation of the two-dimentional growth, expresses itself in a very conspicuous inhibition of the longitudinal growth of the protonema even before the transition to two-dimensional growth takes place.

The action spectrum (Fig. 2) — morphogenetic index as a function of wavelength — shows that the morphogenetic effectiveness is very high below 500 nm. This is taken as an indication that the blue-UV photoreceptor dominates in the light control of development of fern gametophytes. However, the additional structures of the action spectrum which appear at longer wavelengths (> 500 nm) indicate the involvement of phytochrome as well. A straightforward interpretation is that light between 500 and 750 nm exerts a positive effect on protonematal growth. Far-red light (700–750 nm) is the

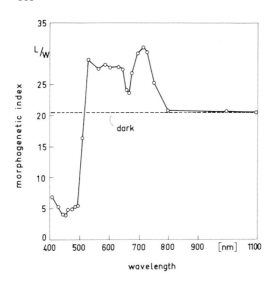

Fig. 2. The morphogenetic index L/W (for definition see text) of the protonema of the fern *Dryopteris filixmas* as a function of wavelength. Notice that a low morphogenetic index is the result of a high morphogenetic effectiveness of the light. Measurements were made after 6 days' culture in continuous monochromatic light at an energy fluence rate of 0.2 Wm^{-2}. Under these conditions and within this period of time the sporelings remained filamentous throughout, even under blue light. (After Mohr 1956b)

most effective. The promotive effect of red light has recently been studied by direct observations at the apex of the protonema in *Adiantum capillus-veneris* by Kadota (quoted after Furuya 1978). We think that this promotive effect of light is due to the formation of relatively small amounts of P_{fr}. Increasing levels of P_{fr} lead then to a relative inhibition of protonematal growth, resulting in the considerable peak of action which appears as a dip in the red part of the action spectrum (Fig. 2).

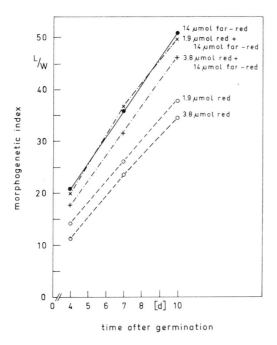

Fig. 3. Results of experiments with dichromatic irradiation: The effect of red light on the morphogenetic index (see Fig. 2) of a fern protonema (*Dryopteris filix-mas*) can be nullified by the simultaneous application of far-red light at a suitable photon fluence fluence rate (the photon fluence rate is given in $\mu mol \cdot m^{-2} \cdot s^{-1}$). (After Schnarrenberger and Mohr 1967)

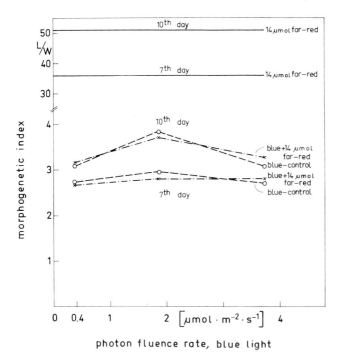

Fig. 4. Results of experiments with dichromatic irradiation: The effect of blue light on the morphogenetic index (see Fig. 2) of a fern protonema (*Dryopteris filix-mas*) cannot be influenced significantly by the simultaneous application of far-red light (the photon fluence rate is always given in μmol $\cdot$ m^{-2} $\cdot$ s^{-1}). (After Schnarrenberger and Mohr 1967)

The explanation of the peak of action in the red spectral part was checked in experiments using dichromatic irradiation. When the protonemata were irradiated simultaneously with red and with far-red light at suitable photon fluence rates ("intensities") the morphogenetic effect of the red light (lowering of the L/W index) is cancelled by the simultaneous far-red light (Fig. 3). On the other hand, the strong morphogenetic effect of the blue light does not show any significant interaction with simultaneously applied far-red light (Fig. 4). All data available from our experiments indicate that the effect of blue light is due to the blue-UV photoreceptor and totally independent of phytochrome. This conclusion received support by recent data obtained in Furuya's laboratory (Furuya 1978). It was found that a brief irradiation with blue light inhibited the apical growth in the protonema of *Adiantum capillus-veneris*. Since this blue light effect could not be modified by subsequently given red or far-red light, it was concluded that the blue light effect was not mediated by phytochrome.

Recently Cooke and Paolillo (1979) have replaced the morphogenetic index L/W by the cross-sectional area of the filament tip. Even though the choice of the cross-sectional area as a measure of the photocontrol over filamentous growth leads to a significant refinement of the argument, the major result remains unchanged, namely "that the pigments responsible for the control of filamentous growth should . . . .be identified as phytochrome and a blue-light-absorbing pigment."

7 Evolutionary Spectulations

It seems that the blue-UV photoreceptor has originated early in evolution. This photo-receptor is still detectable in all classes of plants and is being used as a sensor pigment in a multiplicity of responses. Phytochrome, on the other hand, has so far been detected only in organisms which have originated in the course of evolution from green flagel-lates. Occasional reports about functional phytochrome in fungi, red algae, or brown algae do not meet the critical standards established in phytochrome research; in par-ticular crucial experiments with dichromatic irradiation are lacking.

On the other hand, phytochrome was clearly demonstrated, at least operationally (see Mohr 1972) in all major groups of green plants from *Chlamydomonas* to Douglas fir and duck weed. However, in comparing different groups of green plants we find that the function of phytochrome differs greatly. It is obvious that a conspicuous change of function has occurred during the evolution of green plants. As an example, in green algae phytochrome operates as a sensor pigment for optimum modulation of already established cell functions [e.g., orientation of the chloroplast in *Mougeotia* (Haupt 1973)]; in fern gametophytes, phytochrome is the sensor pigment for spore germination and protonematal phototropism; in the angiosperms, phytochrome's major function is the mediation of photomorphogenesis (= normal development of a seedling) and control of growth in older, green plants. In addition phytochrome is the sensor pigment for photoperiodic responses. There is only an indirect relationship to phototropism, the sensor pigment of which in higher plants is always the blue-UV photoreceptor. As far as the relative importance of phytochrome and blue-UV photoreceptor is concerned, it is obvious that in the course of evolution in higher plants phytochrome became more and more predominant while the significance of the blue-UV photoreceptor decreased (except in phototropism).

Acknowledgments. Supported by Deutsche Forschungsgemeinschaft (SFB 46). I thank Drs. H. Drumm, V.K. Kochhar and S. Sawhney for the permission to use unpublished data. I am grateful to Drs. H. Drumm and E. Schäfer for valuable discussions.

References

Borodin J (1867) Über die Wirkung des Lichts auf einige höhere Kryptogamen. Bull Acad Imp Sci St Petersbourg 12: 432–448
Colomas J, Bulard O (1975) Irradiations à faible énergie et biosynthèse d'amarantine chez des plan-tules d'Amaranthus tricolor L. var. bicolor ruber Hort. Planta 124: 245–254
Cooke TJ, Paolillo DJ (1979) The photobiology of fern gametophytes. II. The photocontrol of filamentous growth and its implications for the photocontrol of the transition to two-dimensional growth. Am J Bot 66: 376–385
Drumm H, Mohr H (1978) The mode of interaction between blue (UV) light photoreceptor and phytochrome in anthocyanin formation of the Sorghum seedling. Photochem Photobiol 27: 241–248
Etzold H (1965) Der Polarotropismus und Phototropismus der Chloronemen von Dryopteris filix-mas (L.) Schott. Planta 64: 254–280
Furuya M (1978) Photocontrol of developmental processes in fern gametophytes. Bot Mag Tokyo Spec Issue 1: 219–242

Gaba V, Black M (1979) Two separate photoreceptors control hypocotyl growth in green seedlings. Nature (London) 278: 51–54

Haupt W (1973) Role of light in chloroplast movement. BioScience 23: 289–296

Köhler KH (1972) Photocontrol of betacyanin synthesis in Amaranthus caudatus seedlings in the presence of kinetin. Phytochemistry 11: 133–137

Mohr H (1956a) Die Beeinflussung der Keimung von Farnsporen durch Licht und andere Faktoren. Planta 46: 534–551

Mohr H (1956b) Die Abhängigkeit des Protonemawachstums und der Protonemapolarität bei Farnen von Licht. Planta 47: 127–158

Mohr H (1972) Lectures on photomorphogenesis. Springer, Berlin Heidelberg New York

Mohr H, van Nes E (1963) Der Einfluß sichtbarer Strahlung auf die Flavonoid-Synthese und Morphogenese der Buchweizen-Keimlinge (Fagopyrum esculentum Moench.) I. Synthese von Anthocyan. Z Bot 51: 1–16

Mohr H, Ohlenroth K (1962) Photosynthese und Photomorphogenese bei Farnvorkeimen von Dryopteris filix-mas. Planta 57: 656–664

Presti D, Delbrück M (1978) Photoreceptors for biosynthesis, energy storage and vision. Plant Cell Environ 1: 81–100

Rau W (1975) Zum Mechanismus der Photoregulation von Morphosen am Beispiel der Carotinoidsynthese. Ber Dtsch Bot Ges 88: 45–60

Raven CW, Spruit CJP (1972) Induction of rapid chlorophyll accumulation in dark grown seedlings. II. Photoreversibility. Acta Bot Neerl 21: 640–654

Schäfer E (1975) A new approach to explain the 'high irradiance response' of photomorphogenesis on the basis of phytochrome. J Math Biol 2: 41–56

Schäfer E (1977) Kunstlicht und Pflanzenzucht. In: Albrecht H (ed) Optische Strahlungsquellen, pp 249–266. Lexica-Verlag, Grafenau

Schäfer E, Lassig T-U, Schopfer P (1975) Photocontrol of phytochrome destruction in grass seedlings. The influence of wavelength and irradiance. Photochem Photobiol 22: 193–202

Scherf H, Zenk MH (1967) Induction of anthocyanin and phenylalanine amminia-lyase formation by a high energy light reaction and its control through the phytochrome system. Z Pflanzenphysiol 56: 203–206

Schnarrenberger K, Mohr H (1967) Die Wechselwirkung von Hellrot, Dunkelrot und Blaulicht bei der Photomorphogenese von Farngametophyten (Dryopteris filix-max (L.) Schott). Planta 75: 114–124

Schneider MJ, Murray BJ (1979) Phytochrome mediation of uredospore germination in the fungus Puccinia graminis. Photochem Photobiol 29: 151–152

Shropshire W (1974) Phototropism. In: Schenk GO (ed) Progress in photobiology, paper No 024. Deutsche Gesellschaft für Lichtforschung, Frankfurt

Stafford HA (1965) Flavonoids and related phenolic compounds produced in the first internode of Sorghum vulgare Pers. in darkness and in light. Plant Physiol 40: 130–138

Stafford HA (1966) Regulatory mechanisms in anthocyanin biosynthesis in first internodes of Sorghum vulgare: effect of presumed inhibitors of protein synthesis. Plant Physiol 41: 953–961

Steiner AM (1967) Action spectra for polarotropism in germlings of a fern and a liverwort. Naturwissenschaften 54: 497–498

Steinitz B, Drumm H, Mohr H (1976) The appearance of competence for phytochrome-mediated anthocyanin synthesis in the cotyledons of Sinapis alba L. Planta 130: 23–31

Sugai M (1971) Photomorphogenesis in Pteris vittata IV. Action spectra for inhibition of phytochrome-dependent spore germination. Plant Cell Physiol 12: 103–109

Sugai M, Furuya M (1967) Photomorphogenesis in Pteris vittata I. Phytochrome-mediated spore germination and blue light interaction. Plant Cell Physiol 8: 737–748

Virgin HI (1961) Action spectrum for the elimination of the lag phase in chlorophyll formation in previously dark grown leaves of wheat. Physiol Plant 14: 439–452

Wildermann A, Drumm H, Schäfer E, Mohr H (1978) Control by light of hypocotyl growth in deetiolated mustard seedlings. I. Phytochrome as the only photoreceptor pigment. Planta 141: 211–216

Sensory Transduction in Phycomyces Photoresponses*

E.D. LIPSON[1]

1 Introduction

Phycomyces blakesleeanus exhibits several quantifiable responses to blue light stimulation. The best characterized responses are those which occur in the mature sporangiophore, namely phototropism and the fundamentally related light-growth response. The sporangiophore is sensitive to a number of other stimulus modes, including gravity, wind, chemicals, and the presence of nearby objects (avoidance response). The mycelium is responsive to blue light in at least two ways, namely control of β-carotene biosynthesis and control of the initiation of sporangiophore development (see chapters by Cerdá-Olmedo and by Russo, this vol.). A comprehensive review on early work on *Phycomyces* is provided by Bergman et al. (1969). A more current review concentrating on behavioral genetics has been written by Cerdá-Olmedo (1977).

As with most blue light-sensitive systems, the identity of the photoreceptor chromophore has long been in question, primarily as a choice between β-carotene and riboflavin. In *Phycomyces* phototropism, the evidence has recently been accumulating in favor of riboflavin. Conversely β-carotene has been ruled out as the receptor pigment for phototropism.

2 Blue Light Responses

2.1 Phototropism and Light Growth Response

Elongation and bending of the sporangiophore are always confined to a growing zone about 3 mm in extent and located beneath the spherical sporangium. With symmetrical illumination or in darkness, the mature sporangiophore grows vertically at about 3 mm per hour. A change in ambient blue light intensity causes a transient change in the growth rate, known as the light-growth response (Foster and Lipson 1973). The long stationary growth phase of the sporangiophore and the availability of an automated tracking machine make this particular response convenient to study. The light-growth response exhibits many properties analogous to those in higher sensory systems including adaptation, rectification, logarithmic transduction and large operating range. System

* Presented at International Conference on the Effects of Blue Light in Plants and Microorganisms, Marburg, West Germany, July 1979
1 Department of Physics, Syracuse University, Syracuse, NY 13210, USA

analysis methods have been applied, primarily to the wild-type response, to help understand these phenomena (Lipson 1975). The method involves a randomized stimulus program for the blue light intensity. Computer analysis using correlation techniques between the stimulus and response permits the numerical evaluation of certain mathematical functions (called kernels) in the context of a theory by Wiener. The result is an accurate determination of the input-output relation for the light-growth response, including nonlinear properties. The kernels may be examined in detail using system analysis methods to model the underlying kinetics of the stimulus-response pathway. The combination of this approach with genetic analysis will be discussed below in Section 3.

The relation between the light-growth response and phototropism has been clarified in recent years (Dennison and Foster 1977; Medina and Cerdá-Olmedo 1977a). Phototropism can be explained largely as a consequence of (a) the properties of the light-growth response (time course, adaptation, latency), (b) the cylindrical lens properties of the sporangiophore, which cause a light beam to focus into a sharp band on the distal side, (c) the azimuthal rotation of the cell wall in the growing zone, in connection with the spiral growth pattern of the sporangiophore and (d) the assumptions that photoreceptors rotate along with the cell wall and that adaptation is confined locally. In this framework phototropism can be explained quantitatively as a summation of local light-growth responses mediated by a "carousel" of photoreceptors moving through the focused band of light.

2.2 Light-Induced Carotene Synthesis

Phycomyces normally has a vivid yellow color due to the presence of β-carotene. Powerful genetics methods have been applied to the problem of regulation of carotene biosynthesis in *Phycomyces* (reviewed by Cerdá-Olmedo and Torres-Martinez 1979). Carotene production may be perturbed by (a) blue light, (b) mutations in several genes (*car*AR, *car*B, *car*S), (c) chemicals (e.g., vitamin A), and (d) mating. The structure of the carotene synthetic pathway has been elucidated by quantitative complementation studies using the *car* mutants. The result is a simple enzyme aggregate model involving a dehydrogenase (*car*B product) and a cyclase (*car*R product) each occurring multiply in a given assembly.

In darkness, *Phycomyces* produces β-carotene at a level of about 50 μg/g (dry weight). Growth under continuous illumination (2 W m^{-2}, blue) increases the concentration about tenfold. Certain mutants with abnormal photoropism (see Sect. 3) are also abnormal for light-induced carotene synthesis. This finding can be explained by the assumption of a common photoreceptor for both processes. Recently mutants have been isolated which are abnormal specifically for light-induced carotene synthesis. These so-called *pic* mutants should be helpful in elucidating the pathway for light control of carotene synthesis (see Cerdá-Olmedo, this vol.).

Another approach has been applied recently to the study of light-induced carotene synthesis (Jayaram et al. 1979). Short-term rather than continuous irradiation was applied. Carotene levels were measured relatively by an in vivo spectrophotometric technique. Studies at various fluence levels indicated that light-induced carotene synthesis is under dual control The low-fluence component saturates at about 20 J m^{-2} while the high-fluence component saturates at about 1000 J m^{-2}. The high-fluence component

alone is sensitive to actinomycin D and cycloheximide, which are inhibitors of transcription and translation. Thus protein synthesis appears to be involved only in the high-
fluence component.

The *pic* mutants were found to be normal at low fluence but deficient at high fluence. These studies show that at least two processes are involved in light-induced carotene synthesis in *Phycomyces* and provide a hopeful basis for the elucidation of these
processes. It will be particularly interesting to determine just which early steps in light-
controlled carotene synthesis are in common with the input stages of the phototropism
channel (see Sect. 3).

3 Light-Induced Sporangiophore Initiation

Bergman (1972) discovered that blue light can control the development of sporangiophores. He exposed mycelia (growing in racing tubes or on petri dishes) to periodic light
programs, e.g., a cycle of 23 h dark and 1 h light. Under these conditions sporangiophores appeared in bands at positions where the mycelial boundary had been during
exposures. No evidence was found for an associated circadian rhythm. Rather light appeared to exert its control directly.

Galland and Russo (1979) have pursued this stimulus-response channel with a different protocol to allow better quantitation. Their procedure was to grow mycelia in
shell vials grouped together in air-tight glass beakers. These conditions inhibit sporangiophore production in the dark. They showed that blue light can reverse this inhibition
and carried out detailed fluence-response measurements for wild-type, *mad* mutants
and *car* mutants. Substantial increases in thresholds (i.e., reduction in sensitivity to blue
light) were found for mutants affected in genes *mad*A and *mad*B, as well as *car*A, *car*B
and *car*R. They interpreted these findings in terms of a model with a common photoreceptor with the phototropism channel (presumably a flavin). They inferred further
that β-carotene or a metabolite played an intermediate role in the pathway, rather than
acting as a secondary receptor. Further details and interpretations of these results may
be found in the contribution by Russo (this vol.).

4 Genetic Analysis of Phototropism Mutants

Mutants with abnormal phototropism are designated by the genotype symbol *mad*. Several hundred *mad* mutants have been isolated using a simple screening procedure (Bergman et al. 1973) after treatment of spores with a chemical mutagen, usually nitroso-
guanidine. A sample of 29 *mad* strains has been examined by phenotypic classification
and by complementation analysis (Bergman et al. 1973; Ootaki et al. 1974; Ootaki et
al. 1977). In the former work, the *mad* mutants were tested for a number of other stimulus-response modes, including avoidance, geotropism, light-induced carotene synthesis
and light-induced sporangiophore initiation. In the latter, the *mad* strains were sorted
into seven complementation groups (genes) denoted as *mad*A, *mad*B, . . ., *mad*G. The
complementation tests were performed using artificial heterkaryons produced by a
microsurgical technique (Ootaki 1973). The results of both of these studies are sum-

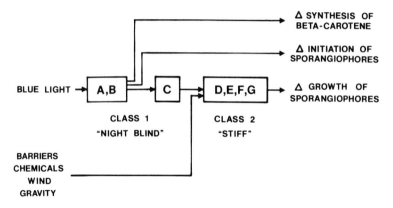

Fig. 1. Sensory pathways inferred from classification of *mad* mutants. The *letters in boxes* refer to the seven genes *mad*A to *mad*G. A mutation in a particular gene leads to abnormalities in all stimulus-response channels which pass through the box for that gene. The central pathway corresponds to phototropism and the light-growth response. (After Bergman et al. 1973)

marized in Fig. 1. Those *mad* mutants which are defective for all tropisms (viz. geotropism and avoidance) are designated as class 2 ("stiff"). Complementation tests have revealed that these class 2 mutants carry mutations in any of the four genes *mad*D, *mad*E, *mad*F or *mad*G. According to the scheme in Fig. 1, these four genes correspond to common output stages of several convergent sensory pathways leading to tropisms and growth responses.

The remaining *mad* mutants studied are normal for tropisms other than phototropism. They are designated phenotypically as class 1 ("night-blind"). Some of the class 1 *mad* strains, in fact those which carry mutations in genes *mad*A or *mad*B, are abnormal for several other blue light responses, namely light-induced carotene synthesis, and light-induced sporangiophore initiation. In these responses the blue light reception occurs in the mycelium rather than the sporangiophores. The fact that a mutation in *mad*A or *mad*B can simultaneously affect several light responses suggests that these responses share common steps at the beginning of their sensory pathways, including the photoreceptor. However it is important to keep in mind that one is comparing sensory pathways in different developmental stages, i.e., mycelium vs. sporangiophores.

Finally, there is one remaining subclass within class 1, corresponding to *mad*C. About half of the mutants studied fall in this group. These mutants appear normal for all stimulus-response channels examined, except for phototropism. Thus, as indicated in Fig. 1, they appear to be affected at an intermediate stage of the sensory pathway.

Additional genetic studies have revealed that the *mad* mutations are strongly recessive to wild type (Medina and Cerdá-Olmedo 1977b) and the genes are mutually unlinked (Eslava et al. 1976). The latter finding reinforces the results from the complementation tests that the seven different complementation groups identified correspond to different genes. Interestingly, none of the existing *mad* mutants, which were isolated after mutagenesis, are totally blind. At sufficiently high intensity all of these *mad* mutants exhibit phototropism. The residual phototropism may be quantitated conveniently by measurements of photogeotropism which are summarized in Fig. 2 (Bergman et

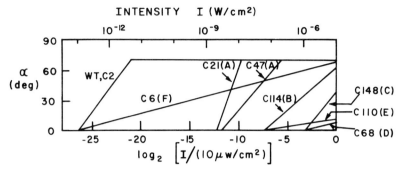

Fig. 2. Composite of threshold curves for *mad* mutants. Sporangiophores in shell vials were exposed for 6 h to a horizontal beam of blue light at the intensity shown on the abscissa. The ordinate α is the mean bending angle towards the light beam. The curve marked *WT,C2* pertains to the wild type and a phototropically normal albino strain. For the other curves, the *letters in parentheses,* after the strain numbers, refer to the *mad* genes as in Fig. 1. The curve for *mad*G (not shown) would be similar to that for C68 (*mad*D). (For original data see Bergman et al. 1973; Ootaki et al. 1974)

al. 1973; Ootaki et al. 1974). In this method one measures the bending angle α, measured from the vertical, of a sporangiophore toward a horizontal light source after a 6-h exposure. The measurements are recorded as a function of light intensity. The wild-type threshold occurs at approximately 10^{-13} W cm^{-2}. Above 10^{-10} W cm^{-2}, the bending reaches an angle of about $70°$ corresponding to an equilibrium between phototropism and geotropism. The thresholds of class 1 strains are shifted by many orders of magnitude. Above their thresholds they behave roughly like wild type. Thus these strains are termed "night-blind". The loss of sensitivity of one or more of the class 1 strains might be due to a corresponding reduction in the concentration of receptor pigment.

Class 2 mutants, to varying degrees, show a reduction in their responsiveness in that their threshold curves generally are shallow over their whole range. Although it is not evident in Fig. 2, their threshold curves seem to extrapolate to the wild-type threshold (for related evidence, see Lipson 1975). For both class 1 and class 2 strains the threshold data are consistent with the scheme of Fig. 1: class 1 strains affected near the input are less sensitive to light; class 2 strains, affected near the output are less responsive.

In my laboratory we have lately been involved in extending the genetic analysis of the *mad* mutants to "second-order" by constructing double *mad* mutants in all 21 combinations for the seven *mad* genes. These strains have been constructed by mating suitable single-mutant strains and then sorting out the progeny with the help of complementation tests. Threshold data have been taken and will be reported elsewhere in the near future (Lipson et al. 1980). Practically all of the double mutants show substantial phototropism in the upper part of the tested range (up to 0.2 mW cm^{-2}). Only the combinations *mad*A *mad*B and *mad*B *mad*C are very blind over the full tested range. All 21 double *mad* strains, as well as the single *mad*'s and wild type are now being studied by the system analysis method described earlier (Lipson 1975) in order to evaluate kinetic interactions among the *mad* gene products.

5 Nature of the Photoreceptor

On the basis of the genetic analysis (Fig. 1) we assume that a common receptor pigment is used for the several blue-light response systems in *Phycomyces*. Most recent work on identification of the photoreceptor has concentrated on phototropism and the related light-growth response. The classic action spectrum of Delbrück and Shropshire (1960) for both of these responses was more consistent with a flavin than a carotene photoreceptor on the basis of peaks at 280 nm and 385 nm.

System analysis of the light-growth response in the high intensity range has led to an estimate of the photoreceptor extinction coefficient of $(1.5 \pm 0.2) \times 10^4$ l mol^{-1} cm^{-1} close to that of oxidized riboflavin, 1.25×10^4 l mol^{-1} cm^{-1} (Lipson 1975). This estimate was derived from a kinetic model for bleaching and regeneration kinetics of the photoreceptor to help explain the loss of sensitivity at very high intensity, as occurs in human vision.

Delbrück et al. (1976) extended the action spectrum of the light growth response to the region from 575 to 630 nm. Here the relative quantum efficiency is down by about 10^8 relative to the blue peak. The very high stimulus intensities needed were obtained with a tunable dye laser. They observed a small shoulder in the action spectrum at about 595 nm. Subtraction of a gaussian extrapolation of the blue action spectrum for phototropism (adjusted to their short wavelength data points), indicated a small peak in the action spectrum at 595 nm. They interpreted this peak as evidence for a direct excitation of a flavin triplet state.

The alternative of β-carotene as the photoreceptor for phototropism has been ruled out by behavioral studies of carotene mutants (Presti et al. 1977). The strains used were blocked in all of the last six steps of β-carotene biosynthesis and were found indeed to have no detectable β-carotene. However their phototropic thresholds and kinetics were identical to those of wild type.

The above evidence together points strongly to a flavin photoreceptor. This by no means constitutes a solution to the problem of identifying the photoreceptor, for we still have little idea which of the numerous flavins and flavoproteins in the cell might be actual photoreceptor. Light-growth response measurements with polarized light (Jesaitis 1974) imply that the photoreceptor is oriented and dichroic, located probably in the plasma membrane. The final identification and localization of the flavin photoreceptor will require a specific assay.

One approach is the study of light-induced absorbance changes. Poff and Butler (1974) reported a flavin-mediated cytochrome photoreduction in *Phycomyces* and *Dictyostelium* which might conceivably involve the same flavin as in phototropism. Then Lipson and Presti (1977) applied this approach to representative *mad* mutants, in particular *mad*A, *mad*B and *mad*C strains. Unfortunately no significant differences in the absorbance changes were observed. Either none of these mutants is affected close to the photoreceptor or else the observable flavin-cytochrome absorbance changes are probably not relevant to the physiological responses. Unless *mad* mutants become available which can show a correlated deficiency in the spectrophotometric effect, it is prudent to remain skeptical. Furthermore, recent work has shown that these flavin-cytochrome effects appear to be rather nonspecific (see Butler, this vol.).

An additional genetics approach toward a specific assay for the flavin photoreceptor is now in progress in my laboratory. We are conducting a hunt for "bright-seeing" mutants of *Phycomyces*, i.e., mutants that are able to respond phototropically at high intensities where the wild type is effectively blinded. At these intensities (2 mW cm^{-2}, blue) the flavin bleaching has presumably overtaken the capacity for regeneration. Mutations affecting the quantum efficiency for bleaching or the regeneration rate constant could permit phototropism to be restored at these high intensities. Thus one could hope that bright-seeing mutants might be affected at or near the photoreceptor.

We have recently obtained candidates from this mutant hunt and are characterizing them to establish the specificity of their bright-seeing character. Then, if favorable, they will be analyzed genetically spectrophotometrically, and biochemically. We hope they will provide the specific assay needed for identification and isolation of the flavin photoreceptor in *Phycomyces*.

6 Biochemical Correlates of Sensory Transduction

Very little is known regarding the molecular components in the sensory pathways. In particular none of the *mad* gene products have been identified, nor is it known whether or not any of the class 1 *mad* mutants are affected in the concentration of the photoreceptor. The long-term goal of explaining sensory transduction in *Phycomyces* at the molecular level will require more effort in this direction.

One hopeful avenue has been provided by the discovery of a light-induced transient decrease in cyclic AMP concentration in sporangiophores (Cohen 1974a). In the same work it was shown that addition of cyclic AMP to sporangiophores could induce a growth response. Cohen (1978) has shown further that the light-induced cyclic AMP effect is much weaker in a *mad*D mutant. This work suggests that cyclic AMP may play a regulatory role, probably toward the output end of the phototropism pathway.

In approaching the sensory pathway from a biochemical standpoint, it is logical to start at the extremes. The input is defined by the flavin photoreceptor. Research concerned with this end has been described already in Section 4. It is not so obvious what molecule(s) correspond to the output of the sensory pathway; it may be that there is a regulatory enzyme which controls the growth rate subject to modulation by sensory signals. In this spirit, investigators have studied chitin synthetase (Jan 1974) and associated proteinases and inhibitors (Fischer and Thomson 1979). In the former work, Jan found an enhancement of chitin synthetase activity after exposure of samples to intense blue light. This finding correlates with a known increase of the steady-state growth rate of sporangiophores at high intensity (Foster and Lipson 1973). In the work on proteinases and inhibitors, specific changes were detected in certain *mad* mutants. More work is needed to establish the significance of these changes. Studies by Gamow and coworkers indicate that cell wall growth involves much more than simple polymerization of chitin. According to their model, the spiral growth of sporangiophores is governed by a transverse-to-longitudinal reorientation of microfibrils in the growing zone (Gamow and Bottger 1979). This reorientation may be coverned in part by lytic enzymes, such as chitinase (Cohen 1974), which in turn may be modulated by sensory signals.

7 Conclusion

Despite its overt structural simplicity, *Phycomyces* exhibits a rich variety of responses to blue light and other modes of stimulation. The three major blue light responses (Sect. 2) are all under intensive study. New methods and approaches have been developed allowing improved quantitation, such as accurate fluence-response relations for the mycelial photoresponses. This quantitation permits careful comparison between mutant and wild-type strains. Studies of *mad, car* and *pic* mutants have already provided considerable insight into the organization of the sensory pathways. In particular the abnormality of *mad*A and *mad*B strains for all blue light responses implies that these responses have their early steps in common, including the photoreceptor.

Several lines of evidence point strongly toward a flavin chromophore for the blue light receptor in *Phycomyces.* The normal phototropism of *car* mutants rules out β-carotene for this role. Hopefully the new bright-seeing mutants will provide a specific genetic assay to permit the positive identification and localization of the flavin photoreceptor.

Beyond the problem of the photoreceptor, the ultimate identification and biochemical characterization of all the *mad* gene products will provide a basis for understanding the molecular mechanisms of sensory transduction in *Phycomyces* phototropism. Similarly identification of the *pic* and *car* gene products will greatly promote the understanding of light-induced carotene synthesis. The extensive physiology and genetics studies to date have raised many specific questions awaiting answers at the molecular level. The mutants hold the key.

References

Bergman K (1972) Blue-light control of sporangiophore initiation in *Phycomyces.* Planta 107: 53–67

Bergman K, Burke PV, Cerdá-Olmedo E, David CN, Delbrück M, Foster KW, Goodell EW, Heisenberg M, Meissner G, Zalokar M, Dennison DS, Shropshire W (1969) *Phycomyces.* Bacteriol Rev 33: 99–157

Bergman K, Eslava AP, Cerdá-Olmedo E (1973) Mutants of *Phycomyces* with abnormal phototropism. Mol Gen Genet 123: 1–16

Cerdá-Olmedo E (1977) Behavioral genetics of *Phycomyces.* Annu Rev Microbiol 31: 535–547

Cerdá-Olmedo E, Torres-Martinez S (1979) Genetics and regulation of carotene biosynthesis. Pure Appl Chem 51: 631–637

Cohen RJ (1974) Cyclic AMP levels in *Phycomyces* during a response to light. Nature (London) 251: 144–146

Cohen RJ (1978) Aberrant cyclic nucleotide regulation in a behavioral mutant of *Phycomyces blakesleeanus.* Plant Sci Lett 13: 315–319

Delbrück M, Shropshire W (1960) Action and transmission spectra of *Phycomyces.* Plant Physiol 35: 194–204

Delbrück M, Katzir A, Presti D (1976) Responses of *Phycomyces* indicating optical excitation of the lowest triplet state of riboflavin. Proc Natl Acad Sci USA 73: 1969–1973

Dennison DS, Foster KW (1977) Intracellular rotation and the phototropic response of *Phycomyces.* Biophys J 18: 103–123

Eslava AP, Alvarez MI, Lipson ED, Presti D, Kong K (1976) Recombination between mutants of *Phycomyces* with abnormal phototropism. Mol Gen Genet 147: 235–241

Fischer EP, Thomson KS (1979) Serine proteases and their inhibitors in *Phycomyces blakesleeanus*. J Biol Chem 254: 50–56

Foster KW, Lipson ED (1979) The light growth response of *Phycomyces*. J Gen Physiol 62: 590–612

Galland P, Russo VEA (1979) Photoinitiation of sporangiophores in *Phycomyces* mutants deficient in phototropism and in mutants lacking β-carotene. Photochem Photobiol 29: 1009–1014

Gamow RI, Bottger B (1979) *Phycomyces:* Modification of spiral growth after mechanical conditioning of the cell wall. Science 203: 268–270

Jan YN (1974) Properties and cellular localization of chitin synthetase in *Phycomyces blakesleeanus*. J Biol Chem 249: 1973–1979

Jayaram M, Presti D, Delbrück M (1979) Light-induced carotene synthesis in *Phycomyces*. Exp Mycol 3: 42–52

Jesaitis AJ (1974) Linear dichroism and orientation of the *Phycomyces* photopigment. J Gen Physiol 63: 1–21

Lipson ED (1975) White noise analysis of Phycomyces light growth response system. Biophys J 15: 989–1045

Lipson ED, Presti D (1977) Light-induced absorbance changes in *Phycomyces* photomutants. Photochem Photobiol 25: 203–208

Lipson ED, Terasaka D, Silverstein P (1980) Double mutants of Phycomyces with abnormal phototropism. Mol Gen Genet, in press

Medina JR, Cerdá-Olmedo E (1977a) A quantitative model of *Phycomyces* phototropism. J Theoret Biol 69: 709–719

Medina JR, Cerdá-Olmedo E (1977b) Allelic interaction in the photogeotropism of *Phycomyces*. Exp Mycol 1: 286–292

Ootaki T (1973) A new method for heterokaryon formation in *Phycomyces*. Mol Gen Genet 121: 49–56

Ootaki T, Fischer EP, Lockhart P (1974) Complementation between mutants of *Phycomyces* with abnormal phototropism. Mol Gen Genet 131: 233–246

Ootaki T, Kinno T, Yoshida K, Eslava AP (1977) Complementation between *Phycomyces* mutants of mating type (+) with abnormal phototropism. Mol Gen Genet 152: 245–251

Poff KL, Butler W (1974) Absorbance changes induced by blue light in *Phycomyces blakesleeanus* and *Dictyostelium discoideum*. Nature (London) 248: 799–801

Presti D, Hsu W, Delbrück M (1977) Phototropism in *Phycomyces* mutants lacking β-carotene. Photochem Photobiol 26: 403–405

Regulation of Cell Growth and Cell Cycle by Blue Light in *Adiantum* Gametophytes

M. FURUYA, M. WADA, and A. KADOTA[1]

1 Introduction

Tissues and organs of higher plants were mainly used as experimental materials in early studies of plant photomorphogenesis. But such multicellular systems are too complex and heterogeneous to be analyzed at the cellular and subcellular levels. In contrast, the developmental processes of haploid generation in ferns, which have been long known to be under photocontrol (Borodin 1867; Life 1907; Klebs 1916, 1917), provide excellent opportunities for microscopic analysis. In fact, spore germination (Mohr 1956a; Sugai and Furuya 1967), timing of cell division (Furuya et al. 1967; Ito 1970; Wada and Furuya 1972, 1974, 1978), orientation of the cell plate (Wada and Furuya 1970, 1971), protonemal growth (Mohr 1956b; Miller and Miller 1967a; Kadota and Furuya 1977; Kadota et al. 1979), phototropism (Mohr 1956b), polarotropism (Etzold 1965; Steiner 1969), apical swelling (Mohr and Holl 1964; Howland 1972; Wada et al. 1978) and antheridium formation (Schraudolf 1967) have been reported to be regulated by light.

Whenever the effect of light on growth and division of protonemal cells is studied, it is noticed that cell division is always accompanied by the cessation of cell growth at an early stage of the cell cycle. The separation of the light effect on cell division from that of apical growth is necessary to study the real effects of light on each process of photomorphogenesis. However, previous work on cell elongation in fern protonemata (Mohr 1956b; Miller and Miller 1961, 1963, 1964, 1967a, b; Miller and Wright 1961; Schnarrenberger and Mohr 1967; Greany and Miller 1976) has not succeeded in separating these two processes.

In the past decade we have attempted, first, to establish an experimental system in which effects of light on elementary processes of protonemal development are separated from each other, then to determine the pigments involved and their intracellular localization and finally to analyze the relationship between diverse target organelles and multiple effects in apical cell of the *Adiantum* gametophyte. This article is not intended to cover the literature fully on the present subject, but to provide a review on the work of this laboratory since 1968.

---

1 Botany Department, Faculty of Science, University of Tokyo, Hongo, Tokyo, Japan 113

2 Instruments

A new instrument always provides us with a new opportunity for advancement. This is particularly true in photobiology. We have thus constructed a spectrograph for mono-chromatic irradiation to determine the action spectra on new photoreactions, a time-lapse video recorder with infra-red sensitive visicon tube for continuous observation under microscope without an acting observing light, and a microbeam irradiator with monochromatic light for searching the intracellular localization of photoreceptive sites. Most of the results described in this article were obtained by using one of these newly built instruments, so that we would like to describe briefly the properties of these in-struments at first.

2.1 Spectrograph

Monochromatic exposure of fern gametophytes was carried out with a spectrograph, which was a combination of a 6.5 kW xenon short-arc lamp, a quartz condensing lens, a running-water filter (10 cm thick) with quartz windows, a knife edge slit of several alternative widths, a concave collimating mirror, and a Bausch and Lomb grating (154 mm X 206 mm; 1200 lines mm^{-1}; blazed at 500 nm). The spectro-irradiator provided a spectrum of monochromatic light in the range of 250–810 nm on its focal plane (4000 mm wide, 100 mm high), in which dispersion was 0.16 nm mm^{-1}. Spectral band width measured at half the peak energy of each wavelength was less than 10 nm. The intensity of monochromatic light was measured with a YSI-Kettering model 65 radio-meter and was adjusted with neutral density filters and carbon-coated glass plates. The duration of monochromatic irradiation was controlled by a shutter at the top window of the dark box, in which a sample dish was kept at 25°C.

2.2 Time-Lapse Video Recorder

In earlier experiments, the development of individual protonema was recorded with photomicrograph using infra-red film and infra-red light provided by a tungsten lamp with an infra-red filter and a heat filter. The developed film was magnified with a pho-tographic enlarger, and the images of protonemata were measured with a ruler.

The above method, however, was not sufficient to record continuously the pro-cessions of cell growth and cell division without any influence of the observing light. Thus, we have built an instrument that will monitor precisely and record cellular and subcellular changes in protonemal development continuously through a Nikon Biophoto microscope with Nomarski interference-contrast optics, coupled with a video camera equiped with an infra-red sensitive tube, connected to a video tape recorder and a video monitor. As shown in Fig. 1, the recorder and the monitor are controlled by a time-lapse controller and a timer. The time is shown on the screen of the monitor in terms of month, day, hour, minute, second and 100 ms (see Figs. 2 and 8).

The sample is set on the stage of microscope in total darkness or under a green safe-light. The microscopic observation is made using light of wavelength longer than 780 nm through a video monitor, and recorded using the same infra-red light. Cell growth

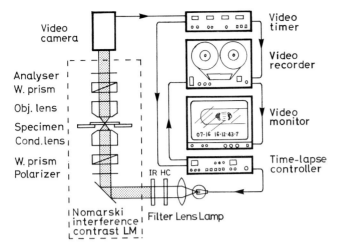

Fig. 1. Diagram showing the time-lapse video recorder coupled to light microscope equiped with Nomarski interference-contrast equipment. *W. prism* Wollaston prism; *Obj. lens* objective lens; *Cond. lens* condenser lens; *IR* infra-red filter; *HC* heat filter

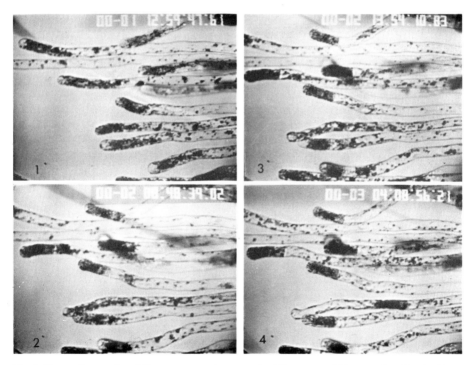

Fig. 2. Photomicrographs of filamentous protonemata in *Adiantum* on TV screen (see Fig. 1, video monitor) showing the time course of apical growth and cell division. Protonemata precultured under continuous red light for 5 days were kept for 1 day under continuous red light and then in the dark for (*1*) 12 h 59 min 47 s 610 ms; (*2*) 1 day 8 h 48 min 39 s 20 ms; (*3*) 1 day 13 h 54 min 10 s 830 ms; (*4*) 2 days 4 h 8 min 56 s 210 ms

is determined so that recorded images of cells on television screen can be measured in a stop-motion mode at chosen intervals of the time (Fig. 2). The progression of M phase in the cell cycle recorded is then observed by replaying the video-tape at normal speed (Fig. 8).

2.3 Microbeam Irradiator

An important problem to resolve is the intracellular localization of photoreceptor pigments in the protonemal cell. In this connection a microbeam irradiator (Fig. 3) was designed in this laboratory and constructed by Hitachi Co. Ltd., Tokyo (Kadota and Furuya 1977; Wada and Furuya 1978). The apparatus has two light source "I" and "II", one of which is used as safe light for observation and the other is used for microbeam irradiation. Light intensity at the sample level was measured by a silicon photocell placed at the bottom of objective (Fig. 3, Sp).

A dish in which protonemata were cultured is placed on the focusing stage (Fig. 3, Fs). Ca. 10 protonemata growing in parallel toward continuous red light (Fig. 4a) were observed under red light provided by light source II (Fig. 3). A narrow beam of red light of higher intensity from light source I was focused on the sample by moving the objective (Fig. 3, C-ob) vertically so that the position of the narrow beam could be adjusted on any regions of the protonemata by moving C-ob horizontally. The protonemata and the position of the narrow beam were photographed (Fig. 3, Ph) under the red safe-light from light source II and a narrow beam of red light from source I just before the treatment with blue or far red light (Fig. 4a). All procedures were carried out at 25°C in the dark or under a dim, green safe-light.

Immediately after the light treatment, the samples were put into a dark box and kept at 25 ± 1°C. The cell development was observed under the microscope at various times thereafter (Fig. 4b), referring to the position of the narrow beam in the photographs taken at 0 h (Fig. 4a).

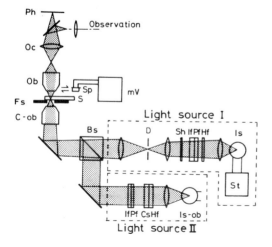

Fig. 3. Diagram showing the microbeam irradiator used. *Ph* photographic camera; *Oc* ocular lens; *Ob* objective lens; *S* specimen; *Fs* focusing stage; *C-ob* condenser objective lens; *Sp* silicon photocell; *mV* millivolt meter; *Bs* beam splitter; *D* diaphragm; *Sh* shutter; *If* interference filter; *Pf* plastic filter; *Hf* heat filter; *Is* irradiation source; *St* stabilizer; *Cs* CuSO$_4$ solution; *Is-ob* irradiation source for observation. (From Wada and Furuya 1978)

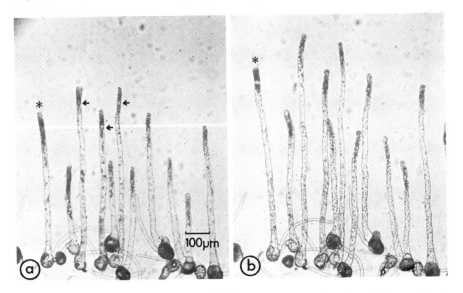

Fig. 4a, b. Microphotographs of a culture of *Adiantum* protonemata immediately before microbeam irradiation (**a**) and 30 h after irradiation (**b**). *Arrows* show nuclear regions. The nuclear region of protonema marked with an *asterisk* was exposed to the narrow beam (10 μm in width) at 0 h (**a**) and showed cell division 30 h thereafter (**b**). (From Wada and Furuya 1978)

3 Blue Light Effects on Cell Growth

3.1 Phytochrome-Mediated Induction of Apical Growth

When spores of *Adiantum* were cultured under continuous red light, the irradiation resulted in filamentous, one-cell protonemata (Figs. 4a and 5 o–o). When the one-cell protonemata were transferred to darkness, the apical growth continued for the subsequent 24 h. Growth ceased within 72 h thereafter (Fig. 5 ●–●), during which cell division took place. The growth in the dark was greatly inhibited by a brief irradiation with far red light given immediately before the dark period (Fig. 5 ■–■), and the effect of far red light was fully reversed by red light. The photoreversibility was repeatedly observed, suggesting the involvement of phytochrome (Kadota and Furuya 1977). The photoreceptive site of this phytochrome system was not localized in any particular region of the protonemata but rather was spread over the entire region of the cell (Kadota and Furuya 1977).

A dichroic orientation of the photoreceptor phytochrome in the protonema was evident so that the linearly polarized far red light vibrating at an angle normal to the developmental axis of protonemata was significantly less effective in preventing the apical growth than all of the other treatments (Furuya 1978).

It is very interesting that nongrowing two-celled protonemata kept in the dark for 3 days resumed the apical growth ca. 6 h after they were reexposed to red light irrespective of the incident energy (Kadota and Furuya 1980). When kept under continuous red light, the resumed apical growth continued at a constant rate thereafter (Fig. 5 □–□), while the growth induced by a brief irradiation with red light, ceased within some hours

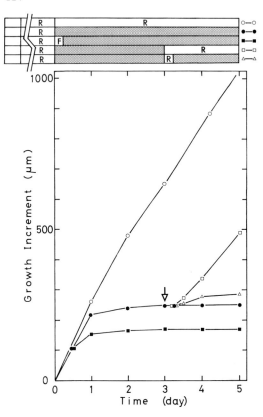

Fig. 5. Effect of red, far red light, and darkness on the apical growth of filamentous protonemata in *Adiantum capillus-veneris*. Filamentous protonemata precultured under continuous red light for 6 days were treated as shown in the top of the figure. *R* red light of 0.5 Wm^{-2}; *F* far red light of 1.8 Wm^{-2}. *Arrow* indicates the onset of red light irradiation

depending upon the dose of red light (Fig. 5 △—△). The results on the wavelength effect on this phenomenon and the red far red reversibility indicated an involvement of phytochrome. A dichroic orientation of the phytochrome was also shown by the exposure of polarized red and far red light.

This red light-induced, single-celled protonema has been used as the experimental material in the works described in this article.

3.2 P$_{b-nuv}$-Mediated Inhibition of Apical Growth

Besides the phytochrome effect described above, it is known that protonemal growth is controlled by either a brief irradiation (Miller and Miller 1964) or a prolonged irradiation (Howland 1972) with blue light. In this connection, action spectra for the effect of brief light treatment on subsequent apical growth in the dark were determined before and after compensation of the phytochrome effect in single-celled protonemata of *Adiantum* grown initially under continuous red light for 6 days. The irradiation lowered the growth rate in subsequent darkness and reduced the final length. The action spectrum before the compensation showed peaks at 450 and 470–480 nm with a shoulder in the near-UV region and another peak at 730 nm, whereas after the compensation the action spectrum was similar in the blue and near-UV region but totally lost

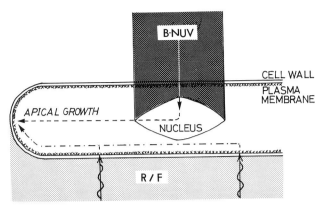

Fig. 6. Schematic illustration of region responsible to red, far red, and blue light which affect the apical growth of protonema in *Adiantum*

the peak at 730 nm (Kadota et al. 1979). Thus it is concluded that the process is under control of two blue light-absorbing pigments such as P_{b-nuv} and phytochrome.

Narrow beam irradiation of the filamentous protonemata with blue light was effective only when the nuclear region of the protonemal cell was exposed. Polarized blue light having different directions of electrical vectors was equally effective in inducing growth inhibition. The experimental evidence supports the conclusion that P_{b-nuv} located in the nuclear region of the filamentous protonema controls a process of growth cessation in the dark, in addition to the phytochrome effect (Kadota et al. 1979). The results in this and the previous sections are schematically summarized in Fig. 6.

3.3 P_{b-nuv}-Mediated Swelling of Protonemal Apex

If blue light is exposed to filamentous protonemata for a longer period of time the response to the prolonged irradiation is quite different from that to brief irradiation, as reported by Drumm and Mohr (1967), Davis (1969) and Howland (1972). When single celled protonemata were continuously exposed to high intensity blue light, apical swelling was brought about within the first few hours (Wada et al. 1978) parallel with the vacuolization in the tip of the protonemata (Wada and O'Brien 1975).

As the photoreceptor pigment controlling the blue light-induced swelling of the protonemal tip has not been identified, action spectra for the induction of apical swelling in red light-grown single-celled protonemata of *Adiantum* were determined by continuous irradiation with monochromatic light for 5 h. The resultant action spectra showed a sharp peak at 480 nm with a broad plateau in the region of blue and near-UV light. Wavelengths longer than 520 nm had no effect, indicating a possible involvement of P_{b-nuv} in this photo-reaction as the photoreceptor (Wada et al. 1978).

To investigate the photoreceptive site of this effect of prolonged irradiation with blue light, a narrow beam of blue light was directed onto various regions of red light-precultured filamentous protonemata for 3 h. When the tip of filamentous protonemata was irradiated with the beam, both apical swelling and inhibition of apical growth ob-

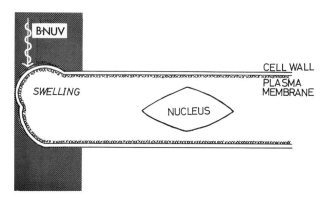

Fig. 7. Schematic illustration of region responsible to blue light which induces the apical swelling in *Adiantum* protonema

viously took place in all protonemata tested, while no significant effect on the swelling was observed when any regions other than the tip were irradiated. Polarized blue light vibrating parallel with the developmental axis of protonemata induced apical swelling and also prevented apical growth as effectively as nonpolarized light, but that vibrating in a normal direction was significantly less effective, indicating a dichroic orientation of P_{b-nuv} in the protonemal tip. The present result is schematically shown in Fig. 7, which has clearly confirmed the result of Etzold (1965) that a blue light-absorbing pigment is dichroic and located at the apex of protonema. But the result is completely different from that induced by a brief irradiation with blue light (Fig. 6).

4 Photocontrol of Cell Cycle

4.1 The Timing of Cell Division

The regulatory mechanism in cell division has not yet been elucidated in both plant and animal cells, but both internal and external factors must control the procession of the cell cycle. Continuous irradiation with red light brings about a marked depression of mitotic activity in cells of fern gametophytes, so spores cultured under red light give rise to single-celled protonemata. When such a single-celled protonema is transferred to the dark, the cell subsequently divides into a new apical cell and a basal cell. The timing of the cell division was first found to be photocontrollable in *Pteris* (Furuya et al. 1967) and later in *Adiantum* (Wada and Furuya 1972).

When the one-celled protonema of *Adiantum* was transferred to the dark, the cell divided sometime during a period between 24 and 36 h thereafter. On the other hand, when red light-precultured protonema was briefly irradiated with far red light and then transferred to darkness, the timing of the resultant cell division was significantly delayed by the far red light treatment. The effect of far red light was repeatedly reversed by a small dose of red light given immediately after the far red light, indicating that the timing of cell division in the dark was regulated by a phytochrome system (Wada and Furuya 1972).

When single-celled protonemata of *Adiantum* were briefly exposed to blue light just before the dark incubation, a cell division occurred after a 17–26 h of the dark period. This blue light induction of synchronous cell division did not require a prolonged irradiation, but was a trigger reaction. The action spectrum for this photoreaction was determined 24 h after irradiations of light between 350 and 750 nm at 10 nm intervals. The above-mentioned effect of phytochrome on the timing of cell division was minimized by a short exposure to red light which was given immediately after the monochromatic irradiation. The result shows a peak in the neighborhood of 460 nm with shoulders and another peak in the near UV region, indicating an involvement of P_{b-nuv} in this photoreaction too (Wada and Furuya 1974). It is now evident that the timing of cell division is controlled by both phytochrome and P_{b-nuv}.

4.2 P_{b-nuv} in Nuclear Region and Phytochrome in Plasma Membrane

Using the technique of local irradiation with a narrow beam (Figs. 3 and 4), the intracellular localization of phytochrome and P_{b-nuv} was determined in single-celled protonemata of *Adiantum*. The result (Wada and Furuya 1978) clearly shows that only irradiation of the region containing the nucleus induced cell division. Beams of 30 μm in width, which corresponds to the diameter of the nucleus, or wider, were equally effective; beams 10 μm wide or less were less effective. It is quite likely that the nucleus, or at least some organelle closely attached to the nucleus, is indeed perceiving the blue light, but it cannot yet be concluded in a definitive manner that the nucleus itself is the photoreceptive site.

The effect of a narrow beam of far red light, which delays the onset of the blue light-induced cell division, was found to be present along the entire length of the protonema cell, including the largely vacuolated basal region of the cell. The possibility of a dichroic orientation of the phytochrome in protonemata was studied by means of irradiation with polarized light (Wada and Furuya 1978), indicating that polarized far red light vibrating parallel with the protonemal axis was significantly less effective than the other treatments which were as effective as nonpolarized far red light. This evidence supports the hypothesis that the phytochrome controlling the timing is localized in the plasma membrane.

4.3 Blue Light Effect on Duration of G_1 Phase

The fact that the timing of cell division in *Adiantum* protonemata is so dependent upon the environmental light condition (Wada and Furuya 1972, 1974) poses the question as to which component phase of the cell cycle is controlled by light absorbed by either P_{b-nuv} or phytochrome in the protonemata. Thus, the duration of each component phase of the cell cycle under various light conditions was determined by the following methods; the initiation and the expiration of S phase were detected autoradiographically by the incorporation of ^{3}H-thymidine into nuclei, and mitotic index and the timing of cell plate formation were examined in the same samples as those used in autoradiography. The duration of M phase was counted by a quick motion picture described in Figs. 1 and 8.

In *Pteris vittata*, protonemata grown under red light were arrested in the early G_1 phase (Ito 1970). Miyata et al. (1979) have found that protonemata of *Adiantum* were also arrested in early G_1 phase under continuous red light irradiation. When single-celled

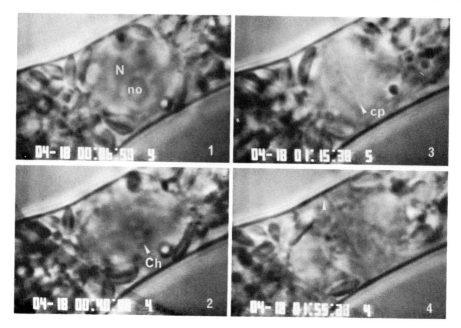

Fig. 8. Photomicrographs on TV screen of the equipment shown in Fig. 1 undergoing cell division induced by blue light irradiation; (*1*) the stage, just before the disappearence of nucleolus, *N* nucleus, *no* nucleolus; (*2*) metaphase, *arrow* shows chromosomes arranged on the equatorial plate, *Ch* chromosome; (*3*) early stage of cell plate formation, *cp* cell plate; (*4*) *arrow* indicates the position in which new cell plate reached to old cell wall

protonemata in G_1 phase were irradiated with different dosages of blue light, only the duration of G_1 phase in cell cycle was significantly reduced by the blue light exposure comparing with that in dark control, and the higher the dosage of blue light irradiation, the shorter the duration of G_1 phase was observed (Table 1). The duration of S, G_2, and M phases was not significantly affected by the irradiation.

Table 1. Effect of blue light on duration of each component phase of cell cycle in dark-incubated protonemata of *Adiantum*

Light Treatment (duration)	Component phase (h)				Total (h)
	G_1	S	G_2	M	
Blue (continuous)[a]	7.5 ± 0.6	3.5 ± 0.6	3.9 ± 1.2	3.4 ± 0.2	(18.3 ± 1.3)
Blue (4 h)[a]	8.0 ± 0.0	2.5 ± 0.7	5.7 ± 0.7	3.8 ± 0.6	(20.0 ± 1.4)
Blue (1 h)[a]	13.0 ± 2.6	5.0 ± 1.4	3.9 ± 1.8	4.6 ± 0.5	(26.5 ± 1.9)
Blue (10 min)[b]	15.5 ± 0.7	4.0 ± 1.4	5.4 ± 0.6	4.6 ± 0.6	(29.5 ± 0.7)
Dark	22.3 ± 0.6	5.7 ± 1.2	7.0 ± 2.4	4.0 ± 0.4	(39.0 ± 1.0)

[a] $0.5\ W\ m^{-2}$, [b] $1.5\ W\ m^{-2}$

4.4 Phytochrome Effect on Duration of G_2 Phase

Finally, the effect of brief irradiation with far red light on the duration of each component phase in cell cycle was estimated with red light-precultured protonemata by the same methods as described in the previous section. The result is presented in Table 2 (Miyata et al. 1979).

It is evident that only the duration of G_2 phase was significantly extended by the irradiation, while the duration of G_1, S, and M phases was not significantly affected by the irradiation. Further, the observation of M phase by the time-lapse video recorder (Figs. 1 and 8) clearly showed that the duration of elementary processes in M phase was not affected at all by the far red light.

The effect of far red light on G_2 phase was reversible by subsequently given red light, and the effect of both lights was repeatedly reversible, while a percent of labeled nuclei was not influenced by the light treatments. The results strongly indicate that a phytochrome system controls the duration of G_2 specifically.

Table 2. Effect of far red light on duration of each component phase of cell cycle in dark-incubated protonemata of *Adiantum*

Light treatment	Component phase (h)				Total (h)
(duration)	G_1	S	G_2	M	
Far red (15 min)[a]	22.5 ± 0.7	3.5 ± 0.7	14.3 ± 2.9	4.2 ± 0.8	(44.5 ± 2.1)
Dark	22.3 ± 0.6	5.7 ± 1.2	7.0 ± 2.4	4.0 ± 0.4	(39.0 ± 1.0)

[a] 30 W m^{-2}

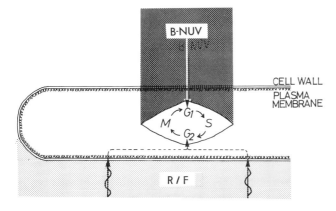

Fig. 9. Schematic illustration of region responsible to red, far red, and blue light which affect the duration of cell cycle in *Adiantum* protonema

5 Concluding Remarks

It is evident that phytochrome and P_{b-nuv}, both of which can absorb blue light, are the photoreceptor pigments controlling elementary processes of development in the *Adian-*

tum gametophytes (Furuya 1978). As far as the phenomena described in this chapter
are concerned, no other pigments are indicated to be involved in the photoreactions,
although some other pigments have been suggested in plant photomorphogeneis (Miller
1968; Mohr 1972; Furuya 1977).

When the protonemal cell is macroscopically irradiated, interaction between the
effects by P_{b-nuv} and those by phytochrome is usually found in elementary processes
of photomorphogenesis in the fern (Table 3). The fact that only the two pigments are
engaged to control such many different responses in this single cell suggests a possibility
that P_{b-nuv}- and phytochrome-induced stimuli may be transduced through different effector systems in each photoreceptive site.

If sample protonemata are briefly irradiated with blue light, the light absorbed by
only nuclear region of the cell causes the photoinhibition of apical growth and the induction of synchronous cell division (Figs. 6 and 9), for the time being. However, if
high fluence of blue light is given to the same sample, the light absorbed by only protonemal apex gives rise to a swelling of the apex (Fig. 7). Therefore, intracellular photoreceptive sites for these two reactions are separated from each other by giving different
fluence of blue light. As the action spectra on the former reaction (Kadota et al. 1979)
and on the latter (Wada et al. 1978) are seemingly very similar, the same or closely related blue light-absorbing pigments may act differently when the pigment is located in
different organelles. In addition, a dichroic effect is observed in the latter effect but
not at all in the former. This evidence indicates that there are at least two target sites
for blue light effects in this cell.

When microbeam irradiation with blue or far red light is done on single-celled
protonemal cell, the light absorbed by P_{b-nuv} in the nuclear region controls the duration
of G_1 phase in cell cycle, while the light absorbed by phytochrome in plasma membrane
regulates that of G_2 phase (Fig. 9). It is important to be pointed out that, when each
photoreceptive site is irradiated with microbeam, the above-mentioned two photocontrolled processes are totally independent so that the previously reported interaction be-

Table 3. Elementary processes of blue light-dependent photomorphogenesis in fern gametophytes

Phenomena	Pigment [Intracellular localization]	
	Phytochrome	P_{b-nuv}
Spore germination	Pfr-dependent germination [?] (Furuya et al. 1980)	Low energy inhibition [?] (Sugai and Furuya 1967)
Apical growth	Pfr dependent induction [PM] (Kadota and Furuya 1980)	Low energy inhibition [NR] (Kadota et al. 1979)
Apical swelling	–	High energy induction [Apex] (Wada et al. 1978)
Cell cycle	Longer G_2 period by Pr [PM] (Miyata et al. 1979)	Shorter G_1 period [NR] (Miyata et al. 1979)
Timing of cell plate formation	Delay by Pr [PM] (Wada and Furuya 1972)	Synchronous division [NR] (Wada and Furuya 1974)

PM: plasma membrane; *NR:* nuclear region

tween the effects by P_{b-nuv} and phytochrome (Wada and Furuya 1974) is no longer detected. This experimental results clearly demonstrate that intracellular localization of target sites is again separatable in the single cell.

The photoreceptive site and the loci of photomorphogenetic responses are also separatable in a single cell in the case of apical growth (Fig. 6) and cell division (Fig. 9), but not in the cases such as apical swelling (Fig. 7) by the technique using microbeam exposure.

Finally we should like to make a comment that one must be most cautious to interpret the photomorphogenetic response caused by macroscopic irradiation with light or under the sunlight, assuming that many photo-induced reactions may progress in parallel in a cell and that their interaction can be very complex. Therefore, it is essential to find and separate multiple effects of one pigment in a cell and diverse target sites of one pigment in a cell.

Acknowledgment. We wish to thank Professor Malcom B.Wilkins for careful reading of the manuscript.

References

Borodin J (1867) Über die Wirkung des Lichtes auf einige höhere Kryptogamen. Bull Acad Imp Sci St Petersbourg 12: 432–448

Davis BD (1969) The transition from filamentous to two-dimentional growth in fern gametophytes II. Kinetic studies on *Pteridium aquilinum.* Am J Bot 56: 1048–1053

Drumm H, Mohr H (1967) Die Regulation der RNS-Synthese in Farngametophyten durch Licht. Planta 72: 232–246

Etzold H (1965) Der Polarotropismus und Phototropismus der Chloronemen von *Dryopteris filix mas* (L.) Schott. Planta 64: 254–280

Furuya M (1977) Photomorphogenesis in plants. Rec Prog Natl Sci Jpn 2: 45–69

Furuya M (1978) Photocontrol of developmental processes in fern gametophytes. Bot Mag Tokyo Spec Issue 1: 219–242

Furuya M, Ito M, Sugai M (1967) Photomorphogenesis in *Pteris vittata.* Jpn J Exp Morphol 21: 398–408

Furuya M, Kadota A, Uematsu H (1980) Percent P_{fr}-dependent germination of spores in *Pteris vittata.* Dev Growth Differ submitted

Greany RH, Miller JH (1976) An interpretation of dose-response curves for light-induced cell elongation in fern protonemata. Am J Bot 63: 1031–1037

Howland GP (1972) Changes in amounts of chloroplast and cytoplasmic ribosomal-RNAs and photomorphogenesis in *Dryopteris* gametophytes. Physiol Plant 26: 264–270

Ito M (1970) Light-induced synchrony of cell division in the protonema of the fern, *Pteris vittata.* Planta 90: 22–31

Kadota A, Furuya M (1977) Apical growth of protonemata in *Adiantum capillus-veneris.* I. Red far-red reversible effect on growth cessation in the dark. Dev Growth Differ 19: 357–365

Kadota A, Furuya M (1980) Apical growth of protonemata in *Adiantum capillus-veneris.* IV. Phytochrome-mediated induction of apical growth in non-growing two cell protonemata. Plant Cell Physiol submitted

Kadota A, Wada M, Furuya M (1979) Apical growth of protonemata in *Adiantum capillus-veneris.* III. Action spectra for the light effect on dark cessation of apical growth and the intracellular photoreceptive site. Plant Sci Lett 15: 193–201

Klebs G (1916/17) Zur Entwicklungs-Physiologie der Farnprothallien. Sitzungsber Heidelberg Akad

Wiss 7B (4): 1–82; 8B (3): 1–138; 8B (7): 1–104

Life AC (1907) Effect of light upon the germination of spores and the gametophytes of ferns. Annu Rep Mo Bot Garden 18: 109–122

Miller JH (1968) Fern gametophytes as experimental material. Bot Rev 34: 361–440

Miller JH, Miller PM (1961) The effect of different light conditions and sucrose on the growth and development of the gametophyte of the fern, *Onoclea sensibilis*. Am J Bot 48: 154–159

Miller JH, Miller PM (1963) Effects of red and far-red illumination on the elongation of fern protonemata and rhizoids. Plant Cell Physiol 4: 65–72

Miller JH, Miller PM (1964) Blue-light in the development of fern gametophytes and its interaction with far-red and red light. Am J Bot 51: 329–334

Miller JH, Miller PM (1967a) Action spectra for light-induced elongation in fern protonemata. Physiol Plant 20: 128–138

Miller JH, Miller PM (1967b) Interaction of photomorphogenetic pigments in fern gametophytes: Phytochrome and a yellow-light-absorbing pigment. Plant Cell Physiol 8: 765–769

Miller JH, Wright DR (1961) An age-dependent change in the response of fern gametophytes to red light. Science 134: 1629

Miyata M, Wada M, Furuya M (1979) Effects of phytochrome and blue-near ultraviolet light-absorbing pigment on duration of component phases of the cell cycle in *Adiantum* gametophytes. Dev Growth Differ 21: 577–584

Mohr H (1956a) Die Beeinflussung der Keimung von Farnsporen durch Licht und andere Faktoren. Planta 46: 534–551

Mohr H (1956b) Die Abhängigkeit des Protonemawachstums und der Protonemapolarität bei Farnen vom Licht. Planta 47: 127–158

Mohr H (1972) Photomorphogenesis. Springer, Berlin Heidelberg New York

Mohr H, Holl G (1964) Die Regulation der Zellaktivität bei Farnvorkeimen durch Licht. Z Bot 52: 209–221

Schnarrenberger C, Mohr H (1967) Die Wechselwirkung von Hellrot, Dunkelrot und Blaulicht bei der Photomorphogenese von Farngametophyten (*Dryopteris filix-mas* (L.) Schott). Planta 75: 114–124

Schraudolf H (1967) Die Steuerung der Antheridienbildung in *Polypodium crassifolium* L. (*Pessopteris crassifolia* Underw. and Maxon) durch Licht. Planta 76: 37–46

Steiner AM (1969) Action spectrum for polarotropism in the chloronema of the fern *Dryopteris filix-mas* (L.) Schott. Photochem Photobiol 9: 507–513

Sugai M, Furuya M (1967) Photomorphogenesis in *Pteris vittata*. I. Phytochrome-mediated spore germination and blue light interaction. Plant Cell Physiol 8: 737–748

Wada M, Furuya M (1970) Photocontrol of the orientation of cell division in *Adiantum*. I. Effects of the dark and red periods in the apical cell of gametophytes. Dev Growth Differ 12: 109–118

Wada M, Furuya M (1971) Photocontrol of the orientation of cell division in *Adiantum*. II. Effects of the direction of white light on the apical cell of gametophytes. Planta 98: 177–185

Wada M, Furuya M (1972) Phytochrome action on the timing of cell division in *Adiantum* gametophytes. Plant Physiol 49: 110–113

Wada M, Furuya M (1974) Action spectrum for the timing of photoinduced cell division in *Adiantum* gametophytes. Physiol Plant 32: 377–381

Wada M, Furuya M (1978) Effects of narrow-beam irradiations with blue and far-red light on the timing of cell division in *Adiantum* gametophytes. Planta 138: 85–90

Wada M, O'Brien TP (1975) Observations on the structure of the protonema of *Adiantum capillus-veneris* L. undergoing cell division following white-light irradiation. Planta 126: 213–227

Wada M, Kadota A, Furaya M (1978) Apical growth of protonemata in *Adiantum capillus-veneris*. II. Action spectra for the induction of apical swelling and the intracellular photoreceptive site. Bot Mag 91: 113–120

Blue Light and Transcription

J. GRESSEL[1]

1 Introduction: The Problems

In a recent lengthy review with the general title of *The Effect of Light on RNA and Protein Synthesis in Plants* [23], only the relationship between light absorbed by the phytochrome system and the synthesis of RNA was covered. Does this mean that the crytochrome[2] and other even less well understood blue light, nonphytochromal acceptor systems, have nothing to do with RNA synthesis? How much can/will this review rectify the situation? I shall try to analyze the direct and indirect evidence that has accumulated where blue light is implicated in affecting transcription. We shall have to ask some questions about relating the blue syndrome with transcription: (1) Is RNA synthesis really necessary? It has been a basic premise of developmental biologists that induced changes in developmental patterns require new transcriptional patterns. But do we need transcription for a transient phototropic response? Are not translational or activity controls sufficient? (2) How good are the data supporting transcriptional changes? Indirect evidence, especially with inhibitors, must not be fully accepted as conclusive, even after weighing the known drawbacks. (3) Do we have evidence for specificity; distinct transcriptional products which should lead to distinct translational products? An increase in growth rate or "step-up" will bring concurrent increases in all RNA species; can the processes be separated? (4) How closely is the photoact linked to transcriptional events? Is transcription many steps down the line from photoreception or are the processes tightly and closely coupled in sequence? (5) Is transcription the "amplifier" in the "black box" or does the "amplifier" or its products induce transcription?

These questions are not just pointed towards blue-induced development and transcription. They can still be pointed towards "classic" well-accepted experiments designed to show relationships between a morphogenetic inducer and transcription. In viewing and reviewing both our own and other contributions to these questions one must heed the warning issued by Thurber: "The conclusion you jump to may be your own".

---

1 Department of Plant Genetics, Weizmann Institute of Science, Rehovot, Israel

2 Cryptochrome is defined herein as that pigment system having an action spectrum somewhat characteristic of flavins and some carotenes. This name refers to its occurrence in cryptogams and its cryptic nature, which is not completely correct. Action spectra analyses implicate the presence of other blue and near UV absorbing pigments as well in many lower organisms

The case I should like to be able to report would be one where a short pulse of low-energy blue light is given; after a minimum of time in the photochemical black box, derepression of genes occurs; new and different discrete mRNA's are transcribed. These mRNA differences can be viewed by RNA/DNA saturation hybridization techniques under the most stringent conditions; new and differently defined known translation products can be visualized both in vitro and in vivo; development then ensues. Proper controls have been carried out throughout that preclude translational and turnover controls, metabolic pool problems, etc. I cannot report any complete success in reaching this utopia, which has been achieved for other inducers; especially in virology and prokaryotic biology. It is doubtful if this goal has been reached in any induced development system in eukaryotes.

Even the classical systems of induced bacterial sporulation where RNA/DNA hybridization was first used in a clearcut manner [23] or induced slime mold sporulation where the de novo synthesis of particular stalk enzymes was claimed (cf. [72]) there are esthetic draw-backs. The inducer was not a "hormone" or some other minuscule factor; the inducer was starvation — a sledgehammer promotor of novel transcription products. The ideal inducer requires a minimum quantity of molecules or quanta to attain a maximal developmental effect. Hartmann has succinctly described the advantage of low flux light over hormones by stating that "light lacks inertia" [36]. Lights can be extinguished, but it is hard to wash out hormones. If we were to pick the ideal photosystem to study light and transcription we would pick phytochrome for its added advantage for control; photoreversibility. We are not addressing ourselves to the ideal photosystem but to the data relating blue light treatment to transcription.

This review is divided into three major parts, each for a group of systems with some unique characters: (1) the fungi which bend or reproduce as a result of blue light. (2) Green plant morphogenetic systems with definable responses via "cryptochrome". Biplanar fern growth is the most studied example. (3) Chloroplast development. White light is usually used by most developmentalists to "green"[3] plastids. In those few cases where light was split to components, interesting results have been obtained. Greening is not only modulated by protochlorophyll and phytochrome but seems to use much of the spectrum to control a complex of processes. Greening will first be discussed in *Euglena* and algae where there are fewer interactions with other photocontrol systems, and finally in other plants where interactions with phytochrome occur.

2 Cryptochrome and Transcription in Fungi

Fungi offer a choice system to work on blue light effects on transcription. From the photobiological point of view fungi are useful as they often lack phytochrome and chlorophyll to complicate matters although they still have many pigments which absorb light. In our discussion here the photoreceptor is more or less irrelevant. Fungi are eukaryotic and thus can be considered a model for higher organisms with many of their

---

3 In this review the more positive terminologies of "green" and "etiolate" will be used for the more accepted "de-etiolate" and "de-green", respectively

control mechanisms. Furthermore, fungi can be cultured axenically and can be handled like bacteria, allowing easy manipulation.

Blue light has a multitude of effects on fungi from changing respiration pathways from the Embden-Meyerhof to the pentose shunt (cf. [28]) to movement, pigment syntheses and morphogenesis. Many of the fungi with photoresponses are soil micro-organisms. They have a variety of evolutionary reasons for "wanting to know" when they reach the soil surface and "see the light". It is then time to reproduce to achieve propagule distribution in the air, and it is time to protect a multitude of enzymes from light. These processes require large energy inputs, new and different enzymes, substrates and products. It would be extremely wasteful, bringing about strong evolutionary unfitness, if all the enzymes involved were constitutive or if all the mRNA's were masked; all awaiting the light, for the RNA to be translated or the enzymes activated. Thus, there is good reason to look for new transcription products in these organisms after illumination, as it is inconceivable that all of these processes are controlled at the level of translation or activity. Considering the extensive genome of fungi and plants, something should occur at the level of transcription.

2.1 Requirement for Transcription in Phototropism

At first glance, one might guess that the rapid movement responses, be they photophobic movements of free floating cells (cf. [22]), phototropism in fungi (cf. [9]), or even the photo-induced rapid changes in the rate of transpiration [81] would not require new transcription products. Sufficient enzymes should be available to provide the necessary components for such transient phenomena. Still, the kinetics of *Phycomyces* phototropism do not preclude transcription; they are even suggestive of the possibility. There is a lag period between illumination and bending of about 4–6 min sufficient time for transcription and translation. There is no direct evidence linking transcription to the tropic response. Still there are "unbending" mutants which do not respond to light [9] but it is not clear what the gene products are, when they are transcribed, or how they fit in.

2.2 Induced Carotenogenesis and Transcription

The ecological importance of carotenoids as photoprotectants in microorganisms is well documented [45]. Carotene induction in terrestrial species is typically modulated by cryptochrome. Still, within the same genus, different species can have different photoacceptors. In *Mycobacterium* sp. cryptochrome perceives, while in *M. marinum* it seems to be a heme (cf. [2]). Nature has more than one way to achieve the same end and picks the best photoacceptors for each ecological niche. The amounts of carotene are rather large, some of the enzymes involved are known, and their cell-free synthesis has been accomplished [14]. Prephytoene synthetase activity was high in the light and undetectable in the dark [41] but this does not constitute evidence for the de novo synthesis which would then suggest de novo transcriptional changes. Evidence for de novo synthesis has been shown in an analogous system; melanin synthesis. Melanin, like carotene, may serve as a photoprotectant. The tyrosinase required for melanogenesis was only found after (chemical) induction in *Neurospora* cultures, by using an immunoassay which even detects enzymatically inactive peptide [42].

All aspects of the photo-induced development[4] of carotene synthesis in micro-organisms have been recently reviewed by Rau [64]. Some aspects not covered in his review were covered earlier [2]. The nature of the carotene development should make it most amenable to studies of transcription. As carotene does not seem to have a rapid turnover there should be a transient increase in the mRNA's for the required enzymes. The first evidence that new RNA synthesis was involved came from work with actinomycin D which only partially blocked carotenogenesis [55, 88]. In *Phycomyces,* the photoinduction of carotene is biphasic. The processes have been separated by light saturation dose and by sensitivity to actinomycin D. The low energy flux response is reported to be insensitive to actinomycin D [40] possibly explaining the previously seen partial inhibitions [55, 88]. A careful study of in vitro and in vivo transcription products may well be of assistance in further separating the two putative photo-processes as well as provide information on the transcriptional and translational controls of the pathway.

One study, bearing direct evidence of a specific increase in transcriptional control during carotenogenesis stands out [76]. Immediately after photoinduction, the relative amount of incorporation into polyadenylated RNA increases (Table 1). The data are most interesting and significant even though the method of presentation does not allow conclusions on the absolute levels of syntheses of all RNA species. Also, the methods involved a perturbation of the steady state by first having a "step down" treatment and then a pretreatment with an inhibitor of cytoplasmic protein synthesis. Still, this seems to be the first demonstration of a change in (presumably) mRNA following a cryptochromal photoinduction of carotene synthesis. One looks forward to seeing a continuation of this work showing metabolism of all RNA species up to the time of carotene synthesis and to see data for transcription and translation of specific mRNA's.

Table 1. Photo-induced incorporation into poly A^+ RNA

RNA fraction	Light : dark
Polysomal RNA	1.00
Poly(A)$^-$ RNA	1.02
Poly(A)$^+$ RNA	1.49

Fusarium aquaeductum cultures were harvested, placed in a step down buffer for 24 h, illuminated for 80 min (with cycloheximide included for the terminal) 30 min. Light and parallel dark controls were labeled separately with both ^{14}C and ^{3}H uridine for the final 10 min. A light and dark culture with opposite isotopes were mixed and RNA fractionated to a polysomal pellet or to total RNA. Total RNA was fractionated in to poly (A)$^+$ and poly(A)$^-$ on a poly U sepharose column. Data are averages of when (^{3}H) uridine and (^{14}C) uridine were used in the light and dark, and are calculated from Table 1 of Schrott and Rau [76]

4 This author believes that it is overstated to call carotene development "photomorphogenesis" [64]. Changes internal to a cell which do not modify its external shape had best be termed "development"

2.3 Transcription Following Photo-Induced Morphogenesis of Reproductive Structures

Photo-induced production of fungal reproductive structures, sexual or asexual spores is rather common and the photobiological aspects have been reviewed rather recently [91]. While it is assumed that there is synthesis of new RNA's and proteins, (or at least new levels of them), the evidence is mainly indirect. There is evidence of new mRNA synthesis, (from RNA/DNA hybridization) during induced spherulation of *Physarum* (cf. [19]). Unfortunately the inducer is not blue light; mannitol or starvation must be used.

Peculiarly there are no reports relating *Phycomyces* sporangiophore induction to nucleic acid metabolism. This should be an interesting system for such studies as some of the nonphototropic mutants do not sporulate, suggesting a common genetic lesion early in the sporulation pathway [8]. It was not reported whether the same mutants can be photo-induced to synthesize carotene.

The relationships between photoinduction and RNA synthesis have been studied in depth in only one system; the photo-induced conidiation of *Trichoderma viride*. This system has many advantages; there are two ways of inducing sporulation; a single pulse of blue light induces conidiation of the young tissue giving rise to a ring of spores (Fig. 1). The light requirement is typically "cryptochromal" and is amongst the most sensitive of the blue phenomena [33]. Conidiation is also induced by starvation, giving rise to diffuse sporulation over the dish. It is not clear how different these processes

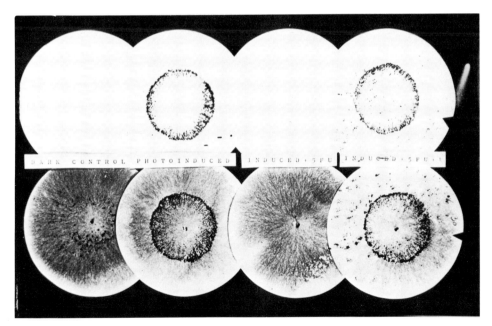

Fig. 1. The suppression of photo-induced sporulation by fluorouracil. The analog was applied at 3×10^{-5} M for 5 h beginning 30 min before a 3-min illumination of *Trichoderma* cultured on filter paper. 24 h later the papers were photographed (as in *top row*) or after staining with cotton blue. Unpublished figure but cf. [31, 32]

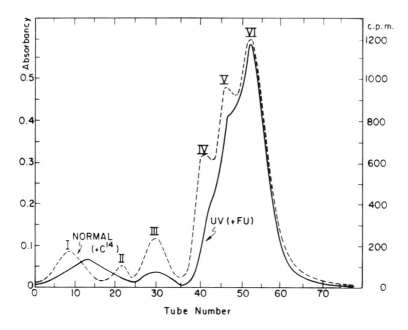

Fig. 2. Nucleic acid pattern of *Trichoderma* cultured with fluorouracil. Fluorouracil was added at 2×10^{-4} M before spore germination and 18 h later mycelia were collected and nucleic acids extracted. The nucleic acids were fractionated on a MAK column after adding a small amount of ^{14}C labeled RNA from a normal culture (as a marker). Note different pattern between the 4S(I) and 5S(TT) RNA peaks and the change in pattern in rRNA region (IV, V, VI). Reproduced from [32] with permission

really are. Formally it could be claimed that light somehow induces a transient starvation of the young tissue. This could be supported by the findings of many groups that cryptochromal blue light stimulates oxygen uptake, which probably results from a change in respiratory pathways [44]. In another system respiration could be precluded; the energy requirements to stimulate respiration are much higher [86].

The first indirect implication of nucleic acid involvement came from studies using the inhibitors 5-fluorouracil and 8-azaguanine [10–12, 31, 32]. They completely suppressed sporulation with hardly measurable effects on growth (Fig. 1). This lack of effect on growth is surprizing as RNA separation profiles of fungi cultivated in fluorouracil are quite modified (Fig. 2) (cf. [31, 32]). Nucleic acids were thus implicated as the DNA, RNA, and protein contents of the colonies vastly increased during the day it takes to make conidia [10]. This finding may be an indirect late result of the overall processes leading to conidiation.

The group that thas taken this work up summarized their working hypothesis as to the sequence of events as follows, mainly based on their inhibitor studies : Light perception → photochemical changes producing primary metabolic responses (ATP) level changes → membrane effects → sporulation specific gene transcription → novel translation → conidiation [11]. So far there are no kinetic data, even with the inhibitors, to support the sequence. The insertion of a "membrane" step before transcription might be supported by data showing that acetylcholine, in the presence of a cholinesterase inhibitor can replace light as an inducer [35].

Inhibitor studies have their drawbacks. Even inhibitors that are considered highly specific can have unknown effects. 8-Azaguanine has been thought to prevent *Trichoderma* condidiation via an effect on transcription or transcription products [12, 30, 31]. But guanine is also involved in flavin synthesis and azaguanine reduced flavin synthesis in a yeast by a far greater extent than it inhibited growth [47]. If an organism lacks a flavin photoacceptor it should have problems in reacting to light.

A genetic approach has been used which possibly contradicts some of the studies with inhibitors of translation. A nonleaky leucine minus mutant of *Trichoderma* was found which immediately stops incorporating exogenous radioactive amino acids when leucine is omitted from the medium. This suggests that there is little protein synthesis, including from internal turnover when there is no leucine. If leucine was removed just prior to induction, full conidiation still commenced if the tissue was returned to a leucine-containing medium within 6 h (Fig. 3) (Aviv and Galun, unpublished results). This may indicate that no protein synthesis is required in those critical 6 h or that the photoact can be "remembered" for quite a while (as it can be in the cold) [30].

The direct approach of trying to find new and different mRNA's during photo-induced sporulation [83] was not successful, despite successes in showing just such changes during starvation induced sporulation of slime molds [19]. Whereas DNA/RNA hybridization techniques could show the presence of different levels of mRNA species between germinating and growing mycelia, no differences between growing and photo-induced mycelia could be detected [83]. The level of detection with the methods used required gross changes, similar to those found with starvation-induced morphogenesis. The results do not preclude transcriptional changes; they indicate that they may be subtle, i.e., quantitatively small.

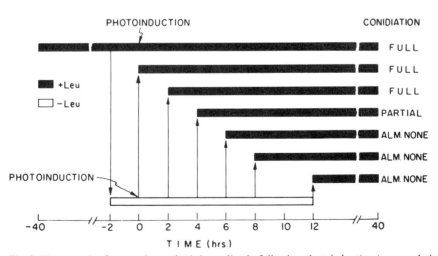

Fig. 3. The necessity for protein synthesis immediately following photoinduction for sporulation. Mutant leu^{-49} which stringently required leucine for protein synthesis was cultivated on filter paper as per [31]. 2 h before photoinduction colonies were washed and placed on a leucine-free medium and left there for various durations. Leucine was later added at 40 μg/ml (*solid bars*) and conidiation measured 40 h after induction (Aviv and Galun, unpublished)

Again, genetic studies provide evidence that different mRNA's are required for conidiation than are required for growth, provided the premise that gene products must be
transcribed to be expressed is accepted. A series of 37 nonconidiating mutants of different complementation groups were isolated. These grow normally but do not conidiate ([99] and Aviv and Galun, unpublished) suggesting that many gene products are
specifically required for conidiation. The lesions may be anywhere from photoaccep-
tion through the multitude of pathways involved in morphogenesis and may occur before the photoact.

3 Blue Light and Transcription in Ferns

Fern gametophyte growth in many species is controlled by at least two pigment systems; phytochrome which initiates germination, and cryptochrome which both suppresses the small amount of dark germination [90] and also initiates biplanar growth
of the germinated sporelings (cf. [27, 49, 51]).

3.1 Transcription Following Blue Suppression of Fern Spore Germination

An attempt was made to measure DNA, RNA, and protein synthesis during germination by radioautography, but the results were neither quantified nor compared to dark
or blue controls [61]. Immediately after imbibition and even before fern spores can be
photo-induced to germinate, radiolabeled adenosine is incorporated into a poly A^+ rich
RNA of low molecular weight (50–150 nucleotides) [101]. It is not clear if this dark
incorporation represents transcription or an enzymatic synthesis of poly A^+ tail-type
material. Immediately after phytochromal induction (with a parallel blue light illumination to suppress dark germination) there is a rapid increase in incorporation, reaching
a doubling of the rate within 16 h (Fig. 4). The initial increase of incorporation is into

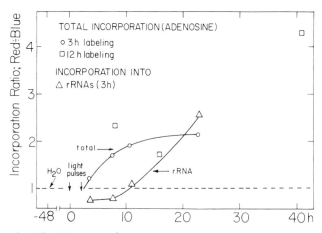

Fig. 4. Rapid increase of adenosine incorporation into RNA during fern spore red light-induced
germination. Dark germination was prevented by treating the control cultures with blue light at the
same time the double pulse of red light was presented. (From Zilberstein, Gressel and Arzee, in
preparation)

the poly A$^+$ material. Within 48 h the enhancement increases by five times and is largely due to incorporation into cytoplasmic rRNA's. It is interesting that isolated mRNA's translate new levels of some polypeptides in vitro long before these changes appear in vivo [102]. These kinds of data suggest that phytochrome sets in motion both controlled transcription and controlled translation.

3.2 Transcription During Photo-Induced Biplanar Growth

The long history of blue light initiation of biplanar prothalli has been summarized in the extensive review of Miller [49] and more recently by Furuya [27]. The protonema of many fern species continue growing unidirectionally indefinitely unless they receive blue (or white) light. Then, either the apical, or one or two subapical cells change the plane of division and a totally new type of growth ensues. Unlike the red/phytochrome treatment which can be a pulse, it seems that the blue light must be given continuously [49, 51]. The morphogenesis is not related to relative photosynthesis rates as it occurs in the presence of sucrose or when light fluxes are adjusted to give equal photosynthesis (cf. [51]).

The early cytological events after induction of biplanar growth are quite suggestive of requirements for both transcription and translation. Within 3 h of transfer to blue light, the apical region becomes bulbous [24], chloroplast diameter increases [4] but only in mature cells [7], nuclear volume increases [5, 57], nucleolus volume decreases [6], the rate of elongation decreases and the rate of cell division increases (cf. [27]).

The photoacceptor(s) for all these events may not be the same. Whereas the action spectrum for division appears typically cryptochromal [95], the spectra published for rapidly inducing apical swelling appear to have a major peak at 480 nm [97] and 6 days later seems ill-defined but blue [50]. Experiments with narrow beam light (30 μm) are highly suggestive of a rapid translational effect. Division was stimulated only when the nucleus was in the area illuminated with blue light. The nucleus is located 60 μm from the tip, but most of the plastids are in the tip region [96].

All of the morphological signs are there to suggest a rapid effect of light, leading to mobilization and division with new transcription and translation products. The field was considerably set back by attempts to demonstrate RNA involvement using inhibitors. A series of nucleic acid analogs and compounds purported to affect transcription were claimed to prevent biplanar growth (cf. [62]). Efforts to verify these findings occupied the efforts of more than two groups. The consensus was that these compounds prevent biplanar division by nonspecifically preventing growth ([16, 48], Gressel and Diamant, unpublished). Once a critical cell number is reached (which may take a considerable time) biplanar growth ensues, even in the presence of the inhibitors.

There was some difficulty in showing that there was an increase in protein and RNA syntheses upon biplanar growth in sporelings developing normally in white light (cf. [3]). Such changes could be demonstrated when blue and red light fluxes were controlled to give equal dry weight addition. The increases in RNA accumulation preceded the protein accumulation (cf. [24]). The measured differences in accumulated RNA were first discernable after 2 days in blue light, but differences in the rate of uridine incorporation were found in 3 h [24].

In a separate attempt to better study the kinetics of RNA accumulation, no definitive differences between red and blue were found. Alas, the species chosen was *Dryopteris borreri* in which biplanar growth occurs at a fixed cell number, irrespective of whether sporelings are cultivated in red or blue light [17].

Gross changes in nucleotide composition have been reported upon change to biplanar growth in unlabeled RNA [37]. These had been attributed to a change in mRNA [37] but this is hard to comprehend; such changes would require immense changes in the mRNA which is usually less than 5% of the total RNA. The changes in base composition were confirmed, but attributed to a different cause [38]; chloroplast rRNA. Chloroplast rRNA increases from 25% of the total RNA of protonemata to 45% in prothalli [38]. This increase in chloroplast rRNA occurs at some time between one day and one week after transfer to blue light [38]. This is much later than the tip swelling which is rapidly visible, suggesting that the chloroplast rRNA effect is hardly primary. Unfortunately, there are no data on base composition of chloroplast and cytoplasmic rRNA's in this species. The base ratios of chloroplast rRNA's have been studied in only a few cases. Most green algae and higher plant chloroplasts reportedly have a G + C of 51%–56%, which is not too different from the G + C of the cytoplasmic rRNA's, and there is much overlapping [25]. Thus one cannot conclusively say whether the change in the total base ratio of the ferns was really due to the increase in chloroplast rRNA.

From the above data it is evident that the fern system should be able to provide much additional evidence on transcription and translation early on after the transfer to blue light with the finer techniques which are now available.

4 Blue Light and Transcription During Greening

Greening is a fascinating complex of morphogenetic systems controlled by light. Proplastids and etioplasts contain small amounts of the major protein component of our universe, ribulose 1,5–bisphosphate carboxylase. In the chloroplast it can be 70% of the soluble protein. There are but few membranous structures in proplastids and etioplasts, and their protein and lipid compositions are very different from that of the chloroplasts. As major and minor enzymes of the chloroplast are coded in both the nucleus or plastid, at least two genomes must be mobilized for plastid development.

Unfortunately, few attempts have been made to separate the different processes into components; the possibility that there may be a multitude of photoacceptors seems to be ignored. Blue light is highly inefficient in converting protochlorophyll [26]. The phytochrome pretreatment which often reduces the lag of greening in white light is often discussed as "the sensor pigment controlling the development of ultrastructure and essential functions of the plastid . . ."[52]. Many lower plants do not contain phytochrome yet normal appearance of chlorophyll and chloroplast rRNA synthesis occurs in low-intensity green light [59, 60, 80] or in blue light as discussed below. There are many processes in which blue light is far more efficient than other wave lengths in affecting processes relating to greening, as will be discussed below.

4.1 Greening in *Euglena*

Euglena must be separated from the plant kingdom not only because of its debatable
evolutionary position; blue light is the best light inducer of normal plastid developmental processes in this genus. A blue light receptor system performs most of the tasks
of the phytochrome system (vis a vis greening) in higher plants. Green algae sometimes
contain phytochrome and their rRNA is like higher plants and quite unlike *Euglena*
(cf. [46]). Thus green algae are treated separately in the following section.

Much of the *Euglena* work is the result of a few groups and their overall findings
have been reviewed recently [56, 73].

Blue light is the best (but not sole) stimulator of transcription of both cytoplasmic
and plastid rRNA's (Fig. 5) [20] and other processes. We will therefore assume that in
cases where white light was used to stimulate greening [18, 18a, 56, 93], it is really a
blue light effect.

The series of events occurring following illumination are well documented and reviewed [18, 18a, 56] and will only be discussed briefly in the context under review. By
experimentally controlling the level of reserves to be degraded, the rate of cytoplasmic
rRNA accumulation in the light varied from negative to rapid; but the turnover rate
(thus rate of synthesis) always increased upon illumination (cf. [56]). This new rate of
synthesis could be detected within 5 min of illumination. The proportion of cytoplasmic ribosomes in polysomes increased within the first hour. The newly appearing
mRNA's have been separated by property into two categories; "photo-induced" (which
continue to appear for quite a while even if the cells are returned to the dark), and
"photo-active" (which are rapidly lost in the dark [56]). It is thought that the rapidly appearing poly A^+ RNA's are the main component of the photo-active group, which may
well be an oversimplification.

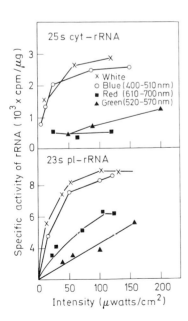

Fig. 5. The effectiveness of various spectral regions in promoting incorporation into dark-grown *Euglena*. Cells exposed to light and $^{32}P_i$ for 3 h before extracting RNA.
Reproduced in modified form from [20] with permission

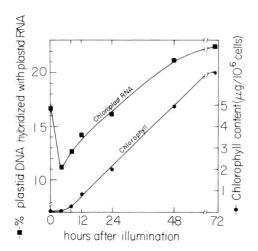

In the plastid, an interesting series of events takes place, as has been ascertained by hybridization of transcribed plastid RNA with plastid DNA [18]. There is a sharp significant initial drop in the proportion of the plastid genome transcribed, followed later by an increase to a much higher percent of genome (Fig. 6). From this it is apparent that new genes must be transcribed in the light and it is possible from the initial drop, that some genes are also being turned off. More recently, this group has shown that there are genes being turned off and new ones being turned on, by hybridizing the plastid mRNA with plastid DNA restriction fragments and a temporal program of transcription was worked out [18a]. It is interesting also that the reannealing kinetics of the plastid RNA are different at different times during the greening [18]. That some of these plastid mRNA's are new is supported by the qualitative and quantitative changes in in vivo translation patterns after illumination [13].

From the above we see some evidence for new transcription of both cytoplasmic and plastid RNA's during the lag phase of greening. How close this occurs to photoperception and how closely they are linked in time and space are yet to be shown.

4.2 Greening in Green Algae

In many respects the greening processes in algae are similar to those in *Euglena*. Considerable work has been done on blue light effects in three genera of green algae; *Chlorella, Scenedesmus,* and *Chlamydomonas.* Phytochrome is not known in them but other wavelengths have some effects on algal development. Blue light (λ_{max} 480 nm), of intensity and distribution equivalent to light penetrating to 10 m in the sea is equivalent to white light in bringing about normal plastid development in marine algae (cf. [94]). Red light effects usually require more radiant energy and are thus (along with evidence presented below) considered to be photosynthesis-driven. There seem to be two separate light reactions which can promote cell division in synchronized *Chlorella* [78]. The blue peak (λ_{max} 485 nm) which is considered to be more morphogenetically specific is not affected by the Hill reaction inhibitor DCMU, whereas the stimulation by the red (λ_{max}

674 nm) is suppressed [78]. It was also possible to separate the red- from the blue-induced stimulation of algal nucleic acid synthesis using DCMU (see below) [20]. Both
red and blue stimulated RNA accumulation in wild-type *Chlorella,* but continuous blue
light alone stimulated RNA accumulation in an achlorophyllous mutant [20]. The action
spectra for various blue-stimulated phenomena in greening are not fully identical [43]
and the spectrum for RNA accumulation does not appear to be cryptochromal [20].

The most rapid blue light effect in algae seems to be enhanced respiration (O_2 consumption). The enhancement, which has a cryptochromal action spectrum [15, 58],
is measurable within minutes, with the kinetics extrapolatable to the time of illumination [15, 74]. This is interesting, as other cryptochrome-controlled events can be induced in the absence of oxygen [30]. The blue stimulation of respiration can be separated from the blue stimulation of RNA synthesis; the enhancement of incorporation
into RNA occurs at much lower energy levels than those effecting increased oxygen uptake [86].

There were/are problems in understanding which RNA species are stimulated by
blue light. There are no kinetic data showing how soon after blue illumination the stimulated incorporation ensues. Does RNA synthesis jump in the 2-h lag preceding chlorophyll synthesis? Presumably it should, as absolute protein increases during this period.
Initially, it was thought that the stimulation of cytoplasmic rRNA was greater than the
stimulation of chloroplast rRNA [84, 85]. This may only be a result of the slower processing of chloroplast rRNA's (cf. [71]). Additionally, fractionations were run under
circumstances where the precursor of the plastid rRNA's cannot be assessed as it remains at the origin of the gel [84–86]. Conditions could have been altered to allow the
precursor to have entered (cf. [34]). They suggest that the chloroplast rRNA synthesis
may be secondary as it can be suppressed by inhibitors without affecting cytoplasmic
rRNA [85], but the converse experiment was not reported.

Blue light has been found to be stimulatory for the synthesis of 4S and 5S RNA's
as well, within 5 min of illumination [87]. The RNA transcribed under these fast kinetic
conditions hybridizes solely to nuclear DNA [87].

4.3 Blue-Induced Transcription During Greening of Phytochrome-Containing Plants

In previous sections it was possible to get some "white" light studies in "under the line"
because of the lack of other pigments. As was pointed out in Sect. 4, different wavelengths may stimulate different partial processes during greening. Thus, it is unfortunate
that so many workers doing excellent biochemistry do not "split" the light to separate
the phenomenon (e.g. [79]).

In higher plants the lag period between illumination and the linear appearance of
chlorophyll can be shortened by (far-red reversible) red light. Thus phytochrome is un-
questionably involved in the preparation for greening. Using 15-s red illumination (far-
red reversible), it was possible to show the appearance in dark-grown barley of mRNA
activity for the in vitro translation of a thylakoid protein of great importance; the
26,000 MW apoprotein of the light harvesting chlorophyll a/b complex [1]. This poly
A^+ mRNA is of nuclear origin and is translated in the cytoplasm. Continuous white light
is required to have this protein inserted into the thylakoids [1]. Blue light was not reported as tested in this system. In isolated dark-grown pea roots, blue light could bring

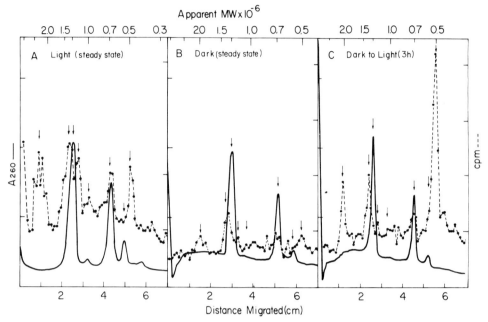

Fig. 7. The effect of light on the rate of incorporation into the 0.5×10^6 MW plastid mRNA. Whole cell RNA was fractionated on polyacrylamide gels. Reproduced in modified form from [69] with permission

about formation of the complete a/b complex while red light stimulated only partial processes (Richter, personal communication).

There is an enigmatic plastid transcription/translation clearly stimulated by blue light. This is a well-characterized plastid mRNA with an apparent molecular weight of 0.5×10^6, which appears as a discrete peak in pulse labeled plants (Fig. 7) [69, 82, 100]. When steady-state dark-grown plants of *Spirodela* are transferred to the light there is a sudden jump in relative incorporation into this mRNA at 3 h, well before stimulation of plastid rRNA synthesis and before the appearance of chlorophyll (Fig. 8) [69]. Continuous blue light is far more effective than red light in stimulating the appearance of this mRNA (Fig. 9), pulsed light is totally ineffective [29]. Various types of correlations have been made between this mRNA and a translation product; a high turnover thylakoid protein of 32,000 MW. The first correlations were temporal; the 32,000 MW protein appeared as a major translation product both in vitro and in vivo (Fig. 10). At the same time there was an increase in 0.5×10^6 MW mRNA levels [65]. It is interesting to note the timing, in line with the phytochrome-induced appearance of the 26,000 MW chlorophyll a/b apoprotein. During white light-induced greening the 32,000 MW protein appears in vivo well before the 26,000 MW protein (Fig. 10). Later a biochemical correlation was shown; the 0.5×10^6 MW mRNA was most efficient in translating the 32,000 MW protein [66]. More recently we have found that blue light is far more effective than red light in stimulating *Spirodela* to incorporate ^{35}S methionine into the 32,000 MW thylakoid protein and its 33,500 MW precursor during regreening. Blue light also hastened the processing of the precursor, in an intensity-dependent process

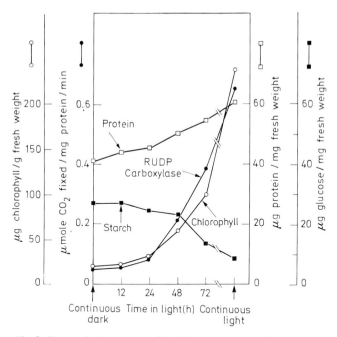

Fig. 8. Changes in the activity of RuDP carboxylase and in the amounts of chlorophyll, soluble proteins and starch after illumination of dark-grown *Spirodela* plants. Reproduced from [69] with permission

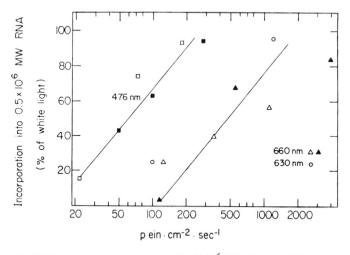

Fig. 9. Enhanced incorporation into 0.5×10^6 MW plastid mRNA as a function of red and blue light. Steady-state dark-grown plants were illuminated with blue or red light and labeled for the terminal h, and the RNA fractionated at 4 h. Data are presented as percent of parallel white light illuminated and labeled dark-grown plants which received 10 W/m^2 under the same conditions. Reproduced in modified form from [29] by permission

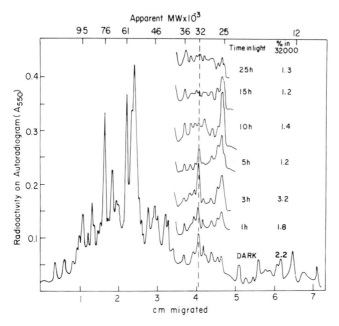

Fig. 10. Fractionation of total membrane proteins following a 1 h ^{35}S methionine incorporation in vivo during photo-induced *Spirodela* greening. Equal amounts of radioactivity were placed in each slot on the polyacrylamide gel. The complete densitometric scan of the autoradiograph is presented only for the membrane proteins of steady-state dark-grown fronds. The inserts are scans of the region containing the 32,000 and 26,000 MW proteins for the other time points. Reproduced in modified form from [65] by permission

(Mattoo, Gressel and Edelman, unpublished). The DCMU-type experiments performed with algae (cf. Sect. 4.2) to try to separate the blue and red effects have yet to be performed in *Spirodela*.

It is yet a problem to suggest a function for this blue-stimulated plastid mRNA synthesis and its thylakoid product. An early hypothesis based on its early appearance during greening suggested that the protein might be related to development. Recent experiments showed that in normal plants the mRNA and its product are synthesized in greater relative amounts in tissue with fully developed plastids than in young developing tissue [93, 100]. Because of the high turnover of the protein it was possible to deplete plastids of most of the 32,000 MW protein with chloramphenical, yet photosynthetic CO_2 fixation could continue. Thus, the protein seems not to be a direct part of photosynthesis.

During greening there is a relative increase in plastid rRNA's [70] due to a change in both the rate of synthesis and in processing [68]. In isolated pea roots this increase was found to be blue-dependent [67]. It was reported that blue light was more effective than red light at equal fluences (Richter, personal communication).

5 Concluding Remarks

A key question that must be asked is about the primacy of transcription; how close is it to the photoact? What processes are affected first? It was pointed out earlier that light is a more esthetic inducer of morphogenesis than starvation. Many of the phenomena induced by light could be also induced by starvation or envisaged as being caused by a transient starvation. Phototropism may be the result of less growth on the lit side. Enhanced respiration, pigmentation, and sporulation can be results of either light or stress conditions. Light can photoactivate enzymes [21, 39] and it can also photoinactivate them [54, 75]. Low [89] and high [92] intensity light inhibit growth of bacteria and yeast. Part of this can be a direct effect on RNA; 4-thiouridine containing tRNA is inactivated by blue light [63]. Many of the enzymes inactivated by blue light are flavin enzymes which have a cryptochromal absorption spectrum. Could cryptochrome be an early evolutionary attempt at photocontrol which utilizes the inactivation of many enzymes? The heterogeneity of action spectra for various controlled events within the same organism may be suggestive of this (cf. Sect. 4.2). Thus, it was pleasant to note that in at least one case, where measurements of fluxes were made, transcription was stimulated with much less light than was required for starvation-simulating processes. Still, precise kinetics are required to show how close to illumination blue-stimulated transcriptions begin in discussing every process.

In inductive processes it is easier to work with pulses of the inducer. Many such phytochromal processes exist. Of all the processes described herein only a few can be pulsed: Photo-induced sporulation (cf. Sect. 2.3), some carotene synthesis [64] and the suppression of dark germination of fern spores (cf. Sect. 3.1). All other processes described seem to perceive continuous blue light.

Of the various photo-induced processes described, only greening is known to require massive changes in protein constituents. Thus, it is almost naive to expect to be able to find specific mRNA's for specific proteins for the other processes discussed. Considering both the huge potential genome size of the nucleus, the proportional change in enzymes required for any process will be minute. In the plastid we have a different situation; a small genome which codes for one protein in massive quantities (the large subunit of ribulose bisphosphate carboxylase) as well as some thylakoid proteins. Thus in the "prokaryotic" part of the genome it should be possible to find more specific mRNA's, as has be done with other prokaryotes.

References

1. Apel K (1979) Phytochrome-induced appearance of mRNA activity for the apoprotein of the light-harvesting chlorophyll a/b protein of barley (*Hordeum vulgare*). Eur J Biochem 97: 183–188
2. Batra PP(1971) Mechanism of light-induced carotenoid synthesis in nonphotosynthetic plants. In: Giese (ed) Photophysiology, vol VI, pp 47–76. Academic Press, London New York
3. Bell PR, Zafar AH (1961) Changes in the level of the protein nitrogen during growth of the gametophyte and the initiation of the sporophyte of *Dryopteris borreri* Newm. Ann Bot 25: 531–546
4. Bergfeld R (1963) Die Wirkung von hellroter und blauer Strahlung auf die Chloroplastenausbildung. Z Naturforsch 18b: 328–331

5. Bergfeld R (1963) Die Beeinflussung der Zellkerne in den Vorkeimen von *Dryopteris filix-mas* durch rote und blaue Strahlung. Z Naturforsch 18b: 557–562

6. Bergfeld R (1967) Kern- und Nucleolusausbildung in den Gametophytenzellen von *Dryopteris filix-mas* (L.) Schott bei Umsteuerung der Morphogenese. Z Naturforsch 22b: 972–976

7. Bergfeld R (1970) Feinstrukturen der Chloroplasten in den Gametophytenzellen von *Dryopteris filix-mas* (L.) Schott nach Einwirkung hellroter und blauer Strahlung. Z Pflanzenphys 63: 55–64

8. Bergman K (1972) Blue-light control of sporangiophore initiation in *Phycomyces.* Planta 107: 53–67

9. Bergman K, Burke PV, Cerda-Olmedo E, David CN, Delbrück M, Foster KW, Goodell EW, Heisenberg M, Meissner G, Zalokar M, Dennison DS, Shropshire W (1969) Phycomyces. Bacteriol Rev 33: 99

10. Betina V, Spisiakova J (1976) Suppression of photo-induced sporulation in *Trichoderma viride* by inhibitors. Folia Microbiol 21: 362–370

11. Betina V, Spisiakova J, Zajacova J (1978) Photoinduced conidiation in *Trichoderma viride* is prevented by inhibitors of RNA and protein synthesis. 12th FEBS Meet, Dresden, July 2–8, Abstr no 0411

12. Betina V, Zajacova J (1978) Inhibition of photo-induced *Trichoderma viride* conidiation by inhibitors of RNA synthesis. Folia Microbiol 23: 460–464

13. Bingham S, Schiff JA (1976) Cellular origins of plastid membrane polypeptides in *Euglena.* In: Mucher TH et al. (eds) Genetics and biogenesis of chloroplasts and mitochondria, pp.79–86. North Holland, Amsterdam

14. Bramley PM, Davies BH (1975) Carotene biosynthesis by cell extracts of mutants of *Phycomyces blakesleeanus.* Phytochemistry 14: 463–465

15. Brinkmann G, Senger H (1978) The development of structure and function in chloroplasts of greening mutants of *Scenedesmus* IV. Blue light-dependent carbohydrate and protein metabolism. Plant Cell PHysiol 19: 1427–1437

16. Burns RG, Ingle J (1968) The induction of biplanar growth in fern gametophytes in the presence of RNA base analogues. Plant Physiol 43: 1987–1990

17. Burns RG, Ingle J (1970) The relationship between the kinetics of ribonucleic acid accumulation and the morphological development of the fern gametophyte, *Dryopteris borreri.* Plant Physiol 46: 423–428

18. Chelm BK, Hallick RB (1976) Changes in the expression of the chloroplast genome of *Euglena gracilis* during chloroplast development. Biochemistry 15: 593–599

18a. Chelm BK, Hallick RB, Gray PW (1979) Transcription program of the chloroplast genome of *Euglena* during chloroplast development. Proc Natl Acad Sci USA 76: 2258–2262

19. Chet I (1973) Changes in ribonucleic acid during differentiation of *Physarum polycephalum.* Ber Dtsch Bot Ges 86: 77–92

20. Cohen D, Schiff JA (1976) Events surrounding the early development of *Euglena* chloroplast 10. Photoregulation of the transcription and translation of chloroplastic and cytoplasmic rRNA. Arch Bioch Biophys 177: 201–216

21. Cohen RJ, Atkinson MM (1978) Activation of *Phycomyces* adenosine 3', 5'-monophosphate phosphodiesterase by blue light. Biochem Biophys Res Commun 83: 616–621

22. Creutz C, Colombetti G, Diehn B (1978) Photophobic behavioral responses of *Euglena* in a light intensity gradient and the kinetics of photoreceptor pigment interconversions. Photochem Photobiol 27: 611–616

23. Doi RH, Igarashi RT (1964) Genetic transcription during morphogenesis. Proc Natl Acad Sci USA 52: 755–762

24. Drumm H, Mohr H (1967) Die Regulation der RNS-Synthese in Farngametophyten durch Licht. Planta 72: 232–246

25. Ellis RJ, Hartley MR (1974) Nucleic acids of chloroplasts. In: Burton K (ed) Biochemistry of nucleic acids, MTP Int Rev Sci Biochem Ser 1, vol VI, pp 1–26. Butterworth, London

26. Frank SR (1946) The effectiveness of the spectrum in chlorophyll formation. J Gen Physiol 29: 157–179

27. Furuya M (1978) Photocontrol of developmental processes in fern gametophytes. Bot Mag Tokyo Spec Issue 1: 219–242

28. Graafmans WDJ (1977) Effect of blue light on metabolism in *Penicillium isariiforme.* J Gen Microbiol 101: 157–161
29. Gressel J (1978) Light requirements for the enhanced synthesis of a plastid mRNA during *Spirodela* greening. Photochem Photobiol 27: 167–169
30. Gressel J, Bar-Lev S, Galun E (1975) Blue light induced response in the absence of free oxygen. Plant Cell Physiol 16: 367–370
31. Gressel J, Galun E (1967) Morphogenesis in *Trichoderma:* Photoinduction and RNA. Dev Biol 15: 575–598
32. Gressel J, Galun E (1970) Sporulation in "Trichoderma": a model system with analogies to flowering. In: Bernier G (ed) Cellular and molecular aspects of floral induction, pp 152–170. Longman, Oxford
33. Gressel J, Hartmann KM (1968) Morphogenesis in *Trichoderma*: Action spectrum of photoinduced sporulation. Planta 79: 271–274
34. Gressel J, Rosner A, Cohen N (1975) Temperature of acrylamide polymerization and electrophoretic mobilities of nucleic acids. Anal Biochem 69: 84–91
35. Gressel J, Strausbauch L, Galun E (1971) Photomimetic effect of acetylcholine on morphogenesis in *Trichoderma.* Nature (London) 232: 648
36. Hartmann KM, Cohnen-Unser I (1972) Analytical action spectroscopy with living systems: Photochemical aspects and attenuance. Ber Dtsch Bot Ges 85: 481–551
37. Hotta Y, Osawa S, Sakaki T (1959) Ribonucleic acid and differentiation of the gametophyte of a polypodiaceous fern. Dev Biol 1: 65–78
38. Howland GP (1972) Changes in amounts of chloroplast and cytoplasmic ribosomal-RNAs and photomorphogenesis in *Dryopteris* gametophytes. Physiol Plant 26: 264–270
39. Hug DH (1978) The activation of enzymes with light. In: Smith KC (ed) Photochemical and photobiological reviews, vol III, pp 1–35. Plenum Press, New York
40. Jayaram M, Presti D, Delbrück M (1979) Light-induced carotene synthesis in *Phycomyces.* Exp Mycol 3: 42–52
41. Johnson JH, Reed BC, Rilling HC (1974) Early photoinduced enzymes of photoinduced carotenogenesis in a *Mycobacterium* species. J Biol Chem 249: 402–406
42. Katan T, Arnon R, Galun E (1975) Immunochemical studies on tyrosinase induction in *Neurospora.* Eur J Biochem 59: 387–394
43. Klein O, Senger H (1978) Biosynthetic pathways to aminolevulinic acid induced by blue light in the pigment mutant C-2A' of *Scenedesmus obliquus.* Photochem Photobiol 27: 203–208
44. Kowallik W, Gaffron H (1967) Enhancement of respiration and fermentation in algae by blue light. Nature (London) 215: 1038–1040
45. Krinsky NI (1968) The protective function of carotenoid pigments. In: Giese AC (ed) Photophysiology, vol III, pp 123: 195. Academic Press, London New York
46. Loening UE (1968) Molecular weights of rRNA in relation to evolution. J Mol Biol 38: 355–365
47. Madia AM, Mattoo AK, Modi VV (1975) 8-Azaguanine and flavinogenesis in *Eremothecium ashbyii.* Biochim Biophys Acta 385: 51–57
48. Miller JH (1968) An evaluation of specific and non-specific inhibition of 2-dimensional growth in fern gametophytes. Physiol Plant 21: 699–710
49. Miller JH (1968) Fern gametophytes as experimental material. Bot Rev 34: 361–440
50. Mohr H (1956) Die Abhängigkeit des Protonemawachstums und der Protonemapolarität bei Farnen vom Licht. Planta 47: 127–158
51. Mohr H (1964) The control of plant growth and development by light. Biol Rev 39: 87–112
52. Mohr H, Oelze-Karow H (1978) Phytochrome and chloroplast development. In: Akoyunoglou G et al. (eds) Chloroplast development, pp 769–779. Elsevier North-Holland Biomedical Press, New York
53. Mohr H, Schopfer P (1978) The effect of light on RNA and protein synthesis in plants. In: Bogorad L, Weil JH (eds) Nucleic acids and protein synthesis in plants, pp 239–260. Plenum Press, New York London
54. Montagnoli G (1974) Biological effects of light on proteins: Enzyme activity modulation (yearly review). Photochem Photobiol 26: 679–683

55. Mummery RS, Valadon LRG (1978) Effects of certain nucleic acid and protein inhibitors on carotenogenesis in *Verticillium agaricinum*. Physiol Plant 28: 254–258
56. Nigon V, Heizmann P (1978) Morphology, biochemistry, and genetics of plastid development in *Euglena gracilis*. Int Rev Cytol 53: 211–290
57. Ootaki T (1963) Modification of the developmental axis by centrifugation in *Pteris vittata*. Cytologia 28: 21–29
58. Pickett JM, French CS (1967) The action spectrum for blue light stimulated oxygen uptake in *Chlorella*. Proc Natl Acad Sci USA 57: 1587–1593
59. Possingham JV (1973) Effect of light quality on chloroplast replication in spinach. J Exp Bot 24: 1247–1260
60. Possingham JV, Cran DG, Rose RJ, Loveys BR (1975) Effects of green light on the chloroplasts of spinach leaf discs. J Exp Bot 26: 33–42
61. Raghavan V (1977) Cell morphogenesis and macromolecule synthesis during phytochrome-controlled germination of spores of the fern, *Pteris vittata*. J Exp Bot 28: 439–456
62. Raghavan V, Tung HF (1967) Inhibition of two dimensional growth and suppression of RNA and protein synthesis in the gametophytes of the fern, *Asplenium nidus,* by chloramphenicol, puromycin and actinomycin D. Am J Bot 54: 198–205
63. Ramabhadran TV, Jagger J (1976) Mechanism of growth delay induced in *Escherichia coli* by near ultraviolet radiation. Proc Natl Acad Sci USA 73: 59–63
64. Rau W (1980) Photoregulation of carotenoid biosynthesis: An example of photomorphogenesis. In: Czygan HCh (ed) Pigments in plants. Fischer, Stuttgart in press
65. Reisfeld A, Gressel J, Jakob KM, Edelman M (1978) Characterization of the 32,000 dalton membrane protein – I. Early synthesis during photoinduced plastid development of *Spirodela*. Photochem Photobiol 27: 161–165
66. Reisfeld A, Jakob KM, Edelman M (1978) Characterization of the 32,000 dalton chloroplast membrane protein. II. The molecular weight of chloroplast messenger RNAs translating the precursor to P-32000 and full-size RUDP carboxylase large subunit. In: Akoyunoglou G et al. (eds) Chloroplast development, pp 669–674. Elsevier/North Holland, Amsterdam
67. Richter G, Dirks W (1978) Blue-light induced development of chloroplasts in isolated seedling roots. Preferential synthesis of chloroplast ribosomal RNA species. Photochem Photobiol 27: 155–160
68. Rosner A, Gressel J (1977) Discoordination of rRNA metabolism during metabolic shifts of *Spirodela* plants. Biochem Biophys Acta 474: 386–397
69. Rosner A, Jakob KM, Gressel J, Sagher D (1975) The early synthesis and possible function of a 0.5×10^6 mRNA after transfer of darkgrown *Spirodela* plants to light. Biochem Biophys Res Commun 67: 383–391
70. Rosner A, Porath D, Gressel J (1975) The distribution of plastid ribosomes and the integrity of plastid rRNA during the greening and maturation of *Spirodela* fronds. Plant Cell Physiol 16: 891–902
71. Rosner A, Posner HB, Gressel J (1973) RNA synthesis in *Lemna*: Characterization by gel electrophoresis. Plant Cell Physiol 14: 555–564
72. Roth A, Ashworth JM, Sussman M (1968) Periods of genetic transcription required for the synthesis of three enzymes during cellular slime mold development. Proc Natl Acad Sci USA 59: 1235–1242
73. Schiff JA (1978) Photocontrol of chloroplast development in *Euglena*. In: Akoyunoglou G et al.(eds) Chloroplast development, pp 747–768. Elsevier/North-Holland Biochemical Press, Amsterdam
74. Schmid GH (1969) Blue light enhanced respiration in a colorless *Chlorella* mutant. Z Physiol Chem 350: 1513–1520
75. Schmid GH (1970) The effect of blue light on some flavin enzymes. Z Physiol Chem 351: 575–578
76. Schrott EL, Rau W (1977) Evidence for a photoinduced synthesis of poly (A) containing mRNA in *Fusarium aquaeductuum*. Planta 136: 45–48
77. Senger H, Bishop NI (1968) An action spectrum for nucleic acid formation in an achlorophyllous mutant of *Chlorella pyrenoidosa*. Biochim Biophys Acta 157: 417–419

78. Senger H, Schoser G (1966) Die spektralabhängige Teilungsinduktion in mixotrophen Synchronkulturen von *Chlorella*. Z Pflanzenphysiol 54/4: 308–320
79. Siddell SG, Ellis RJ (1975) Protein synthesis in chloroplasts: characteristics and products of protein synthesis in vitro in etioplasts and developing chloroplasts from pea leaves. Biochem J 146: 675–685
80. Simonova EI, Kudinova LI, Novikova NS (1977) Regulation of chlorophyll biosynthesis by light of different spectral compositions. Fiziol Rast 24: 1154–1158
81. Skaar H, Johnsson A (1978) Rapid, blue-light induced transpiration in *Avena*. Physiol Plant 43: 390–396
82. Speirs J, Grierson D (1978) Isolation and characterization of 14-S RNA from spinach chloroplasts. Biochim Biophys Acta 521: 619–633
83. Stavy (Rodeh) R, Galun E, Gressel J (1972) Morphogenesis in *Trichoderma:* RNA DNA hybridization studies. Biochim Biophys Acta 259: 321–329
84. Steup M (1975) Die Wirkung von blauem und rotem Licht auf die Synthese ribosomaler RNA bei *Chlorella*. Arch Microbiol 105: 143–151
85. Steup M (1977) Blue light-dependent regulation of cytoplasmic ribosomal RNA synthesis in *Chlorella*. Arch Microbiol 112: 277–282
86. Steup M, Ssymank V (1978) Photoregulation of ribosomal RNA synthesis in a mutant of *Chlamydomonas reinhardii*. Ber Dtsch Bot Ges 91: 243–250
87. Steup M, Ssymank V, Winkler U, Glock H (1977) Photoregulation of transfer and 5S ribosomal RNA synthesis in *Chlorella*. Planta 137: 139–144
88. Subden RE, Bobowski G (1973) Evidence for actinomycin D inhibition of transcription of carotenoid loci in *Neurospora*. Experientia 29: 965–967
89. Sulkowski E, Guerin B, Defaye J, Slonimski PP (1964) Inhibition of protein synthesis in yeast by low intensities of visible light. Nature (London) 202: 36–39
90. Sugai M (1971) Photomorphogenesis in *Pteris vittata* IV. Action spectra for inhibition of phytochrome-dependent spore germination. Plant Cell Physiol 12: 103–109
91. Tan KK (1978) Light-induced fungal development. In: Smith JE, Berry DR (eds) The filamentous fungi, vol III, p 9. E Arnold, London
92. Taylor BL, Koshland Jr DE (1975) Intrinsic and extrinsic light responses of *Salmonella typhimurium* and *Escherichia coli*. J Bacteriol 123: 557–569
93. Verdier G (1975) Synthesis and translation site of light-induced mRNAs in etiolated *Euglena gracilis*. Biochim Biophys Acta 407: 91–98
94. Vesk M, Jeffrey SW (1977) Effect of blue-green light on photosynthetic pigments and chloroplast structure in unicellular marine algae from six classes. J Physiol 13: 280–288
95. Wada M, Furuya M (1974) Action spectrum for the timing of photo-induced cell division in *Adiantum* gametophytes. Physiol Plant 32: 377–381
96. Wada M, Furuya M (1978) Effects of narrow-beam irradiations with blue and far-red light on the timing of cell division in *Adiantum* gametophytes. Planta 138: 85–90
97. Wada M, Kadota A, Furuya M (1978) Apical growth of protonemata in *Adiantum capillus-veneris* II. Action spectra for the induction of apical swelling and the intracellular photoreceptive site. Bot Mag 91: 113–120
98. Weinbaum SA, Gressel J, Reisfeld A, Edelman M (1979) Characterization of the 32000 dalton chloroplast membrane protein III. Probing its biological function in *Spirodela*. Plant Physiol 64: 828–832
99. Weinman-Greenshpan D, Galun E (1969) Complementation in nonconidiation mutants of *Trichoderma*. J Bacteriol 99: 802–806
100. Wollgiehn R, Lerbs S, Munsche D (1978) Eigenschaften einer Chloroplasten RNA vom Molekulargewicht 0.5×10^6 aus *Nicotiana rustica*. Biochem Physiol Pflanz 173: 60–66
101. Zilberstein A, Gressel J, Arzee T (1976) RNA synthesis at the onset of fern spore germination. Isr J Bot 25: 97
102. Zilberstein A, Gressel J, Arzee T, Edelman M (1979) Differentiation between control events in fern spore germination. Plant Physiol 63S: 18

Photoreceptors and Primary Reactions

Spectroscopic and Photochemical Characterization of Flavoproteins and Carotenoproteins as Blue Light Photoreceptors

P.S. SONG[1]

1 Introduction

One of the most fundamental processes in nature involves interactions between organisms and light ranging in wavelength from 260 nm to 900 nm. Photosensory transductions in many organisms exemplify such interactions in terms of direct utilization of light energy into biological responses. In particular, blue light is widely perceived by a number of organisms in their photosensory behaviors (e.g., phototaxis and phototropism). The aim of the present paper is to elucidate the excited states of blue light photoreceptors (mainly flavoproteins and carotenoproteins) in terms of their spectroscopy and photoreactivity. Thus, this paper represents an up-dated version of our previous work (Song et al. 1972). We also suggest various molecular mechanisms possible with these photoreceptors on the basis of spectroscopic and photochemical properties of their electronic excited states. To do so, we start by reviewing some of the molecular relaxation processes of photoreceptors which may be coupled to the primary trigger system of photosensory responses of organisms toward blue light.

A photoreceptor bound to a protein and/or membrane absorbs a quantum of blue light. The light-energized photoreceptor may then relax to the ground state via several modes of transition. Such relaxation modes can be coupled to dynamic as well as static conformation changes of protein and/or membranes to which the photoreceptor is bound. The coupled relaxation may then affect the membrane potential, structural specificity of the photoreceptor relative to its effector site, and regulation of growth regulators or hormones, etc. Several modes of coupling between photoreceptors' relaxation processes and the macromolecular conformation can be envisioned in terms of the following schemes.

Photophysical Processes. Internal conversion, intersystem crossing and radiative emission from the lowest electronic excited state are not by themselves useful for effecting conformation changes of protein or membrane. However, heat produced from the radiationless transitions can be used for "local heating" of the photoreceptor proper, depending on the thermal conductivity of the photoreceptor binding environment. In some instances, it is feasible to use the local heating to induce reorientation of the photoreceptor itself. For example, Albrecht (1958) has shown theoretically that local heating may cause local orientation of the solute molecule due to thermally induced rotational dif-

---

1 Department of Chemistry, Texas Tech University, Lubbock, TX 79409, USA

fusion. However, there is no evidence for or against the possibility of local heating being a factor in the blue light-induced photoresponses (vide infra).

Polarizability. The polarizability, both isotropic and anisotropic, of photoreceptor changes upon excitation by light. Because of its instantaneous evolution and disappearance in concert with the light-absorption process, the utility of polarizability change as a driving force for conformational change is probably very limited, although anisotropic polarizability may have a significant effect on the microenvironmental dynamics of the chromophore binding site.

Excited State Dipole Moment. The permanent dipole moment of a photoreceptor usually changes upon light excitation, entailing reorientation of the photoreceptor and/ or surrounding residue dipoles of the protein or membrane.

Acid-Base Equilibria. The pK_a of acidic and basic centers of photoreceptor can change dramatically upon light excitation. Aside from the obvious implication in the light-induced proton flux and transmembrane potential associated with such a change, the pK_a change due to pK_a^* may also induce static and dynamic conformational changes of interacting proteins and/or membranes. In order to generate a proton gradient, it is assumed that an efficient proton-conducting network is established in association with the acid-base groups, as found in *Halobacterium halobium* (Danon and Stockenius 1974), and as has been suggested for the photophobic response of Stentor (Walker et al. 1979). Phototautomerism involving proton transfer in a photoreceptor molecule can also be considered in the same context as in the acid-base equilibria.

Conformational Change and Isomerization. Electronic excitation can bring about conformational changes in the photoreceptor chromophore. Photoisomerization in rhodopsin is well known. These chromophore changes are often accompanied by conformation changes of the interacting protein and/or membrane. Photo-induced viscosity changes of a polymer solution arising from the photoisomerization of bound azo dyes serve as a good illustration of this effect (Lovrien 1967).

Photodissociation. Photodissociation equilibria of molecular complexes are well known. These include not only dissociation of molecular complexes upon light excitation, but also the formation of excimers and exciplexes during the excited state lifetime. The former may temporarily dissociate in the excited state or in the metastable state, resulting in conformational changes in the protein or membrane. The reverse is also possible (e.g., binding of phytochrome P_{fr} produced from the excitation of Pr; Song et al. 1979). The dissociation may be brought about after photoreactions, as in the photoisomerization of 11-cis retinal rhodopsin to all-trans retinal and release of opsin.

Photoredox Reactions. It is possible that a photoredox reaction may have a direct role in producing light-induced membrane potentials, as found in the photosynthetic electron transport.

2 Materials and Methods

Various spectroscopic equipments such as subnanosecond lifetime and time-resolved spectrometers and ORD-CD spectrometer have been used, as described in our previous publications (Fugate and Song 1976; Song et al. 1976, 1979; Walker et al. 1979).

The isolation procedure for the plasma membrane bound flavoprotein(s) from seedlings of *Zea mays* L. (WF9xBear 39) was based on differential centrifugation as described by Brain et al. (1977), with modifications described elsewhere (Song et al. 1980). The photosynthetic antenna complexes of dinoflagellates were isolated as described previously (Song et al. 1976; Koka and Song 1977).

3 Summary of Results

3.1 Flavoprotein Photoreceptor

In contrast to the well-characterized photoreceptors such as chlorophylls, phytochrome, and rhodopsin, the identity of blue light photoreceptors is neither certain nor unique in terms of the two likely chromophore candidates, flavin and carotenoid (Song and Moore 1974). However, in many systems such as *Neurospora* (Muñoz and Butler 1975), *Phycomyces* (Delbrück et al. 1976; Presti et al. 1977), *Euglena* (Diehn and Kint 1970) and others, flavin now seems to be the logical photoreceptor, based on various types of information including spectroscopic and photochemical considerations (Song et al. 1972; Song and Moore 1974; Song 1977). The difficulty in the isolation of the flavoprotein photoreceptor lies in the fact that flavin is ubiquitous in living organisms and is the coenzyme for numerous flavoenzymes, thus making the task of finding a unique photoreceptor flavin extremely difficult. We report herein an attempt to find a flavin photoreceptor from corn coleoptiles.

Coleoptiles have long been known to be the photoactive portion of plant phototropism, as the coleoptiles with their tips shielded from light do not phototropically respond to blue light. In *Avena* coleoptiles, the phototropic response is maximal for the upper 50 μm and it is ca. 4000 times less sensitive 2 mm further down the tip (Briggs 1963).

The photoreceptor for phototropic response in *Avena* and other plants seems to be plasma membrane-bound (Brain et al. 1977). This is suggested by the fact that auxin is laterally translocated across the coleoptile on irradiation with blue light, presumably as a result of light-mediated transport across cell membranes, and that in *Phycomyces* the photoreceptor is thought to be oriented with respect to the axes of the organism (Jesaitis 1974).

In spite of the apparent implication for the involvement of a flavoprotein as the blue light photoreceptor in corn and other organisms, no flavoprotein photoreceptors have ever been isolated and characterized. There are two formidable difficulties which cannot be readily overcome; these are, (a) the concentration of flavin photoreceptors is likely to be quite low and (b) once isolated, the assay for the flavin photoreceptor function is not available, unlike other photoreceptor systems such as phytochrome and rhodopsin.

Table 1. Protein and flavin content in various fractions from differential centrifugation of corn coleoptiles

Fraction	Protein[a] (mg/ml)	Flavin[b] (mol/mg)$\times 10^{10}$	NADPH dehydrogenase activity[c] (ΔA_{600}/s/mg) $\times 10^3$
Crude extract	2.91	7.11	—
9 kp	1.91	5.86	36
21 kp	2.72	4.78	36
50 kp	2.78	3.13	62
50 ks	1.42	7.89	76

[a] Determined with the Bio-Rad protein assay reagent
[b] Determined by the lumiflavin method (Yagi 1972). This determination does not include covelently bound flavins
[c] Dichlorophenolindophenol as electron acceptor, with absorbance changes measured at 600 nm

Characterization of a Membrane-Bound Flavoprotein from Corn Coleoptile. We have studied flavins specifically to ascertain wether or not their excited state dynamics are compatible with the primary photoreactivity required for initiating the phototropic transduction mechanism. We have also examined a preparation from the corn coleoptile tips that is enriched in what is probably plasma membrane (Brain et al. 1977) and photoresponsive with respect to the light-induced reduction of a specific b type cytochrome, which has been proposed to be involved in the blue light response (Schmidt and Butler 1976; Brain et al. 1977). We describe results on the corn coleoptile preparation in this section.

Table 1 shows protein and flavin analyses of the corn coleoptile preparations from the differential centrifugation (see Fig. 1 for notation). The 9 kp, 21 kp and 50 kp fractions are presumably mitochondrial, plasma membrane and endoplasmic reticulum enriched fractions, respectively, as determined by various markers (Brain et al. 1977). Several flavoenzymes were assayed for in these preparations utilizing 2,6-dichlorophenolindophenol as the artificial electron acceptor. The only enzyme found in significant quantity among those assayed for was NADH dehydrogenase. This flavoprotein accounts for the major fraction of the flavin content in 21 and 50 kp preparations shown in Table 1. The NADH dehydrogenase (DH) activity on SDS gel was shown by a Tetrazolium Blue assay at the gel origin (> 175,000 MW) (Song et al. 1980).

We have solubilized 21 and 50 kp membrane fractions with SDS and examined them utilizing SDS gel electrophoresis according to Fairbanks et al. (1971). The solubilized membranes of 21 and 50 kp gave essentially the same pattern of 5.6% acrylamide gels with respect to protein band, indicating that they were predominantly of the same plasma membrane composition. Since 50 kp has been shown to contain substantial endoplasmic reticulum (Brain et al. 1977), the plasma membrane is represented by the solubilized fractions. The SDS gel electrophoresis also showed a yellow band at Rf 0.7 in both 21 and 50 kp fractions. This corresponds to mol wt of 15,000 ± 2,000 daltons. This flavoprotein is not NADH dehydrogenase. Is it the phototropic receptor flavoprotein or some unknown protein (including proteolytic degradation product)? The only known flavoprotein with MW in this region is flavodoxin (ca. 17,000) found in anaerobic bacteria. A proteolytic product can

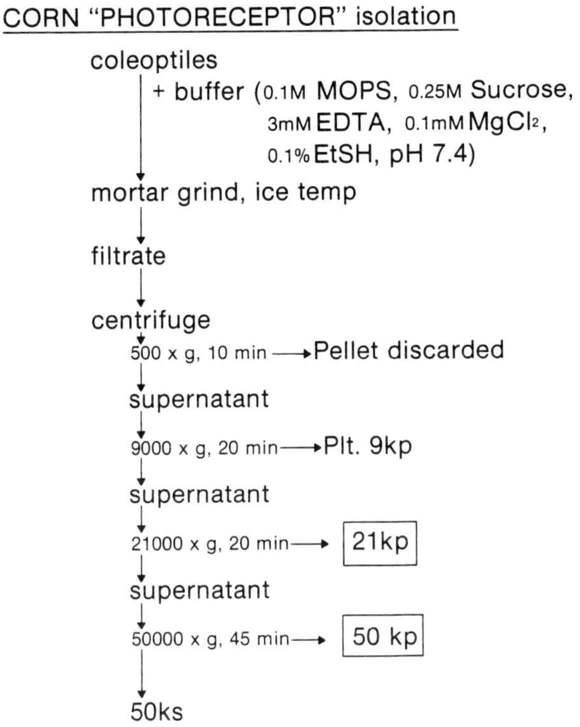

CORN "PHOTORECEPTOR" isolation

Fig. 1. The isolation protocol for
21,000 and 50,000 dalton membrane preparations from corn coleoptiles. *p* pellet, *s* supernatant

coleoptiles

| + buffer (0.1M MOPS, 0.25M Sucrose,
| 3mM EDTA, 0.1mM MgCl₂,
| 0.1% EtSH, pH 7.4)

mortar grind, ice temp

filtrate

centrifuge
    500 x g, 10 min ⟶ Pellet discarded

supernatant

    9000 x g, 20 min ⟶ Plt. 9kp

supernatant

    21000 x g, 20 min ⟶ ⎢ 21kp ⎥

supernatant

    50000 x g, 45 min ⟶ ⎢ 50 kp ⎥

50ks

Fig. 1. The isolation protocol for 21,000 and 50,000 dalton membrane preparations from corn coleoptiles. *p* pellet, *s* supernatant

probably be eliminated since the PMSF treatment and all steps in the procedure were carried out at $4°C$, but other possibilities must await further work, including elution from the gel for further characterization. However, it has not escaped our attention that the yellow flavin chromophore on the SDS gel is likely to be covalently bound. This is of considerable interest since other well-characterized photoreceptors of sensory transducing organisms possess covalently linked chromophores, presumably to achieve the maximum efficiency and stability in their photochemical and photobiological functions (Song 1977). Periodic acid-Schiff reagent staining for carbohydrates of the SDS gel revealed numerous glycoproteins, but whether or not the yellow band is a glycoprotein remains to be established. There was faint staining corresponding to the yellow band, however.

The absorption spectra of 21 and 50 kp showed peaks at 422, 449, and 481 nm, consistent with previously reported absorption spectra (Briggs 1974). These spectra are in reasonable agreement with the light-minus-dark difference spectrum reported by Galston et al. (1977) for the analogous preparations from pea epicotyls.

We have recorded the uncorrected fluorescence emission spectrum of 21 kp in a triangular cuvette obtained by front-surface excitation and detection on a single photon counting spectrofluorometer (Song et al. 1980). Both the fluorescence emission and excitation spectra are attributable to the flavin. The same spectrum was obtained for 50 kp.

Table 2 shows the fluorescence lifetimes of various flavins. Table 3 shows the fluorescence lifetimes for the corn coleoptile preparations. It can be seen that both 21 and 50 kp fractions contain fluorescence components of lifetime ≤ 1 ns within experimental errors, while 9 kp and 50 kp show somewhat longer lifetime ~ 1.4 ns. All fractions are

Table 2. Fluorescence lifetimes of flavins and flavoenzymes at room temperature measured on the phase-modulation fluorometer (average of 20–30 values)

Flavin	Medium	τ_F, ns
Riboflavin tetrabutyrate	EtOH	5.65
	CCl$_4$	6.94
	CCl$_4$ + 3 mM TCA[a]	4.27
	CCl$_4$ + 20 mM TCA	2.55
FMN	H$_2$O	4.65
	90% Glycerol:H$_2$O	5.40
	H$_2$O	2.35
	33% Sucrose:H$_2$O	3.40
	90% Glycerol:H$_2$O	5.00
Lipoamide dehydrogenase	30 mM Pi Buffer, 0.3 mM EDTA, pH 7.2	3.50
D-Amino acid oxidase	PB Buffer, pH 8.5	2.10[b]
	+ 10 mM ϕ-pyruvate	0.44
	+ 2 mM Indole-carboxylic acid	2.65

[a] Trichloroacetic acid
[b] The steady-state polarization degree of fluorescence for this enzyme (absorbance at 450 nm = 0.9) was 0.36, indicating that the major fluorescence intensity was due to the bound flavin. The high polarization may also be due to shorter lifetimes (40 PS for dimer and 130 PS for monomer) (Nakashima et al. 1980). The longer lifetime then represents free FAD

Table 3. Fluorescence lifetimes of various fractions of the corn coleoptile (see Fig. 1) at room temperature and at 30 MHz modulation frequencies. The excitation and emission wavelengths were 441.6 nm and 534.2 nm, respectively (10 nm bandpass). The samples were excited from the front surface of a triangular cell. The protein concentration was ca. 4 mg/ml

Centrifugal fraction	Fluorescence lifetime (ns)[a]	
	By phase shift[b]	By demodulation
9 kp	1.357 ± 0.425	2.177 ± 0.414
21 kp	1.018 ± 0.349	2.572 ± 0.381
50 kp	0.767 ± 0.262	1.544 ± 0.381
50 ks	1.400 ± 0.164	3.020 ± 0.395

[a] The phase and modulation modes yield two different sets of lifetimes, indicating the heterogeneity of the flavin fluorescence
[b] Dilution of the sample yielded even shorter lifetimes. For example, 10 × dilution of 50 kp gave 0.63 ns, indicating that the longer component contributed to these lifetimes to some extent

seen to be heterogeneous, indicating presence of more than one fluorescent flavin in these preparations. It should be noted that the lifetimes shown in Table 3 were measured at 30 MHz which is best suited for short lifetime components. Independent measurements on a nanosecond pulse fluorometer (box car averager) also suggested that 21 and 50 kp preparations show a fluorescence lifetime < 1 ns as the major component (Fig. 2). Auxin showed a 30% fluorescence quenching of 50 kp and 21 kp at 10 mM, again indicating that the singlet excited state of the flavin chromophore in the plasma membrane preparations undergoes an efficient reaction with auxin. This quenching is par-

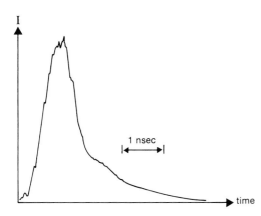

Fig. 2. The fluorescence decay curve (not deconvoluted) of 21 kp recorded on a box-car averaging nanosecond time-resolved spectrofluorometer

1 nsec

time

Table 4. Quenching constants of flavins with various quenchers

Flavin	Quencher	Solvent	K_Q, M^{-1} [a]	$k_Q, M^{-1}s^{-1} \times 10^{-9}$
Riboflavin	H^+	H_2O-Perchlorate	43.4 77.4	7.9
Riboflavin	Azide	H_2O	24.6 35.5	5.1
RFTB	Auxin	Pyridine	24.9 46.6[a]	3.6
Riboflavin	Auxin	H_2O	49.1 54.7[b]	10.0

[a] First value as determined from lifetime data; second values from $\emptyset_F$ data
[b] Upward deviations due to static quenching

ticularly significant, since auxin does not quench the fluorescence of D-amino acid oxidase in a similar concentration range.

Primary Photoreactivity of Flavins. Table 4 shows quenching data for flavin in the presence of various quenchers. The quenching by indole acetic acid is not linear (Fig. 3), indicating that static (complexation) quenching occurs at higher concentrations of auxin. From the lifetime decrease measurements (Table 4 and Fig. 3), it is clear that the dynamic quenching of flavin by auxin is essentially diffusion-controlled. Azide and other quenchers listed in Table 4 also showed dynamic quenching. Tryptophan showed a slight dynamic quenching, as the static quenching predominated.

In an attempt to learn more about the singlet excited state reactions of flavin in solution and plasma membrane preparations, the photoreaction was followed both aerobically and anaerobically by measuring photobleaching of riboflavin by auxin.

From Stern-Volmer kinetic results, the maximum quantum yield was found to be 0.72, indicating an efficient photoreaction between the excited flavin and auxin. The photoreaction was affected by neither oxygen nor KI which are both excellent quenchers for the flavin triplet. This suggests that the photoreaction of flavin with auxin

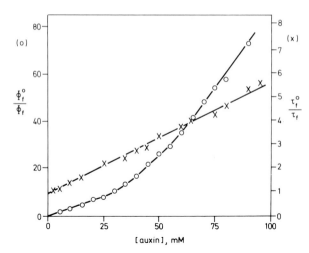

Fig. 3. The Stern-Volmer plot of the riboflavin fluorescence (−o−) and lifetime (−x−) in water at room temperature as a function of the indole acetic acid concentration

proceeds via the singlet state of the former, consistent with the quenching data shown in Table 4 and Fig. 3. However, oxygen clearly participates in secondary reactions, since the tlc autoradiogram patterns for the aerobic and anaerobic photolysis are different (Kurtin 1969).

Finally, we have shown that a special b-type cytochrome is present in the membrane preparations (21 and 50 kp) of corn coleoptiles, as monitored by dithionite-induced and exogenous methylene blue-sensitized absorbance changes. No absorbance change due to flavin reduction was evident, possibly because of an equilibrium amount of bisulfite in the dithionite treatment which reoxidizes the reduced flavin (Mayhew 1980).

3.2 Carotenoprotein Photoreceptor

A number of action spectra for photoresponses of organisms resemble the absorption spectra of carotenoids. The most recent identification of the action spectrum in terms of a carotenoid is for the photo-induced carotenoid biosynthesis in *Neurospora crassa* (De Fabo et al. 1976). In the case of phototaxis in *Euglena* (Diehn and Kint 1970) and phototropism in *Phycomyces* (Delbrück et al. 1960, 1976; Presti et al. 1977) and *Avena* coleoptile (Thimann and Curry 1961), the photoreceptor is likely to be a flavoprotein. In our previous analysis (Song and Moore 1974), it was concluded that carotenoids were not suited as photoreceptors in these photoresponses. Among other reasons, the short lifetime due to efficient internal conversion from the lowest excited singlet state was regarded as the most critical barrier for a carotenoid or carotenoprotein to overcome as a primary photoreceptor. This viewpoint still holds, as far as the primary photoreceptor for sensory transduction is concerned. Nonetheless, it is worthwhile to reexamine carotenoid as a possible photoreceptor in light of recent findings. First, we summarize results on carotenoproteins which act as the secondary (or antenna) photoreceptor.

The problem of the short lifetime of the excited state in carotenoids can be partly overcome by resonance (exciton) interactions among more than one carotenoid mole-

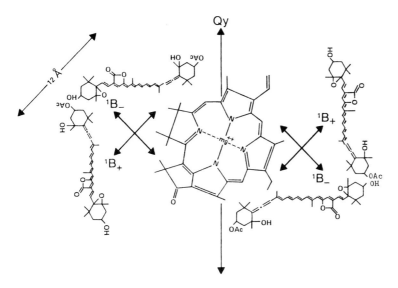

Fig. 4. The molecular topography of antenna photoreceptor complex, peridinin-chlorophyll a-protein, isolated from marine dinoflagellates. Two peridinin molecules form exciton paris, resulting in two exciton states, B$_+$ and B$_-$

cule and between the carotenoid and acceptor molecule (primary photoreceptor), as shown recently in the photoreceptor pigment complex composed of 4 peridinin, 1 chlorophyll a and 1 protein isolated from marine dinoflagellates (Song et al. 1976; Koka and Song 1977). The molecular topography of the dimeric exciton pairs of peridinin is shown in Fig. 4. Energy transfer from the exciton state to the excited state of the primary photoreceptor molecule (i.e., chlorophyll a) can account for the phototactic action spectra of several dinoflagellates (Haupt 1975). For example, the action spectrum for phototaxis of *Gyrodinium dorsum* (Forward 1973) is consistent with the organism's in situ absorption spectrum which is largely attributable to the carotenoprotein and chlorophyll a. We have also shown that the organism photoaccumulates within the lighted area of the optical cuvette illuminated through the single beam spectrophotometer slit even after pretreatment of the algae at 77 K for 30 min (Song, unpublished observation). An approximate action spectrum of this photoaccumulation includes contribution from the chlorophyll absorption band at 660 nm. Thus, accurate action spectra of dinoflagellates containing peridinin-chlorophyll proteins may coincide with the fluorescence excitation spectra which reveal a high efficiency (approaching 100%) of energy transfer from the carotenoid to chlorophyll a in vitro (Song et al. 1976; Koka and Song 1977).

There are not definitely established carotenoproteins as the primary photoreceptors which have been isolated and characterized. Retinal Schiff's base in rhodopsin and in bacteriorhodopsin acts as the primary photoreceptor chromophore. However, these are photoreceptors of the short chain polyene. Although it is possible for the protonated Schiff's base being responsible for blue light response, as the protonated Schiff's base in solution absorbs maximally at 450 nm region, there is no evidence that this type of polyene is actually involved in blue light responses.

The absorption spectrum of β-apo-8'-carotenal Schiff's base has the most likely absorption spectrum in fitting several phototropic and phototactic action spectra, with resolution in the blue light region, except for the lack of a strong near-UV peak (Moore and Song 1974). The latter difficulty can be resolved if the carotenoid is isomerized to the cis form. The protonated Schiff's base of apo-carotenal shows its λ_{max} shifted too far to the long wavelength region to be comparable with blue light action spectra. Reticulataxanthin and citranaxanthing Schiff's bases show similar spectral results (Moore and Song 1974).

Carotenals and their Schiff's bases were found to be nonfluorescent, suggesting that the excited state lifetimes are extremely short as in other carotenoids (Moore and Song 1974). However, the permanent dipole moment of the Schiff bases of carotenals increases by a factor of 4 upon excitation of the Schiff's bases to their ^{1}B excited state. The pK_a^* of the Schiff's bases nitrogen is also predicted to increase upon light excitation.

4 Discussion

Photophysical Requirements of Flavin as the Photoreceptor. With the notable exceptions of D-amino acid oxidase and lipoamide dehydrogenase, most flavoproteins are nonfluorescent or only very weakly fluorescent. This is most probably due to strong interactions between FMN or FAD and aromatic residues, particularly trp and tyr. Flavoproteins of this type are not suitable as the photoreceptor, since the excited state is effectively quenched via static and/or dynamic quenching processes. Furthermore, the highly efficient intersystem crossing in free flavins is almost completely suppressed in flavoproteins (McCormick 1977).

From the brief consideration given above, it is suggested that flavins in the blue light receptors can not be bound at a site where the excited state is quenched via static charge transfer interactions. Thus, the flavoprotein photoreceptor is likely to be fairly fluorescent, particularly when the photoreceptor is functionally disrupted (e.g., isolated or frozen). It is anticipated that the photoreceptor in a fully operational form in vivo is nonfluorescent or very weakly fluorescent owing to the efficient primary photoprocesses responsible for triggering blue light responses. In this connection, a covalently bound flavin is of special interest (vide supra).

What is the Most Likely Mode of the Primary Photoprocess in the Blue Light Receptor? The phototropic and phototactic response of certain organisms is an energy conversion process. Light absorbed by the photoreceptor is transduced to a chemical or membrane potential which can trigger phototropism. The membrane-bound photoreceptor is excited by light. The excited photoreceptor relaxes via several modes of relaxation processes which can be coupled to dynamic as well as static conformational changes of protein and/or membranes to which the flavin photoreceptor is bound. We have discussed several coupling modes between the photoreceptor's electronic relaxation and the macromolecular conformation change (vide supra). Light excitation changes acidicity and basicity of photoreceptor chromophores dramatically. Thus, it is conceivable that a transient proton gradient, transmembrane potential and dynamic and static conformational changes of interacting proteins and/or membrane can be induced by changes in pK_a upon excitation (pK_a^*). The pK_a^* of riboflavin was estimated to be ca. -4 from changes in the absorption spectrum for the neutral and protonated species, since the proton quenching prevented the direct titration using the fluorescence lifetime method (Lasser and Feitelson 1973). An approximate pK_a^* was calculated from the lifetime data, however, assuming that the N_1-protonated flavin does not fluoresce, yielding pK_a^* of -2. The pK_a^* of N_3 was determined from lifetime measurements, which yielded a value of 10.52 for the fluorescent state compared to $\sim$10 for the ground state. It seems unlikely that these extreme pK_a^* values would be particularly effective as a primary photoprocess of the flavin photoreceptor at physiological pH.

The change in dipole moment of a photoreceptor upon light excitation entails re-orientation of the photoreceptor and/or surrounding residue dipoles of the protein or membrane. From the solvent Stokes' shifts of absorption and fluorescence spectra of flavin, the fluorescent state of riboflavin tetrabutyrate was found to be 1.75 times greater than that of the ground state. Again, this does not seem particularly significant to be used as an efficient mode of the primary photoprocess of the flavin photoreceptor. In fact, several flavoproteins including D-amino acid oxidase, lipoamide dehydrogenase and coleoptile plasma membrane preparations showed no significant degree or rate of relaxation in the nanosecond time-resolved spectroscopic measurements (Song et al. 1980).

It is possible that the bound photoreceptor dissociates upon excitation. However, the degree of dissociation and its role in the excited flavin photoreceptor processes are uncertain.

From the above considerations and the fact that the plasma membrane-bound flavo-protein (of 21 and 50 kp) possess a short fluorescence lifetime compared to other fluores-cent flavins and flavoproteins, we come to the conclusion that the photochemical role of flavin in triggering the phototropic event is a photoreduction (Song et al. 1980). Flavin is readily photoreduced by auxin (Galston 1974; vide supra). The action spectrum for the photooxidation of auxin in an etiolated pea homogenate is consistent with flavin as the photoreceptor (Galston and Baker 1949).

Auxin is apparently laterally translocated in coleoptiles from the lighted to the shaded side without significant photooxidation of auxin (Briggs et al. 1957; Thimann and Curry 1961; Pickard and Thimann 1964). Previously, we proposed a photochemical scheme whereby auxin bound to a macromolecule is released for translocation induced by the photoexciation of the flavin photoreceptor (Song et al. 1972). This model ac-counts for the lateral auxin transport in coleoptiles without photodecomposition. We now attempt to further refine this model by considering specific photoprocesses which may be responsible for the phototropic transduction of blue light energy.

Let us assume that auxin is either bound to a protein (receptor) or membraneous compartment analogous to the acetylcholine receptor sites on synaptic membrane. This assumption is experimentally supported (Goldsmith and Thimann 1962; Hertel et al. 1972). To trigger release of the bound auxin, the excited flavin photoreceptor changes the conformation of the auxin-bound protein and/or membrane, resulting in the release of the bound auxin in a nonstoichiometric ratio (i.e., the number of auxin released is much greater than the number of photoreceptor excited, and the photosignal is thus amplified). This scheme is analogous to the rhodopsin-induced release of the visual ex-citation transmitter (Ca^{2+}) from the rod outer segment discs (Hagins 1972).

The auxin release can be triggered by the photoreduction of the flavin photoreceptor itself by a hydrogen donor. In this respect, it is interesting that indole acetic acid
itself is an excellent donor. Furthermore, the flavin photoreduction by auxin is nearly
oxygen-independent (Kurtin 1969), in contrast to other hydrogen donors examined.
The photoreaction by either singlet or triplet flavin is, therefore, faster than or comparable in its rate to the quenching of the singlet or triplet flavin by oxygen. For these
reasons, the auxin release from the membrane or membraneous compartment may be
triggered by the photoreduction of flavin. This mechanism does not result in significant
photodecomposition of auxin, since the local concentration of flavin photoreceptor
can be much lower than that of auxin at the primary photoreaction site where auxin
is accumulated. This speculative model also accounts for the necessary amplification
factor to effect the phototropic curvature, as discussed previously (Thimann and Curry
1961; Song et al. 1972).

It is noteworthy that phototropic auxin transport in coleoptiles is oxygen-dependent
(von Guttenberg 1959; Thimann and Curry 1961; Goldsmith and Thimann 1962; Thimann 1967). This could be accounted for by the reoxidation of the photoreduced flavin photoreceptor by oxygen, which could also give rise to the post-auxin transverse
electrical potential.

Similar photochemical mechanism(s) are possible for other blue light responsive
organisms (fungus and algae, etc.), as long as a good hydrogen donor is available for
the primary photoreaction, i.e., photoreduction of flavin photoreceptor, for phototropism and phototaxis.

*Can the Plasma Membrane-Bound Flavoprotein From Coleoptiles be the Phototropic
Receptor?* Most flavoproteins are nonfluorescent or only very weakly fluorescent due
to static quenching at their binding sites. Thus, their fluorescence lifetimes are not necessarily short compared to those of free flavins. In fact, even weakly fluorescent D-amino
acid oxidase, lipoamide dehydrogenase and several other flavoproteins show fluorescence
lifetimes considerably longer than those expected from the dynamic quenching and from
the relationship between the fluorescence lifetime and quantum yield ($\phi_f = \tau_f/\tau_f^0$, where
τ_f^0 is the radiative lifetime). In view of this observation, it is safe to conclude that the
plasma membrane-bound flavin with $\tau_f \leqslant 1$ ns represents a unique flavoprotein, which
may well be the phototropic receptor for corn. It is clear that the flavin in these preparations is subject to an efficient photoprocess due to its binding environment. Thus, the
fluorescence lifetime of plasma membrane-bound flavin is given by $\tau_f = [k_f + k_r + k_{ic} + k_{isc}]^{-1} \leqslant 1$ ns where k_r represents the primary photoreaction via the singlet excited state,
and k_f, k_{ic}, and k_{isc} are rate constants for fluorescence, internal conversion and inter-
system crossing, respectively. Such a photoprocess is likely to be a reaction with an endo-
genously bound donor in order to shorten the fluorescence lifetime substantially. The
identity of the reacting donor is not known, but several model reactions (Table 4) indicate that the singlet excited state of flavin can undergo an efficient photoreaction.
Xe did not specifically inhibit phototropic responses of corn, suggesting that the triplet
flavin is probably not involved (R. Vierstra, K. Poff and P.S. Song, unpublished results).
The short fluorescence lifetime of the plasma membrane-bound flavin reflects an efficient dynamic photoreaction of the flavin in its binding environment. The plant growth
hormone, auxin, involved in the phototropically coupled lateral transport in vivo serves
as an excellent donor model.

In this connection, it is intersting to note that the plasma membrane-bound flavin is photochemically reactive, as shown previously (Song et al. 1980). On the other hand, there was no comparable quenching of D-amino acid oxidase by auxin. The nature of the fast photoprocess in the singlet excited state of flavins and plasma membrane-bound flavoprotein is yet to be elucidated mechanistically. It is possible that an encounter complex between auxin and the excited flavin (exciplex) contributes to the quenching, followed by nucleophilic addition of the former to the N_5 or C_{4a} position of the excited flavin. The primary photoreaction product may well be of a covalent type described by Hemmerich et al. (1967), thus reducing the flavin.

Although 21 and 50 kp preparations contained b-type cytochrome, it is not possible to ascertain whether the photoreduced flavin produced above is the electron donor to cytochrome. However, it is noteworthy that some type of efficient photoreduction of the flavin is the most likely mode of the primary photoprocess, because other modes of electronic relaxation processes were found to be rather sluggish, making it difficult to account for the efficiency of phototropism in general and the short fluorescence lifetime of the plasma membrane-bound flavin in particular.

Light Harvesting by Carotenoproteins as the Blue Light Photoreceptor. To overcome the most critical kinetic difficulty due to the short lifetime of the excited state of carotenoid, we have shown that the excited state of peridinin can be stabilized by dimeric resonance interaction (exciton) with another molecule of peridinin (vide supra). This is illustrated below (Scheme 1):

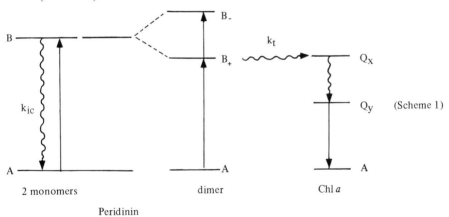

(Scheme 1)

Further resonance stabilization is possible by exciton interactions among degenerate or near-degenerate excited states of peridinin and chlorphyll a. It was shown by the picosecond spectroscopy that an efficient energy trasfer occurs from the resonance stabilized dimeric state of peridinin to chl a, while the transfer is halted when the dimeric arrangement is destroyed (Koka and Song 1977). The rate constant for the transfer was found to be $k_t \geqslant 10^{12}$ l mol^{-1} s^{-1}.

Many blue light response action spectra of photosynthetic algae and dinoflagellates can be explained in terms of energy transfer from the secondary (peridinin) to the primary (chl a) photoreceptors (Scheme 1). For those action spectra lacking red sensitivity, it is necessary to assume that only a fraction of the photosynthetic apparatus (the primary photoreceptor) is utilized for the sensory transduction. To test the validity of this assumption, dye laser excitation can be used for recording weak action spectra from 550–700 nm, eliminating any contamination of blue light in the action spectral measurement.

Can a Carotenoid Function as the Primary Photoreceptor? The answer to this question can be affirmative only if the excited states can be stabilized to the extent that the excitation energy can be utilized directly (rather than transferring to another primary photoreceptor) for the primary photoprocess of sensory transduction. One possibility invokes the role of carotenal Schiff's base which functions as a proton uptaker (Song 1977). This can be made feasible if the excited state of Schiff's base can be resonance-stabilized by exciton coupling with other Schiff's base molecules.

The theory of local heating and rotational reorientation of molecules around the excited molecule which locally releases its energy as heat has been worked out (Albrecht 1958). It may be possible that carotenoids absorbing blue light dissipate the energy as heat which activates sensory response systems such as carotenogenesis. In this case, the short excited state lifetime is an advantage as the blue light photoreceptor. However, this conjecture remains highly speculative at the present, since action spectrum based on the local heating could have higher sensitivity in the near-UV and UV regions than the blue maximum region.

5 Concluding Remarks

It is clear that the current status of our understanding of blue light photoreceptors leaves much to be desired. It is ironic that part of the difficulty in identifying flavin or carotenoid as the primary photoreceptor is the wide diversity of photoreactivity and the apparent lack of it, respectively. In this report, we presented arguments that fluorescence lifetime can be taken as a reflection of the flavin photoreactivity in the attempted isolation and characterization of flavoprotein photoreceptor. We have also shown that the carotenoid excited state can be resonance-stabilized to enhance its probability of transferring the excitation energy to primary photoreceptor. Whether carotenoid can function as the primary photoreceptor remains an open question.

Acknowledgments. This work was supported by the R.A. Welch Foundation (D-182) and NSF (PCM 79-06806). Technical assistance of Drs. R.D. Fugate and P. Koka is greatly appreciated.

References

Albrecht AC (1968) J Chem Phys 1144–1145
Brain RD, Freeburg JA, Weiss CF, Briggs WR (1977) Plant Physiol 59: 948–952
Briggs WR (1963) Annu Rev Plant Physiol 14: 311–352
Briggs WR (1974) Carnegie Inst Yearb 74: 807–809
Briggs WR, Tocher RD, Wilson JF (1957) Science 125: 210
Danon A, Stockenius W (1974) Proc Natl Acad Sci USA 71: 1234–1238
De Fabo EC, Harding RW, Shropshire W Jr (1976) Plant Physiol 57: 440–443

Delbrück M, Shropshire W, Jr (1960) Plant Physiol 35: 194
Delbrück M, Katzir A, Presti D (1976) Proc Natl Acad Sci USA 73: 1969–1973
Diehn B, Kint B (1970) Physiol Chem Phys 2: 483–488
Fairbanks G, Steck TL, Wallach DFH (1971) Biochemistry 10: 2606–2617
Forward RB (1973) Planta 111: 167–178
Fugate RD, Song PS (1976) Photochem Photobiol 24: 479–481
Galston AW (1974) Plant Physiol 54: 427–436
Galston AW, Baker RS (1949) Am J Bot 36: 773
Galston AW, Britz SJ, Briggs WR (1977) Carnegie Inst Yearb 76: 293–295
Goldsmith MHM, Thimann KV (1962) Plant Physiol 37: 492
Guttenberg H von (1959) Planta 53 412
Hagins WH (1972) Annu Rev Biophys Bioeng 1: 131
Haupt W (1975) General review on phototactic microorganisms. In: Colombetti G (ed) Proc Biophys Photoreceptors and Photobehavior of Microorganisms, 4–23
Hemmerich P, Massey V, Weber G (1967) Nature (London) 213: 728–730
Hertel R, Thomson KS, Russo V (1972) Planta 107: 325
Jesaitis AJ (1974) J Gen Physiol 63: 1–21
Koka P, Song PS (1977) Biochim Biophys Acta 495: 220–222
Kurtin WE (1969) PhD Dissertation, Texas Tech University, Lubbock
Lasser N, Feitelson J (1973) J Phys Chem 77 1101–1105
Lowrien R (1967) Proc Natl Acad Sci USA 57: 236
Mayhew SG (1980) In: Yagi K (ed) Flavins and flavoproteins, pp 189–197. University Park Press, Baltimore
McCormick DB (1977) Photochem Photobiol 26: 169–177
Moore TA, Song PS (1974) J Mol Spectrosc 52: 209–214
Muñoz V, Butler WL (1975) Plant Physiol 55: 421–426
Nakashima N, Yoshihara K, Tanaka F, Yagi K (1980) J Biol Chem, in press
Pickard BG, Thimann KV (1964) Plant Physiol 39: 341
Presti D, Hsu WJ, Delbrück M (1977) Photochem Photobiol 26: 403–405
Schmidt W, Butler WL (1976) Photochem Photobiol 24: 71–80
Song PS (1977) J Agric Chem Soc Korea; CA Lee 60th Birthday Commemorative Issue 20: 10–25
Song PS, Moore TA (1974) Photochem Photobiol 19: 435–441
Song PS, Moore TA, Sun M (1972) In: Chichester CO (ed) The chemistry of plant pigments, pp 33–74. Academic Press, London New York
Song PS, Koka P, Prezelin B, Haxo FT (1976) Biochemistry 15: 4422–4427
Song PS, Chae Q, Gardner J (1979) Biochim Biophys Acta 568: 479–497
Song PS, Fugate RD, Briggs WR (1980) In: Yagi K (ed) Flavins and flavoproteins, pp 443–453. University Park Press, Baltimore
Thimann KV (1967) Compr Biochem 27: 1 and references therein
Thimann KV, Curry GM (1961) In: McElroy WD, Glass B (eds) Light and life, pp 665–666. Johns Hopkins University Press, Baltimore
Walker EB, Lee TY, Song PS (1979) Biochim Biophys Acta 587: 129–144
Yagi K (1972) Methods Enzymol 18B: 290–296

Carotenoids as Primary Photoreceptors in Blue-Light Responses

W. SHROPSHIRE, JR[1]

1 Introduction – Advocatus Diaboli

The preponderance of evidence for the identification of the so-called "blue light" photoreceptor (cryptochrome) is clearly in favor of a flavoprotein (Presti and Delbrück 1978). However, there remain some photobiological responses to blue light whose mediation cannot be explained readily by a flavoprotein photoreceptor. Therefore, the organizers of this First International Conference on the Effects of Blue Light in Plants and Microorganisms have requested that I serve as an advocatus diaboli for carotenoids, since their possible function as primary photoreceptors for some responses has not been excluded.

In moot court proceedings or debate, such an advocatus diaboli presents evidence against the prevailing common opinion. Therefore, I will limit my arguments primarily to those in favor of carotenoids since the putative role of flavoproteins has been extensively presented by others at this meeting. In addition, there have been a number of recent reviews summarizing this long-standing controversy. Among the best which specifically compare the relative merits of flavins and carotenoids are the reviews by Song et al. (1972), Song and Moore (1974), Bensasson (1975), McKellar et al. (1975), Briggs (1976), Presti and Delbrück (1978), Tan (1978), and Nultsch and Häder (1979). The reader is referred to these excellent reviews for details of the arguments and experimental evidence.

Since most of the previous speakers in this meeting have emphasized flavins and especially light-induced absorbance changes (LIAC) mediated by flavins, I wish to begin by reminding you that such LIAC have been known for some time. In fact, interest in such systems declined because of the lack of specificity of action brought about by the photooxidations that occur in homogenates where flavins are present. Interest also declined when it was discovered that auxin is not oxidized in coleoptile tips in phototropism but rather is redistributed. Galston (1977) has recently summarized this history which is not as well known as it should be by current workers in the blue light field.

1 Smithsonian Radiation Biology Laboratory, 12441 Parklawn Drive, Rockville, MD 20852, USA

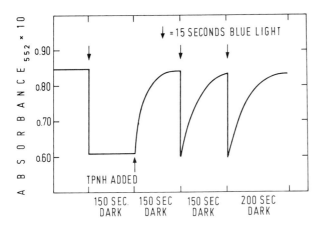

Fig. 1. Light-induced absorbance changes in extracts from *Euglena*. First exposure to blue light bleaches system. Addition of TPNH(NADPH) regenerates system for subsequent cycling by blue light. (Redrawn from Lewis et al. 1961)

2 Light-Induced Flavin-Mediated Cytochrome Absorbance Changes

A clear example of such blue light-induced absorbance changes in partially purified homogenates was described 18 years ago by Lewis et al. (1961). Using an in vitro system obtained from *Euglena* cells they found that exposure to 15 s of broad-band blue light (400–500 nm) produced a significant reduction in absorbance measured on a Cary 14 spectrophotometer at 552 nm (Fig. 1). No further change in absorbance occurred in the bleached system in the dark unless TPNH (NADPH) were added to the reaction mixture. Following the addition there was a gradual increase in absorbance in the dark and subsequent blue light exposures could recycle the reaction. Examination of the components indicated that cytochrome 552 was being oxidized in the presence of a light-activated flavoprotein and re-reduced in the dark by an enzymatic reaction requiring TPNH.

Lewis et al. (1961) proposed a reaction scheme to explain these light-induced absorbance changes (Fig. 2).

In general, it is concluded that such flavin systems can occur readily. I only wish to remind this audience that they have been known for some time (Galston 1977). The central problem, of course, is how to verify that such reactions are responsible for the mediation of the blue light-induced physiological response which is being observed, and are not simply unspecific photooxidations catalyzed by flavins present in the homogenates.

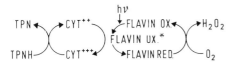

Fig. 2. Hypothetical reaction scheme proposed for absorbance changes described in Fig. 1. Cytochrome absorbs maximally at 552 nm. Blue light is absorbed by oxidized flavin. Reduced cytochrome is regenerated by an enzyme utilizing TPNH (NADPH). (Redrawn from Lewis et al. 1961)

2.1 Carotenoids Localized Near Light-Sensitive Regions

But let us return to my principal task. Phototropism has long been known to be a blue light response (Darwin 1896). Professor Erwin Bünning at the University of Tübingen noted the high concentration of carotenoids in the tips of photosensitive grass coleoptiles. He suggested that carotenoids might be the photoreceptors for phototropism (Bünning 1937) and for the last 40 years the pendulum has swung between riboflavin and β-carotene (Fig. 3) as the principal candidate. Almost every major worker in this field has changed his opinion at least once between these two as new data have been published.

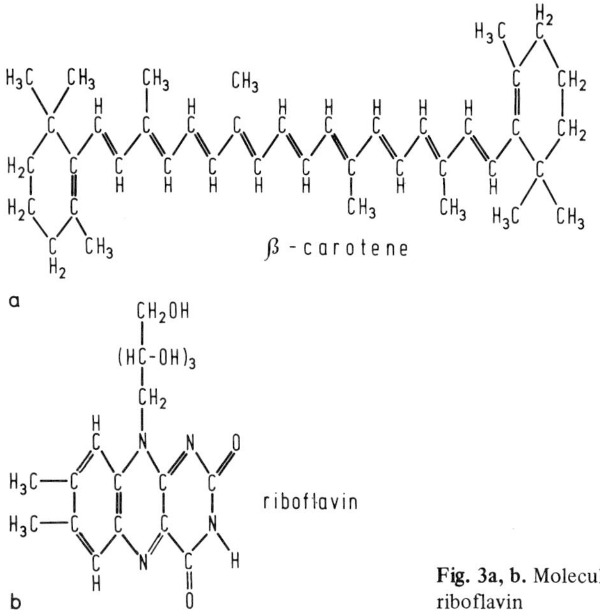

Fig. 3a, b. Molecular structure of β-carotene and riboflavin

2.2 Large Extinction Coefficients

Any chromophore for the mediation of a photoresponse would be expected to have a large value for the extinction coefficient in order to absorb quanta maximally. The value given for most carotenoids ranges between 10^4 and 10^5 dm^3 mol^{-1} cm^{-1}. The $E \frac{1\%}{1\,cm}$ for β-carotene is approximately 2500 at 450 nm depending upon the solvent in which it is measured. In nonpolar solvents such as petroleum ether β-carotene has a value of $\epsilon = 1.3 \times 10^5$ dm^3 mol^{-1} cm^{-1} (Davies 1976). Thus, carotenes are efficient absorbers of blue light.

2.3 Widely Distributed Among Living Organisms

Carotenoids are found in almost every living system. Plants and microorganisms are capable of synthesizing their own carotenoids and for most the regulation is under light control (Rau 1976; Harding and Shropshire 1980). Animals are unable to synthesize their own carotenoids and must ingest them both for obtaining carotenoids and for obtaining the raw materials for related molecules such as the retinals of rhodopsins needed for vision.

Carotenoids are lipid-soluble and as such are associated in fungi, for example, with lipid droplets that may be observed moving about the cell in cytoplasmic streaming. In higher plants capable of synthesizing chlorophyll, carotenoids are found in plastids in the lipophillic portion of the plastid envelope and in photosynthetic lamellae. The biosynthesis of carotenoids apparently takes place entirely in plastids for ready insertion into photosynthetic membranes. The intracellular localization of carotenoid biosynthesis in fungi is not known. It is also not known whether carotenoids may be present in other membranes of cells such as the plasmalemma. Since carotenoids have very low solubility in aqueous systems it is generally assumed that they do not occur in the cytosol. However, it is interesting to note that the inhibition of enzyme activity (Schmid, this vol.) has been reported for reaction mixtures in which a thin layer of hexane containing β-carotene has been layered over an aqueous solution of enzymes. Apparently, enough β-carotene is partitioned into the aqueous fraction to give significant reductions in some enzyme activities. An alternative explanation could be the transfer of carotenoids to the enzymes at the interface. In any case, many more data are needed about the intracellular localization of carotenoids and the site of biosynthesis.

2.4 Absorption Spectra Congruent with Some Blue Light Action Spectra

A number of action spectra for blue light responses have been presented during the first two days of this conference (see Presti and Delbrück 1978 for a representative sampling). However, it is to be noted that for a number of these action spectra the effectiveness between 330 nm and 400 nm is relatively low as compared to the effectiveness of the maxima between 400 nm and 500 nm. This is especially true for action spectra of photomovement in microorganisms (see Nultsch and Häder 1979).

Oftentimes, the argument is made that inert screening external to the photoreceptor by other near-ultraviolet absorbers could reduce the height of the action spectrum. This suggestion is certainly valid. However, in light-induced carotenogenesis in *Neurospora* (Fig. 4) the absorbance through a single layer of mycelial pad was measured. Even if most of the absorbance in such a mycelial pad was external to the photoreceptor as a screen, the low height of the action spectrum in this wavelength region could not be accounted for (De Fabo et al. 1976).

Such is the case for photomovement of microorganisms where the action spectra sometimes have maxima between 500 nm and 600 nm. Some have fine structure with several maxima while others have a smooth single maximum reminiscent of visual response curves for invertebrates and vertebrates.

Some authors studying photomovements have concluded on the basis of careful analysis of the action spectra and screening that the photoreceptor is a carotenoprotein. If screening effects are removed by measuring in *Volvox* the stop response alone, the best fit for the action spectrum found is a carotenoprotein. An excellent example of such an analysis is to be found in the work of Schletz (1975) on the phototactic orientation of *Volvox*. The photoreceptor is believed to be localized in the plasmalemma covering the concave side of the stigma (Schletz 1975) and as the organism rotates is

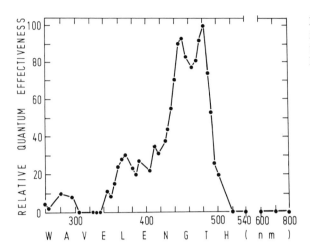

Fig. 4. Action spectrum of photo-induction of carotenoid biosynthesis in *Neurospora crassa*. (De Fabo et al. 1976)

alternately screened by the carotenoids of the stigma. In this case, the action spectrum for the stop response without screening has a single smooth maximum centered near 485 nm and extending from 400 nm to 600 nm (Fig. 5).

    Thus, we are forced to conclude that for some blue light responses the action spectra can best be matched by the absorption spectra for carotenoproteins or, in fact, by retinal systems.

    For phototropism, both in higher plants and fungi, a flavin system appears most likely. For example, Presti et al. (1977) have observed a maximum for the light-growth response action spectrum of *Phycomyces* near 595 nm, a maximum corresponding to direct optical excitation of a flavin. The comparable triplet state for β-carotene is in the infra-red (Land et al. 1971). Bensasson et al. (1976) give triplet values for bacterial polyenes with maxima occurring between 400 nm and 600 nm depending upon the number of conjugated double bonds. However, it should be noted that singlet to triplet intersystem crossing efficiencies of these polyenes are very low, a few percent or less. These authors concluded that these triplet states cannot be populated intramolecularly but only via energy transfer from another excited state, usually a triplet, except in the case of singlet oxygen.

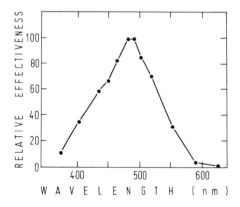

Fig. 5. Action spectrum of the stop response in *Volvox*. Relative effectiveness values are normalized to 482 nm. (Redrawn from Schletz 1975)

2.5 Carotenoids as Accessory Pigments

The role of carotenoids as accessory pigments in photosynthesis is well established. In this case quanta are absorbed by the carotenoids and energy is transferred to chlorophyll on its way to a reaction center for photosynthesis. The ability of carotenoids to transfer energy is very efficient and may in both in vivo and in vitro experiments approach 100% (Bensasson 1975). Thus, the question of low reactivity in redox reactions when compared to the flavins may not be important if there are other mechanisms for the transfer of energy from an excited molecule into a reaction sequence.

2.6 Carotenoid Derivatives

Action spectra of most blue light responses have an action spectrum maximum between 260 nm and 280 nm. This maximum is probably due to absorption by an aromatic amino acid, indicating that the photoreceptor is bound to a protein. Carotenoid derivatives (especially of β-carotene), such as retinals, are attached by Schiff base linkage to the ϵ-amino groups of lysine in opsin proteins and have been recognized as the photoreceptors for vision. In man, the chromophore of rhodopsin is an 11-cis-hindered retinal (Fig. 6) having an extinction coefficient $\epsilon_{500} = 4 \times 10^4$ dm^3 mol^{-1} cm^{-1} (Wald 1968). Such rhodopsins were thought to occur only in animals until a few years ago when the photoreceptor of the purple bacterium, *Halobacterium* was discovered by Oesterhelt and Stoeckenius (1971) to be also a retinal-protein. This bacteriorhodopsin functions as a light-driven proton pump converting light into chemical energy stored as ATP. However, recent data (reviewed by Hildebrand 1977) suggest that two photoreceptors are involved in the motor responses of *Halobacterium*. One of these is bacteriorhodopsin. The other is either a flavoprotein (Hildebrand and Dencher 1975) or, as more recent data indicate (Hildebrand 1977), it is more likely a retinylidene protein which is probably a precursor for bacteriorhodopsin.

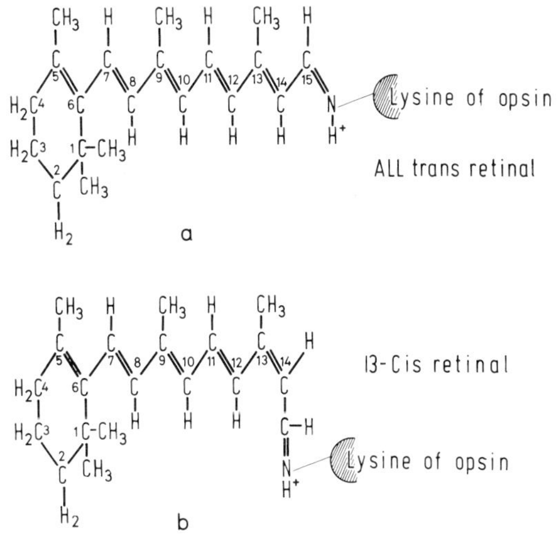

Fig. 6a, b. Molecular structure of retinal chromophores attached by a Schiff base to the ϵ-amino group of a lysine residue in opsin molecule. In vertebrates the all trans form is released from the opsin. (Redrawn and modified from Henderson 1977)

In vertebrates the all trans form of retinal hydrolyzes from the opsin. In invertebrates and *Halobacterium* the retinal remains attached to the opsin. In *Halobacterium* the retinal establishes a photoequilibrium between the 13-cis form and the all trans (Fig. 6). It is not clear whether the initial visual event after light absorption is a cis-trans isomerization or whether large polarization changes occur in the electronically excited polyenes. The isomerization may result in a sudden flow of charge which is perhaps coupled to a chemiosmotic proton pump.

For our purposes, the point is that such retinal systems are very light-sensitive and rapidly mediate light signals. Such molecules are derived from carotenoids and have been shown to be present in at least one bacterial system. They may be present in cryptogams.

2.7 Membrane-Associated Carotenoproteins

Carotenoids have been implicated also in nonvisual responses of animals. In the sea slug *Aplysia californica* abdominal and circumesophageal ganglia have cytoplasm which contains masses of yellow-orange membrane-bound granules (lipochondria) (Krauhs et al. 1977). The pigmented portion of such ganglia comprises 20% of their weight. Extraretinal potential changes are induced by blue-green light which increases a calcium-activated membrane potassium conductance. The involvement of calcium in the light-generated potentials is clear since the potentials may be generated also by the addition of the calcium ionophore A23187.

The absorption spectrum of this orange pigment in the lipochondria has maxima at 413, 433, and 459 nm. It was first identified by Raman spectroscopy to be β-carotene (Sordahl et al. 1976). However, recent data (Sordahl, personal communication) by mass spectroscopy indicate the photoreceptor pigment is not β-carotene nor a retinol, but consists rather of "exotic carotenoids".

2.8 Cis-Trans Isomerization; May Lead to Photochromism

One of the characteristics of carotenoids is the large number of isomeric forms which may exist. For such conjugated double-bond-rich molecules the number of isomers is given by 2^n, where n equals the number of double bonds. For β-carotene (Fig. 3) with 9 double bonds, 512 isomers are theoretically possible. Many fewer actually exist because some are prohibited by steric hindrance. However, it may be that where fine tuning might be required, either energetically or for a specific photoequilibrium value for a process, the availability of a large number of isomers would be advantageous.

In *Euglena gracilis,* blue light-induced absorbance changes have been observed in vivo (Fong and Schiff 1978). These absorption changes occur in zeta-carotene. They are due to differences in absorption between cis and trans zeta-carotene. Interestingly, the action spectrum for inducing these cis-trans isomerizations has action maxima between 370–390 nm and 420–448 nm with no activity past 500 nm. This unknown photoreceptor was suspected to be a flavoprotein (Fong and Schiff 1979). The flavoprotein might sensitize changes in a membranous organelle which are then expressed by absorption changes in zeta-carotene. However, flavins and flavoproteins appear to be excluded since iodide ions and redox agents do not inhibit the isomerizations of zeta-carotene. Spectroscopic data favor a two-pigment system but the initial sensitizer, which is clearly not zeta-carotene, remains unknown.

Stavenga (this vol.) clearly summarizes in his chapter how photochroism occurs in visual pigments. Such photochromism, in which at least two absorption forms exist, allows for a photostationary state to be established in the light as a function of the spectral distribution of the light. In invertebrates where the retinal is not released from the opsin, a photoequilibrium is established between 11-cis retinal and the all trans form. Such a mechanism confers in extra-retinal systems the ability to measure precisely the spectral quality of the light being received.

Extra-retinal systems have been described for many years in amphibians for such behavioral responses as orientation and motion (Parker 1903). Fast photovoltages and slow electrophysiological responses have also been observed from isolated pieces of frog skin. Wald and Rayport (1974) measured the potentials generated by skin photoreceptors in the leopard frog (electrodermograms) and found them to be mediated by a photochromic retinal pigment. They proposed (Rayport and Wald 1978) the following scheme to explain their action spectra data (Fig. 7).

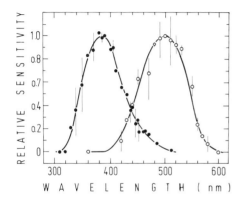

Fig. 7. Action spectrum for leopard frog electrodermogram. (Rayport, personal communication)

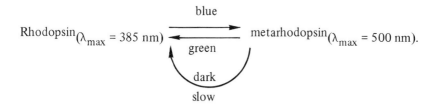

The amplitude of the electrodermogram potentials varies linearly with intensity. The photoresponse of the rhodopsin conversion is approximately twice as effective as the photorecovery system. In the dark the metarhodopsin reverts slowly with a half time of more than 70 min to the rhodopsin form (Rayport, personal communication).

These photochromic extra-retinal systems in amphibians are analogous to the phytochrome photochromic system of higher plants (Shropshire 1972, 1978).

2.9 Carotenoids as Quenchers

Carotenoids are known to be effective quenchers of energetic molecules (Krinsky 1968, 1979). They are quenchers of free radicals, triplet excited sensitizers such as chlorophyll, and of singlet oxygen. Takahama (1978) found that lipid peroxidation in the chloroplasts of spinach leaves was suppressed by 50% when extracted chloroplasts were reconstituted with β-carotene. As a control, β-carotene added to nonextracted chloroplasts did not affect the normally occurring lipid peroxidation. He concluded that β-carotene functions as an efficient quencher of singlet oxygen generated by visible light which in turn induces the lipid peroxidation.

2.10 Photoresponses Inhibited by Carotenoid-Specific Chemical Inhibitors

A number of photoresponses have been reported to be inhibited by chemical inhibitors of carotenoid synthesis. For example, the inhibitor diphenylamine which blocks conversion of phytoene to phytofluene by phytoene dehydrogenase also inhibits photoinduced sporulation in the fungus *Trichoderma viride* (Kumagai and Oda 1969). Phenylacetic acid, a specific inhibitor of flavins, had no effect on the photoresponses and the flavin inhibitor quinacrine also was without effect.

In higher plants the nor-fluorozon herbicide 9789 markedly inhibits carotenoid synthesis. White seedlings of herbicide-treated barley, oats, and corn which are dark-grown have normal phototropic responses to blue light stimuli. Their sensitivity and fluence-response curves are identical to normally pigmented dark-grown seedlings (Shropshire, unpublished). In addition, the well-known Briggs and Chon effect of red and far red light exposures on phototropic sensitivity operates normally through the phytochrome system which is unaffected by the herbicide (Jabben and Deitzer 1979).

2.11 Some Mutants Lacking Carotenoids Lose Photoresponses

The mutant diatom *Nitzschia alba* is colorless and does not contain carotenoids (Nultsch and Wenderoth 1973). It has no phototactic reactions (Wenderoth 1975), whereas pigmented *Nitzschia* strains show normal phototactic responses. Nevertheless, double mutant sporangiophores of *Phycomyces* (Presti et al. 1977) in which the β-carotene content has been reduced to 4×10^{-5} the level of that contained in the wild type have unaltered phototropic responses.

2.12 Sensitivity of Some Photoresponses Are a Function of Carotenoid Content

The formation of sporangiophores in *Phycomyces* is inhibited if the mycelium is grown in a closed container. Blue light exposure of the mycelium reverses the inhibition. Galland and Russo (1979a) measured the irradiance threshold for mutants deficient in β-carotene (Fig. 8). Those mutants deficient in genes *car*A, *car*B and *car*R have a threshold increased by a factor of 100 to 2000 as compared to the wild-type NRRL-1555. *car*A mutants are affected in the synthesis of phytoene, *car*B mutants are defective in the four dehydrogenation steps from phytoene to lycopene, and *car*R mutants are defective

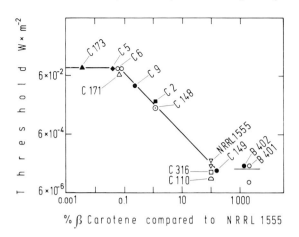

Fig. 8. Response threshold for photoinitiation of sporangiophores in *Phycomyces*. Wild type is NRRL-1555. The β-carotene concentration of the mutant C173 is below detectability and the indicated value is an upper estimate. (Redrawn from Galland and Russo 1979)

in the two cyclization steps from lycopene to β-carotene (Cerdá-Olmedo 1977; Cerdá-Olmedo and Torres-Martinez 1979).

Russo (this vol.) cautions that the threshold relationship to β-carotene content is a complex one and also may well be dependent upon the concentration of inhibiting gas.

Oxygen is required for the light-induced synthesis of β-carotene. Galland and Russo (1979b) have observed that the oxygen threshold for these *Phycomyces* mutants is proportional to the cube root of the carotene content. These are the only two responses of *Phycomyces* for which there is evidence at present that the sensitivity is a function of the β-carotene content.

Singlet oxygen has been postulated as a possible effector molecule for the avoidance response and some of the gaseous inhibition of differentiation in the fungus *Phycomyces* (R. Cohen, personal communication). Gaseous singlet oxygen, which surprisingly has a half life of about 45 min at low concentrations (Kasha and Kahn 1970; Kahn 1978) could be quenched effectively by carotenoids.

2.13 Photomotile Reactions of Some Microorganisms

As reviewed by Nultsch (this vol.) carotenoids have been implicated as the photoreceptors for a number of organisms. I will only remind you of some of these systems (Table 1) from the review of Lenci and Colombetti (1978).

2.14 Biphasic Fluence-Response Curves for Light-Induced Carotenogenesis

Harding (1974) first observed a biphasic fluence dependence of the photocontrol of carotenogenesis in *Neurospora*. Relatively low fluences saturate the synthesis for short time exposures but several hours' exposure to higher fluences results in increased synthesis.

Recent work on *Neurospora* has confirmed this observation (Schrott, this vol.). Jayaram et al. (1979) have observed a similar biphasic fluence dependence for caroteno-

Table 1. Pigments involved in the photomotile reactions of some microorganisms. (After Lenci and Colombetti 1978)

Microorganism	Researcher	Photoreceptive pigments
Chlamydomonas	Halldal	Carotenoids
Cryptomonas	Watanabe Furuya	Phycoerythrin
Gyrodinium	Forward	Carotenoprotein (plus a sensitivity-controlling phytochrome)
Gymnodinium	Forward	Carotenoprotein
Volvox	Schletz	Carotenoprotein (carotenoids as screening pigments)
Euglena	Diehn	Flavins (carotenoids in the stigma as screening pigments)
Halobacterium	Stoeckenius	Bacteriorhodopsin (step-down photophobic response)
		Retinylidineprotein (step-up photophobic response)
Phormidium	Nultsch Häder	Photosystem I (under Photosystem II light background illumination)
		Photosystem II (under Photosystem I light background illumination)

genesis in *Phycomyces* (Fig. 9). The biphasic nature of the response in *Phycomyces* is also confirmed by inhibitor studies in which the response to high fluence levels alone is inhibited by transcription and translation inhibitors such as actinomycin D and cycloheximide (Jayaram et al. 1979; cf. Sandmann and Hilgenberg 1978). The suggestion is made that the low-fluence response may be mediated by β-carotene while the high-fluence response may be mediated by a flavin.

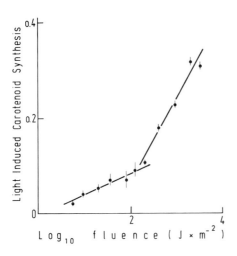

Fig. 9. Biphasic fluence-response curves for light-induced carotenoid synthesis. Values were measured in vivo spectrophotometrically for *Phycomyces* NRRL-1555. (Data replotted from Jayaram et al. 1979)

I have confirmed the biphasic fluence-response curve for carotenogenesis in *Phycomyces*. Preliminary action spectra measurements at 380, 405, and 455 nm for both fluence ranges indicate a greater effectiveness at 380 nm for the high-fluence range than for the low-fluence range in support of two photoreceptors.

However, caution must be exerted and careful measurements of the reciprocity relationships for both of these ranges determined. The possibility exists that these biphasic curves simply result from rapid utilization of a precursor material which results in low fluence saturation. More precursor may be synthesized during continuous irradiation over extended time periods and result in additional accumulation of β-carotene (Harding 1974; Schrott, this vol.).

2.15 Photochromic Photoreceptor in Phycomcyes

Evidence for a long-suspected and searched-for photochromic photoreceptor in *Phycomyces* (Shropshire 1975) has been obtained by Löser and Schäfer (this vol.) in support of the theoretical prediction of Hartmann (1977). If these exciting data are confirmed and possible subtle artifacts of methodology excluded, a retinal photoreceptor for *Phycomyces* may once again be suspected. A search was made for retinal a number of years ago in *Phycomyces* (Meissner and Delbrück 1968) and was considered unsuccessful. The extremely low amounts reported were thought to be due to in vitro reactions occurring after extraction. These new data may once again stimulate a re-examination, especially in light of the recent discoveries in *Halobacterium*.

In any case, the existence of photochromic absorption changes in *Phycomyces* may well lead to the development of extraction and purification techniques in an analogous fashion to the successful methods, developed for phytochrome (Shropshire 1977).

3 Conclusion

Finally, I wish to close this review with a return to the setting of a moot trial between two protagonists and conclude with a Mulla Nasrudin tale by Idries Shah (1971). These tales have become enormously popular among physicists in the United States. I think you will find this one instructive.

"The Mulla was made a magistrate. During his first case the plaintiff argued so persuasively that he exclaimed:

'I believe that you are right!'

The clerk of the court begged him to restrain himself, for the defendant had not been heard yet.

Nasrudin was so carried away by the eloquence of the defendant that he cried out as soon as the man had finished his evidence:

'I believe you are right!'

The clerk of the court could not allow this.

'Your honour, they cannot both be right.'

'I believe you are right!' said Nasrudin."

Therefore, if you were to ask me today if the blue-light photoreceptor (cryptochrome) is a flavin or a carotenoid, I would be tempted to answer as Nasrudin:

'Yes, I believe that you are right!'

Seriously, the data may ultimately force us to acknowledge that there is not just a single photoreceptor capable of explaining all blue light responses as attractive and appealing as that generalization may be. Responses requiring flavins and carotenoids may remain. Careful examination may disclose the presence of retinals in additional organisms. However, if the blue light photoreceptor proves to be photochromic, then the possibility for a single cryptochrome photoreceptor remains open. Many differently shaped action spectra are capable of being generated by a photochromic system. Only future research will resolve the issue.

Acknowledgments. Appreciation is expressed for travel support from the Smithsonian Institution Fluid Research Fund 1236F902, to Tom Roman, Physics Department, Syracuse University, who introduced me to the Mulla Nasrudin stories, and to Prof. Dr. H. Senger who kindly provided assistance in the preparation of the figures.

References

Bensasson RV (1975) Spectroscopic and biological properties of carotenoids. In: Colombetti G (ed) Biophysics of photoreceptors and photobehavior of microorganisms, pp 146–163. Proc Int School Badia Fiesolana, Lito Felici, Pisa

Bensasson RV, Land EJ, Maudinas B (1976) Triplet states of carotenoids from photosynthetic bacteria studied by nanosecond ultraviolet and electron pulse irradiation. Photochem Photobiol 23: 189–193

Briggs WR (1976) The nature of the blue light photoreceptor in higher plants and fungi. In: Smith H (ed) Light and plant development, pp 7–18. Butterworths, Boston

Bünning E (1937) Phototropismus und Carotinoide. II. Das Carotin der Reizaufnahmezonen von *Pilobolus, Phycomyces* und *Avena.* Planta 27: 148–158

Cerdá-Olmedo E (1977) Behavioral genetics of *Phycomyces.* Annu Rev Microbiol 31: 535–547

Cerdá-Olmedo E, Torres-Martinez S (1979) Genetics and regulation of carotene biosynthesis. Pure Appl Chem 51: 631–637

Darwin C (1896) The power of movement in plants, p 592. D Appleton and Company, New York

Davies BH (1976) Analysis of carotenoid pigments. In: Goodwin TW (ed) Chemistry and biochemistry of plant pigments, vol II, p 154. Academic Press, London New York

DeFabo EC, Harding RW, Shropshire W Jr (1976) Action spectrum between 260 and 800 nanometers for the photoinduction of carotenoid biosynthesis in *Neurospora crassa.* Plant Physiol 57: 440–445

Fong F, Schiff JA (1978) Blue-light absorbance changes and phototaxis in *Euglena.* Plant Physiol 59: 74–405

Fong F, Schiff JA (1979) Blue light-induced absorbance changes associated with carotenoids in *Euglena.* Planta 146: 119–128

Galland P, Russo VEA (1979a) Photoinitiation of sporangiophores in *Phycomyces* mutants deficient in phototropism and in mutants lacking β-carotene. Photochem Photobiol 29: 1009–1014

Galland P, Russo VEA (1979b) Regulation of sporangiophorogenesis in *Phycomyces:* joint control by oxygen, retinol and blue light. Planta 146: 257–262

Galston AW (1977) Riboflavin retrospective or deja-vu in blue. Photochem Photobiol 25: 503–504

Harding RW (1974) The effect of temperature on photoinduced carotenoid biosynthesis in *Neurospora crassa.* Plant Physiol 54: 142–147

Harding RW, Shropshire W Jr (1980) Photocontrol of carotenoid biosynthesis. Annu Rev Plant Physiol 31: in press

Hartmann KM (1977) Aktionsspektrometrie. In: Hoppe W, Lohmann W, Markl H, Ziegler H (eds) Biophysik, ein Lehrbuch, pp 197–222. Springer, Berlin Heidelberg New York

Henderson R (1977) The purple membrane from *Halobacterium holobium.* Annu Rev Biophys Eng 6: 87–109

Hildebrand E (1977) What does *Halobacterium* tell us about photoreception? Biophys Struct Mech 3: 69—77

Hildebrand E, Dencher N (1975) Two photoreceptors controlling behavioral responses of *Halobacterium halobium*. Nature (London) 257: 46—48

Jabben M, Deitzer GF (1979) Effects of the herbicide san 9789 on photomorphogenic responses. Plant Physiol 63: 481—485

Jayaram M, Presti D, Delbrück M (1979) Light-induced carotene synthesis in *Phycomyces*. Exp Mycol 3: 42—52

Kasha M, Kahn AU (1970) The physics, chemistry and biology of singlet molecular oxygen. Ann NY Acad Sci 171: 5—23

Khan AU (1978) Activated oxygen: Singlet molecular oxygen and superoxide anion. Photochem Photobiol 28: 615—627

Krauhs JM, Sordahl LA, Brown AM (1977) Isolation of pigmented granules involved in extra-retinal photoreception in *Aplysia californica* neurons. Biochim Biophys Acta 471: 25—31

Krinsky NI (1968) The protective function of carotenoid pigments. In: Giese AC (ed) Photophysiology, vol III, pp 123—195. Academic Press, London New York

Krinsky NI (1979) Carotenoid protection against oxidation. Pure Appl Chem 51: 649—660

Kumagai T, Oda Y (1969) An action spectrum for photoinduced sporulation in the fungus *Trichoderma viride*. Plant and Cell Physiol 10: 387—392

Land EJ, Sykes A, Truscott TG (1971) The invitro photochemistry of biological molecules. II. The triplet states of β-carotene and lycopene excited by pulse radiolysis. Photochem Photobiol 13: 311—320

Lenci F, Colombetti G (1978) Photobehavior of microorganisms. Annu Rev Biophys Bioeng 7: 341—361

Lewis SC, Schiff JA, Epstein HT (1961) Photooxidation of cytochromes by a flavoprotein from *Euglena*. Biochem Biophys Res Commun 5: 221—225

McKellar JF, Phillips GO, Checucci A, Lenci F (1975) Excited states of flavins in photoreception processes. In: Birks JB (ed) Excited states of biological molecules, pp 92—105. Wiley, New York

Meissner G, Delbrück M (1968) Carotenes and retinal in *Phycomyces* mutants. Plant Physiol 43: 1279—1283

Nultsch W, Häder D-P (1979) Photomovement of motile microorganisms. Photochem Photobiol 29: 423—437

Nultsch W, Wenderoth K (1973) Phototaktische Untersuchungen an einzelnen Zellen von *Navicula peregrina* (Ehrenberg) Kützing. Arch Microbiol 90: 47—58

Oesterhelt D, Stoeckenius W (1971) Rhodopsin-like protein from the purple membrane of *Halobacterium halobium*. Nature New Biol 233: 149—152

Parker GH (1903) The skin and eyes as receptive organs in the reactions of frogs to light. Am J Physiol 10: 27—33

Presti D, Delbrück M (1978) Photoreceptors for biosynthesis, energy storage and vision. Plant Cell Environ 1: 81—100

Presti D, Hsu W-J, Delbrück (1977) Phototropism in *Phycomyces* mutants lacking β-carotene. Photochem Photobiol 26: 403—405

Rau W (1976) Photoregulation of carotenoid biosynthesis in plants. Pure Appl Chem 47: 237—243

Rayport S, Wald G (1978) Frog skin photoreceptors. Program Abstr 6th Annu Meet Am Soc Photobiol, June 11—15, pp 94—95

Sandmann G, Hilgenberg W (1978) Förderung der β-carotin-synthese durch Licht bei *Phycomyces blakesleeanus* Bgff. Biochem Physiol Pflanz 172: 401—407

Schletz K (1975) Phototaxis bei *Volvox*-Pigmentsysteme der Lichtrichtungsperzeption. Z Pflanzenphysiol 77: 189—211

Shah I (1971) The pleasantries of the incredible Mulla Nasrudin, p 71. EP Dutton, New York

Shropshire W Jr (1972) Phytochrome, a photochromic sensor. Photophysiology 7: 32—72

Shropshire W Jr (1975) Phototropism. In Schenck GO (ed) Progress in photobiology, Proc 6th Int Congr Photobiol Drsch Ges Lichtforsch eV, Frankfurt 1974, Abst 024, pp 1—5. Springer, Berlin Heidelberg New York

Shropshire W Jr (1977) Photomorphogenesis. In: Smith KC (ed) The science of photobiology, pp 281—312. Plenum Press, New York London

Shropshire W Jr (1978) Photochromic systems, program and abstracts 6th Annu Meet Am Soc Photobiol, June 11–15, p 91

Song P-S, Moore TA (1974) On the photoreceptor pigment for phototropism and phototaxis: Is a carotenoid the most likely candidate? Photochem Photobiol 19: 435–441

Song P-S, Moore TA, Sun M (1972) Excited states of some plant pigments. In: Chichester CO (ed) The chemistry of plant pigments, pp 33–74. Academic Press, London New York

Sordahl LA, Lewis A, Pancurak J, Brown AM, Perreault G (1976) Identification of the photosensitive pigment in the giant neuron of *Aplysia californica.* Biophys J 16: 146a

Takahama U (1978) Suppression of lipid peroxidation by β-carotene in illuminated chloroplast fragments: Evidence for β-carotene as a quencher of singlet molecular oxygen in chloroplasts. Plant Cell Physiol 19: 1565–1569

Tan KK (1978) Light induced fungal development. In: Smith JE, Berry DR (eds) The filamentous fungi, vol III, pp 334–357. John Wiley and Sons, New York

Wald G (1968) Molecular basis of visual excitation. Science 162: 230–239

Wald G, Rayport S (1974) Skin photoreceptors in the leopard frog. Biol Bull 147: 503

Wenderoth K (1975) Untersuchungen der photo-phobotaktischen Reaktionen einzelner Diatomeenzellen, pp 1–72. Inaugural-Dissertation (Dr. rer. nat.) der Phillips Universität Marburg/Lahn. Görich und Weiershäuser, Marburg (Lahn)

On the Nature of the Blue Light Photoreceptor: Still an Open Question

E. DE FABO[1]

1 Introduction

For many years the literature has been replete with investigations by scientists fascinated with the diverse effects of light on organisms. Curiously enough, when one analyzes the "type" of light involved in many of these responses, one discovers that much of the living world shows a predilection for blue light. That this predilection exists is well-documented [5, 6, 7, 8, 9, 29, 36, 55, 57, 60].

In many cases, action spectra have been determined in an attempt to identify the chromophore(s) involved in the absorption of this blue light [13, 15, 17, 24, 28, 39, 40, 45, 56, 59, 62]. If these action spectra are compared with the absorption spectra of possible photoreceptors, two different groups of compound are usually indicated. These are the carotenoids and the flavins. Presently, flavins seem to be the favored candidates, but as we shall attempt to point out, the question is far from settled (see also Shropshire, this vol.).

Both carotenoids and flavins are similar in their ability to absorb in the blue and near-ultraviolet. A major difficulty in choosing between the two groups is that both show absorption spectra that match closely the action spectra of many of the blue light photoresponses. That this difficulty exists is attested to by the diverse review articles on this point [10, 12, 23, 29, 42, 52].

For carotenoids, the fine structure in the blue generally shows three distinctive peaks or two peaks and a shoulder, whereas flavins tend not to show such detailed fine structure, although under certain conditions, fine structure seems to be determinable [53, 54]. Also, in the 350–400 nm range and especially around 370 nm, the flavins generally show a much broader and more effective type of absorption [54] as compared to the carotenoids [63]. This is an important point since many investigators use this as a major criterion in concluding which photoreceptor is involved. However, it should be noted that flavins are not the only compounds which show this type of absorption. For example, carotenoproteins also show very efficient absorption in this wavelength region [11]. Because of the importance some investigators attach to this region in interpreting action spectra, more detailed comments regarding this "370 nm peak" will be made later.

Since much of the evidence often quoted in favor of either flavin or carotenoid as the blue light photoreceptor is action spectra data, it might be worthwhile to review

1 Stratospheric Impact Research and Assessment Staff (RD 683), U.S. Environmental Protection Agency, Washington, D.C. 20460, USA
Present address: Cancer Biology Program, NCI-Frederick Cancer Research Center, Frederick, MD 21701, USA

some of the basic problems and assumptions associated with action spectroscopy. Several excellent reviews on this topic already exist [1, 26, 50]. Therefore, we shall dwell only on those points which we feel are the most critical and which so often seem to be overlooked by individuals when interpreting (and obtaining) action spectrum data.

2 Action Spectroscopy

Any response of an organism to light must involve at least two processes: (1) direct light absorption by a pigment (the photoreceptor) and (2) following this absorption, a series of events (dark reactions) which lead to some observable response. It is the purpose of action spectroscopy to obtain information regarding the identification of the photoreceptor.

2.1 Definition

An action spectrum can be most simply defined as a plot of the reciprocal of the number of quanta required to produce a given effect versus wavelength [26, 50]. The actual procedure involves measuring the dose response relationships for effective wavelengths and then choosing some response level common to all the wavelengths used. If the dose response curves for each wavelength are similar, this implies that the mechanism of action is the same for all wavelengths [26].

The amount of energy needed to produce a particular response varies with wavelength with the most effective wavelength producing the response at the lowest dose and vice versa. This difference in energy requirement is due to the fact that the absorbing pigment, i.e., the photoreceptor, has for each wavelength a different probability of absorbing the photons of that wavelength.

This concept can easily be seen after one derives a mathematical expression describing the relationship between the response measured, the number of incident photons needed to produce that response, and the probability of absorption (the absorption coefficient) [50].

Simply stated: Response = number of absorbed quanta = kI_0 bc

where I_0 = incident number of quanta, b = the beam pathlength through some absorbing substance (cm), c = the concentration of absorber (molecules · cm^{-3}), k = arbitrary proportionality constant, R = response.

The absorption coefficient, μ, is directly proportional to the reciprocal of the number of incident photons. That is

$$\mu = \frac{kR}{I_0 bc},$$

and since K = kR/bc which is itself a constant when the same response is measured for each wavelength, then

$$\mu = \frac{K}{I_0}.$$

The absorption coefficient is thus mathematically defined and can be thought of as a cross-sectional area which expresses the probability that the photoreceptor will, in fact, absorb a photon. It has the units $cm^2 \cdot molecule^{-1}$, i.e., the larger its value, the greater the probability of absorption. More importantly, this probability is characteristic for the absorbing substance and is a function of wavelength.

2.2 Distortions of the Action Spectrum

A major problem in interpreting an action spectrum is that the effect may not accurately describe the absorption spectrum of the photoreceptor. This is primarily due to the fact that some basic assumptions need to be made [50], all of which must be valid to fully minimize any distortions to the action spectrum. These assumptions are:

1. The photoreceptor is evenly distributed throughout the system studied.
2. The concentration of the photoreceptor is very small so that not all of the incident radiation is absorbed by the sample (self-screening).
3. Screening by other pigments is negligible, or if not, then it is either uniform throughout the wavelength region used or it can be corrected for by direct transmission measurements.
4. The reciprocity relationship between irradiance and time holds under the experimental conditions used so that the response is determined to be a function only of the total dose. (Irradiance × Time = Dose).
5. The quantum efficiency is the same for all wavelengths.
6. For some constant response, the number and rate of formation of intermediate photoproducts is the same for all wavelengths used.
7. Little or no fluorescence occurs in the system.

If all of the above hold true, then the action spectrum should be exactly congruent to the in vivo absorption spectrum of the photoreceptor [50].

For some types of experiment, namely, the determination of growth and phototropic action spectra, the perturbations due to internal and external screening can be corrected for by careful measurement and calculations, and it may be possible to show that the photoreceptor is not screened or at most very slightly screened, as will be discussed shortly. Also, any variability in sensitivity (e.g., biological rhythms) should be noted. If detected, this variability can be eliminated by using the "null response" technique [17]. However, this technique may not be applicable to all types of action spectrum determinations. If the null technique cannot be used, a correction for sensitivity changes may be made by measuring the response to a standard dose for each experiment and then correcting the experimental values appropriately [14].

2.3 Dose Response Curves

In order to determine accurately the number of incident quanta that will produce some constant response, a dose response curve must be determined for each wavelength. If linearity is noted, then an equation can be easily determined and used to calculate any level of response desired. This constant response level should arbitrarily be a value near the 50% of maximum level. This tends to minimize any errors in linearity of the dose response curves. The incident dose values should be expressed in SI units such as J/m^2, i.e., the total incident energy striking the organism per unit area. However, since the response is proportional to the number of incident quanta per unit area, the incident dose values must be converted to moles of quanta [50].

In determining dose response curves, the Bunsen-Roscoe Law of Reciprocity needs first to be validated. If this law holds, it means that the effect is a function only of total dose and independent of the irradiation time or irradiance. It is very important that this parameter be checked since, if it is found not to hold over the time limits used to determine the dose response curves, misinterpretation of the action spectrum derived could occur. This would imply that a single photochemical act may not be the only rate-limiting step [33].

3 On the Nature of the Blue Light Photoreceptor: Carotenoids vs. Flavins

Classically, the arguments usually presented against carotenoids as being the photoreceptor for various blue light responses are: (1) the presence of screening pigments which could lower the response in the ultraviolet (UV) (this makes the action spectrum look less like flavin absorption), (2) flavins show a much greater versatility as biological photosensitizers [53], (3) in some studies with albino mutants, photoresponses occur although carotenoid pigments appear to be absent [34, 42, 43], and (4) altering environmental conditions can induce carotenoid-like fine structure in flavin absorption spectra [53, 57].

More recent arguments cited against a carotenoid-type photoreceptor include: (1) blue light -induced absorbance changes in fungi [37, 38, 41], (2) inhibition of some physiological responses by flavin inhibitors [2, 35, 39, 48], (3) long-wave induction (595 nm) of blue light responses in *Phycomyces* [16], and (4) dye-mediated photooxidations as the primary reaction upon illumination of the photoreceptor [31].

While the above arguments are formidable, we believe that careful scrutinizing of each will show none is conclusive. For example, in regards to point 1 if it can be shown that the photoreceptor is generally external to some (or most) screening compounds then the effect of such a screen on lowering the action spectrum would be greatly modified (if not entirely eliminated) [15]. One way to check this is to measure the transmission through the tissue being tested. As an example, a transmission curve through one nonirradiated mycelial pad of *Neurospora* was determined [15]. This curve showed that below about 300 nm, light passing through the pad is absorbed so strongly that any interpretation of the action spectrum below this point is questionable. However, above this wavelength it can be seen that transmission through the pad does occur such that in the blue, the transmission is not so greatly different relative to that in the near-ultraviolet. From the transmittance at 300 nm, which is approximately 10%, and assuming homogeneous screening with the photoreceptor located within the upper 10% of the mycelium, it can be calculated that 80% of the incident energy reaches the photoreceptor. If the photoreceptor is in a layer 25% below the surface, this uniform screen would decrease the transmittance to about 60%. It would appear, then, from these arguments that screening probably did not have a significant distorting effect on the determination of this action spectrum for all wavelengths greater than 300 nm [15]. Such an approach should be carried out whenever an action spectrum shows a low UV response. Point 2: It has been known for some time that flavins have a wide range of photoreactivity [53]. Recently, it was reported that the life time of excited single-state β-carotene is very short (on the order of 10^{-14} s) and that intersystem crossing of carotenoid to the triplet state is very inefficient. Based mainly upon these arguments, it was proposed that flavin would more likely be a better candidate for the role of a biological photoreceptor than would carotenoids, at least for phototropism [52].

Notwithstanding the cogent nature of point two, it has been pointed out that these observations, too, are not conclusive [42]. In any event, the point we wish to make here is that despite the versatility of flavins as biological photosensitizers, other mechanisms exist whereby energy input may stimulate some effector system. For example, steric alterations in bio-molecules which bring about activation are known to occur. The best example of this is the cis-trans isomerization in vertebrate vision [61].

Therefore, as an alternative proposition to the high photoreactivity of flavins one could conceive that photoisomerizations might mediate blue light responses [14, 15]. For example, several investigators have proposed enzyme induction as the primary event following blue light activation of carotenoid synthesis [3, 4, 20, 21, 45, 46]. One can envision enzyme *induction* as being coupled directly to enzyme *activation* mediated by carotenoids or carotenoid-protein complexes [14, 15]. For example, Prins [44] provides data which supports the suggestion that plants contain photochemical moieties associated with membrane protein and that signal transduction can occur through photoregulation of the conformation and/or assembly of these chromoproteins.

In another report [60] Voskresenskaya states ". . .therefore, one cannot deny the possibility of photoregulation of non-photosensitive enzymes in living cells with the intervention of chromophore molecules absorbing light selectively in different regions of the spectrum. . ." While there is no direct evidence presently available to support it, the hypothesis that blue light-induced enzymatic processes may be regulated by the photoisomerization of a membrane-bound, carotenoid-protein photoreceptor should be considered valid until proven otherwise [14, 15]. Whatever the mechanism eventually turns out to be there seems to be no apparent reason, a priori, why carotenoids cannot act as a blue light photoreceptor by an isomerization mechanism as presented [14]. In fact, it has been reported that, surprisingly, little attention was being paid to this possible mechanism [53]. Furthermore, light activation of enzymes by a variety of mechnisms is an area of study which seems to be developing rapidly [25].

Point 3: Relative to "carotenoid-less" mutants, it may be surmised that in some mutant albino studies, carotenoids may still be present in such small amount as to be beyond spectrophotometric detection but still be effective in producing the photoresponse [8, 47]. Furthermore, this is consistent with the idea that regulatory molecules tend naturally to be in low concentrations. Even in the case of the "carotenoid-less" double mutant studies of *Phycomyces* [43], in which this organism showed normal phototropic responses to blue light, and in which no β-carotene could be detected, the data are not foolproof. One could argue that the carotenoid might be either (a) linked to a protein and/or (b) membrane-bound (see point 2 above). These conditions could make the detection of such a compound very difficult; to be conclusive, such conditions must be ruled out.

In reference to point 4, flavin absorption spectra are very sensitive to the external environment associated with them [53]. Consequently, special conditions are required to resolve their fine structure. For example, in the case of riboflavin tetrabutyrate (RFTB), when this compound was embedded in polymethylmethacrylate, or placed in ethanol at 77K, fine structure in the ultraviolet and blue regions could be detected which resembled the action spectrum of *Avena* phototropism [53]. While the argument is made that embedding in plastic simulates a tightly bound condition of the photoreceptor in vivo, no direct evidence supporting this idea is presented. However, one observation has been made linking a similarity between absorption spectra of chromophores in frozen solution at low temperatures to absorption spectra of bound chromophores in solution at room temperature [52]. Unfortunately, such an observation does not allow for a definitive conclusion to be made.

Concerning the more recent arguments supporting flavins, one should note the following:

Point 1: With regard to the blue light-induced absorbance changes, whereby a b-type cytochrome was shown to be reduced by a flavin [37, 38, 41], the physiological relevance of this response has been questioned by a number of investigators [15, 22]. Besides this, it still remains to be shown just what this photoreduction means in terms of the various blue light responses. None of the reported observations has been experimentally determined to be involved directly with the initial light reaction. Several recent review articles [22, 32, 42] address this problem in some detail and point out the need for further clarification.

Point 2: As to the abrogation of a blue light response by flavin inhibitors, the major drawback in interpreting these results is the same as for all inhibitor studies. That is, one can never be sure about the specific site of action of the inhibitor. Until it can be shown that the inhibitor is acting specifically at the level of the photoreceptor, arguments based on these types of study cannot be considered as conclusive.

Point 3: Dye-mediated photooxidations using red light to bring about induction normally requiring blue light indicate clearly that these dyes can act as artificial photoreceptors [31]. Since carotenoids cannot bring about photooxidations, this would tend to argue against them as possible photoreceptors. Yet, as was reported [31], it is still an open question whether these dyes are able to replace the natural photoreceptor or act via some other mechanism which transfers the energy to the in vivo photoreceptor. Interestingly enough, energy transfer was reported to occur from these dyes to carotenoids [31].

Point 4: With regards to the long wave (ca. 595 nm) induction of light-growth response [16] in *Phycomyces,* the argument is made that this peak corresponds with the expected transition for riboflavin from the ground state directly to the first excited triplet state.

This is perhaps the most formidable argument yet presented in favor of a flavin as the blue light photoreceptor. These elegant experiments by Delbrück et al. [16] showed what appears to be the direct optical excitation from the ground state to the lowest triplet state of riboflavin. The action spectrum for this response shows a peak at 595 nm. Notwithstanding the elegant nature of these experiments, it is nevertheless important to note that the observed peak at 595 nm is arrived at by subtracting the observed values for a growth response (575 nm $< \lambda <$ 630 nm) from a curve which is an extrapolation of the sum of two tropic responses at two different intensities, both assumed to be following a gaussian distribution, one centered at 450 nm, the other centered at 483 nm.

The authors claim that, based on the excellent fit in slope and approximate fit in the height of the experimental values, ". . .it seems reasonable to trust Eq. (1) further out as representing the long wavelength tail of the strong transition of the receptor pigment". ["Equation (1) is the expression of the photon cross section as the sum of the two gaussians centered as described above".]

While this may seem reasonable, it does not constitute unambiguous proof that this assumption is correct. This is a very important point since if the assumption is not valid, then the peak given as being at 595 nm would in fact be shifted elsewhere and may no longer correspond to the direct optical excitation of the lowest triplet state of riboflavin.

4 The Near Ultraviolet (370 nm) Dilemma

Finally, in addition to the specific points outlined above, another point which must be considered in some detail is the following one.

Many investigators favor a flavin or flavoprotein photoreceptor primarily because of the high near-ultraviolet (ca. 370 nm) response often seen in many of the so-called "blue light" action spectra. It is true that because of this high response peak, many of these "blue light" action spectra do indeed more closely resemble the absorption spectra of flavin or flavoproteins [42, 53] than they do carotenoids. However, this argument, too, is equivocal.

In a recent publication Hager [19] pointed out that the fine structure of the characteristic three-peak absorption spectrum for carotenoids diminishes and eventually disappears, concomitant with the appearance of a new peak in the near ultraviolet region, when water is added to a polar (alcohol) solution of carotenoids. In addition, it was also shown that water-soluble complexes, isolated from spinach chloroplasts, show strong light absorption in the ultraviolet and a one-peak absorption curve in the blue. After transfer of this complex to polar solutions the maximum in the ultraviolet disappears and a characteristic three-peak carotenoid curve appears in the blue region of the spectrum. It is concluded that ". . .carotenoid complexes which are bound to membrane or particles in the intact cells may have a four-peaked absorption curve similar to the pigments which are dissolved in the water-containing alcohols. . .". Furthermore, ". . .It is conceivable that those carotenoids which do not form ultraviolet peaks in the dissolved state are able to do so under conditions under which carotenoids are bound to membrane or particle." This study further states "on the basis of the described abilities of the carotenoids to form an absorption peak in the long wave ultraviolet the appearance of such a maximum in an action spectrum (in the region about 370 nm) can no longer be considered to be sufficient proof for the participation of a flavin as light acceptor."

Since many investigators continue to interpret the high response peak in the near-ultraviolet around 370 nm as indicative of a flavin photoreceptor [53, 57], it is important to note these arguments carefully. However, some counterarguments to this observation by Hager have been noted [52]. Still another observation [49] provides data that seem to support Hager's position. The absorption spectrum of a carotenoid protein complex isolated from the photosynthetic bacterium *Rhodospirillum rubrum,* showed only a single peak at 370 nm when dissolved in 0.01 M phosphate buffer. However, after acetone extraction, the spectrum of spirilloxanthin appeared showing maximum absorption at approximately 499 nm. Spirilloxanthin (or one of its derivatives) is the end product of carotenoid biosynthesis in these photosynthetic bacteria [30].

Hence, a carotenoid-protein complex in vivo could also show absorption characteristics similar to flavoprotein especially in regards to the near-ultraviolet (370 nm) peak observed in the various spectra [11, 58].

A second point about the 370 nm peak is the following: Some investigators [8, 51] point out that certain compounds might actually be excited by the near-ultraviolet and fluoresce in the blue. These blue quanta could then be re-absorbed by the photoreceptor and enhance whatever response was being measured in vivo. The end result of this would be a higher response peak in the action spectrum located at the peak of the exciting wavelength. This would, of course, lead to an erroneous interpretation of the action spectrum and hence greatly confuse the issue when using the action spectrum to indentify the photoreceptor. So many different types of compounds can be excited by UV light and fluoresce in the blue [18] that unless strong fluorescence is ruled out, it is possible that some of the high response peaks at 370 nm noted in nearly all of the blue light action spectra are due to re-absorption of blue fluorescence. In any case, the dilemma posed by the 370 nm peak remains unresolved.

5 Conclusions

An attempt has been made to compare and evaluate the major arguments concerning the nature of the blue light photoreceptor. While many arguments have been used to support a flavin hypothesis, we conclude that, at present, no unequivocal data exist to rule out carotenoid (or any other compound for that matter) as the physiological blue light photoreceptor. Undoubtedly, many other arguments exist both pro and con other than the ones emphasized here, and we sincerely regret the ommissions. Nevertheless, our objective was to try to point out the open nature of this question of the identity of the blue light photoreceptor.

To further emphasize this view, we cite a very recent publication which presents data supporting the hypothesis that, in fact, two photoreceptors can exist, at least for carotenoid induction in *Phycomyces* [27]. One photoreceptor is postulated to be a carotenoid and the other a flavin. If the idea of more than one blue light photoreceptor gains momentum, as it seems to be doing (see also [35]), then we may hope that future searching for the photoreceptor(s) by "blue light scientists" will be at least twice as interesting, and as fascinating as has been the search for a single blue light photoreceptor the last four decades.

6 Recommendations

1. Action Spectra: When presenting action spectrum data, one should provide all the dose response curves as well as reciprocity, transmission, and fluorescence data. This would allow one to determine to what extent distortions to the action spectrum might exist. These data are invaluable if one wishes to maximize the value of determining and interpreting an action spectrum.

2. Newer Approaches: Since an action spectrum alone cannot provide conclusive data as to the nature of the photoreceptor, other approaches to this problem are needed. Genetic and biochemical approaches have been used for years and have contributed much information. What is needed now, it seems, is the application of some of the more modern and sophisticated techniques available, for example, to vision photochemists. Some of these are enumerated and described by Stavenga (this vol.).

3. Bi (or Tri)-ennial "Blue Light" Conferences: If one is to maintain the momentum stimulated by the wonderful meeting at Marburg, superbly organized by Dr. Senger and his staff, a meeting on the blue light "syndrome" every two or three years would certainly be scientifically profitable and highly desirable.

Acknowledgments. The author wishes to acknowledge Drs. Roy Harding and Walter Shropshire, Jr., of the Smithsonian Institute's Radiation Biology Laboratory, Rockville, Maryland, U.S.A. for very helpful comments and suggestions. He also thanks Dr. Margaret Kripke, of the Frederick Cancer Research Center, (FCRC), Frederick, Maryland, U.S.A. for the use of her laboratory's facilities during the preparation of this manuscript.

He is indebted especially to Jan Jenkins, Donna Follin, and Sandy Kimberly, of FCRC, for making available to him their very fast and efficient typing skills.

This report has not been reviewed for approval by the Agency and hence its contents do not represent the views and policies of the U.S. Environmental Protection Agency, nor does mention of trade names or commercial products constitute endorsement or recommendation for use.

References

1. Allen MB (1964) Absorption spectra, spectrophotometry and action spectra. In: Giese AC (ed) Photophysiology, vol I, pp 83–110. Academic Press, London New York
2. Bara M, Galston AW (1968) Experimental modification of pigment content and phototropic sensitivity in excised *Avena* coleoptiles. Physiol Plant 21: 109–118
3. Batra PP (1971) Mechanisms of light induced carotenoid synthesis in nonphotosynthetic plants. In: Giese AC (ed) Photophysiology, vol I, pp 47–76. Academic Press, London New York
4. Batra PP (1964) On the mechanism of photoinduced carotenoid synthesis: Aspects of the photo-induced reaction. Arch Biochem Biophys 107: 485
5. Beck SD (1968) Insect photoperiodism, pp 18–19. Academic Press, London New York
6. Benoit J, Ott L (1944) External and internal factors in sexual activity. Effect of irradiation with different wavelength on the mechanism of photostimulation of the hypophysis and on testicle growth in the immature duck. Yale J Biol Med 17: 27
7. Bergman K, Burke PV, Cerdá-Olmedo E, David CN, Delbrück M, Foster KW, Goodell EW, Heisenberg M, Meissner G, Zalokar M, Dennison DS, Shropshire W Jr (1969) Phycomyces. Bacteriol Rev 33: 99–157
8. Briggs WR (1964) Phototropism in higher plants. In: Giese AC (ed) Photophysiology, vol I, pp 223–268. Academic Press, London New York
9. Briggs WR (1963) The phototropic responses of higher plants. Annu Rev Plant Physiol 14: 311–352
10. Cerdá-Olmedo E (1977) Behavioral genetics of *Phycomyces*. Annu Rev Microbiol 31: 535–547
11. Cheesman DF, Lee WL, Zagalsky PF (1967) Carotenoproteins in invertebrates. Biol Rev Cambridge Philos Soc 42: 131

12. Curry GM (1969) Phototropism. In: Wilkins MB (ed) Physiology of plant growth and development, pp 245–273. McGraw-Hill, New York

13. Curry GM, Gruen HE (1959) Action spectra for the positive and negative phototropism of *Phycomyces* sporangiophores. Proc Natl Acad Sci USA 45: 797–804

14. De Fabo EC (1974) PhD Thesis George Washington University, Washington DC

15. De Fabo EC, Harding RW, Shropshire W Jr (1976) Action spectrum between 260 and 800 nanometers for the photoinduction of carotenoid biosynthesis in *Neurospora crassa*. Plant Physiol 57: 440–445

16. Delbrück M, Katzir A, Presti D (1976) Responses of *Phycomyces* indicating optical excitation of the lowest triplet state of riboflavin. Proc Natl Acad Sci USA 73: 1969–1973

17. Delbrück M, Shropshire W Jr (1960) Action and transmission spectra of *Phycomyces*. Plant Physiol 35: 194–204

18. Goodwin RH (1953) Fluorescent substances in plants. Annu Rev Plant Physiol 4: 283–304

19. Hager A (1970) Ausbildung von Maxima in Absorptionsspektrum von Carotenoiden im Bereich um 370 nm: Folgen für die Interpretation bestimmter Wirkungsspektren. Planta 91: 38–53

20. Harding R (1968) PhD Thesis. California Institute of Technology, Pasadena, California

21. Harding RW, Mitchell HK (1968) The effect of cycloheximide on carotenoid biosynthesis in *N. crassa*. Arch Biochem Biophys 128: 814–818

22. Harding RW, Shropshire W Jr (1980) Photocontrol of carotenoid biosynthesis. Annu Rev Plant Physiol 31: in press

23. Hartmann KM, Cohnen-Unser I (1973) Carotenoids and flavins versus phytochrome as controlling pigments for blue-UV mediated photoresponses. Z Pflanzenphysiol 69: 109–124

24. Howes CD, Batra PP (1970) Mechanism of photoinduced carotenoid synthesis. Further studies on the action spectrum and other aspects of carotenogenesis. Arch Biochem Biophys 137: 175

25. Hug DH (1978) The activation of enzymes with light. In: Smith KC (ed) Photochemical and photobiological reviews, vol III, pp 1–33. Plenum Press, New York

26. Jagger J (1967) Introduction to research in ultraviolet photobiology. Prentice-Hall, Englewood Cliffs

27. Jayaram M, Presti D, Delbrück M (1979) Light-induced carotene synthesis in *Phycomyces*. Exp Mycol 3: 42–52

28. Kowallik W (1967) Action spectrum for an enhancement of endogenous respiration by light in *Chlorella*. Plant Physiol 42: 672–676

29. Krinsky N (1971) Function. In: Isler O (ed) Carotenoids, pp 669–716. Birkhauser, Basel

30. Liaaen-Jensen SL, Andrewes AG (1972) Microbial carotenoids. Annu Rev Microbiol 26: 225–248

31. Lang-Feulner J, Rau W (1975) Redox dyes as artificial photoreceptors in light-dependent carotenoid synthesis. Photochem Photobiol 21: 179–183

32. Lipson ED, Presti D (1977) Light-induced absorbance changes in *Phycomyces* photomutants. Photochem Photobiol 25: 203–208

33. Mancinelli AI, Rabino I (1979) The "high-irradiance responses" of plantmorphogenesis. Bot Rev 44: 129–181

34. Meissner G, Delbrück M (1978) Carotenes and retinal in *Phycomyces* mutants. Plant Physiol 43: 1279–1283

35. Mikolajczyk E, Diehn B (1975) The effect of potassium iodide on photophobic responses in *Euglena*: evidence for two photoreceptor pigments. Photochem Photobiol 22: 269–271

36. Mohr H (1972) Lectures on photomorphogenesis, pp 190–204. Springer, Berlin Heidelberg New York

37. Muñoz V, Brody S, Butler WL (1974) Photoreceptor pigment for blue light responses in *Neurospora crassa*. Biochem Biophys Res Commun 58: 322–327

38. Muñoz V, Butler WL (1975) Photoreceptor pigment for blue light in *Neurospora crassa*. Plant Physiol 55: 421–426

39. Page RM, Curry GM (1966) Studies on phototropism of young sporangiophores of *Pilobilus kleinii*. Photochem Photobiol 5: 31–40

40. Pickett JM, French CS (1967) The action spectrum for blue light stimulated uptake in *Chlorella*. Proc Natl Acad Sci USA 57: 1587–1593

41. Poff KL, Butler WL (1974) Absorbance changes induced by blue light in *Phycomyces blakesleeanus* and *Dictyostelium discoideum*. Nature (London) 248: 799–801

42. Presti D, Delbrück M (1978) Photoreceptors for biosynthesis, energy storage and vision. Plant Cell Environ 1: 81–100
43. Presti D, Hsu W-J, Delbrück M (1977) Phototropism in *Phycomyces* mutants lacking β-carotene. Photochem Photobiol 26: 403–405
44. Prins W (1973) Model studies on biological photoregulation. In: Programs and abstracts, p 183. Am Soc Photobiol, Washington DC
45. Rau W (1967) Untersuchungen über die lichtabhängige Carotinoidsynthese. I. Das Wirkungsspektrum von *Fusarium aquaeductuum*. Planta 72: 14–28
46. Rilling HC (1964) On the mechanism of photoinduction of carotenoid synthesis. Biochim Biophys Acta 79: 464
47. Sargent ML, Briggs W (1967) The effects of light on circadian rhythm of conidiation in *Neurospora*. Plant Physiol 43: 1504–1510
48. Schmidt W, Hart J, Filner P, Poff K (1977) Specific inhibition of phototropism in corn seedlings. Plant Physiol 60: 736–738
49. Schwenker U, Gingras G (1973) A carotenoprotein from chromatophores of *Rhodospirillum rubrum*. Biochem Biophys Res Commun 51: 94
50. Shropshire W Jr (1972) Action spectroscopy. In: Mitrakos K, Shropshire W Jr (ed) Phytochrome, pp 161–181. Academic Press, London New York
51. Shropshire W Jr, Withrow RB (1958) Action spectrum of phototropic tip curvature of *Avena*. Plant Physiol 33: 360–365
52. Song P-S, Moore TA (1974) On the photoreceptor pigment for phototropism and phototaxis: is a carotenoid the most likely candidate? Photochem Photobiol 19: 435–441
53. Song P-S, Moore TA, Sun M (1972) Excited states of some plant pigments. In: Chichester CO (ed) The chemistry of plant pigments, pp 33–74. Academic Press, London New York
54. Sun M, Moore TA, Song P-S (1972) Molecular luminescence studies of flavins. I. The excited states of flavins. J Am Chem Soc 94: 1730–1740
55. Thimann KV, Curry GM (1960) Phototropism and Phototaxis. In: Flurkio M, Mason S (ed) Comparative biochemistry, vol I, pp 243–309. Academic Press, London New York
56. Thimann KV, Curry GM (1961) Phototropism. In: McElroy WD (ed) Light and life, pp 646–672. Johns Hopkins Press, Baltimore
57. Thomas JB (1965) Primary photoprocesses in biology. North Holland Publ Co, Amsterdam
58. Thommen H (1971) Metabolism. In: Isler O (ed) Carotenoids, pp 637–668. Birkhauser, Basel
59. Virgin HI (1952) An action spectrum for the light induced changes in the viscosity of plant protoplasm. Physiol Plant 5: 575–582
60. Voskresenskaya NP (1972) Blue light and carbon metabolism. Annu Rev Plant Physiol 23: 219–234
61. Yoshizawa T, Wald G (1963) Prelumirhodopsin and the bleaching of visual pigments. Nature (London) 197: 1279–1286
62. Zalokar M (1955) Biosynthesis of carotenoids in *Neurospora*. Action spectrum of photoactivation. Arch Biochem Biophys 56: 318–325
63. Zechmeister L (1962) Cis-trans isomeric carotenoids, vitamins A and arylpolyones. Academic Press, London New York

Conformational Changes Caused by Blue Light

G.H. SCHMID[1]

Photoregulation of metabolic reaction sequences in plants by blue light is possible by a control of enzyme syntheses or by a direct effect of blue light on the enzyme activity itself. Possible photoreceptors for these effects are either phytochrome or yellow pigments such as flavins or carotenoids. A typical property of most of these blue light effects seems to be that light of very low intensities which is absorbed by very small amounts of a yellow pigment leads to such amplified consequences as enzymic activation or inactivation. Both the absorbing pigment as well as blue light are ultimarely catalysts in an enzymic or chemical reaction sequence. This can easily be demonstrated by the blue light sensitivity of a number of enzymes under in vitro conditions. In all these cases blue light sensitivity is induced by a chromophore which is attached to the enzyme protein. If the chromophore is naturally occurring on that enzyme the system might have a certain probability for photoregulation also under in vivo conditions.

The blue light effects described in the following are all mediated by FMN (flavin mononucleotide) as the photoreceptor. In all these cases FMN is either the natural coenzyme of the light-sensitive enzyme or FMN can be attached to the enzyme, which means that a binding site for FMN exists somewhere on the enzyme without the FMN having a recognizable coenzyme function. The binding site might be naturally occupied by flavin (e.g., lactate dehydrogenase from baker's yeast EC 1.1.2.3) or merely a potential one. Such blue light effects lead either to an enzyme inactivation or to an enhancement of activity.

If FMN is the photoreceptor, the effect of light on the enzyme is generally mediated by one of the three mechanisms described in the following or by any possible simultaneous combination or mixture of these three reaction types:

Firstly: Triplet FMN is directly quenched by the apoprotein leading to an inhibition or stimulation of enzyme activity. This reaction type does not involve oxygen, proceeds consequently also under anaerobic conditions and has rather tendencies to be absent under higher oxygen partial pressures. The light effect might be fully reversible.

The second and third mechanism both involve oxygen.

Secondly: Triplet FMN reacts with oxygen forming a FMN peroxide which then via oxygen transfer oxidizes the substrate, that is the apoenzyme in our case. The light effect is in general an irreversible inactivation of the enzyme.

1 Centre d'Etudes Nucleaires de Cadarache, Service de Radioagronomie, B.P. No. 1
  13115 Saint-Paul-Lez-Durance, France

Thirdly: In this case triplet FMN leads to singlet oxygen formation which then as the reactive species alters the apoprotein, thus also leading in general to an irreversible inactivation.

Glycolate oxidase of higher plants (EC 1.1.3.1) is a typical FMN enzyme which oxidizes its substrate glycolate by using oxygen as hydrogen acceptor. The enzyme, together with RuDP-oxygenase, is thought to be one of the key enzymes of photorespiration (Lorimer et al. 1973; Zelitch 1975). The enzyme from *N. tabacum* is blue light-sensitive. Very low intensities inhibit the enzyme activity under optimal reaction conditions to a very large extent (Fig. 1). This inactivation is caused by a direct interaction between the photoexcited coenzyme and the apoenzyme. This is demonstrated by the fact that the blue light sensitivity of the enzyme is decreased by lowering the reaction temperature (Table 1) and by the fact that the light reaction proceeds also in the absence of oxygen (Schmid 1969) which means that triplet FMN is directly quenched by the apoenzyme. As the interaction appears to be irreversible, a stable chemical or conformational alteration of the apoenzyme structure must be considered.

Glycine oxidase of the colorless *Chlorella* mutant 125 (Schwarze and Frandsen 1960) is stimulated by blue light. The stimulatory effect of blue light is immediate and shows no lag period after the onset of light (Fig. 2). Upon switching off the light the

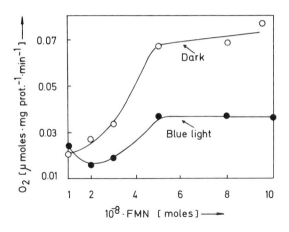

Fig. 1. Dependence of the activity of dialyzed tobacco glycolate oxidase on blue light and flavin mononucleotide. Enzyme of *N. tabacum* var. Su/su. The assay was illuminated with 550 erg $\cdot$ s^{-1} $\cdot$ cm^{-1} of blue light (380 nm $<\lambda<$ 575 nm). Temp. 30°C

Table 1. Temperature dependence of the photoinactigation of tobacco glycolate oxidase in blue light

Temperature (°C)	Light condition	Spec. enzyme activity μl O$_2$ uptake $\cdot$ mg protein^{-1} $\cdot$ h^{-1}
12.7 ± 0.5	Dark	19.6
12.7 ± 0.5	Blue light	21.0
26.5 + 1	Dark	55.5
26.5 ± 1	Blue light	27.9

Enzyme preparation of *N. tabacum* var. Su/su. Light intensity 2070 erg $\cdot$ s^{-1} $\cdot$ cm^{-2} of blue light 380 nm $<\lambda<$ 575 nm

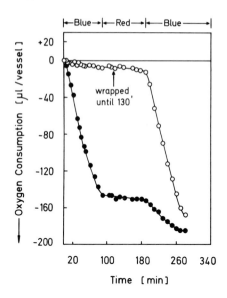

Fig. 2. Dependence of the glycine oxidase activity in the colorless *Chlorella* mutant 125 on blue light. The reaction mixture contained in 3 ml: 1.5 ml 0.1 M K_2HPO_4, 1 ml crude enzyme aquiv. 19.8 mg protein, and 0.1 ml 10^{-3} M FMN. The reaction mixture was illuminated with either 2070 erg · sec · cm^{-1} of blue light 380 nm $<\lambda<$575 nm or 39 000 erg · sec · cm^{-1} of red light 575 nm $<\lambda<$750 nm transmitted through blue or red Plexiglas filters. Temp. 30°C

enzyme activity disappears immediately also without any sign of overshoot, clearly indicating a reversible conformational change in the enzyme protein. The isolated enzyme depends for activity on the addition of FMN and functions apparently only in the light as shown in Fig. 2. The enzyme was discovered by Schmid and Schwarze (1969) in the above-described colorless *Chlorella* mutant. The mutant contains no detectable amounts of carotenoids and has apparently a block in carotenoid synthesis on the phytoen level (Schmid and Schwarze 1969). Under starved conditions the mutant also shows a blue light-enhanced respiration (Kowallik 1967; Kowallik and Gaffron 1967) and excretes FMN into the medium, which clearly speaks in favor of FMN as the photoreceptor. Later Lee et al. (1971) observed in another colorless *Chlorella* mutant G-27, which in contrast to mutant 125 is FMN-deficient, the same type of a blue light and FMN requiring glycine oxidase.

It has been conclusively shown that this glycine oxidase does not play a role in Kowallik and Gaffron's blue light-enhanced respiration (Kowallik and Ryters 1976).

NAD-Malate dehydrogenase (Ec 1.1.1.37) from *Euglena gracilis, N. tabacum,* and *Zea mays,* which is a key enzyme of the Krebs cycle, is inhibited by blue light in the presence of FMN (Codd 1972a). The effect of blue light on this enzyme is the result of two mechanisms occurring simultaneously. First a direct inactivation of the enzyme via FMN as the sensitizer and via an inactivation involving singlet oxygen as the reactive species. This is shown by the fact that inactivation of the enzyme is reduced by only 50% in oxygen-free nitrogen (Codd 1972a) and that typical singlet oxygen quenchers like β-carotene (Foote and Denny 1968; Khan and Kasha 1970) and certain amines (Quannes and Wilson 1968) like TMPD in the reaction assay effectively quenched the blue light effect on the enzyme. This example shows the effect of blue light on a respiratory enzyme.

TrZnsketolase (EC 2.2.1.1) of *N. tabacum* is involved in the Calvin-Bassham cycle of photosynthesis and participates in the formation of glycolate which is thought to be a key substrate of photorespiration in higher plants. The isolated enzyme is inactivated by low blue light intensities in the presence of low concentrations of FMN which acts as the sensitizer (Codd 1972b). FAD in the tenfold concentration or thiamine pyrophosphate did not act as sensitizers.

Table 2. Dependence of the photoinhibition of tobacco tranksketolase on the presence of oxygen

Conditions	Blue light	Relative rates
Normal aerobic	Dark	100
	Dark + FMN	100
	Blue light	100
	Blue light + FMN	35.8
Gassed with oxygen	Blue light	92.1
	Blue light + FMN	66.1
Gassed with oxygen-free nitrogen	Blue light	95.4
	Blue light + FMN	100

The reaction mixture contained in 3 ml when indicated 0.5 μmol FMN. The blue light intensity was 1542 erg $\cdot$ s^{-1} $\cdot$ cm^{-2} 380 nm $<\lambda<$575 nm. Temp. 25°C

The transketolase FMN system is a typical example for an inactivation mechanism which proceeds via singlet oxygen formation. If the reaction assay is gassed with oxygen-free nitrogen, partically no blue light effect is observed (Table 2). In the presence of a normal oxygen partial pressure the blue light effect is quenched by substances which deactivate triplet FMN by collision such as tryptophane (Radda 1966) or which affect triplet FMN by complex formation such as CMU (Homann and Gaffron 1964) (Table 3) but also by substances such as β-carotene or hydroquinone which are thought to act as singlet oxygen quenchers (Foote and Denny 1968; Foote et al. 1970). Thus, the mechanism of the transketolase inactivation by blue light is clearly different from that of glycolate oxidase which proceeded also in the absence of oxygen and which was only quenched upon addition of substances with paramagnetic properties such as oxygen itself or certain heavy metal ions (Schmid 1969).

In the preceding paragraphs it was shown that FMN-enzymes or enzymes that are able to bind FMN such as uricase (EC 1.7.3.3) (Schmid 1970), NAD-malate dehydrogenase or transketolase (Codd 1972a, b) are inhibited by blue light. The next obvious question to ask is whether blue light would affect a typical FAD-enzyme. A characteristic enzyme of this class is

β-D-Glucose oxidase (EC 1.1.3.7). In context with photorespiration, the enzyme is of special interest because glycolate may be produced by the pentose phosphate cycle. The native FAD-enzyme shows no blue light sensitivity at all and can be irradiated in the entire pH range of its activity for hours, which is certainly of physiological significance. This fits into the observation mentioned above, namely that FAD is no blue light sensitizer under light conditions in which FMN affects a variety of enzyme activities. If the concentration of FAD in the assay medium is increased far beyond the concentration which is effective with FMN, no blue light effect is produced either, nor is

Table 3. Effect of various compounds on the blue light inactivation of tobacco transketolase

Additions	Blue light	Relative rates
None	Dark	100
	Dark + FMN	100
	Blue light	100
	Blue light + FMN	35.8
Tryptophane	Dark	95
	Dark + FMN	92
	Blue light	90
	Blue light + FMN	81
CMU[a]	Dark	93
	Dark + FMN	92
	Blue light	93.4
	Blue light + FMN	83.6
Hydroquinone	Dark	100
	Dark + FMN	95
	Blue light	100.2
	Blue light + FMN	84.4
β-carotene	Dark	93.3
	Dark + FMN	93.3
	Blue light	89.4
	Blue light + FMN	65.1

[a] CMU, N-[p-Chlorophenyl]N,N'-dimethyl urea. The reaction mixture contained where indicated either 1 μmol tryptophane, CMU, hydroquinone, or β-carotene in 3 ml. 0.5 μmol FMN/3 ml assay mixture when indicated. Light intensity 1542 erg $\cdot$ s^{-1} $\cdot$ cm^{-2} 380 nm $< \lambda < 575$ nm. Temp. 25°C

the native enzyme sensitive to blue light in the presence of FMN in the assay medium. The FAD can be removed at least partially from the enzyme by dialysis which leads in the known manner to a decrease in the enzyme activity (Table 4). Readdition of the coenzyme FAD restores that activity. If the dialyzed enzyme is supplemented with FMN instead of FAD, no significant restoration of the enzyme activity is observed, which simply means that FMN is not able to substitute for the natural coenzyme FAD. However, in the presence of blue light FMN restores the enzyme activity, which means that the enzyme has now become light-sensitive. FMN plus blue light appear to substitute for the original coenzyme FAD. It is quite obvious that this activation is caused by a conformational change of the apoenzyme induced by excited FMN (Schmid 1971).

I think I have shown that in the presence of FMN blue light can activate or inhibit a variety of enzymes. A photoregulation of these enzymes by blue light under in vivo conditions could well take place and would depend on the presence of substances in the cell compartment or in the immediate vicinity of the enzyme which affect or quench excited FMN and/or singlet oxygen.

Table 4. Blue effect of FMN and blue light on the activity of β-D-glucose oxidase

Enzyme condition	Blue light $erg \cdot s^{-1} \cdot cm^{-2}$	Specific enzyme activity $\mu mol\ O_2$ uptake $\cdot mg\ protein^{-1} \cdot h^{-1}$
Native	Dark	251
Native + FMN	Dark	290
Native	2100	255
Native + FMN	2100	243
Dialyzed	Dark	130
Dialyzed + FMN	Dark	162
Dialyzed + FAD	Dark	260
Dialyzed + FMN	2100	248

When indicated the assay mixture contained 0.1 μmol FMN or FAD. The assay contained 50 μg of enzyme protein. Temp. 30°C

References

Codd GA (1972a) The photoinhibition of malate dehydrogenase. FEBS Lett 20: 211−214

Codd GA (1972b) The photoinactivation of tobacco transketolase in the presence of flavin mono-nucleotide. Z Naturforsch 27b: 701−704

Foote CS, Denny RW (1968) Chemistry of singlet oxygen. XII. Quenching by β-carotene. J Am Chem Soc 90: 6233−6235

Foote CS, Denny RW, Weaver L, Chang Y, Peters J (1970) Ann NY Acad Sci 171: 139

Homann P, Gaffron H (1964) Photochemistry and metal catalysis: studies on a flavin sensitized oxidation of ascorbate. Photochem Photobiol 3: 499−519

Khan AU, Kasha M (1970) Chemiluminescence arising from simultaneous transitions in pairs of singlet oxygen molecules. J Am Chem Soc 92: 3293−3300

Kowallik W (1967) Action spectrum for an enhancement of endogenous respiration by light in *Chlorella*. Plant Physiol 42: 672−676

Kowallik W, Gaffron H (1967) Enhancement of respiration and fermentation in algae by blue light. Nature (London) 215: 1038−1040

Kowallik W, Ryters G (1976) Über Aktivitätssteigerungen der Pyruvatkinase durch Blaulicht oder Glucose bei einer Chlorophyll-freien *Chlorella* Mutante. Planta 128: 11−14

Lee D, Sargent DF, Taylor CPS (1971) Blue light stimulated oxygen uptake mediated by FMN in a colorless *Chlorella* mutant. Can J Bot 49: 651−655

Lorimer GH, Andrews TJ, Tolbert NE (1973) Ribulose diphosphate oxygenase. II. Fruther proof of reaction products and mechanism of action. Biochemistry 12: 18−23

Quannes C, Wilson T (1968) Quenching of singlet oxygen by tertiary aliphatic amines. Affect of DABCO. J Am Chem Soc 90: 6527−6528

Radda GK (1966) The chemistry of flavins and flavoproteins. II. Inhibition of photoreduction of flavin nucleotides and analogues. Biochim Biophys Acta 112: 448−458

Schmid GH (1969) The effect of blue light on glycolate oxidase of tobacco. Hoppe Seylers Z Physiol Chem 350: 1035−1046

Schmid GH (1970) The effect of blue light on some flavin enzymes. Hoppe Seylers Z Physiol Chem 351: 575−578

Schmid GH (1971) Photoregulation of β-D-glucose oxidase by blue light. Phytochemistry 10: 2041−2042

Schmid GH, Schwarze P (1969) Blue light enhanced respiration in a colorless *Chlorella* mutant. Hoppe Seylers Z Physiol Chem 350: 1513−1520

Schwarze P, Frandsen N (1960) Herstellung von *Chlorella* Farbmutanten mit Hilfe von radiaktiven
   Isotopen. Naturwissenschaften 47: 47–48
Zelitch J (1975) Photosynthesis, photorespiration and plant productivity. Academic Press, London
   New York

Interactions of Flavins with Cytochrome C and Oxygen in Excited Artificial Systems

R.J. STRASSER[1] and W.L. BUTLER[2]

1 Introduction

Flavin-mediated photoreactions are found in many biological systems. Some of these photoreactions can be simulated in artificial systems comprised of flavin light and a suitable electron donor and acceptor. Such artificial systems have been reported earlier (Schmidt and Butler 1976) showing that excited flavins are able to photoreduce cytochrome c under aerobic and anaerobic conditions. Semi-biological or semi-artificial systems consisting of biological membrane fractions suspended in a flavin solution have also been reported (Ninnemann et al. 1976). In these experiments the redox changes of membrane-bound cytochromes could be mediated by the excited exogeneous flavins.

The complexity of all possible reactions even in a simple artificial system becomes apparent as soon as flavins are present in the reaction mixture. The purpose of these investigations was to follow simultaneously several parameters in an artificial system and to compare the model reactions with photocontrolled reactions in biological systems.

2 Material and Methods

To compose the reaction mixture the following stock solutions in phosphate buffer 0.1 mM pH 7 were made: EDTA 50 mM, riboflavin 0.5 mM, cytochrome c 1 mM, NBT 100 mM. The reaction mixture always contained 50 μl EDTA, 50 μl riboflavin, 50 μl cytochrome c or/and 25 μl NBT when indicated. Phosphate buffer was added to a final volume of 250 μl.

The reaction mixture was placed in a multibranched fiber optic system as described earlier (Ninnemann et al. 1976). The technical set-up allowed the simultaneous measurements of two absorption kinetics, the fluorescence emission and the variations of the oxygen concentration while the sample was excited with blue light or not.

The excitation beam was a broad blue band of 450 ± 30 nm of 3 W/m^2 at the surface of the sample (tungsten lamp plus the Corning filters CS 9782 and two CS 5543). The excitation beam was chopped in 1 s light and 2 s dark cycles. One measuring beam was set at $550 \pm$ nm (tungsten lamp, Corning filter CS 9782, monochromator, fibersystem, sample, fibersystem, Corning filter CS 3486 interference filter 549 nm, Hamamatsu photomultiplyer).

1 Present address: Institute of Plant Physiology, University of Geneva, 1211 Geneva, Switzerland
2 University of California at San Diego, La Jolla, CA 92037, USA

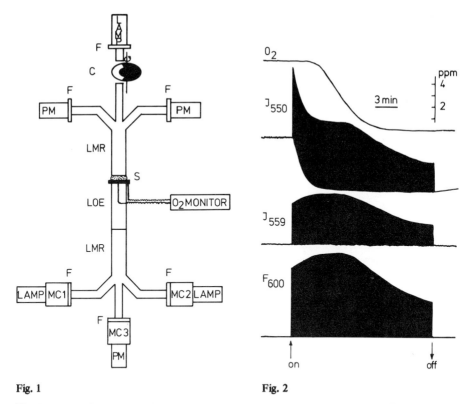

Fig. 1   **Fig. 2**

Fig. 1. Fiber optic set-up allowing to measure simultaneously two absorptions, one fluorescence, and an oxygen signal in the dark or during blue light excitation. *MC* monochromator, *F* filtercombination, *PM* photomultiplyer, *LMR* light-mixing rod, *LOE* light-guiding oxygen electrode, *S* sample, *C* chopper

Fig. 2. Simultaneous recordings of oxygen concentration, higher and lower envelope curves of the transmissions at 550 nm and 559 nm and the fluorescence emission at 600 nm. The reaction mixture contained: EDTA 10 mM, riboflavin 0.1 mM, cytochrome c 0.2 mM in phosphate buffer 0.1 mM pH 7.0

   The fluorescence was filtered through Corning blocking filters CS 2418 and 2424 and a monochromator set at 600 nm and it was measured by a Hamamatsu photomultiplyer).

   To have homogeneous illumination all light beams passed through a light mixing rod placed on both sides of the sample. The oxygen concentration was measured with a Clark-type oxygen electrode which was built on fiber optics. (LOE: Light guiding oxygen electrode, Strasser 1973, 1974). The set-up is shown schematically in Fig. 1. The slow chopping of the exciting beam made it possible to follow the photocurrents at 550 nm and at 559 nm in the dark (pure measuring beams passing through the sample) and during excitation with the exciting blue light (measuring beams superimposed with the fluorescence at these wavelengths). The lower envelope curves of the J 550 nm and the 559 nm signals indicate therefore the transmission of the sample at these wavelengths. Figure 2 shows such an experiment (discussed as experiment 2 in the Results section). A mixture of EDTA riboflavin (RF) and cytochrome c was used. The lower envelope curve of the J 550 nm signal shows the reduction of cytochrome c. The unvaried lower envelope of the J 559 nm signal shows that the transmission signal at this wavelength is not influenced by the reduction of cytochrome c. Adding another electron acceptor e.g., NBT (nitroblue tetrazolium) makes it possible to measure cytochrome c and NBT reduction simultaneously in the same reaction mixture.

3 Results

3.1 The Reaction Mixtures

Four sets of experiments are shown in Fig. 3. In all experiments the electron donor and the flavin concentration were the same and all experiments were started under

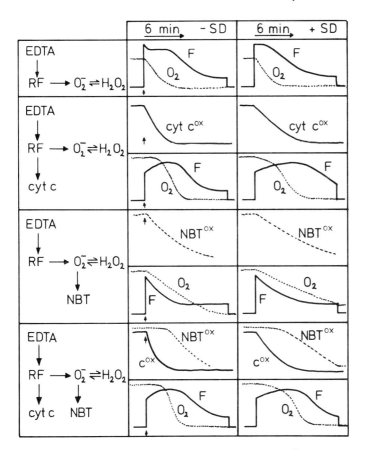

Fig. 3. Four times the same EDTA-riboflavin-system with different combinations of electron acceptors, such as oxygen (Exp. 1, top), oxygen and cytochrome c (Exp. 2), oxygen and NBT (Exp. 3), oxygen, cytochrome c and NBT (Exp. 4). The reaction mixture was as indicated in Fig. 2. NBT was 10 mM when added. O_2 oxygen concentration, F fluorescence emission at 600 nm, $Cyt\ c^{ox}$ transmission at 550 nm, NBT^{ox} transmission at 550 nm, $+SD$ superoxide dismutase added, *arrow* light cycles on

aerobic conditions. As indicated, the experiments were performed in presence of cytochrome c (cyt c), nitroblue tetrazolium salt (NBT) or cyt c plus NBT, or without any added electron acceptor. The addition of superoxide dismutase (SD) did not change the principal behavior of the measured parameters, it only affected the velocities of the reactions due to the decrease of the O_2^- concentration in the reaction mixture.

3.2 The Reduction of Electron Acceptors

Whenever an electron acceptor was present it was reduced immediately upon illumination. This was the case for cytochrome c as well as for NBT or oxygen. However if cytochrome c and NBT were added simultaneously, cytochrome c was reduced first followed by the reduction of NBT.

3.3 The Oxygen Measurements

The reduction of NBT was always paralled with an oxygen uptake, while for the reduction of cytochrome c no or only very little oxygen consumption was observed. Without NBT or cytochrome c present, oxygen was consumed immediately upon illumination.

3.4 The Fluorescence Measurements

The fluorescence signal can mainly be attributed to the oxidized form of the flavin. Under anaerobic conditions the fluorescence is very low due to the accumulation of fully reduced flavin FlH_2. The decrease of fluorescence even in the presence of oxygen may indicate the presence of the flavin radical FlH. The reduction of NBT was always paralleled by a decrease of the flavin fluorescence. If, however, cytochrome c was the electron acceptor the flavin fluorescence was quenched partially at the beginning and increased, while the cytochrome was becoming reduced. Similar behavior was observed also when the experiment was started under anaerobic conditions (data not shown).

4 Discussion

Due to the reported results we propose that for the different sets of experiments the following reactions are predominant:

4.1 Primary Reactions

For all combinations we can assume the following fast reactions:

$$Fl \xrightarrow{\text{blue light}} Fl^*$$
$$Fl^* \longrightarrow Fl \quad + \quad \text{fluorescence}$$
$$O_2 \quad + \quad {}^3Fl \longrightarrow Fl \quad + \quad O_2^*$$
$$EDTA^{red} \quad + \quad {}^3Fl \longrightarrow FlH_2 \quad + \quad EDTA^{ox}$$
$$FLH_2 \quad + \quad Fl \rightleftharpoons 2 \, FlH$$

The established equilibrium between the oxidized and the reduced flavin and the flavin radical will determine the different groups of reactions shown in the experiments 1 to 4 in Fig. 3.

4.2 Oxygen as the Only Electron Acceptor (Exp. 1)

In addition to the primary reactions

$$O_2^* \quad + \quad FlH_2 \longrightarrow FlH \quad + \quad O_2^-$$
$$O_2 \quad + \quad FlH_2 \longrightarrow Fl \quad + \quad H_2O_2$$
$$O_2 \quad + \quad FlH \longrightarrow Fl \quad + \quad O_2^-$$
$$\overline{EDTA^{red} + O_2 + Fl \xrightarrow{\;+ light\;} FlH + O_2^- + EDTA^{ox}}$$

The intermediate decrease of flavin fluorescence can be attributed to the small turn-over concentration of the flavin radical FlH. The oxygen consumption is mainly due to the autoxidation of FlH_2 directly forming H_2O_2.

4.3 Cytochrome c as Electron Acceptor (Exp. 2)

In addition to the primary reactions

$$\overline{Cyt\ c^{ox} + FlH \longrightarrow Fl \quad + \quad Cyt\ c^{red}}$$
$$EDTA^{red} \quad + \quad Cyt\ c^{ox} + Fl \xrightarrow{+ light} Fl \quad + \quad Cyt\ c^{red} + EDTA^{ox}$$

As cytochrome c is directly reduced by the flavin molecule, the rate of photochemistry leading to the reduction of cytochrome c competes with the fluorescence emission. Therefore the fluorescence is partially quenched at the onset of illumination and increases while cytochrome c is being reduced. The presence of oxidized cytochrome c keeps the concentration of reduced flavin FlH_2 close to zero so that only little O_2 is consumed and little O_2^- and H_2O_2 are formed.

4.4 NBT and Oxygen as Electron Acceptors (Exp. 3)

In addition to the primary reactions

$$O_2 \quad + \quad FlH_2 \longrightarrow Fl \quad + \quad H_2O_2$$
$$O_2^* \quad + \quad FlH_2 \longrightarrow FlH \quad + \quad O_2$$
$$O_2 \quad + \quad FlH \longrightarrow Fl \quad + \quad O_2^-$$
$$O_2^- \quad + \quad NBT^{ox} \longrightarrow NBT^{red} \quad + \quad O_2$$
$$\overline{EDTA^{red} + NBT^{ox} + O_2 + Fl \xrightarrow{+ light} FlH + H_2O_2 + NBT^{red} + EDTA^{ox}}$$

The trapping of O_2^- radicals by NBT speeds up the oxidation of FlH_2. The oxidation of FlH is slower than the oxidation of FlH_2. Therefore the concentration of FlH increases and the fluorescence decreases even in presence of oxygen.

4.5 Cytochrome c, NBT and Oxygen as Electron Acceptors (Exp. 4)

This experiment shows at the beginning a reaction type of experiment 2 as long as cytochrome c is present in its oxidized form, followed by a reaction type of experiment 3
as soon as cytochrome c is reduced. The experiment shows clearly that under the chosen
conditions cytochrome c and NBT are reduced by different electron donors.

4.6 General Remarks

Once the reaction mixture is anaerobic, FlH_2 can be accumulated in all four experiments.
Under these conditions FlH_2 will act as a strong reducing agent for the remaining still
reducible molecules. Only the predominant reactions have been mentioned. Under different conditions other combinations of reactions may become important such as, e.g.,
the reduction of cytochrome c by the radical O_2^- as reported elsewhere (Schmidt and
Butler 1976). The reported model system suports several proposed reactions observed
in biological membranes (Fig. 4) e.g., (a) O_2^- reduces cytochrome b in the respiratory
chain of mitochondria (Ninnemann et al. 1976). (b) Flavin mediates the direct reduction of a b-type cytochrome in the nitratereducrase complex (Ninnemann and Klemm-
Wolfgramm, this vol.).
    Whenever a flavin system is excited the four following overall reactions have to be
considered:

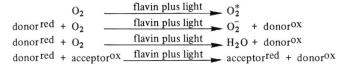

$$O_2 \xrightarrow{\text{flavin plus light}} O_2^*$$
$$donor^{red} + O_2 \xrightarrow{\text{flavin plus light}} O_2^- + donor^{ox}$$
$$donor^{red} + O_2 \xrightarrow{\text{flavin plus light}} H_2O + donor^{ox}$$
$$donor^{red} + acceptor^{ox} \xrightarrow{\text{flavin plus light}} acceptor^{red} + donor^{ox}$$

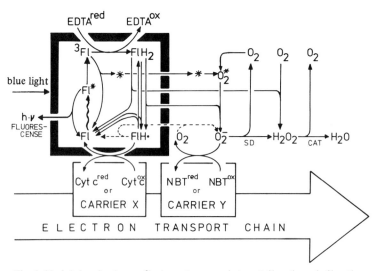

Fig. 4. Model showing how a flavin system may interact directly or indirectly over oxygen, with
electron carriers in biological systems. The biological electron carriers seem to be often b-type cytochromes

In Fig. 4 the discussed reactions are put together and are shown in one scheme. It proposes especially that a flavin system may potentially interfere with biological electron transport chains at different sites. First, flavins interact directly with electron carriers in a biological system and second, they interact with oxygen to form oxygen radicals or excited oxygen, which afterwards react with the biological electron carriers in a membrane. This dual capacity of interference of flavins with biological systems presents a very subtle regulation mechanism for many phenomena triggered by blue light.

Acknowledgment. This work was supported by a grant from the National Institute of Health, GM-20648.

References

Ninnemann H, Klemm-Wolfgramm (1980) In: Senger (ed) this volume
Ninnemann H, Strasser RJ, Butler WL (1976) Photochem Photobiol 26: 41−47
Schmidt W, Butler WL (1976) Photochem Photobiol 24: 71−75
Strasser RJ (1973) Arch Int Physiol Biochim 81: 935−955
Strasser RJ (1974) Experientia 30: 320

Artificial Flavin/Membrane Systems; a Possible Model for Physiological Blue Light Action

W. SCHMIDT[1]

1 Introduction

The blue light photoreceptor responsible for a great variety of physiological responses is most likely a flavin, bound in a highly dichroic manner to a membrane moiety [10, 11, 28]. The mechanism termed *sensory transduction,* by which the impinging blue light is transformed into a biochemical equivalent, appears to be a flavin-mediated photo-redox reaction. This is suggested by several observations. (1) Molecular oxygen is a necessary prerequisite for the primary blue light action [9, 21, 27]. (2) Oxidizing agents are capable of simulating physiological blue light action [26] and consequently reducing agents are capable of suppressing it [1, 26]. (3) The primary step of blue light action is temperature-independent [6, 7, 16, 20, 21]. (4) Photodynamically active dyes act as (artificial) photoreceptors, photodynamically inactive dyes do not [3, 13]. (5) Inhibitors of light-driven, flavin-mediated electron flow inhibit physiological blue light responses specifically [18, 23]. (6) The blue light-induced, dark-reversible photoreduction of a b-type cytochrome, as observed in several blue light-sensitive organisms, appears to be correlated to physiological primary reaction of blue light [19]. Redox reactions of carotenoids appear to be physiologically irrelevant.

In addition, several organisms provide good evidence for the assumption that the blue light receptor flavin is localized in the plasma membrane, and that the three components flavin, b-type cytochrome, and membrane interact in the sensory transduction process on the basis of redox reactions [4, 10, 11, 22, 28].

Due to the lack of a clear-cut assay for the blue light photoreceptor and its small concentration compared to the "bulk pigments" in the cell (cf. page 136 in [2]), the primary blue light reaction can scarcely be further elucidated in vivo. In addition, very little is known about the only physiologically applicable "anisotropic flavin chemistry" (i.e., the chemistry of bound flavins), in contrast to the well-understood "isotropic flavin chemistry" (i.e., the chemistry of the free flavin [8]).

As a first approach for a model of the blue light photoreceptor and anisotropic flavin chemistry, three "amphiflavins" bearing C_{18}-hydrocarbon chains at positions 3, 7, and 10 (cf. Fig. 1) have been anchored within artificial vesicle membranes made from three different saturated lecithins. On this basis we hope to learn more about anisotropic flavin chemistry which in turn is mandatory for the understanding of the primary blue light action.

---

1 Biologisches Institut der Universität Konstanz, Postfach 7733, 7750 Konstanz, FRG

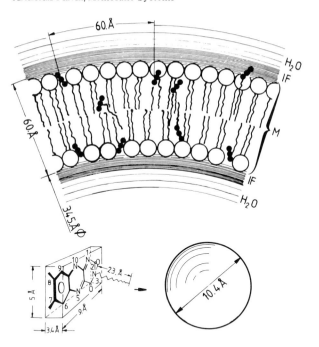

Fig. 1. *Top* idealized sketch of the vesicle-bound amphiphilic flavins. The average distance between the flavins is about 60 Å. In the crystalline state of the membrane (*M*) they are mainly localized near the hydrophilic head groups of the phospholipids, but sink deeper into the more hydrophobic parts of the membrane upon phase transition. The permeability of the interface (*IF*) for exogenous molecules solubilized in the water phase (H_2O) depends on the membrane phase and the charge of the molecules. *Bottom* crude dimensions of the amphiphilic flavin "AFl 3". The other amphiflavins employed bear the aliphatic chain in pos. 7 and 10, respectively. For a first quantitative approach of the rotational mobility of the flavin within the membrane, it was crudely taken as a sphere of 10.4 Å diameter [24]

In the first paper of a planned series on this subject, the microenvironment and rotational motion of the flavins as a function of the specific flavin and the membrane phase (crystalline-liquid crystalline) has been studied by fluorescence analysis [24], (cf. Fig. 1). These investigations provide a good starting point to study the physiologically more relevant *photochemistry* of membrane-bound flavins. With this prospect, further fluorescence properties of membrane-bound flavins are elucidated below. Preliminary photochemical results are presented.

2 Materials and Methods

The amphiphilic flavins 7,8,10-trimethyl-3-octadecyl-isoalloxazin (AFl 3), 3,8,10-trimethyl-7-octadecyl-isoalloxazin (AFl 7), 3,7,8-trimethyl-10-octadecyl-isoalloxazin (AFl 10), 3-methyl-10-octadecyl-isoalloxazin (AFl 10,7-8-H) were synthesized by Dr. W.R. Knappe. The crude structure and size of AFl 3 is depicted in Fig. 1. The phospholipid L-β-γ dimyristoyl-α-lecithin (DML,N 42803, melting point 23°C), L-β-γ dipalmitoylα-lecithin (DPL,N 42556, melting point 41°C) and L-β-γ distearoyl-α-lecithin (DSL,N 43698, melting point 55°C), puriss., were purchased from Fluka, Neu-Ulm. The detailed preparation of the flavin-loaded vesicles has been described previously [24]. The experiments were generally performed in 0.01 M phosphate buffer at pH 8.0, containing 0.01 M NaCl. The pH's for the investigation of the pH dependence of the fluorescence quantum efficiency and the pK determination were adjusted as follows: pH 0/1, HCl; pH 2/3, sulfate; pH 4/5 acetate; pH 6/7/8 phosphate; pH 9/10/11 borate; pH 12, phosphate; pH 13, NaOH/KCl. All suspension mediums are photochemically inert.

Fluorescence measurements were performed with the fluorimeter model MPF 3 (Perkin Elmer) under aerobic conditions. Cuvette temperature was controlled by circulating thermostated water through a copper cell holder. Photoreduction experiments were performed with the Cary 118 spectrophotometer under anaerobic conditions. For details see [24].

3 Results and Discussion

3.1 Phase Dependence of Fluorescence

The dependence of the fluorescence of vesicle-bound flavins on temperature (i.e., on the membrane phase, "crystalline" or "liquid crystalline") has recently been monitored kinetically near the fluorescence maximum at 520 nm [24]. An increase of the flavin fluorescence at the phase transition was interpreted as delocalization of the flavin chromophore from the hydrophilic head group into the more hydrophobic hydrocarbon region of the membrane. This is further supported by the corresponding fluorescence spectra. "Model spectra" of isotropically dissolved flavin in solvents of quite different polarity are shown in Fig. 2A. Clearly, with decreasing polarity of the solvent, the fluorescence quantum yield increases, the fluorescence peak undergoes a hypsochromic shift, and the vibrational structure is better resolved. Analogous fluorescence characteristics are exhibited by the vesicle-bound flavins as a function of the membrane phase, as shown for the DSL/AFl 3 system (Fig. 2B). This feature is even more clearly seen by the corresponding difference spectra (dotted lines in Fig. 2).

Utilizing the formulas by *Perrin* and *Stoke/Einstein*, from fluorescence polarization measurements the rotational relaxation time of the membrane-bound flavin has been shown to decrease from about 75 ns in the crystalline state to about 10 ns in the liquid crystalline state [24]. This decrease has a twofold origin. (1) The fluidity of the membrane in the liquid crystalline state is higher than in the crystalline state, (2) the chromophores are delocalized upon phase-transition (vide supra); it is well known that, independent of phase, motion increases from the glycerol backbone of the lipid towards the terminal methyl group [15]. The corresponding micro-viscosities, as encountered by the flavin, are about 160 cp below and about 30 cp above phase transition [24].

A closer inspection of the single components of polarized fluorescence reveals more differentiated information, additional to the integral values of viscosity, polarity, and mobility as obtained from fluorescence polarization, averaged over all flavins present. For example, Fig. 3 shows the temperature dependence of the extreme fluorescence polarization components at 0° and 90°, including some intermediate angles, of AFl 10 (7-8-H), bound to DML-, DPL-, and DSL-vesicles. Clearly, the curves are not propor-

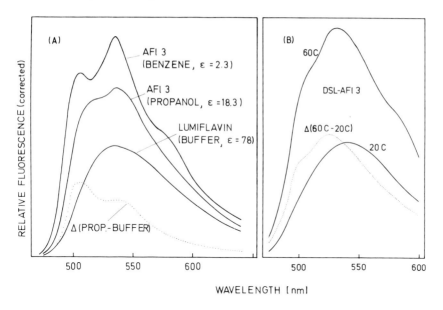

Fig. 2. A Dependence of the flavin fluorescence emission spectrum on the polarity of the solvent. For the nonpolar solvents benzene and propanol the amphiphilic flavin AFl 3 was used and for the aqueous solution, lumiflavin. **B** Fluorescence emission spectrum of the amphiphilic flavin AFl3, bound to DSL-vesicles, in the crystalline ($20°$C) and liquid crystalline ($60°$C) state of the membrane. All flavin concentrations are 4 μM, λ_{ex} = 470 nm

tional to each other, as expected for a unique flavin population and/or flavin environment. The $0°$-components show stronger temperature quenching and a scarcely pronounced fluorescence change ("dip") near the phase transitions, in contrast to the $90°$-components. For comparison, under the same conditions an aqueous solution of lumiflavin does not show (apparent) fluorescence polarization, indicating that the rotational relaxation is much faster than the fluorescence life-time (Fig. 3). The straightforward interpretation encompasses two distinct flavin moieties, differing in their localization. Due to the rotational relaxation of the membrane-bound flavins, increasing angles between polarizer and analyzer are equivalent to increasing time gaps between excitation and observation.

$90°$-component: A relatively long-living component is observed. It exibits only modest quenching by temperature and shows a marked fluorescence change ("dip") at the phase transitions. This indicates a deeper embedding in the more hydrophobic phase of the membrane. Phase transitions mainly involve the hydrocarbon chains, rather than the polar head groups [14].

$0°$-component: This component shows an even stronger fluorescence quenching by temperature than equeous lumiflavin. It is relatively short-lived and observed right from the beginning of excitation. Since a fluorescence "dip" at the phase transitions is barely seen, this flavin moiety appears to be mainly localized in the polar head group area (cf. Fig. 1).

This interpretation in terms of two distinct flavin environments holds for a previously badly understood finding as well. The fluorescence polarization of membrane-

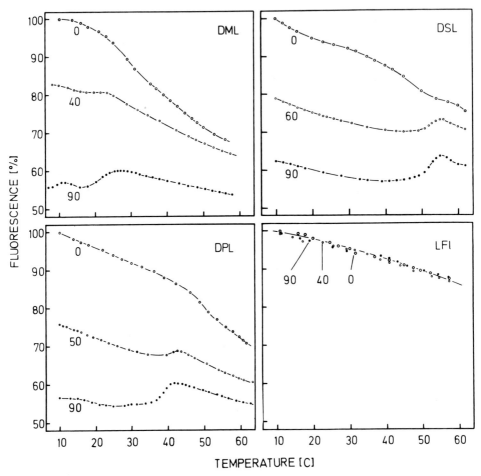

Fig. 3. Fluorescence polarization of AFl 10 (7-8-H) in DML-, DPL- and DSL-vesicles as function of temperature. For comparison, aqueous lumiflavin was measured under the same conditions. The flavin concentrations are 4 μM. The fluorescence was excited at 470 nm with vertically polarized light. The analyzing polarization filter was set vertically (90°), horizontally (0°) or to intermediate angles as indicated (for details of the measuring procedure cf. [24])

bound flavins depends on the suspension viscosity (as achieved by different glycerol concentrations) in a biphasic way (Fig. 9 in [24]), whenever the membrane is in the liquid crystalline state (in the crystalline state of the membrane the glycerol has not access to the membrane at all, Fig. 9 in [24]). The flavins localized in the head group area of the membrane are about ten times more efficiently hindered in their rotational motion than those localized in the deeper hydrophobic fatty acid chain area [24]. For further elucidation of this feature, direct fluorescence life-time measurements are projected.

3.2 Phase Dependence of Flavin Accessibility

The access of exogenous molecules to the membrane-bound flavins depends strongly
on the membrane phase and the specific charge of the molecules. Iodide, azide [25],
and EDTA (ethylenediaminetetraacetic acid, vide infra) as anions have easier access to
the flavins in the crystalline state. Glycerol [24] and protons, however, appear to pene-
trate the membrane/water interface more easily in the liquid crystalline state of the
membrane. This feature is probably correlated with the zwitterionic lipid head groups.
The polar head group dipole of the lecithin molecule is supposed to be oriented parallel
to the membrane surface in the crystalline state and orthogonal to it in the liquid crystalline state of the membrane [14]. Figure 4 shows the quantum efficiency of flavin
fluorescence as a function of the pH (i.e., proton concentration) and the phase for the
DPL/AFl 10 system (dotted curve: fluorescence dependence on pH of aqueous riboflavin [12]). All curves are normalized to 100%. Clearly, in the liquid crystalline state
and above pH = 2, the flavin fluorescence is sensitive to proton concentration, but not
in the crystalline state. The curve for riboflavin falls off at approximately pH3, whereas
those for the vesicle-bound flavin decrease at a ten times higher proton concentration,
independent of phase. The latter feature remains unexplained in molecular terms.

The fluorescence decrease of aqueous riboflavin in the alkaline range is due to the pK
of about 10 of the N3-hydrogen (cf. [8], Fig. 2). Unfortunately, the corrresponding
(apparent) pK of the membrane-bound amphiflavin (AFl 10) cannot be observed since
it is blocked at position N3 by a methyl group (in contrast to normal isoalloxazines car-
rying a hydrogen at N3), ensuring full fluorescence up to pH 10/pH 13.

That pK's of flavins can change dramatically upon binding to membranes is demonstrated in detail for the pK of the N1-hydrogen of AFl 3 in DPL-vesicles. The dependence of the absorption spectrum of the reduced system on the pH can be taken as convenient assay (Fig. 5). As a control, the known pK for aqueous (lumi-) flavin (6.25,
[17]) was verified (Fig. 6). Despite the large measuring error, the pK is more increased
below phase transition (10.1) than above phase transition (9.4).

EDTA is known as a suitable substrate for flavin photoreduction [5]. Preliminary
experiments with membrane-bound flavins reveal an unexpected feature which, however,
is in good consistency with our previous results. The quantum efficiency Φ of flavin

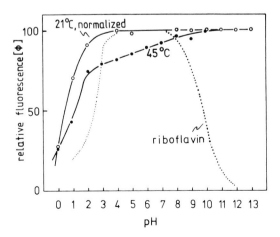

Fig. 4. Dependence of the fluorescence
quantum yield on the pH and the membrane phase. *Dotted line* aqueous riboflavin [12] *Open circles* AFl 10 in
DPL-vesicles with the membrane in
the crystalline state. *Closed circles*
AFl 10 in DLP-vesicles with the membrane in the liquid crystalline state.
For comparison, all curves are normalized to 100%

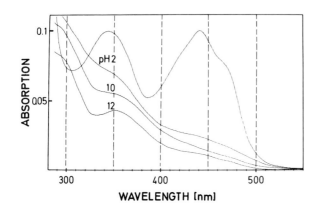

Fig. 5. Absorption spectra of AFl 3 bound to DPL-vesicles. The spectrum of the oxidized form is mainly independent of the pH. Due to the pK of the N1-hydrogen, the reduced form changes significantly with the pH, which was used for its determination. Because of strong scattering of the vesicle suspensions, the absorption measurements were performed with flavin-free vesicles as reference. The flavin concentration was 4 μM

photoreduction as a function of EDTA concentration and temperature is depicted in Fig. 7 for the DPL/AFl 10-system (other systems are currently being investigated and will be described in a forthcoming paper [25]). Below phase transition temperature, significant photoreduction occurs even in the absence of EDTA, indicating that (part of) the lipid itself seves as electron donor. Additional EDTA concentrations promote the speed of photoreduction even more. Above phase transition temperature it is necessary to have a high concentration of EDTA in order for any photoreduction to take place. This, again, is consistent with our previous conclusion of (1) a flavin displacement upon phase transition and (2) the easier access of anions to the flavins in the crystalline state.

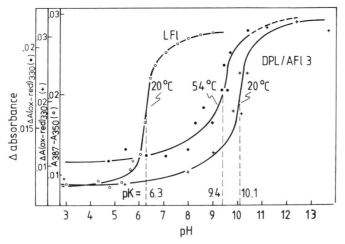

Fig. 6. Determination of the pK of the N1-hydrogen from spectra as shown exemplarily in Fig. 5. *Open circles* the absorption difference between 387 and 350 nm is taken as assay for the change of the spectrum of the reduced aqueous lumiflavin (LFl) with the pH. This assay cannot be used for the membrane-bound flavin, since the relatively small differences are completely obscured by the unavoidable variations in scattering. This uncertainty, however, can be mainly bypassed if the absorption change between the oxidized and the reduced form at one wavelength (330 nm) is measured as a function of pH of the same individual sample. Reduction was achieved by anaerobic irradiation with blue light in the presence of EDTA. Subsequently the samples were autoxidized

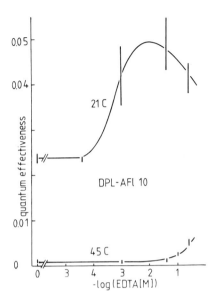

Fig. 7. Quantum efficiency of photoreduction of DPL-AFl 10 as a function of temperature and exogeneous EDTA-concentration. For details cf. [25]

Which part of the lipid serves as electron donor when the membrane is in the crystalline state? Considering the binding energies of hydrogen, we can mainly exclude the fatty acid residues, and there remain the choline group and the C-H-bonds of the glycerol backbone of the phosphatidylcholin as contenders. This assumption is currently being investigated with corresponding model substrates under isotropic conditions [25].

4 Summary and Prospects

Amphiphilic flavins can be bound in reasonable concentrations ([lipid]/[flavin] $\approx$ 100/1) to artificial membrane vesicles. Their orientation, localization and mobility can be controlled by the specific attachment of the hydrocarbon chain ("anchor") to the flavin, and the membrane phase [24]. Based on polarized fluorescence measurements, two flavin moieties, differing in the microenvironment within the membrane, have been postulated. The flavins are highly demobilized by binding to the membranes, the access of exogenous molecules ("substrates") to the flavin is largely diminished and depends on the charge of the molecule and the phase of the membrane. The currently used membrane-bound flavins provide efficient but simple enough model systems to study anisotropic flavin chemistry. Based on the experience obtained from these systems, we intend for our future work to switch to vesicle-bound flavoproteins as more realistic models for the blue light photoreceptor. Recently we succeeded in binding a lipophilic flavoenzyme (monoamineoxidase) to artificial membrane vesicles made from egg-lecithin, under preservation of its activity.

Acknowledgments. The author wishes to thank Prof. Hemmerich for his continued interest in aspects of this work. Mr. G. Gleichauf is thanked for excellent technical assistance. The author is indebted to Dr. W.R. Knappe for the synthesis of the different amphiflavins. Dr. F. Armstrong is thanked for reading the manuscript. This work was financially supported by the Deutsche Forschungsgemeinschaft (Sonderforschungsbereich 138, "Biologische Frenzflächen und Spezifität", A1).

References

1. Batra PP, Rilling HC (1964) Arch Biochem Biophys 107: 485
2. Bergman K, Burke PV, Cerdá-Olmedo E, David CN, Delbrück M, Foster KW, Goodell EW, Heisenberg M, Meissner G, Zalokar M, Dennison DS, Shropshire Jr W (1969) Bacteriol Rev 33: 99
3. Blum HF, Scott KG (1933) Plant Physiol 8: 525
4. Brain R, Freeberg JA, Weiss CV, Briggs WR (1977) Plant Physiol 59: 948
5. Fife DJ, Moore WM (1979) Photochem Photobiol 29: 43
6. Guttenberg H v (1959) Planta 53: 412
7. Haupt W (1957) Planta 49: 61
8. Hemmerich P (1976) The present status of flavin and flavoenzyme chemistry. Prog Chem Org Nat Prod 33: 451
9. Howes CD, Batra PP (1970) Arch Biochem Biophys 137: 175
10. Jaffe L, Etzold H (1962) J Cell Biol 13: 13
11. Jesaitis AJ (1974) J Gen Physiol 63: 1
12. Kavanagh R, Goodwin RH (1949) Arch Biochem 20: 315
13. Lang-Feulner J, Rau W (1975) Photochem Photobiol 21: 179
14. Lee AG (1975) Prog Biophys Mol Biol 29: 5
15. Levine YK, Birdsall NJM, Lee AG, Metclfe JC (1972) Biochemistry 11: 1416
16. Mathews MM (1963) Photochem Photobiol 2:1
17. Michaelis L, Schwarzenbach G (1938) J Biol Chem 123: 538
18. Mikolajczyk E, Diehn B (1975) Photochem Photobiol 22: 269
19. Ninnemann H, Klemm F (1976) Plant Physiol 58 Abstr: 112
20. Poff KL, Butler WL (1974) Photochem Photobiol 20: 241
21. Rilling HC (1962) Biochem Biophys Acta 60: 548
22. Schmidt W, Thompson KS, Butler WL (1977) Photochem Photobiol 26: 407
23. Schmidt W, Hart J, Filner P, Poff KL (1977) Plant Physiol 60: 736
24. Schmidt W (1979) J Membr Biol 47: 1–25
25. Schmidt W, Hemmerich P (1979) J Membr Biol in preparation
26. Theimer RR, Rau W (1970) Planta 92: 129
27. Zalokar M (1954) Arch Biochem Biophys 50: 71
28. Zurzycki J (1972) Acta Protozool 11: 189

Effects of UV and Blue Light on the Bipotential Changes in Etiolated Hypocotyl Hooks of Dwarf Beans

E. HARTMANN[1] and K. SCHMID[1]

1 Introduction

One of the most complex topics to study in biology is the ability of organism to perceive, code, transmit, and integrate environmental information, which is used to direct the cellular metabolism and developmental processes occurring. The detection of different wavelengths of light by specific mechanism plays a key role in plant development. Although great progress has been made in the study of plant photoreceptor pigments, some pigment systems are understood better than others.

More is known about the phytochrome pigment system, which works in the red and far red regions of the visible light spectrum (Smith 1975) than about the pigments which operate in the short wavelength regions. The main reason for this situation seems to be that there are problems involved in detecting the pigment system (or systems) responsible for the blue light effects. Virtually nothing is known about the events which occur between the reception of the light stimulus and the early intracellular transduction events. Light-induced biopotential changes may provide the opportunity to study these little-understood processes. We have investigated the induction of biopotential changes in etiolated bean hypocotyl hooks using short-wavelength light. Under natural occurring conditions the hypocotyl hook is the part of the seedling which has the first contact with light. After the perception of light a hook opening growth reaction quickly follows (Klein et al. 1956). It is known that the hook region contains high amounts of phytochrome (Furuya and Hillman 1964) but nothing is known about the blue light photoreceptor.

2 Material and Methods

Dwarf bean seedlings (*Phaseolus vulgaris* cv. *St. Andreas,* Erste Ernte, grünhülsig) were grown in vermiculite in absolute darkness with a constant room temperature of $20 \pm 1°C$. The moisture of the substrate is an important factor for the conditions of the seedlings. It is always necessary to use the same amount of distilled water to wet the dry vermiculite. In the experiments where inhibitors were used the control and the test seedlings were grown in separate plastic boxes side by side in a third large box, to maintain similar microenvironmental conditions for both cultures.

1 Institut für Allgemeine Botanik der Universität Mainz, Saarstraße 21, 6500 Mainz, FRG

The carotenoid content of the etiolated seedlings was manipulated by the following inhibitors which were applied in the water used for wetting the substrate: Amitrole (3-amino-1.2.4-triazole), SAN 6706 [4-chloro-5(dimethylamino)-2-(trifluoro-m-tolyl)-3(2H)pyridotinone], SAN 9774, and SAN 97854 (Sandoz, Switzerland). The most potent inhibitor from the SAN groupe was SAN 6706, which was used in all experiments described. The inhibitors were dissolved in water by an ultrasonic treatment.

Six to seven day-old bean seedlings were collected, selected for a hook angle of 180°, and cut 5 cm below the hoop apex. As bathing solution we used: KCl 10.0; $CaCl_2 \times 2H_2O$ 0.1; $MgSO_4 \times 7H_2O$ 0.05; KH_2PO_4 1.0; Fe(III)citrate 0.008 mM, pH 5.2. All amipulations with the etiolated seedlings were performed under dim green light (0.2 W m^{-2}).

The electrical contacts were established with surface electrodes of the silver-silver chloride type in 0.3 M KCl. The biopotentials were measured with a KEITHLY electrometer Mod. 602 and a self-built high impedance preamplifier. A highly stabilized voltage source was used to compensate the offset voltage. The signals were recorded by a micrograph BD 5 recorder (Kipp and Zonen). We used different combinations of instruments to make the measurements. One combination was equipped to channel the biopotential changes from one bean (Fig. 1) and another was designed to measure a greater number of hypocotyl hooks (Fig. 2). It was possible by the second instrument to overcome the problems of variation in biopotential responses typical for individual bean seedlings. The hooks (10 or 18 in one vessel) were electrically connected in series. This way of connecting had many advantages over connecting them in parallel. There were no problems with high compensating currents typical for parallel connections. The potentials of the different beans could change without mutual interaction. In this way a highly stabilized biopotential with little noise was obtained. This signal represented the mean value of the potential for the number of hooks used in the experiment.

The measurements were made in a Faraday cage which was installed in a thermoconstant room at 20 ± 1°C. It was necessary to allow the hook-preparations to equilibrate after the cutting until a stabilized potential was reached. After 3–4 h the drift of the biopotential was negligible. We always observed a typical shape of the transient change

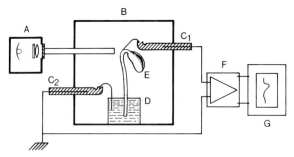

Fig. 1. Assemblage for the measurement of light-induced biopotential changes at the hook of dwarf bean seedlings. *A* Light projector with a glass fiber cable and interference filters, *B* black sealed lucite box, *C* silver-silver-chloride-electrodes in 0.3 M KCl solution, contact via KCl soaked cotton threads, *D* small lucite vessel containing the bathing solution and *E* the excised *Phaseolus* seedlings, *F* Keithly 602 electrometer, *G* chart recorder

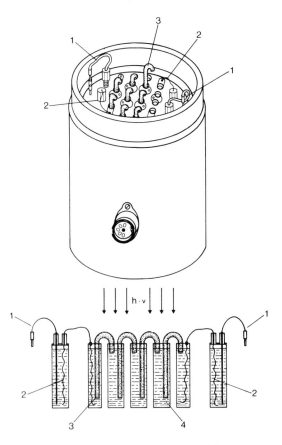

Fig. 2A, B. Assemblage for the measurement of light-induced biopotential changes at 10 or 18 hooks of dwarf bean seedlings. **A** shows the vessel with 10 or 18 hooks, **B** the electrical connection of the bean hooks and the electrodes. *1* Silver-silver-chloride-electrodes in 0.3 M KCl solution, *2* contacts between the two cutting sites of the hypocotyl hooks and the electrodes. The contacts were KCl soaked cotton threads, *3* hypocotyl hooks, *4* bathing solution

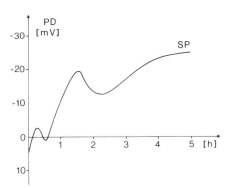

during the equilibration time (Fig. 3). The bean hooks were irradiated using a projector (150 W) and a light-conducting glass fiber cable producing a light spot of about 6 mm². The instrument equipped for 10 or 18 hooks could be fixed directly on the projector, and the hooks were irradiated from above on their apices. The high-intensity irradiations were performed with a Zeiss Xenosol III projector equipped with an Osram xenon bulb XB2500 W. Interference filters (half band with 15 nm) were placed in the light paths of the different projectors. Gray filters were used to alternate the light intensity.

Fig. 3. Characteristic biopotential change of etiolated hypocotyl hooks of beans during the equilibration time. The measurement started just after the cutting of the seedlings. After about 3 h a stabilized potential difference (SP) of about −25 mV was reached

Polarized light was produced with polarizing foils (type PW 44, Käsemann, FRG) which were put directly over the hooks. The irradiance was measured with a YSI-radiometer model 65 (Yellow Springs Instruments).

Electrical stimulation of the hooks also gave a biopotential response. A constant current stimulator model ELS-CS-1 (Electronics for Life Science) was used for this purpose. The stimulus was transmitted as a square wave of 10 ms using platinum electrodes and different currents were tested. The extraction of carotenoids was performed after Nielsen and Gough (1974). Ten hooks were deep-frozen in liquid nitrogen. The samples were homogenized with an Ultra-Turrax in 10 ml ice-cold methanol and small amount of $CaCO_3$. The homogenate was centrifuged with 4000 g for 5 min and the pellet was exhaustively extracted with ethanol (3–5 times). The supernatants were combined and shaken with hexane (Uvasol, Merck). The hexane phase was shaken with water to separate the carotenoids. After complete separation of both layers the hexane phase was used for carotenoid determinations. The absorption spectra or concentrations of the carotenoids were determined with a SHIMADZU spectralphotometer UVS-210. The statistical evaluations used are described in the text or in the legends of the figures. All experiments reported were reproduced more than five times independently. Comparable control samples were always used, so that differences in developmental age or equilibration times could be monitored. All data are corrected for those control determinations.

3 Results

3.1 The Effect of Blue Light (437 nm)

The general biopotential transient change after a blue light stimulus of 60 s is shown in Fig. 4 and represents a hyperpolarization. The reaction starts with a lag-time between 1 and 5 s depending on the light intensity used for the stimulus. The maximum of the biopotential change is reached after about 2 min. There is no significant difference between a saturating short blue light stimulus or continuous irradiation of a comparable intensity; in both cases we observed the shown transient change. A light stimulus given during the period of potential change had little or no effect on the transient. There is a refractory period of about 30–40 min before a new light stimulus can induce a full effect.

The dose response relationship for a wavelength of 437 nm is demonstrated in the Figs. 5–8. In the plots of Figs. 5 and 6 the saturation effect of doses between 10^{-7} and 10^{-3} Mol m^{-2} is clearly demonstrated. With higher intensities like 28 W m^{-2} a remarkable increase in the amplitude of the biopotential change was observed.

Another possibility to describe dose and response correlations is by varying the irradiation time and keeping the irradiance constant. These results are shown in Fig. 7. We observed again different curves for the high light intensity and the lower light intensities. With the high light intensity a saturation of the maximum biopotential change is reached after much shorter irradiation times than in other cases. The semi-logarithmic plot of the data (Fig. 8) shows this result still more impressively. The Bunsen-Roscoe reciprocity law could only be established for the linear part of the dose response curves determined with 8.5 and 2 W m^{-2}. Using higher doses the law does not hold (Fig. 9).

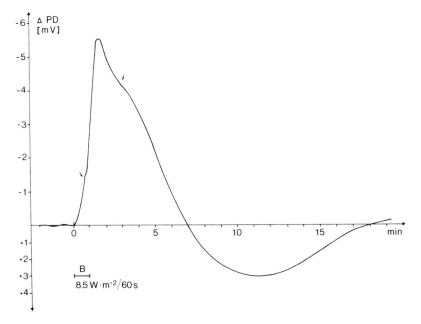

Fig. 4. Typical biopotential transient change after an irradiation with blue light (437 nm). The effect was a hyperpolarization. There were two typical shoulders in the curve labeled with the *arrows*

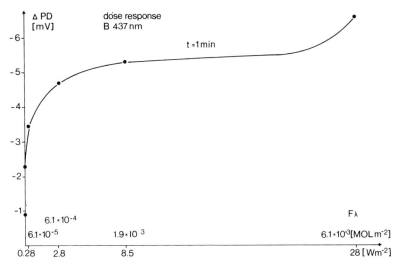

Fig. 5. Dose response curve of the effect of different quantum flux densities of blue light (437 nm) on the light-induced biopotential change

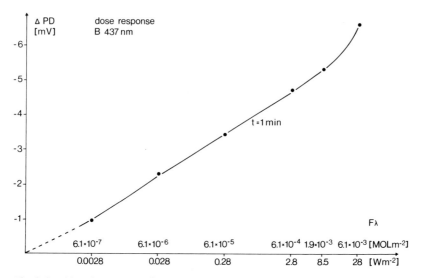

Fig. 6. Semi-logarithmic plot of the dose response curve shwon in Fig. 5

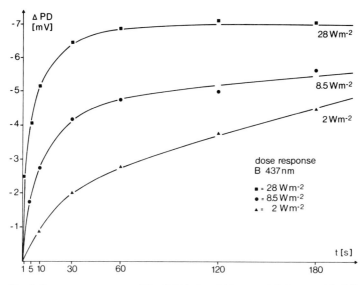

Fig. 7. Dose response curve of the light-induced biopotential change with different irradiances and different irradiation times of blue light (437 nm). The values represent means from six measurements

3.2 The Effect of UV Light (365 nm and 254 nm)

Besides the effects of the blue light (437 nm) we have tested the influence of UV light. We had no instrument to produce monochromatic UV light therefore we used light sources with a relatively broad UV light-emitting spectrum. We had two different sources, one which produced UV light with a wavelength maximum of 365 nm, and another with a maximum around 254 nm.

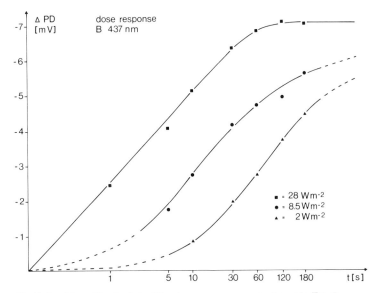

Fig. 8. Semi-logarithmic plot of the dose response curves shown in Fig. 7

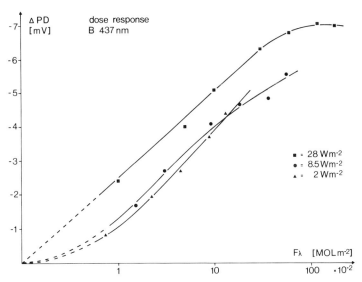

Fig. 9. Comparison of different quantum flux densities of blue light (437 nm). The reciprocity law holds for moderate intensities but not for the high irradiance of 28 W m^{-2}

The biopotential changes induced by the long-wavelength UV light (365 nm) were fully comparable with the blue light effects. A dose of 1.46 Mol m^{-2} (calculated for 365 nm) gave a biopotential change of -1.4 mV which is about 20% less than a blue light dose of the same amount.

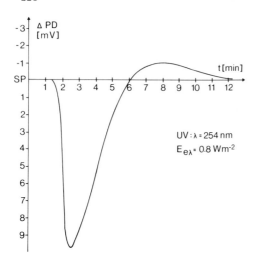

Fig. 10. Typical recording of a biopotential change induced by a far UV light stimulus. The effect represents a depolarization

A completely different biopotential change occurred after a short-wavelength UV light stimulus (Fig. 10). Instead of a hyperpolarization typical for blue light and long UV light (365 nm) a strong depolarization was induced. The lag-phase of the beginning of the photoelectrical effect (25 s) was significantly lower than that found with blue light (1 s). The slope of the resulting biopotential change was very fast. The dose response curve for short UV light showed a saturation after exposure times longer than 90 s (Fig. 11).

As mentioned earlier in this paper, the hypocotyl hooks needed a refractory period after a light stimulus before they could respond again. This was found also for repeated pulses of short UV light (Fig. 12A). A blue light stimulus (437 nm) which was given about 10 min after the first irradiation with short UV light was as effective as the irradiation of nonpreirradiated hooks (Fig. 12B). This result showed that the UV light dose used in this experiments did not injure the plant tissue, and that the UV light-absorbing substances or pigments had to reconstitute, but the blue light-absorbing pigments were not influenced by these processes.

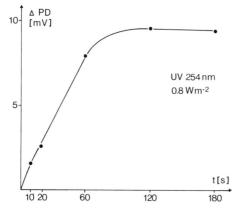

Fig. 11. Dose response curve of far UV light-induced biopotential changes. The curve represents the mean values of four measurements

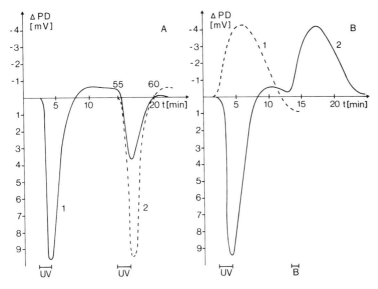

Fig. 12A, B. Recordings demonstrating the necessity of a refractory period for a repeated far UV light stimulus in order to reach the full biopotential change of the first stimulation (**A**). Such a re-fractory period was not necessary for a blue light stimulus following the far UV light irradiation (**B**). The peaks of the nonpreirradiated hooks (*1*) and of the sample preirradiated with far UV light (*2*) were identical

3.3 Carotenoid Concentrations and Light-Induced Biopotential Changes

The distribution of carotenoids in the etiolated hypocotyl hooks showed a gradient (Fig. 13). The highest amount was found in the part of the hook where the hypocotyls are inserted. The carotenoid concentrations decreased progressively toward the base, but remained more or less constant further than 3 cm away from the elbow parts of the hook. The carotenoid content of different parts of the hypocotyl hook was not correlated with the sensitivity to a blue light stimulus (UV light has not been tested so far). We found a relatively low carotenoid concentration in the region where the highest light effect was observed (cf. Fig. 13 with Fig. 19).

There was also no correlation between the carotenoid content of the seedlings and their age-dependent sensitivity to light-induced biopotential changes (Fig. 14). Six-day-old seedlings showed a maximum sensitivity to photostimulation. The light-induced responses decreased with the age of the bean seedlings and the carotenoid content of the seedlings increased during the whole experimental time of nine days. In contrast to the data reported so far, the experiments with inhibitors indicated that carotenoids are important for the biopotential changes. The carotenoid content of the tissue could be altered over a wide range by the inhibitor SAN 6'06. It was not possible to block the carotenoid synthesis completely by all the tested SAN-inhibitors or amitrole in the dark. The lowest concentration which we could find in the etiolated seedlings using the highest inhibitor concentration which could be dissolved (10^{-4} M) was about 15% of the untreated control seedlings.

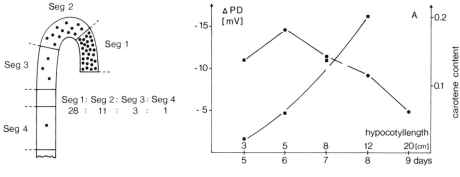

Fig. 13   **Fig. 14**

Fig. 13. Distribution profile of carotenoids in etiolated hypocotyl hooks of dwarf beans. The different segments (*Seg*) which were extracted are labeled with *1, 2, 3,* and *4*. The highest sensitivity to the blue light induced biopotential change was found in segment *3*

Fig. 14. Comparison between the age-dependent sensitivity of bean seedlings to the light-induced biopotential change and the corresponding carotenoid content of the seedlings

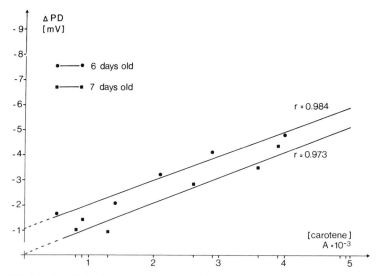

Fig. 15. Correlation between the carotenoid content of 6- and 7-day-old etiolated bean seedlings treated with SAN 6706 and the biopotential change induced by blue light (437 nm). The light effect decreased with decreasing carotenoid concentrations. The curves represent the mean values of eight measurements. The linear regression was calculated with a Texas Instrument calculator TI 59. The regression coefficients (*r*) are shown in the graph

The positive correlation between the carotenoid concentration and the blue light (437 nm)-induced biopotential changes is demonstrated in Fig. 15. The response was strong with high carotenoid concentrations and weak with low ones. The differences in the sensitivities between six- and seven-day-old seedlings was also evident under con-

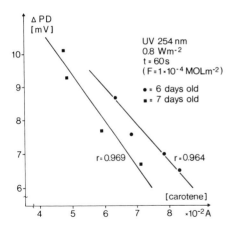

Fig. 16. Correlation between the carotenoid content of 6- and 7-day-old etiolated bean seedlings treated with SAN 6706 and the biopotential change induced by far UV-light (254 nm). The light effect increased with decreasing carotenoid concentrations. The curves represent the mean values of eight measurements. The linear regression was calculated with a Texas Instrument calculator TI 59. The regression coefficients (r) are shown in the graph

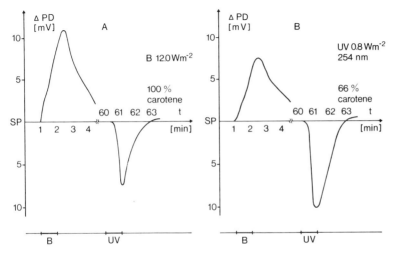

Fig. 17A, B. Comparison between the effects of blue light and far UV light on the biopotential changes in control seedlings and seedlings reduced in their carotenoid content. The hook preparations were first irradiated with blue light and after the biopotential had stabilized again an irradiation with far UV light followed

ditions of inhibited carotenoid synthesis. Just the opposite reaction occurred with short UV light irradiation (Fig. 16). We found a high negative correlation. The UV light effect increased with decreasing carotenoid concentrations. These correlations are again shown with the original recordings from two batches of beans with different carotenoid concentrations irradiated first with blue light and then with short UV light (Fig. 17). We did some preliminary studies to prove the reactability of the tissues with a constant current stimulation. The results seem to indicate that there were no differences in response behavior between the control seedlings and the inhibitor-treated seedlings, but it is necessary to continue this investigation in the future in order to improve the methods.

Amitrole was much less effective at inhibiting the carotenoid synthesis in etiolated seedlings. We found a tendency for a reduced light reaction but these effects were not statistically significant.

4 Sensitivity of Different Regions of the Hypocotyl Hook to Blue Light and the Effect of Polarized Light

Using polarized light for the induction of biopotential changes we found a significant dependence on the vibration plane of the polarized light. The effect of an electrical vector perpendicular to the hook apex was set to 100%. The difference of an irradiation with an electrical vector parallel to the hook apex was 7.3% less. After four independent measurements with different batches of beans, the Student t-test showed that the results were significantly different ($p < 0.5$).

The result is shown in Fig. 18. We have recorded beside the 100% situation the difference transient between both vibration planes of polarization. The irradiation was performed with two identical reacting bean samples which were irradiated at the same time with two different vibration planes of polarized light. The difference between both conditions is clear to recognize.

The sensitivity of different regions of the hypocotyl hook to light stimuli was tested using glass fiber optics. The direction of incident light and the position of the spots irradiated are shown in Fig. 19; irradiating the side of the hook was more or less ineffective. Irradiation of the outer part of the hook always gave a biopotential change, but with significantly different amplitudes between different regions. The construction of the vessel did not permit us to irradiate the hook from below into the inside of the hook region. In order to obtain information about this part of the hook it will be necessary to improve the measuring system.

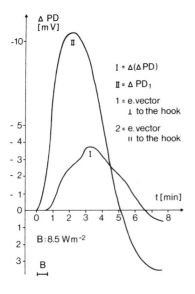

Fig. 18, Demonstration of a difference-measurement with different vibration planes of polarized blue light. *Curve II* shows the response of an irradiation with an E vector of the polarized light perpendicular to the hook elbow. *Curve I* is the difference transient change directly recorded between situation II and an E vector of the polarized light parallel to the hook elbow (the representative curve of this biopotential change could not be registered). This experiment was performed with two batches of hook preparations which reacted to light stimuli with perfectly comparable reactibility

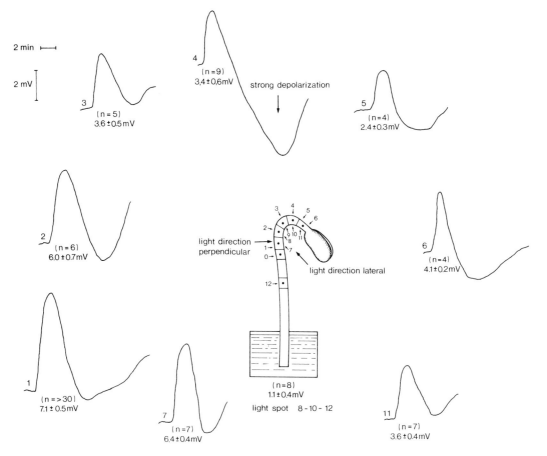

Fig. 19. Sensitivity of different parts of the bean hook to blue light-induced biopotential changes. The hook was irradiated with a light spot of about 6 mm^2. The light direction was perpendicular to the hook elbow and lateral from the side to the elbow. The light spots are labeled by *numbers*

5 Discussion

This investigation was initiated to obtain more information about the effects of blue light, especially regarding the induction of biopotential changes in multicellular plant tissue as reported by Briggs (1976) and Hartmann (1975). The first step was to improve the method used for the measurement of the biopotential changes of one bean hook in order to reduce the variations and other problems causing difficulties. The newly developed instrument allowed us to measure ten or more hooks in one experiment, and the highly reproducible and stable signals were a great advantage for routine determinations of biopotential changes.

It was also very important to confirm that the results reproted by Hartmann (1975) were fully reproducible with the new technique. The light-induced responses had some characteristics of an action potential. The maximum amplitude of the biopotential change was obtained once a threshold dose of light was given. Increasing the quantum flux density in a moderate range did not alter the response of Δ PD. A new light stimulus had no effect unless the stabilized potential was reached again. The refractory period for a new full effective light stimulus was about 40 min. The dose response curves for a wavelength of 437 nm showed that the reaction is saturated in a wide range of moderate light intensities. With high light intensities ($>$ 20 W m^{-2}) the effect was remarkably increased. We do not know the reasons for this effect but it is characteristic that blue light reactions can always be separated in low- and high-energy reactions (Briggs 1976). The results about the Bunsen-Roscoe reciprocity law were consistent with this general observation. The reciprocity law was fulfilled only with lower blue light intensities and never with higher light intensities. The same relationship has been found for other blue light-regulated processes like the geotropic curvature of avena coleoptiles (reviewed by Dennison 1979).

Hartmann (1975) published a preliminary action spectrum for the blue light-induced biopotential changes in a wavelength range from 400–520 nm (longer wavelengths were ineffective, shorter ones could not be measured). This action spectrum showed three peaks (438 nm, 468 nm, and 499 nm). The wavelength of 438 nm gave a much more pronounced peak than the others. We have no information about the function of the wavelengths shorter than 400 nm, therefore it will be necessary to repeat the action spectrum for the light-induced biopotential change for the whole spectrum with the newly designed instrument. It is also well known that the action spectra from blue light-induced reactions give only limited information about the pigment system involved in the reaction. The action spectra are similar so that in most of the published data either flavins or carotenoids could be the photoreceptors. Progress has been made both in the use of highly sophisticated spectrophotometrical techniques and in the methods used to calculate the fourth derivate for the data (Poff et al. 1973; Poff and Butler 1974; Munoz and Butler 1975). These results gave strong evidence that a flavin is the photoreceptor, but nevertheless other evidence indicates that carotenoids are the responsible photoreceptors for blue light (Shropshire and Withrow 1958).

If carotenoids are not the photoreceptor we have to ask what the function of carotenoids is? A rather old assumption is that the carotenoids have a screening function (Bünning 1938), but besides some evidence for this function (Jacob 1964) there are data which do not coincide with this screening function of carotenoids (Curry and Thimann 1961; Page and Curry 1966). The present data demonstrate an involvement of carotenoids in the light-induced biopotential changes. We found a significant correlation between the intensity of the blue light effect increased proportionally to the carotenoid concentrations. This result was only observed in those bean seedlings in which the carotenoid content was altered by inhibitors. In untreated seedlings there was no correlation between the carotenoid content and the sensitivity to blue light.

Therefore we have to prove that the inhibitors only alter the carotenoid content of the tissues and do not effect the general reactibility of the seedlings by inhibiting other cellular processes. We tried to show this by electrical stimulation of the hook preparations. The preliminary evidence of these tests was that there were no differences in reactibility between control and inhibitor-treated seedlings. The main question which then arises is whether or not it is valid to use this sort of test since we do not know if the same ways of sensory transduction are used for the response to electrical and light stimuli. Other evidence for a comparable reactibility in both inhibitor-treated and untreated tissues is that the age dependent sensitivity of the tissues was not influenced by the inhibitors. Schmidt et al. (1977) used geotropism as a control for the specifity of potential inhibitors of phototropism by corn coleoptiles. They found three categories of inhibitor: (1) ineffective, (2) inhibition of both reactions, and (3) specific inhibitors for phototropisms. These inhibitors were KJ, NaN_3, and phenylacetic acid. These inhibitors show special reactions with flavins. We must test in future the effects of those inhibitors at the blue light-induced biopotential changes. There is only very little information about what is happening during sensory transduction in a plant organism after the perception of a light stimulus. The question whether the alterations in membrane permeability demonstrated by biopotential changes are primarily caused by the light transduction reaction or are only a consequence of other processes triggered by light remains open. Therefore the electrophysiological studies must be accompanied by biochemical studies of the metabolic reactions occuring during those processes.

In order to obtain some preliminary information about the influence of wavelengths shorter than 400 nm, broad long and short UV light was tested. The effects of long UV light (about 365 nm) were fully comparable with the effects of blue light (437 nm). Short UV light (about 254 nm) produced completely different reactions. Instead of a hyperpolarization a strong depolarization occurred and although the lag-phase for the beginning of the biopotential change increased more than 20 times. The resulting transient change showed a very fast slope. The effect of short UV light increased with decreasing carotenoid concentration of the tissue.

The dose of short UV light used for the induction of the described effects obviously did not injure the tissue. A blue light stimulus had the same effectiveness with or without UV preirradiation. In the UV part of the spectrum still more substances are presumed to be photoreceptors for UV light. It is not possible to speculate about the system which might be activated by UV light, but it is important to mention that short UV light effects showed typical saturating dose response relationship and a refractory period for a repeated stimulus. These data are fully comparable with the effects reported for blue light. In the case of blue light a special photoreceptor pigment is postulated and the question arises if it will be necessary to claim also a special photoreceptor for UV light. This question is stressed by the results which showed that the refractory periods for far UV light and blue light did not coincide. The question cannot be answered at the moment but it may be better to consider that the plant cell is responding as whole, without the involvement of a specific receptor. More work is necessary to come to a conclusion about the function of short UV light in mediated biopotential changes.

The correlation between the effectiveness of short UV light and the carotenoid concentration should be interpreted as a screening function of the carotenoids.

Another important question for the understanding of the function of a photoreceptor is its localization and orientation in the tissue or the cell.

From many studies (algae, fungi, mosses, and ferns) it is known that the photoreceptor molecules are dichroically orientated. The situation in higher plants is still obscure. Marmé and Schäfer (1972) reported data from avena coleoptiles which demonstrated a dichroic orientation of phytochrome. The effects of polarized light in complex organs are influenced much more by the structure of the cuticula and the cell layers where the photoreceptor is localized than by single cells. It is possible that the structures themself have a polarizing effect on nonpolarized light.

Using etiolated bean seedlings we found a difference in the effectiveness of blue light depending on the vibration plane of polarization. Transversely polarized light (the E vector perpendicular to the elbow of the hook, 437 nm) was 7.3% more effective than longitudinally polarized light (the E vector parallel to the elbow of the hook). This result is in the range which was described for avena coleoptiles (15%–20%) for phytochrome effects depending on the E vector of the polarization (Marmé and Schäfer 1972). The polarized light effect in *Phycomyces* is not very strong and also comparable with our data (Haupt and Buchwald 1967). The special function of the hook of the beans in light absorption is stressed by the data which demonstrated a dependence of the blue light-induced biopotential on the light direction. A light stimulus given perpendicular to the hook always induced a response (but with significant differences depending on the position of the light spot on the hook). A lateral illumination of the side of the hook was in most cases ineffective and showed only a response at the most sensitive part of the hook. It is difficult to explain this phenomenon with the data so far produced. We have to consider the different possibilities such as absorption structures of the hook surface, light channeling in the tissue, or localized perception regions of the organ.

The special function of the hypocotyl hook as an organ for coding, transmitting, and integrating information was also reported by Rubinstein (1971) and Gee and Vince-Prue (1976). They showed a different sensitivity of different regions of the hook for the red light-induced hook opening and De Greef et al. (1976) demonstrated a relay system in phytochrome-mediated interorgan cooperativity for the same reaction.

The experiments presented show that the plant organ used and the electrophysiological methods are good tools for furthering the investigation of sensory transduction in plant organs.

Acknowledgments. The investigation was supported by a grant from the Deutsche Forschungsgemeinschaft to E.H. We thank N. Grimsley for reading the manuscript.

References

Briggs WR (1976) The nature of the blue light photoreceptor in higher plants and fungi. In: Smith H (ed) Light and plant development, pp 7–18. Butterworths, London

Bünning E (1938) II. Das Carotin der Reizaufnahmezone von Pilobolus, Phycomyces und Avena. Planta 27: 148–158

Curry GM, Thimann KV (1961) Phototropism: The nature of the photoreceptor in higher and lower plants. In: Christensen BC, Buchmann B (eds) Progress in photobiology, pp 127–134. Elsevier Publishing Co, New York

De Greef JA, Cauberg R, Verbelen JP, Moerells E (1976) Phytochrome-mediated inter-organ dependence and rapid transmission of the light stimulus. In: Smith H (ed) Light and plant development, pp 295–316. Butterworths, London

Dennison DS (1979) Phototropism. In: Pirson A, Zimmermann MH (eds) Encyclopedia of plant physiology, New Ser, vol VII, pp 506–560. Springer, Berlin Heidelberg New York

Furuya M, Hillman WS (1964) Observations on spectrophotometrically assayable phytochrome in vivo in etiolated Pisum seedlings. Planta 63: 31–42

Gee H, Vince-Prue D (1976) Control of the hypocotyl hook angle in Phaseolus mungo L.: The role of parts of the seedling. J Exp Bot 27: 314–323

Hartmann E (1975) Influence of light on the bioelectric potential of the bean (Phaseolus vulgaris) hypocotyl hook. Physiol Plant 33: 266–275

Haupt W, Buchwald M (1967) Die Orientierung der Photoreceptor-Moleküle im Sporangienträger von Phycomyces. Z Pflanzenphysiol 56: 20–26

Jacob F (1964) Über die Funktion eines Karotin-Lichtschirmes bei dem Phototropismus von Sporangienträger chromosporer Pilobolus-Arten. Flora 155: 209–222

Klein WH, Withrow RB, Elstad VB (1956) Response of the hypocotyl hook of bean seedlings to radient energy and other factors. Plant Physiol 31: 289–294

Marmé D, Schäfer E (1972) On the localization and orientation of phytochrome molecules in corn coleoptiles (Zea mays L.). Z Pflanzenphysiol 67: 192–194

Munoz V, Butler WL (1975) Photoreceptor pigment for blue light in Neurospora crassa. Plant Physiol 55: 421–426

Nielsen OF, Gough S (1974) Macromolecular physiology of plastids. XI. Carotenes in etiolated tigrina and xantha mutants of barley. Physiol Plant 30: 246–254

Page RM, Curry GM (1966) Studies on phototropism of young sporangiophores of Piloboluskleinii. Photochem Photobiol 5: 31–40

Poff KL, Butler WL (1974) Absorption changes induced by blue light in Phycomyces blakesleeanus and Dictyostelium discoideum. Nature (London) 248: 799–801

Poff KL, Butler WL, Loomis WF Jr (1973) Light-induced absorbance changes associated with phototaxis in Dictyostelium. Proc Natl Acad Sci USA 70: 813–816

Rubinstein B (1971) The role of various regions of the bean hypocotyl on red light-induced hook opening. Plant Physiol 48: 183–186

Schmidt W, Hart I, Filner Ph, Poff KL (1977) Specific inhibition of phototropism in corn seedlings. Plant Physiol 60: 736–738

Shropshire W Jr, Withrow RB (1958) Action spectrum of phototropic tip-curvature of Avena. Plant Physiol 33: 360–365

Smith H (1975) Phytochrome and photomorphogenesis. McGraw Hill, London

Blue Light-Controlled Conidiation and Absorbance Change in *Neurospora* are Mediated by Nitrate Reductase

H. NINNEMANN[1] and E. KLEMM-WOLFGRAMM[1]

After our knowledge has expanded somewhat on the nature of photoreceptors involved in certain blue light-regulated biological responses, I would like to draw your attention to the question of how the physical signal light may be transformed in an organism into a biochemical information. We tried to approach this problem with the rhythmic albino mutant band al-2, bd of *Neurospora crassa*. At least two blue light-controled responses are known in this organism: (1) phase shifts of the circadian rhythm of conidiation and (2) promotion of conidiation under certain preconditions. We worked with in vivo and in vitro systems, using the light-induced absorbance change (ΔA) first reported in *Neurospora* by Muñoz et al. (1974) as an assay system.

Absorbance changes are no simple parameters to work with. At various times after harvesting mycelium and transferring it into starvation conditions a number of different ΔA's may be observed, some of them due to reduction of mitochondrial cytochromes. These seem to be unrelated to the problem we are concerned with. The typical, "specific" absorbance increase we see in light-minus-dark difference spectra occurs at 423 and 557 nm and is due to a flavin-mediated reduction of a nonmitochondrial b-type cytochrome. Under physiological conditions, this ΔA_{423} is never observed in fresh samples but occurs only in starved mycelium. Concomitantly an absorbance decrease at around 460 nm can be seen due to flavin reduction in the light.

1 Correlation Experiments in Vivo

Our results on trials to establish possible correlations between the blue light-induced ΔA_{423} and blue light-induced physiological responses of *Neurospora* have already been published (Klemm and Ninnemann 1978). The set-up for these experiments included (a) growing *Neurospora* in petri dishes on an $(NH_4)NO_3$-containing agar medium where the circadian rhythm of conidiation is entrained and the mycelium allowed to grow in darkness until part of it is used, (b) transferring the growth front of this mycelium without medium into special optical cuvettes (starvation conditions) where ΔA's are monitored and where inductive blue or white light can be applied, and (c) inoculating this mycelium after variable time spans in the optical cuvette into growth tubes with solid medium (second transfer) so that the effects of handling the mycelium in darkness and of the inductive irradiation, respectively, can be observed.

1 Institut für Chemische Pflanzenphysiologie der Universität, Corrensstraße 41, 7400 Tübingen, FRG

1.1 Correlation Between ΔA_{423} and Phase Shifts?

Conditions can be found where a phase shift of the circadian rhythm of conidiation can be induced without a concomitant absorbance increase at 423 and 557 nm in the mycelium; also, with appropriate timing, ΔA_{423} can be brought about without a corresponding phase shift. Thus ΔA_{423} does not correlate with light-induced phase shifts of conidiation — possibly because no cytochrome b might be involved in this reaction chain.

1.2 Correlation Between ΔA_{423} and Light-Promoted Conidiation

The absorbance increase at 423 nm, however, correlates well with the formation of a light-promoted (first) conidiation band if the mycelium had been starved in the optical cuvette for more than 30–60 min. With shorter starvation times, the mycelium is able to conidiate without light, and under these conditions no ΔA_{423} can be found. Therefore we concluded that prolonged starvation (1–5 h if the mycelium had been grown on solid medium) is a precondition for light-induced conidiation and for the light-induced absorbance change ΔA_{423}. At this point we began to search for key enzymes which are important for conidiation and could at the same time show the light-induced ΔA.

2 In Vivo Experiments on Nitrate Reductase

A number of observations during experimentation with *Neurospora* convinced us that the nitrogen metabolism is an essential determinant for conidiation to occur and for the rhythm of conidiation to become expressed. The N-source of the medium (NO_3^-, NH_4^+, certain amino acids) influence conidia formation on the medium and after starvation. Our choice fell on the hemoflavoprotein nitrate reductase, the synthesis of which can be induced e.g., by NO_3^- and repressed by NH_4^+; NH_4^+ also inhibits nitrate reductase activity. Nitrate reductase constitutes a small electron transport chain in itself and its subunit activities can be monitored separately beside the total NADPH $-\!\!\rightarrow NO_3^-$ activity:

$$\text{NADPH} \begin{array}{c} \text{FADH}_2 \\ \downarrow \\ \nearrow \text{FAD} \rightarrow \text{Cyt. b}_{557} \\ \\ \searrow \text{FAD} \rightarrow \text{Cyt. b}_{557} \\ \text{Cyt. c} \end{array} \begin{array}{c} \text{MVH} \\ \downarrow \\ \rightarrow \text{Mo} \rightarrow 2e \end{array} \begin{array}{c} \nearrow NO_3^- \\ \\ \searrow NO_2^- \end{array} \qquad \text{(Pan and Nason 1978)}$$

The following experiments were performed with mycelium grown in liquid culture; then about 16 h are necessary to starve the mycelium sufficiently so that upon irradiation it shows ΔA. Following nitrate reductase activities as a function of starvation time showed that total NADPH $-\!\!\rightarrow NO_3^-$ activity (cytoplasmic) decreased, but the subunit activites FADH$_2$ $-\!\!\rightarrow NO_3^-$ and MVH (reduced methyl viologen) $-\!\!\rightarrow NO_3^-$ increased with prolonged (up to 20 h) starvation of the mycelium (Klemm and Ninnemann 1979). The increased subunit activities became increasingly membrane-bound, mainly in the plasma membrane-enriched fraction (21,000 g) and somewhat less in the mitochondria-enriched fraction (9000 g). The starvation effect on nitrate reductase activities could be mimicked by a brief acetone treatment of the mycelium before homogenizing and fractionating it. After the acetone treatment, ΔA became visible in freshly harvested mycelium without delay. Irradiation of the starved mycelium reactivated total NADPH $\rightarrow NO_3^-$ activity in the cytoplasm and led to conidiation.

Thus starvation or acetone treatment or elevated temperature (experiments not shown) might create conformational changes in the membranes or local changes of the subunits with respect to each other which then changes the nitrate reductase activities. According to our working hypothesis, the cytoplasmic (or lightly membrane-associated) nitrate reductase chain is more or less reduced under growth conditions. Upon starvation, the chain must become oxidized, the subunits change position relative to each other or separate so that active sites for exogenous substrates become more exposed, and they become membrane-bound. At this time total nitrate reductase activity (NADPH $\rightarrow NO_3^-$) is impaired. Brief irradiation of the mycelium now leads to ΔA_{423}; light also causes reactivation of total nitrate reductase activity: the subunits detach from the membranes and recombine or rearrange to form a functional nitrate reductase electron transport chain.

3 In Vitro Experiments on Nitrate Reductase

Isolated and partially purified nitrate reductase (Garrett and Nason 1969) of *Neurospora* mutant band grown on nitrate or ammonia-nitrate medium can be photoreduced (see Fig. 1: absorbance increases at 423 and 557 nm, decreases at about 460 and 480 nm) if an electron donor for the endogenous flavin (like EDTA) is added to the sample (Ninnemann, in preparation). Irradiation of the samples with white or blue light (Fig. 2) was performed at room temperature, but spectra were taken at $-196°C$, since the reoxidation of cytochrome b-557 in the absence of nitrate in darkness at room temperature is very fast (b-557 reoxidizes totally within the thawing time of the 220 μl sample, i.e., within 20–30 s; this is in contrast to purified and photoreduced yeast lactate dehydrogenase (= cytochrome b_2) the reoxidation of which is slow with $t_{1/2}$ = 20 min; Ninnemann, unpublished). The peak positions of the photo-reduced nitrate reductase are identical with the NADPH-reduced maxima (Fig. 2). The FAD of isolated *Neurospora* nitrate reductase seems to be lightly bound to the protein, since 2 h of dialysis in the cold remove virtually all of the flavin. Photoreducibility and nitrate reductase activity, however, can be reconstituted with exogenous FAD (4 μM). The flavin-mediated photoreduction (with either endogenous FAD or with added FAD to FAD-depleted nitrate reductase) of cytochrome b-557 of nitrate reductase does not proceed via O_2^- (in contrast to results with mitochondrial b-type cytochromes, see Ninnemann et al. 1977) since additon of superoxide dismutase does not decrease the amount of absorbance increase at 423 and 557 nm (it prevents, however, the photoreduction of the component showing a shoulder in the light-minus-dark difference spectrum at around 440 nm which is probably due to contamination with some cytochrome oxidase).

Photoreduction can proceed in presence of NADPH: after prereduction with 2.5 mM NADPH, no additional reduction occurred with time in darkness. A following brief ir-

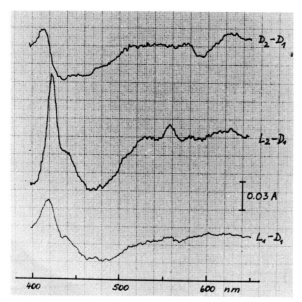

Fig. 1. Photoreduction of partially purified nitrate reductase with white light. *Lower curve* 1 min of 2×10^6 ergs/cm^{2}s white light minus dark sample before irradiation; *middle curve* 3 min of white light minus dark; *upper curve* reoxidation in darkness within 20 s. D_1 dark sample before photoreduction D_2 = dark sample after photoreduction and thawing. Sample: 200 μl nitrate reductase, 24 mg protein/ml, + 20 μl EDTA 250 mM pH 7.2, irradiated at room temp., frozen in light, spectra taken at $-196°$C. Nitrate reductase from mycelium grown for 14 h on NO_3^+-medium

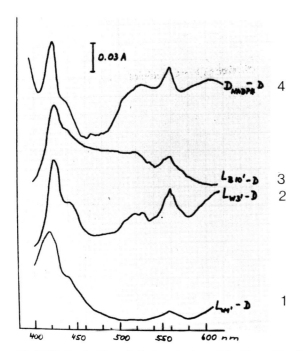

Fig. 2. Photoreduction of nitrate reductase with white or blue light compared to NADPH-reduced nitrate reductase. Difference spectra ($-196°$) from bottom to top: *1* 1-min white light (2×10^6 ergs/cm^{2}s) minus dark, *2* 3-min white light minus dark, *3* 10-min blue light (Corning 5562, 4×10^5 ergs/cm^{2}s) minus dark, *4* sample + 4.3 mM NADPH, 2 min reduced minus without NADPH. Sample constituents as in Fig. 1, protein content 21.5 mg/ml

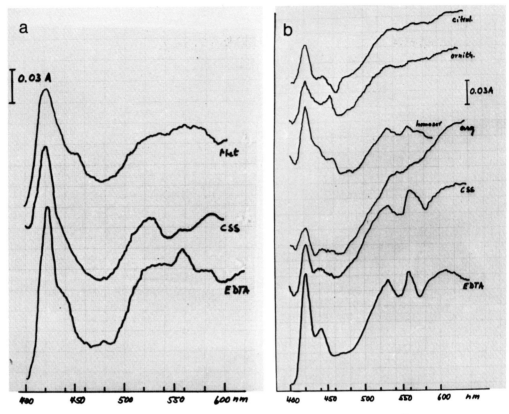

Fig. 3a, b. Electron donors for irradiated flavin of nitrate reductase. Light-dark difference spectra (3 min white light 2×10^6 ergs/cm^2 s). **a** From bottom to top 22.7 mM EDTA pH 7.2, 18.2 mM cysteine sulfinic acid (CSS) pH 8.9, 22.7 mM methionine pH 7.0. **b** From bottom to top: 22.7 mM EDTA pH 7.2, 18.2 mM CSS pH 8.9, 22.7 mM arginine pH 6.5, 18.2 mM homoserine pH 5.6, 22.7 mM ornithine pH 6.9, 22.7 mM citrulline pH 7.8. Spectra taken at $-196°$C

radiation (3 min, 2×10^6 ergs/cm^{2}s white light) caused strong reduction of cytochrome b-557 and of the FAD. In darkness cytochrome b-557 was reoxidized faster than the FAD. When the same sample was repeatedly photo-reduced and reoxidized in darkness, the NADPH became oxidized, while most of the nitrate reductase remained intact. But of greater importance for starved *Neurospora* cells might be the possibility to find electron donors different from NADPH for the nitrate reductase electron transport chain in light: the irradiated FAD component of nitrate reductase can be re-reduced by EDTA and by some amino acids from the cysteine, methionine, or arginine metabolism: cysteine sulfinic acid, methionine and homoserine, ornithine, citrulline, and less well arginine (Fig. 3a, b) were successfully applied (Ninnemann, unpublished), though all these amino acids are less suitable donors for the flavin than is EDTA (Frisell et al. 1959). The possible natural donor for irradiated FAD of nitrate reductase is still unknown. But the possibility exists that the specificity of this donor determines the specificity of a flavo(hemo)protein as photoreceptor molecule in a given blue light-regulated process.

References

Frisell WR, Chung CW, Mackenzie CG (1959) J Biol Chem 234: 1297–1302
Garrett RH, Nason A (1969) J Biol Chem 244: 2870–2882
Klemm E, Ninnemann H (1978) Photochem Photobiol 28: 227–230
Klemm E, Ninnemann H (1979) Photochem Photobiol 29: 629–632
Muñoz V, Brody S, Butler WL (1974) Biochem Biophys Res Commun 58: 322–327
Ninnemann H (1980) in preparation
Ninnemann H, Strasser RJ, Butler WL (1977) Photochem Photobiol 26: 41–47
Pan SS, Nason A (1978) Biochim Biophys Acta 523: 297–313

Phototropism in *Phycomyces:*
a Photochromic Sensor Pigment?

G. LÖSER[1] and E. SCHÄFER[1]

1 Introduction

The blue light reactions of *Phycomyces* have been investigated by several authors during the past years. A broad review was given by Bergman et al. (1969). Light-induced absorption changes in *Phycomyces* and their possible implications for a blue light receptor were investigated by Poff and Butler (1974), Lipson and Presti (1977) and, for *Neurospora,* by Klemm and Ninnemann (1978). Galston (1977) gave a short historical review of the search for a blue light receptor in *Phycomyces* and other species. New evidence against β-carotene as the blue light receptor in *Phycomyces* has been presented by Presti et al. (1977). The absolute extinction coefficient for the receptor pigment in *Phycomyces* was estimated by Lipson (1975) and proved to be of the same size as that of riboflavin. Delbrück et al. (1976) supported the candidacy of riboflavin as the receptor pigment by measuring the relative quantum efficiency for the light growth reaction of *Phycomyces* up to 630 nm wavelength.

On the basis of a study of analytical action spectroscopy theory, Hartmann (1977) concluded that the blue light receptor action spectra in general could be represented in principle as the action spectra of a photochrome. Detailed analysis of the fluence response curves for polarotropism of *Dryopteris* protonemata (Steiner 1969) indicates that the ineffective form of this hypothetical photochrome should have an absorption maximum at 450 nm and a shoulder at 480 nm. The active form should absorb predominantly at 370 nm and 420 nm (Hartmann 1977).

Here we report experiments using dichromatic irradiation that give further evidence for a photochromic photoreceptor pigment in *Phycomyces.*

2 Materials and Methods

The experiments were carried out using wild-type *Phycomyces* NRRL 15555 (−) obtained from V.E.A. Russo (Max-Planck-Institut für molekulare Genetik, Berlin) and cultured on bacto potato dextrose agar (DIFCO Labs, Detroit, MI, USA) in small glass vials of 1 cm diameter each. Culture conditions are as described elsewhere (Bergmann et al. 1973).

---

1 Biologisches Institut II, Universität Freiburg, Schänzlestr. 1, 7800 Freiburg, FRG

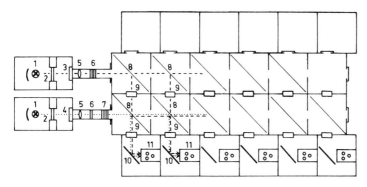

Fig. 1. Scheme of the experimental arrangement (top view). *1* light source; *2* heat filter glass; *3* interference filter; *4* interference filter; *5* collimating lenses; *6* neutral density glasses; *7* cut-off-glasses; *8* 50% transmission mirrors; *9* neutral density glasses; *10* surface mirrors; *11* adjustable tables with specimens in glass vials

The experimental arrangement is a modified and extended version of the one used by Varjú et al. (1961) and Bergman et al. (1973) and is shown in Fig. 1. 250 Watt Osram halogen lamps were used as the light sources. After passing a heat filter (5 mm thickness) and collimating lenses monochromatic light of 450 nm was defined by an interference filter (PAL 450, Schott Mainz, FRG). The 573 nm, 605 nm, and 621 nm beams originated from interference filters AL 573, AL 605, and Al 621 resp. (Schott), all of about 20 nm half-width. For these wavelengths considerable care was taken to suppress undesired faint light of shorter wavelengths using a stack of cut-off glasses (Schott), e.g., OG 550 plus two OG 590, 3 mm thickness each for the 605 nm beam, producing a nominal blue light suppression of about $10^{-5} \cdot 10^{(-5)} \cdot 3 = 10^{-20}$ together with the AL 605 filter.

In order to check experimentally for blue contamination of the 605 nm beam described, BG 12 (4 mm thickness) and BG 28 (2 mm) glasses (Schott) were added with 20% of the residual blue light passing. Under these conditions *Phycomyces* sporangiophores did not show phototropic curvature, whereas without the additional glasses phototropic curvature was measurable, i.e., due to action of 605 nm light.

For orange red light fluence rates between 150 W/m² and 1600 W/m² at the object site and for horizontal beams only, a 2.5 kW Xenonol-III-source with water-cooled filters and cut-off-glasses was used. The monochromatic and the Xenosol-III experiments were performed — before the apparatus shown in Fig. 1 was available — with a single chamber apparatus for the specimens, i.e., without mirrors.

The slightly divergent beams had to pass a set of 50% transmission mirrors and surface mirrors (Spindler and Hoyer, Göttingen, FRG), the latter could be used to produce an inclined beam. The beams had a cross section of about $4 \cdot 4$ cm² at the object sites and were checked for sufficient homogeneity. In all cases stray light was reduced by blackening surfaces, adding additional apertures (cf. Fig. 1) and introducing additional barriers against light from outside.

The symmetric apparatus (Fig. 1) can be used simultaneously for two monochromatic fluence response experiments or for one unilateral dichromatic experiment. In the latter

case the 50% transmission mirrors were placed as in Fig. 1, the 450 nm fluence rate at
equivalent positions of each object table being adjusted to equal values by inserting
neutral density glasses between the chambers followed by tests for equal photogeotropic
equilibrium bending angle of the sporangiophores. The 605 nm fluence rate decreased
in steps of two to four from one object chamber to the following one.

For the one-chamber monochromatic experiments, polarized light, if used, was generated by a polarization foil (type PW 44, Käsemann, 8203 Oberaudorf, FRG). According to Castle (1934) "transversely" polarized light is 10% to 15% more effective for
phototropic bending of *Phycomyces* than "longitudinally" polarized light, thus creating
only small differences in the fluence response curve. The same holds for 605 nm light
as shown by our observations (Fig. 2). The terms "longitudinal" and "transverse" are
used according to Jesaitis (1974).

Stage 4b-specimens – one to three in each vial – were placed on adjustable tables
in the chambers. The bending angle was measured after six hours, according to Varjú
et al. (1961). Each point in the figures (see Sect. 3) represents the average deviation
from vertical of typically 15 sporangiophores after about 6 h of exposition. The errors
indicated are standard errors of the arithmetical means.

3 Results

3.1 Fluence Rate Response Curves

Using light beams inclined by $20°$, the apparent saturation level is significantly lower
for 605 nm light than for 450 nm light (Fig. 2 and Table 1). For 605 nm light the fluence
rate response curve shows oscillatory behavior in the apparent saturation region. These

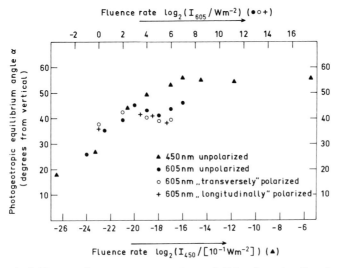

Fig. 2. Phototropic response to monochromatic light of wavelength and polarization as indicated
in the figure. The light incident on the sporangiophores is inclined by $20°$ down from the horizontal.
For "longitudinal" and "transverse" see Sect. 2

Table 1. Saturation levels of the photogeotropic equilibrium angle α for monochromatic light of different wavelengths λ and of different inclinations β from horizontal. Mean values obtained from averages weighed with the number of individuals[d]

λ / nm β	$0°$	$20°$	Polarization
450	69.3 ± 1	55.8 ± 1	Unpolarized
450	–	57.0 ± 1	Polarized
573[a]	–	56.0 ± 1	Unpolarized
605[a]	65.5 ± 2	44.4 ± 3	Unpolarized
605[b]	–	39.9 ± 2	"Transversal"
605[b]	–	39.3 ± 2	"Longitudinal"
621	67.1 ± 4[c]	$\geqslant 46.0 \pm 2$[a]	Unpolarized

[a] Fluence rate up to $\log_2 I = 7$
[b] Fluence rate up to $\log_2 I = 6$
[c] With Xenosol-3 as light source, $\log_2 I = 7$ to 11
[d] The error indicated represents the angle difference between the mean value and the highest or lowest avarage value at different fluence rates considered in the apparent saturation region

effects were not observed for 450 nm light and could not be observed for 621 nm light because of experimental limitations.

Polarized light of 605 nm wavelength produced slightly smaller saturation levels than did nonpolarized light (Fig. 2 and Table 1).

Using horizontal irradiation the differences between the blue saturation level and the apparent saturation levels for 605 nm and 621 nm light are small (Table 1). Oscillatory behavior could also be observed at the beginning of the saturation region for 605 nm and 621 nm light.

The slope of the fluence rate response curves has not yet been investigated in detail. However, at $20°$ inclination of the incident light, the slopes in the ascendant part of the curves measured did not differ significantly.

The effects observed are difficult to explain on the basis of a single photoreceptor. Similar behavior has been reported by Steiner (1969) for the fluence response curves of *Dryopteris* protonemata especially after relatively short exposure times.

3.2 Dichromatic Reversion Experiments

As a test for a possible photochromic nature of the photoreceptor for phototropism in *Phycomyces* we used dichromatic experiments. Light was given unilaterally, the fluence rate of 605 nm light varied and the nonsaturating 450 nm fluence rate held constant, cf. methods section. Both beams were inclined by $20°$. Figures 3 and 4 show our results for two different fluence rates of blue light.

The shape of the dichromatic curves shows a clear minimum at 605 nm fluence rates not far from the phototropic threshold for this wavelength. The minima are significantly below the reference reaction for blue light given alone and can be interpreted as partial reversion of the phototropic reaction by 605 nm light. At very low 605 nm fluence rates, the reaction corresponds to 450 nm light given alone, whereas at the beginning

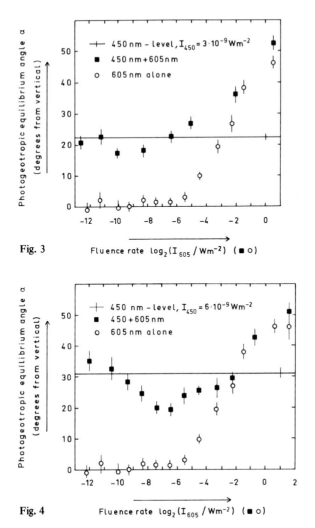

Fig. 3

Fluence rate $\log_2(I_{605}/Wm^{-2})$ ($\blacksquare$ o)

Fig. 4

Fluence rate $\log_2(I_{605}/Wm^{-2})$ ($\blacksquare$ o)

Figs. 3 and 4. Phototropic response to unilateral dichromatic irradiation of 450 nm and 605 nm wavelength. Experimental procedure see Sect. 2. The reference phototropic response for 605 nm monochromatic light is also given. The response for constant 450 nm fluence rate given alone is indicated by the straight line evaluated from more than 60 sporangiophores

of the 605 nm saturation zone, the dichromatic irradiation produces a slightly larger response than 605 nm light alone.

A fact of interest is that the minimum point of the dichromatic curve is situated at higher 605 nm fluence rate if a higher 450 nm fluence rate is used (cf. Figs. 3 and 4).

At higher blue light fluence rates close to saturating the response – with bending angles around 45° for beams inclined by 20° – we could observe minimum structures, too, but surprisingly with no significant reversion of the response obtained so far.

The order of magnitude of the relative quantum efficiency of blue and orange light for the induction of phototropism, for an equal photogeotropic equilibrium angle of

$30°$, expressed as n_{450}/n_{605} is about 10^{-8} according to our observations (Figs. 2, 3 and 4).

4 Discussion

Our data, i.e., (a) a lower apparent saturation level for the photogeotropic equilibrium angle at 605 nm compared with 450 nm light, (b) partial reversion of the blue light photogeotropic effect when using unilateral dichromatic irradiation of 450 nm and 605 nm wavelength simultaneously, indicate that phototropism in *Phycomyces* is not mediated by a single photoreceptor type.

Complex fluence rate response behavior for phototropism of *Phycomyces* was reported earlier by Varjú et al. (1961), see Fig. 4 in their report, where 380 nm light shows deviations from blue light (480 nm) saturation behavior: lower apparent saturation level and partially different slope.

Also variable and complex fluence response patterns were reported by Steiner (1969) for polarotropism of *Dryopteris* protonemata and by Shropshire and Withrow (1958) for the phototropic tip curvature of *Avena*.

Our dichromatic curves (450 nm and 605 nm light) with partial reversion correspond to those presented by Hartmann (1967) for the inhibition of hypocotyl growth of *Lactuca* for simultaneous irradiation with 658 nm and 768 nm light. This was interpreted as due to the action of the photochromic photoreceptor phytochrome (Hartmann 1967; Gruber and Schäfer 1977).

The question arises as to what mechanism could explain our observations (a) and (b).

1. The partial reversion of the 450 nm light-induced photogeotropic equilibrium curvature by 605 nm light given simultaneously could be explained on the basis of an antagonistically acting photoreceptor. This is in contrast to our finding that the relative quantum efficiency for the positive phototropism using 605 nm light alone has the same order of magnitude as observed by Delbrück et al. (1976) for the light growth reaction. Although there are of course very complex photoreceptor models possible to explain our observations, we prefer at the moment the following interpretation.

2. In agreement with Hartmann (1977) we suggest a photochromic system of at least two photoreceptors $A \underset{h\nu'}{\overset{h\nu}{\rightleftharpoons}} B$ with possibly overlapping absorption bands and having different quantum efficiencies at different wavelengths to be the mediator of the responses we observed. The very weak action of 605 nm light therefore should represent the long wavelength tail of a photoreceptor absorption band. At 605 nm, the apparent photoconversion cross section of the active form B should be larger than that of the inactive form A resulting in partial reversion under dichromatic irradiation conditions at relevant fluence rate ratios.

These observations and interpretations are in contrast to those of Delbrück et al. (1976). For the induction of the light growth response in the *Phycomyces* albino mutant C2, they attributed the action of light with wavelengths around 600 nm to an excitation of the lowest triplet state of riboflavin.

Acknowledgments. We are grateful to K.M. Hartmann and R. Hertel for initiating this work and R. Hertel for his continuous interest and support. We thank Ch. Beggs for his help in the revision of the English text. The support of the Evangelisches Studienwerk Villigst e.V. by a grant to G. Löser and of the Deutsche Forschungsgemeinschaft (SFB 46) by funds given to E. Schäfer is appreciated.

References

Bergman K et al (1969) Phycomyces. Bacteriol Rev 33: 99–157

Castle ES (1934) The phototropic effect of polarized light. J Gen Physiol 17: 41

Delbrück M, Katzir A, Presti D (1976) Responses of phycomyces indicating optical excitation of the lowest triplet state of riboflavin. Proc Natl Acad Sci USA 73: 1969–1973

Galston AW (1977) Riboflavin retrospective or déjà-vu in blue. Photochem Photobiol 25: 503–504

Gruber R, Schäfer E (1977) Pre-steady state analysis of HIR reaction. In: Annu Eur Symp Photomorphogen, Bet Dagan, Israel, Abstr 39

Hartmann KM (1967) Photoreceptor problems in photomorphogenic responses under high-energy-conditions (UV-Blue-Far-Red). Eur Photobiol Symp, Hvar, Jugoslavia, Abstr 29

Hartmann KM (1977) Aktionsspektrometrie. In: Hoppe E et al. (eds) Biophysik. Springer, Berlin Heidelberg New York

Jesaitis AJ (1974) Linear dichroism and orientation of the phycomyces photopigment. J Gen Physiol 63: 1–21

Klemm E, Ninnemann H (1978) Correlation between absorbance changes and a physiological response induced by blue light in neurospora. Photochem Photobiol 28: 227–230

Lipson ED (1975) White noise analysis of phycomyces light growth response system, II. Biophys J 15: 1013–1031

Lipson ED, Presti D (1977) Light-induced absorbance changes in phycomyces photomutants. Photochem Photobiol 25: 203–208

Poff KL, Butler WL (1974) Absorbance changes induced by blue light in phycomyces blakesleeanus and dictyostelium discoideum. Nature (London) 248: 799–801

Presti D, Hsu W-J, Delbrück M (1977) Phototropism in phycomyces mutants-lacking β-carotene. Photochem Photobiol 26: 403–405

Shropshire W, Withrow RB (1958) Action spectrum of phototropic tip-curvature of avena. Plant Physiol 33: 360–365

Steiner AM (1969) Dose response behaviour for polarotropism of the chloronema of the fern dryopteris filix-mas (L.) Schott. Photochem Photobiol 9: 493–506

Varjú D, Edgar L, Delbrück M (1961) Interplay between the reactions to light and to gravity in phycomyces. J Gen Physiol 45: 47–58

Blue and Near Ultraviolet Reversible Photoreaction in Conidial Development of Certain Fungi

T. KUMAGAI[1]

1 Introduction

Since the end of the 19th century numerous studies on the effect of light in development of fungi have been made. Among them, Barnett and Lily (1950) confirmed that *Choanephora cucurbitarum* fails to form conidia in continuous bright light and in continuous total darkness. However, it produces conidia in profusion in cultures incubated under bright light followed by darkness, but not under a reverse light-dark regime. Therefore, they proposed the hypothesis that two metabolic reactions are involved in the sporulation of *C. cucurbitarum,* one needing light and the other being inhibited by light. Thereafter, Leach (1967) investigated the effects of light and temperature on the sporulation of certain fungi and placed them into two categories termed "diurnal sporulators" (*Alternaria dauchi, A. Tomato* and *Stemphylium botryosum*), and "constant-temperature sporulators" (*Helminthosporium catenarium, Fusarium nivale* and *Cerocosporella herpotrichoides*). Photosporogenesis in "diurnal sporulators" has two distinct phases, an "inductive phase" which leads to the formation of conidiophores, and a "terminal phase" which results in the formation of conidia. The "inductive phase" is stimulated by near-UV light, whereas the "terminal phase" is inhibited by near-UV or blue light. The "constant-temperature sporulators" show no clear separation of photosporogenesis into two distinct phases; sporulation occurs abundantly under continuous near-UV light although it is more abundant under light exposure followed by darkness.

On the other hand, it has been found that when *A. tomato* (Kumagai and Oda 1969), *Botrytis cinerea* (Suzuki et al. 1977) and *H. oryzae* HA_2 (Honda et al. 1968) were grown on a potato-dextrose-agar medium in total darkness, no conidiophores and conidia were formed in any of the cultures, but when dark-grown colonies were exposed to near-UV or black light followed by darkness, conidiation occurred in profusion in the narrow region of the colony produced just prior to illumination. When the cultures were incubated under continuous blue light after irradiation with near-UV light, conidiophores formed dedifferentiated into longer, slender hyphae and conidiation was completely inhibited. Hence, it was concluded that these fungi belong to "diurnal sporulators".

Furthermore, conidiation occurred successively in cultures under continuous near-UV light containing no wavelengths above 400 nm. However, when cultures were grown under continuous black light, no conidia were formed for a long time, although conidiophores were produced in profusion. It is interesting to note that conidiophores formed

---

1 Institute for Agricultural Research, Tohoku University, 980 Sendai, Japan

remained in that stage and neither dedifferentiated into longer, slender hyphae nor lost the capacity to conidiate under black light. Why do conidiophores remain at a definite stage under black light?

2 Opposing Effects of Blue and Near-UV Radiations in Conidial Development

The effects of blue light, near-UV light and black light appear to differ from each other, and near-UV and blue regions included in black light seemed to interact. Thus, opposing effects of these two lights in conidial development were studied. There was no conidiation when cultures were continuously irradiated with blue light for 10–12 h immediately following inductive irradiation with near-UV light. Furthermore, 2-h blue light irradiation was applied at different times during the darkness after inductive irradiation. The phase most sensitive to suppression by blue light was at 6–8 h in *A. tomato* and 4–8 h in *B. cinerea* after the beginning of the dark period. However, suppression by blue light was almost completely reversed by near-UV light applied immediately after blue light. Furthermore, it was found that blue light and near-UV light acted antagonistically in conidial development of these fungi (Table 1).

Table 1. Conidiation of certain fungi after exposure to blue (B) and near-UV (NUV) radiation in sequence

Irradiation	Conidiation (%)			
	A. tomato	*H.oryzae HA₂*	*B.cinerea*	*H.oryzae KU-13*
None (dark control)	100	100	100	100
NUV	105	95	107	–
B	20	11	28	39
B + NUV	100	93	95	91
B + NUV + B	45	38	52	58
B + NUV + B + NUV	115	86	81	83
B + NUV + B + NUV + B	58	34	55	60
B + NUV + B + NUV + B + NUV	–	91	–	–

6th h (*A. tomato:* Kumagai et al. 1969), 8th h (*H. oryzae HA₂*: Honda et al. 1968), and 5th h (*B. cinerea:* Suzuki et al. 1977) after the beginning of darkness following inductive irradiation with near-UV light, conidiophores were irradiated alternatively with blue light (1.5 W m^{-2}) for 1 h and near-UV light (0.6 W m^{-2}) for 1 h. *H. oryzae KU-13* (Yamamura et al. 1978) was treated at 9th h after the beginning of darkness following irradiation with black light for 5 days

   Which stage of conidiophore development is controlled by the blue and near-UV reversible photoreaction? We cannot detect the precise stage of conidiophore development in *Alternaria* and *Helminthosporium,* since both are morphologically too similar. However, developmental stages of conidiophores during darkness after inductive irradiation in *B. cinerea* can be divided into six numbered stages as shown in Fig. 1. Cultures with a selected conidiophore at one of the numbered stages were exposed to blue light for 1 h. At stage 1, conidiation was not inhibited by blue light but completed normally

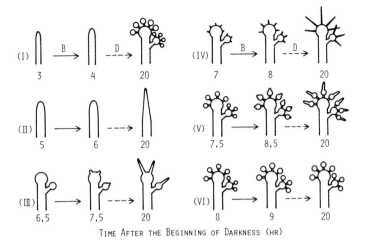

TIME AFTER THE BEGINNING OF DARKNESS (HR)

Fig. 1. Sterile conidiophores dedifferentiated by blue light irradiation at various stages of conidiophore development of *Botrytis cinerea*. Cultures with selected conidiophores at various stages, following inductive exposure to black light for 12 h, were irradiated with blue light (*B*) for 1 h. *D* darkness. *Stage I* initial conidiophore; *stage II* immature conidiophore; *stage III* spherical ampulla at tip
of conidiophore; *stage IV* denticles blown out from the ampulla; *stage V* conidium initials at the
tip of the denticles; *stage VI* mature conidia. (Suzuki et al. 1977)

10 h after the beginning of darkness following inductive irradiation. However, conidiophores from stages 2 to 5 were inhibited by the 1-h blue light break and conidiophores
already formed in these stages dedifferentiated into longer, slender hyphae, that is,
"sterile" conidiophores. Furthermore, when near-UV light was applied immediately
following exposure to blue light at stages 2 to 5, conidiation proceeded normally just
as in the case without blue light inhibition. When near-UV light was applied immediately
following exposure to blue light, recovery was most effective, indicating the longer the
dark period inserted after irradiation with blue light, the weaker the reversal effect by
near-UV light. Thus, it was confirmed that the inhibitory effect of blue light could be
completely reversed by near-UV irradiation, and the effect of near-UV light acts only
at the same developmental stage as that inhibited by blue light.

On the other hand, another isolate of *H. oryzae KU-13* formed conidiophores with
conidia when cultured under continuous darkness (Yamamura et al. 1978). However,
many of the conidiophores formed under continuous black light remained in the conidiophore stage and neither produced conidia nor lost the capacity to conidiate. Hence,
when cultures were placed in darkness after incubating under black light, conidiophores
in any aged loci of the colony began to form conidia simultaneously and matured during
a subsequent dark period. Thus, the inhibitory effect of blue light could be critically
detected at a definite stage in conidial development when black light was used. From
microscopic examination, this stage also seemed to correspond with the conidiophore
maturation stage. Furthermore, the suppression of conidial development by blue light
could be reversed by immediate near-UV light and the effects of the two lights were
alternatively reversible (Table 1).

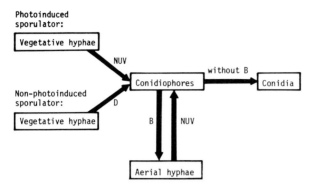

Fig. 2. Schematic explanation of the conidial development in two types of fungi, photo-induced and nonphoto-induced sporulators. "Photo-induced sporulators" and "nonphoto-induced sporulator" indicate the fungi *H. oryzae HA*$_2$, *A. tomato,* and *B. cinerea* which required near-UV light for conidiophore induction and the fungus *H. oryzae KU-13* which did not require near-UV light for conidiophore induction, respectively. *NUV, B,* and *D* indicate the cultural conditions under near-UV light, blue light, and darkness, respectively (Yamamura et al. 1978)

It is noteworthy that the blue and near-UV reversible photoreaction functioning in the conidial development of *H. oryzae KU-13* was very similar to that mentioned previously in other fungi, i.e., requires near-UV irradiation for conidiophore induction. From these results, it was inferred that a pigment system, "mycochrome", plays an important role in blue and near-UV reversible photoreaction in conidial development of these fungi and the response is determined by the final quality of light received. When the final light is near-UV, a conidium develops; when it is blue, conidiation is inhibited, "sterile" conidiophore being formed. The effect of light on conidiation of these fungi is schematized in Fig. 2. The difference between these fungi can be recognized merely in the induction stage of conidiophores, that is, near-UV light is required in *H. oryzae HA*$_2$, *A. tomato* and *B. cinerea,* termed "photo-induced sporulators", but it is not required in *H. oryzae KU-13,* which is termed "nonphoto-induced sporulator".

3 Significance of Near-UV Radiation and Certain Natures of P$_B$ in Mycochrome of *Alternaria tomato*

Blue and near-UV photoreaction has also been found in a light-minus-dark difference spectrum of cell extracts isolated from dark-grown mycelia of *A. tomato;* two pigments, P$_{NUV}$ (near-UV-absorbing pigment) and P$_B$ (blue-absorbing pigment) are involved in that photoreaction (Kumagai et al. 1976). That is, a supernatant fraction obtained by centrifugation at 1000 g for 10 min of the cell extracts was fractionated by a two-layer sucrose density gradient centrifugation consisting of 68% and 30% (W/W) sucrose. After centrifugation at 40,000 g for 120 min, three fractions located at each discontinuity of sucrose were obtained. The uppermost fraction was further treated by ammonium sulfate fractionation. The supernatant fraction obtained by saturated ammonium precipitation was used as a P$_{NUV}$ fraction. When this fraction was irradiated with 305 nm, a difference spectrum showed decrease only at about 300 nm and was not influenced by a subsequent irradiation with blue light. The active principle contained in P_{NUV} fraction was lost by dialysis, recovered in the fraction that passed through the cellophane membrane and was not precipitated by the addition of perchloric acid (final volume of 5%). Hence it was assumed that P_{NUV} is a protein-free, low molecular material and it is either soluble in the cytoplasm or very loosely bound to the particulate fraction. On the other hand, the fraction located at 68% layer of sucrose was further dialyzed against 0.1 M phosphate buffer (pH 7.0) with cellophane membrane and this dialyzed particulate fraction was used as a P_B fraction. A difference spectrum of this fraction was almost uninfluenced by light irradiation. However, when the P_B fraction plus P_{NUV} fraction was exposed to 305 nm, the light-minus-dark difference spectrum showed a peak absorbance near 400 nm. Such a light-induced absorbance change occurred only under coexistence of P_{NUV} and P_B fractions.

P_B may be considered to be tightly bound to the particulate fraction since it could not be solubilized by sonication or detergent treatment. In this connection, the intracellular distribution of P_B was investigated. That is, hyphae of *A. tomato* grown in the presence of polyoxin, which is known to have an effect on the synthesis of cell wall chitin, form protoplast-like loosely shaped structures which are easily disrupted by a Potter-type homogenizer (Yoshioka et al. 1975). Polyoxin below 10 μg/ml medium has very little effect on photosporulation and photoreaction of this fungus. Distribution of P_B and several marker enzymes were investigated using such hyphae as shown in Table 2. The maximum level of succinic dehydrogenase was recovered in the fraction located on 55% sucrose; hence it was considered that most mitochondria were contained in that fraction. However, the maximum level of near-UV light-induced absorbance change near 400 nm and ATPase activity were recovered in the fraction located on 68%, and the distribution of those activities was very similar. Thus, it is likely that P_B mostly localizes on membranous structures.

When the P_B plus P_{NUV} fraction was irradiated with 400 nm after exposure to 305 nm, the difference spectrum partially dissipated. When molecular oxygen, rather than blue light irradiation, was introduced after irradiation with 305 nm photo-induced peak

Table 2. Specific activities of succinic dehydrogenase, ATPase, and photoresponse

Conc. of sucrose (%)	Succinic dehydrogenase (mV/min)	ATPase (μmol/H)	Response relative
30	0.4	1.21	0.026
55	2.0	0.87	0.025
68	0.5	1.6	0.048

The mycelia obtained as reported elsewhere (Yoshioka et al. 1975) were suspended in a medium containing 0.02 M Tris-HCl (pH 7.2), 0.5 M mannitol, 2.5 mM EDTA and 1 mg/ml bovine serum albumin and then disrupted by a Potter-type homogenizer in ice-water bath. Supernatant fraction of homogenate after centrifugation at 1000 g for 10 min was further centrifuged at 40,000 g for 120 min. The 40,000 g pellet was resuspended in the same buffer and fractionated by using three-layer sucrose density gradient (30%, 55%, 68%) centrifugation at 40,000 g for 120 min. Three fractions located at each sucrose discontinuity were extracted. Activities of succinic dehydrogenase, ATPase and protein concentration were assayed by the methods of Veeger et al. (1969), Nombela et al. (1974), Lowry et al. (1951) (Yoshioka, unpublished) absorbance near 400 nm rapidly disappeared. Following irradiation with 305 nm after introduction of molecular oxygen, peak absorbance near 400 nm increased again. The antagonistic action of near-UV irradiation and introduction of molecular oxygen could be reversibly repeated. When sodium hydrosulfite was added to the sample cuvett containing only P_B fraction, a reduced-minus-oxidized difference spectrum showed a main peak at 395 nm, sub-peak at 495 nm and a broad dip at 580 nm; those disappeared during subsequent introduction of molecular oxygen. Those Em, 7 values were +160 mV (n = 1), +58 mV (n = 2) and +48 mV (n = 2), respectively (Kumagai 1978). Thus, it was considered that the active principle showing light-induced peak absorbance near 400 nm may correspond with peak absorbance at 395 nm in the reduced-minus-oxidized difference spectrum, which is a reduced form of P_B; consequently, P_B in mycochrome is an oxidation-reduction pigment, autooxidizable and its Em, 7 value is +160 mV with n = 1. Also, the mycochrome system suggested in photo-conidiation is consistent with substances showing these light-induced absorbance changes which are caused by photo-oxidation reduction reaction.

The significance of near-UV light in P_{NUV}-mediated photoreductions of P_B and cytochrome c, which was chosen for its similar mode of action to that of P_B, was further investigated (Kumagai, in press). Measured wavelengths in the dual mode were set at 395 and 460 nm for reduction of P_B, since the light-induced difference spectrum showed a peak near 395 nm and an isobestic point near 460 nm, and 548 and 540 nm for reduction of cytochrome c. As shown in Fig. 3, photoreduction of P_B under anaerobic condition occurred at a rate several times larger than under aerobic condition. Similarly, when cytochrome c was exposed to a wavelength of 305 nm in the presence of P_{NUV} fraction, it was reduced under anaerobic condition at a faster rate than under aerobic condition. Here, a gas-tight cuvett was bubbled continuously for 30 min with nitrogen gas deoxygenated by passage over hot copper wire for measurement under

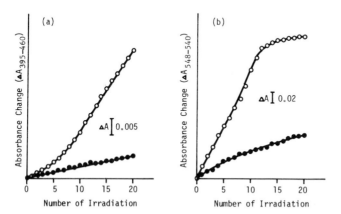

Fig. 3a, b. P_{NUV}-mediated photoreduction of P_B (a) and cytochrome c (b) by near-UV light. a Sample cuvett contained 1 ml of P_B fraction, o.5 ml of P_{NUV} fraction and 1.5 ml of 0.1 M phosphate buffer (pH 7.0). Reduction of P_B is shown as $\Delta A_{395-460}$. b Sample cuvett contained 50 μl of 1% cytochrome c, 0.5 ml of P_{NUV} fraction and 2.5 ml of 0.1 M phosphate buffer (pH 7.0). Reduction of cytochrome c is shown as $\Delta A_{548-540}$. Measurement was done every 1 min of irradiation with 305 nm (0.8 W m^{-2}). *Black and blank circles* show the values obtained under aerobic and anaerobic conditions, respectively (Kumagai, in press)

anaerobic condition. Under aerobic condition, both P_{NUV}-mediated photoreductions of P_B and cytochrome c by near-UV light were inhibited by superoxide dismutase, but not under anaerobic condition. It was concluded that photoreaction under aerobic condition mediated through P_{NUV} and near-UV light produced superoxide anion, which reduced P_B and cytochrome c. In fact, both P_B and cytochrome c could be reduced by superoxide anion produced by xanthine-xanthine oxidase, which was also inhibited by either superoxide dismutase or tiron. Furthermore, both photoreductions of P_B and cytochrome c under anaerobic condition were not inhibited by superoxide dismutase; consequently, some photoproduct other than superoxide anion formed by the cooperation of P_{NUV} and near-UV light may have directly reduced either P_B or cytochrome c.

As described above, P_B is an oxidation-reduction pigment with an oxidation-reduction potential of +160 mV with n = 1. P_B also showed a mode of action similar to cytochrome c in P_{NUV}-mediated photoreduction by near-UV light. Therefore, an investigation of the effect of NADH on P_B and its possible correlation with cytochrome c was attempted. When NADH was added to a sample cuvett containing only P_B fraction, P_B was not reduced. However, if NADH was added to a sample cuvett containing P_B fraction and the soluble fraction obtained by two-layer sucrose density gradient centrifugation, P_B was reduced as shown in Fig. 4. Similarly, cytochrome c was reduced by adding NADH in the presence of crude soluble supernatant fraction. In these reductions, NADH was about three times more active than NADPH. These reductions were not influenced by amytal and antimycin A. Soluble supernatant fraction boiled at 100 C for 5 min did not show NADH-dependent-reduction activity; hence, it was concluded that NADH-dependent cytochrome c reductase was contained in the soluble fraction obtained by

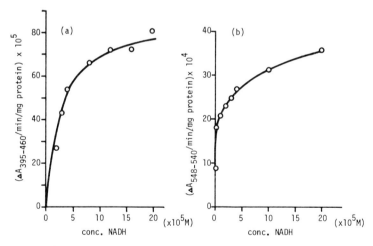

Fig. 4a, b. NADH-dependent reductions of P_B in mycochrome (a) and cytochrome c (b). a Sample cuvett contained 2 ml of P_B fraction and 1 ml of soluble supernatant fraction obtained by two-layer sucrose density gradient centrifugation. *Vertical line* shows the initial reduction rate of P_B ($\Delta A_{395-460}$/min/mg protein) and *horizontal line* shows the concentrations of NADH added. b Sample cuvett contained 50 μl of 1% cytochrome c, 1 ml of soluble supernatant fraction obtained by two-layer sucrose density gradient centrifugation and 2 ml of 0.1 M phosphate buffer (pH 7.0). *Vertical line* shows the initial reduction rate of cytochrome c ($\Delta A_{548-540}$/min/mg protein) and *horizontal line* shows the concentration of NADH added (Kumagai, in press)

two-layer sucrose density gradient centrifugation. Since reduction of cytochrome c by NADH occurred by dialyzed P_B fraction, bound-type NADH-dependent cytochrome c reductase may also be involved. Similarly, it was presumed that some type of enzyme, which functioned on NADH-dependent reduction of P_B, was contained in the crude soluble fraction. When cytochrome c was added to a sample cuvett after irradiation with 305 nm, photo-reduced P_B was easily oxidized. The crude enzyme fraction was not required for cytochrome c-dependent oxidation of reduced P_B. Hence, it was concluded that reduced P_B could be oxidized directly by cytochrome c itself.

4 Discussion

As previously stated, P_B was reduced by NADH under the existence of a certain enzyme fraction, designated "NADH-dependent P_B reductase" in comparison to "NADH-dependent cytochrome c reductase" and its reduction was not inhibited by amytal or antimycin A. A reduced form of P_B caused by NADH or near-UV light was easily oxidized by either cytochrome c or molecular oxygen. Hence, it is considered that the mycochrome system functions in close relationship to mitochondrial electron flow system since they have two connecting points; one is NADH as an entrance, and the other cytochrome c which finally leads to molecular oxygen. From these results, the mycochrome system may function as a kind of "photorespiration" similar to "alternative respiration" in cyanide-resistant respiration in higher plants (Solomos 1977); mycochrome-mediated photoreaction is closely associated with respiration, ultimately causing conidiation. Furthermore, blue light may function in the same area as near-UV light, but this is still unclear.

P_B was reduced by superoxide anion which was produced by cooperation of P_{NUV} and near-UV light under aerobic condition. This photoreduction appeared to be small since P_B once reduced was rapidly oxidized by molecular oxygen. On the other hand, it may be considered that photoreduction under anaerobic condition was much larger compared to that under aerobic condition, and this photoreduction was not influenced by superoxide dismutase. Hence the unknown, $P_{NUV}*$, may be produced by the cooperation of P_{NUV} and near UV light, which directly reduces P_B. In this connection, Briggs (1976) has indicated that a flavin-mediated photo-oxidation reduction reaction of cytochromes may relate to the effect of blue light in phototaxis of *Dictyostelium*, phototropism of *Phycomyces*, and photoinhibition of circadian rhythm of *Neurospora*. The photoresponsive system of *A. tomato* perhaps differs from those, however. It is noteworthy that the appearance of P_{NUV}-mediated photoreduction of P_B occurs via a similar mechanism of flavin-mediated photoreduction of cytochrome c by blue light as in the model system by Schmidt and Butler (1976) if P_{NUV}, P_B, and near-UV light are replaced by flavin, cytochrome c, and blue light, respectively. In any case, free radicals such as superoxide anion or the yet unknown $P_{NUV}*$, produced by light irradiation, may drive the vegetative growth of fungi into an adverse environmental condition; it may finally cause conidia formation considered as one of the bodies resistant to adverse environmental condition. Through such a process, P_B may function as a scavenger or protection against free radicals; consequently, photo-oxidation reduction once begun may be closely associated with respiration, which in turn is connected to photosporogenesis. This would explain the difference in near-UV requirements for conidiophore induction between "photo-induced sporulators" and "nonphoto-induced sporulator".

5 Summary

A photoreceptor system, "mycochrome", is involved in a blue and near-UV reversible photoreaction which in turn plays an important role in the photocontrol of conidial development in "photo-induced sporulators" and "nonphoto-induced sporulator". Near-UV irradiation is required for conidiophore induction of photo-induced sporulators such as *Alternaria tomato, Botrytis cinerea* and *Helminthosporium oryzae HA$_2$*, but not for a nonphoto-induced sporulator such as *H. oryzae KU-13*. However, conidial development of both sporulators is suppressed by blue light applied at a definite stage of conidiophore development, and the conidiophores dedifferentiate into longer, slender hyphae. The effect of blue light can be reversed by immediate exposure to near-UV light and the effects of these two spectral regions are alternatively reversible.

Blue and near-UV photoreaction has also been found in a light-minus-dark difference spectrum of cell extracts isolated from dark-grown mycelia of *A. tomato*, and two pigments, P_B (blue-absorbing type) and P_{NUV} (near-UV-absorbing type) are involved in that photoreaction. P_{NUV} is either soluble in the cytoplasm or very loosely bound to the particulate fraction, while P_B is tightly bound to the particulate fraction. Blue and near-UV photoreaction occurs only under coexistence of P_B and P_{NUV}. In this article, near-UV irradiation and the significance the photoresponsive system, "mycochrome", in *A. tomato* is shown. Near-UV irradiation caused photoreduction of both P_B and cytochrome c, which was chosen for its similar mode of action to that of P_B, mediated through P_{NUV}. Both P_{NUV}-mediated photoreductions of P_B and cytochrome c were larger under anaerobic condition than aerobic condition. Both photoreductions under aerobic condition could be inhibited by superoxide dismutase; consequently, it is presumed that a superoxide anion was produced when P_{NUV} was exposed to near-UV light under aerobic condition which also reduced P_B and cytochrome c. Under anaerobic condition, a certain photoproduct resulting from the cooperation of P_{NUV} and near-UV light (with the exception of a superoxide anion) may directly reduce either P_B or cytochrome c, since these photoreductions were not inhibited by superoxide dismutase. Furthermore, P_B and cytochrome c were reduced by NADH independent of light, if some crude enzyme fraction was added to that system. Reduced P_B was easily oxidized by the addition of cytochrome c or the introduction of molecular oxygen. Thus, it is concluded that photoreaction mediated through the mycochrome system may be linked to electron flow.

Acknowledgment. This work was supported in part by a grant from the Ministry of Education of Japan.

References

Barnett HL, Lily VG (1950) Influence of nutritional and environmental factors upon asexual repro-
    duction of *Choanephora cucurbitarum* in culture. Phytopatholgy 40: 80–89
Briggs WR (1976) The nature of the blue light photoreceptor in higher plants and fungi. In: Smith
    H (ed) Light and plant development, pp 7–18. Butterworths, London
Honda Y, Sakamoto M, Oda Y (1968) Blue and near ultraviolet reversible photoreaction on the
    sporulation of *Helminthosporium oryzae.* Plant Cell Physiol 9: 603–607
Kumagai T (1978) Mycochrome system and conidial development in certain fungi imperfecti. Photo-
    chem Photobiol 27: 371–379
Kumagai T (1979) Near ultraviolet light-induced absorbance changes and mycochrome system in
    the fungus, *Alternaria tomato.* Agric Biol Chem in press
Kumagai T, Oda Y (1969) Blue and near ultraviolet reversible photoreaction in conidial development
    of the fungus, *Alternaria tomato.* Dev Growth Differ 11: 130–142
Kumagai T, Yoshioka N, Oda Y (1976) Further studies on the blue and near ultraviolet reversible
    photoreaction with an intracellular particulate fraction of the fungus, *Alternaria tomato.* Biochim
    Biophys Acta 421: 133–140
Leach CM (1967) Interaction of near-ultraviolet light and temperature on sporulation of the fungi
    Alternaria, Cercosporella, Fusarium, Helminthosporium, and *Stemphylium.* Can J Bot 45: 1999–
    2016
Lowry OH, Rosebrough NJ, Farr AL, Randall JR (1951) Protein measurement with the folin phenol
    reagent. J Biol Chem 193: 265–275
Nombela C, Uruburu F, Villanueva JR (1974) Studies on membranes isolated from extracts of *Fusa-
    rium culmorum.* J Gen Microbiol 81: 247–254
Schmidt W, Butler WL (1976) Flavin-mediated photoreactions in artificial systems: A possible model
    for the blue-light photoreceptor pigment in living systems. Photochem Photobiol 24: 71–75
Solomos T (1977) Cyanide-resistant respiration in higher plants. Annu Rev Plant Physiol 28: 279–
    297
Suzuki Y, Kumagai T, Oda Y (1977) Locus of blue and near ultraviolet reversible photoreaction in
    the stages of conidial development in *Botrytis cinerea.* J Gen Microbiol 98: 199–204
Veeger C, DerVartanian DV, Zeylemaker WP (1969) Succinate dehydrogenase. In: Lowenstein JM
    (ed) Methods in enzymology, vol VIII, pp 81–90. Academic Press, London New York
Yamamura S, Kumagai T, Oda Y (1978) Mycochrome system and conidial development in a non-
    photoinduced isolate of *Helminthosporium oryzae.* Can J Bot 56: 206–208
Yoshioka N, Kumagai T, Oda Y (1975) Effect of polyoxin on the morphogenesis of the fungus,
    Alternaria tomato, with special reference to photosporulation. Dev Growth Differ 17: 281–286

Blue Light Responses in the Siphonaceous Alga *Vaucheria**

W.R. BRIGGS[1] and M.R. BLATT[1]

During the past two decades, there has been a great deal of work on the influence of light on chloroplast movement (Haupt and Schönbohm 1970; Haupt 1973). Indeed much has been learned about the photoreceptor phytochrome from studies of light-regulated chloroplast movement in the alga *Mougeotia* (Haupt 1970). The regulation of chloroplast movement by light acting through phytochrome, however, seems to be the exception; more commonly regulation is achieved via a blue light photoreceptor system. Such is the case with the siphonaceous alga *Vaucheria* (Fischer-Arnold 1963; Blatt and Briggs 1980). A blue-absorbing pigment not only regulates chloroplast movement in this alga, but in addition induces phototropic curvature (Kataoka 1975a, 1977b) and lateral branching (Kataoka 1975b). The present paper provides a brief review of these responses, with emphasis on the response mechanism at the cellular level when possible.

1 Phototropic and Straight Growth Responses

Virtually all that we know of the growth responses of *Vaucheria* to blue light comes from work by Kataoka, of Osaka University in Japan. Kataoka (1975a) first studied filament elongation under a variety of light regimes. Under a daily cycle of 12 h light/12 h dark, growth occurred predominately during the dark period, coming to a halt within 2 or 3 h of the onset of the light period. Upon transfer to continuous light, however, growth began again after about 12 h, and, after a transient decrease in growth at 24 h, persisted at a relatively constant rate for 5–7 days. A morphological change (a slight swelling), occurring with the onset of growth, provided a fixed intrinsic marker for determining growth rate, commonly about 200 μm/h. By studying the relative movements of adhering resin particles or distribution of a fluorescent dye, Kataoka (1975a) verified that filament elongation was caused by true tip growth, without elongation below the hemispherical tip region (Kataoka 1975a).

Kataoka next investigated the various phototropic responses of *Vaucheria* filaments. *Vaucheria* phototropism is complex, exhibiting curvature responses comparable to first-positive (Kataoka 1975a), first-negative, and second-positive (Kataoka 1977b) curvatures

* C.I.W. – D.P.B. Publication No. *681*

1 Department of Plant Biology, Carnegie Institution of Washington, 290 Panama St., Stanford, CA 94305, USA

in *Avena* (see Briggs 1963). For most phototropic experiments, Kataoka illuminated the tip of a horizontally growing filament from above using an ingenious split field illumination that allowed each of the two halves of the tip to receive a different wavelength of light (Kataoka 1975a). Normally one was a reference wavelength to which there was no significant response (e.g., in the red). When the other wavelength was in the blue, tip growth became reoriented so that the filament grew into the blue field. Note that the response in this case is not phototropism in the strictest sense since the direction of growth is not precisely controlled by the direction of incident blue illumination. In other experiments, however, Kataoka (1975b) did demonstrate true phototropic growth toward incident blue light, by illuminating filaments unilaterally from the side. The response obtained with split field illumination will nevertheless be considered as phototropic in the broad sense in this paper.

Just beneath the apical wall of the growing tip of a *Vaucheria* filament is a region of clear cytoplasm which Kataoka refers to as a hyaline cap. When a phototropic response occurred, it was always preceded by a shift of the hyaline cap so that its axis moved away from the longitudinal axis of the filament (Kataoka 1975b). The number of degrees by which it shifted provided an accurate index of the magnitude of the subsequent phototropic response. The curvature mechanism was thus very similar to that described earlier by Page (1962) for the phototropic response of the young sporangiophore of the fungus *Pilobolus* and that described by Etzold (1965) for polarotropic curvature of the young filamentous gametophyte of the fern *Dryopteris*.

The phototropic response of *Vaucheria* was rapid, beginning within 2 min of the onset of illumination and continuing for an additional 10–20 min depending upon dose (Kataoka 1975a). The dose of light at 450 nm needed to obtain a threshold response was about 30 W-s m^{-2} and from this dose to doses two orders of magnitude higher, the reciprocity law was valid. The action spectrum showed maximal photosensitivity near 450 nm, with shoulders near 415 and 495 nm. There was no significant photoactivity for wavelengths beyond 550 nm. Unfortunately no measurements were made below 400 nm, so comparison with the action spectra for phototropism in *Avena* or *Phycomyces* (see Curry and Thimann 1961) or other blue-ultraviolet processes (see Presti and Delbrück 1978) is not possible, though all show similar features in the blue.

In addition to demonstrating that the curvature response in *Vaucheria* was brought about by a localized reorientation of tip growth, Kataoka (1975a) also showed that only the apical half of the apical dome was photosensitive (Kataoka 1975a). Although such localized photosensitivity might lead one to expect different effects of polarized light with the E vector in different planes, Kataoka reported no such effects with polarized blue light (1975a). The problem merits further investigation, however. Kataoka was watching a lateral curvature response to illumination from above, rather than true directed phototropic curvature, as mentioned above. The effects of plane polarization of unilateral illumination on curvature have not been determined.

In his investigation of the mechanism of light-induced growth reorientation, Kataoka (1975b) also demonstrated a positive light-growth reaction (straight growth) in *Vaucheria*. When tips were illuminated uniformly with blue light, their growth rate was 30% or more higher than when they were provided with an equal energy of red light. Furthermore, the response occurs rapidly (within a few minutes as nearly as one can judge), and could be fast enough to play a role in phototropic curvature. However, in a split field experiment, such a blue light-induced growth acceleration would lead to curvature of the tip out of the blue field and into the red, just the opposite of what actually occurs. Thus the light-growth reaction itself can only be antagonistic to the development of positive phototropic curvature, and cannot be directly involved in the events which bring about curvature in the first place.

As mentioned above, Kataoka did a few experiments with unilateral illumination to demonstrate genuine phototropic curvature toward the actinic light source. In these experiments, he attempted to reverse the direction of curvature by varying the refractive index of the medium. The curvature response remained positive however, and schematic diagrams of the path of light through the tip under the various conditions suggest that under these conditions the side of the hemispherical tip toward the light would always receive more light than the side away, whether the refractive index of the surrounding medium was greater, the same as or less than that of the filament tip.

In another study, Kataoka (1977b) extended the phototropic dose response curve to far higher light doses. Between 10 and 10^3 J m^{-2} of blue light, the response increased log-linearly, and the reciprocity law was valid. At higher doses, administered with relatively high intensity actinic light (46 W m^{-2}) almost no curvature was seen near 10^5 J m^{-2}. The reciprocity law was not valid in this range of doses, however, since if the actinic light intensity was lowered, substantial curvatures occurred. When actinic intensity was increased even more, clear negative curvature was observed. This entire phototropic response system is remarkably similar in its properties to that described by Zimmerman and Briggs (1963) for *Avena* and by Briggs (1960) for *Zea* coleoptiles. All three systems show a range of positive curvature response within which the reciprocity law is valid, a higher range within which with high intensity light curvature is first either lacking or is actually negative, and then becomes positive with increase in dosage, and within which the reciprocity law is not valid. In all three systems, the positive curvature within this range shows some sort of dependence on duration of actinic illumination, though the precise relationship is not worked out for *Vaucheria*. Whether *Vaucheria* shows phytochrome-mediated alterations in these phototropic dose response relationships, as was the case with *Avena* (Zimmerman and Briggs 1963) and *Zea* (Chon and Briggs 1966) is unknown.

Recently, Kataoka (1979) has studied the effect of intermittent light on phototropic responses in *Vaucheria*. For small responses in the first-positive curvature range, the alga could tolerate a separation of 1 s light pulses by as much as 40 s darkness without any decrease in the magnitude of the response obtained. Experiments with higher doses, and much shorter light and dark intervals indicated that with light pulses of 10 ms or longer, and dark periods between 15 and 150 ms, intermittent light was more effective in inducing curvature than was continuous irradiation of the same energy, suggesting restriction of the response by a limiting dark reaction. The results are in some ways similar to those obtained for *Zea* coleoptiles (Briggs 1960). Briggs (1960), however, was studying the second positive curvature response, and working under conditions of reciprocity failure, whereas Kataoka was clearly working with doses below the maximum for first-positive curvature, doses for which the reciprocity law holds for *Vaucheria*. Overall, the photobiological similarities between *Vaucheria* and coleoptiles of *Avena* or *Zea* are nonetheless remarkable, particularly in view of the very different bending mechanisms involved.

Finally, Kataoka (1977a) reported that cyclic AMP and its analog dibutyryl cyclic AMP both strongly inhibited the phototropic response of *Vaucheria* at concentrations of 10^{-4} M without affecting the growth rate. Both caffeine and theophylline also inhibited phototropism but not growth. This latter result would be expected if the two compounds inhibit a cyclic AMP phosphodiesterase in *Vaucheria* as they are reported in animals (Cehovic et al. 1972) since their action would presumably lead to a higher internal concentration of cyclic AMP. However, in view of the difficultires encountered by others in demonstrating cyclic AMP unequivocally in higher plants (Amrhein 1977), the preliminary nature of the results with *Vaucheria*, and the reliance on undetermined specificity of action of inhibitors, the conclusions should be viewed with caution.

2 Branching

Brief mention should be made of the influence of blue light on branching in *Vaucheria*. Kataoka (1975b) illuminated a small and relatively mature region of a *Vaucheria* filament with low intensity blue light and chronicled the events leading to branching. Within seconds, chloroplast aggregation in the light began (see below) and by 1 h, the thickness of the cytoplasmic layer adjacent to the cell wall had increased about tenfold. The cytoplasm itself in this region was packed with chloroplasts. After about 4–5 h of illimination, a small hyaline area, similar to that at the filament apex, could be seen. This hyaline area marked the point at which a proturberance ultimately appeared with the onset of branching. Large doses of blue light were required (over 7000 J m^{-2}) and a minimum of 5 h from the onset of illumination was needed before the protuberance was visible. Despite the far greater light dose required for branching than for phototropism, and the much greater lag period, cyclic AMP apparently inhibited both processes (Kataoka 1977a).

3 Chloroplast Movement

The final process in *Vaucheria* to be considered in this brief review is the regulation of chloroplast movement by blue light. By 1908, Senn was able to describe two distinct responses of *Vaucheria* chloroplast movement to light. The first was accumulation of chloroplasts in illuminated parts of the filament when the light intensity was low. The second was migration of the chloroplasts from the illuminated to the lateral flanks of the filament when the light intensity was high. It was then over 50 years before *Vaucheria* was again the subject of careful photobiological investigation. Haupt and Schönfeld (1962) then obtained an action spectrum for the migration of chloroplasts to the flanks of strongly illuminated cells. They found action maxima near 450 and 370 nm, with a distinct minimum near 400 nm. There was also no activity at wavelengths longer than 500 nm. Haupt and Schönfeld alse report examining this chloroplast response with polarized light, stating that the response was independent of the plane of polarization. Like Kataoka (1975a), however, they present no details or data.

A year later, Fischer-Arnold (1963) presented results of studies both of chloroplast aggregation in weak, and avoidance in strong blue light. The action spectrum for the weak light response was very similar to that obtained by Haupt and Schönfeld (1962) for the strong light response, suggesting a similar photoreceptor pigment (flavins were favored in both articles). Fischer-Arnold also investigated the effect of red light. While inactive by itself, red light increased the subsequent responses to blue. This red effect was presumably not mediated by phytochrome, since it was not reversible by far red light. The light perception process for chloroplast aggregation was not affected by temperature between $10°$ and $30°C$, although the subsequent rate of chloroplast accumulation showed a pronounced temperature dependency with a Q_{10} near 2. The avoidance reaction in strong light appeared insensitive to temperature over the same range.

Fischer-Arnold (1963) next investigated the two chloroplast responses in *Vaucheria* in somewhat more detail. The avoidance reaction in strong light had a lag period of approximately 10 min. Thereafter, the chloroplasts followed a somewhat complicated pathway to the cytoplasm along the lateral walls. Upon cessation of irradiation, the chloroplasts became redispersed, a process which started almost immediately, and was completed within 10 min. The aggregation of chloroplasts in weak light occurred far more rapidly. Fischer-Arnold reported a detectable reaction within 10 s of the onset of illumination under certain conditions. Blatt and Briggs (1980) report responses within 20 s with relatively high intensity light, and longer lag periods with decreasing light intensity for this same aggregation reaction. Redispersal was rather slower, with both Fischer-Arnold (1963) and Blatt and Briggs (1980) reporting the process to last about 10 min.

Finally, Fischer-Arnold elegantly demonstrated that the photoreceptor for the accumulation response was located not in the chloroplasts themselves but rather in a peripheral cytoplasmic region close to the plasma membrane (the experiments did not exclude the plasma membrane itself as the site of the photoreceptor). If she illuminated a region of cytoplasm devoid of chloroplasts, then chloroplasts which entered this region became trapped. On the other hand, if a filament of *Vaucheria* was moved carefully so that a streaming group of chloroplasts remained in the light spot, then no aggregation occurred. Both Fischer-Arnold (1963) and Senn (1908) attributed chloroplast aggregation in dim light to a localized increase in cytoplasmic viscosity. Indeed, Virgin (1954) and references cited therein reports both increases in cytoplasmic viscosity in *Helodea densa* leaves in response to weak light and decreases in response to strong.

Most recently, Blatt and his colleagues (Blatt and Briggs 1980; Blatt et al. 1980) have reinvestigated the cellular basis for the aggregation response to weak light. Using Nomarski Differential Interference Contrast optics, they were able to resolve a series of longitudinal cables visible just beneath the cell wall by light microscopy. Chloroplast movement could be seen to occur along the cables, with chloroplasts occasionally appearing to move in opposite directions along the same cable, bumping each other, and then passing on their way. When a small region of cytoplasm was irradiated with blue light, the cables became rapidly destabilized, forming a reticulum-like structure within which branches were repeatedly formed and broken. When a chloroplast moving along a cable in an unilluminated protion of the filament entered the illuminated area, it encountered the reticulum and remained in that area.

By using single fiber optics, Blatt and Briggs (1980) could simultaneously illuminate a small region of a filament with blue light and monitor the transmission of red light through the same region. The decrease in transmission occurring as chloroplasts became

aggregated allowed quantitative investigation of the kinetics of reticulum formation
and aggregation under a variety of light intensities and wavelengths. As mentioned above,
aggregation occurred within 20 s when the light intensity was high (400 mV m^{-2}). The
reciprocal of the lag period increased linearly with intensity. Regardless of intensity,
however, reticulation of the cables always preceded actual chloroplast aggregation by
at least 10 s. A study of wavelength dependence of reticulation showed similarity to the
action spectrum of Fischer-Arnold in the blue region of the spectrum. The use of fiber
optics precluded measurements in the ultraviolet.

In a second study, Blatt et al. (1980) then investigated the nature of the cables,
using a combination of light microscopy, electron microscopy, and experiments with
inhibitors. They first made protoplasts by cutting *Vaucheria* filaments and extruding
small droplets of cytoplasm surrounded by plasma membrane. These droplets showed
both cytoplasmic streaming, and, under Nomarski optics, cables. Furthermore, they retained their sensitivity to weak blue light. The protoplasts could then be burst on electron microscope grids while kept under observation in the light microscope. If they were
then rinsed with a solution containing ATP, much of the cellular debris was removed,
but groups of cables were still clearly visible. Williamson (1975) used this ATP technique
to clear perfused internodal cells of *Chara* of organelles. Most of the organelles remaining within the perfused cell then streamed out one of the open ends. Blatt et al. (1980)
then treated the preparation with subfragment 1 of rabbit muscle myosin, prior to preparation for electron microscopy. This subfragment forms characteristic arrowhead
patterns when it reacts with actin (see Spudich et al. 1972). Under examination by electron microscopy, the cables visible under Nomarski optics appear to be composed of
bundles of fine filaments on which arrowheads are visible in some cases. Treatment of
the preparation with ATP just after treatment with the subfragment 1 preparation completely removes the decoration, leaving filaments about 5–7 nm in diameter, in the correct size range for actin (Spudich et al. 1972). Britz (1979) has recently reviewed the
small but growing literature on the occurrence of actin in algae.

That an actin-based system played a role not only in chloroplast movement but also
in the aggregation response to weak blue light was suggested by results with inhibitors.
Cytochalasin B, a strong inhibitor of actomyosin-related cell movements in animals and
plants (Wessells et al. 1971; Williamson 1976), completely stopped cytoplasmic streaming in *Vaucheria,* as one might have expected. Phalloidin, a potent toxin from the fungus
Amanita phalloides, favors equilibration in the direction of filamentous actin, away
from the unassociated globular subunits (see Wieland 1977), and might be expected to
stabilize actin filaments against factors leading to their disruption. Phalloidin was completely without effect on cytoplasmic streaming in *Vaucheria.* However, it completely
prevented any response to weak light: chloroplasts moved unhindered through the illuminated area. Addition of cytochalasin B in the presence of phalloidin no longer stopped streaming. However, it did allow reticulation and chloroplast aggregation in blue
light, apparently antagonizing the phalloidin sufficiently to permit a light-induced destabilization of the cables to occur.

From this brief overview of the photobiology of *Vaucheria,* it is clear that significant progress has been made in understanding at least one of the reactions, chloroplast
aggregation in dim blue light. This progress could well lead to advances in our understanding of other photoresponses as well. We are left, however, with a number of ques-

tions. First, in connection with the aggregation response, how does photoreceptor excitation lead to action filament destabilization? Are ionic mechanisms involved? Second, could light effects on cable stability be related to the chloroplast avoidance reaction in strong light? Third, is the shift in the orientation of the hyaline cap related in any way to light-induced changes in the integrity of actin filaments? Fourth, is some underlying disturbance of polarity of cables of actin filaments an initial event in the chain of reactions leading to branch formation at an illuminated region of the cell? Fifth, what is the fine structure of the cables themselves, particularly as regards the polarity of the individual actin filaments and the presence or absence of membrane? Finally, what is the precise subcellular location and chemical nature of the photoreceptor pigment? All of these questions are susceptible to experimental attack, and within the next few years, we may expect to learn a great deal more about the way in which this organism responds to blue light. Whether the mechanisms unearthed by such studies are of more general applicability to blue light responses remains to be seen, but they appear to involve both pigments and cellular systems which are widespread in green plants and fungi.

References

Amrhein N (1977) The current status of cyclic AMP in higher plants. Annu Rev Plant Physiol 28: 123–132

Blatt MR, Briggs WR (1980) Blue light-induced cortical fiber reticulation concomitant with chloroplast aggregation in the alga *Vaucheria sessilis*. Planta 147: 355–362

Blatt MR, Wessells NK, Briggs WR (1980) Actin and cortical fiber reticulation in the siphonaceous alga *Vaucheria sessilis*. Planta 147: 363–375

Briggs WR (1960) Light dosage and the phototropic responses of corn and oat coleoptiles. Plant Physiol 35: 951–962

Briggs WR (1063) The phototropic responses of higher plants. Annu Rev Plant Physiol 14: 311–352

Britz SJ (1979) Chloroplast and nuclear migration. In: Haupt W, Feinleib ME (eds) Encyclopedia of plant physiology, New Ser, vol 7, Physiology of movements, pp 170–205. Springer, Berlin Heidelberg New York

Cehovic G, Posternak T, Charollais E (1972) A study of the biological activity and resistence to phosphodiesterase of some derivatives and analogues of cyclic AMP. In: Greengard P, Robinson GA, Paoletti R (eds) Advances in cyclic nucleotide research, vol I, pp 521–540. Raven Press, New York

Chon HP, Briggs WR (1966) The effect of red light on the phototropic sensitivity of corn coleoptiles. Plant Physiol 41: 1715–1724

Curry GM, Thimann KV (1961) Phototropism; the nature of the photoreceptor in higher and lower plants. In: Christensen B, Buchmann B (eds) Progress in photobiology, pp 127–134. Elsevier, Amsterdam

Etzold H (1965) Der Polarotropismus und Phototropismus der Chloronemen von *Dryopteris filix-mas* (L.) Schott. Planta 64: 254–280

Fischer-Arnold G (1963) Untersuchungen über die Chloroplastenbewegung bei *Vaucheria sessilis*. Protoplasma 56: 495–520

Haupt W (1970) Localization of phytochrome in the cell. Physiol Veg 8: 551–563

Haupt W (1973) Role of light in chloroplast movement. Bioscience 23: 289–296

Haupt W, Schönbohm E (1970) Light-oriented chloroplast movements. In: Halldahl P (ed) Photobiology of microorganisms, pp 283–307. Wiley Interscience, London

Haupt W, Schönfeld I (1962) Über das Wirkungsspektrum der „negativen Phototaxis " der *Vaucheria*-Chloroplasten. Ber Dtsch bot Ges 75: 14–23

Kataoka H (1975a) Phototropism in *Vaucheria geminata*. I. The action spectrum. Plant Cell Physiol 16: 427–437

Kataoka H (1975b) Phototropism in *Vaucheria geminata*. II. The mechanism of bending and branching. Plant Cell Physiol 16: 439–448

Kataoka H (1977a) Phototropic sensitivity in *Vaucheria geminata* regulated by 3',5'-cyclic AMP. Plant Cell Physiol 18: 431–440

Kataoka H (1977b) Second positive- and negative phototropism in *Vaucheria geminata*. Plant Cell Physiol 18: 473–476

Kataoka H (1979) Phototropic responses of *Vaucheria geminata* to intermittent blue light stimuli. Plant Physiol 63: 1107–1110

Page RM (1962) Light and the asexual reproduction of *Pilobolus*. Science 138: 1238–1245

Presti D, Delbrück M (1978) Photoreceptors for biosynthesis, energy storage and vision. Plant Cell Environ 1: 81–100

Senn G (1908) Die Gestalts- und Lageveränderung der Pflanzenchromatophoren. Engelmann, Leipzig

Spudich J, Huxley H, Finch J (1972) Regulation of skeletal muscle contraction II. Structural studies of the interaction of the tropomyosin-troponin complex with actin. J Mol Biol 72: 619–632

Virgin HI (1954) Further studies of the action spectrum for light-induced changes in the protoplasmic viscosity of *Helodia densa*. Physiol Plant 7: 343–353

Wessells N, Spooner B, Ash J, Bradley M, Ludena M, Taylor E, Wrenn J, Yamada K (1971) Microfilaments in cellular and developmental processes. Science 171: 135–143

Wieland T (1977) Modification of actins by phallotoxins. Naturwissenschaften 64: 303–309

Williamson R (1975) Cytoplasmic streaming in *Chara*: a cell model activated by ATP and inhibited by cytochalasin B. J Cell Biol 17: 655–668

Williamson R (1976) Cytoplasmic streaming in characean algae. In: Wardlaw I, Paffioura J (eds) Transport and transfer processes in plants, pp 51–58. Academic Press, London New York

Zimmerman BK, Briggs WR (1963) Phototropic dosage-response curves for oat coleoptiles. Plant Physiol 38: 248–253

Cis to *Trans* Photoisomerization of ζ-Carotene in *Euglena gracilis* Var. *bacillaris* W₃BUL: Further Purification and Characterization of the Photoactivity

Y.L. STEINITZ[1], J.A. SCHIFF[2], T. OSAFUNE[2], and M.S. GREEN[2]

1 Introduction

We have previously described a blue light-induced photoactivity in intact cells of certain *Euglena* mutants which accumulate ζ carotene (Fong and Schiff 1979). Blue light brings about absorption changes in intact cells and in extracts which are consistent with those expected for a cis to trans photoisomerization of ζ carotene. In this paper we describe the further purification and characterization of this system.

2 Materials and Methods

2.1 Organisms

Mutant strain W₃BUL derived from *Euglena gracilis* Klebs var. *bacillaris* Cori (Fong and Schiff 1979; Schiff et al. 1979) was grown on Hutner's medium, pH 3.5, as described previously (Greenblatt and Schiff 1959; Lyman et al. 1961). Strain W₃BUL lacks detectable plastid DNA and green plastid pigments (Edelman et al. 1965; Schiff 1979). Cells were grown in darkness to a density of $2-3 \times 10^6$ cells ml^{-1} unless otherwise noted, and were prepared under green safelights as previously described (Schiff 1972; Fong and Schiff 1979). The method of Zeldin and Schiff (1967) was used for determining cell number.

2.2 Measurement of Absorption Spectra and Absorbance Changes

Isolation of cells and absorption measurements with a computerized Biospect model 61 Spectrophotometer were carried out as previously described by Fong and Schiff (1979) including illumination for 120 s with blue light from a tungsten lamp filtered by a Corning No. 5433 broad-band blue filter (420 nm maximum) at fluence of 140 W m^{-2}. When the absorption of samples in organic solvents was measured, a 1.0 ml aliquot was used. Some absorption spectra were measured with a Cary model 14 spectrophotometer.

---

1 Present address: Department of Microbiology, Faculty of Agriculture, Hebrew University, Rehovoth, Israel
2 Institute for Photobiology of Cells and Organelles, Brandeis University, Waltham, MA 02254, USA

2.3 Preparation of Cell Extracts and Purification of the Photoactivity

All steps were carried out in the absence of light or under green safelights (Schiff 1972). To prepare extracts 4-day-old cells ($1.5-2.0 \times 10^6$ cells ml^{-1}) were centrifuged from 2 l of medium at 500 g for 1 min at room temperature and were resuspended in a buffer solution 0.25 M with respect to sucrose and 0.05 M with respect to potassium phosphate pH 7.0. The highest photoactivity was found at cell densities from 1.5 to 2.5 × 10^6 cells ml^{-1} representing cultures at the end of the exponential phase of growth and in the early part of the stationary phase (from about 4 to 8 days after innoculation under our conditions). The cells were then re-sedimented at 500 g for 5 min and were resuspended in 20 ml of fresh buffer at 4°C. (From this point on all manipulations were performed at 4°C.) This suspension was passed through a chilled French pressure cell at 4000 lb in^{-2} and was then centrifuged at 1000 g for 10 min. Enough disodium ethylene diamine tetraacetate was added to the supernatant fluid to yield a final concentration of 0.005 M. The supernatant fluid was recentrifuged at 43,000 g for 20 h. The floating orange-colored, lipid-like material close to the surface of the tube was collected in 1 ml of fresh sucrose phosphate buffer and exhibited most of the original photoactivity. This suspension was used for the experiments to be described where 0.1 ml was mixed with 0.1 ml of buffer and the photoactivity measured in a 10.5 mm diameter cuvette. This floating fraction in sucrose-phosphate buffer could be stored in darkness at −18°C for at least 7 days without loss of activity.

The following additions to this purified fraction were made in some experiments (final concentration shown): disodium ethylene diamine tetraacetate 0.001−0.010 M; cetyltrimethyl-ammoniumbromide 0.001 M; magnesium chloride 0.015 M−0.1 M; sucrose 0.05 M−1.0 M; HCl 0.001 M−0.1 M.

In some experiments the floating material was heated at 100°C for 10 min, cooled, and the photoactivity measured.

In some experiments, the purified floating material was extracted with 90% acetone (v/v). In others a transfer from acetone to fresh diethyl ether followed as described in Vaisberg and Schiff (1976).

2.4 Chromatographic Analysis of Extracts of the Photoactive Fraction

The ether fraction was taken to dryness in vacuo, resuspended in ether and chromatographed on Eastman (Rochester, N.Y.,U.S.A.) precoated thin layer silica gel on plastic sheets (No. 13953) without fluorescent indicator, using 20% ethyl acetate in methylene chloride (v/v) (Vaisberg and Schiff 1976) as the solvent. Individual spots were eluted in ether and their absorption was determined in a Cary Model 14 spectrophotometer.

The appropriate extinction coefficient was used to determine the concentration of ζ and β carotene (Jensen and Jensen 1971; Fong and Schiff 1979). The ultraviolet (UV) spectrophotometric method of Shaw and Jefferies (1953) was used for identification of ergosterol.

2.5 Electron Microscopy

Samples were fixed with 2.5% (v/v) glutaraldehyde in the sample buffer (0.25 M sucrose, 0.05 M potassium phosphate pH 7.0) for 90 min at room temperature. After withdrawing this solution from the particles with a pasteur pipet, 2% osmium tetroxide in 0.1 M sodium cacodylate buffer pH 7.2 was added and postfixation proceeded for 12 h at 4°C. After washing once with cacodylate buffer, the particles were dehydrated successively with a 50% to 100% ethanol series and with 100% acetone. The samples were embedded according to the methods of Spurr (1969) and thin sections were cut on a Reichert Ultramicrotome OmU2 with a diamond knife and were placed on Formvar filmed grids. The sections were stained with 2% uranyl acetate overnight, washed with deionized water, and stained with lead salt solution for several minutes (Sato 1968; Osafune et al. 1976). They were then examined with a Phillips-300 electron microscope.

2.6 Chemicals and Sonication

All chemicals used were of analytical grade. The membrane filters used in the indicated experiments were Millipore HA 0.45 uM pore size. For sonication the Sonifier Cell Disrupter Model 200 (Branson Sonic Power Company, Danbury, Conn., U.S.A.) was used; the sample was cooled with ice water.

3 Results

If whole cells of *Euglena gracilis* variety *bacillaris* W₃ BUL are irradiated with blue light (Fig. 1), the blue light minus dark difference spectrum shows positive peaks at about 445 nm, 412 nm, and 385 nm (Fong and Schiff 1979). The absorption spectrum of these cells shows contributions from carotenoids, cytochromes, and other pigments. After cell breakage and centrifugation at 1000 g, the supernatant fraction shows a more carotenoid-like absorption and the blue light minus dark difference spectrum indicates that most of the original activity is present. EDTA was added at this point to dissociate the ribosomes present and release trapped photoactivity; the concentration of EDTA used (5 mM) was found to have no effect on the photoactivity in control experiments. After centrifugation at 43,000 g for 20 h the orange lipid-like floating fraction shows an absorption spectrum with clear peaks and a good recovery of photoactivity as judged from the difference spectrum. The peaks at 385 nm, 409 nm, and 434 nm are those expected for ζ carotene (Fong and Schiff 1979) in cells and cell fractions and the peaks in the blue light minus dark difference spectrum at 390 nm, 415 nm, and 441 nm correspond to the differences to be expected on conversion of cis to trans ζ carotene. Small amounts of β carotene which accompany the ζ carotene during purification are still present to some extent as judged from the augmented absorption at 434 nm where both ζ and β carotenes absorb and from the small absorptions at somewhat longer wavelengths; however as previously observed (Fong and Schiff 1979) no photoactivity attributable to β carotene is evident in the difference spectra. The clear peaks attributable to ζ carotene in the floating material isolated in darkness, the increase in absorption of these peaks on illumination together with the small shifts of the absorption peaks to

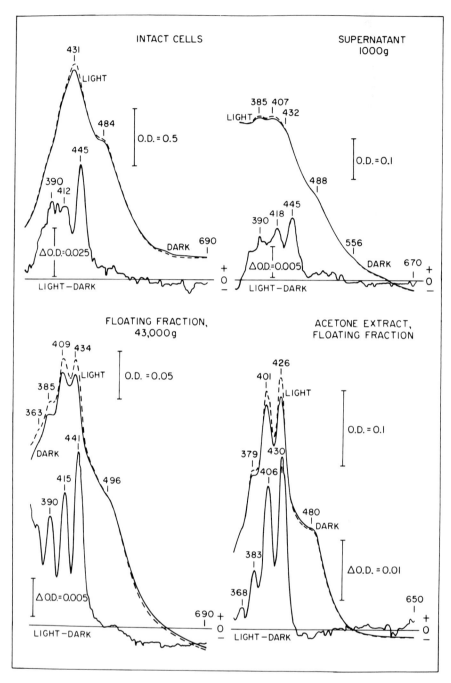

Fig. 1. Representative absorption and blue light minus dark difference spectra of various preparations from *Euglena gracilis* var. *bacillaris* W₃BUL. These include whole cells, the 1000 g supernatant fluid prepared from cell extracts, floating material from 43,000 *g* centrifugation and a 90% (v/v) acetone extract of the floating material

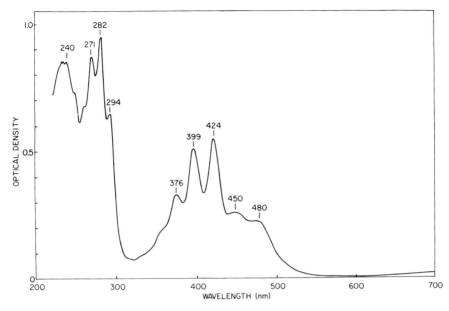

Fig. 2. Absorption spectrum of the diethyl ether eluate of the chromatographic spot moving near the front on silica gel thin layer chromatography, determined in a Cary Model 14 recording spectrophotometer. The acetone extract of the purified floating material was transferred to ether, the ether solution was applied to the silica gel, and the chromatogram was developed with ethyl acetate-methylene chloride

longer wavelengths expected for the cis to trans isomerization all indicate the considerable degree of purification achieved using these techniques of separation.

Extraction of the highly purified floating fraction with acetone, transfer to ether, and chromatography of the ether extracts provides further evidence of purity. While the whole cells contain small amounts of xanthophylls in addition to ζ and β carotene (Fong and Schiff 1979), the highly purified floating fraction contains only ζ carotene and smaller amounts of β carotene among the colored carotenoids. All of the xanthophylls and a significant portion of the β carotene are lost during purification. Thus the highly purified floating fraction contains no xanthophylls and has a ratio of ζ carotene to β carotene of about 3 compared with the whole cells where the ratio is about 1.

Elution with ether of the chromatographic spot moving near the front which contains the carotenes, and determination of the absorption spectrum yields the data of Fig. 2. The spectrum in the violet and blue regions shows the absorptions to be expected from ζ carotene and the smaller amounts of β carotene. The spectrum in the ultraviolet, however, below 300 nm shows the characteristic absorption spectrum of ergosterol, one of the major sterols of the *Euglena* cells (Shaw and Jeffries 1953; Stern et al. 1960; Avivi et al. 1967; Anding and Ourisson 1973). Sterols are thought to be membrane constituents in eukaryotic cells.

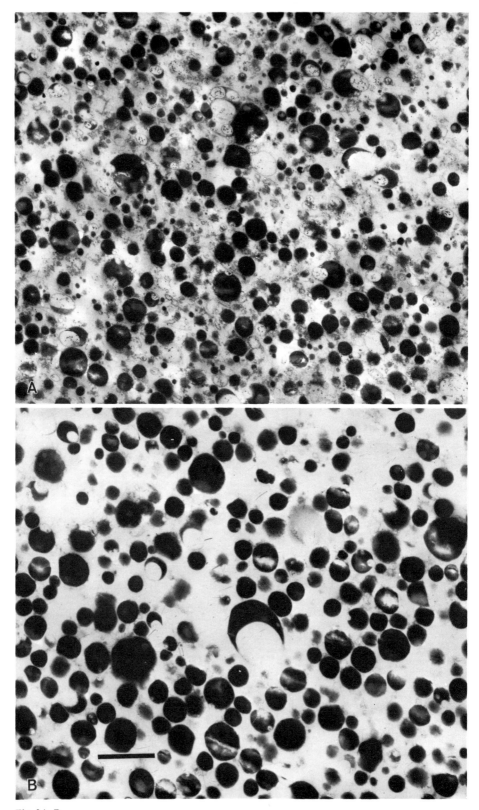

Fig. 3A, B.

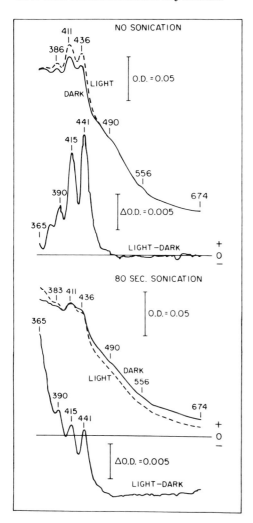

Fig. 4. Absorption and blue light minus dark difference spectra of purified floating material from W$_3$BUL before and after sonication for 80 s at ice water temperature

Consistent with the particulate nature of the photoactivity (Fong and Schiff 1979) the photoactivity is retained by and can be measured directly on a Millipore filter of 0.45 um pore size.

Figure 3 shows that the highly purified floating fraction is composed largely of fairly homogeneous densely staining particles of various sizes when viewed in the electron microscope. Although varying amounts of contaminating materials are seen, it seems likely that the photoactivity and, therefore, the ζ carotene is associated with these particles.

Fig. 3A, B. Electron micrographs of thin sections through the purified floating material. The *marker* indicates 1.0 um for both halves of the figure. A Purified floating material from dark-grown W$_3$BUL; B purified floating material from dark-grown W$_3$BUL exposed to white fluorescent lamps at a distance of 1.0 cm for 5 min before fixation

We have previously found (Fong and Schiff 1979) that additions of various inhibitors and redox reagents had no effect on the photoactivity. In the present work this list can be extended to include cetyltrimethylammonium bromide (1.0 mM), EDTA (0.001–0.01 M), magnesium chloride (0.025–0.1 M), sucrose (0.05–1.0 M), HCl (0.001–0.1 M), and 10 min at 100°C or freezing and thawing, none of which affected the photoactivity in the highly purified floating fraction.

Sonication for 80 s delivered as equally spaced pulses changed the shape of the absorption spectrum and decreased the photoactivity shown as the blue light minus dark difference spectrum (Fig. 4). A sharp rise in the difference spectrum toward shorter wavelengths is seen in the sonicated material suggesting that a new peak in the blue light minus dark difference spectrum may exist below 350 nm, the lower limit of the present measurements.

Much to our surprise, the acetone and ether extracts of the highly purified floating material prepared in darkness retained photoactivity. Figure 1 shows that in the 90% acetone extract the peaks of absorption are now in the expected positions for ζ carotene in solvents. The blue light minus dark difference spectra are similarly shifted but the increase in absorption and small shift to longer wavelengths to be expected from photo-isomerization of cis to trans ζ carotene on illumination are still observed. Essentially the same results were obtained on transfer of the pigments from acetone to diethyl ether (data not shown).

4 Discussion and Conclusions

The results reported here show that a particulate fraction of low density can be purified from dark-grown cells of *Euglena gracilis* var. *bacillaris* bleached strain W₃BUL which contains ζ carotene and exhibits a photoconversion of cis ζ carotene to trans ζ carotene on illumination. Extraction of the purified particles from dark-grown cells with acetone and transfer to ether yields solutions which still show the expected absorptions and spectral shifts expected for the photoisomerization of cis ζ carotene to the trans form.

A comparison of the absorption bands of ζ carotene in the particles, in the acetone and ether extracts, and in other solvents (Table 1) shows that the ζ carotene molecules in the particles (and in the intact cells, see Fig. 1) are in an environment which is differ-

Table 1. Wavelength maxima of ζ carotene in various environments

	Peak wavelengths (nm)			Reference
Hexane (Pet Ether)	380,	400,	425	Davis et al. (1966); Davis (1970)
Toluene	383,	406,	431	Davis (1976)
Chloroform	384,	405,	430	Davis (1976)
Ether	377,	399,	424	Fong and Schiff (1979); this paper
Acetone (90% in H₂O)	379,	401,	426	This paper
Purified floating fraction	385,	409,	434	This paper
Carbon disulfide:Chloroform 3:1	389,	414,	442	Fong and Schiff (1979)

ent from acetone, ether, and other common solvents. Only in mixtures of chloroform and carbon disulfide (Table 1), (Fong and Schiff 1979) are the absorption bands of ζ carotene greatly shifted to longer wavelengths as in the particles and cells. However, in acetone, ether, and in particles illumination with blue light causes the small shifts in absorption to longer wavelengths together with an increase in the optical density to be expected from isomerization of cis ζ carotene to the trans form (Fong and Schiff 1979).

The spectroscopic properties of the ζ carotene in the particles and in solvents compared with information available in the literature (Claes 1961; Davis et al. 1966; Davis et al. 1969; Raymundo and Simpson 1972; Qureshi et al. 1974; Simpson et al. 1976) suggest that it is the 15 cis isomer. It has been suggested that the biosynthetic pathway of carotenoids may involve 15 cis isomeric intermediates up to and including ζ carotene (Davis et al. 1966). Thus the photoisomerization reaction we have described here appears to involve the conversion of 15 cis ζ carotene to trans ζ carotene.

The following information is consistent with the photoreaction being a cis to trans photoisomerization of ζ carotene taking place in a lipid environment: the low density of the particles containing the activity, the lack of influence of added redox reagents (Fong and Schiff 1979), ions, chelating agents and pH changes on the activity, and the variable effects of detergents (some have no effect, others such as triton inhibit) (Fong and Schiff 1979). The spectroscopic information, however, indicates that the environment in the particles is somewhat different from that provided by nonpolar solvents (Table 1). Consistent with the nature of the reaction, heating the system briefly before assay does not destroy the activity. Although some cis isomers of carotenoids are unstable to heat, others are not (Moss and Weedon 1976); the cis isomer of ζ carotene studied here does not appear to be affected by brief heating in its normal environment.

The work of Claes (1961) and our own previous work (Fong and Schiff 1979) showed that the cis to trans photoizomerization of ζ carotene in solvents takes place in the presence of iodine. It is well known that the photoisomerization of carotenoids in general is catalyzed by iodine (Zechmeister 1962). In the present work, however, we have shown that the photoisomerization of cis ζ carotene to trans in solvent extracts of particles does not require the addition of iodine. Either this photoisomerization does not require iodine catalysis or some catalyst is extracted from the particles along with the cis ζ carotene. We favor the former explanation, that no catalyst is required, since photoisomerization of many other carotenoids has been observed to take place in solvents in the absence of iodine catalysis (Zechmeister and Polgar 1944; Zechmeister 1962; Kropf and Hubbard 1970; Weedon 1971; Moss and Weedon 1976; Maeda et al. 1978).

The electron micrographs of the purified particles indicate that the major constituents of this fraction are homogeneous densely staining particles of a range of sizes with a varying background of contaminating vesicles. It is likely that the photoactivity is associated with the densely staining particles but more work is necessary to be certain of this. A strong possibility for the origin of these densely staining particles is the stigma material of the cells. In dark-grown cells of W_3 BUL the stigma is an amorphous irregularly shaped mass of lipoidal material which becomes organized into the characteristic stigma granules on exposure to light (Osafune and Schiff 1979). It has often been suggested that the hematochrome particles found throughout *Euglena* cells originate from a breaking up of the stigma (Leedale 1967; Buetow 1968). In red *Euglena* which contain large amounts of hematochrome, the expansion and contraction of the cellular hematochrome distribution which results in a change of the color of the organism from green to red on an increase in illumination is known to be a blue light-induced phenomenon (Johnson and Jahn 1942). The range of particle sizes found in the electron micrographs suggest that the ζ carotene photoactivity may be associated with stigma granules and/or hematochrome and perhaps the light-induced changes in hematochrome placement and stigma morphology in *Euglena* cells are related to the pigments they contain. Bartlett et al. (1972) isolated a similar particle fraction from wild-type *Euglena* cells.

We have shown in this paper that sonication alters the absorption spectrum of the purified particles and destroys the usual photoisomerization reaction seen in the blue light minus dark difference spectrum. Since the ζ carotene is in an environment somewhat different from nonpolar solvents, sonication may expose the ζ carotene to a more polar aqueous environment by breaking up the particles and providing a greater surface to volume ratio. Perhaps this more polar environment does not provide conditions which allow the usual photoisomerization reaction to take place.

We have previously shown that the action spectrum for photoisomerization of cis ζ carotene to trans does not correspond to the absorption spectrum of ζ carotene (Fong and Schiff 1979). The action spectra have rather sharp peaks at 370 nm, 390 nm, 423 nm, and 448 nm, and no effectiveness at wavelengths longer than about 460 nm. Thus we suggested that another molecule might be present which serves to absorb the light and to photosensitize the conversion of cis ζ carotene to the trans form. With the finding in the present work that solvent extracts of purified particles exhibit the photoreaction, it is possible that no other molecule than cis ζ carotene is necessary to allow the photoreaction to take place. If ζ carotene absorbs the light, the action spectrum for its own isomerization could represent absorption by both cis and trans ζ carotene, leading to different equilibrium mixtures at each wavelength depending on the absorption of each of the two forms at any particular wavelength. This is reminiscent of the treatment of the kinetics of phytochrome conversion by Hartmann and Unser (1972), where different steady-state conversions of P_r to P_{fr} are observed depending on the wavelength used to elicit the reaction. It remains for future work to derive the necessary kinetic relationships to permit a test of this model by fitting these relationships to the observed dose response curves and the action spectra (Fong and Schiff 1979).

The most likely function of the photoconversion of cis ζ carotene to trans is in carotenogenesis where it may serve as one path of formation of trans carotenoids as previously suggested (Fong and Schiff 1979). The more tenuous relation of this reaction to phototaxis and other blue light reactions in bleached strains of *Euglena* is also discussed by Fong and Schiff (1979).

Acknowledgments. We thank Nancy O'Donoghue for her assistance with the electron microscopy. This work was supported by grant GM-14595 from the National Institute of Health. Jerome A. Schiff is the Abraham and Etta Goodman Professor of Biology.

References

Anding C, Ourisson G (1973) Presence of ergosterol in light-growth and dark-grown *Euglena gracilis*. Eur J Biochem 345–346

Avivi L, Laron O, Halevy S (1967) Sterols of some algae. Comp Biochem Physiol 321–326

Bartlett CJ, Walne PL, Schwarz OJ, Brown DH (1972) Large scale isolation and purification of eye-spot granules from *Euglena gracilis* var. *bacillaris*. Plant Physiol 49: 881–885

Buetow DE (1968) Morphology and ultrastructure of Euglena. In: Buetow DE (ed) The biology of *Euglena,* vol I, pp 146–149. Academic Press, London New York

Claes H (1961) Energieübertragung von angeregten chlorophyll auf C_{40}-Polyene mit verschiedenen chromophoren Gruppen. Z Naturforsch 166: 445–454

Davies BH (1970) A novel sequence for phytoene dehydrogenation in *Rhodospirillum rubrum*. Biochem J 116: 93–99

Davies BH (1976) Carotenoids. In: Goodwin TW (ed) Chemistry and biochemistry of plant pigments, vol II, pp 38–165. Academic Press, London New York

Davies BH, Holmes EA, Loeber DE, Toube TP, Weedon BCL (1969) Carotenoids and related compounds XXIII. Occurence of 7,8,11,12-tetrahydrolycopene, spheroidene 3,4,11',12'-tetrahydro-speroidene and 11',12'-dihydrospheroidene in *Rhodospirillum rubrum*. J Chem Soc c 1266–1268

Davis JB, Jackman LM, Siddons PT, Weedon BCL (1966) Carotenoids and related compounds XV. The structure and synthesis of phytoene, phytofluene, ζ carotene and neurosporene. J Chem Soc 2154–2165

Edelman M, Schiff JA, Epstein HT (1965) Studies of chloroplast development in *Euglena* XII. Two types of satellite DNA. Mol Biol 11: 769–776

Fong F, Schiff JA (1979) Blue light induced absorbance changes associated with carotenoids in Euglena. Planta 146: 119–127

Greenblatt CL, Schiff JA (1959) A pheophytin like pigment in dark adapted *Euglena gracilis*. J Protozool 6: 23–28

Hartmann KM, Cohnen Unser L (1972) Analytical action spectroscopy with living systems: Photochemical aspects and attenuance. Ber Dtsch Bot Ges 85: 481–551

Jensen SL, Jensen A (1971) Quantitative determination of carotenoids in photosynthetic tissues. Methods Enzymol 23: 586–602

Johnson LP, Jahn TL (1942) Cause of the green-red color change in *Euglena rubra*. Physiol Zool 15: 89–94

Kropf A, Hubbard R (1970) The photoisomerization of retinal. Photochem Photobiol 12: 249–260

Leedale GF (1967) Euglenoid flagellates, p 25. Prentice Hall Inc, New Jersey

Lyman H, Epstein HT, Schiff JA (1961) Studies of chloroplast development in *Euglena* I. Inactivation of green colony formation by u.v. light. Biochem Biophys Acta 50: 301–309

Maeda A, Schichida Y, Yoshizawa T (1978) Formation of 7-cis retinal by the direct irradiation of all-trans retinal. J Biochem 83 (3): 661–663

Moss GP, Weedon BCL (1976) Chemistry of the carotenoids. In: Goodwin TW (ed) Chemistry and biochemistry of plant pigments, vol I, pp 149–224. Academic Press, London New York

Osafune T, Schiff JA (1979) Light induced changes in a proplastid remnant in dark-grown resting *Euglena gracilis* var. *bacillaris* W_3BUL. Plant Physiol 63: 5–27

Osafune T, Mihara S, Hase E, Ohkuro I (1976) Electron microscope studies of the vegetative cellular life cycle of *Chlamydomonas reinhardi* Dangeard in synchronous culture IV. Mitochondria in dividing cells. J Electron Microsc 25: 261–269

Qureshi Asaf A, Qureshi N, Kanok K, Porter JW (1974) Isolation, purification, and characterization of cis ζ carotene and the demonstration of its conversion to acyclic, monocyclic, and dicyclic carotenes by a soluble enzyme system obtained from the plastids of tangerine tomato fruits. Arch Biochem Biophys 162: 117–125

Raymundo C, Sympson KL (1972) The isolation of poly cis ζ carotene from the tangerine tomato. Phytochemistry 11: 397–400

Sato T (1968) A modified method for lead staining thin sections. J Electron Microsc 17: 158–159

Schiff JA (1972) A green safelight for the study of chloroplast development and other photomorphogenetic phenomena. Methods Enzymol 24: 321

Schiff JA (1979) Blue Light and the photocontrol of chloroplast development in *Euglena.* This volume, pp 495–511

Schiff JA, Lyman H, Russell UG (1979) Isolation of mutants from *Euglena gracilis.* In: San Pietro A (ed) Methods in enzymology, vol 23, pp 143–162. Academic Press, London New York Revision and updated version: Isolation of mutants from *Euglena gracilis.* Ibid 59: in press

Shaw WH, Jefferies JP (1953) The determination of ergosterol in yeast. Analyst 78: 509–528

Simpson KL, Lee TC, Rodriguez DB, Chichester CO (1976) Metabolism in scenescent and stored tissues. In: Goodwin TW (ed) Chemistry and biochemistry of plant pigments, vol I, pp 806–807. Academic Press, London New York

Stern AI, Schiff JA, Klein HP (1960) Isolation of ergosterol from *Euglena gracilis:* distribution among mutant strains. J Protozool 7: 52–55

Vaisberg AJ, Schiff JA (1976) Events surrounding the early development of *Euglena* chloroplasts 7. Inhibition of carotenoid biosynthesis by the herbicide SAN 9789 (4-chloro-5-(methylamino)-2-(α,α,α-trifuluoro-m-tolyl-3(2H pyridazinone) and its developmental consequences. Plant Physiol 57: 260–269

Weedon BCL (1971) Stereochemistry. In: Isler O (ed) Carotenoids, pp 267–319. Birkhauser, Basel

Zechmeister L (1962) Cis-trans isomeric carotenoids, vitamins A and arylpolyenes, p 50. Academic Press, London New York

Zechmeister L, Polgar A (1944) On the occurence of a fluorescing polyene with a characteristic spectrum. Science 100: 317–318

Zeldin MH, Schiff JA (1967) RNA metabolism during light induced chloroplast development in *Euglena.* Plant Physiol 42: 922–932

Spurr AR (1969) A low-viscosity epoxy resin embedding medium for electron microscopy. J Ultrastruct Res 26: 31–43

Carotenogenesis

Blue Light-Induced Carotenoid Biosynthesis in Microorganisms

W. RAU[1]

1 Introduction

Among the mass pigments occurring in the plant kingdom, carotenoids are obviously the most widespread ones; not only all photosynthetic organisms contain carotenoids but also coloring of many fungi and bacteria are caused by these pigments. Carotenoids have many functions in plant life: they act as protective substances against intensive irradiation, they are accessory pigments in photosynthesis, they are attractants for animals in flowers and fruits, and they play a role in sexual reproduction of fungi, to mention only the most important functions.

In contrast to animals, plants and bacteria have the capacity for de novo biosynthesis of carotenoids and in many species carotenogenesis is under photocontrol. Especially during seedling development of angiosperms, the formation of carotenoids is stimulated by light, but this regulation is only one part of the general photomorphogenetic transformations and may therefore be not entirely independent of them. In algae photoregulation of carotenoid biosynthesis has been reported only for a few species mainly in connection with the development of chloroplasts and the photosynthetic apparatus; *Euglena gracilis* as well as mutants of *Chlorella* and *Scenedesmus* have been studied extensively in this respect. In a number of fungi – such as *Phycomyces* or *Mucor* – light increases the rate of carotenoid synthesis, and it is only in a few species, which have been listed previously (Rau 1975), that biosynthesis is strictly photoregulated. In bacteria such strict photo-induction has only been detected and thoroughly investigated in species of three genera: *Mycobacterium, Myxococcus,* and *Flavobacterium* (listed by Batra 1971).

Since carotenogenesis in angiosperms is under phytochrome control and since blue light-mediated changes of carotenoid production in algae will be the topic of other papers in this book, the present article will be restricted to results reported for microorganisms; however, for comparison analogous data from other organisms have to be considered in part. Fungi and bacteria in which illumination is obligatory for massive carotenogenesis seem to be particularly suited for investigations into the sequence of reactions that intervene between absorption of the irradiation and response. The purpose of this article will therefore be to summarize our present knowledge of these processes and to discuss problems that are still open. It is hoped that results on this special blue light-induced reaction may contribute to the knowledge of blue light responses in general.

1 Botanisches Institut der Universität München, Menzinger Straße 67, 8000 München 19, FRG

2 The Phenomenon

Fungi and bacteria which obligatorily need light for the massive production of carotenoids synthesize only trace amounts of pigment when grown in the dark. Already a
brief exposure to light – i.e., a few seconds – induces substantial carotenogenesis although higher doses of irradiation are necessary for bulk production. Light saturation
of the response has been reported for *Mycobacterium* sp. (Howes and Batra 1970),
Mycobacterium marinum (Batra 1971) and *Neurospora crassa* (Zalokar 1955), but no
saturation could be achieved in *Fusarium aquaeductuum* (Rau 1967a); in these organisms it has also been found that for the photo-induction reaction the reciprocity law
(I X t = const.) is valid. When mycelia of *Fusarium* were illuminated with white light
for various periods, the data revealed a logarithmic relationship between the amount
of pigment produced and the illumination time. Recent results concerning saturation
and dose relationship in *N. crassa* will be presented and discussed in another paper
(Schrott, this vol.).

The kinetics of photo-induced accumulation of carotenoids are very similar in all
species studied although the absolute lengths of characteristic periods are different. As
an example the results obtained for *Fusarium* (Fig. 1) show that after a lag period the
amount of pigment increases rapidly for a certain period of time and thereafter carotenoid synthesis ceases. However, continuous illumination leads to a prolonged accumulation which seems to be linear with time. From the data gathered from experiments
with brief illumination periods it is obvious that photoreactions and subsequent dark
reactions are involved. The dark reactions are consistently a strict consequence of photo-
induction indicating that in the sequence of events a very early "point of no return"
exists. Therefore photo-induced carotenogenesis in microorganisms exhibits all features
of a "classical" induction mechanism.

As a consequence of photo-induction the organisms synthesize a set of carotenoids
characteristic for the particular species (for ref. see Isler 1971). *M. marinum* and *Phycomyces blakesleeanus* produce β-carotene as main pigment, the intermediates – members
of the Porter-Lincoln series – are found only in trace amounts. On the contrary in *N.
crassa, F. aquaeductuum,* and *Mycobacterium* sp., these intermediates are accumulated

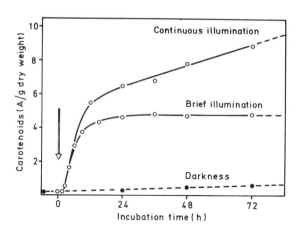

Fig. 1. Accumulation of carotenoids
in mycelia of *Fusarium aquaeductuum* in the dark, following a brief
illumination and in continuous
light. The *arrow* indicates the period of brief illumination

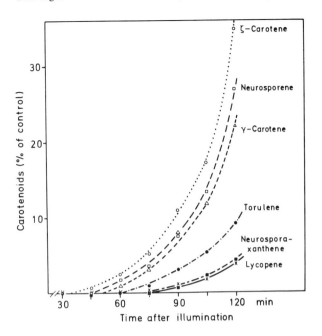

Fig. 2. Time course of synthesis of individual carotenoids in *Fusarium aquaeductuum* after a brief (10 min) exposure to light. Carotenoid content of control samples was determined when carotenogenesis has finished (36 h). (From Bindl et al. 1970)

to some extent; in these organisms the individual carotenoids are synthesized after photo-induction in a sequence closely following the proposed biosynthetic pathway (Fig. 2). A typical end product in both fungal species as well as in *Verticillium agarici- num* seems to be the C_{35}-carotenoid acid neurosporaxanthin, which so far has been detected only in a few fungi. Results and problems of the biosynthesis of carotenoids have been reviewed in extenso in the past years (Goodwin 1971, 1979; Britton 1976; Davies and Taylor 1976; Davies 1979).

3 The Photo-Induction

Since review articles and papers in this book are especially dedicated to various aspects of this topic, only results on blue light-induced carotenogenesis will be summarized and only conclusions from these results will be discussed.

3.1 The Photoreceptor(s)

Zalokar (1955) was the first to determine an action spectrum of carotenogenesis in the spectral region between 400 and 500 nm for mycelia of *N. crassa* using nonconidiating cultures; prevention of conidiation is important because in conidia of this species bio- synthesis of carotenoids is not light-dependent. A more detailed action spectrum has been established later for this fungus as well as for *F. aquaeductuum* and *Mycobacterium* sp. These spectra are presented in Fig. 3; they all show characteristic features also known for other blue light responses: A maximum at 370–380 nm and three peaks or at least

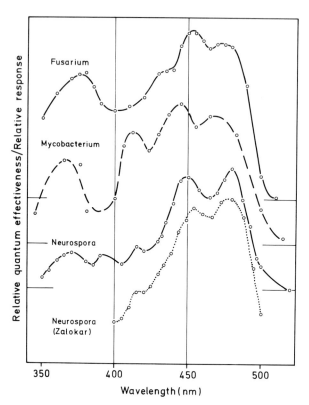

Fig. 3. Action spectra of photo-induced carotenoid biosynthesis in *Neurospora crassa* (calculated from Zalokar 1955 and from DeFabo et al. 1976), *Fusarium aquaeductuum* (from Rau 1967a) and *Mycobacterium* sp. (from Howes and Batra 1970). For convenience the base lines have been shifted as indicated

shoulders between 400 and 500 nm; light with a wave length longer than approximately 520 nm is ineffective. Comparing the shapes of the action spectra one has to bear in mind that they are plotted in different ways. In two bacterial species the shapes of the spectra are different (Fig. 4), suggesting that in these cells the photoreceptor could be a porphyrin. Additional support for this conclusion comes from the fact that porphyrins were extracted from the cells that have absorption spectra corresponding more or less with the action spectra obtained. Recently for the fungus *Verticillium agaricinum* phytochrome-mediated regulation of carotenogenesis has been reported (Valadon et al. 1979); the photoreaction shows the typical induction by red light and subsequent reversion by far-red light. It should be emphasized at this point that so far phytochrome was assumed to be absent from fungal cells.

Although for the cases illustrated by Fig. 3 the nature of the photoreceptor — flavin or carotenoid — will not be discussed further, a few results obtained by the use of inhibitors should be mentioned here. Hydrosulfite (Dithionite) inhibits photoinduction in *Mycobacterium* sp. but not in *M. marinum* (Batra and Rilling 1964). This observation is of particular interest since hydrosulfite is known to reduce flavins, thus eliminating their excitability by blue light and only in *Mycobacterium* sp. is a flavin assumed to be the photoreceptor. The photoreaction is inhibited by azide in both *Mycobacterium* species (Batra and Rilling 1964) as well as in *F. aquaeductuum* (Lang-Feulner and Rau 1975). Since this substance is known to complex transition metals, Batra and Rilling discussed the possibility that such a metal is part of the photoreceptor system; however, since photo-induction by red light via mythelene blue is also inhibited by azide (see

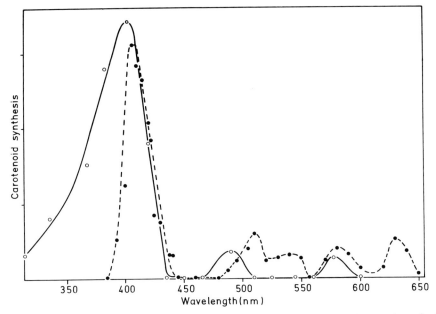

Fig. 4. Action spectra of photo-induced carotenoid synthesis in *Mycobacterium marinum* ○—○ (from Batra and Rilling 1964) and in *Myxococcus xanthus* ●— – —● (from Burchard and Hendricks 1969)

Sect. 3.2) their assumption does not seem to be likely. In *F. aquaeductuum* a few chemicals known to interfere with photoreactions of flavins proved to be completely or nearly ineffective in suppression of light-induced carotenogenesis: Rotenon up to concentrations of 20 μmol/l, amytal up to 5 mmol/l and also dichloromethylurea (DCMU) — when present only during illumination — up to 10 mmol/l. However, the lack of action could also be due to insufficient uptake of the drugs by the fungal cells.

3.2 "Primary Reactions"

3.2.1 Temperature Dependence

The photoreaction was found to be independent of temperature in all organisms investigated (Zalokar 1955; Rau 1962; Rilling 1962; Matthews 1963; Valadon and Mummery 1971; Seviour and Codner 1973; Weeks et al. 1973; Harding 1974); it was therefore concluded that light induction is a photochemical process.

3.2.2 Oxygen Requirement

For optimal photo-induction the presence of oxygen is essential, although minor induction in the absence of O_2 has been observed for *N. crassa* (Zalokar 1954; Rau 1969), for *Aspergillus giganteus* (Trinci and Banbury 1969) and for *F. aquaeductuum* (Rau 1969) but not for the *Mycobacteria* (Rilling 1964; Howes et al. 1969). Accordingly, the latter species did not produce carotenoids in N_2-atmosphere when proper electron acceptors were present during illumination. In *Fusarium* saturation of the photoreaction under anaerobic conditions is reached at a relatively low light dosage. It proved to be independent of both light intensity and time of illumination. Moreover, when mycelia were illuminated in N_2-atmosphere with saturating dosages, subsequently supplied with oxygen in the dark and then illuminated a second time in the presence or absence of oxygen they were susceptible to an additional photo-induction also under anaerobic conditions. Obviously the photoreceptor system can be reactivated by oxygen in the dark; this regeneration does not occur in the absence of O_2, is completed after approximately 10 min, and is independent of the temperature. In contrast to Rilling (1964) and Howes et al. (1969), we therefore concluded that oxygen is not involved directly in the photochemical reaction of the photoreceptor but rather functions as an electron acceptor keeping the photoreceptor in a proper state of oxidation. It should be emphasized, however, that from the results so far available the possibility of different mechanisms in the presence or absence of oxygen cannot be ruled out.

3.2.3 Effect of Dithionite and H_2O_2

As mentioned before, dithionite inhibits the presumed flavin-mediated photo-induction in *Mycobacterium* sp.; corresponding results have been obtained for *F. aquaeductuum* (Fig. 5). Moreover in *Fusarium* dithionite prevents photo-induction specifically when applied to the mycelia before or immediately after illumination; the effect decreases gradually with the time elapsed between illumination and application of the drug. When added approximately 20 min after the onset of irradiation dithionite is completely ineffective. Essentially the same results were obtained when hydroxylamine was used as a reducing agent (Theimer and Rau 1970). This inhibitory effect acting after illumination needs a different interpretation than inhibition during the photoreaction. On the other hand incubation of the mycelia with buffered hydrogen peroxide solution in the dark substitutes, at least to a certain degree, for light in inducing carotenoid synthesis.

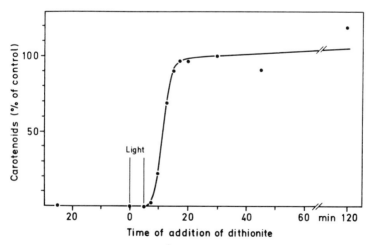

Fig. 5. Effect of dithionite (5×10^{-3} M) applied to mycelia of *Fusarium aquaeductuum* at various times before or after photo-induction on carotenogenesis. The inhibitor was removed 30 min after addition by rinsing the mycelia with buffer. (From Theimer and Rau 1970)

3.2.4 Artificial Photoreceptors

Biosynthesis of carotenoids is also induced when mycelia of *Fusarium* are incubated
with methylene blue, toluidine blue, or neutral red (dyes which all absorb red light),
and simultaneously are illuminated with red light that is ineffective in photo-induction
via the natural photoreceptor. In accord with the results found for photo-induction
with white light as a convenient and effective source for blue light, a logarithmic relationship between the amount of pigment synthesized and illumination time was observed
also with red light in the presence of methylene blue. Photo-induction by red light is
triggered only in the presence of these photodynamically active redox dyes, whereas
dichlorophenol-indophenol and malachite green show no effect (Lang-Feulner and Rau
1975). Although the question is still open whether the redox dyes are by-passing the
natural photoreceptor or whether they act as an additional receptor transferring the
energy to the natural photoreceptor, recent results obtained by Britz et al. (1979) tend
to support the former possibility.

3.2.5 Conclusions

For some other blue light-sensitive systems direct evidence for redox reactions as a consequence of irradiation has been accumulated in the last years; the results are presented
and discussed in another contribution in this book. Summarizing the data for photo-
induced carotenogenesis and also considering the results for the other phenomena we
may draw a hypothetical scheme illustrating the events during photo-induction via the
blue light photoreceptor (Fig. 6). In this scheme we assume that irradiation of the photoreceptor causes by its own reduction a concomitant oxidation of a yet hypothetical
"substance" X_{red}; the resulting X_{ox} is stabilized rapidly – in *Fusarium* presumably
within 20 min (Fig. 5) – by subsequent reactions yielding a "photo-oxidation product".
The reduced photoreceptor may be re-oxidized by transferring electrons either to a
cytochrome and finally to O_2 or directly to oxygen. For further induction steps two
"key substances" may be considered as "triggers": the or one of the reduced components
(cytochrome?) or the oxidized component (X_{ox}). Taking into account the data on the

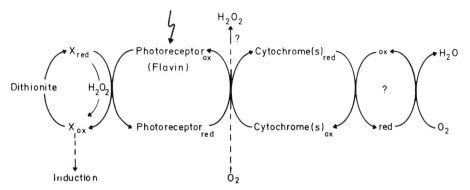

Fig. 6. Scheme illustrating the proposed events during the "primary reaction" of the "blue light-
photoreceptor"

effects of dithionite and H_2O_2 we consider a stabilized X_{ox} to be the more likely candidate.

3.3 "Chemoinduction"

In a few organisms blue light as inducing agent of carotenogenesis can be replaced by application of certain chemicals. In *F. aquaeductuum* incubation of the mycelium with p-chloro- or p-hydroxymercuribenzoate — both known as inhibitors of SH-groups — causes an abundant synthesis of carotenoids in the dark (Rau 1967b; Rau et al. 1967). Accumulation of pigment starts after a lag period of 4–6 h, increases until 24 h after addition of the compound and then continues for some days at a constant rate. Other SH-group blocking substances do not induce carotenoid synthesis; other mercuric compounds and benzoic acid are also ineffective. In *M. marinum*, which probably uses a porphyrin as photoreceptor, application of antimycin A, an inhibitor of electron transport, leads to carotenogenesis in the dark (Batra 1967; Batra et al. 1969, 1971); mercuribenzoate proved to be ineffective in this organism. The regulation of carotenoid biosynthesis in *Phycomyces blakesleeanus* is very complex including influence of mating, stimulation by blue light and also a "chemostimulation" to different extents by vitamin A or β-ionon (Eslava et al. 1974); it should be emphasized that these substances are not incorporated into the carotenoid molecules, and therefore the stimulating effect is not due to an additional supply of precursors.

In all three organisms, the effect of "chemo-induction" is additive to photo-induction under various experimental conditions tested (Batra 1971; Theimer and Rau 1972). Therefore, it seems to be doubtful that "chemo-induction" may contribute to the understanding of the photoreaction. However, "chemo-induction" has to be considered for the understanding of the further sequence in the induction mechanism of carotenogenic enzymes.

3.4 Regulatory Mutants

Mutants which carry genetic blocks for different steps in the biosynthetic pathway of carotenoids are well known, especially in *N. crassa* and *Phycomyces blakesleeanus*. For the elucidation of the mechanism of photoregulation another type of mutant, i.e., regulatory mutants, might be useful, that are either blocked in the photo-induction of carotenogenesis or able to synthesize carotenogenic enzymes without photo-induction. A mutant of *Phycomyces* has been described and examined which produces high amounts of β-carotene when grown in the dark and which therefore is supposed to be a regulatory mutant (Murillo et al. 1978). Using UV-irradiation of spores of *F. aquaeductuum* we obtained several mutant strains which synthesize more pigment in the dark than the wild type after illumination with the highest light dosages available (Theimer and Rau 1969). From additional data we concluded that these mutants have lost photoregulation and that carotenogenic enzymes have became constitutive. Recent reinvestigations indicate that in addition to the high rate of carotenoid production in the dark, photo-induction of minor amounts of pigments is still possible.

4 Acetylcholine as a Possible "Inducer"

One of the primary events in rapid phytochrome-mediated responses (i.e., leaf movements) was considered to be a change in the permeability of membranes. Jaffe (1970)
found that the mammalian neurohumor, acetylcholine, is also present in plant tissues
and that its level is controlled via phytochrome; subsequently it was shown that this
compound regulates a number of photomorphoses (for ref. see Briggs and Rice 1972;
Kandeler 1972). For the blue light-dependent sporulation of the fungus *Trichoderma*,
Gressel et al. (1971) reported on the substitution of acetylcholine for light; but the
acetylcholine esterase had to be inhibited by simultaneous application of eserin. We
therefore tested the effect of acetylcholine on the induction of carotenoid biosynthesis
in *F. aquaeductuum*. The substance was applied in the dark to mycelia (in submerged
cultures or on solid media) up to a concentration of 10 mmol/l both with or without
eserin (conc. up to 1 mmol/l). In the various combinations tested acetylcholine turned
out to be ineffective in inducing any pigment synthesis in the mycelia.

5 De Novo Synthesis of Carotenogenic Enzymes

As a consequence of photo-induction dark reactions are induced which finally lead to
an accumulation of carotenoids. To explain this accumulation two alternative mechanisms may be taken into consideration: "photoactivation" of preformed enzymes or
de novo synthesis of the enzyme molecules. Photoactivation of many enzymes – though
not carotenogenic enzymes – by blue light has been reported previously and will be
described and discussed in the contribution by Schmidt in this book. However, in all
organisms so far tested except *Flavobacterium dehydrogenans* pigment accumulation
after photo-induction starts only after a lag period.

5.1 Inhibitor Studies

Photo-induced carotenoid accumulation is completely blocked by cycloheximide, when
applied prior to or immediately after illumination of the fungi *F. aquaeductuum* (Rau
1967b; Bindl et al. 1970), *N. crassa* (Harding and Mitchell 1968; Rau et al. 1968; Harding 1974) and *Verticillium agaricinum* (Mummery and Valadon 1973); corresponding
results have been obtained using chloramphenicol with the *Mycobacteria* (Rilling 1962;
for ref. see Batra 1971) and *Flavobacterium dehydrogenans* (Weeks et al. 1973). From
these results it was concluded that light induces de novo production of carotenogenic
enzymes. Although the relevance of this kind of conclusion has been criticized (e.g.,
Schopfer 1977) additional data confirm this interpretation.

The inhibition of carotenogenesis in *Fusarium* by cycloheximide (Rau 1967b) and
in *M. marinum* by chloramphenicol (Batra 1967) is reduced with the time elapsed between illumination and addition of the inhibitor; application a few hours after the onset
of illumination, when the rate of pigment accumulation is highest, proved to be nearly
ineffective, indicating that the inhibitor does not interfere with the catalytic activity
of the enzymes.

5.2 Lag Phases

The kinetics of carotenoid accumulation following a brief illumination show (Fig. 1) that synthesis of pigments probably ceases after a certain time. These results indicate that the activity of carotenogenic enzymes is lost either due to inactivation or to degradation of the enzyme molecules. In phytochrome-mediated accumulation of anthocyanins (Lange et al. 1967) and also that of carotenoids in *Sinapis* seedlings (Schnarrenberger and Mohr 1970) no lag phase was found after a second illumination. The lack of a second lag phase observed for several different photoresponses seems to exclude that a second illumination leads again to de novo synthesis of enzymes. In mycelia of *Fusarium* and *Neurospora* (Rau and Rau-Hund 1977) the time courses of the carotenoid accumulation after a brief first and a second illumination are identical, except for the velocity of the reaction; therefore only the results for *Neurospora* are shown in Fig. 7. Renewed pigment accumulation after a second light treatment is similar to that after a first illumination period since it starts only after a lag phase. Such a second lag phase was not only observed when carotenoid synthesis induced by the first illumination is completed, but also when the second light treatment was given during the period of rapid pigment accumulation after a first illumination. Furthermore, the realization of the second photo-induction can also be blocked by cycloheximide; the gradual loss of inhibitory power with time occurs in an identical pattern after both the first and the second illumination (Fig. 8).

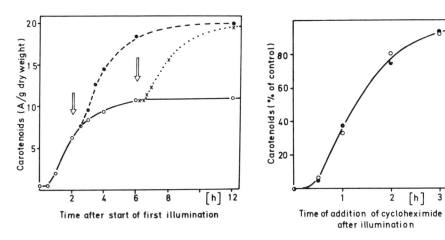

Fig. 7   Fig. 8

Fig. 7. The course of carotenoid accumulation in *Neurospora crassa* after first and second illumination (10 min, white light). O——O first illumination only; ●——● second illumination after 2 h; x. . .x second illumination after 6 h. *Arrows* indicate time of second illuminations. (From Rau and Rau-Hund 1977)

Fig. 8. Inhibition of carotenoid synthesis in *Fusarium aquaeductuum* by cycloheximide (6 · 10⁻⁵ M), applied various times after start of first (O) or second (●) illumination. Amount of pigments was estimated after 36 h of incubation; control: no cycloheximide. (From Rau and Rau-Hund 1977)

5.3 Cell-Free Systems for Carotenoid Biosynthesis

For direct evidence that irradiation induces de novo synthesis of carotenogenic enzymes absent in dark-grown organisms both the preparation of crude or purified enzyme solutions and a functional in vitro assay system for the accurate estimation of enzyme activities are required. Unfortunately only few cell-free systems from different organisms have yet been established and moreover, in the organisms which served as sources for the cell-free systems (e.g., tomato plastids or *Phycomyces*) biosynthesis of carotenoids is not strictly photoregulated (Yokoyama et al. 1962; Porter 1969; Bramley and Davies 1975; Kushwaha et al. 1976). The only cell-free system derived from an organism with strict photoregulation so far examined in detail is that from *Mycobacterium* sp. (Johnson et al. 1974); with this system the authors demonstrated that in homogenates of dark-grown cells the activity of the enzyme required for the conversion of geranylgeranyl-pyrophosphate to prephytoene-pyrophosphate is totally absent and that this enzyme is probably synthesized de novo in photo-induced cells. Furthermore, although low activity of geranylgeranyl-pyrophosphate synthetase present in dark-grown cultures was increased substantially after light treatment of intact cells, the activity of isopentenyl-pyrophosphate isomerase was not changed. Only recently Spurgeon et al. (1979), as well as our group (Mitzka-Schnabel and Rau 1978, and to be published), elaborated a cell-free system obtained from *N. crassa,* that converts precursors to phytoene and, to a much lower extent, to carotenoids.

The first step of the biosynthetic pathway of carotenoids that is under photoregulation is still a matter in dispute. From earlier investigations using *N. crassa* it has been assumed that only enzymes following phytoene in the biosynthetic pathway are regulated by light. However, evidence has recently accumulated that preceding reactions are photoregulated, although phytoene is synthesized to some extent also in dark-grown cultures. This problem has been reviewed elsewhere in more detail (Rau 1976).

5.4 "Organization" of the Carotenogenic Enzyme System

From genetic and physiological results on the carotene biosynthesis in *Phycomyces,* Cerda-Olmedo deduced a model for the "organization" of the carotenogenic enzymes (for ref. see Cerda-Olmedo and Torres-Martinez 1979). In this model it is assumed that the enzymes — dehydrogenases and cyclases — are organized in an enzyme assembly; one might also use the terms "enzyme aggregate" or "multi-enzyme complex" for this form of organization. Since the author presents the results and conclusions in his own article in this book, only two additional results should be pointed out briefly. Our recent data using a cell-free system from *N. crassa* (Mitzka-Schnabel and Rau 1978, and to be published) show that activity of carotenogenic enzymes is enriched in membrane fractions indicating that the enzymes are membrane-bound; this result might support the model of an "enzyme aggregate". On the other hand when cycloheximide is applied to mycelia of *Fusarium* different times after irradiation and subsequently sufficient time is given to complete carotenoid synthesis in the dark, the individual amount of each of the different carotenoids produced may be taken as a relative indicator of the particular enzyme activity already present at the time of addition of the inhibitor. The results of such an experiment (Fig. 9) show that the different enzyme activities of the

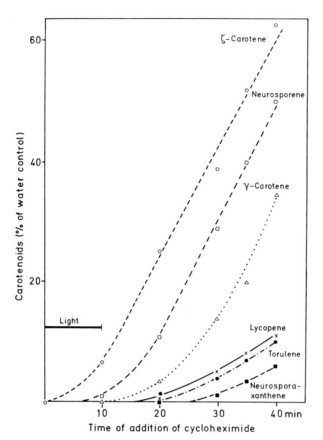

Fig. 9. The amounts of individual carotenoids synthesized in *Fusarium aquaeductuum* during a 36-h incubation as a function of time of addition of cycloheximide after illumination. (From Bindl et al. 1970)

biosynthetic pathway appear in sequence; whether such sequential enzyme formation as well as the differential inhibition by diphenylamin of the carotenoid synthesis in *Fusarium* (Bindl et al. 1970) contradict the concept of an enzyme aggregate has to await further examination.

5.5 Conclusion

In summary, it can be stated that the results so far available are consistent with the assumption that as a consequence of photo-induction the enzymes of the biosynthetic pathway of carotenoids are produced de novo. However, for unequivocal evidence isolation, biochemical characterization and direct proof of enzyme synthesis in undoubtedly necessary.

6 The Level and Possible Mechanism of Photo-Induction of Carotenogenic Enzymes

As mentioned before, photo-induced pigment synthesis is completely blocked by cyclo-heximide applied prior to illumination. However, in *Fusarium* this inhibition is reversible: cycloheximide was removed by rinsing the mycelium at various times after illumination during the subsequent dark period, then permitting sufficient time for completion of carotenoid synthesis; thereafter the amount of pigment was estimated (Fig. 10). From these results we drew two conclusions: (1) Even under conditions when protein synthesis is blocked photo-induction takes place; this is demonstrated by the carotenoid production in the dark simply by removing the block. Therefore, irradiation must induce the formation of an "induction product" which ranges prior to translation and enzyme synthesis in the sequence of photo-induced reactions. (2) The "induction product" appears to be rather stable, which is indicated by the finding that after withdrawal of cycloheximide the same amount of carotenoids was synthesized as had been produced by mycelia that were illuminated after the treatment with the inhibitor (cf. both types of point in Fig. 10).

Does this hypothetical "induction product" regulate the synthesis of carotenogenic enzymes at the transcription or at the translation level? Is there any evidence for the synthesis of specific messenger-RNA via gene de-repression? Since this problem will be discussed in a separate article (Gressel, this vol.) for blue light-regulated phenomena and has also been recently reviewed in detail for phytochrome-mediated responses (Schopfer 1977), the following report is restricted to a brief summary of results and conclusions derived from investigations on photo-induced carotenogenesis in microorganisms.

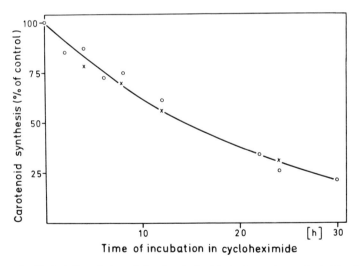

Fig. 10. Stability of photo-induction during blocked protein synthesis (O--O): Cycloheximide applied prior to illumination, washed out at different times prior (control sample) or after illumination, and then amount of carotenoids determined after completion of pigment accumulation. "Ability for synthesis" of the mycelium (x—x): Illumination only after removing cycloheximide at different times. (From Rau 1971)

Inhibition of carotenoid biosynthesis by compounds known as inhibitors of transcription has been reported for a few organisms. In *Mycobacterium* sp. proflavin was found to inhibit light-mediated pigment production; investigations with *Fusarium* yielded results indicating that the effect of proflavin is unspecific (Schrott 1973). Actinomycin D prevented synthesis of carotenoids in *Flavobacterium dehydrogenans* (Weeks et al. 1973) but in *Verticillium agaricinum* (Mummery and Valadon 1973) and in *N. crassa* (Subden and Bobowski 1973) it inhibits photo-induced pigment production only partly. A few years ago we succeeded in finding a potent inhibitor of carotenogenesis in *Fusarium;* incubation of the mycelia with distamycin A, added prior to illumination, in concentration of $3 \cdot 10^{-4}$ mol/l completely blocked light-induced carotenoid formation. The specificity of the effect of distamycin A was tested by examination of the incorporation of radioactively labeled uridine and leucine (Felbermeir and Rau 1974; Jochum-Felbermeir and Rau, in prep.). Very recently we found an antimetabolite of RNA synthesis — tubercidine, an analog of adenosine produced by *Streptomyces tubercidicus* — which also completely inhibits carotenoid synthesis in *Fusarium.*

Direct evidence for an involvement of de novo synthesis of mRNA in photoregulated biosynthesis of carotenoids has also been obtained from investigations with *Fusarium* (Schrott and Rau 1975, 1977). Using short periods of illumination, labeling with specific precursors and subsequent affinity chromatography of double-labeled polysomal RNA on poly(U)-sepharose we were able to demonstrate different ^{3}H/^{14}C ratios for the adsorbed poly(A)RNA-containing fraction when compared to the unbound fraction: when ^{3}H-uridine was present in the light-induced sample and ^{14}C-uridine in the dark control the ^{3}H/^{14}C ratio in the bound material was found to be higher than that of the unbound fraction, and vice versa. No such shift was observed when both sample and control were kept in darkness.

These results and the data from the inhibitor experiments are consistent with the hypothesis that photo-induction of carotenogenesis, at least in fungi, is regulated at the level of transcription. However, unequivocal evidence in support of this hypothesis as well as proof for the proposal that photoregulation might either cause the inactivation of a repressor or the production of an inducer (Rilling 1964; Batra 1971), requires further detailed investigations.

Acknowledgments. I thank most warmly all co-workers who have contributed ideas and results to work from our laboratory and also all colleagues for helpful and stimulating discussions. In particular I should like to mention Prof. Dr. Theimer, Dr. Schrott, and my wife Dr. Rau-Hund, and thank them also for their help in preparing this manuscript. My thanks are also due to Deutsche Forschungsgemeinschaft for generous financial support of our work.

References

Batra PP (1967) J Biol Chem 242: 5630–5635
Batra PP (1971) In: Giese AC (ed) Photophysiology, vol VI, Mechanism of light induced carotinoid synthesis in non-photosynthetic plants, pp 47–76
Batra PP, Rilling HC (1964) Arch Biochem Biophys 107: 485–492
Batra PP, Gleason Jr RM, Jenkins J (1969) Biochim Biophys Acta 177: 124–135
Batra PP, Harbin TL, Howes CD, Bernstein SC (1971) J Biol Chem 246: 7125–7130
Bindl E, Lang W, Rau W (1970) Planta 94: 156–174

Bramley PM, Davies BH (1975) Phytochemistry 14: 463–465
Briggs WR, Rice HV (1972) Annu Rev Plant Physiol 23: 293–334
Britton G (1976) Pure Appl Chem 47: 223–236
Britz SJ, Schrott E, Widell S, Briggs WR (1979) Photochem Photobiol 29: 359–362
Burchard RP, Hendricks SB (1969) J Bacteriol 97: 1165–1168
Carlile MJ (1956) J Gen Microbiol 14: 643–654
Cerda-Olmedo E, Torres-Martinez S (1979) Pure Appl Chem 51: 631–637
Davies BH (1980) In: Czygan F-C (ed) Pigments in plants. Fischer, Stuttgart
Davies BH, Taylor RF (1976) Pure Appl Chem 47: 211–221
DeFabo EC, Harding RW, Shropshire Jr W (1976) Plant Physiol 57: 440–445
Eslava AP, Alvarez MI, Cerda-Olmedo E (1974) Eur J Biochem 48: 617–623
Felbermeir M, Rau W (1974) Zusammenfassung der Vorträge. Dtsch Bot Ges Ver Angew Bot Tag Würzburg, p 147
Goodwin TW (1971) In: Isler O (ed) Carotenoids, pp 577–636. Birkhäuser-Verlag, Basel Stuttgart
Goodwin TW (1979) Annu Rev Plant Physiol 30: 369–404
Gressel J, Strausbauch L, Galun E (1971) Nature (London) 232: 648
Harding RW (1974) Plant Physiol 54: 142–147
Harding RW, Mitchell HK (1968) Arch Biochem Biophys 128: 814–818
Howes CD, Batra PP (1970) Arch Biochem Biophys 137: 175–180
Howes CD, Batra PP, Blakeley CF (1969) Biochim Biophys Acta 189: 298
Isler O (ed) (1971) Carotenoids. Birkhäuser Verlag, Basel Stuttgart
Jaffe MJ (1970) Plant Physiol 46: 768–777
Johnson JH, Reed BC, Rilling HC (1974) J Biol Chem 249: 402–406
Kandeler R (1972) Fortschr Bot 34: 211–226
Kushwaha SC, Kates M, Porter JW (1976) Can J Biochem 54: 816–823
Lange H, Bienger I, Mohr H (1967) Planta 76: 359–366
Lang-Feulner J, Rau W (1975) Photochem Photobiol 21: 179–183
Matthews MM (1963) Photochem Photobiol 2: 1–8
Mitzka-Schnabel U, Rau W (1978) In: Abstr Contrib Pap 5th Int Symp Carotenoids, p 41
Mummery RS, Valadon LRG (1973) Physiol Plant 28: 254–258
Murillo FJ, Calderon IL, Lopez-Diaz I, Cerda-Olmedo E (1978) Appl Environ Microbiol 36: 639–642
Porter JW (1969) Pure Appl Chem 20: 449
Rau W (1962) Planta 59: 123–137
Rau W (1967a) Planta 72: 14–28
Rau W (1967b) Planta 74: 263–277
Rau W (1969) Planta 84: 30–42
Rau W (1971) Planta 101: 251–264
Rau W (1975) Ber Dtsch Bot Ges 88: 45–60
Rau W (1976) Pure Appl Chem 47: 237–243
Rau W, Rau-Hund A (1977) Planta 136: 49–52
Rau W, Feuser B, Rau-Hund A (1967) Biochim Biophys Acta 136: 589–590
Rau W, Lindemann I, Rau-Hund A (1968) Planta 80: 309–316
Rilling HC (1962) Biochim Biophys Acta 60: 548–556
Rilling HC (1964) Biochim Biophys Acta 79: 464–475
Schnarrenberger C, Mohr H (1970) Planta 94: 296–307
Schopfer P (1977) Annu Rev Plant Physiol 28: 223–252
Schrott EL (1973) Ph D Thesis, Univ München
Schrott EL, Rau W (1975) Ber Dtch Bot Ges 88: 233–243
Schrott EL, Rau W (1977) Planta 136: 45–48
Seviour RJ, Codner RC (1973) J Gen Microbiol 77: 403–415
Spurgeon SL, Turner RV, Harding RW (1979) Arch Biochem Biophys 195: 23–29
Subden RE, Bobowski G (1973) Experientia 29: 965–967
Theimer RR, Rau W (1969) Biochim Biophys Acta 177: 180–181
Theimer RR, Rau W (1970) Planta 92: 129–137

Theimer RR, Rau W (1972) Planta 106: 331–343
Trinci APJ, Banbury GH (1969) Trans Br Mycol Soc 52: 1–14
Valadon LRG, Mummery R (1971) Microbios 4: 227–240
Valadon LRG, Osman M, Mummery RS (1979) Photochem Photobiol 29: 605–607
Weeks OB, Saleh FK, Wirahadikusumah M, Berry RA (1973) Pure Appl Chem 35: 63–80
Yokoyama H, Nakayama TOM, Chichester CO (1962) J Biol Chem 237: 681–686
Zalokar M (1954) Arch Biochem Biophys 50: 71–80
Zalokar M (1955) Arch Biochem Biophys 56: 318–325

Photokilling and Protective Mechanisms in *Fusarium aquaeductuum*

A. HUBER[1] and E.L. SCHROTT[1]

1 Introduction

Light and air are the dominant factors affecting organisms in their natural environment. Much work has been done to elucidate their influence on plants and microbes.

In some nonphotosynthetic bacteria and fungi blue and near-UV light (UV-A) induces carotenogenesis, as illustrated by the action spectra, e.g., for *Fusarium aquaeductuum* (Rau 1967a). Depending on the energy fluence rate light of the same wavelengths inducing carotenogenesis is also capable of reducing viability and even causing destruction of the organism; this has been called "photokilling" (for refs. see Webb 1977). Both optimum carotenoid production as well as photokilling require the presence of oxygen. The protective role of carotenoids has already been demonstrated with *Corynebacterium poinsettiae* (Kunisawa and Stanier 1958; for further ref. see Krinsky 1976). For fungi it has so far been demonstrated only in single cells of *Sporodiobolus johnsonii* (Goldstrohm 1964) and *Dacryopinax spathularia* (Goldstrohm and Lilly 1965). Blanc et al. (1976) investigated the harmful effects of visible light, UV-A and far-UV on suspensions of *Neurospora crassa* conidia and concluded that carotenoids can protect conidia against UV-A and visible light. In a most recent paper of Ramadan-Talib and Prebble (1978) the photosensitivity of respiration in *Neurospora* mitochondria and the protective role of carotenoids were tested. Hence it was of interest to examine the effect of light on mycelia and the protective function of carotenoids against injurious effects of irradiation.

2 Results

2.1 Loss of Ability for Carotenoid Production Caused by UV-A Treatment

2.1.1 Materials and Methods

The same strain of *Fusarium aquaeductuum* was used as in previous investigations (Rau 1967a; Schrott and Rau 1977). The growth medium was inoculated with 25 ml of exponential phase cultures and grown for 18 h; accommodation to the minimal medium was reduced to 30 min unless stated otherwise. Samples were illuminated using banks of fluorescent tubes (Osram 65 W/19, providing 40 W m^{-2} within the spectral range of 400–520 nm) for white light illumination and black light fluorescent tubes (Osram L

1 Botanisches Institut der Universität, Menzinger Straße 67, 8000 München 19, FRG

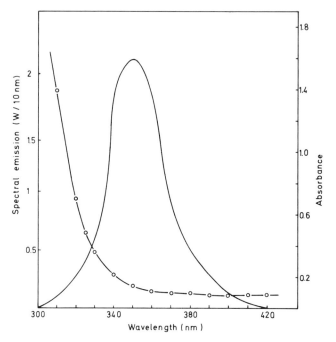

Fig. 1. Spectral emission of black light lamps (Osram L 40 W/73; *solid line*) and the spectral absorption of the glass of gas wash bottles used in the experiments (O—O)

40 W/73, 56 W m^{-2}) for UV-A irradiation. The irradiance was varied by grids and was measured with a Quantum spectrometer (QSM – 2500, Techtum Instrument, Umea, Sweden) or a Thermopile (Type E 1, Kipp, Delft, Holland) connected to a Multiflex Galvanometer "MGF O" (Lange, Berlin).

The spectral emission of the black light lamps and the absorption of the gas wash bottles used for illumination of the submerged mycelia are shown in Fig. 1.

A flow of oxygen or air (4 l/h) was passed through the samples (1 g fresh weight/ 100 ml) during exposure to light or during dark incubation, respectively. Samples were kept at $22° \pm 2°C$. Extraction and quantitative determination of carotenoids were performed as described elsewhere (Rau 1967a). Data presented are the average of at least three independent experiments.

2.1.2 *Effect of Light Source and Intensity*

Dose response curves of carotenoid production induced by different light sources and energy fluence rates are presented in Fig. 2 indicating that white light and low fluence rates of UV-A are able to induce carotenogenesis proportional to the incident energy, whereas maximum fluence rate of UV-A is effective only when short illumination periods are used; continued irradiation causes the mycelium to lose its capability of pigment production rather quickly. Formation of conidia can account for the presence of a basic amount of pigment although respiration is blocked completely as shown in

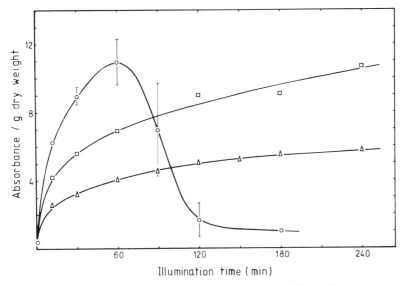

Fig. 2. Dose response curves for carotenogenesis induced by different light sources and fluence rates
○—○ UV-A (56 W m^{-2}); △—△ UV-A (1 W m^{-2}); □—□ white light (40 W m^{-2})

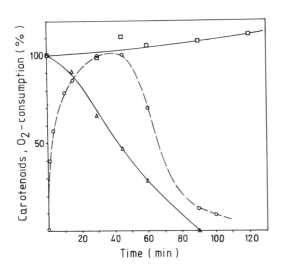

Fig. 3. Oxygen consumption of mycelia of *Fusarium* at the end of the illumination period (△—△; dark control: □—□), and ability for carotenoid production during the following 24 h dark period (○— — —○). Mycelia were cultivated for 40 h prior to harvesting

Fig. 3. The negligible growth rate of the fungus on agar plates after treatment with UV-A would indicate photokilling of the organism.

The loss of ability for carotenoid production occurs only when oxygen is present and is insignificant in a nitrogen atmosphere. No exogeneous sensitizer is required. For artificial light sources maximum energy fluence rate of black light is necessary for apparent photokilling. No killing was observed with low fluence rates of black light (1 W m^{-2}) or white light (40 W m^{-2}) even after an illumination period of 24 h. On the other hand bright sunlight causes photokilling within 30—60 min, but as yet no statement

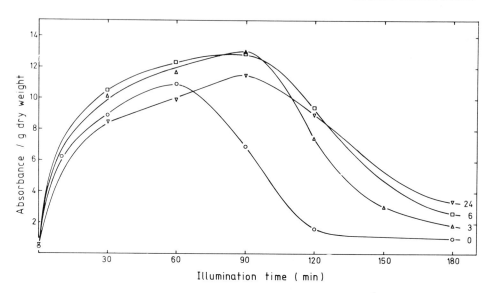

Fig. 4. Dose response curves for carotenogenesis induced by UV-A (56 W m^{-2}) following different dark periods between harvesting and onset of light. Dark periods elapsed: ○—○ 0 h; △—△ 3 h; □—□ 6 h; ▽—▽ 24 h

can be made whether the UV-A portion of the sunlight is responsible for killing and/or the high fluence rates of other spectral regions. Wavelength dependence of photoinhibition of transport in *E. coli* (Sprott et al. 1976) and the action spectrum for inactivation of *E. coli* cells (Webb and Brown 1976) demonstrate that wavelengths up to 650 nm may still be effective.

2.1.3 Influence of the Physiological State of the Organism

As described above, photokilling is competing with photoinduction and therefore leads to an optimum curve regarding the amount of carotenoids formed. The position of the peak depends on the physiological age of the organism (Fig. 3), i.e., sensitivity increases with culture time.

As shown in Fig. 4, peak position also depends on the time elapsed between harvesting and onset of light. After an intervening dark incubation period of 3 h the peak position is shifted from 60 min to 90 min of irradiation time; identical results are obtained when the incubation period is 2 h (data not shown). Dark incubation periods exceeding 3 h do not alter the position of the optimum for carotenoid induction even though resistance to irradiation still increases slightly. Furthermore, it is found that with increasing resistance against the injurious effect of irradiation, carotenoid synthesis also increases except for the 24 h incubation period. We assume that during the dark incubation period the cells are recovering from the stress caused by the harvesting procedure and accommodate to the minimal medium which does not permit further growth due to the lack of nitrogen compounds.

2.2 Photokilling in Terms of Inhibition of Leucine Incorporation and Membrane Damage

2.2.1 Materials and Methods

For pulse experiments the amount of mycelia was reduced to 150 mg fresh weight/ 100 ml of buffered glucose solution containing 100 μg/ml chloramphenicol and 0.1 mmol/l of unlabeled L-leucine. After 30 min of adaptation 2.5 μCi (= 9.25×10^4 s^{-1}) of L-[U-^{14}C]leucine (354 mCi/mmol = 1.309×10^{10} s^{-1} mmol^{-1}) were added to each sample. Incubation was continued for additional 30 min and then the mycelia was collected by vacuum filtration, rinsed and resuspended in buffered glucose for 5 min, collected and resuspended again in the medium omitting leucine. One third of this suspensuion was kept in the dark throughout the experimental period, the other parts were illuminated continuously either with white light (40 W m^{-2}) or UV-A (56 W m^{-2}).

For determination of the trapped (i.e., uptake plus incorporation) leucine 5 ml aliquots were withdrawn, collected on Whatman GF/C filter discs, and rinsed thoroughly with cold deionized H$_2$O. Incorporation was determined by precipitating a 5 ml aliquot with 5 ml of cold 20% TCA (w/v) 5 min prior to collection on filter discs, rinsed twice with 10 ml of cold 10% TCA (w/v) followed by 10 ml of cold 70% EtOH (v/v). The discs were dried and the radioactivity measured as described previously (Schrott and Rau 1977).

2.2.2 Results

The effect of UV-A and white light on trapped or incorporated leucine compared to the dark control is demonstrated by Fig. 5. The amount of incorporation is reflected by the data obtained after precipitation of the protein by TCA. It is inhibited to some extent by illumination with white light and is blocked completely upon irradiation with UV-A. The amount of trapped leucine represented by the results which were obtained when H$_2$O was used as a collecting medium remains unaffected for approximately 45 min. But when incubation is continued efflux of labeled leucine was observed that depends on the type of treatment: time course of the efflux under white light coincides with that of the dark control, whereas for UV-A treated samples, non-incorporated leucine is released completely by the cell within the following 75 min. This indicates that the diffusion barrier must have been removed, presumably reflecting membrane damage.

2.3 Protecting Systems Against Photokilling

2.3.1 Carotenoids

Experiments were designed to investigate whether the carotenoids present in *Fusarium aquaeductuum* are able to protect the organism against photokilling. As this organism is photochromogenetic (Eberhard et al. 1961) we are able to produce mycelia containing different amounts of carotenoids by varying the lengths of pre-illumination time followed by a constant dark period, or by using a constant pre-illumination time followed by different periods of dark incubation. The pigmented mycelia is then exposed to the high energy fluence rate of UV-A which is capable of causing photokilling.

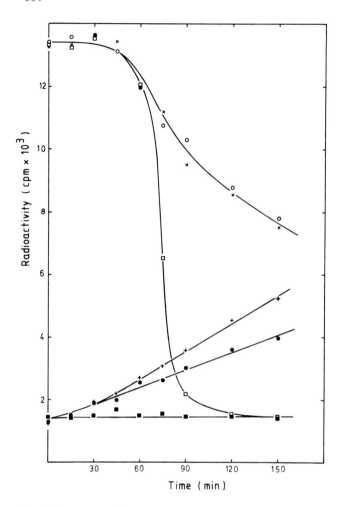

Fig. 5. Time course of incorporation and release of trapped leucine as a function of the treatment during incubation. Incorporated leucine: +——+ dark control; ●——● white light (40 W m^{-2}); ■——■ UV-A (56 W m^{-2}). Trapped leucine: x——x dark control; O——O white light (40 W m^{-2}); □——□ UV-A (56 W m^{-2})

Figure 6 shows the results from experiments carried out to compare the susceptibility of pigmented mycelia with that of nonpigmented. Two hours after harvesting, initial illumination with white light (40 W m^{-2}) for 1, 30, and 100 min, followed by 4 h dark incubation, produced mycelia that contained different amounts of carotenoids at the start of UV-A irradiation (56 W m^{-2}). The results demonstrate that in pigmented mycelia the ability for carotenoid synthesis is retained even after extended periods of irradiation, as indicated by the shift of the peak position. When irradiation is continued the extent of photokilling is inversely proportional to the illumination time used for providing the protection potential.

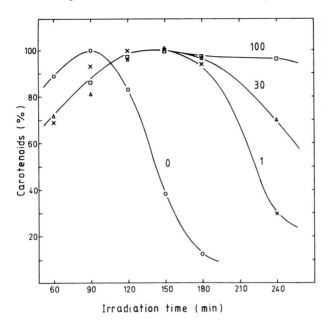

Fig. 6. Dose response curves for carotenogenesis induced by UV-A irradiation (procedure described in the text) as a function of carotenoid content at the start of irradiation. Time of initial illumination: x——x 1 min; △——△ 30 min; □——□ 100 min; ○——○ dark control

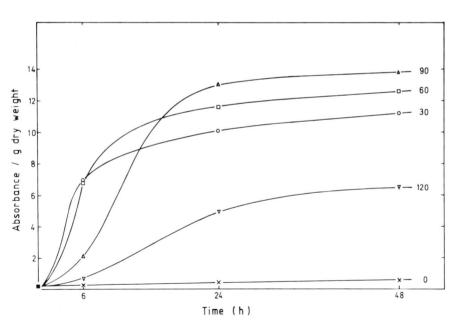

Fig. 7. Time course of accumulation of carotenoids induced by different UV-A irradiation periods, following 2 h of adaptation. ○——○ 30 min; □——□ 60 min; △——△ 90 min; ▽——▽ 120 min; x——x dark period

2.3.2 "Repair System"

A second protecting system is proposed partly based on the following experimental data (Fig. 7): The time course of carotenoid accumulation after induction with different UV-A irradiation periods shows, as one would expect, that increasing length of irradiation time results in enhanced pigment formation (except for 120 min). But concomitantly the rate of synthesis is remarkably reduced, predominantly during the initial phase. This indicates that mycelia first hampered in carotenoid production are able to recover with increasing incubation time in the dark.

Comparison of the dose response curve of carotenoid induction with the time course of the inhibition of respiration by continuous UV-A irradiation (see Fig. 3) shows that mycelia produce maximum amounts of carotenoids in the subsequent dark period even though respiration is reduced to 50% at the end of the irradiation period. This could also indicate recuperation of the organism during the dark phase. But stronger evidence for a repair process such as measurements of the recovering rates of respiration during the dark incubation period has yet to be obtained.

3 Discussion

The data clearly demonstrate that in *Fusarium aquaeductuum* the effect of white and near-UV light not only triggers induction of carotenoid biosynthesis but also inhibits physiological processes and even causes damage of cell structures.

Upon illumination mycelia respond to increasing irradiation times with the synthesis of increasing amounts of carotenoids. We assume that the harmful effect of light, although present (Fig. 5), is not visibly apparent on carotenoid synthesis as long as energy fluence rate is relatively low. High energy fluence rate of UV-A (56 W m^{-2}) as well as bright sunlight are far more effective in the induction of carotenogenesis which, however, depends on the developmental stage of the cell population used for the experiment at least when UV-A is used (Figs. 2 and 3). Induction also is effected by the time elapsed between harvesting and the onset of light (Fig. 4) and on pre-illumination (Fig. 6).

Extension of the irradiation time leads to the loss of ability for carotenoid formation during the subsequent dark period. Therefore, the dose response curve is the result of two counteracting processes that lead to an optimum curve. When irradiance is sufficiently low, no optimum curve was yet observed even when energy fluence was as high as 7×10^5 J m^{-2} for *Fusarium aquaeductuum* (Theimer 1968) or 5×10^6 J m^{-2} for *Fusarium oxysporum* (Schrott, this vol.).

On the basis of our observations indicating an immediate inhibition of leucine incorporation upon irradiation (Fig. 5), a decrease in respiration (Fig. 3) and an initial delay in carotenoid synthesis during the subsequent dark period (Fig. 7), as well as the findings of Blanc et al. (1976) and of Ramadan-Talib and Prebble (1978), we propose one or more "repair systems" capable of restoring photodamaged cell components. When energy fluence rate (or energy fluence) is low, the "repair system" is efficient enough to prevent photokilling. The increasing susceptibility to light with age may then be explained by the decreasing physiological activity that includes, of course, the activity of the "repair system(s)". In contrast to *Fusarium,* susceptibility to light of *E. coli* decreases with age (Peak 1970).

The resistance of mycelia to the injurious effect of light is enhanced mainly by pre-illumination. Since the time of pre-illumination is correlated to the amount of carotenoids present within the cells at the start of the UV-A irradiation one may conclude that carotenoids play a protective role (for ref. see Krinsky 1976).

Apart from lipid globules, carotenoids of *Neurospora* are mainly localized in membrane fractions (Mitzka-Schnabel and Rau 1980). Preliminary results indicate that in *Fusarium* the subcellular distribution of carotenoids seems to be essentially the same (Schrott, unpublished data). Since irradiation also seems to cause membrane damage (Fig. 5) carotenoids may also act as a stabilizer for membranes. As an alternative, the possibility cannot yet be ruled out that "repair system(s)" is induced or activated by light, which thus may, at least partly, be responsible for the protection effects observed.

Since the resistance of mycelia to photokilling events depends on the physiological state of the cells (apart from the relevance of this fact concerning the efficacy of "repair"), the content or distribution of (UV-A) light-absorbing compounds other than carotenoids could vary with the developmental stages. Such compounds could be endogenous sensitizers or protective molecules which are increased/decreased quantitatively or become more/less effectively distributed in the cells with age, respectively.

In summary, *Fusarium* mycelia are susceptible to photodestruction. For protection from these lethal events the organism appears to activate at least two different protection mechanisms, namely carotenoids and "repair system(s)".

Acknowledgments. This work was supported by grants of the Deutsche Forschungsgemeinschaft to Dr. W. Rau. We are indebted to Dr. W. Rau for generous help and encouragement. The skilful technical assistance of Miss T. Driver as well as her and Dr. R.R. Theimer's support in preparing the manuscript is gratefully acknowledged.

References

Blanc PL, Tuveson RW, Sargent ML (1976) Inactivation of carotenoid-producing and albino strains of *Neurospora crassa* by visible light, blacklight, and ultraviolet radiation. J Bacteriol 125: 616–625

Eberhard D, Rau W, Zehender C (1961) Über den Einfluß des Lichts auf die Carotenoidbildung von *Fusarium aquaeductuum.* Planta 56: 302–308

Goldstrohm DD (1964) The effect on intense light on pigmented and nonpigmented cells of *Sporidiobolus johnsonii.* Proc West Va Acad Sci 36: 17–21

Goldstrohm DD, Lilly VG (1965) The effect of light on the survival of pigmented and nonpigmented cells of *Dacryopinax spathularia.* Mycologia 57: 612–623

Krinsky NI (1976) Cellular damage initiated by visible light. In: Gray TGR, Postgate JR (eds) The survival of vegetative microbes, pp 209:239. Cambridge University Press, Cambridge

Kunisawa R, Stanier RY (1958) Studies on the role of carotenoid pigments in a chemoheterotrophic bacterium, *Corynebacterium poinsettiae.* Arch Microbiol 31: 146–156

Mitzka-Schnabel U, Rau W (1980) The subcellular distribution of carotenoids in *Neurospora crassa.* Phytochemistry in press

Peak MJ (1970) Some observations on the lethal effects of near-ultraviolet light on *Escherichia coli,* compared with the lethal effects of far-ultraviolet light. Photochem Photobiol 12: 1–8

Ramadan-Talib Z, Prebble J (1978) Photosensitivity of respiration in *Neurospora* mitochondria. A protective role for carotenoid. Biochem J 176: 767–775

Rau W (1967a) Untersuchungen über die Lichtabhängige Carotinoidsynthese. I. Das Wirkungsspektrum von *Fusarium aquaeductuum.* Planta 72: 14–28

Schrott EL, Rau E (1977) Evidence for a photoinduced synthesis of poly(A) containing mRNA in *Fusarium aquaeductuum*. Planta 136: 45–48

Sprott GD, Martin WG, Schneider H (1976) Differential effects of near-UV and visible light on active transport and other membrane processes in *Escherichia coli*. Photochem Photobiol 24: 21–27

Theimer RR (1968) Untersuchungen zum Wirkungsmechanismus von Licht und Mercuribenzonat bei der Carotinoidsynthese in *Fusarium aquaeductuum*. Ph D Thesis, Universität München

Webb RB (1977) Lethal and mutagenic effects of near-ultraviolet radiation. In: Smith KC (ed) Photochemical and photobiologic reviews, vol II, pp 169–261. Plenum Press, New York

Webb RB, Brown MS (1976) Sensitivity of strains of *Escherichia coli* differing in repair capability for far UV, Near UV and visible radiations. Photochem Photobiol 24: 425–432

Dose Response and Related Aspects of Carotenogenesis in *Neurospora crassa*

E.L. SCHROTT[1]

1 Introduction

For red light, biphasic dose response curves have been reported for the growth of both *Avena* mesocotyl and coleoptiles (for ref. see Blaauw et al. 1968), for anthocyanin synthesis in mustard seedlings (Blaauw-Jansen 1974) and for the suppression of mesocotyl elongation of maize seedlings (Vanderhoef and Briggs 1977). For blue light effects such biphasic dose response relationships are only scarcely documented in the literature. Most recently Jayaram et al. (1979) claim a biphasic dose response curve for light-induced carotenogenesis in *Phycomyces*. Another example seems to be the phototropic curvature of *Avena* coleoptiles described by Zimmerman and Briggs (1963), if one interpolates the curves obtained by irradiation with low intensity blue light of 1.4 and 0.14 $pEcm^{-2}s^{-1}$ following red light pretreatment.

Biosynthesis of carotenoids in *Neurospora* is photoregulated. In the dark only a small amount of carotenoids is produced. The photoinduction is independent of the temperature, while formation of carotenoids in the subsequent dark reaction is temperature-sensitive (for further details on the mechanism of photo-induced carotenogenesis in microorganisms see Rau, this vol.).

According to the results of Zalokar (1955) and of Rau et al. (1968) the photoreaction of *Neurospora* is saturated at low dosages. In contrast, dose response curves for induction of carotenoid synthesis in *Fusarium aquaeductuum* (Theimer 1968), *F. coeruleum* (Schrott, unpublished data) and in *F. oxysporum* (data presented in Fig. 1) do not show saturation up to an energy fluence of 5×10^6 J m^{-2} and, in addition, are biphasic.

Harding (1974) has shown that the amount of pigment produced by *Neurospora* is increased further when long irradiation times are used, i.e., the period of irradiation within a 24-h incubation is varied from 0 to 24 h. But this type of experiment only yields an equivocal dose response curve. Since both dose response curves published cover only short illumination times and, on the other hand, the second phase of the dose response curves of the *Fusarium* species is only observed when illumination times are longer than approximately 15 min, we tried to complete the *Neurospora* dose response picture as far as carotenoid synthesis in the wild type is concerned. At this point it should be mentioned that in the band strain of *Neurospora* a second blue light-mediated phenomenon was observed, namely the inhibition of periodic formation of conidia (Sargent and Briggs 1967).

1 Botanisches Institut der Universität, Menzinger Straße 67, 8000 München 19, FRG

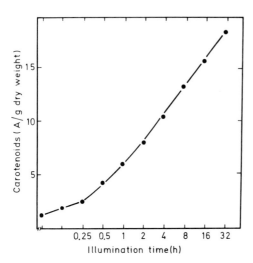

Fig. 1. Dose response curve for carotenogenesis in *Fusarium oxysporum* (40 W m^{-2})

The method used for our investigations is a modification of the procedure given by Rau (1967a) and Rau et al. (1968) and will be detailed elsewhere (Schrott 1980). Culture conditions were worked out that made use of the advantage provided by submerged growth over mycelial pads: Submerged mycelia consist of a homogeneously distributed network of short thread-like hyphae, the cells of which are of similar age and development and thus furnish samples of high conformity and experimental reproducibility.

Although only the blue portion of white light is active in carotenogenesis, white light fluorescent tubes were chosen as a light source for easy illumination of a number of samples at the same time. Fluorescent tubes emitting blue light only turned out to be less effective.

2 Results and Discussion

When dark-grown mycelia of *Neurospora* are irradiated with white light, saturation of photo-induced carotenogenesis is reached with an energy fluence of approximately 300 J m^{-2} when the cells are exposed to fluence rates between 0.3 and 40 W m^{-2} for up to 16 min. Extension of the illumination time beyond 16 min results in an additional increase in the amount of carotenoids synthesized during the subsequent dark period (Fig. 2), whereby the slope of this second phase of response depends on the energy fluence rate. In contrast, the slope of the first phase of the dose response curves is essentially constant irrespective of the fluence rate applied (also see Rau et al. 1968). But due to the saturation by a constant energy fluence the dose response curves all meet in the "transition point" to the second phase (which is the second of the "saddle points", characteristic for the course of the curve; the first saddle point is represented by the lower end of the plateau).

Energy fluence rates below 0.3 W m^{-2} are obviously insufficient for first-phase saturation. Experiments designed to investigate whether the second phase may also be saturated yielded equivocal results due to photodestruction of pigments formed during

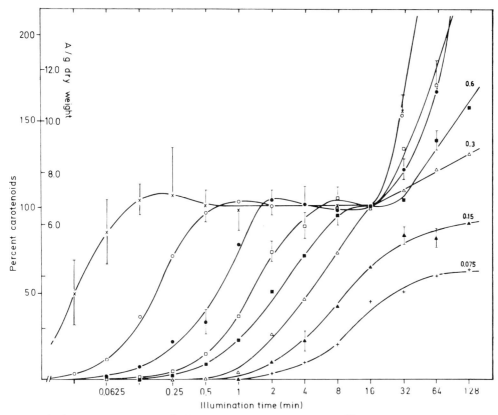

Fig. 2. Dose response curves for light-dependent carotenogenesis in *Neurospora* when mycelia are illuminated with different fluence rates at 20°C ± 1°. Each point represents the mean value of at least three independent experiments (each with duplicate), computed as percent of the 16 min value (100%). For reasons of scale values higher than 200% are not given in this diagram. Standard deviations of the mean exceeding ± 5% are presented by the error bars. x—x 40 W m^{-2}; ○—○ 8 W m^{-2}; ●—● 2 W m^{-2}; □—□ 1.1 W m^{-2}. Lower fluence rates are indicated above each curve

illumination and to photokilling events. But data reported by Harding (1974) suggest such a final saturation, which is conceivable if one takes into account the limited space within the cell for carotenoid accumulation.

The effect of temperature during the entire length of the experiment on the shape of the dose response curve is shown in Fig. 3b. The data clearly demonstrate that the "transition point" is shifted to longer illumination times when temperature is reduced. In addition, the slope of the second phase of the dose response curve tends to become more gentle.

The curves of the absolute amounts formed at different temperatures are presented in Fig. 3a. Maximum production of carotenoids is obtained at 10°C. In contrast to the findings of Harding (1974) the amount of carotenoids produced at 6°C was lower than that formed at 10°C, which could be explained by the differences in the culture and irradiation procedures.

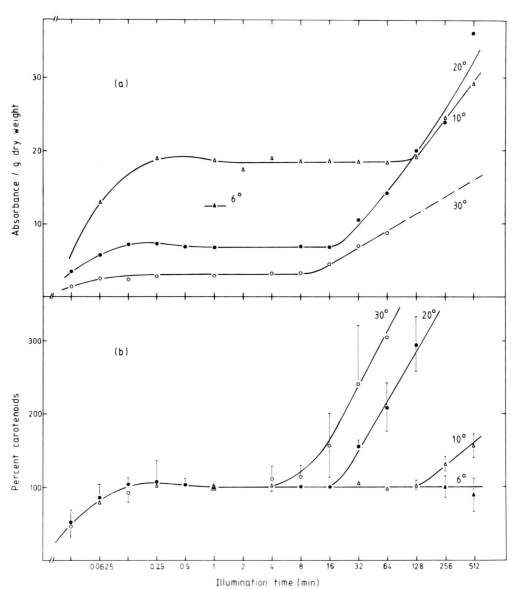

Fig. 3a, b. Effect of temperature during the experiment on dose response. Fluence rate used was 40 W m^{-2}. Standard deviations of the mean exceeding ± 5% are presented. In **b** the amount of carotenoids is normalized with respect to the 1 min value (100%)

The results indicate that the reciprocity law holds only for a definite temperature; change of the temperature yields changed amounts of carotenoids formed after induction with a given energy fluence. Reciprocity that is usually thought to be temperature-independent is probably modified in *Neurospora* during one (or more) of the subsequent dark reactions. Harding (1974) proposed that the temperature sensitivity "is due to the

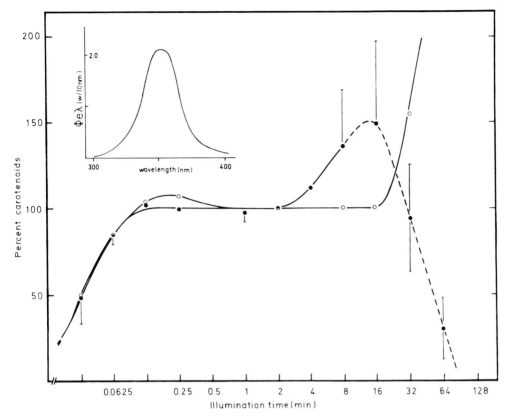

Fig. 4. Dose response curves for the effect of near UV (UV-A) illumination on carotenogenesis (●——●) in comparison to the dose response curve (○——○) obtained after illumination with 40 W m^{-2} white light. Temperature throughout the experiment was 20°C. The values are normalized with respect to the 2 min value (100%). The emission spectrum of the black light fluorescent tubes is shown in the insert. Standard deviations of the mean exceeding ± 5% are presented

degradation of the product of the light reaction (or a compound derived from it) in a temperature-dependent reaction". His results would suggest that a product closely associated with the light reaction is effected by the change in temperature rather than half-life or activity of carotenogenic enzymes.

On the other hand, the common origin of the second phase of the dose response curves (the "transition point") which is independent of fluence rate used but dependent on temperature indicates that essential elements of the photoreception sequence itself become depleted and therefore cause saturation of the first phase of the response (i.e., the plateau); the temperature-dependent process responsible for the shift of the "transition point" could then be the regeneration or replacement of the exhausted components of the photoreception chain.

Figure 4 illustrates the dose response curve obtained when mycelia of *Neurospora* are irradiated with blacklight fluorescent tubes that emit near-UV radiation of predominately 350 nm (UV-A) also capable of inducing carotenogenesis (Zalokar 1955; DeFabo et al. 1976). Since blacklight fluorescent tubes were shown to be more effective than white light in the induction of carotenoid formation in *Fusarium* (Huber and Schrott, this vol.), they were chosen also in these experiments to substitute for high fluence rates of blue light.

The data clearly demonstrate that UV-radiation results in the same dose response relationship in the first phase of the curve as illumination with white light. But the start of the second phase is triggered already after an illumination time of approximately 2 min. Thus, de novo synthesis of depleted photoreceptive material is probably not involved. In the following, the process of regaining photosensitivity either by regeneration or by replacement via de novo synthesis is therefore termed "restoration".

Since the second phase of the dose response curves is observed only after a distinct period of illumination, i.e., 16 min after onset of illumination (at 20°C), the question arises whether illumination necessarily has to be continuous or if a dark period can be intercalated into the plateau without loss in carotenoid production. The following experiment was designed: Samples were illuminated for different lengths of time followed by darkness up to the 16th min after onset of light. Subsequently exposure of all samples was continued for another 16 min. After completion of carotenoid formation the amount of pigment was compared to that of the control which had been illuminated continuously for 32 min. Fluence rate was chosen to be 8 W m^{-2} at which saturation of carotenogenesis is reached after about 1 min of illumination (see Fig. 2) allowing a sufficient number of samples to be exposed for shorter periods of time.

The results shown in Fig. 5 indicate that continuous illumination is not obligatory as long as illumination is resumed at least at the transition point to the second phase.

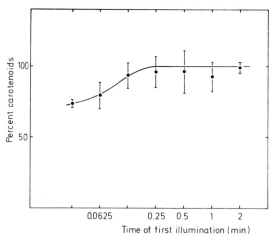

Fig. 5. The amount of carotenoids induced by the light regime described in the text as compared to controls illuminated continuously for 32 min (100%). The amount of carotenoids induced by the illumination period between the 16th and 32nd minute was previously substracted from those produced by the samples and by the control. Four experiments were carried out in triplicate including controls for the position of the "saddle points"

The amount of carotenoids specifically induced by the illumination between the 16th and 32nd minute was substracted from the total amount present in the samples and the control. Thus the data mathematically represent the amount of pigment induced by the initial illumination. One therefore would expect a dose response curve similar to that shown in Fig. 2 achieved by illumination with the same fluence rate. However, the results suggest that it is not even necessary to illuminate the length of time required to reach the plateau; an initial illumination period of approximately 15 s followed by illumination between the 16th and 32nd minute after the first onset of light are as effective for induction of carotenogenesis as continuous illumination for 32 min. Also, the very short illumination periods are far more efficient compared to the single-illumination experiment.

Extending the time elapsed between first and second illumination to 2 h leads to results reported previously by Rau and Rau-Hund (1977). They show that an additional amount of carotenoids is induced by the second illumination only after a second lag-phase, indicating that the carotenoid-producing mechanism is restarted.

Apart from the length of illumination time (see below) the amount of carotenoids formed after a second illumination depends on the time elapsed after first exposure to light: The longer the time period between first and second illumination, the more effective is the second induction; after approximately 2 h maximum competence for an additional induction is restored (Fig. 6).

Compared to the preceding experiment only time delay is extended. Therefore, the mechanism of this restoration is assumed to be identical to the restoration reactions during the plateau. Restoration of inactivated material probably starts immediately but becomes effective only at and past the "transition point". Therefore light is not per-

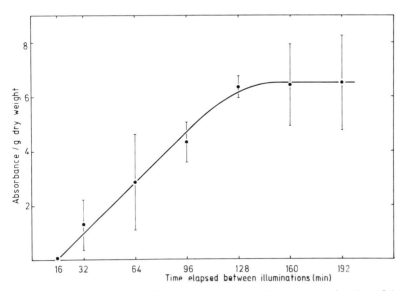

Fig. 6. The amount of carotenoids induced by a second illumination as function of time elapsed between first (1 min) and second (8 min) illumination (8 W m^{-2}). Mean values of 3 experiments (run in duplicates) are shown

ceived and consequently not necessary between depletion and occurrence of the first restored molecules, i.e., during the plateau.

If depletion is not quantitative, which presumably happens when the illumination time is shorter than 1 min, remaining material is still available during the subsequent exposure in addition to the regenerated components which then would permit to produce the same amount of carotenoids as in the control although only initially illuminated for 15 s instead of 1 min, the length of time required for saturation when a fluence rate of 8 W m^{-2} is used in a single-illumination experiment (see Fig. 2).

After an initial saturating induction, restoration is completed when maximum competence for a second illumination is achieved. The fact that the amount of carotenoids maximally induced by the second illumination is about the same as by the first exposure is in favor of regeneration of depleted material rather than of a de novo synthesis. During continuous illumination in the second phase consumption and restoration is assumed to reach a steady state corresponding to a constant rate of induction as long as no final saturation mechanism is superimposed.

The interpretation of the results is also confirmed by the following experiment: Samples were illuminated using again a fluence rate of 8W m^{-2} at 20°C either for 1 min or for 16 min representing the two "saddle points" of the dose response curve. After 2 h in the dark all samples were illuminated again for the periods of time specified. As shown in Table 1 the light administered throughout the stage of the plateau is not accounted for in the second illumination. On the other hand, the amount of carotenoids induced depends on the length of the second exposure, i.e., the illumination time period represented by the saddle points characteristic for the dose response curve of the first illumination are no more equal in their inducing capability in a second illumination. As a consequence dose response curves of the second illumination must exhibit a different shape.

In fact, longer illumination times are necessary for a saturating induction, whereas the second of the "saddle points" remains unaffected (Fig. 7). The slope of the second phase of the dose response curve appears to be less steep, suggesting that sensitivity to light is reduced after an initial illumination.

Table 1. The effect of the light regime, described in the text, on the amount of carotenoids induced. Values are the average of 6 experiments.☐ Illumination period,▤ dark period

Illumination regime	Percent carotenoids
	100
	96.1 ± 16.1
	192.8 ± 23.8
	186.7 ± 34.3

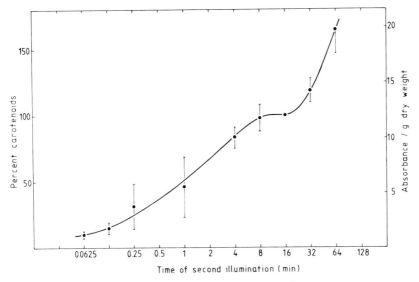

Fig. 7. Dose response curve for a second illumination (8 W m^{-2}) 2 h after an initial saturating exposure (1 min). Standard deviations exceeding ± 5% of mean values of 4 experiments are presented

A biphasic regulation mechanism seems to be ecologically ingenious taking into account the protective function of the carotenoids (for details see Huber and Schrott, this vol.). Organisms lacking photoregulation of carotenoid synthesis produce sufficient amounts to enable the species to survive. Also, photoregulated organism grown in the dark synthesize at least small amounts of carotenoids as a basic protection potential. In addition, some organisms have obviously developed rather sophisticated regulatory mechanisms that permit to economize the metabolism which then has to respond only to the extent absolutely necessary.

Acknowledgments. This work was supported by a grant from the Deutsche Forschungsgemeinschaft to Dr. W. Rau. We are indebted to Dr. W. Rau for valuable support and stimulating discussions, to Dr. R.R. Theimer for his help during the preparation of the manuscript and to Miss C. Hopf for skilful technical assistance.

References

Blaauw OH, Blaauw-Jansen G, van Leeuwen WJ (1968) An irreversible red-light-induced growth response in *Avena.* Planta 82: 87–104

Blaauw-Jansen G (1974) Dose response curves for phytochrome-mediated anthocyanin synthesis in the mustard seedling (*Sinapis alba* L.). Acta Bot Neerl 23: 513–519

DeFabo EC, Harding RW, Shropshire W Jr (1976) Action spectrum between 260 and 800 nanometers for the photoinduction of carotenoid biosynthesis in *Neurospora crassa.* Plant Physiol 57· 440–445

Harding RW (1974) The effect of temperature on photo-induced carotenoid biosynthesis in *Neurospora crassa.* Plant Physiol 54: 142–147

Jayaram M, Presti D, Delbrück M (1979) Light-induced carotene synthesis in *Phycomyces*. Exp Mycol 3: 42–52

Rau W (1967a) Untersuchungen über die lichtabhängige Carotinoidsynthese. I. Das Wirkungsspektrum von *Fusarium aquaeductuum*. Planta 72: 14–28

Rau W, Rau-Hund A (1977) Light-dependent carotenoid synthesis. X. Lag-phase after a second illumination period in *Fusarium aquaeductuum* and *Neurospora crassa*. Planta 136: 49–52

Rau W, Lindemann I, Rau-Hund A (1968) Untersuchungen über die lichtabhängige Carotinoidsynthese. III. Die Farbstoffbildung von *Neurospora crassa* in Submerskultur. Planta 80: 309–316

Sargent ML, Briggs WR (1967) The effects of light on a circadian rhythm of condidiation in *Neurospora*. Plant Physiol 42: 1504–1510

Schrott EL (1980) Manuscript in preparation

Theimer RR (1968) Untersuchungen zum Wirkungsmechanismus von Licht und Mercuribenzoat bei der Carotinoidsynthese in *Fusarium aquaeductuum*. Ph D Thesis, Universität München

Vanderhoef LN, Briggs WR (1977) Kinetic and spectral studies on red light-induced suppression of mesocotyl elongation in maize. Carnegie Inst Wash Yearb 76: 283–286

Zalokar M (1955) Biosynthesis of carotenoids in *Neurospora*. Action spectrum of photoactivation. Arch Biochem 56: 318–325

Zimmerman BK, Briggs WR (1963) Phototropic dosage-response curves for oat coleoptiles. Plant Physiol 38: 248–253

Carbon Metabolism and Respiration

Effects of Blue Light on Respiration and Non-Photosynthetic CO_2 Fixation in *Chlorella vulgaris* 11 h Cells

SHIZUKO MIYACHI[1], AKIO KAMIYA[2], and SHIGETOH MIYACHI[1,3]

1 Introduction

With yellow mutant cells of *Chlorella vulgaris* (Mutant 211-11 h) which has no chloro-phyll, Kowallik and Gaffron (1966) reported that the rate of respiration declined when these cells were kept in the dark in phosphate buffer. Blue light counteracted this de-cline and high respiratory rate was maintained. Kowallik (1969) further showed that O_2 uptake in DCMU-poisoned *Chlorella pyrenoidosa* was also greatly enhanced by blue light illumination.

On the other hand, Krotkov (1964) and Ogasawara and Miyachi (1969, 1970) found that blue light enhances $^{14}CO_2$ incorporation into aspartate, glutamate, malate, and fumarate, while red light stimulates its incorporation into sucrose and starch in *Chlorella* and other green algae. The finding by Ogasawara and Miyachi that the above-mentioned blue light effect on CO_2 fixation takes place under the experimental conditions in which normal photosynthesis is completely inhibited by CMU, and that extremely low light intensity suffices to saturate the effect indicated that this effect is operated by a mech-anism independent of normal photosynthetic CO_2 fixation. They further showed that *Chlorella* cells incubated in the dark in phosphate medium longer than 12 h showed pronounced blue light effect on CO_2 fixation, whereas those kept under continuous light showed only a slight blue light effect, if any. During the dark starvation, phospho-enolpyruvate (PEP) carboxylase activity and the capacity for dark $^{14}CO_2$ fixation de-creased significantly (Ogasawara and Miyachi 1971).

With colorless mutant cells of *Chlorella vulgaris* (Mutant 125), Kamiya and Miyachi (1974) found that the rates of dark $^{14}CO_2$ fixation and respiration decreased by dark starvation and recovered by illumination with blue light. The main CO_2 fixation product under blue light was aspartate. Aspartate and other compounds into which $^{14}CO_2$ was mostly incorporated under blue light were presumed to be derived from oxalacetate, formed as a result of a C_1-C_3 carboxylation reaction. They concluded that PEP carbo-xylase activity which decreased during dark starvation is enhanced by blue light illumina-tion, and this causes the blue light effect on CO_2 fixation. Kamiya and Miyachi further showed that the enhancing effects of blue light (456 nm) were saturated at intensities as low as 400–800 erg/cm^2 s^{-1}. The action spectra for these effects were also similar to

1 Radioisotope Centre, University of Tokyo, Bunkyo-ku, Tokyo 113, Japan
2 Present address: Department of Chemistry, Faculty of Pharmaceutical Sciences, Teikyo University, Sagamiko, Kanagawa 199-01, Japan
3 Institute of Applied Microbiology, University of Tokyo, Bunkyo-ku, Tokyo 113, Japan

each other; both showed peaks at 460 and 380 nm. These results indicate that the same mechanism underlies the effects of blue light on CO_2 fixation and respiration in *Chlorella vulgaris* 125 cells.

The present investigation was carried out to see whether the similarities in the two kinds of blue light effects observed with colorless mutant cells of *Chlorella* could also be observed in wild-type cells of *Chlorella vulgaris*. The positive results are described in this paper.

2 Material and Methods

2.1 Algal Culture

Chlorella vulgaris 11 h (Algal collection of the Plantphysiological Institute of Göttingen University) was grown at $23°-25°C$ in a flat oblong vessel containing 500 ml of inorganic culture medium (Ogasawara and Miyachi 1970) with a light-dark rhythm of 16 and 8 h. Air enriched with ca. 2% CO_2 by volume was constantly bubbled through the algal suspension. During the light period, the vessel in the water-bath was illuminated with an incandescent lamp. The light intensities were increased stepwise as follows: 500 lx (1st day), 1800 lx (2nd–3rd day), 3000 lx (4th–5th day), 6000 lx (6th–7th day) and 10,000 lx (8th day). Cells were harvested 3 h after the start of the last light period (8th or 9th day). The light intensity during the last 1 h was increased to 20,000 lx. The cells were washed twice with 2×10^{-3} M K_2SO_4 and suspended in 2.5×10^{-3} M phosphate buffer (pH 7.6). The final cell density was 10 ml packed cell volume per liter. In some experiments, the flask containing the algal suspension was covered with aluminum foil and horizontally shaken in the dark with a reciprocal motion (length of motion, 3 cm; frequency, 120–130 returns/min) at $25°C$. In other experiments, the cells were immediately subjected to the determination of $^{14}CO_2$ fixation or oxygen gas exchange.

2.2 Determination of Oxygen Gas Exchange and $^{14}CO_2$ Fixation under Varied Illumination Conditions

The rate of oxygen gas exchange was determined with 6 ml of the algal suspension placed in the glass cylinder equipped with a Clark-type oxygen probe (Rank Brothers, London). After the rate of oxygen uptake was determined, the vessel was illuminated with red or blue monochromatic light at varied intensities from one side. The monochromatic light (half-band width, 10 nm) was obtained by the combined use of an interference filter and an appropriate color filter (Miyachi et al. 1978). The light intensity was adjusted by changing the distance and/or by inserting the neutral density filters (Shonan Komaku Kenkyusho, K.K., Tokyo) between the oxygen probe and the light source (500 w-xenon lamp). The light intensity was determined with a Radiometer (YSI-Kettering, Model 65). $^{14}CO_2$ fixation was started by adding $NaH^{14}CO_3$ solution (final concentration, 0.63 mM). Determination of the rate of $^{14}CO_2$ fixation and two-dimensional paper chromatographic analysis of photosynthetic products were carried out according to the procedures described elsewhere (Miyachi et al. 1978).

3 Results and Discussion

3.1 Effects of Blue Light on Respiration

When *Chlorella vulgaris* 11 h cells which had been shaken overnight in phosphate buffer in darkness (1-day-starved cells) were illuminated with blue light of low intensity (456 nm, 340 erg/cm^2 s^{-2}), the rate of oxygen uptake started to increase after a lag period which lasted for ca. 3 min (Fig. 1). No such enhancement in oxygen uptake was observed in non-starved algal cells. As shown in the figure, oxygen uptake was sometimes slightly slowed down upon blue light illumination. Figure 1 further shows that the rate of dark respiration was greatly decreased during dark starvation of the algal cells (see also Table 1).

Figure 2 shows that the rate of oxygen uptake in 1-day-starved cells of *Chlorella vulgaris* 11 h was enhanced by illumination of blue light at low intensities. The rate attained the maximum at ca. 550 erg/cm^2 s^{-1}. The further rise of the intensity caused the decrease in the rate of oxygen uptake and oxygen was evolved at intensities higher than ca. 3000 erg/cm^2 s^{-1}. On the other hand, the blue light-induced oxygen uptake was not observed in nonstarved cells.

Figure 3 shows that red light (660 nm) at intensities 0–10,000 erg/cm^2 s^{-1} did not show enhancement of oxygen uptake irrespective of whether the algal cells had been starved or not.

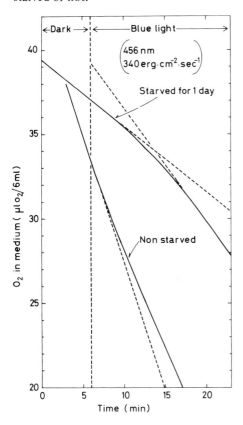

Fig. 1. Effects of blue light on respiration in nonstarved and 1-day-starved cells of *Chlorella vulgaris* 11 h

Table 1. ^{14}C-Incorporation into various compounds during dark $^{14}CO_2$ fixation in *Chlorella vulgaris* 11 h cells

Compounds	^{14}C incorporated (cpm)				
	Cells which were used without dark-precincubation (A)		Cells which had been shaken in the dark for 1 day (B)		(A-B)/A
Insolubles	503	(5.5)	517	(12.5)	
Phosphate esters	351	(3.8)	276	(6.7)	
Aspartate	5464	(59.7)	1513	(36.7)	
Glutamate	1380	(15.1)	743	(18.0)	
Malate	834	(9.1)	452	(11.0)	
Others	628	(6.9)	619	(15.0)	
Total ^{14}C fixed	9160	(100.1)	4120	(99.9)	0.55
Rate of O_2 uptake ($-\mu lO_2/\mu l$ cells h^{-1})	1.85		0.62		0.66

The figures in parentheses show percent incorporation of ^{14}C. Time of $^{14}CO_2$ fixation, 10 min

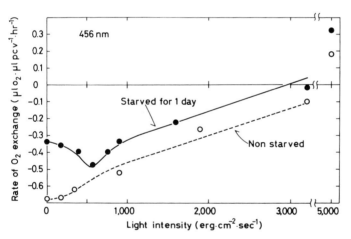

Fig. 2. Rate of oxygen exchange vs. blue light intensity in nonstarved and 1-day-starved cells of *Chlorella vulgaris* 11 h

In the presence of DCMU, oxygen uptake in the starved cells was enhanced in parallel with the intensity of blue light and attained the stationary level at ca. 2000 erg/cm^2 s^{-1} (Fig. 4). In another experiment which is not shown here, the rate of oxygen uptake at 1000 erg/cm^2 s^{-1} was the same as that at 2000 erg/cm^2 s^{-1}. These results show that the enhancement of respiration by blue light (456 nm) attains the maximum level at intensity as low as 1000 erg/cm^2 s^{-1}. On the other hand, red light (660 nm) did not show any enhancement in oxygen uptake in the presence of DCMU (Fig. 4).

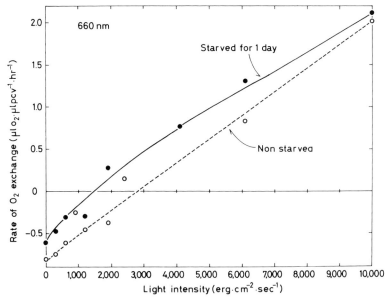

Fig. 3. Rate of oxygen exchange vs. red light intensity in nonstarved and 1-day-starved cells of *Chlorella vulgaris* 11 h

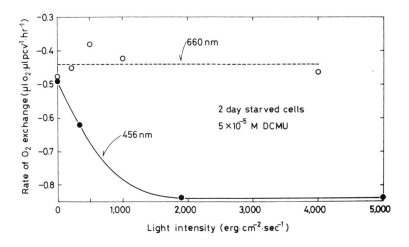

Fig. 4. Effects of red and blue lights at varied intensities in the presence of DCMU on oxygen exchange in 2-day-starved cells of *Chlorella vulgaris* 11 h

Thus, it is clear that, as was found in colorless mutant cells of *Chlorella vulgaris* (Mutant 125) (Kamiya and Miyachi 1974), the respiratory activity in starved cells of *C. vulgaris* 11 h was specifically enhanced by blue light and there was a lag period which lasted at least for several minutes before respiration was enhanced by illumination with short wavelength light. Essentially the same results have been obtained with DCMU-poisoned *Chlorella pyrenoidosa* (Kowallik 1969).

3.2 Effects of Blue Light on Nonphotosynthetic Carbon Metabolism

Table 1 shows that similar to oxygen uptake, the capacity of dark $^{14}CO_2$ fixation decreased during dark starvation of the algal cells. In starved as well as nonstarved cells, the radioactivities were mostly found in aspartate, glutamate, and malate, indicating that dark $^{14}CO_2$ fixation was mediated by a C_1-C_3 carboxylation reaction and that the capacity for the C_1-C_3 carboxylation reaction decreased during starvation. Table 2 shows that in the presence of DCMU, blue light significantly enhanced $^{14}CO_2$ fixation in 1-day-starved cells (50.3% increase), while the effect of red light was relatively small (17.6% increase). The distribution of radioactivity among $^{14}CO_2$ fixation products revealed that blue light especially enhanced $^{14}CO_2$ fixation into phosphate esters and aspartate. $^{14}CO_2$ incorporation into phosphate esters was also enhanced by red light. These results suggest that the enhancement in $^{14}CO_2$ incorporation into phosphate esters is due to photosynthesis. [Although photosynthesis was inhibited by DCMU, the remaining activity (about 3% of normal) is still comparable to that of dark $^{14}CO_2$ fixation.] Other than aspartate, $^{14}CO_2$ incorporation into glutamate, fumarate, citrate and malate was enhanced by blue light, while red light did not show significant enhancement.

Table 2. $^{14}CO_2$ incorporation into products in 1-day-starved *Chlorella vulgaris* 11 h cells under various illuminating conditions in the presence of DCMU (5×10^{-5} M)

Compound	Dark (A)	Blue light (B)	(B-A)/A × 100	Red light (C)	(C-A)/A × 100
Insolubles	900 (13.6)	835 (8.4)	−7.2	930 (12.0)	3.3
P-esters	960 (14.6)	1970 (19.8)	105.2	1830 (23.5)	90.6
Sucrose	595 (9.0)	290 (2.9)	−51.2	950 (12.2)	59.7
Aspartate	2100 (31.7)	3800 (38.2)	81.0	1940 (24.9)	−7.6
Glutamate	650 (9.9)	865 (8.7)	33.1	690 (8.8)	6.1
Malate	790 (12.0)	955 (9.6)	20.9	660 (8.5)	−16.5
Citrate	450 (6.8)	590 (5.9)	31.1	475 (6.1)	5.6
Fumarate	170 (2.6)	230 (2.3)	35.3	110 (1.4)	−35.9
Others	0	410 (4.1)		195 (2.5)	
Total ^{14}C fixed	6615 (100.2)	9945 (99.9)	50.3	7780 (99.9)	17.6

The figures in parentheses show the percent incorporation of ^{14}C. Time of $^{14}CO_2$ fixation, 15 min. Light intensity: 1900 erg cm^{-2} s^{-1} for blue light (456 nm) and 2000 erg cm^{-2} s^{-1} for red light (660 nm), respectively

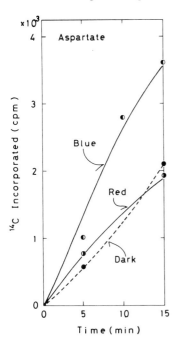

Fig. 5. Effects of red and blue lights on $^{14}CO_2$ incorporation into aspartate in the presence of DCMU in 1-day-starved cells of *Chlorella vulgaris* 11 h. The data were derived from the experimental results shown in Table 2

Table 3. $^{14}CO_2$ incorporation into products in nonstarved *Chlorella vulgaris* 11 h cells under various illuminating conditions in the presence of DCMU (5×10^{-5} M)

Compound	Dark (A)		Blue Light (B)		(B-A)/A $\times$ 100	Red light (C)		(C-A)/A $\times$ 100
Insolubles	2160	(13.7)	3330	(18.0)	54.2	3290	(20.0)	52.3
P-esters	2780	(17.6)	3320	(17.9)	19.4	3110	(18.9)	11.9
Sucrose	320	(2.0)	600	(3.2)	87.5	340	(2.1)	6.3
Aspartate	3170	(20.1)	3900	(20.8)	21.9	2900	(17.6)	−8.5
Glutamate	2790	(17.7)	2560	(13.8)	−8.6	2160	(13.2)	−22.6
Malate	1460	(9.3)	1400	(7.6)	−4.1	1070	(6.5)	−26.7
Citrate	2200	(13.9)	2220	(12.0)	0.0	2530	(15.4)	15.0
Fumarate	350	(2.2)	270	(1.4)	−22.9	220	(1.4)	−37.1
Others	560	(3.6)	960	(5.2)	71.4	815	(5.0)	45.5
Total ^{14}C fixed	15,790	(100.1)	18,560	(99.9)	17.5	16,435	(100.1)	4.1

The figures in parentheses show the percent incorporation of ^{14}C. Time of $^{14}CO_2$ fixation, 15 min. Light intensity: 1900 erg cm^{-2} s^{-1} for blue light (456 nm) and 2000 erg cm^{-2} s^{-1} for red light (660 nm), respectively

Figure 5 shows that in the presence of DCMU, $^{14}CO_2$ incorporation into aspartate in 1-day-starved *Chlorella* cells was greatly enhanced by blue light, while it was not enhanced by red light. In nonstarved cells, neither blue nor red light showed significant effect on $^{14}CO_2$ incorporation in the presence of DCMU (Table 3).

These results show that the starvation of *Chlorella vulgaris* 11 h cells decreases the activity for a C_1-C_3 carboxylation reaction (possibly PEP carboxylase reaction), and hence the capacity of dark $^{14}CO_2$ fixation. When the starved cells are illuminated with blue light, the C_1-C_3 carboxylation is greatly enhanced. In contrast, red light does not show an appreciable effect on a C_1-C_3 carboxylation reaction in the starved cells. In contrast to the starved cells, C_1-C_3 carboxylation reaction in the nonstarved cells was not affected by blue light illumination.

4 Summary

The rates of respiration and dark CO_2 fixation in *Chlorella vulgaris* 11 h cells were suppressed when these cells were starved in darkness. Both rates were recovered by illumination with blue light, while red light did not show any effect. Enhancing effect of blue light (456 nm) on respiration was saturated at light intensities as low as ca. 1000 erg $cm^2\ s^{-1}$. Under the conditions in which photosynthesis in the starved cells was inhibited by DCMU, the main $^{14}CO_2$ fixation product under blue light was aspartate. These findings indicate that the regulation of respiration and non-photosynthetic CO_2 fixation by blue light as observed in a colorless *Chlorella vulgaris* (Mutant 125) can also occur in photosynthesizing wild type cells of *Chlorella*.

Acknowledgments. This research was supported by grants from the Japanese Ministry of Education, Science, and Culture, and the Ministry of Agriculture, Forestry, and Fisheries (GEP-54-II-I-11).

References

Kamiya A, Miyachi S (1974) Effects of blue light on respiration and carbon dioxide fixation in colorless *Chlorella* mutant cells. Plant Cell Physiol 15: 927–937

Kowallik W (1969) Der Einfluß von Licht auf die Atmung von *Chlorella* bei gehemmter Photosynthese. Planta 86: 50–62

Kowallik W, Gaffron H (1966) Respiration induced by blue light. Planta 69: 92–95

Krotkov G (1964) The influence of wavelength of incident light on the path of carbon in photosynthesis. Trans R Soc.Can 4th Ser 2: 205–215

Miyachi S, Miyachi Sh, Kamiya A (1978) Wavelength effects on photosynthetic carbon metabolism in *Chlorella*. Plant Cell Physiol 19: 277–288

Ogasawara N, Miyachi S (1969) Effect of wavelength on $^{14}CO_2$-fixation in *Chlorella* cells. In: Metzner H (ed) Progress in photosynth Res, vol III, pp 1653–1661. Verlag C Lichtenstern, München

Ogasawara N, Miyachi S (1970) Regulation of CO_2-fixation in *Chlorella* by light of varied wavelengths and intensities. Plant Cell Physiol 11: 1–14

Ogasawara N, Miyachi S (1970) Effects of disalicylidenepropandiamine and near far red light on $^{14}CO_2$-fixation in *Chlorella* cells. Plant Cell Physiol 11: 411–416

Ogasawara N, Miyachi S (1971) Effects of dark preincubation and chloramphenicol on blue light-induced CO_2 incorporation in *Chlorella* cells. Plant Cell Physiol 12: 675–682

Effect of Blue Light on CO_2 Fixation in Heterotrophically Grown *Scenedesmus obliquus* Mutant C-2A′

SHIGETOH MIYACHI[1,2], SHIZUKO MIYACHI[2], and HORST SENGER[3]

1 Introduction

It has been established that blue light enhances respiration in various algae (for literature, see Voskresenskaya 1972). With colorless mutant and wild type cells of *Chlorella vulgaris* as well as *Chlorella ellipsoidea,* Miyachi and coworkers found that blue light also enhanced CO_2 incorporation into aspartate (Ogasawara and Miyachi 1970; Kamiya and Miyachi 1974; Miyachi et al. 1979). The characteristics of both blue light effects were very similar, indicating that the same mechanism underlies blue light-induced respiration and CO_2 fixation. With heterotrophically-grown wild type and mutant C-2A′ cells, Watanabe et al. (1979) found that $^{14}CO_2$-incorporation was enhanced upon illumination with fluorescent lamps. In this paper it was proved that blue light caused an enhancement of $^{14}CO_2$-incorporation in the mutant cells of *Scenedesmus.* In the case of *Chlorella* cells, $^{14}CO_2$-fixation was enhanced immediately the blue light was switched on, while respiration was enhanced only after a lag period of several minutes. However, in mutant cells of *Scenedesmus,* both respiration and $^{14}CO_2$ fixation were enhanced immediately blue light illumination commenced. A possible explanation for such difference was discussed.

2 Material and Methods

Mutant C-2A′ cells of *Scenedesmus obliquus* were shaken in a dark incubator (Gyrotary Shaker G25, New Brunswick) at 30°C in a 500 ml Erlenmeyer flask containing 250 ml of inorganic "basal medium" (Senger and Bishop 1972) enriched with glucose (0.5%) and yeast extract (0.5%). After 3 days, the cells (equivalent to 10 μl PCV) were transferred to 250 ml of fresh medium and grown for another 3 days. Then, the cells were washed and suspended in 4 mM phosphate buffer, pH 8.0, containing KNO_3 (8 mM) and $MgSO_4$ (1 mM). The cells were shaken in the dark for 1 h and then illuminated with blue light (399 nm , half-band width, 20 nm) at 500 erg/cm² s⁻¹. Monochromatic light was obtained with an interference filter. For details on light source and interference filter etc., see Brinkmann and Senger (1978). At intervals, aliquots of the algal suspension were taken out to determine the rates of oxygen uptake and $^{14}CO_2$ fixation. Rate of respiration was determined polarographically with a Gilson Medical Electronics Oxygraph (Middelton, Wis.). To determine rates of $^{14}CO_2$-fixation, $NaH^{14}CO_3$ (2 mM) was added to the algal suspension. Aliquots were taken out at 2-min intervals to the

1 Institute of Applied Microbiology, University of Tokyo, Bunkyo-ku, Tokyo 113, Japan
2 Radioisotope Center, University of Tokyo, Bunkyo-ku, Tokyo 113, Japan
3 Fachbereich Biologie-Botanik, Universität Marburg, Lahnberge, 3550 Marburg, FRG

planchets which contained acetic acid. Sampling lasted for 8 min under continued stirring. Samples were dried under an infra-red lamp and the radioactivity of fixed ^{14}C was
determined with a gas-flow counter.

3 Results and Discussion

Figure 1 shows that both oxygen uptake and $^{14}CO_2$ fixation were enhanced by blue
light of low intensity (399 nm, 500 erg/cm^2 s^{-1}). The magnitude of the enhancement
on $^{14}CO_2$ fixation was similar to that observed when the cells were illuminated with
fluorescent lamps at 10 W m^{-2}. We therefore concluded that the effects of illumination
with fluorescent lamps on $^{14}CO_2$ fixation during an early stage of greening of *Scenedesmus obliquus* (Watanabe et al. 1980) were caused by blue light.

With colorless mutant cells of *Chlorella vulgaris* #125, Kamiya and Miyachi (1974)
found that $^{14}CO_2$ fixation was enhanced immediately upon illumination with blue light,
while respiration was enhanced only after the lag period of several minutes. Since the
main $^{14}CO_2$ fixation product was aspartate, they assumed that the C_1-C_3 carboxylation reaction is increased by blue light and that this causes an enhancement in the formation of oxalacetic acid (OAA) and then aspartate. An enhancement in the supply of
OAA by this route will accelerate the tricarboxylic acid (TCA) cycle and consequently
the oxygen uptake. The lag period observed in the blue light enhancement of respiration
will therefore correspond to the time required for the blue light-induced increase in the

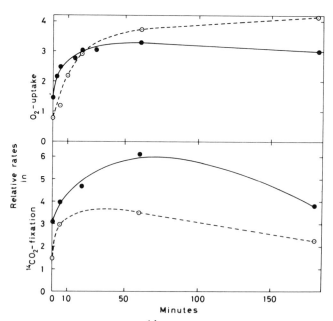

Fig. 1. Effects of blue light on $^{14}CO_2$-fixation and respiration in *Scenedesmus obliquus* mutant
C-2A' cells. *Solid lines:* the algal cells which had been kept in phosphate buffer containing KNO$_3$
and MaSO$_4$ for 1 h in the dark were irradiated with 399 nm-light at 500 erg cm^{-2} s^{-1}. *Dotted lines:*
the algal cells suspended in the above medium were illuminated with fluorescent lamps at 10 W m^{-2}.
O$_2$-uptake and $^{14}CO_2$-fixation shown by *dotted lines* were measured with the algal cells which had
been starved in the dark for 4 h and immediately after the algal cells were suspended in the inorganic medium, respectively

levels of intermediates of the TCA cycle. According to their reasoning, the initial changes in the carbon metabolism induced by blue light illumination is an enhancement of the C_1-C_3 carboxylation reaction possibly catalyzed by phosphoenolpyruvate (PEP) carboxylase. They assumed that the enhancements in oxygen uptake and carbohydrate breakdown are induced secondarily as a result of a blue light-induced increase in PEP carboxylase activity.

However, in heterotrophically grown mutant cells of *Scenedesmus obliquuus*, not only $^{14}CO_2$-fixation but also oxygen uptake was enhanced immediately upon the commencement of blue light illumination, indicating that the rate-determining step is different from that in *Chlorella* cells. (No lag period in oxygen uptake was observed irrespective of the starvation period to which the algal cells were subjected.) One possible explanation for this is that the blue light-induced breakdown of carbohydrate is far greater in heterotrophically grown *Scenedesmus* cells than in starved *Chlorella* cells, so that the enhanced supply of the substrate such as PEP and acetyl-CoA immediately enhances both respiration and PEP carboxylation reaction.

Figure 1 also shows that the rate of $^{14}CO_2$ fixation was decreased when the algal cells were illuminated for 180 min. When these algal cells were then bubbled with CO_2-free air for 10 min before a second addition of $NaH^{14}CO_3$, the rate of $^{14}CO_2$ fixation was greatly enhanced (5.4 times increase in one experiment). We therefore assume that the apparent decrease during the initial 180 min is due to the production of unlabeled CO_2 by respiration causing a dilution of the labeled CO_2-leading, in turn, to a decrease in incorporation of $^{14}CO_2$.

4 Summary

Blue light enhances not only O_2 uptake but also $^{14}CO_2$ fixation in heterotrophically-grown *Scenedesmus obliquus* mutant C-2A'. Unlike *Chlorella* cells, there was no lag period in the blue light-enhancement of O_2 uptake of this alga.

Acknowledgments. The authors wish to thank Frau I. Koss for her skillful technical assistance. Thanks are also due to Deutsche Akademischer Austauschdienst for the financial support to S. Miyachi.

References

Brinkmann G, Senger H (1978) The development of structure and function in chloroplasts of greening mutants of *Scenedesmus* IV. Blue light-dependent carbohydrate and protein metabolism. Plant Cell Physiol 19: 1427–1437

Kamiya A, Miyachi S (1974) Effects of blue light on respiration and carbon dioxide fixation in colorless *Chlorella* mutant cells. Plant Cell Physiol 15: 927–937

Miyachi Sh, Kamiya A, Miyachi S (1980) Effects of blue light on respiration and non-photosynthetic CO$_2$ fixation in *Chlorella vulgaris* 11 h cells. This volume, pp 321–328

Ogasawara N, Miyachi S (1970) Regulation of CO$_2$-fixation in *Chlorella* by light of varied wavelengths and intensities. Plant Cell Physiol 11: 1–14

Senger H, Bishop NI (1972) The development of structure and function in chloroplasts of greening mutants of *Scenedesmus* I. Formation of chlorophyll apparatus. Plant Cell Physiol 13: 633–649

Voskresenskaya NP (1972) Blue light and carbon metabolism. Annu Rev Plant Physiol 23: 219–234

Watanabe M, Oh-hama T, Miyachi S (1980)Light-induced carbon metabolism in an early stage of greening in wild type and mutant C-2A' cells of *Scenedesmus obliquus*. This volume, pp 332–343

Light-Induced Carbon Metabolism in an Early Stage of Greening in Wild Type and Mutant C-2A′ Cells of *Scenedesmus obliquus*

MASAYO WATANABE[1], TAMIKO OH-HAMA[1], and SHIGETOH MIYACHI[1]

1 Introduction

It has been reported that the first event which occurs during the light-induced green-ing process of heterotrophically grown mutant cells of *Scenedesmus obliquus* (mutant C-2A') is the enhancement of respiration and blue light is effective for the enhancement (Senger and Bishop 1972).

Blue light-induced respiration has been observed with various algae (Kowallik and Gaffron 1967; Pickett and French 1967; Schmid and Schwarze 1969; Kamiya and Mi-yachi 1974; for literature, see Jackson and Volk 1970). With colorless mutant and wild-type cells of *Chlorella,* Miyachi et al. (Ogasawara and Miyachi 1970; Kamiya and Miya-chi 1974; Miyachi et al. 1980b) found that blue light also enhanced $^{14}CO_2$ incorporation into aspartate. Based on the findings that characteristics for the blue light-induced re-spiration and $^{14}CO_2$-fixation were very similar to each other, they concluded that the same mechanism underlies the two kinds of blue light effect.

In this article, we report that $^{14}CO_2$ fixation into aspartate is enhanced when hetero-trophically grown mutant as well as wild-type cells of *Scenedesmus obliquus* are illumi-nated. The radioactivity is then found in RNA and the protein fraction. Mechanism of the effects of 6-MP, CH and the removal of nutrient are also discussed.

2 Materials and Methods

2.1 Algal Culture

The strain of *Scenedesmus obliquus* D_3 and mutant C-2A' which had been obtained by X-ray irradiation were kindly provided by Professor H. Senger, Philipps University, Mar-burg. The culture medium containing 0.5% glucose and 0.25% yeast extract in the basal medium (0.4 g $NaH_2PO_4 \cdot H_2O$; 0.015 g $CaCl_2 \cdot 2H_2O$, 0.178 g $Na_2HPO_4 \cdot 2H_2O$; 0.47 g NaCl; 0.81 g KNO_3; 0.246 g $MgSO_4 \cdot 7H_2O$; 0.003 g $FeSO_4 \cdot 7H_2O$; 1 ml Arnon's A 5 microelement solution in 1 l) were inoculated with dark-grown algal cells in concentra-tion of 0.1% packed cell volume (PCV), and were shaken for 4—8 days in the dark at 30°C.

1 Institute of Applied Microbiology, University of Tokyo, Bunkyo-ku, Tokyo 113, Japan

2.2 Experiments of $^{14}CO_2$ Fixation in the Light and in Darkness

Algal cells were washed with 4 mM phosphate buffer, pH 7.6 and suspended either in the same buffer or in the buffer containing KNO_3 and $MgSO_4$ at the same concentrations in basal medium. The suspension was also provided with penicillin (1.3×10^5 units per l) and shaken in the dark for 20 h at $30°C$. The suspension (4–6 ml) was placed in a flat-bottomed reaction vessel equipped with a rubber stopper (Miyachi 1959). $NaH^{14}CO_3$ (specific activity, 50–60 mCi/mmol, final concentration, 0.5 mM) was then added and the vessel was placed in the acrylic plastic incubator containing water at $30°C$. The vessel was shaken under illumination from below (10 klx) or in the dark. At intervals, portions of the cell suspension (0.6–1.0 ml) were transferred into methanol (final conc., 80%, v/v) chilled with NaCl and ice. The total ^{14}C fixed in algal cells was determined according to Ogasawara and Miyachi (1970).

2.3 Analysis of $^{14}CO_2$ Fixation Products

Algal cells were extracted successively with 80% methanol, 20% methanol and water for 15 min each at $65°C$. Extracts were combined (Alcohol-water-soluble fraction). The residue was further extracted with methanol (4 times), ethanol-ether (3:1) at $60°C$ and with ether. Dried residue was fractionated according to the method of Schneider (1945) with modifications as follows. The material was washed with cold 5% perchloric acid (PCA). Then the residue was extracted with 5% PCA for 30 min at $65°C$ and then washed with cold 5% PCA. Washings and extract were combined (Hot PCA soluble fraction = nucleic acid fraction). The residue contained protein (protein fraction). The radioactivity in both fractions was determined with a liquid scintillation spectrophotometer (Beckman LS-230) using Tritosol (Fricke 1975) as scintillation cocktail. The protein and nucleic acid fractions contained large amounts of polysaccharide. However, in the mutant cells, the radioactivity was not detected in polysaccharide.

Alcohol-water-soluble fraction was analyzed using two-dimensional paperchromatography as described by Ogasawara and Miyachi (1970). When necessary, electrophoresis (Camag, Switzerland) was carried out using a solution consisting of formic acid-acetic acid-water (26:120; 1000), pH 1.9 at 4000 V for 20 min. Nucleic acid was hydrolyzed either with N HCl or with 70% PCA according to Oh-hama and Hase (1976). Hydrolysates were separated by paperchromatography with solvent consisting of isopropanol (130 ml): conc. hydrochloric acid (33.2 ml): water (total volume, 200 ml). Temperatures during the development of PCA-hydrolysate (purine and pyrimidine bases) and HCl-hydrolysate (purine bases and pyrimidine ribophosphates) were $10°C$ and $25°C$, respectively. Protein fraction was hydrolyzed with 6 N HCl at $100°C$ for 18 h. Amino acids thus liberated were determined by two-dimensional paperchromatography (Ogasawara and Miyachi 1970). The radioactivity was determined with a liquid scintillation spectrophotometer or G.M. counter. The values thus obtained were normalized to those determined with a gas flow counter.

2.4 Light-Induced Greening and Chlorophyll Analysis

For greening experiment, algal suspensions were illuminated with daylight fluorescent lamp (10 klx), and aerated with air containing 1% CO_2 at $30°C$. Chlorophyll was extracted with hot methanol containing $MgCO_3$ and its concentration was determined with the use of the specific absorption coefficients of chlorophylls a and b given by Ogawa and Shibata (1965).

2.5 6-MP and CH

One mM 6-MP or 0.1 μg/ml CH (final conc. of methanol, 0.05%) was added into algal suspension unless otherwise described.

3 Results

3.1 Effect of Light on CO_2 Fixation in Mutant Cells Suspended in Phosphate Buffer

Figure 1 shows that $^{14}CO_2$ fixation in heterotrophically grown yellow cells of *Scenedesmus obliquus* C-2A' was enhanced by illumination. In alcohol-water soluble fraction, ^{14}C was mostly incorporated into aspartate (Fig. 2). The radioactivity was also found in glutamate, malate, and citrate in the decreasing order. The radioactivity incorporated into phosphate esters was very small. Time course of the percent incorporation of ^{14}C into aspartate showed significant negative slope, while that into nucleic acid-plus-protein fraction showed positive slope (Fig. 3). Figure 4 shows that ^{14}C-incorporation into protein as well as nucleic acid fraction was greatly enhanced by illumination.

3.2 Effects of 6-MP and Nutrients on Respiration and Light-Induced $^{14}CO_2$ Fixation in Mutant Cells

With respect to the inhibition of CH on respiration during the greening process of the mutant cells, Oh-hama and Senger (1975) found that the light-induced respiration consists of two successive steps; an initial CH-insensitive and the following CH-sensitive step. Similarly, when the mutant cells were illuminated in the presence of 6-MP, an inhibitor of nucleic acid synthesis, the light enhancement in respiration stopped after 5–10 min (Fig. 5A). On the other hand, when the mutant cells suspended in the basal medium were illuminated, there occurred a gradual increase following an initial rapid enhancement in respiration. Enhancement in respiration was also stopped after 5–10 min when the mutant cells suspended in the nitrate-free medium were illuminated (Fig. 5B). Different from nitrate, the removal of $MgSO_4$ from basal medium did not suppress the light-induced respiration. These results indicate that the initial enhancement is 6-MP-insensitive, while the later phase is sensitive to this inhibitor and requires the addition of N-source in the medium.

Figure 6A shows the time courses of $^{14}CO_2$ fixation in the mutant cells which were suspended in phosphate buffer and that in which nitrate and magnesium had been added (N + Mg medium). When $NaH^{14}CO_3$ was added simultaneously with turning on the light, the addition of nitrate and magnesium did not show significant effect. However, when $NaH^{14}CO_3$ was added 35 min after the start of illumination, $^{14}CO_2$ fixation in the cells suspended in N + Mg medium was significantly faster than in those suspended in phosphate medium. Figure 6B shows that ^{14}C-incorporation into protein as well as nucleic acid fraction was higher in N + Mg medium than in phosphate medium, and the difference was greater in the later period than immediately after the start of illumination.

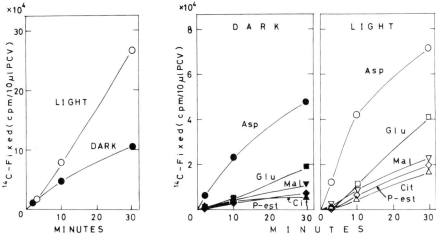

Fig. 1   **Fig. 2**

Fig. 1. Time courses of $^{14}CO_2$ fixation in *Scenedesmus obliquus* C-2A' in the light and in darkness. Algal cells were suspended in 4 mM phosphate buffer, pH 7.6. Cell density, 11 μl PCV/ml. Light intensity, 10 klx. $NaH^{14}CO_3$, 0.5 mM. The data shown in Figs. 1–4 were the results of the same experiment

Fig. 2. Time courses of ^{14}C-incorporation in alcohol-water-soluble products. *Asp* aspartate; *Glu* glutamate; *Cit* citrate; *Mal* malate; *P-est* phosphate esters

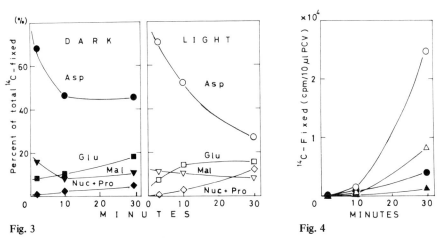

Fig. 3   **Fig. 4**

Fig. 3. Percent incorporation of ^{14}C into products. *Nuc* nucleic acid, *Pro* protein

Fig. 4. Time courses of ^{14}C-incorporation into nucleic acid and protein fraction in the light and in darkness. *Triangles* nucleic acid; *circles* protein. *Open symbols* light incubation; *solid symbols* dark incubation

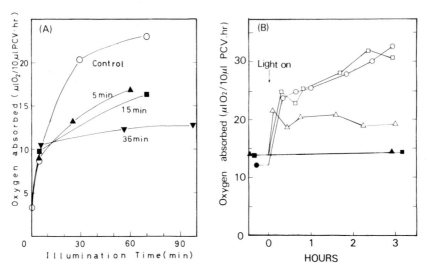

Fig. 5A, B. Effects of 6-MP (**A**) and nitrate and magnesium (**B**) on light-induced respiration. Rate of oxygen uptake was determined with Clark type electrode (Rank Brothers, London) at 30°C. **A** Time indicated at each curve shows the duration of preincubation with 1 mM 6-MP in the dark. Eight-day old culture was illuminated without medium change. **B** ○ basal medium; △ basal medium minus nitrate; □ basal medium minus magnesium sulfate. *Open symbols* light incubation; *solid ones* dark incubation

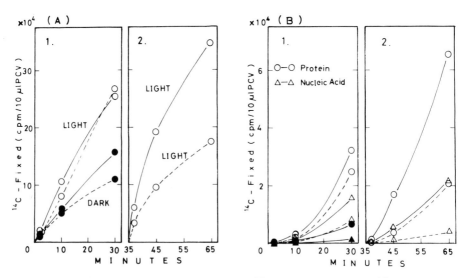

Fig. 6A, B. Effect of nitrate-plus-magnesium on total $^{14}CO_2$ fixation (**A**) and in ^{14}C-incorporation into nucleic acid and protein fraction (**B**). − − − phosphate buffer; ⎯⎯ phosphate buffer plus 8 mM nitrate and 1 mM magnesium sulfate. 0.5 mM NaH^{14}CO$_3$ was added at 0 min (A-1, B-1) or 35 min after the start of illumination (A-2, B-2) (for symbols see Fig. 4)

3.3 Effects of 6-MP and CH on Light-Induced $^{14}CO_2$ Fixation in Mutant Cells

The addition of 6-MP did not show an appreciable effect, while CH showed slight but reproducible stimulation on the light-induced $^{14}CO_2$ fixation in mutant cells which were suspended in phosphate medium (Fig. 7A). Figure 7B shows that 6-MP greatly decreased ^{14}C-incorporation into nucleic acid fraction. ^{14}C-incorporation into protein fraction was greatly inhibited by CH. 6-MP also inhibited ^{14}C-incorporation into pro-

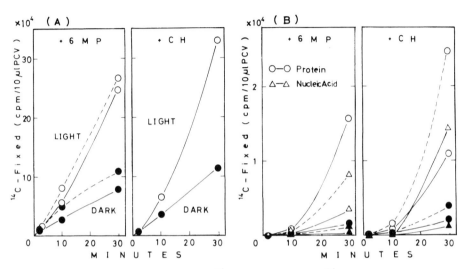

Fig. 7A, B. Effects of 6-MP and CH on total $^{14}CO_2$ fixation (**A**) and ^{14}C-incorporation into nucleic acid and protein fraction (**B**) in mutant cells kept in phosphate buffer. Cells were preincubated with 6-MP or CH for 10 min in the dark. – – – Control without inhibitor; —— plus inhibitor. Concentration of NaH$^{14}CO_3$, 0.5 mM; *open symbols* light incubation; *solid ones* dark incubation

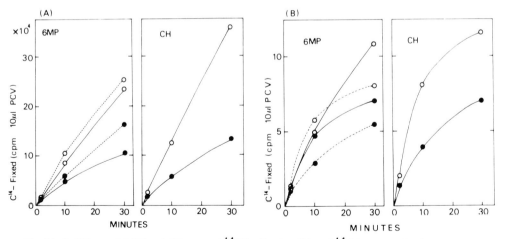

Fig. 8A, B. Effects of 6-MP and CH on total $^{14}CO_2$ fixation (**A**) and ^{14}C-incorporation into aspartate (**B**) in mutant cells suspended in nitrate and magnesium medium (for explanation see Fig. 7)

tein fraction, while CH did not inhibited ^{14}C-incorporation into nucleic acid fraction. Table 1 shows that the radioactivity in aspartate was greatly increased by the addition of 6-MP and CH.

The similar effects of 6-MP and CH were observed in the mutant cells which were suspended in N + Mg medium: Total ^{14}CO$_2$ incorporation in light was enhanced by CH, but 6-MP did not show appreciable affect (Fig. 8). ^{14}C-incorporation into aspartate in light was also enhanced by 6-MP and CH.

Table 1. Effects of 6-MP and CH on incorporation of radioactivity in products of ^{14}CO$_2$ fixation in mutant cells kept in phosphate buffer in the light or in darkness

	Radioactivity incorporated ($\times 10^3$ cpm/10 μl PCV · 30 min)					
	Light			Dark		
	Control	6-MP	CH	Control	6-MP	CH
Aspartate	71.1 (26.8)	100.5 (40.6)	114.0 (35.4)	47.9 (45.6)	38.5 (49.7)	50.2 (49.4)
Malate	22.5 (8.5)	12.4 (5.0)	18.7 (5.8)	11.6 (11.0)	7.6 (9.8)	7.9 (7.4)
Citrate	16.2 (6.1)	8.7 (3.5)	14.4 (4.5)	5.7 (5.4)	4.4 (5.7)	4.8 (4.7)
Glutamate	40.2 (15.3)	28.8 (11.6)	31.6 (9.8)	19.3 (18.4)	13.1 (16.9)	15.4 (15.2)
P-esters	19.8 (7.5)	6.2 (2.5)	42.3 (13.1)	7.3 (7.0)	1.6 (2.1)	5.8 (5.7)
Nucleic acid	8.2 (3.1)	3.4 (1.4)	14.3 (4.4)	1.3 (1.2)	0.4 (0.5)	1.3 (1.3)
Protein	24.7 (9.3)	15.6 (6.3)	10.9 (3.4)	3.9 (3.7)	1.6 (2.1)	2.1 (2.1)
Others	61.8 (23.3)	71.9 (29.1)	75.9 (23.6)	8.0 (7.6)	10.3 (13.3)	14.1 (13.9)
Total	264.9	247.5	322.1	105.0	77.5	101.6

Figures in parentheses show percent incorporation of radioactivity

3.4 Effects of 6-MP and Nutrients on Chlorophyll Synthesis in Mutant Cells

Figure 9A shows that the addition of MgSO$_4$ and KNO$_3$ to phosphate medium was necessary for the light-induced chlorophyll formation. Figure 9B shows that 1 mM 6-MP which inhibited the secondary phase of light-induced respiration almost completely suppressed chlorophyll synthesis.

3.5 Effects of Light in the Presence of CMU on ^{14}CO$_2$ Fixation and Respiration in Wild-Type Cells

Figure 10A, B shows that illumination of dark-grown wild-type cells (which contained chlorophyll at 4 μmol/ml PCV) in the presence of CMU greatly enhanced respiration and ^{14}CO$_2$ fixation. The main initial ^{14}CO$_2$ fixation product was aspartate and the radioactivity incorporated into phosphate esters was very small, if any (Fig. 10C). ^{14}C-incorporation into nucleic acid and protein fraction was also enhanced by illumination in the presence of CMU (data not shown). These results indicate that the photoinduced carbon metabolism as observed in mutant cells is also occurring in the wild-type cells.

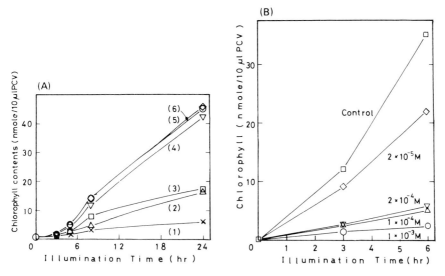

Fig. 9A, B. Effects of magnesium and nitrate (**A**) and 6-MP (**B**) on light-induced chlorophyll formation. **A:** *1* 4 mM phosphate buffer (pH 7.6) only; *2* plus 3 mM nitrate; *3* plus 1 mM magnesium sulfate; *4* plus nitrate and magnesium sulfate; *5* basal medium; *6* the medium in which algal cells had been grown. Light intensity, 10 klx. **B** Six-day-old culture was illuminated without medium change. Concentration of 6-MP is indicated in the figure

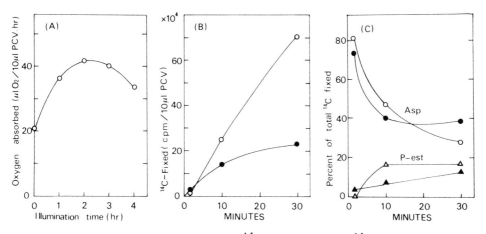

Fig. 10A -C. Effects of light on respiration (**A**) and $^{14}CO_2$ fixation (**B**) and ^{14}C-incorporation into aspartate and phosphate esters (**C**) in *Scenedesmus obliquus* wild-type cells grown in the dark. Algal cells were suspended in a medium containing 4 mM phosphate buffer (pH 7.6), 8 mM nitrate, 1 mM magnesium sulfate and 10 μM CMU (for other conditions see Fig. 1)

3.6 Light-Induced ^{14}C-Incorporation into Purine and Pyrimidine Bases in Nucleic Acid Fraction as Well as into Amino Acids in Protein Fraction

When mutant cells were fed with $NaH^{14}CO_3$ for a short period (30–60 min) in the light in the presence or absence of KNO_3-plus-$MgSO_4$, ^{14}C was incorporated into purine and pyrimidine bases as well as pyrimidine ribophosphate in nucleic acid fraction (Table 2). No radioactivity was detected in thymine, and the radioactivity incorporated into uracil was higher than that incorporated into cytosine. Similar results were obtained when dark-grown wild-type cells were illuminated in the presence of CMU. We, therefore, concluded that, under illumination, ^{14}C was incorporated into RNA in the nucleic acid fraction.

6-MP inhibited ^{14}C-incorporation into all kinds of bases, while CH did not show such inhibition.

Paperchromatographic analysis of the hydrolysates of protein fraction obtained after 60 min of $^{14}CO_2$ fixation in mutant cells revealed that the percent radioactivities in aspartate and glutamate were 35% and 20%, respectively. The same values obtained with wild-type cells (in the presence of CMU) were 28% and 14%, respectively.

Table 2. Incorporation of radioactivity among the purine and pyrimidine bases in nucleic acid fraction in *Scenedesmus* cells which were incubated with $NaH^{14}CO_3$ in the light

Expt.	Time of $^{14}CO_2$ fixation (min)	CMU	CH	6-MP	KNO$_3$+ MgSO$_4$	Uracil	(Uridylic acid)	Cytosine	(Cytidylic acid)	Adenine	Guanine	Total
							Radioactivity incorporated ($\times 10^3$ cpm/10 μl PCV) into					
	30	–	–	–	–	1.9		1.1		3.6	1.5	8.1
	30	–	–	–	–		(2.5)		(1.0)	2.7	1.9	8.1
1	30	–	+	–	–	3.0		1.1		3.3	3.2	10.6
	30	–	+	–	–		(3.3)		(1.4)	3.7	6.2	14.6
	60	–	–	–	+	11.8		6.2		9.4	7.5	34.9
	60	–	–	–	+		(13.7)		(8.3)	6.6	8.9	37.5
2	60	–	+	–	+	6.3		2.7		5.3	5.8	10.1
	60	–	–	+	+	3.8		1.6		2.3	1.7	9.4
3	60	+	–	–	+	37.6[a]		18.1		14.2	4.6	–
	60	+	–	–	+		(26.5)		(14.6)	10.7	5.5	57.3

Expt. 1 and 2: mutant cells; Expt. 3: wild-type cells
[a] The radioactivity due to glucose is also included

4 Discussion

We showed that the primary product in the light-induced $^{14}CO_2$ fixation in mutant as well as dark-grown wild-type cells of *Scenedesmus obliquus* C-2A' is aspartate. With *Chlorella* cells, Kamiya and Miyachi (1974) found that blue light enhanced not only respiration but also $^{14}CO_2$ fixation into aspartate. According to their conclusion phosphoenolpyruvate (PEP) carboxylase activity is enhanced by the short wavelength light and the resulting product, OAA, is converted to aspartate. They also showed evidence indicating that blue light effects on $^{14}CO_2$ fixation and respiration are closely related with each other. It has been shown that blue light is effective for the enhancement of respiration in the mutant cells (see Introduction). With the same mutant cells of *Scenedesmus*, Miyachi et al. (1980a) recently found that blue light is also effective for $^{14}CO_2$ fixation. We therefore concluded that the light-induced $^{14}CO_2$ fixation observed with mutant and wild-type cells of *Scenedesmus* is caused by blue light. Possibly, the initial $^{14}CO_2$ fixation was due to the blue light-enhanced PEP carboxylase reaction. ^{14}C would then be incorporated into RNA and protein fraction. Analyses of these macromolecules showed that about half of total radioactivity was in aspartate and glutamate of protein fraction and both purine and pyrimidine bases in RNA had radioactivity (Table 2). There-fore, at least half of ^{14}C incorporated into protein fraction in light would be derived from OAA formed as a result of PEP carboxylation reaction (Fig. 11). The radioactivity in pyrimidine bases in RNA will be derived from aspartate and carbamoylphosphate. On the other hand, only the nitrogen atom of the aspartate molecule is transferred to purine bases and radioactivity is increased by the direct $^{14}CO_2$-incorporation (Schulman 1961). Figure 11 shows that OAA and aspartate formed as a result of light-induced PEP carbo-xylation reaction is essential to the light-induced syntheses of protein and RNA in mutant cells of *Scenedesmus*.

Steup et al. (1977) reported the evidence for blue light-induced synthesis of cyto-plasmic RNA in *Chlorella*. Schrott and Rau (1977) found Poly-(A)-containing RNA in *Fusarium*, whose carotenoid synthesis is known to be blue light-dependent. The form of RNA produced under illumination of *Scenedesmus* cells is yet to be studied. In this connection, we found that when ^{14}C-incorporation into RNA was inhibited by 6-MP,

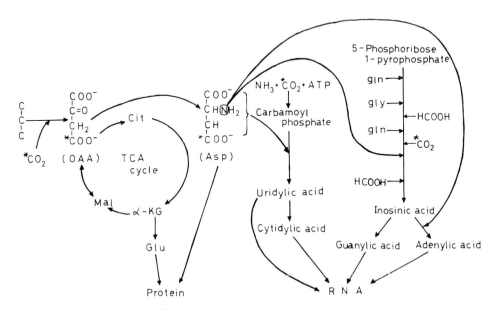

Fig. 11. A possible pathway of $^{14}CO_2$ fixation stimulated by light

[14]C-incorporation into protein (Fig. 7) and the secondary phase of light-induced respiration (Fig. 5A) were also suppressed. However, 6-MP did not affect total $^{14}CO_2$ fixation. This may be the reason why ^{14}C-aspartate was accumulated in the presence of this inhibitor (Table 1, Fig. 8). Gerhart and Pardee (1962) reported that cytidine triphosphate inhibits aspartate transcarbamylase which catalyzes the reaction between carbamoyl-phosphate and aspartate (Fig. 11). It is therefore also possible that this type of feedback inhibition which leads to the accumulation of aspartate is controlling pyrimidine bio-synthesis of this alga. It has been known that citrate synthase is inhibited by ATP and limits the turnover rate of the TCA cycle. It is also possible that suppression of respiration coupled with the inhibition of RNA synthesis by 6-MP is caused by the above type of inhibition.

The addition of CH which inhibits (cytoplasmic) protein synthesis did not suppress ^{14}C-incorporation into RNA (Fig. 7B and Table 2). CH rather enhanced total $^{14}CO_2$ fixation and ^{14}C-incorporation into aspartate (Fig. 8). One of the possibilities which caused the enhancement of $^{14}CO_2$ fixation is the acceleration of PEP carboxylase reaction by CH. It is also possible that the supply of PEP to the site of carboxylase reaction is enhanced by CH. It has been shown that blue light enhances carbohydrate breakdown (Miyachi et al. 1977; Brinckmann and Senger 1978). Schwartzbach et al. (1975) found that illumination of *Euglena* in the presence of CH accelerated the breakdown of para-mylon. It was also found that CH transiently enhanced respiration (Oh-hama and Senger 1975). All these data are in accord with the inference that the enhanced supply of PEP due to the CH-induced breakdown of starch enhances PEP carboxylation reaction, and hence formation of ^{14}C-aspartate.

5 Summary

Fixation of $^{14}CO_2$ in heterotrophically grown wild-type and mutant C-2A' cells of *Scene-desmus obliquus* was enhanced by light. Analysis of the early $^{14}CO_2$-fixation products revealed that radioactivity was mostly localized in aspartate, glutamate and malate in decreasing order.

[14]C-incorporation into purine and pyrimidine bases in RNA as well as aspartate and glutamate in protein fraction was also stimulated by light. The addition of 6-methyl purine (6-MP) inhibited ^{14}C-incorporation not only into RNA but also into protein fraction. Cycloheximide (CH) inhibited ^{14}C-incorporation into protein fraction, while that into RNA was not suppressed for the initial 30 min. The amount of ^{14}C-aspartate was increased by both inhibitors.

Enhanced respiration occurring during several minutes after the start of illumination was insensitive to 6-MP. However, the later increase was suppressed by this inhibitor. Removal of nitrate from the suspending medium also suppressed this secondary increase in respiration.

Based on the above results, possible regulatory effects of light on carbon metabolism during the early period of greening was discussed.

Acknowledgments. This research was supported by grants from Japanese Ministry of Education, Science and Culture, and Ministry of Agriculture, Forestry and Fisheries (GEP-54-II-I-12).

References

Brinckmann G, Senger H (1978) Light dependent formation of thylacoid membranes during the development of photosynthetic apparatus in pigment mutant C-2A' of *Scenedesmus obliquus*. In: Akoyunoglou et al.(ed) Chloroplast development, pp 201–206. Elsevier/North-Holland Biochemical Press, Amsterdam

Fricke U (1975) Tritosol: A new scintillation cocktail based on Triton X-100. Anal Biochem 63: 555–558

Gerhart JC, Pardee AB (1962) The enzymology of control by feedback inhibition. J Biol Chem 237: 891–896

Jackson WA, Volk RJ (1970) Photorespiration. Annu Rev Plant Physiol 21: 385–432

Kamiya A, Miyachi S (1974) Effects of blue light on respiration and carbon dioxide fixation in colorless *Chlorella* mutant cells. Plant Cell Physiol 15: 927–937

Kowallik W, Gaffron H (1967) Enhancement of respiration and fermentation in algae by blue light. Nature (London) 215: 1038–1040

Miyachi S (1959) Effect of poisons upon the mechanism of photosynthesis as studied by the pre-illumination experiments using carbon-14 as a tracer. Plant Cell Physiol 1: 1–15

Miyachi S, Kamiya M, Miyachi Shizuko (1977) Wavelength effects of incident light on carbon metabolism in *Chlorella* cells. In: Mitsui A et al (eds) Biological solar energy conversion, pp 167–182. Academic Press, London New York

Miyachi S, Miyachi Shizuko, Senger H (1980a) Effect of blue light on CO_2 fixation in heterotrophically grown *Scenedesmus obliquus* mutant C-2A', pp 329–331. This volume

Miyachi Shizuko, Kamiya A, Miyachi S (1980b) Effects of blue light on respiration and non-photosynthetic CO_2 fixation in *Chlorella vulgaris* 11 h cells, pp 321–328. This volume

Ogasawara N, Miyachi S (1970) Reguration of CO_2-fixation in *Chlorella* by light of varied wavelengths and intensities. Plant Cell Physiol 11: 1–14

Ogawa T, Shibata K (1965) A sensitive method for determining chlorophyll *b* in plant extracts. Photochem Photobiol 4: 193–200

Oh-hama T, Hase E (1976) Enhancing effects of CO_2 on chloroplast regeneration in glucose-bleached cells of *Chlorella protothecoides* I. Role of non-photosynthetic CO_2-fixation in chloroplast regeneration. Plant Cell Physiol 17: 45–53

Oh-hama T, Senger H (1975) The development of structure and function in the chloroplasts of greening mutants of *Scenedesmus* III. Biosynthesis of δ-aminolevulinic acid. Plant Cell Physiol 16: 395–405

Pickett JM, French CS (1967) The action spectrum for blue-light-stimulated oxygen uptake in *Chlorella*. Proc Natl Acad Sci USA 57: 1587–1593

Schmid GH, Schwarze P (1969) Blue light enhanced respiration in a colorless *Chlorella* mutant. Hoppe Seylers Z Physiol Chem 350: 1513–1520

Schneider WC (1945) Phosphorus compounds in animal tissues I. Extraction and estimation of deoxypentose nucleic acid and pentose nucleic acid. J Biol Chem 161: 293–303

Schrott EL, Rau W (1977) Evidence for a photoinduced synthesis of poly(A) containing mRNA in *Fusarium aquaeductum*. Planta 136: 45–48

Schulman MP (1961) Purines and pyrimidines. In: Greenberg DM (ed) Metabolic pathways, vol II, pp 389–457. Academic Press, London New York

Schwartzbach SD, Schiff JA, Goldstein NH (1975) Events surrounding the early development of *Euglena* chloroplasts. V. Control of paramylum degradation. Plant Physiol 56: 313–317

Senger H, Bishop NI (1972) The development of structure and function in chloroplasts of greening mutants of *Scenedesmus* I. Formation of chlorophyll. Plant Cell Physiol 13: 633–649

Steup M, Ssymank V, Winkler U, Glock H (1977) Photoregulation of transfer and 5S ribosomal RNA synthesis in *Chlorella*. Planta 137: 139–144

Enhancement of Carbohydrate Degradation by Blue Light

W. KOWALLIK[1] and S. SCHÄTZLE[1]

Following the observation of Russian scientists of a greater accumulation of nitrogenous substances and of a smaller piling-up of carbohydrates in different higher plants under blue light exposure (Voskresenskaya 1952), it could be shown that short-wave visible radiation caused an increased loss in reserve carbohydrates in several species of unicellular green algae. In the upper line of Table 1 data for autotrophically grown *Chlorella* cells, in which photosynthesis had been inhibited by application of 10^{-5} mol/l DCMU to their normal growth medium are given. Such cells consumed about one half of their endogenous carbohydrates when kept in darkness, but lost two thirds of it when exposed to blue light for 5 h. In the lower line comparable data are listed for equally poisoned cell samples, this time, however, suspended in plain phosphate buffer instead of in growth medium. During the same time these cells used up less carbohydrates; but again lost about 30% more on blue illumination. These results make the enhancement of carbohydrate degradation by short-wave visible radiation as a consequence of biosynthesis of new proteins unlikey; they rather point to the catabolism of carbohydrate itself as the target for blue light action.

This assumption is supported by the fact that CO_2-output and O_2-uptake of DCMU-poisoned wild-type *Chlorella* cells are clearly enhanced on blue light treatment, even when the cells are suspended in buffers without any nitrogen source. Figure 1 shows that using phosphate buffer the exchange of both gases is equally increased resulting in an unchanged RQ of around unity.

For a closer characterization of this phenomenon its intensity and wavelength dependences have been measured. Figure 2 shows in the upper half the need of only very small amounts of blue light for half-saturation, which is already achieved at about 10 μW

Table 1. Loss in carbohydrates of DCMU-poisoned (10^{-5} mol/l) wild-type *Chlorella* (211-8b) suspended in different media measured after 5 h in darkness or in blue light (30°C)

	Blue	Dark	Blue / Dark
Growth medium (pH 6.5)	67.5%	54.8%	1.23
Phosphate buffer (0.1 mol/l; pH 6.5) + NO_3	57.2%	44.2%	1.29
Phosphate buffer (0.1 mol/l; pH 6.5)	38.5%	30.4%	1.27

1 Fakultät für Biologie, Universität Bielefeld, Universitätsstraße, 4800 Bielefeld, FRG

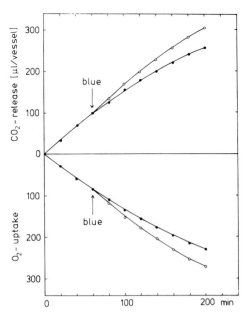

Fig. 1. Oxygen uptake and carbon dioxide release of DCMU-poisoned (10^{-5} mol/l) *Chlorella* cells suspended in 0.1 mol/l phosphate buffer (pH 6.5) in darkness (●) or in blue light (○) ($\lambda = > 380 - < 500$ nm, max. 435 nm, ~ 300 μW cm^{-2}) (29.5°C)

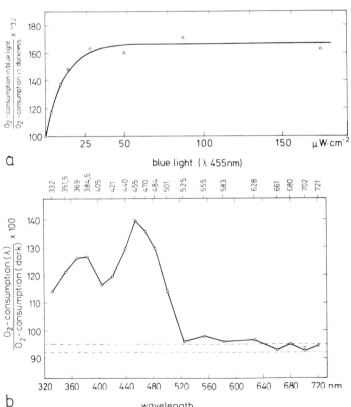

Fig. 2. a Intensity dependence, **b** wavelength dependence of the light enhanced O_2-uptake of DCMU-poisoned (10^{-5} mol/l) *Chlorella* cells, expressed as the percentage of O_2-uptake in light of that in preceding darkness. 0.1 mol/l phosphate buffer pH 7.95 (26°C). *Broken lines* indicate the variance of the dark control

cm^{-2}. It shows in the lower half that wavelengths around 460 nm and around 370 nm yield greatest effects. All wavelengths greater than 520 nm – that means the yellow, orange, red, and far red region of the visible spectrum – prove to be ineffective (Kowallik 1969a).

For further analysis of this action of blue quanta a nonphotosynthetic mutant of *Chlorella vulgaris* proved to be useful. By exposure to radioactive phosphate this organism has become unable to synthesize chlorophyll, so that no photosynthetic O_2-production occurs under illumination (Schwarze and Frandsen 1960). The cells are bright yellow because of their rather high content in carotenoids (Kowallik 1966).

This organism has to be grown heterotrophically. For that purpose glucose is added to the normal inorganic growth medium to a final concentration of 1%. The static culture is held at 30°C in darkness; it is continuously aerated with air. As a measure for growth packed cell volume, dry weight, cell number and optical density can be used (Fig. 3). They all exhibit similar trends: A steep rise to about 20-fold the initial value during the first three days is followed by a leveling-off thereafter. Reaching the plateau of propaga-

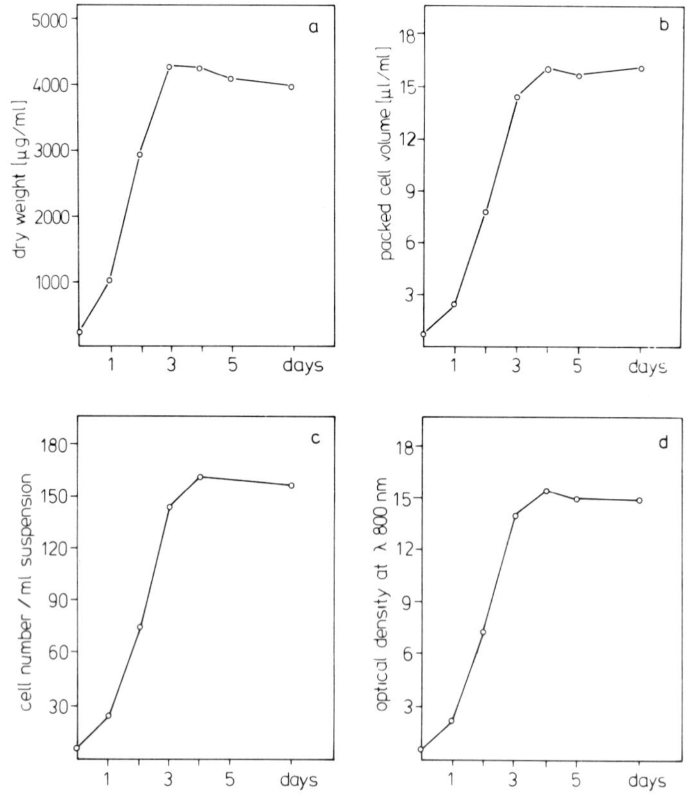

Fig. 3a-d. Dry weight (**a**), packed cell volume (**b**), cell number (**c**) and optical density (**d**) per ml cell suspension in the course of a static culture of the chlorophyll-free *Chlorella* mutant 211-11h/20 in darkness. Aeration with air (30° ± 1°C)

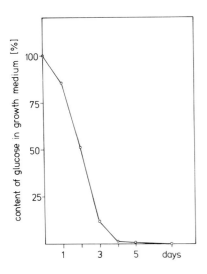

Fig. 4. Decrease of glucose in the medium of a static culture of the chlorophyll-free *Chlorella* mutant 211-11h/20 in darkness. Aeration with air ($30° \pm 1°C$)

tion coincides with an almost complete exhaustion of the medium in its organic carbon source (Fig. 4). According to the rapid uptake of exogenous glucose the amount of cell-bound carbohydrates rises initially quite drastically (Fig. 5). It declines when exogenous glucose is lacking, initially fast, after 2–3 days increasingly more slowly because of respiratory action. Cells from about seven-day-old cultures have reached a level, which only drops very little further with time (Georgi 1974).

The respiratory O_2-uptake of these mutant cells depends largely on the level of reserve carbohydrates (Fig. 6). Its rate drops with the decrease in substrate (Kirst 1974). Because of their low and only slowly changing metabolic activity with time, cells from about seven-day-old cultures are called "resting" cells in the following.

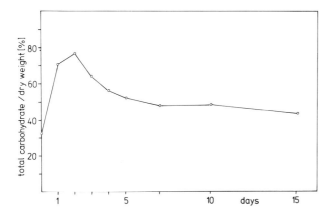

Fig. 5. Changes in the content of total carbohydrates in cells of the chlorophyll-free *Chlorella* mutant 211-11h/20 in the course of a static culture in darkness. Aeration with air ($30° \pm 1°C$)

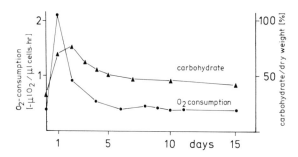

Fig. 6. Total carbohydrates (▲) and O_2-uptake (●) of cells of the chlorophyll-free *Chlorella* mutant 211-11h/20 in the course of a static culture in darkness (30° ± 1°C)

This yellow *Chlorella* mutant, too, responds to blue illumination with an increased oxidative breakdown of its carbohydrate reserves (Kowallik and Gaffron 1966). The light effect is most pronounced in gently starved "resting" cells (Fig. 7) (Kirst 1974). Figure 8 proves that their blue light-enhanced loss in carbohydrates (a) is accompanied by an equally enhanced output of CO_2 (b) and by an increased uptake of oxygen (c). Because of the greatest light effect of such "resting" cells we have preferentially used samples from 8–14-day-old cultures for further experiments.

At first we had to confirm further the equality of the mutants' response to blue light with that of the wild-type cell. Intensity and wavelength dependences have therefore been measured. Figure 9 demonstrates both these characteristica for the extra loss in carbohydrates (Georgi 1974), Fig. 10 for the simultaneous extra O_2-uptake (Kowallik 1967). In both cases the data resemble those of the green cells very closely and thus present no objection to using the yellow mutant for further experiments aimed to the elucidation of the blue light action.

It should be mentioned here that another pigment-mutant of *Chlorella*, which besides being devoid of chlorophyll is also devoid of detectable carotenoids, and which therefore appears "white" to the eye, responds quite similarly to visible radiation (Schmid and Schwarze 1969). The above characteristics of its light-enhanced O_2-uptake have been determined independently by Miyachi's group (Kamiya and Miyachi 1974) and in our laboratory (Feindler 1974) with basically identical results.

For an enhanced breakdown of carbohydrate two ways are easily imaginable: Blue light might either speed up the dark-type respiration or it might bring about a different

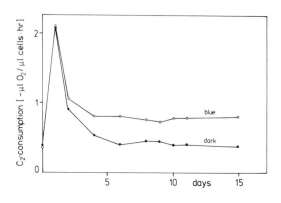

Fig. 7. O_2-uptake of cells of the chlorophyll-free *Chlorella* mutant 211-11h/20 at different phases of a static dark culture in darkness (●) or in blue light (○) ($\lambda >$ 380 nm – $<$500 nm, max. 435 nm, $\sim$ 300 μW cm^{-2}) (30° ± 0.1°C)

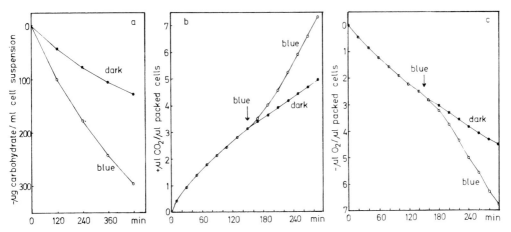

Fig. 8a–c. Loss of carbohydrate (**a**), production of CO_2 (**b**), and uptake of O_2 (**c**) of "resting" cells of the chlorophyll-free *Chlorella* mutant 211-11h/20 in darkness (●) or in blue light (○) ($\lambda > 380$ nm – < 500 nm, max. 435 nm; ~ 300 μW cm^{-2})

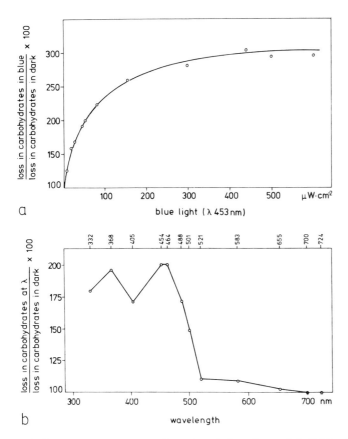

Fig. 9. a Intensity dependence, **b** wavelength dependence of the light-enhanced loss in carbohydrates of "resting" cells of the chlorophyll-free *Chlorella* mutant 211-11h/20; within 4 h (30°C)

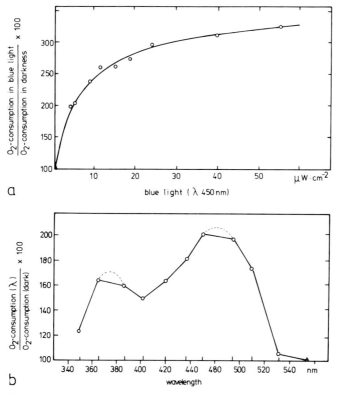

a

b

Fig. 10. a Intensity dependence, **b** wavelength dependence of the light-enhanced O_2-uptake of "resting" cells of the chlorophyll-free *Chlorella* mutant 211-11h/20. 0.1 mol/l phosphate buffer pH 6.5 $(30° \pm 0.1°C)$

way of carbohydrate degradation, replacing or superimposing the well-known mitochondrial catabolism. Such a different way is known generally in photorespiration. Its key reaction is the oxidation of glycolate. In green cells this substrate most likely derives from splitting of ribulosediphosphate and in this way connects photorespiration with the reductive pentosephosphate cycle of photosynthesis. It is, however, not definitely decided yet if it cannot also arise from other sources.

There are mainly three criteria which are widely used to discriminate between dark-type and photorespiration: First, glycolate oxidase is a flavine enzyme, and as such insensitive to cyanide; the final oxidase in dark respiration is a cytochrome, which is highly sensitive to that inhibitor. Second, glycolate oxidase is allosterically inhibited by hydroxymethanesulfonates, substances which have no effect on cytochromes at all. And third, the flavin-dependent glycolate oxidase has a very low affinity to oxygen, while cytochromes on the contrary exhibit a high affinity.

Using cyanide with *Chlorella* cells it must be kept in mind that their endogenous dark respiration cannot be suppressed completely by that poison, that the oxidation of exogenous glucose, however, is strongly affected. Figure 11 shows that KCN abolishes

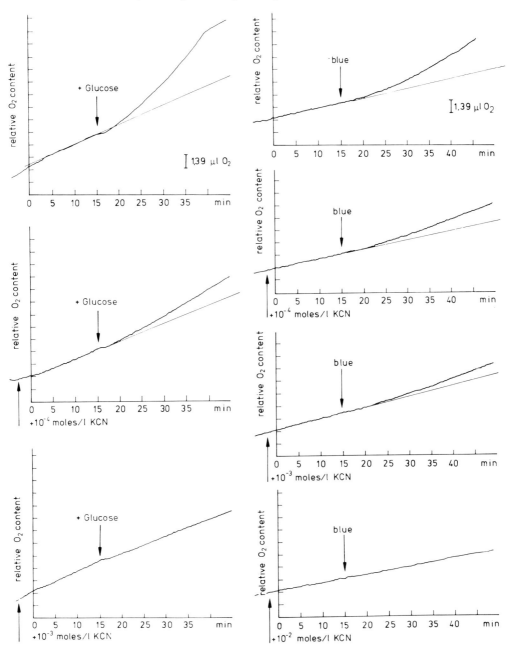

Fig. 11. Influence of different concentrations of cyanide on the O_2-uptake of "resting" cells of the chlorophyll-free *Chlorella* mutant 211-11h/20 in absence or in presence of glucose and in darkness or in blue light (λ 455 nm, $\sim$ 25 μW cm^{-2}). 0.1 mol/l phosphate buffer pH 6.5 (30° ± 0.1°C)

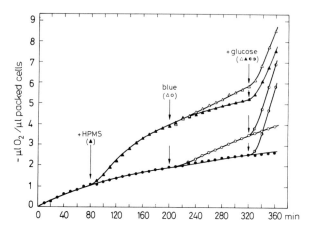

Fig. 12. Influence of hydroxypyridylmethanesulfonate ($2 \cdot 10^{-3}$ mol/l) on the O_2-uptake of "resting" cells of the chlorophyll-free *Chlorella* mutant 211-11h/20 in darkness (●▲) or in blue light (○△) ($\lambda > 380$ nm– < 500 nm, max. 435 nm; ~ 300 μW cm^{-2}) ($30° \pm 0.1°C$)

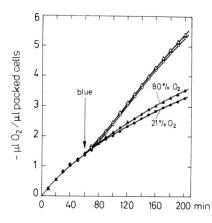

Fig. 13. Influence of different partial pressures of O_2 on the O_2-uptake of "resting" cells of the chlorophyll-free *Chlorella* mutant 211-11h/20 in darkness (●▲) or in blue light (○△) ($\lambda > 380$ nm– < 500 nm, max. 435 nm; ~ 300 μW cm^{-2}) ($30° \pm 0.1°C$)

the blue light-dependent enhancement in O_2-uptake just as it does the extra O_2-consumption brought about by exogenous glucose (Schwarzmann 1973). From Fig. 12 it can be seen that HPMS does not eliminate the enhancement in O_2-uptake by light, and Fig. 13 proves that raising the O_2-content from 21% up to 80% neither alters the cells' O_2-uptake in darkness nor changes the increased uptake in the light remarkably (Kowallik 1971). Thus none of these experiments gives rise to the idea of photorespiration being the pathway leading to greater O_2-uptake in blue light. The additional observation of a light-induced increase in the ATP-level, following intensity and wavelength dependences as the extra O_2-uptake does (Fig. 14) is no support for that assumption either (Kowallik and Scheil 1976). It rather points to an enhanced dark-type respiration. This appears also most likely from the result of chasing the products of externally applied ^{14}C-labeled glucose. Its metabolites detectable some seconds up to some minutes after application are not different in cells respiring in darkness from those in blue light (Georgi 1974).

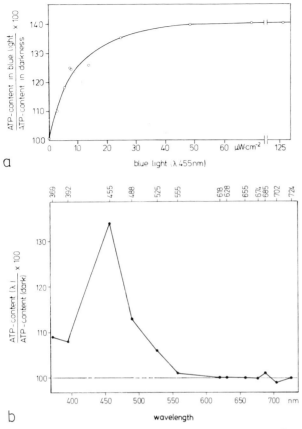

Fig. 14. a Intensity dependence, **b** wavelength dependence of the light-enhanced ATP content of "resting" cells of the chlorophyll-free *Chlorella* mutant 211-11h/20 (30°C). ATP determined enzymatically

Presuming, therefore, an increase in the well-known catabolic pathway of carbohydrates by light the question after the mechanism of such an effect of visible radiation arises. Looking for the answer it appeared rewarding to narrow down the point of attack of the effective light quanta.

Unicellular green algae are no obligate aerobes. They are able to degrade carbohydrates under anaerobiosis, too, producing different organic acids, such as acetic acid, lactic acid, formic acid, and CO_2 under exclusion of oxygen. The amounts and the ratios of the fermentation products depend largely on the algal strain and on the environmental conditions. The yellow mutant also proved to be able to produce organic acids from reserve carbohydrates under an atmosphere of nitrogen (Kowallik and Gaffron 1967). It thus offered the opportunity to check a response to blue light of glycolysis. Figure 15 shows that shining blue light on fermenting mutant cells results in an increase in acid output. This enhancement also is brought about by rather small amounts of effective wavelengths, which again are those of the near ultraviolet and the blue part of the visible spectrum (Fig. 16) (Kowallik 1969b).

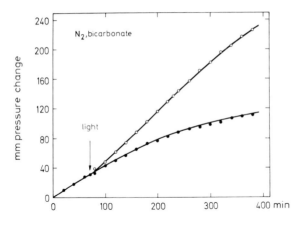

Fig. 15. Acid production of "resting" cells of the chlorophyll-free *Chlorella* mutant 211-11h/20 in darkness (●) or in blue light (○) ($\lambda > 380$ nm$- <500$ nm, max. 435 nm; ~ 300 μW cm^{-2}) under nitrogen; $30° \pm 0.1°$C. Expressed by CO_2 liberated from the acidified suspension medium (0.1 mol/l $NaHCO_3$)

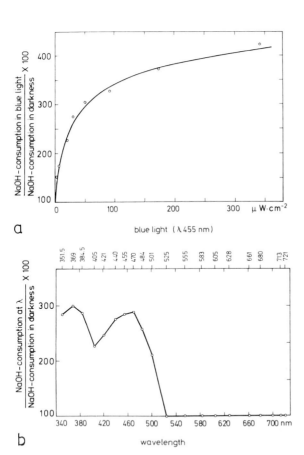

Fig. 16. a Intensity dependence, **b** wavelength dependence of acid excretion of "resting" cells of the chlorophyll-free *Chlorella* mutant 211-11h/20 under an atmosphere of nitrogen. The acid produced was measured by automatic titration with 0.01 mol/l NaOH to a steady pH. 0.002 mol/l phosphate buffer ($30°$C)

Table 2. Specific activity of *phosphofructokinase* in crude extracts of "resting" cells of a chlorophyll-free *Chlorella* mutant (211-11h/20) kept in phosphate buffer (0.1 mol/l; pH 7.5) in darkness or under blue light for different periods of time (30°C)

Time	Enzyme activity (nmol/min · mg protein)		
	Dark	Blue	$\dfrac{\text{Blue}}{\text{Dark}} \times 100$
1 h	80.5	79.8	99
4 h	82.5	84.1	102
24 h	88.2	89.3	101

Table 3. Specific activity of NAD-dependent *glyceraldehyde-3-phosphate dehydrogenase* in crude extracts of "resting" cells of a chlorophyll-free *Chlorella* mutant (211-11h/20) kept in bicarbonate (0.1 mol/l; pH 7.6) in darkness or under blue light for different periods of time (30°C)

Time	Enzyme activity (nmol/min · mg protein)		
	Dark	Blue	$\dfrac{\text{Blue}}{\text{Dark}} \times 100$
1 h	1510	1522	101
4 h	1490	1400	94
8 h	1448	1411	97

These results localize the point of attack of blue light within the non strictly oxygen-dependent part of carbohydrate breakdown. Consequently we next looked for changes in the activity of some glycolytic enzymes. While hexokinase, glucose-6-phosphate-dehydrogenase, and aldolase are still under examination, there is already some information available on phosphofructokinase, on NAD-dependent glyceraldehydephosphate-dehydrogenase, and on pyruvate-kinase.

Phosphofructokinase, which prepares its substrate molecule for splitting into two triosephosphates by phosphorylation, is well known as a regulatory enzyme in animal cells. Its action is especially affected allosterically by ATP, thus regulating demand and supply of useable energy. In crude extracts of yellow mutant cells obtained from slightly starved cultures, phosphofructokinase converts 80–90 nmol of fructose-6-phosphate to fructose-1,6-diphosphate per mg protein per minute under optimal conditions (Table 2). Extracts of cells exposed to blue light for different times do not exhibit any difference in their phosphofructokinase activity (Schätzle 1978).

NAD-dependent glyceraldehydephosphate-dehydrogenase, oxidizing glyceraldehydephosphate to phosphoglycerate, can operate with a much greater turnover rate. In vitro about 1500 nmol of substrate can be converted by one mg protein per minute. These rates, too, are not changed in crude extracts of previously blue light exposed "resting" cells (Table 3) (Conradt 1976). Because of the high turnover potency this is, however, not unexpected.

Pyruvate kinase finally, which by transferring the phosphate group of phosphoenol-pyruvate to ADP, thus forming pyruvate on the one side and fixing parts of the oxidation

Table 4. Specific activity of *pyruvate kinase* in crude extracts of "resting" cells of a chlorophyll-free *Chlorella* mutant (211-11h/20) kept in bicarbonate (0.1 mol/l; pH 7.5) in darkness or under blue light for different periods of time (30°C)

Time	Enzyme activity (nmol/min · mg protein)		
	Dark	Blue	$\dfrac{\text{Blue}}{\text{Dark}} \times 100$
1 h	42.7	54.1	127
4 h	47.0	69.0	147
8 h	52.1	88.3	170

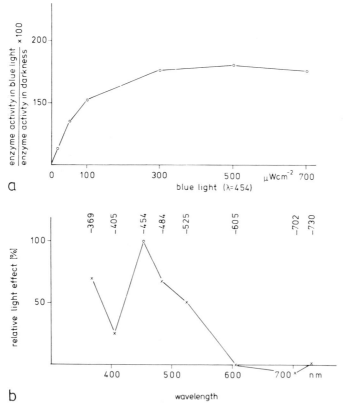

Fig. 17. a Intensity dependence, **b** wavelength dependence of the light-enhanced activity of pyruvate kinase in crude extracts from "resting" cells of the chlorophyll-free *Chlorella* mutant 211-11h/20 (30°C). Enzyme activity has been determined photometrically

energy in ATP on the other side, exhibits a maximal in-vitro-turnover rate in starved cells of 40–50 nmol per mg protein per minute only. With this low rate this enzyme could be a major bottle neck of the complete pathway and thus be privileged for its

Table 5. Specific activity of *pyruvate kinase* in crude extracts of "resting" cells of a chlorophyll-free *Chlorella* mutant (211-11h/20) kept in bicarbonate (0.1 mol/l; pH 7.5) ± cycloheximide (10^{-3} mol/l) or ± actinomycin D (150 μg/ml) in darkness or in blue light for 3 h (30°C)

| | Enzyme activity (nmol/min · mg protein) | | | |
| | Cycloheximide | | Actinomycin | |
	−	+	−	+
Dark	48.2	35.0	48.6	41.7
Blue	75.8	37.7	75.9	45.7
Increase in blue	57.3%	7.7%	56.2%	9.6%

regulation. In crude extracts of blue illuminated cells pyruvate kinase converts almost twice as much substrate as in extracts from cells in darkness (Table 4) (Ruyters 1976). This effect appears directly connected with the light-increased oxidative carbohydrate breakdown: Its intensity and wavelength dependences (Fig. 17) both resemble those for light-enhanced O_2-uptake quite closely (Ruyters and Kowallik, 1980).

From this it appears possible that blue light takes part in the regulation of the degradative carbohydrate flow by influencing the turnover rate of bottle neck enzymes.

To achieve that two major possibilities can easily be seen, first an activation of enzyme protein present, and second biosynthesis of new protein appear to be possible. To obtain first information on this, yellow mutant cells have been exposed to blue light either in the presence or in the absence of inhibitors of protein biosynthesis. Table 5 shows the results obtained with cycloheximide and actinomycin D: While in the unpoisoned controls there are enhancements in activity by blue light of 57% and 56%, in the poisoned samples increases of less than 10% are recorded (Kowallik and Ruyters 1976).

(It may be noted that on poisoning the pyruvate kinase activity per protein drops also in dark-kept cells by 20%–30%; this may be taken as information on the life-time of the enzyme. We did not go into this any further.)

These results are certainly a strong indication for a blue light-induced biosynthesis of pyruvate kinase. This conclusion obtains additional support by two more observations: First the increase in pyruvate kinase activity by light is still greater in cells supplied with exogenous nitrogen sources, and second, there is no measurable enhancement in the activity of that enzyme when crude extracts, prepared from previously dark-kept cells, are illuminated for up to two hours. The definite proof of light-induced protein-synthesis, however, is still missing. Neither attempts to produce a heavier enzyme fraction by applying fitting isotopes during the period of illumination, nor experiments to separate different species of pyruvate kinase from illuminated and from dark-kept cells have been performed yet.

Assuming at this point the induction of new protein synthesis as the only response of the enzyme to blue light — an assumption which will partly by questioned in the following Chapter by Ruyters — we have to ask for the mechanism of that light action. From the literature it is well known that in animal tissues, but also in some plants like yeast and *Euglena,* the activity of pyruvate kinase is — besides allosterically — also regulated by synthesis of the enzyme. Its amount depends largely on the level of carbohy-

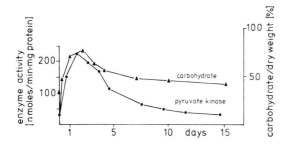

Fig. 18. Changes in the content of total carbohydrates (▲) and in the activity of pyruvate kinase (●) in the course of a static dark culture of the chlorophyll-free *Chlorella* mutant 211-11h/20 (30°C)

Table 6. Specific activity of *pyruvate kinase* in crude extracts of "resting" cells of a chlorophyll-free *Chlorella* mutant (211-11h/20) kept in bicarbonate (0.1 mol/l; pH 7.5) in absence or in presence of 1% glucose for 3 h in the dark (30°C)

	− Glucose	+ Glucose	Increase by glucose
Enzyme activity (nmol/min · mg protein)	50.1	86.8	73.3%

drate in the organism's food. Since in a static culture the level of exogenous glucose drops continuously with time, it provides the chance to check this possibility by looking for the turnover rate of the enzyme in one and the same culture in the course of about 15 days. It can be seen from Fig. 18 that the in-vitro-activity of pyruvate kinase is rather high during the first 2–3 days, but that it declines subsequently. At about the 10th day it reaches a rate which only slowly changes later on (Ruyters 1976). This trace resembles quite closely that of the content of endogenous carbohydrates. It thus nourishes the possibility of a substrate-dependent biosynthesis of that enzyme for the yellow *Chlorella* mutant, too. To gain additional information on this we added exogenous glucose to 10–14 days old – i.e., carbohydrate-depleted – cells. Table 6 shows that the activity of pyruvate kinase rose by more than 70% within 3 h. This increase, too, is largely prevented by poisoning with cycloheximide or with actinomycin D (Table 7) (Kowallik and Ruyters 1976): While the controls show increases in activity per protein by 69% and by 75% on addition of glucose, the samples plus inhibitor exhibit much smaller effects. They come about largely, however, by the drop in activity in the glucose-free poisoned sample which is probably due to the rather short life-time of the enzyme. If one assumes that glucose stabilizes the enzyme present, a better comparison might be (minus glucose – minus poison, which is the control) against (plus glucose – plus poison). In that case there is only an enhancement of around 8% observable.

The similarity in the behavior of pyruvate kinase activity of the yellow *Chlorella* mutant to exogenous glucose on the one side and to blue light on the other leads to the idea that short-wave visible radiation might provide a substrate and in this way induce the synthesis of new enzyme protein. The assumption is supported by the observation that cells in which the oxidation of triosephosphate has been inhibited by monoiodoacetate contain roughly 50% more free sugars and sugarphosphates when kept under blue light than those held in darkness (Kowallik and Ruyters 1976).

Table 7. Specific activity of *Pyruvate kinase* in crude extracts of "resting" cells of a chlorophyll-free *Chlorella* mutant (211-11h/20) kept in bicarbonate (0.1 mol/l; pH 7.5) ± cycloheximide (10^{-3} mol/l) or ± actinomycin D (150 μg/ml) in absence or in presence of exogenous glucose (1%) for 3 h in the dark (30°C)

| | Enzyme activity (nmol/min · mg protein) | | | |
| | Cycloheximide | | Actinomycin | |
	−	+	−	+
− Glucose	44.2	33.2	43.3	34.3
+ Glucose	74.7	47.9	76.0	46.8
Increase by glucose	69.1%	44.2%	75.5%	36.4%

The question of how blue light builds up a higher level in respirable molecules cannot yet be answered. It can be shown electronmicroscopically that the cell's carbohydrate reserves are deposited inside its rudimentary chloroplast. This organelle looks like an almost empty bag, containing only some single thylacoid membranes. Thus it can be speculated that blue light somehow increases the outflow of carbohydrates from their storage compartment into the cytosol. This event might give rise to a substrate-induced biosynthesis of bottle neck enzymes, which could result in an increase in the total flow of carbohydrates through their catabolic pathway.

References

Conradt W (1976) Über die Wirkung kurzwellig-sichtbarer Strahlung auf die Aktivität von Atmungsenzymen bei *Chlorella* (PEP-Carboxylase, NAD- bzw. NADP-abhängige GAP-Dehydrogenase). Staatsexamensarbeit, Köln

Feindler U (1974) Über den Einfluß von Licht auf den Atmungsgaswechsel einer farblosen *Chlorella*-Mutante. Energie- und Wellenlängenbedarf. Staatsexamensarbeit, Köln

Georgi M (1974) Über blaulichtbedingte Veränderungen im Kohlenhydrat- und Proteingehalt einer chlorophyllfreien Mutante von *Chlorella vulgaris*. Dissertation, Köln

Kamiya A, Miyachi S (1974) Effects of blue light on respiration and carbon dioxide fixation in colorless *Chlorella* mutant cells. Plant Cell Physiol 15: 927–937

Kirst R (1974) Versuche zur Charakterisierung der Atmung von *Chlorella* im Dunkel und im Licht. Temperatur- und pH-Abhängigkeit. Diplomarbeit, Köln

Kowallik W (1966) Chlorophyll-independent photochemistry in algae. Energy conversion by the photosynthetic apparatus. Brookhaven Symp Biol 19: 467–477

Kowallik W (1967) Action spectrum for an enhancement of endogenous respiration by light in *Chlorella*. Plant Physiol 42: 672–676

Kowallik W (1969a) Der Einfluß von Licht auf die Atmung von *Chlorella* bei gehemmter Photosynthese. Planta 86: 50–62

Kowallik W (1969b) Eine fördernde Wirkung von Blaulicht auf die Säureproduktion anaerob gehaltener Chlorellen. Planta 87: 372–384

Kowallik W (1971) Light stimulated respiratory gas exchange in algae and its relation to photorespiration. In: Hatch MD, Osmond CB, Slatyer RO (eds) Photosynthesis and photorespiration, pp 514–522. Wiley-Interscience, New York London Sydney Toronto

Kowallik W, Gaffron H (1966) Respiration induced by blue light. Planta 69: 92–95

Kowallik W, Gaffron H (1967) Enhancement of respiration and fermentation in algae by blue light. Nature (London) 215: 1038–1040

Kowallik W, Ruyters G (1976) Über Aktivitätssteigerungen der Pyruvatkinase durch Blaulicht oder Glucose bei einer chlorophyllfreien *Chlorella*-Mutante. Planta 128: 11—14

Kowallik W, Scheil I (1976) Lichtbedingte Veränderungen des ATP-Spiegels einer chlorophyllfreien *Chlorella*-Mutante. Planta 131: 105—108

Ruyters G (1976) Über Aktivitätsänderungen kohlenhydratabbauender Enzyme von *Chlorella* im Dunkel und im Blaulicht. Dissertation, Köln

Ruyters G, Kowallik W (1980) Further studies of the light-mediated change in the activity of pyruvate kinase of a chlorophyll-free *Chlorella* mutant. Z Pflanzenphysiol 96: 29—34

Schätzle S (1978) Die Wirkung verschiedener Faktoren auf die Aktivität der Phosphofructokinase bei *Chlorella*. Diplomarbeit, Bielefeld

Schmid GH, Schwarze P (1969) Blue light enhanced respiration in a colorless *Chlorella* mutant. Z Hoppe Seylers Z Physiol Chem 350: 1513—1520

Schwarze P, Frandsen NO (1960) Herstellung von *Chlorella*-Farbmutanten mit Hilfe von radioaktiven Isotopen. Naturwissenschaften 47: 47

Schwarzmann A (1973) Vergleichende Untersuchungen zum lichtgesteigerten Atmungsgaswechsel und zur Glykolat-Oxidation einer chlorophyllfreien *Chlorella*-Mutante. Diplomarbeit, Köln

Voskresenskaya NP (1952) Der Einfluß der spektralen Zusammensetzung des Lichts auf das Verhältnis der Photosyntheseprodukte. Dokl Akad Nauk SSSR 86: 429—435

Blue Light-Effects on Enzymes
of the Carbohydrate Metabolism in *Chlorella*
1. Pyruvate Kinase

G. RUYTERS[1]

In analyzing the effects of blue light on the metabolism of a chlorophyll-free, carotenoid-containing mutant of *Chlorella vulgaris* Kowallik and Ruyters (1976) demonstrated an enhancement in the activity of pyruvate kinase (EC 2.7.1.40). While the intensity- and wavelength-dependences show good agreement with corresponding data of other blue light responses of that alga, the duration of illumination does not (Ruyters and Kowallik 1980). An analogous discrepancy has been reported for the blue light-caused stimulation of phosphoenolpyruvate carboxylase activity and of CO_2 fixation in a colorless *Chlorella* mutant by Kamiya and Miyachi (1975). These authors therefore presumed a dual effect of blue light.

To test the validity of this hypothesis for pyruvate kinase of our yellow *Chlorella* mutant we compared some important kinetic properties of that enzyme from blue-irradiated and from dark-kept cells. For this purpose the cells were broken by shaking with glass beads in a Vibrogen cell mill for 5 min, the resulting homogenate was centrifuged at $20,000\,g$ for 30 min. The supernatant, i.e., the crude extract, was taken for the determination of enzyme activity and soluble protein. Fractionation with $(NH_4)_2SO_4$ and following dialysis caused no fundamental changes in the results obtained, so that we renounced a partial purification of pyruvate kinase at this stadium of our investigations.

Pyruvate kinase catalyzes the formation of pyruvate and ATP, phosphoenolpyruvate (PEP) and ADP being the substrates. In crude extracts from dark-kept cells the enzyme exhibits normal Michaelis-Menten kinetics towards ADP, but shows sigmoidicity towards PEP. Figure 1 shows the effect of increasing concentrations of PEP in the presence of saturating levels of ADP on the initial velocity of the reaction at pH 7.5. The $S_{0.5}$ value for PEP as determined from Hill plots is approx. 2.6 mmol/l (Fig. 2); the Hill coefficient given by the slope of the line is nearly 1.5, thus pointing to at least two binding sites for PEP. According to Koshland et al. (1966) and Atkinson (1966), the term $S_{0.5}$ is used instead of K_m for an enzyme that does not follow simple Michaelis kinetics.

Blue light treatment of the cells causes a clear change in some kinetic properties (Figs. 3 and 4): Firstly, after a 16-h blue illumination the maximal velocity (V_{max}) is enhanced by some 35%, secondly the $S_{0.5}$-value is decreased by approx. 40% to 1.6 mmol/l. This indicates a higher affinity of the "blue light enzyme" for PEP. The Hill coefficient for PEP is not affected by blue light, and neither are the kinetic properties of the enzyme towards ADP.

1 Lehrstuhl für Stoffwechselphysiologie, Fakultät für Biologie der Universität Bielefeld, Universitätsstraße, 4800 Bielefeld 1, FRG

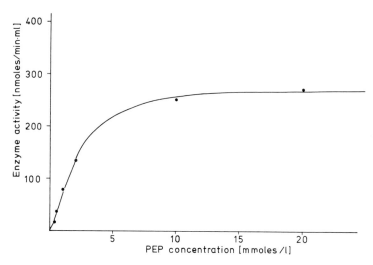

Fig. 1. Activity of pyruvate kinase of the dark-grown *Chlorella* mutant 211-11h/20 as a function of PEP concentration at saturating ADP concentration

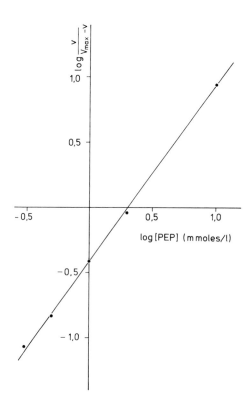

Fig. 2. Hill plot of the data of Fig. 1

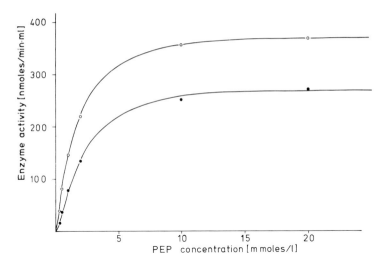

Fig. 3. Activity of pyruvate kinase of the yellow *Chlorella* mutant 211-11h/20 as a function of PEP concentration at saturating ADP concentration; comparison of dark-kept (*closed circles*) and 16 h blue-irradiated cells ($\lambda < 550$ nm; approx. 250 μW cm^{-2}; *open circles*)

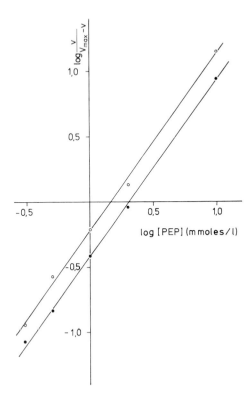

Fig. 4. Hill plot of the data of Fig. 3

It should be stressed that the $S_{0.5}$-values reflect intrinsic properties of the enzyme molecule, whereas the V_{max}-values are a function of both turnover number of the individual enzyme molecule and the amount of enzyme. Therefore, the differences between pyruvate kinase from dark-kept and from blue-irradiated *Chlorella* cells described above may reflect both enzyme properties and quantities; they might thus be the result of only one or of two blue light-dependent processes.

To decide this question we investigated the dependences of both these altered kinetic data on the duration of light treatment and on the light intensity.

In the first case cell suspensions were irradiated with blue light of $\lambda < 550$ nm at a quantum flux of approx. 250 μW cm^{-2} for different times. Figure 5 shows the increase of V_{max}, Fig. 6 the decrease of $S_{0.5}$(PEP) dependent on the time of illumination. For better comparison both these data are plotted together in Fig. 7. It is evident that (1) both processes need a relatively long period of illumination before reaching a maximal effect, (2) the time courses are different: the V_{max} increase has a lag period of about 1/2 to 1 h and is half-saturated after approx. 5 h of blue light, whereas the $S_{0.5}$(PEP) decrease is measurable after only 1/2 h, being half-saturated after 2 h already.

This discrepancy is much more pronounced when the *Chlorella* mutant cells are somewhat depleted in carbohydrate reserves by suspending them in plain buffer for 16 h prior to illumination (Fig. 8). While the time course of the V_{max} increase is not

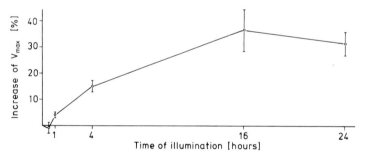

Fig. 5. Blue light-mediated increase of V_{max} of pyruvate kinase of the yellow *Chlorella* mutant 211-11h/20 in dependence on the duration of light treatment ($\lambda < 550$ nm; approx. 250 μW cm^{-2}) expressed as $\dfrac{blue - dark}{dark} \times 100$ (%)

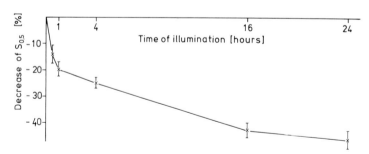

Fig. 6. Blue light-mediated decrease of $S_{0.5}$(PEP) of pyruvate kinase of the yellow *Chlorella* mutant 211-11h/20 in dependence on the duration of light treatment ($\lambda < 550$ nm; approx. 250 μW cm^{-2}) expressed as $\dfrac{blue}{dark} \times 100 - 100$ (%)

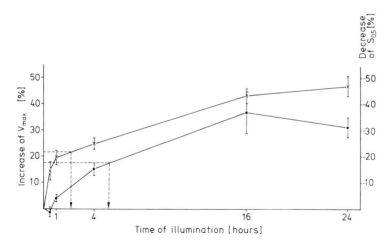

Fig. 7. Comparison of the changes in V_{max} (*closed circles*) and $S_{0.5}$(PEP) (*crosses*) of pyruvate kinase of the yellow *Chlorella* mutant 211-11h/20 mediated by blue light treatment of different duration ($\lambda < 550$ nm; approx. 250 μW cm^{-2}); same data as Figs. 5 and 6

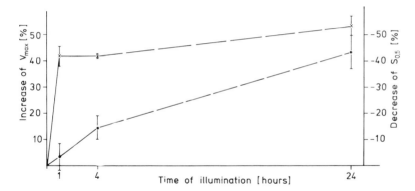

Fig. 8. Blue light-mediated changes in V_{max} (*closed circles*) and $S_{0.5}$(PEP) (*crosses*) of pyruvate kinase of the yellow *Chlorella* mutant 211-11h/20 in dependence on the duration of light treatment ($\lambda < 550$ nm; approx. 250 μW cm^{-2}); cells had been starved by suspending in plain buffer for 16 h prior to illumination

affected by this treatment the $S_{0.5}$(PEP) decrease reaches its maximal value after only 1 h of irradiation.

Corresponding differences are found in the dependence on light intensity (Fig. 9). Approximately 300 μW cm^{-2} lead to saturation of the blue light effect on both V_{max} and $S_{0.5}$; half-saturation, however, of the V_{max} increase is achieved at approximately 70 μW cm^{-2}, whereas the $S_{0.5}$ decrease needs only 30 μW cm^{-2} for half-saturation.

Thus, the requirement of light energy and the influence of the duration of light treatment point to a dual effect of blue light, which may consist in a change of enzyme conformation and in enzyme amount. To prove the latter we tried to inhibit protein

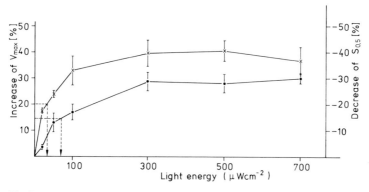

Fig. 9. Blue light-mediated changes in V_{max} (*closed circles*) and $S_{0.5}$(PEP) (*crosses*) of pyruvate kinase of the yellow *Chlorella* mutant 211-11h/20 in dependence on light energy (16 h illumination with λ 454 nm)

synthesis by application of cycloheximide (Table 1). The cells were given 10^{-3} mol/l of poison 1 h before a 4-h illumination. This leads to a complete inhibition of the enhancement in maximal velocity; it might reduce but definitely does not prevent the $S_{0.5}$ decrease.

Anaerobic conditions for 4 h — reducing the ATP production in the cells (Kobr and Beevers 1971) — lead to the same result (Table 2). This might indicate an energy requirement of the blue light-mediated V_{max} increase, and is thus in accordance with the assumption of a de novo enzyme synthesis.

The results presented here support the conception of a dual effect of blue light on pyruvate kinase. It might consist first in a change of the enzyme conformation resulting in a higher PEP affinity, and second in a de novo synthesis of enzyme protein.

Table 1. Influence of 10^{-3} mol/l cycloheximide on the blue light-mediated changes in V_{max} and $S_{0.5}$(PEP) of pyruvate kinase of the yellow *Chlorella* mutant 211-11h/20

	− Cycloheximide		+ Cycloheximide	
	Dark	Blue	Dark	Blue
Relative V_{max}	100	113.8 ± 1.5	100	98.8 ± 1.9
Relative $S_{0.5}$	100	63.3 ± 1.8	100	84.2 ± 1.8

Table 2. Influence of anaerobic conditions on the blue light-mediated changes in V_{max} and $S_{0.5}$ (PEP) of pyruvate kinase of the yellow *Chlorella* mutant 211-11h/20

	Air		N_2	
	Dark	Blue	Dark	Blue
Relative V_{max}	100	115.2 ± 2.3	100	99.7 ± 1.2
Relative $S_{0.5}$	100	75.1 ± 2.0	100	87.1 ± 2.6

References

Atkinson DE (1966) Regulation of enzyme activity. Annu Rev Biochem 35: 85–124

Kamiya A, Miyachi S (1975) Blue light-induced formation of phosphoenolpyruvate carboxylase in colorless *Chlorella* mutant cells. Plant Cell Physiol 16: 729–736

Kobr MJ, Beevers H (1971) Gluconeogenesis in the castor bean endosperm 1. Changes in glycolytic intermediates. Plant Physiol 47: 48–52

Koshland DE, Nemethy G, Filmer D (1966) Comparison of experimental binding data and theoretical models in proteins containing subunits. Biochemistry 5: 365–385

Kowallik W, Ruyters G (1976) Über Aktivitätssteigerungen der Pyruvat-Kinase durch Blaulicht oder Glucose bei einer chlorophyllfreien *Chlorella*-Mutante. Planta 128: 11–14

Ruyters G, Kowallik W (1980) Further studies of the light-mediated change in the activity of pyruvate kinase of a chlorophyll-free *Chlorella* mutant. Z Pflanzenphysiol 96: 29–34

Blue Light-Effects on Enzymes
of the Carbohydrate Metabolism in *Chlorella*
2. Glyceraldehyde 3-Phosphate Dehydrogenase
(NADP-Dependent)

W. CONRADT[1] and G. RUYTERS[1]

In higher plants, as well as in green algae, blue light has been found to develop and to maintain functioning chloroplasts (for review see Voskresenskaja 1972). In trying to understand this action of blue quanta mutants of unicellular green algae have been used in some cases (Dresbach 1973; Brinkmann and Senger 1978). The results obtained lead to the assumption of an involvement of enhanced carbohydrate breakdown in the development of the photosynthetic apparatus. In this context we have studied the influence of short-wave visible radiation on a typical chloroplast enzyme, the glyceraldehyde 3-phosphate dehydrogenase (NADP-GAPDH, EC 1.2.1.13). To avoid interactions by photosynthetic reactions we used a chlorophyll-free, carotenoid-containing mutant of *Chlorella vulgaris* as object for our investigations.

This organism has only a rudimentary plastid, but nevertheless NADP-dependent GAPDH-activity is measurable in crude extracts of dark-grown cells prepared as described by Kowallik and Ruyters (1976) (for further methods see Conradt 1980). Its level varies from approximately 150 to 300 nmol/min · mg protein dependent on the age of a static dark culture (Fig. 1). When cell suspensions are irradiated with blue light for 4 h GAPDH-activity is enhanced. The effect is most pronounced in cultures somewhat depleted in carbohydrate reserves (Figs. 1 and 2).

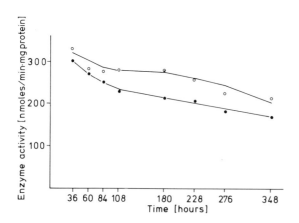

Fig. 1. Activity of NADP-dependent GAPDH (nmol/min mg protein) of dark-kept (*closed circles*) and blue-irradiated cells ($\lambda < 550$ nm; approx. 300 μW cm^{-2}; 4 h; *open circles*) of the yellow *Chlorella* mutant 211-11h/20 in the course of a static dark culture

1 Lehrstuhl für Stoffwechselphysiologie, Fakultät für Biologie der Universität Bielefeld, Universitätsstraße, 4800 Bielefeld 1, FRG

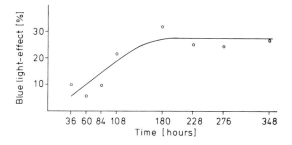

Fig. 2. Effect of blue light ($\frac{\text{blue-dark}}{\text{dark}} \times 100$) on the activity of NADP-dependent GAPDH (data computed from those of Fig. 1)

The intensity-dependence of the blue light-effect on GAPDH-activity is shown in Fig. 3. Saturation is achieved at intensities $\geqslant 300 \ \mu\text{W cm}^{-2}$, half-saturation at approx. $90 \ \mu\text{W cm}^{-2}$.

Red light of different wavelengths from 665 to 730 nm and of different intensities up to $1000 \ \mu\text{W cm}^{-2}$ proved to be completely ineffective. Thus the light effect on GAPDH studied here is quite different from that described by Ziegler et al. (1969). Our results all resemble corresponding data for the blue light-stimulated O_2-uptake and carbohydrate degradation of the yellow *Chlorella* mutant (Georgi 1974; Kowallik 1967), thus pointing to a possible correlation of these processes.

The enhancement of GAPDH-activity, mediated by blue light, could be brought about either by an increase in enzyme level or by an activation of pre-existing enzyme molecules. In a first approach to elucidate the mechanism involved some kinetic properties of GAPDH from blue-irradiated and from dark-kept cells have been compared. Figure 4 shows the dependences of the initial velocity of the enzyme reaction on the concentrations of G-1,3-P_2 or NADPH at fixed concentrations of the second substrate, respectively, plotted according to Hanes (1932): While the apparent K_m-values for G-1,3-P_2 and NADPH remain unchanged the maximal velocity of GAPDH is increased by blue light treatment. These data are compatible with the assumption of an enhanced enzyme level, though other explanations are possible, too (e.g., dissociation of enzyme aggregates, see Melandri et al. 1970).

In further experiments we therefore tried to inhibit protein synthesis by application of 10^{-3} mol/l cycloheximide. As can be seen from Table 1 the inhibitor completely prevents the blue light-caused activity enhancement of GAPDH. This cannot be taken as a proof but is a strong indication of an involvement of de novo enzyme synthesis.

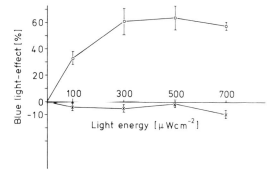

Fig. 3. Dependence of the increase in NADP-dependent GAPDH-activity on the intensity of blue light (λ 454 nm; 16 h; *open circles*). The blue light effect is expressed as $\frac{\text{blue dark}}{\text{dark}} \times 100$. For comparison the respective effect on the content of soluble proteins of the crude extracts is shown (*crosses*)

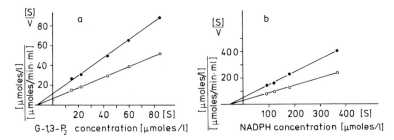

Fig. 4a, b. Velocity of NADPH-dependent reduction of G-1,3-P_2 as a function of substrate concentrations. **a** Influence of increasing G-1,3-P_2 concentrations at a fixed concentration of NADPH of 360 μmol/l. **b** Influence of increasing NADPH concentrations at a fixed concentration of G-1,3-P_2 of 43 μmol/l. Cells of the yellow *Chlorella* mutant were kept in darkness (*closed circles*) or under blue light ($\lambda < 550$ nm; approx. 300 μW cm^{-2}; *open circles*) for 24 h

Table 1. Influence of 10^{-3} mol/l cycloheximide on NADP-dependent GAPDH-activity of dark-kept or blue-irradiated cells ($\lambda < 550$ nm; approximately 300 μW cm^{-2}; 3 h light treatment) of the yellow *Chlorella* mutant 211-11h/20

	– Cycloheximide		+ Cycloheximide	
	Dark	Blue	Dark	Blue
Enzyme activity nmol/min · ml	294.9	356.7	261.0	247.4
Protein mg/ml	2.1	2.1	1.9	1.8
Activity/prot. nmol/min · mg	140.4	170.0	137.4	137.4
Effect of blue light in %	+ 21.1		± 0	

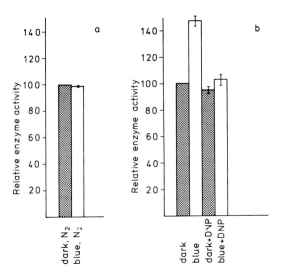

Fig. 5. a Relative activities of NADP-dependent GAPDH in crude extracts of the yellow *Chlorella* mutant 211-11h/20 kept under anaerobic conditions in blue light (λ 454 nm; approx. 300 μW cm^{-2}) or darkness for 4 h. **b** Relative activities of NADP-dependent GAPDH in crude extracts of 2,4-dinitrophenol-treated and untreated cells of the yellow *Chlorella* mutant 211-11h/20 kept in blue light ($\lambda <$ 550 nm; approx. 300 μW cm^{-2}) or darkness for 4 h

Two more results support this probability: First, *Chlorella* cells, kept under anaerobic conditions and thereby lowered in their ATP level (Kobr and Beevers 1971), do not exhibit an increase in enzyme activity on blue illumination for 4 h (Fig. 5a). Second, poisoning *Chlorella* cells with 5×10^{-4} mol/l 2,4-dinitrophenol results in a markedly suppressed light effect on GAPDH-activity (Fig. 5b).

The above findings support the view that the stimulation of NADP-GAPDH-activity is connected with an energy (and C-skeleton) supply via the light-enhanced oxidative carbohydrate breakdown. A similar mechanism is discussed for the blue light-dependent formation of carotenoids in the yellow *Chlorella* mutant and for the formation of chlorophyll in *Scenedesmus* C-2A' (Dresbach 1973; Brinkmann and Senger 1978). Thus there might be a basic master reaction leading to the activation of GAPDH and other known processes involved in the formation and maintenance of a functioning photosynthetic apparatus.

References

Brinkmann G, Senger H (1978) The development of structure and function in chloroplasts of greening mutants of *Scenedesmus*, IV. Blue light dependent carbohydrate and protein metabolism. Plant Cell Physiol 19: 1427–1437

Conradt W (1980) Über die Wirkung von kurzwellig-sichtbarer Strahlung auf die Aktivitäten plastidärer und cytoplasmatischer Enzyme einer chlorophyllfreien *Chlorella*-Mutante. Dissertation, Bielefeld

Dresbach C (1973) Lichtwirkungen auf die Bildung von Photosynthesepigmenten bei Mutanten von *Ankistrodesmus* und *Chlorella*. Dissertation, Köln

Georgi M (1974) Über die blaulichtbedingten Veränderungen im Kohlenhydrat- und Proteingehalt einer chlorophyllfreien Mutante von *Chlorella vulgaris*. Dissertation, Köln

Hanes CS (1932) CLXVII. Studies on plant amylases. I. The effect of starch concentration upon the velocity of hydrolysis by the amylase of germinated barley. Biochem J 26: 1406–1421

Kobr MJ, Beevers H (1971) Gluconeogenesis in the castor bean endosperm, 1. Changes in glycolytic intermediates. Plant Physiol 47: 48–52

Kowallik W (1967) Action spectrum for an enhancement of endogenous respiration by light in *Chlorella*. Plant Physiol 42: 672–676

Kowallik W, Ruyters G (1976) Über Aktivitätssteigerungen der Pyruvat-Kinase durch Blaulicht oder Glucose bei einer chlorophyllfreien *Chlorella*-Mutante. Planta 128: 11–14

Melandri BA, Pupillo P, Baccarini-Melandri A (1970) D-Glyceraldehyde 3-phosphate dehydrogenase in photosynthetic cells, I. The reversible light-induced activation in vivo of NADP-dependent enzyme and its relationship to NAD-dependent activities. Biochim Biophys Acta 220: 178–189

Voskresenskaja NP (1972) Blue light and carbon metabolism. Annu Rev Plant Physiol 23: 219–234

Ziegler H, Ziegler I, Schmidt-Clausen HJ, Müller B, Dörr I (1969) Activation of NADP-dependent glyceraldehyde 3-phosphate dehydrogenase in relation to photosynthetic electron transport. In: Metzner H (ed) Progr Photosynth Res, vol III, pp 1636–1645. Tübingen

Blue Light-Induced Enhancement in Activity of Certain Enzymes in Heterotrophically Grown Cultures of *Scenedesmus obliquus*

G. KULANDAIVELU[1] and G. SAROJINI[1]

1 Introduction

An environmental factor such as light quality was shown to affect the distribution of absorbed carbon into various products of photosynthesis (Cayle and Emerson 1957; Das and Raju 1965; Zak 1965). It is reported that blue light stimulates the incorporation of CO_2 into amino acids, especially alanine, but the red light into sugars. Hauschild et al. (1964) found a similar accumulation of fixed carbon atoms in amino acids and organic acids, particularly in aspartate and malate. Pirson and Kowallik (1964) have reported that plants grown under blue light had a higher protein content, while those under red light were relatively rich in carbohydrates. Increase in the rate of protein synthesis in fern gametophytes after blue light irradiation was also reported (Payer et al. 1969).

Ogasawara and Miyachi (1969, 1970, 1971) showed that in *Chlorella ellipsoidea* cells the amount of $^{14}CO_2$ incorporated into aspartate, glutamate, malate and fumarate under monochromatic blue light (457 nm) was larger than under monochromatic red light (675 nm). They further revealed that this blue light effect on CO_2 fixation was observed even in the presence of DCMU at concentrations which suppress the photosynthetic CO_2 fixation. Low intensities (300 ergs cm^{-2} s^{-1}) of blue light were shown to induce the PEP case activity in starved *Chlorella* cells and this induction was suggested to be due to an increased supply of free sugars. Till this date it is not clear whether the blue light enhances the carbohydrate breakdown, thereby inducing the synthesis of PEP case or whether blue light first induces the PEP case synthesis which eventually brings about the breakdown of the glucose polymer. We have here made an attempt to study the mechanism of the blue light enhancement and the regulation of carbon metabolism in fully greened photosynthetic *Scenedesmus* cells.

Abbreviations. *DCMU:* 3-(3,4-dichlorophenyl)-1,1-dimethylurea; *NAD/NADP-G-3P dehydrogenase:* NAD or NADP-dependent glyceraldehyde-3-phosphate dehydrogenase; *NAD-MDH:* NAD-dependent malate dehydrogenase; *PEP case:* phosphoenolpyruvate carboxylase

1 School of Biological Sciences, Madurai Kamaraj University, Madurai 625 021, India

2 Materials and Methods

2.1 Algal Cultures

The unicellular green alga *Scenedesmus obliquus* D_3 was grown heterotrophically in the dark in liquid culture medium containing 0.5% glucose and 0.1% yeast extract. For growth and ageing conditions see Kulandaivelu and Senger (1976a). Cells were collected at a known period of growth, washed twice and resuspended in 0.1 M PO_4 buffer, pH 6.8.

2.2 Preparation of Chloroplast Particles

Cells at 0.5 mg/ml of chlorophyll concentration were mixed with glass beads (0.55 mm ϕ), packed in a stainless steel container and oscillated at a frequency of 80 Hz in a vibrating cell mill. Temperature was maintained at $0°-2°C$. The homogenate was centrifuged at $3000\,g$ for 5 min and the green supernatant was used as the source of chloroplast particles.

2.3 Preparation of Crude Enzyme Extracts

Cells suspended in the enzyme extraction medium (100 mM Tris-Cl, pH 7.6, 5 mM $MgCl_2$, 1 mM EDTA, 10 mM mercaptoethanol) were broken in the cell mill and the homogenate was clarified by centrifuging at $20,000\,g$ for 30 min. The supernatant was used as the source of enzymes.

2.4 Light Activation

For the light activation, the starved cells were transferred to glass vessels maintained at $25°C$ and illuminated with either broad band blue light (380–500 nm, Corning C.S. 5–61) or red light (620–740 nm, Schott). Irradiance was maintained at 2.5 W m^{-2}. Monochromatic blue light was obtained by using interference filters (Schott, Half Band Width 11 nm). Irradiance was measured with a Licor LI-188 Quantum/Radiometer (Lambda Inst. Corp., USA).

2.5 Measurements

2.5.1 $^{14}CO_2$ Fixation. Dark CO_2 fixation was followed in whole cells after injecting known amount of $NaH^{14}CO_3$. Aliquots were removed at regular time intervals, transferred to Whatman No. 3 filter paper, acidified and the radioactivity was measured in a liquid scintillation counter.

2.5.2 Enzymes. PEP case (EC 4.1.1.31) activity was measured using the method of Kamiya and Miyachi (1975). Decrease in the absorbance at 340 nm to measure the oxidation of NADH was followed at $25°C$. Non-specific oxidation was checked in assays without the substrate. NAD-G-3-P dehydrogenase (EC 1.2.1.12) was assayed spectrophotometrically in the direction of reduction of 3-PGA. NADP-G-3-P dehydrogenase (EC 1.2.1.13) was assayed as NAD dehydrogenase except that 0.6 μmol of NADPH was used instead of NADH. NAD-MDH (EC 1.1.1.37) activity was assayed in the direction of oxidation of NADH at 25°C.

2.5.3 Oxygen Uptake. Respiratory O_2 uptake was continuously followed using a Clark-type electrode in whole cells at different times of light activation. Temperature was maintained at 28°C.

2.5.4 Estimations. Glucose content was determined by anthrone method (Scott and Melvin 1953). Protein was estimated by the method of Lowey et al. (1951). The total chlorophyll in methanol extracts was calculated using the formula of MacKinney (1941).

3 Results

3.1 Heterotrophic Cultures

When the *Scenedesmus* cells were inoculated from a 5-day-old culture to the fresh medium and grown heterotrophically, the cell mass, measured as packed cell volume (PCV) increased rapidly up to 5 days and then slowly (Fig. 1). No significant decrease in its level was noticed even after 60 days. The glucose content of the medium declined rather sharply during the initial phase of the growth and reached a 10% level on the 3rd day and on the 10th day found only in traces. The soluble protein in the cells increased only up to 10 days and then declined slowly reaching ultimately a 5% level on the 70th day.

3.2 Enzymes in Heterotrophic Cultures

The activity of PEP case, the primary heterotrophic carboxylase enzyme, increased rapidly during the initial period similar to the pattern of growth (up to 5 days) and declined drastically thereafter. The glycolysis enzyme, NAD-G-3-P dehydrogenase activity decreased steadily from the early phase of growth. The TCA cycle is found to be more active at least up to 30 days, as indicated by the stable higher level of NAD-MDH activ-

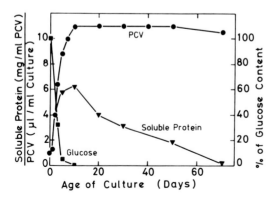

Fig. 1. Changes in the packed cell volume (PCV), glucose content in the medium and soluble protein in ageing heterotrophic cultures of *Scenedesmus.* The age of the culture was calculated from the date of inoculation. Other conditions as under Sect. 2

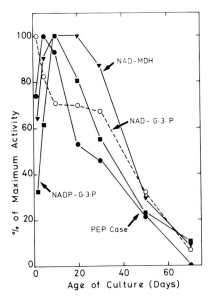

Fig. 2. Changes in the level of a few enzymes in the heterotrophically grown *Scenedesmus* cells. The 100% levels of the enzymes (in μmol NAD/NADPH oxidized mg sol. protein^{-1} h^{-1}) are: NAD-G-3-P dehydrogenase, 5.6; NADP-G-3-P dehydrogenase, 15.5; NAD-MDH, 46.5; PEP case, 3.8. For details see Sect. 2

Table 1. Changes in PEP case, NAD/NADP-G-3-P dehydrogenase activity in *Scenedesmus* cells as a function of the starvation period. For all the experiments 10-day-old culture was used. Figures in parentheses are percentage with reference to the non-starved cells

Enzymes	Enzyme activity μmol NADH/NADPH ox. mg sol protein^{-1} h^{-1}	
	Starvation period	
	0 h	24 h
PEP case	2.23 (100)	1.16 (52)
NAD-G-3-P dehydrogenase	12.00 (100)	0.24 (2)
NADP-G-3-P dehydrogenase	6.10 (100)	5.60 (92)

ity. The level of the Calvin cycle enzyme NADP-G-3-P dehydrogenase, although it reached the maximum level after 10 days, also declined sharply similar to the PEP case (Fig. 2).

3.3 Light Activation

To study the light activation of some of the key enzymes, 10-day-old culture which shows relatively higher level of enzyme activity was selected. As the initial activity was very high, the cells were starved in PO_4 buffer, pH 6.8. Changes in the level of a few important enzymes after 24 h of starvation are given in Table 1. Though PEP case demonstrated only about 50% loss from its original level, the extent of inactivation was over 97% in the case of NAD-G-3-P dehydrogenase. Contrary to the above two enzymes the NADP-G-3-P dehydrogenase was found to be more stable in retaining more than 80% of the original activity after 24 h of starvation.

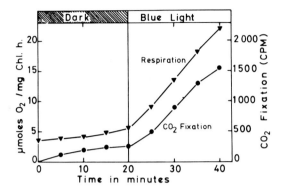

Fig. 3. Effect of weak blue light (2.5 W/m^2) on the overall rate of dark CO_2 fixation and respiration in 10-day-old heterotrophically grown cells of *Scenedesmus*. Cells were starved for 24h in 0.1 M PO_4 buffer, pH 6.8 in the dark at 25°C. For details see Sect. 2

For activation, the starved cells were transferred to blue or red light (2.5 W m^{-2}) and as an initial step the changes in their CO_2 fixation and respiratory activities have been followed. The O_2 uptake rate was enhanced several-fold within 5 min after turning the blue light on, and continued linearly for several minutes (Fig. 3). Similar to the respiration CO_2 fixation which was almost in its minimal level in the dark, enhanced to great extent. Linear rate of CO_2 fixation was seen at least up to 20 min.

To study in detail the nature of activation and also to check the effect of red light, the levels of the enzyme activity were followed in the starved cells after exposing to blue and red lights. Exposure to blue light was found to stimulate specifically the NAD-G-3-P dehydrogenase activity while the NADP-G-3-P dehydrogenase was inhibited up to 30% in 3 h. A higher level of inhibition of the latter enzyme was seen upon exposure for longer duration (data not shown). Red light did not bring any significant effect on the NAD-G-3-P dehydrogenase, while a slight enhancement of the NADP form was noticed (Table 2). A more detailed picture of the time course of enhancement of the PEP case activity on exposure to red and blue light is given in Fig. 4. Blue light stimulated the PEP case activity several-fold after an initial lag for a period of 2 h. Linear rate of activation was seen up to about 12 h which then levels off. Red light in contrast to the NAD/NADP-G-3-P dehydrogenases, marginally inhibited the PEP case activity.

Having established that the blue light has a definite role in the activation of PEP case and NAD-G-3-P dehydrogenase, an attempt was also made to study the nature of the

Table 2. Effect of blue and red light on NAD/NADP dependent G-3-P dehydrogenase activity in heterotrophically grown *Scenedesmus*. Cells were collected from a 10-day-old culture, washed twice and resuspended in 0.1 M PO_4 buffer, pH 6.8 and starved for 24 h in the dark at 25°C. Portions of cells suspension were exposed to either red or blue light. For other details see Methods. Figures in parentheses are percentage activity with reference to the dark level

Enzymes	Time of illumination (h)	Enzyme activity μmol NADH/NADPH ox mg protein^{-1} h^{-1}		
		Dark	Red	Blue
NAD-G-3-P dehydrogenase	3	0.21 (100)	0.22 (102)	0.32 (152)
NADP-G-3-P dehydrogenase	1	4.25 (100)	4.42 (104)	3.92 (92)
	3	6.00 (100)	6.78 (113)	4.09 (68)

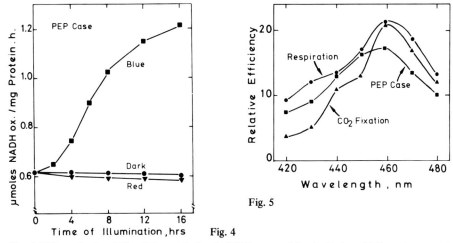

Fig. 4

Fig. 5

Fig. 4. Effect of red and blue light on the level of PEP case activity in 10-day-old *Scenedesmus* cultures. Cells were starved for 24 h in 0.1 M PO$_4$ buffer, pH 6.8 in the dark at 25°C. Other conditions as under Sect. 2

Fig. 5. Action spectra for dark CO$_2$ fixation, respiration and PEP case activity in 10-day-old culture of *Scenedesmus* grown heterotrophically. Monochromatic blue light was obtained using Schott interference filters (half band width, 11 nm). Irradiance was maintained at 0.25 W/m^2. Cells were exposed to monochromatic lights for 3 h. Other details as under Sect. 2

photoreceptor in such greened cells by following the action spectra for various reactions. All the reactions studied viz., the overall CO$_2$ fixation and respiration and the enzyme PEP case activity demonstrated maximal enhancement at wavelengths around 460 nm (Fig. 5).

3.4 Effect of Cycloheximide on Blue Light Induction

The fact that the blue light brings about a two-fold stimulation in the activity of PEP case and NAD-G-3-P dehydrogenase indicates the possibility of de novo synthesis of the enzyme. To check this PEP case induction was followed in the presence of 6 to 8 µg/ml cycloheximide. Apart from inhibiting totally the blue light enhancement it also brought the original level down to about 70% (Table 3).

4 Discussion

The heterotrophically grown and starved cells of *Scenedesmus,* when transferred to low intensity of blue light demonstrate increased rates of CO$_2$ fixation and respiration which is in many respects similar to the observations in the chlorophyll-less mutant of *Chlorella* (Kowallik and Gaffron 1966; Kowallik 1967; Kamiya and Miyachi 1975). Blue light at such intensities increased preferentially the respiratory activity which was reflected also on the rate of dark CO$_2$ fixation. Since such enhancement was observed even in

Table 3. Effect of cycloheximide on the blue light-induced PEP case activity in *Scenedesmus*. Cyclo-heximide was added as methanolic solution. The final concentration of methanol was about 0.2%. Controls received the same amount of methanol. Figures in parentheses are percentage with reference to the dark activity. For other details see Sect. 2

Age of culture	Conditions	PEP case activity μmol NADH ox. mg protein^{-1} h^{-1}
8 days	Dark	1.02 (100)
	Blue light	1.31 (128)
	Blue light + cycloheximide (6 μg/ml)	0.91 (89)
11 days	Dark	1.04 (100)
	Blue light	1.60 (154)
	Blue light + cycloheximide (8 μg/ml)	0.74 (71)

the presence of inhibitory concentration of DCMU it was concluded that blue light en-hances the dark carboxylase activity (Ogasawara and Miyachi 1969).

Studies by Kowallik and his associates in a chlorophyll-less mutant of *Chlorella* showed that transfer of dark-adapted cells to blue light for a longer period induced pyru-vate kinase activity. More detailed and sequential analysis by Miyachi and his co-workers in colourless mutant and partially greened cells of *Chlorella* indicate that blue light en-hances greatly the PEP case, as evidenced by the higher labelling of aspartate in time periods as short as 20 s after the exposure to blue light. Based on such observations they have concluded that blue light induces the PEP case activity which could be due to the increased de novo synthesis. However, the question as to why the PEP case becomes activated under blue light illumination was not fully explained.

In our studies we have selected three key enzymes, viz., NAD-G-3-P dehydrogenase, NADP-G-3-P dehydrogenase and PEP case in order to check the activation/suppression of respectively the glycolysis, Calvin cycle and dark CO_2 fixation. The starved *Scenedes-mus* cells when exposed to red light demonstrated activation of NADP-G-3-P dehydro-genase indicating enhancement of the Calvin cycle. Activation of photosynthetic O_2 evolution in such aged cells by both red and blue light was already known (Kulandaivelu and Senger 1976c). Only marginal activation of the NAD-G-3-P dehydrogenase and PEP case was observed under such conditions. In contrast, under low intensity of blue light the Calvin cycle was found to be preferentially inhibited with concomitant activation of both NAD-G-3-P dehydrogenase and PEP case (Fig. 6). Simultaneous analysis of the time course of induction of the latter two enzymes shows that PEP case enhancement was preceded by the activation of NAD-G-3-P dehydrogenase. Although the cells used in such experiments were initially completely devoid of glucose and are used in glucose-free medium, the necessary substrates for the enhancement of glycolysis might have been obtained through the breakdown of the reserve materials. Ultrastructural studies of such aged cells showed that cells in the late log phase contain a high level of starch reserve which are completely utilized during the subsequent early stationary phase of growth (Kulandaivelu and Senger 1976b). Hence it is probable that blue light in such cells first activates the NAD-G-3-P dehydrogenase, resulting in the increased breakdown of reserved

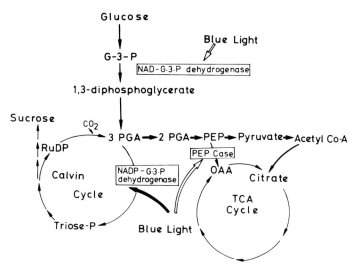

Fig. 6. Scheme showing the probable enzymes that are suppressed (*crossed solid arrow*) and activated (*open arrows*) by low intensity of blue light in *Scenedesmus* cells. For details see text

glucose polymer and accumulate PGA. As the Calvin cycle was blocked under such conditions, further conversion PGA takes place towards the TCA cycle resulting in an increase of the PEP pool and consequent to this de novo synthesis of the PEP case could probably have occurred. Since pyruvate carboxylase-mediated CO_2 fixation is absent in *Scenedesmus* cells, the dark CO_2 fixation is possible only through PEP case. Experiments carried out in the presence of cycloheximide also support the view of substrate-induced de novo synthesis of PEP case as the blue light induction was completely suppressed by this inhibitor.

Our results indicate that the blue light irradiation (1) brings about similar changes in the carbon metabolism in green heterotrophic cells of *Scenedesmus* which additionally possess, in contrast to the chlorophyll-less mutant of *Chlorella,* a functionally active Calvin cycle and (2) inhibits preferentially the Calvin cycle and activates the dark carboxylase through the activation of the glycolysis.

Acknowledgements. The authors thank the University Grants Commission and ICAR for financial support, the latter in the form of Senior Research Fellowship to one of us (G.S.) and Dr. A. Gnanam for reading the manuscript.

References

Cayle T, Emerson R (1957) Effect of wavelength on the distribution of carbon-14 in the early products of photosynthesis. Nature (London) 179: 89–90

Das VSR, Raju PV (1965) Photosynthetic $^{14}CO_2$ assimilation by rice under the influence of blue light. Ind J Plant Physiol 8: 1–5

Hauschild AHW, Nelson CD, Krotkov G (1964) Concurrent changes in the products and the rate of photosynthesis in *Chlorella vulgaris* in the presence of blue light. Naturwissenschaften 51: 274

Kamiya A, Miyachi S (1975) Blue light induced formation of PEP case in colourless *Chlorella* mutant cells. Plant Cell Physiol 16: 729–736

Kowallik W (1967) Action spectrum for an enhancement of endogenous respiration by light in *Chlorella*. Plant Physiol 42: 672–676

Kowallik W, Gaffron H (1966) Respiration induced by blue light. Planta 69: 92–95

Kowallik W, Scheil I (1976) Lichtbedingte Veränderungen des ATP-Spiegels einer chlorophyllfreien *Chlorella*-Mutante. Planta 131: 105–108

Kulandaivelu G, Senger H (1976a) Changes in the reactivity of the photosynthetic apparatus in heterotrophic ageing cultures of *Scenedesmus obliquus*. I. Changes in the photochemical activities. Physiol Plant 36: 157–164

Kulandaivelu G, Senger H (1976b) Changes in the reactivity of the photosynthetic apparatus in heterotrophic ageing cultures of *Scenedesmus obliquus*. II. Changes in ultrastructure and pigment composition. Physiol Plant 36: 165–168

Kulandaivelu G, Senger H (1976c) Changes in the reactivity of the photosynthetic apparatus in heterotrophic ageing cultures of *Scenedesmus obliquus*. III. Recovery of the photosynthetic capacity in aged cells. Physiol Plant 36: 169–173

Lowry OH, Rosenbrough NJ, Farr AL, Randall RJ (1951) Protein measurement with the folin phenol reagent. J Biol Chem 193: 265–275

MacKinney GJ (1941) Absorption of light by chlorophyll solutions. J Biol Chem 140: 315–322

Ogasawara N, Miyachi S (1969) Effect of wavelength on $^{14}CO_2$ fixation in *Chlorella* cells. In: Metzner H (ed) Progr Photosynth Res, vol III, pp 1653–1661. Lichtenstern, München

Ogasawara N, Miyachi S (1970) Regulation of CO_2 fixation in *Chlorella* by light of varied wavelength and intensities. Plant Cell Physiol 11: 1–14

Ogasawara N, Miyachi S (1971) Effects of dark preincubation and chloramphenicol on blue light induced CO_2 incorporation in *Chlorella* cells. Plant Cell Physiol 12: 675–682

Payer HD, Storiffer U, Mohr H (1969) Die Aufnahme von $^{14}CO_2$ und die Verteilung des ^{14}C auf freie Aminosäuren und auf Proteinaminosäuren im Hellrot und im Blaulicht (Objekt: Farnvorkeime von *Dryopteris filix-mas* (L) Schoot). Planta 85: 270–283

Pirson A, Kowallik W (1964) Spectral response to light by unicellular plants. Photochem Photobiol 3: 489–497

Scott Jr TA, Melvin EH (1953) Determination of dextran with anthrone. Anal Chem 25: 1656–1661

Zak EG (1965) Influence of molecular oxygen on the formation of aminoacids during photosynthesis in *Chlorella* under various conditions of illumination. Fiziol Rast 12: 263–269

Effect of 360 nm Light on RuBPCase Products in Vitro – Role of Copper in the Reaction

L.S. DALEY[1], H.F. TIBBALS[2], and L.J. THERIOT[2]

1 Introduction

We will present data supporting: a UV-activated carboxylative biosynthesis of phospho-hydroxypyruvate in vitro by an RuBPCase preparation from tobacco; the presence of more copper in RuBPCase than is generally measured; that the copper is the form of spin-coupled dimeric units of Cu+2- - -Cu+2, and that copper influences the long wavelength UV stimulation of RuBPCase. To facilitate interpretation of this information, some previously published material will be presented. This material includes the action spectra of UV-activated RuBP carboxylation, and the finding that phosphohydroxypyruvate is an early intermediate of carbon fixation in plants. This material will be evaluated in terms of a possible mechanism of photorespiration.

Preliminary data suggest that the products of UV-stimulated RuBPCase activation included 14-C labeled products other than 3-PGA. There are at least four possible explanations for this:

1.1 An Impurity Other Than RuBP was Carboxylated by an Enzyme Other Than RuBPCase

While our enzyme is quite pure (Fig. 1F), unusual carboxylation products have been inferred or reported in plant tissues or extracts [3, 7, 11, 12, 26, 27, 34, 36, 37, 41, 44, 51] and are little understood.

1.2 A Substrate Other Than RuBP is Acted Upon by RuBPCase to Form a Carboxylated Product

Such a reaction is unknown, however it is necessary to take into account, that while the RuBP used here was prepared in a standard fashion [18, 19], RuBP is a labile product [18, 35], and thus could contain a degradation product that could be carboxylated.

Abbreviations. *RuBP:* ribulose-bis-phosphate; *RuBPCase:* ribulose-bis-phosphate carboxylase; *DMSO:* dimethyl sulfoxide; *6C*:* 2-carboxy-3-keto ribitol-1,5 bisphosphate

1 Department of Biological Sciences, North Texas State Universtiy, Denton, Texas 76203, USA
2 Department of Chemistry, North Texas State University, Denton, Texas 76203, USA

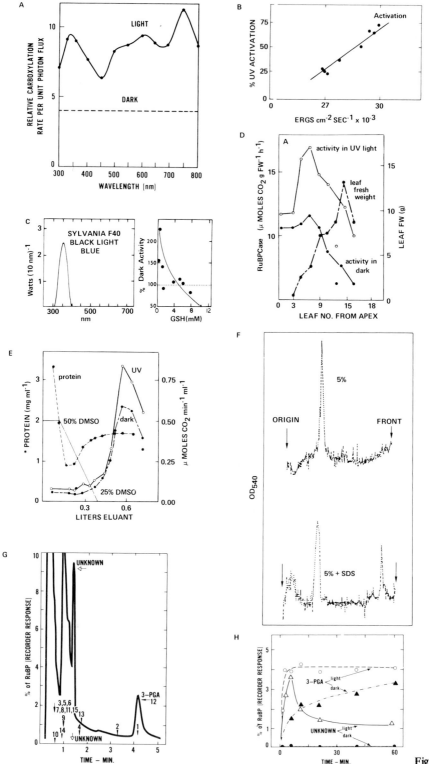

Fig. 1A–H

1.3 The New Products Found Were Derived From 3-PGA (Previously Made by RuBP-Case Action on RuBP and CO_2)

This could be mediated through 3-PGA dehydrogenase present in the chloroplast [31], but it is unlikely due to the unfavorable kinetic parameters of the enzyme (Fig. 2F) [38, 48].

1.4 The Products Found Were Different From 3-PGA, the Enzyme Responsible was RuBPCase, and the Substrate Carboxylated was RuBP

All four possibilities are of interest, thus the nature of the products of the reaction was examined. To evaluate the information presented, several additional matters need to be considered:

1. The spectral response of this system to different wavelengths of light [46, 47] (Fig. 1A).

2. In the field, in the chloroplast stroma, both substrate and enzyme are subject to even more intense UV light than used experimentally [29, 30].

3. Column profiles (Fig. 1E) and time courses of $^{14}CO_2$ fixation (Fig. 2B) demonstrate that in the absence of the enzyme $^{14}CO_2$ fixation did not occur.

Fig. 1A–H. Light activation of RuBPCase. **A** Action spectra for the light activation of RuBPCase (adapted from [46]). **B** Effect of light on percent activation [14]. **C** Spectral distribution of lamps used and effect of reduced glutathion on reaction. Samples were irradiated for the duration of the assay through the open tops of the vials [14]. **D** Extraction of light-activatable RuBPCase from *Nicotiana tabacum*. Extraction volume is proportional to leaf weight [14]. **E** Elution of RuBPCase from tobacco leaf homogenates with DMSO containing buffers. Lyophylized, neutralized tobacco leaf material and mixed bed resin were suspended in 50% DMSO solution and eluted with buffers containing DMSO [14]. **F** Polyacrylamide gel electrophoresis of RuBPCase purified by DMSO extraction in 5% gels with and without sodium dodecylsulfate [14]. **G** Gas liquid chromatographic analysis of the silylated products of the UV-activated reaction of RuBPCase. The substrate concentrations and other conditions are as in [11 and 14]. RuBPCase plus substrates was allowed to react for 5 min under UV irradiation, rapidly frozen, lyophilized, silylated with BSTFA and analyzed by GLC, on a 5% SE column. Retention times are labeled as follows: *1* citric acid, *2* dihydroxyacetone phosphate, *3* glyceraldehyde-3-phosphate, *4* glycoaldehyde phosphate, diethyl acetal, (phosphoglyco-aldehyde diethylacetal), *5* glucose-6-phosphate, *6* phosphoenolpyruvate, *7* 6-phosphogluconate, *8* ribose-5 phosphate, *9* RuBP, *10* sedoheptulose-1,7-bisphosphate, *11* xylose-1-phosphate, *12* 3-PGA, *13* phospho-glycolic acid, *14* fructose-1,6-bisphosphate, *15* inorganic phosphate. Phosphohydroxypyruvate derivatives were prepared in two ways: (1) the dimethyl ketal of phosphohydroxypyruvate was hydrolyzed with 9% formic acid at $100°C$ in the presence of BSTFA, and the mixture was neutralized, or (b) by transamination reaction (glutamic oxaloacetic, plus glutamic pyruvic transaminases) from phosphoserine, and then silylated. The retention time of the unknown UV-irradiated RuBPCase reaction product is indicated by the *arrow*. This retention time corresponds to the phosphohydroxy-pyruvate reaction product. **H** Time course of appearance of the RuBPCase reaction products with and without (dark) UV treatment during reaction. Reaction conditions as in **F** [11, 14]. Substrate concentrations were 8 mM and 20 mM bicarbonate. The reaction was stopped by freezing in dry ice-acetone mixture, the products were lyophylized, derivatized with BSTFA and separated on a 3 mm-diameter GLC column. The RuBPCase assay is completed before RuBP is significantly depleted. In this experiment without 14-C label the concentration of 3-PGA was corrected for 3-PGA derivatives that were present at zero time or in absence of the enzyme

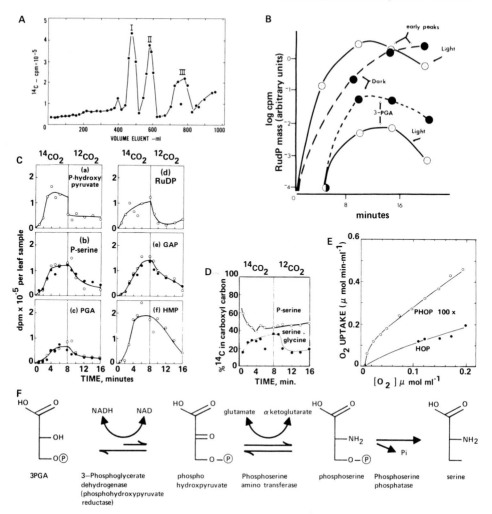

PHOSPHORYLATED PATHWAY TO SERINE

Fig. 2A–F. $^{14}CO_2$ products in vivo and in vitro. **A** $^{14}CO_2$ labeled products of the UV-activated RuBP-Case reaction eluted from a cellulose column. The reaction was carried out for 5 min using 8 mM RuBP and 40 mM C-14 labeled sodium bicarbonate. The reaction was quenched and eluted with butanol: propionic acid: water; 10:5:7; v:v:v. The column, 15 cm by 6 cm in diameter, contained 33 g of Munktell 410 cellulose. Each 20 ml fraction was collected into 5 ml of 1 M semicarbazide hydrochloride in 1 M ammonium-acetic acid buffer pH 5.0. **B** 14-C CO_2 labeled products of light-activated RuBPCase, time course. The reaction conditions were as in Fig. 1. RuBP concentration was 15 mM. The 14-C detector limited the resolution of the early products, thus phosphohydroxy-pyruvate was resolved from 3-PGA but not from what we now believe to be the six carbon inter-mediate (6C*). A 6 mm column GLC was used. Products of the RuBPCase reaction with and with-out UV irradiation are shown. The total counts (7293, at 5 min) are normalized relative to the mass of RuBP and are expressed in arbitrary units. Mass was estimated with a thermal detector. **C** $^{14}CO_2$ pulse chase-labeled metabolites in vivo [12]. **D** Percentage of radioactivity in carboxyl groups of phos-phoserine and serine+ glycine fractions [12]. **E** The effect of oxygen concentration on the nonenzymic oxidation of phosphohydroxy pyruvate (PHOP) or hydroxypyruvate (HOP). Note that the curve for PHOP is graphed × 100 [15]. **F** Pathways of phosphohydroxypyruvate metabolism

4. Phosphohydroxypyruvate will be proposed as the newly found product of the light-activated carboxylation reaction, phosphohydroxypyruvic aldehyde has previously been suggested for this role [44]. Phosphohydroxypyruvate is both easily decarboxylated under physiological conditions (Fig. 2E) [15, 43], and cochromatographs in most systems with 3-PGA [11, 12, 13, 40, 41, 51]. These properties of phosphohydroxypyruvate make it a likely candidate for a role in photorespiration (Fig. 4).

5. Enzymic conversion of 3-PGA to phosphohydroxypyruvate (Fig. 2C) requires the use of additives to trap the phosphohydroxypyruvate since the 3-PGA dehydrogenase reaction strongly favors the production of 3-PGA [38, 48].

6. Chemical procedures described here, while quite harmful to enzymic systems, do not destroy the carboxylation products and these methods do not vary greatly from those used by others [4, 26, 32, 36, 40, 41, 51].

7. The in vitro assay conditions [11, 14] differ from the usual for carboxylase, in that they are carried out for longer times, with more RuBP and bicarbonate. The short time interval in the usual assay helps assure high rates and low Km bicarbonate and approximates the in vivo conditions [22, 23]. However, these conditions are approximated in the type of assay used here [11, 14]. These concentrations of RuBP are close to those found in the chloroplast stroma under certain conditions [24], and aid in the detection of 6C* [6, 11, 20, 34, 40, 41, 51, 52, 53], the six-carbon intermediate of the RuBPCase reaction. However, as a result of these concentrations of RuBP, it is possible to postulate a carbon dioxide or bicarbonate acceptor molecule as an impurity of the preparation. Such an impurity has not been reported in the literature.

8. In vivo increases in carboxylation may follow long-wavelength UV irradiation under certain circumstances and with high flux of photosynthetically active light [42].

The presence of copper (Fig. 3) in the enzyme makes interpretations of oxidative activity more complex [8, 49]. Wishnick and Lane (1969, 1970) reported that there was approximately one atom of copper per molecule of enzyme, since then a considerable number of reports [5, 9, 25, 33] suggest that there is less copper than that in the enzyme. In these reports, except some of those which deal with highly crystallized enzyme, significant amounts of copper are still found, approximately 0.2 atoms copper per molecule enzyme. Since copper is closely associated with the enzyme in vivo [39], this constant low value suggests a systematic error in the technique used. For this reason, we searched for and found a matrix effect [28] in the copper determinations which would give lower detector response than would be expected for the amount of copper present (Fig. 3B).

2 Materials and Methods

References to UV light activation of RuBPCase include: [1, 10, 11, 14, 46, 47]. The materials and methods for extraction, separation, and identification of phosphohydroxypyruvate can be found in [11, 12, 13]. The preparation and assay of light-activatable RuBPCase are described in [10, 11, 14]. The labile nature of phosphohydroxypyruvate and other matters of this sort in [15, 43]. Chromatographic information on phosphohydroxypyruvate can be found in [11, 12, 13]. Determination of related enzymes and metabolites can be found in [11, 14, 31]. Unpublished material can be found in the figure legends.

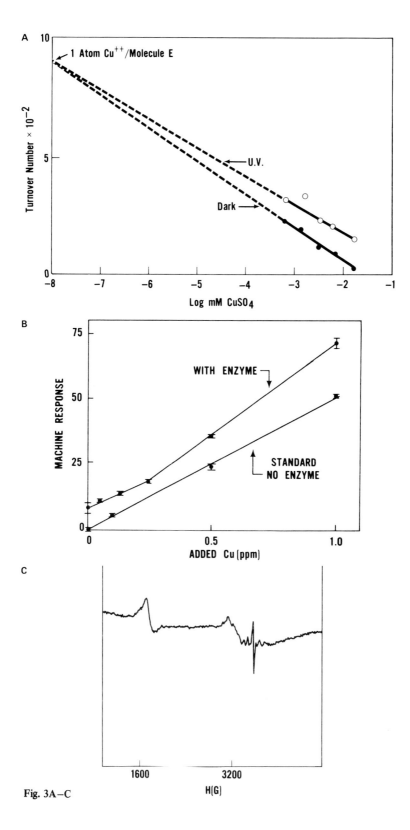

Fig. 3A–C

3 Results and Discussion

Tentative identification of phosphohydroxypyruvate as a carboxylation product of the UV-irradiated reaction mixture (Fig. 1G) [11], permits examination of the four explanations outlined in the introduction.

   1. Is phosphohydroxypyruvate a result of the carboxylation of a substrate other than RuBP by an enzyme other than RuBPCase? One enzymic mechanism [36] is a heat-stable phosphoenolpyruvate carboxylase that adds CO_2 to phosphoenolpyruvate, the resulting adduct being the restructured to form phosphohydroxypyruvate. The possibility of this mechanism being of importance in the UV-activated system is greatly diminished by its low pH optima but cannot be entirely ruled out because of its high activity in tobacco extracts. In addition Fig. 1F indicates that the light-activated RuBPCase is essentially free of contamination by other proteins.

   2. Is a substrate other than RuBP carboxylated by RuBPCase to give phosphohydroxypyruvate? RuBPCase has two known activities, the carboxylation and oxidation of RUBP, therefore, a third activity is not inconceivable, but this has not yet been reported.

   3. Is phosphohydroxypyruvate derived from 3-PGA? This is unlikely for the following reasons: it is not consistent with the time courses presented in Fig. 2B (in vitro) and 2C (in vivo); the 3-PGA dehydrogenase reaction strongly favors 3-PGA production (Fig. 2F) [38, 48]; and it is not consistent with the effect of the 3-PGA dehydrogenase inhibitor bromopyruvate [11].

   4. Is phosphohydroxypyruvate a result of the carboxylation of RuBP by RuBPCase. The possibility of a second carboxylation product (other than 3-PGA) was suggested [3] to explain several kinetic and other discrepancies. The methods used to determine 3-PGA by chromatography are now known to be equivocal for phosphohydroxypyruvate [13, 40]. Earlier enzymic determinations of 3-PGA [18, 21, 45] are currently realized to be much less specific and less pure than were previously thought [17]. Chemical tests such as periodate cleavage yield oxidation products that could have been derived from phosphohydroxypyruvate or 3-PGA [11, 13, 40]. Kinetic studies attempting to determine whether the resulting products of RuBPCase were two 3-PGA molecules or one 3-PGA plus X, are ambiguous if X is phosphohydroxypyruvate [16], because of the presence in the chloroplast of 3-PGA dehydrogenase [31]. Deuterium labeling experiments [20] are short almost exactly one deuterium per mol for two 3-PGA's, but almost exactly correct for one 3-PGA plus one phosphohydroxypyruvate. This is even more apparent when corrections are made for the small amount of deuterium exhanged into

---

Fig. 3A–C. Interactions of copper with RuBPCase. **A** Effect of copper sulfate on UV activation of RuBPCase. Reaction conditions as in Fig. 1 [11, 14]. **B** Copper levels of spinach RuBPCase, apparent matrix effect, that gives low readings if copper is not added. Copper was determined by flame atomic absorption spectrophotometry using Perkin-Elmer Model 370. Instrument response was linear from 0.1 μg/ml to 2.0 μg/ml. Samples were alternated with distilled water blanks and appropriate standards to check on instrument drift. Solvents had no detectable copper. The data points reflect maximum and minimum instrument response for triplicate analysis. The spinach RuBPCase was purchased (Sigma). **C** Electron spin resonance spectra of spinach RuBPCase as a powder obtained with a Varian 4502-12 X-band spectrometer using 100 Kc modulation and a 24-cm magnet. The microwave frequency was 9524 billion Hz. The internal reference was diphenylpicrylhydrazide. The spectrum was scanned from 0–10,000 Gauss, at room temperature

the phosphate group. Phosphohydroxypyruvate has a mass spectra on decomposition that is the same as products of the RuBPoxygenase reaction [43], if dephosphorylation precedes silylation [2, 4, 32]. If we consider alternative (1) unlikely because of the pH of the reaction, alternative (2) unlikely because such reactions are unknown, and alternative (3) less probable because of the reasons expressed above, alternative (4) becomes the most acceptable. Thus, we put forward the hypothesis that under certain conditions such as light activation, RuBP can be carboxylated to yield 3-PGA plus phosphohydroxypyruvate. We may interpret the light activation of RuBPCase by UV and by inference the complete action spectra of this effect (Fig. 1A) in the following fashion: (a) Phosphohydroxypyruvate is produced from RuBP (Figs. 1G, 1H, 2B) perhaps through a redox cleavage reaction of the six-carbon intermediate (6C*) (W.W. Cleland, pers. commun.) (Fig. 4). (b) This reaction is favored by high intensity UV light (Fig. 1B), and inhibited by sulphydryl reagents.(Fig. 1C). (c) Phosphohydroxypyruvate is also an early

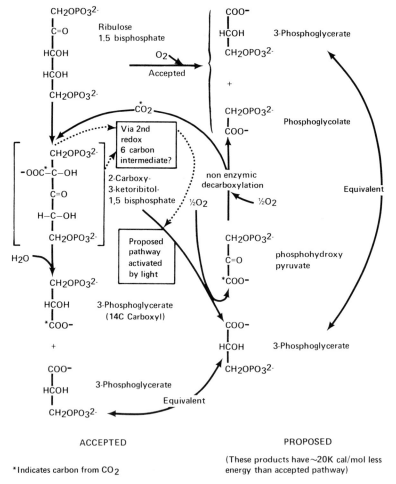

Fig. 4. Proposed mechanism of RuBPCase/oxygenase mode of action with phosphohydroxypyruvate as an intermediate

product of photosynthesis in vivo (Figs. 2C, 2D). (d) This reaction may be mediated by dimeric spin-coupled copper found on the enzyme (Fig. 3C). In the presence of exogenous nondimeric spin-coupled copper the formation of this redox intermediate would be inhibited; perhaps by mass action of copper sulfate. The effect of copper sulfate is partially reversed by high intensity light (Fig. 3A).

Following this view (Fig. 4) the first event of photorespiration, the RuBP oxygenase reaction [52, 53], would be a function of the pool size of the readily oxidizable product phosphohydroxypyruvate. The products of this nonenzymic reaction have been shown to yield the same products on silylation and GLC-mass spectroscopic analysis as phosphoglycolate and carbon dioxide [43]. Thus photorespiration would depend, at least in part, on the ability of the plant to reduce the size of the phosphohydroxypyruvate pool, either via phosphoglycerate dehydrogenase to 3-PGA or via a transaminase reaction to phosphoserine (Fig. 2F). The amount of photorespiratory losses via the serine to glycine pathway [52, 53] would then depend on the ability of the plant to process phosphoserine (Fig. 2F) or 3-PGA via other routes than the serine to glycine pathway.

This hypothesis is presented schematically in Fig. 4. In this figure phosphohydroxypyruvate is produced by the RuBPCase reaction under natural light (UV) conditions or via a carefully preserved activated state of RuBPCase. The phosphohydroxypyruvate that is not converted to phosphoserine or 3-PGA (Fig. 2F) breaks down spontaneously to give carbon dioxide which is recycled, and phosphoglycolate which is processed in the accepted glycine to serine pathway [52, 53]. In the process oxygen is consumed as in the RuBPoxygenase reaction. Phosphohydroxypyruvate then can be considered as an intermediate of the RuBPoxygenase reaction. Phosphoxydroxypyruvate is usually not detected for four reasons: it is readily converted by the chloroplast 3-PGA dehydrogenase [31] to the favored product 3-PGA [38], or by transaminases to phosphoserine; it is labile and readily breaks down to the expected products of the oxygenase reaction [15, 43]; it is not easily resolved by chromatographic means from 3-PGA [13], and it is produced only when RuBPCase is in a particular active state. This active state may be "quenched" by paramagnetic magnetic metals such as copper (Fig. 3A) as may be inferred from data presented at this meeting (G. Schmitt), or may be related to blue light stimulation of respiration (this vol.).

The authors hope that this easily tested hypothesis will lead to applied use, in breeding of food crops, and further investigations in this area of short wavelength light effects on plants [29].

Acknowledgments. This work was supported by Grant GM 10017 from the U.S. Public Health Service, the Welch Foundation and by Lionel Daley.

References

1. Andersen WR, Wildner GF, Criddle RS (1970) Ribulose diphosphate carboxylase. III. Altered forms of ribulose diphosphate carboxylase from mutant tomato plants. Arch Biochem Biophys 137: 84–90
2. Andrews TJ, Lorimer GH, Tolbert NE (1973) Ribulose diphosphate oxygenase. I. Synthesis of phosphoglycolate by fraction-1 protein of leaves. Biochemistry 12: 11–18
3. Bassham JA, Kirk M (1960) Dynamics of the photosynthesis of carbon compounds. Biochem Biophys Acta 43: 447–464

4. Berry JA, Osmond CB, Lorimer GH (1978) Fixation of 18-02 during photorespiration. Plant Physiol 62: 954–967

5. Branden R (1978) Ribulose-1,5-diphosphate carboxylase and oxygenase from green plants are two different enzymes. Biochem Biophys Res Commun 81: 539–546

6. Calvin M (1962) The path of carbon in photosynthesis. Science 135: 879–889

7. Chapman DJ, Leech RM (1976) Phosphoserine as an early product of photosynthesis in isolated chloroplasts and leaves of *Zea mays* seedlings. FEBS Lett 68: 160–164

8. Chiai EIO (1977) Bioinorganic chemistry - An introduction, Ch 9. Allyn and Bacon, Boston

9. Chollet R, Anderson LL, Hovsepian LC (1975) The absence of tightly bound copper, iron and flavin mononucleotide in crystalline ribulose 1,5-bisphosphate carboxylase-oxygenase from tobacco. Biochem Biophys Res Commun 64: 97–107

10. Dailey FA (1977) Model explaining the regulation of RUBP carboxylase and oxygenase activity by various effectors. Ph D Dissertation, University of California

11. Daley LS (1975) Light activation of ribulose diphosphate carboxylase: conditions and products of the reaction. Ph D Dissertation, University of California

12. Daley LS, Bidwell RGS (1977) Phosphoserine and phosphohydroxypyruvic acid. Plant Physiol 60: 109–114

13. Daley LS, Bidwell RGS (1978) Separation of phosphohydroxypyruvate,3-phosphoglyceric acid and O-phosphoserine by paper chromatography and chemical derivitizatation. J Chromatogr 147: 233–241

14. Daley LS, Dailey F, Criddle RS (1978) Light activation of ribulose bisphosphate carboxylase. Purification and properties of the enzyme from tobacco *Nicotianum tabacum.* Plant Physiol 62: 718–722

15. Daley LS, Vines HM, Bidwell RGS (1979) Oxidation of phosphohydroxypyruvate and hydroxypyruvate: physiological implications in plants. Can J Bot 57: 1–3

16. Galmiche JM (1973) Studies on the mechanism of glycerate-3-phosphate synthesis in tomato and maize leaves. Plant Physiol 51: 512–519

17. Heinz F, Bartelsen K, Lamprecht W (1962) D-glycerate dehydrogenase from liver in relation to serine metabolism Z Physiol Chem 329: 222–240

18. Horecker BL, Hurwitz J, Weissbach A (1958) Ribulose diphosphate. Biochem Prep 6: 83–90

19. Huffaker RC, Obendorf RL, Keller CJ, Kleinkopf GE (1966) Effect of light intensity on photosynthetic carboxylation phase enzymes and chlorophyll synthesis in greening leaves of *Hordeum vulgare* L. Plant Physiol 41: 913–918

20. Hurwitz J, Jacoby WB, Horecker BL (1956) On the mechanism of CO_2 fixation leading to phosphoglyceric acid. Biochem Biophys Acta 22: 194–195

21. Jacoby WB, Brummond DO, Ochoa S (1956) Formation of 3-phosphoglyceric acid by carbon dioxide fixation with spinach leaf enzymes. J Biol Chem 218: 811–821

22. Jensen RG, Bahr JT (1974) Properties of ribulose diphosphate carboxylase as observed upon lysis of spinach chloroplasts. In: Proc 3rd Int Congr Photosynth Sept 2–6, pp 1411–1420. The Weizmann Institute of Science. Rehovot, Israel. Elsevier Scientific Publishing Co, Amsterdam

23. Jensen RG, Bahr JT (1977) Ribulose 1,5-diphosphate carboxylase-oxygenase. Annu Rev Plant Physiol 28: 379–400

24. Jensen RG, Sicher RC, Bahr JT (1978) Regulation of ribulose 1,5 bisphosphate carboxylase in the chloroplast. In: Siegelman HW, Hind G (eds) Photosynthetic carbon assimilation, pp 95–112. Plenum Publ Co, New York

25. Johal S, Bourque DP (1979) Crystalline ribulose 1,5-bisphosphate carboxylase-oxygenase from spinach. Science 204: 74–77

26. Kennedy RA, Laetsch WM (1974) Formation of 14-C-labeled alanine from pyruvate during short term photosynthesis in a C-4 plant. Plant Physiol 54: 608–611

27. Kent SS, Pinkerton FD, Strobel GA (1974) Photosynthesis in the higher plant, *Vicia faba.* Plant Physiol 53: 491–495

28. Kirkbright GF, Sargent M (1974) Atomic absorption and fluorescence spectroscopy. Academic Press, London New York

29. Klein RM (1978) Plants and near-ultraviolet radiation. Bot Rev 44: 1–127

30. Kreithen M, Eisner T (1978) Ultraviolet light detection by the homing pigeon. Nature (London) 272: 347–348

31. Larsson S, Albertsson E (1979) Enzymes related to serine synthesis in spinach chloroplasts. Physiol Plant 45: 7–10
32. Lorimer GH, Andrews TJ, Tolbert NE (1973) Ribulose diphosphate oxygenase. II. Further proof of reaction products and mechanisms of action. Biochemistry 12: 18–23
33. McCurry SO, Hall NP, Pierce J, Paech C, Tolbert NE (1978) Ribulose-1,5-bisphosphate carboxylase/oxygenase from parsley. Biochem Biophys Res Commun 84: 895–900
34. Moses V, Calvin M (1958) The path of carbon in photosynthesis XXII. The Identification of carboxyketopentitol diphosphates as products of photosynthesis. Proc Natl Acad Sci USA 44: 260–277
35. Paech C, Pierce J, McCurry S, Tolbert NE (1978) Inhibition of ribulose-1,5-bisphosphate carboxylase/oxygenase by ribulose-1,5-bisphosphate epimerization and degradation products. Biochem Biophys Res Commun 83: 1084–1092
36. Pan D, Waygood ER (1971) A fundamental thermostable cyanide sensitive phosphenolpyruvate acid carboxylase in photosynthetic and other organisms. Can J Bot 49: 631–643
37. Punnett T, Kelley JH (1976) Environmental control over C-3 and C-4 photosynthesis in vascular plants. Plant Physiol 57: Abstr no 305
38. Sallach HJ (1966) D-3-Phosphoglycerate dehydrogenase. Methods Enzymol 9: 216–220
39. Salisbury JL, Floyd GL (1978) Molecular enzymic and ultrastructure characterization of the pyrenoid of the scaly green monad *Micromonas squamata*. J Phycol 14: 362–368
40. Siegel MI, Lane MD (1973) Chemical and enzymatic evidence for participation of 2-carboxy-3-keto-ribitol-1,5-diphosphate intermediate in the carboxylation of ribulose 1,5-diphosphate. J Biol Chem 248: 5486–5498
41. Sjodin B, Vestermark A (1973) The enzymatic formation of a compound with the expected properties of carboxylated ribulose 1,5 diphosphate. Biochem Biophys Acta 297: 165–173
42. Teramura AH, Kossuth SV, Biggs RH (1978) Effects of UV-B enhancement under contrasting PAR growth regimes on NCE, dark respiration, and growth in soybeans (*Glycine max*). Plant Physiol 61: Abstr no 408
43. Tournier P, Espinasse A, Gerster R (1978) Decarboxylation par H_2O_2 de ceto-acides en relation a vec le metabolisme de la photorespiration. CR Acad Sci 287: 729–732
44. Weaver RH, Lardy HA (1961) Synthesis and some biochemical properties of phosphohydroxypyruvic aldehyde and of 3-phosphoglyceryl glutathione thiol ester. J Biol Chem 236: 313–316
45. Weissbach A, Horecker BL, Hurwitz J (1956) The enzymic formation of phosphoglyceric acid from ribulose diphosphate and carbon dioxide. J Biol Chem 218: 795–810
46. Wildner GF, Criddle RS (1969) Ribulose diphosphate carboxylase. I. A factor involved in light activation of the enzyme. Biochem Biophys Res Commun 37: 952–960
47. Wildner GF, Zilg H, Criddle RS (1971) Light effect on isolated ribulose diphosphate carboxylase activity. 2nd Int Congr Photosynth, pp 1825–1830. Stresa
48. Willis JE, Sallach HJ (1964) The occurence of D-3-phosphoglycerate dehydrogenase in animal tissues. Biochem Biophys Acta 81: 39–54
49. Wishnick M, Lane MD, Scrutton MC, Midlvan AS (1969) The presence of tightly bound copper in ribulose diphosphate carboxylase from spinach. J Biol Chem 244: 5761–5763
50. Wishnick M, Lane MD, Scrutton MC (1970) The interaction of metal ions with ribulose 1,5-diphosphate carboxylase from spinach. J Biol Chem 245: 4939–4947
51. Vestermark A, Sjodin B (1972) Isotachoelectrophoresis used alone or in two dimensional combination with zone electrophoresis for the small scale isolation of labelled ribulose-1,5-diphosphate. J Chromatogr 71: 588–592
52. Zelitch I (1975) Pathways of carbon fixation in green plants. Annu Rev Biochem 44: 123–145
53. Zelitch I (1979) Photosynthesis and plant productivity. Chem Eng News 5: 28–48

The Photoinactivation
of Micro-Algal Ribulose Bisphosphate Carboxylase; its Physiological and Ecological Significance

G.A. CODD[1] and R. STEWART[1]

1 Introduction

In limnology, the term photoinhibition is used to describe the phenomenon of a marked decrease in the rate of photosynthesis by phytoplankton at or near the water surface, compared with rates observed at lower levels in the water column. The phenomenon has been frequently observed during in situ depth profiles of primary productivity in both fresh and marine waters throughout the world. The factors involved in the development of photoinhibition are several and their interelations complex. The intensity and duration of incident surface radiation, light penetration through the water, which is in turn affected by the presence of suspended detritus and self-shading by the algae, and the nutritional status of the algae, have all been implicated in the manifestation of photoinhibition (for a review, see Harris 1978).

Although the deleterious effects of UV light on algal photosynthesis have been intensively studied (e.g., Halldal 1967; Halldal and Taube 1972) few attempts have been made to distinguish between the effects of UV and visible wavelengths in photoinhibition in the natural environment. The role of UV in this effect has been indicated in some in situ productivity studies using glass and quartz containers (see Harris 1978). We have carried out studies on the growth and primary productivity of phytoplankton in a Scottish freshwater lake and have found photoinhibition in the surface waters on several occasions throughout the year. Evidence is presented here for the role of visible light in the effect. In vivo and in vitro studies on algal species isolated from the lake have revealed a blue light-dependent photoinactivation of ribulose 1,5-bisphosphate (RuBP) carboxylase. This may account in part for photoinhibition in natural waters.

2 The Photoinhibition of Photosynthesis in Loch Balgavies

Loch Balgavies, near Forfar, Angus, Scotland, has an area of 21 ha, a maximum depth of 9.6 m, a mean depth of 3 m and is surrounded by very productive agricultural land. The values of abiotic and biotic parameters studied with season in Loch Balgavies are typical of a temperate, dimictic, eutrophic lake (Stewart 1978). Figure 1 shows the depth profiles of phytoplankton primary productivity rates measured from May 1975

---

1 Department of Biological Sciences, University of Dundee, Dundee DD1 4HN, Great Britain

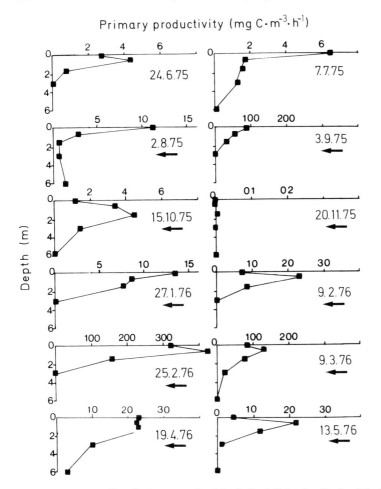

Fig. 1. Vertical profiles of primary productivity in Loch Balgavies, Scotland during 1975–1976. (The *arrows* denote the depth of the euphotic zone)

to May 1976 using $NaH^{14}CO_3$ and glass bottles. The rates of primary productivity were markedly affected by season, temperature, light intensity, light penetration, and phytoplankton standing crop densities. Analyses of the influence of these parameters are given in Stewart (1978). The range in photosynthetic rates observed was from 0 to 970 mg inorganic carbon fixed m^{-3} h^{-1}. A decrease in primary productivity was found on several occasions in the surface layers, compared with 0.5–1.0 m, whether expressed in terms of per unit water volume or as specific rates of photosynthesis (mg carbon fixed mg chlorophyll a^{-1} h^{-1}).

The incubations detailed in Fig. 1 were performed using glass bottles with walls greater than 3 mm in thickness which did not allow the penetration of UV light, and the results therefore indicated that visible wavelengths were to some extent responsible for the photoinhibition observed. The degree of photoinhibition in the surface waters was not increased in the presence of UV light, as evidenced by incubations performed in

Table 1. Effects of incubation at the surface of Loch Balgavies on the photosynthetic activity of a phytoplankton population dominated by *Anabaena* spp. Water samples were incubated for 3 h around midday on 13.5.76

Container	Specific rate of photosynthesis (mg C fixed mg chl a^{-1} h^{-1})
Clear glass	0.063
Clear Plexiglas[a]	0.066
Red Plexiglas[a]	0.377

[a] Constructed from 3 mm thick Plexiglas sheets, supplied by Röhm, Darmstadt, ref. nos. 233 and 501 respectively

clear Plexiglas containers which transmitted both UV and visible wavelengths. Table 1 shows the specific rate of photosynthesis by buoyant *Anabaena* spp. at the lake surface on 13.5.76. Photoinhibition was observed on this occasion (see Fig. 1) but the rates of photosynthesis in thick glass bottles and clear Plexiglas containers were virtually identical. Table 1 also shows the surface rate of specific photosynthesis in red Plexiglas containers, which effectively excluded visible wavelength below ca. 560 nm, but transmitted the same intensities of red light as the clear Plexiglas containers. The specific rate of photosynthesis at the lake surface was ca. sixfold higher in the absence of blue light than in its presence. Diurnal primary productivity studies with depth using red Plexiglas versus glass containers have shown that red light itself contributes to the appearance of surface photoinhibition in Loch Balgavies (Stewart 1978). However, the marked difference between rates of photosynthesis by algae in a photoinhibition state when incubated in clear versus red Plexiglas containers also implicates blue spectral wavelengths in the development of photoinhibition.

The most pronounced surface photoinhibition of photosynthesis occurred when blooms of gas vacuolate cyanobacteria (blue-green algae) accumulated in the surface waters of the lake (principally *Microcystis aeruginosa* and *Anabaena* spp.).

3 An in Vivo Effect of Blue Light on Phytoplankton Ribulose Bisphosphate Carboxylase

The depression of the photosynthetic production of particulate carbon (Harris 1978; Fig. 1) by high light intensities, together with the stimulatory effects of high light on the excretion of photosynthate by phytoplankton notably of glycollic acid (Al-Hasan et al. 1975; Fogg 1975) suggest a direct, or indirect, photoregulation of the key enzyme of photosynthesis and photorespiration, namely RuBP carboxylase/oxygenase. Paradoxically, RuBP carboxylase has been neglected in attempts to identify the targets affected during the photoinhibition of photosynthesis, and we sought evidence for the effect of light quality on this enzyme in phytoplankton species.

Since inflated gas vacuoles in cyanobacteria preferentially scatter blue as compared to red light (see Fogg et al. 1973) and photoinhibition in Loch Balgavies was most severe when buoyant gas vacuolate cyanobacteria were dominant, the effects of irradiating

Table 2. The effect of irradiation of *Microcystis aeruginosa* cells upon levels of extractable RuBP carboxylase activity

Whole cell incubation conditions[a]	RuBP carboxylase activity			
	Cells with inflated gas vacuoles		Cells with collapsed gas vacuoles	
	Sp. Act[b]	Rel. Act[c]	Sp. Act	Rel. Act
Zero time	1.50	100	1.50	100
1 h dark	1.52	101	1.40	93
1 h red light	1.32	88	1.25	83
1 h blue light	0.51	34	1.20	80

[a] Cultures (1 mg dry wt ml^{-1}) were incubated in 0.01 M HEPES buffer, pH 7.5 and incubated at 25°C in darkness, or in 5000 μW cm^{-2} red light or blue light, provided by a Leitz Pradovit tungsten filament bulb plus red Plexiglas (see legend to Table 1) or blue Plexiglas (Röhm ref. no. 627) and a 5% (w/v) solution of copper sulphate
b Measured according to Stewart et al. (1977) using a 20,000 g X 10 min supernatant of cells disrupted immediately after incubation Sp. act., μmol CO_2 fixed mg protein^{-1} h^{-1}. Values are the mean of duplicate determinations
c Relative enzyme activity expressed as a percentage of that measured at zero time

M. aeruginosa cells upon rates of extractable RuBP carboxylase were determined. The significance of gas vacuoles in determining the susceptibility of RuBP carboxylase to in vivo photoregulation was determined by irradiating cells containing inflated gas vacuoles versus cells with gas vacuoles collapsed by brief ultrasonication of the culture. The cultures were irradiated by either red or blue light at intensities which were saturating for carbon dioxide fixation but which were also considerably lower than encountered in the surface layers of Loch Balgavies during photoinhibition in the summer. As shown in Table 2, levels of extractable RuBP carboxylase activity were not significantly affected by 1 h of pre-incubation of buoyant *M. aeruginosa* cells, with inflated gas vacuoles, in red light versus darkness. However, two thirds of the enzyme activity was lost after exposure for 1 h to blue light (400–520 nm). Similar pre-illumination of cells with collapsed vacuoles did not reveal a significant loss of RuBP carboxylase activity. Similar results were obtained using cells of the gas vacuolate cyanobacterium *Anabaena flos-aquae*, also isolated from Loch Balgavies.

4 The Photoinactivation of Purified Microalgal Ribulose Bisphosphate Carboxylase

It is not known whether the observed changes in the levels of extractable RuBP carboxylase (Table 2) were due to the effects of blue light on the enzyme per se, or were caused via an indirect route involving the effects of blue light on related metabolic or photo-oxidative processes (see Bassham 1971; Halliwell 1978a). In order to examine the susceptibility of RuBP carboxylase to inhibition, or inactivation, by blue light, the enzyme was purified from the two Loch Balgavies gas vacuolate cyanobacteria and also from unialgal, axenic cultures of the green algae *Sphaerocystis Schroeteri* and *Scenedesmus quadricauda* isolated from Loch Balgavies.

RuBP carboxylase, homogeneous by the criterion of yielding a single band in polyacrylamide gel electrophoresis, was purified by treating French pressate to high-speed
centrifugation, precipitation by ammonium sulphate and sedimentation through linear
density gradients of sucrose. Full details are given elsewhere (Codd and Stewart 1977;
Stewart et al. 1977). All of the purified enzymes exhibited RuBP oxegenase activity as
evidenced by RuBP-dependent O_2 uptake. Since blue light-absorbing cofactors, namely
flavins, are frequently associated with oxygenases in both prokaryotes and eukaryotes
(Hayaishi et al. 1975) the absorption spectra of the phytoplankton enzymes were measured. Each enzyme showed an absorption maximum around 280 nm but no indication
of a coloured chromophore was apparent, as evidenced by the absence of absorption
above 300 nm.

In keeping with the absorption spectra, the activities of the purified enzymes, when
incubated alone, were not affected by visible wavelengths (Table 3). However, a photoinactivation of each enzyme tested was found in the presence of blue light (500 μW cm^{-2})
plus FMN (0.2 mM final concentration), which resulted in a loss of 68%–82% of enzyme
activity during the experiments. Activities were not affected by similar exposure to red
light either with or without FMN. The blue light photoinactivation was apparently irreversible; no restoration of activity being found after 3 h of storing the irradiated enzyme in the dark.

Further studies on the mechanism of the blue light-dependent inactivation of RuBP
carboxylase have been performed with the comparatively stable *Sph. Schroeteri* enzyme.
As detailed in Table 4, the inclusion of the triplet FMN quenchers potassium iodide and
tryptophan (see Schmid 1969, 1970) protected the enzyme almost completely from
photoinactivation. These data indicate, in common with several other flavin-sensitized
photooxidations of plant enzymes (Schmid 1969, 1970; Codd 1972a, 1972b), that the

Table 3. Effects of irradiation upon ribulose bisphosphate carboxylase purified from Loch Balgavies
microalgal isolates

Incubation conditions[c]	RuBP carboxylase activity[a] Source of enzyme[b]		
	A	B	C
Dark	84	100	79
Dark plus FMN	85	75	82
Red light	84	100	81
Red light plus FMN	86	100	80
Blue light	86	75	83
Blue light plus FMN	18	30	32

[a] Enzyme activities after incubation for 1 h, expressed as percent of activity at zero time
[b] A: *Microcystis aeruginosa.* Enzyme activity at zero time, 0.3 μmoles CO_2 fixed mg protein^{-1} min^{-1}.
B: *Sphaerocystis Schroeteri.* Enzyme activity at zero time, 0.25 μmoles CO_2 fixed mg protein^{-1}
min^{-1}. C: *Scenedesmus quadricauda.* Enzyme activity at zero time, 1.5 μmoles CO_2 fixed mg pro-
tein^{-1} h^{-1}
c As in the legend to Table 2 except that enzymes were exposed to 500 μW cm^{-2} light. FMN was
included at a final concentration of 0.02 mM. All incubations performed under aerobic conditions

Table 4. The effect of various substances on the inactivation by blue light of purified *Sphaerocystis Schroeteri* ribulose bisphosphate carboxylase

Incubation conditions[b]	RuBP carboxylase activity[a]
Dark plus FMN	100
Blue light plus FMN	48
Dark plus FMN plus KI	109
Blue light plus FMN plus KI	82
Dark plus FMN plus tryptophan	100
Blue light plus FMN plus tryptophan	82
Dark plus FMN plus β-carotene	100
Blue light plus FMN plus β-carotene	55
Dark plus FMN minus O_2	100
Blue light plus FMN minus O_2	85

[a] Expressed as percent of that measured at zero time, when a specific activity of 0.28 μmol CO_2 fixed mg protein^{-1} min^{-1} was recorded
[b] As in legend to Table 2. FMN present throughout. Enzyme exposed to 500 μW cm^{-2} blue light. Final concentration of FMN, KI, tryptophan and β-carotene, ca. 0.2 mM. All incubations were performed under air unless stated otherwise

blue light-dependent inactivation of RuBP carboxylase proceeds mainly via triplet FMN. Subsequent energy transfer from triplet FMN directly to the substrate can occur (Foote 1968), but this is unlikely in the case of algal RuBP carboxylase photoinactivation since the blue light, FMN-dependent loss of activity of the *Sph. Schroeteri* enzyme was almost entirely dependent on the presence of oxygen (Table 4). The photo-oxidation of RuBP carboxylase in crude extracts. of *M. aeruginosa* and *Sc. quadricauda* is completely oxygen-dependent (R. Stewart and G.A. Codd, unpublished observations). β-Carotene, a well-established singlet oxygen quencher (Foote et al. 1970) did not significantly prevent the photo-oxidation of the *Sph. Schroeteri* enzyme (Table 4) and no conclusions can yet be drawn about the species of oxygen involved in the inactivation mechanism.

The rate of photoinactivation of the *Sph. Schroeteri* enzyme was lower when RuBP, the enzyme's substrate, was present in the incubation mixture during irradiation (Fig. 2).

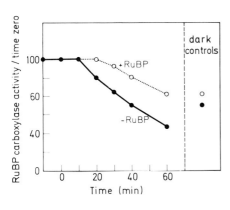

Fig. 2. The effect of the presence of RuBP on the photoinactivation of purified *Sphaerocystis Schroeteri* RuBP carboxylase by blue light plus FMN. The enzyme was incubated in the absence of CO_2 with and without RuBP (1 mM). Other conditions were as detailed in the legends to tables. Aliquots were removed at the time intervals indicated and rapidly assayed for RuBP carboxylase activity. Activities are expressed as percent of activity at zero time (0.3 μmol CO_2 fixed mg protein^{-1} min^{-1}). Dark controls represent activities measured after 1 h in the dark

Enzymic carboxylation of the RuBP during irradiation was minimized by incubation in CO_2-free buffer. Although the possibility of RuBP oxygenation cannot be excluded, since the presence of oxygen is required for photoinactivation to occur (Table 4), the persistence of RuBP throughout the incubation period can be concluded since the formation of acid-stable ^{14}C-labelled material was observed upon the addition of ^{14}C-labelled bicarbonate to aliquots of the incubation mixture. The protective effect of RuBP indicates that a site of photoinactivation may exist at, or near the active centre of the enzyme. These results also have implications about the susceptibility of RuBP carboxylase to photoinactivation in vivo under photo-oxidative conditions.

5 Discussion

The damaging or lethal effect of visible light on cells has been known for about a century (see Krinsky 1976) and although the photoinhibition of primary productivity in natural waters is well documented (Harris 1978), most of the studies on the wavelenghts involved and intracellular processes affected have been performed in the laboratory with pure cultures or in vitro systems. The dynamics of photo-oxidative damage and identification of the toxic species involved, including excited sensitizers, hydrogen peroxide and superoxide and singlet oxygen radicals have been extensively studied under laboratory conditions. Furthermore, the positive roles of carotenoids, catalase, and notably of superoxide dismutase in affording protection against photo-oxidative damage have been clearly shown (see Krinsky 1976; Fridovich 1978; Halliwell 1978b). Shilo's group, working on the surface layers of ponds in Israel have obtained evidence for the role of superoxide dismutase in the protection of cyanobacteria against photo-oxidative death (Abeliovich and Shilo 1972; Abeliovich et al. 1974; Eloff et al. 1976). Gross effects of damaging light, namely on cell death, colony morphology, chlorophyll and protein levels were observed, although the particular metabolic reactions and enzymes affected were not identified (Eloff et al. 1976).

    The present studies in Loch Balgavies have implicated blue light in the surface photo-inhibition of phytoplankton photosynthesis. Although the damaging action of UV wavelengths has been recognized elsewhere (Harris 1978) no evidence for this has been obtained in the surface layers of Loch Balgavies. We have not measured cell mortality, but the pronounced inhibition of photosynthesis in the upper half metre of the water column (Fig. 1), which correlates with the penetration of blue wavelengths and the accumulation of high dissolved oxygen (which in the daytime reaches values up to 150% saturation, Stewart and Codd, in preparation) together indicate that photo-oxidative conditions often occur in Scottish waters (Loch Balgavies lies at latitude ca. 57°N).

    Eloff et al. (1976) significantly point out that considerable differences in sensitivity to photo-oxidation are found between laboratory strains and pond strains of cyanobacteria tested, and stress that caution be used in making ecological conclusions from laboratory studies using strains from culture collections. We have continued to study the effects of light quality on photosynthesis in the laboratory using strains of phytoplankton isolated from the natural populations which were found to be susceptible to photoinhibition. The RuBP carboxylases purified from the two cyanobacteria and from the two green algae isolated were all found to be inactivated by blue light in the presence of FMN and, only in the gas vacuolate cyanobacteria *M. aeruginosa* and *A. flos-aquae* did this occur under mild blue light conditions in vivo. Teleologically, the photoinactivation of the key enzyme of photosynthesis must be viewed as an undesirable event. Measures to protect the enzyme from photoinactivation in vivo, whether by means of carotenoids or superoxide dismutase would clearly enhance the survival prospects of the organism and the apparent absence of photoinactivation of RuBP carboxylase in *Sph. Schroeteri* and *Sc. quadricauda* in vivo may be due to the participation of such protective mechanisms. The loss of RuBP carboxylase activity during whole cell irradiation of the cyanobacteria required the presence of inflated gas vacuoles. From the positive advantage conferred upon the cyanobacterial cell by gas vacuoles, i.e., buoyancy and thus the ability to rise in the water column into the photic zone (Walsby 1975), must therefore be substracted the influence of these inclusions in rendering RuBP carboxylase susceptible to photoinactivation in vivo.

Both the mechanism of photoinactivation in vitro and the natural significance of RuBP photoinactivation in vivo require further elucidation. Data presented herein have shown that FMN acts as a sensitizer in vitro. Although flavins are ubiquitous cellular components (Lehninger 1975) other blue light-absorbing compounds may act as sensitizers of the photoinactivation in vivo. From the studies performed so far, many factors may be expected to interact to determine the extent of photoinactivation in vivo, namely light intensity, quality and duration, endogenous sensitizer and quencher levels, oxygen tension, and the nutritional status of the cell which may influence the RuBP pool size. Also to be taken into account may be the indirect activation of RuBP carboxylase by light, via light-dependent increases in the concentration of Mg^{2+} and the alkalization of the stroma in chloroplasts (see Heldt et al. 1978).

From laboratory studies, it is possible that several light-induced effects, including pigment bleaching and the inactivation photosynthetic electron transport reactions, may contribute to the surface photoinhibition of primary productivity (Harris 1978). Requirements for the photo-inactivation of RuBP carboxylase identified in whole cell and pure enzyme experiments with Loch Balgavies isolates, namely, blue light, oxygen, a low nutrient status (i.e., RuBP pool size), and in cyanobacteria, the presence of inflated gas vacuoles, can apparently be satisfied in the surface waters of the lake. It is therefore suggested that the surface photoinhibition of photosynthesis in Loch Balgavies, particularly when dominated by buoyant gas vacuolate cyanobacteria, may be partly due to the photo-inactivation of RuBP carboxylase by blue light.

Acknowledgment. Research support from the Natural Environment Research Council is gratefully acknowledged.

References

Abeliovich A, Shilo M (1972) Photo-oxidative death in blue-green algae. J Bacteriol 111: 682–689
Abeliovich A, Kellenberg D, Shilo M (1974) Effect of photo-oxidative conditions on levels of superoxide dismutase in *Anacystis nidulans*. Photochem Photobiol 19: 379–382
Al-Hasan RH, Coughlan SJ, Pant A, Fogg GE (1975) Seasonal variations in phytoplankton and glycollate concentrations in the Menai Straits, Anglesey. J Mar Biol Assoc UK 55: 557–565
Bassham JA (1971) The control of photosynthetic carbon metabolism. Science 172: 526–534

Codd GA (1972a) The photoinhibition of malate dehydrogenase. FEBS Letts 20: 211–214

Codd GA (1972b) The photoinactivation of tobacco transketolase in the presence of flavin mononucleotide. Z Naturforsch 276: 701–704

Codd GA, Stewart WDP (1977) D-Ribulose 1,5-diphosphate carboxylase from the blue-green alga *Aphanocapsa* 6308. Arch Microbiol 113: 105–110

Eloff JN, Steinitz Y, Shilo M (1976) Photo-oxidation of cyanobacteria in natural conditions. Appl Environ Microbiol 31: 119–126

Fogg GE (1975) Biochemical pathways in unicellular plants. In: Photosynthesis and productivity in different environments, Int Biol Progr, vol III, pp 437–457. Cambridge University Press, Cambridge

Fogg GE, Stewart WDP, Fay P, Walsby AE (1973) The blue-green algae. Academic Press, London New York

Foote CS (1968) Mechanisms of photosensitized oxidation. Science 162: 963–970

Foote CS, Denny RW, Weaver L, Chang Y, Peters J (1970) Quenching of singlet oxygen. Ann N Y Acad Sci 171: 139–148

Fridovich I (1978) The biology of oxygen radicals. Science 201: 875–880

Halldal P (1967) Ultraviolet action spectra in algology. A review. Photochem Photobiol 6: 445–460

Halldal P, Taube O (1972) Ultraviolet action spectra and photoreactivation in algae. In: Giese AC (ed) Photophysiology, pp 163–188. Academic Press, London New York

Halliwell B (1978a) The chloroplast at work. A review of recent developments in our understanding of chloroplast metabolism. Prog Biophys Mol Biol 33: 1–54

Halliwell B (1978b) Biochemical mechanisms accounting for the toxic action of oxygen on living organisms: The key role of superoxide dismutase. Cell Biol Int Rep 2: 113–128

Harris GP (1978) Photosynthesis, productivity and growth: The physiological ecology of phytoplankton. Arch Hydrobiol Ber Ergeb Limnol 10: 1–171

Hayaishi O, Nozaki M, Abbott MT (1975) Oxygenases: Dioxygenases. In: Boyer PD (ed) The enzymes, vol 12B, pp 119–189. Academic Press, London New York

Heldt HW, Chon CJ, Lorimer GH (1978) Phosphate requirement for the light activation of ribulose-1,5-biphosphate carboxylase in intact spinach chloroplasts. FEBS Letts 92: 234–240

Krinsky NI (1976) Cellular damage initiated by visible light. In: Gray TGR, Postgate JR (eds) The survival of vegetative organisms. Soc Gen Microbiol Symp, vol 26, pp 209–239. Cambridge University Press, Cambridge

Lehninger AL (1975) Biochemistry, 2nd edn. Worth Publishers Inc, New York

Schmid GH (1969) The effect of blue light on glycolate oxidase of tobacco. Hoppe Seylers Z Physiol Chem 350: 1035–1046

Schmid GH (1970) Photoregulation von Flavinenzymen durch Blaulicht. Ber Dtsch Bot Ges 83: 399–415

Steeman-Nielsen E (1952) The use of radio-active carbon (C^{14}) for measuring organic production in the sea. J Cons Int Explor Mer 18: 117–140

Stewart R (1978) The effects of light quality on phytoplankton carbon metabolism. PhD Thesis, University of Dundee

Stewart R, Auchterlonie CC, Codd GA (1977) Studies on the subunit structure of ribulose-1,5-diphosphate carboxylase from the blue-green alga *Microcystis* aeruginosa. Planta 136: 61–64

Walsby AE (1975) Gas vesicles. Annu Rev Plant Physiol 26: 427–439

A Rhythmic Change in the Enhancement
of the Dark Respiration of *Chlorella fusca* Induced
by a Short Blue-Light Exposure of Low Intensity

B. REINHARDT[1]

Chlorella fusca which has been kept in darkness for several hours shows after a short exposure to blue light even of one second only and of low intensity in the following dark a transitory enhancement of the respiratory O_2-uptake which has its maximum between the 6th and 8th min. In favorable circumstances (using autospores and after a sufficiently long period of darkness) the amount of oxygen additionally taken up can be about 500 times as great as that evolved photosynthetically during the short illumination. The maximal enhancement of the O_2-uptake caused by blue light, however, appears only once. Another blue light illumination after the decline of the transitory enhancement of the O_2-uptake is followed by an obviously smaller enhancement of the O_2-uptake.

The reproduction of such a transitory enhancement of the respiratory O_2-uptake caused by blue light and measured on a rate electrode as described by Ried (1965, 1966, 1968, 1969) is represented in Fig. 1. The first part of the curve represents the rate of the dark respiration after a dark interval of several hours. Then the oxygen evolution during the blue light exposure of 1 s follows, and finally the continual rise of the O_2-uptake. Starting from the maximum the effect continually declines to a rate of respira-

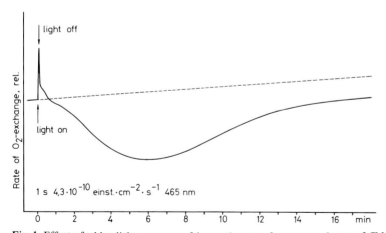

Fig. 1. Effect of a blue light exposure of 1 s on the rate of oxygen exchange of *Chlorella fusca*

1 Fachbereich Biologie der Johann Wolfgang Goethe-Universität (Botanik), Siesmayerstraße 70, 6000 Frankfurt, FRG

tory O_2-uptake which is slightly higher than the dark respiration. The theoretical course of the dark respiration is indicated by the dotted line.

These results are the basis of my investigations. In the following presentation I shall report on attempts to characterize the conditions for occurrence of this type of blue light effect and its dependence on some experimental factors.

All investigations are carried out with autospores of *Chlorella fusca*. The *Chlorella* cells are synchronized in a light-dark cycle of 16 to 8 h, harvested for experiments in each case at the end of the light phase, taken up in fresh growth medium and thereby adjusted to such a cell concentration that during the measurement in the cuvette for which only a limited amount of O_2 is available the respiration can be observed for about 1 h. After the harvest the *Chlorella* is kept in algal cuvettes in continuous darkness at 30°C and aerated with compressed air.

The determination of the O_2-concentration or their change is made polarographical-ly. In order to avoid the limitation of the rate electrode and also for measuring the O_2-exchange in absolute values, as well as carry out a repeated sample change in absolute darkness, a concentration electrode is used which has been developed specially for this purpose. To keep constant all the factors which possibly influence the oxygen measure-ment and to carry out experiments over several days, all measurements, as well as the necessary manipulations, are automated. The measuring method will be demonstrated by the scheme of the experimental setup in Fig. 2. In a Plexiglas cuvette a platin-silver-silver-chloride elecrode is installed which contains KCl as electrolyte and is covered with a Teflon membrane. The voltage which is necessary for polarization of the electrode is provided by a polarograph. The depolarization current which is conducted to the polaro-graph is directly proportional to the O_2-concentration. This current is linearly amplified, changed into voltage and simultaneously recorded on a recorder and printed digitally on a printer after an analog/digital transformation in the multitester. The values printed by the printer are calculated by a Siemens computer.

The sampling of the *Chlorella* out of the algal reserve cuvette and filling into the Plexiglas cuvette with a plunger pump are controlled in such a way that the respiration

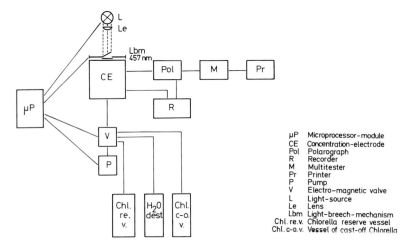

Fig. 2. Scheme of the experimental setup

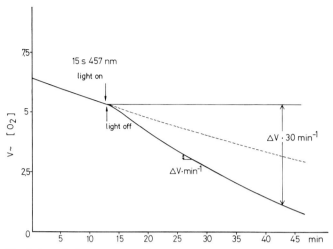

Fig. 3. A typical record curve of the O_2-exchange of *Chlorella fusca* obtained by a concentration-electrode

activity of *Chlorella* is not influenced by the manipulation. After every measurement the sample of *Chlorella* is thrown away and the cuvette, the pump, and the valves which allow a flowrate of liquids and gases is cleaned with distilled water. As light source a halogen lamp of 1000 W is used, from this light 457 nm is filtered out with a Schott interference filter with a half-width of 13 nm. A light breach mechanism controls the illumination time. All these function elements are controlled by a microprocessor module which also allows the programming of an exact experiment programm. All function elements with which *Chlorella* comes into contact are exactly tempered and additionally during the O_2-measurement in the Plexiglas cuvette the temperature of the *Chlorella* sample in the cuvette is measured directly by a thermocouple and is recorded. The temperature deviation does not exceed a tenth degree.

In Fig. 3 the course of a typical curve of the O_2-exchange of *Chlorella* obtained by a concentration electrode is shown. In the left part of the curve the O_2-uptake in the dark (always named dark respiration) and the enhancement of the O_2-uptake which follows the blue light illumination — in this case for 15 s — is presented. The gradient of the curve gives the speed of the O_2-uptake. For the presentation of the kinetics the change of O_2-concentration per minute and for investigations of the effect in dependence on time of dark pretreatment the amount of oxygen taken up during the 30 min after the preceding blue light exposure is determinated. The dotted line shows the theoretical course of dark respiration, namely that which is measured by another sample directly after this blue light experiment, when the blue light effect and the dark respiration are approximately constant.

The kinetics (Fig. 4) which have been calculated from such measurements with the concentration electrode show a maximum which is shifted into the fourth to the sixth minute, but apart from that they correspond to the curve measured with the rate electrode. The experiments presented here differ from each other only in the dark period which precedes each measurement. This period lasts 9 h for the upper kinetic, 21 h for

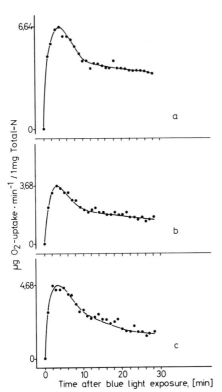

Fig. 4a−c. Kinetics of the respiratory O_2-uptake (measured with a concentration-electrode) after a blue light exposure of 15 s 4,75 · 10^{-9} einstein cm^{-2} s^{-1} 457 nm. The kinetics are measured at different times of dark-pretreatment: **a** after 9h, **b** after 21 h, **c** after 52 h

the middle one and 52 h for the lower one. The light intensity and the illumination time are the same in all cases. It is to be noticed that the curves which are measured at each given point of time are kinetically more or less the same but are different in the absolute quantity of the blue light effect. Measuring the quantity of the effect every hour with increasing time of dark pretreatment the variation of the quantity shows a rhythmical course. Figure 5 shows the quantity of oxygen which is taken up in the first 30 min after the preceding blue light illumination as a function of the time of dark pretreatment which precedes each measurement. The measurement starts 8 h after the light phase of the light-dark-cycle, this is the moment when the normal light starts in the culture cycle. At this time the sporulation is completed to 90%. At this moment the absolute quantity of the O_2-uptake after the blue light illumination reaches a maximum. After that the quantity of the blue light effect decreases to a minimum in the sixteenth hour of the dark period. This is followed by a plateau which covers several hours and an increase of the quantity of the effect to a maximum about 24 h later. Then the quantity of the effect decreases again and after about 24 h it reaches another maximum.

The interruption of the course of the curve is caused by a passing defect of the membrane which is covered over the electrode.

Such regular variations of the quantity of the blue light effect can be observed over several days. The start of the measurement of the series of experiments which is pre-

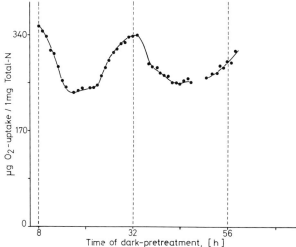

Fig. 5. Quantity of O_2-uptake during 30 min in the dark after a blue light exposure of 15 s in dependence on the time of dark-pretreatment

sented in Fig. 6 is preceded by a dark period of 95 h. At the beginning of the fifth theoretical light-dark cycle here again a rhythmical variation of the quantity of the effect is to be noticed. Compared to the first result the phase position is shifted for several hours after five days which, as well as the running experiments for characterization of the rhythm, might be due to a cicardian rhythm. The curves presented here, as well as the following ones, are all curves of presentative experiments from a series of which the result corresponds to what is shown here in all essential points. It is, however, to be taken into consideration that with 10% of the series of the experiments an enhancement of the respiration caused by blue light is to be observed but no rhythmical variation of the quantity of the effect. Microscopic observation of *Chlorella*, which shows such a devia-

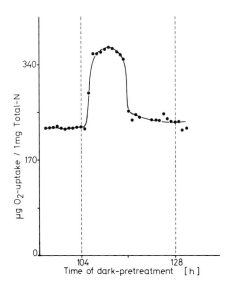

Fig. 6. For legend see Fig. 5

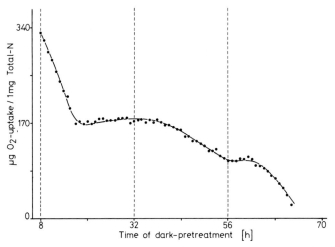

Fig. 7. Quantity of O_2-uptake during 30 min in the dark without a blue light exposure in dependence on the time of dark pretreatment

tion in its physiological behavior, shows an intense vacuolization, which is normally not found in young autospores.

When determining the quantity of the dark respiration of *Chlorella* in exactly the same way as the rhythm of this blue light effect, a step-by-step decline of the respiration activity in continuous darkness is to be observed in the first days of the experiment. Like the preceding figures, Fig. 7 shows the amount of oxygen taken up in 30 min, dependent on the time of dark pretreatment. In the first part the variation of the quantity of the dark respiration shows a course which is very similar to the variation of the quantity of the blue light effect. The decline to a minimum in the 15th to 16th h is followed by another decline to a second slightly less distinct plateau, and finally another decline of O_2-uptake. In spite of the initially parallel course of the magnitude of the dark respiration and the quantity of the blue light effect, the further course of both curves shows clearly that the rhythmical variation of the absolute quantity of the blue light effect does not reflect a constant proportional enhancement of a rhythmically oscillating dark respiration. Instead, it seems that the relative enhancement of the O_2-uptake caused by blue light also shows a rhythmical course.

It is difficult to identify or even to correlate this rhythmical effect with others of the various described effects of blue light on respiration and on connected metabolic activities.

At first it was assumed that these phenomena could possibly be classified according to their dependence on light intensity. This dependence was determined for the effect described here after a time of dark pretreatment after which the quantity of the blue light effect is approximately constant for several hours. Figure 8 shows the quantity of O_2-uptake 10 min after the blue light illumination of 15 s as a function of the light intensity. The effect is half-saturated at 6×10^{-10} einstein cm^{-2} s^{-1} and reaches its saturation in the range of 4.1×10^{-9} einstein cm^{-2} s^{-1}. However, in the recent past it has emerged from experiments still running that the effect reported here appears to be dose-dependent, at least within a limited range.

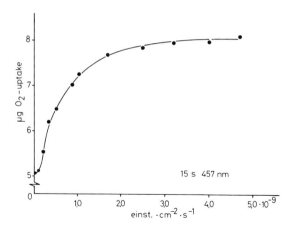

Fig. 8. Quantity of O_2-uptake during 10 min in the dark after a blue light exposure of 15 s in dependence on the light intensity

Therefore a comparison with other effects on the basis of the dependence on intensity does not seem conclusive.

References

Ried A (1965) Carnegie Inst Wash Yearb 64: 399–406
Ried A (1966) Ber Dtsch Bot Ges 79: 112–115
Ried A (1968) Biochim Biophys Acta 153: 653–663
Ried A (1969) Planta 87: 333–346

Interaction Between Blue Light
and Nitrogen Metabolism

Regulation by Monochromatic Light of Nitrate Uptake* in *Chlorella fusca*

F. CALERO[1], W.R. ULLRICH[2], and P.J. APARICIO[1]

1 Introduction

Regulatory effects of monochromatic light in the overall processes of photosynthesis and respiration are widely known in higher plants and algae. Blue light plays an important role by changing the pattern of photosynthetic products and by stimulating mitochondrial respiration even under conditions of active photosynthesis (Voskresenskaya 1972). This blue light effect can be due to an enhancement in the flow of electrons from the noncyclic electron transport chain toward oxygen, promoting an increase in pseudocyclic photophosphorylation, or to a stimulation of NAD(P)H consumption leading to a higher level in C_3 compounds, in glycolate and in amino acids, together with a higher level of ATP (Steup and Pirson 1974; Eichhorn and Augsten 1977). Besides, it certainly also consists in a direct interference of blue light with the activities of various enzymes involved in carbon and amino acid metabolism (Ries and Gauss 1977; Miyachi et al. 1978). The stimulation of protein and amino acid synthesis by blue light requires higher rates of ammonium production, as suggested or already found by several authors (Voskresenskaya and Grishina 1962). As the first steps in the utilization of nitrate, its uptake and/or its subsequent reduction to ammonia could be key reactions to be regulated by monochromatic light.

Many short-term effects of blue light are due to photoreactions of flavins or carotenoids (Voskresenskaya 1972). A rather direct interference of blue light with nitrate reduction in vitro was shown recently (Aparicio et al. 1976; Roldán et al. 1978): nitrate reductase inactivated in vitro or in vivo could be reactivated in vitro by light, mainly blue light, and this photoreactivation was highly accelerated by addition of flavins.

The present paper deals with the short-term effects of monochromatic light on the utilization of inorganic nitrogen compounds in *Chlorella fusca,* especially on nitrate uptake and nitrate reductase activity. Special attention was also paid to the relation of the photosynthetic and respiratory carbon metabolism with this photochromic regulation (cf. Calero 1978).

---

* Dedicated to Prof. W. Simonis on the occasion of his 70th birthday
1 Departamento de Bioquímica, Facultad de Biologia y CSIC, Universidad de Sevilla, Sevilla, Spain
2 Institut für Botanik der Technischen Hochschule, Schnittspahnstr. 3, 6100 Darmstadt, FRG

2 Materials and Methods

Chlorella fusca Shihira et Krauss, strain 211-15 from the collection of algae in Göttingen was cultured under continuous light of 11 klx at 27°C and with 2% CO_2, in the medium of Pirson and Ruppel (1962). The culture was diluted daily to a cell density of 3 to 5 μg chlorophyll per ml and used at a cell density of 10 to 20 μg chl/ml. For the nitrate uptake experiments the algae were collected on millipore filters, washed twice with nitrate-free medium buffered to pH 8.0 with 5 mM Tris-HCl, resuspended in the same nitrate-free medium to a final cell density of 25 to 35 μg chl/ml, and transferred to an experimental Plexiglas cuvette at 28°C. Illumination with white light and aeration with air + 2% CO_2 were started immediately. Light sources were 250 W tungsten-halogen projector lamps, one of them with focusing lens, and a 200 W mercury-xenon lamp (Oriel Co., Stamford, Conn., USA). Filters for monochromatic light were either broad-band interference filters for 450 ± 50 nm, 550 ± 50 nm, and 650 ± 50 nm (K 45, K 55, K 65 from Balzers, Liechtenstein) or narrow-band interference filters (B-1, X-ray and far-infrared blocked, from Baird-Atomic, Boston, Mass., USA). Nitrate, nitrite or ammonia was added after 20 min adaptation of the algae in the experimental cuvette. Samples of 2 ml were taken out every 5 or 10 min and were filtered immediately through paper filters or cellulose membranes. Nitrate concentration was measured by optical density differences between 203 and 250 nm, similarly to Cawse (1967), using 0.2 ml of the sample + 0.8 ml water, 1.0 ml of 2% sulfamic acid and 3.0 ml of 6.67% $HClO_4$. Nitrite was assayed according to Snell and Snell (1948) and was measured spectrophotometrically at 540 nm, ammonia with Nessler's reagent.

   For estimation of the nitrate reductase activity the cells were collected by centrifugation at 5000 g, washed with distilled water (or 10 mM Tris-HCl, pH 7.5) and disintegrated at 0°C with glass beads in a cell mill (Vibrogen-Zellmühle, Bühler, Tübingen, Germany) or ground in a mortar with alumina powder. The homogenate was then filtered through a glass fritte and centrifuged at 27,000 g for 15 min, the clear supernatant serving as enzyme preparation. The standard nitrate reductase assay mixture contained in 1 ml: Tris-HCl pH 7.5, 100 μmol; KNO_3, 10 μmol; NADH, 0.3 μmol, and an adequate volume of the enzyme preparation. After 5 min incubation at 30°C, the nitrite formed was assayed according to Snell and Snell (1949).

   Oxygen evolution experiments were carried out with an oxygen electrode (Rank Brothers, Bottisham, England) also at 28°C, pH 6.3, with carbonic anhydrase, and at the same light intensities as described for the nitrate uptake experiments. For maximum photosynthesis $KHCO_3$ was added during the experiments to give a final concentration of 2 mM.

   The inhibitor DCMU [3-(3,4-dichlorophenyl)-1,1-dimethylurea] was dissolved in ethanol and added to the experiments to give a final concentration of 30 μM in 1% ethanol. The figures show single representative experiments out of a series of varying size, the tables show mean values of various experiments, the number of experiments being indicated in brackets.

3 Results

3.1 Nitrate Uptake

Nitrate uptake and nitrate reduction are strongly light-dependent in algae. Light saturation in white light is achieved at about the same light intensity as that of photosynthetic
oxygen evolution (Ullrich, unpublished). At saturating intensities red or blue light sustained a lower nitrate consumption than white light (Figs. 1 and 2). While at these high
intensities red or blue light was similarly effective, they yielded only about one half of
the nitrate uptake rate obtained in saturated white light. Switching from one light quality to another caused an immediate change in the nitrate consumption rate, the technique employed not allowing the detection of any lag phase, which suggests that the
transition was complete after 2 to 3 min at the latest. This applies also to a change from
light to dark, although the dark uptake rate in the absence of glucose or other reduced
carbon sources is only 0.5 to 1.0 μmol mg^{-1} chl h^{-1}, which is one tenth to one fiftieth
of the rate in saturating white light (Fig. 1A). All experiments were carried out at saturating nitrate concentrations, and it must be emphasized that the rate of nitrate consumption, once established under the particular conditions, remains constant until almost
all nitrate has been consumed (Fig. 1A).

   Addition of light of the complementary wavelength to a saturating red or blue light
produced a significant increase in the nitrate uptake rates (Fig. 2A, B). Even weak red
light (Fig. 2B) in combination with saturating blue light was sufficient to yield 87% of

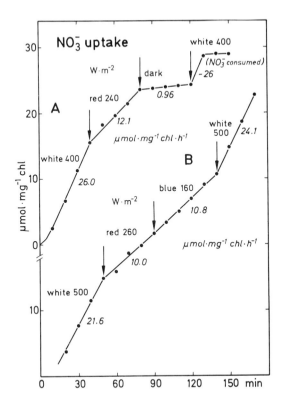

Fig. 1. Nitrate uptake in white, red,
and blue light at high light intensities,
and in the dark, in *Chlorella fusca*. pH
8.0. A and B different experiments
showing different light saturation rates
and the immediate changes upon
changing the illumination

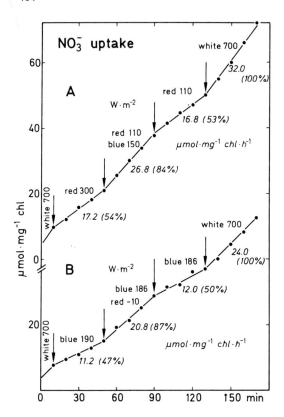

Fig. 2A, B. Complementary effect of red with blue light, compared with only red or blue light, and white light, on nitrate consumption by *Chlorella fusca*. pH 6.3. **A** Addition of strong blue and red light, **B** addition of weak red to saturating blue light

the saturation rate in white light. Green light of a similar intensity did not produce any stimulatory effect (data not shown)

Saturation of nitrate uptake was obtained with different intensities of red and blue light: in blue light 180 W m^{-2} were sufficient, while in red light almost 300 W m^{-2} were necessary (Fig. 3). At lower light intensities of about 60 W m^{-2}, blue light proved to be much more efficient than red light, leading to more than the double relative yield for nitrate uptake (Fig. 3, Table 2).

In the absence of carbon sources as CO_2, the nitrate uptake rates under blue or red light were comparable to those sustained by white light (Table 1), perhaps because the rates were too low to be further limited by monochromatic light. Under these conditions

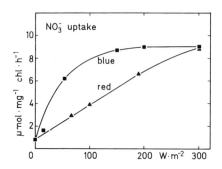

Fig. 3. Light saturation curves with blue and red light of nitrate consumption in *Chlorella fusca*. pH 6.3. Saturation rate in white light at an average of 19 μmol mg^{-1} chl h^{-1}

Table 1. Nitrate consumption by *Chlorella fusca* cells in the absence of CO_2 in white, blue, and red light of 200 W m^{-2}. CO_2-free air, 100% = 11 μmol mg^{-1} chl h^{-1}. 2 mM KNO$_3$. (4 experiments)

Illumination	NO$_3^-$ Uptake (%)
White	100
Blue	99
Red	102

Table 2. Relative yield of photosynthetic O_2 evolution and nitrate uptake in blue related to red light at equal light intensities. Air + 2% CO_2 pH 8.0, 2 mM KNO$_3$. (5 experiments)

Light intensities W m^{-2}	O_2 Evolution Blue/Red	NO$_3^-$ Uptake Blue/Red
60	1.8	2.3
190	1.4	1.4
300	1.0	1.0

nitrate is reduced and high amounts of nitrite and ammonia are excreted to the culture medium, accounting for the total nitrate taken up (Ullrich 1974; Eisele and Ullrich 1975).

3.2 Photosynthetic Oxygen Evolution

The different nitrate uptake rates in white, blue, and red light could also be due to a different efficiency in photosynthesis. In contrast to the much lower rates of nitrate uptake in saturating monochromatic light, oxygen evolution reaches the same 200 μmol mg^{-1} chl h^{-1} in blue or red as in white light (Fig. 4). At low light intensities, there was, indeed, a reasonably higher yield in oxygen evolution in blue light compared with red light, but the efficiency ratios are lower for O_2 evolution than for nitrate consumption (Table 2).

3.3 Nitrite and Ammonia Uptake

Nitrite uptake, under the same conditions as for nitrate uptake, was only slightly modified by a change in the light quality, although the intensities of red and blue light were not fully saturating (Fig. 5). This clearly shows that neither nitrite uptake nor its reduction or further metabolic steps of ammonium assimilation can be directly involved in the effects of monochromatic light. Nitrite uptake rates were higher than nitrate uptake rates and, therefore, its uptake and/or its reduction cannot be rate-limiting to nitrate

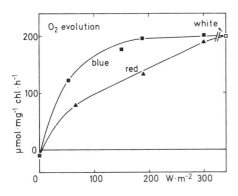

Fig. 4. Light saturation curves with blue and red light of CO_2-dependent oxygen evolution by *Chlorella fusca*. 2 mM HCO$_3^-$ added at pH 6.3 in presence of carbonic anhydrase. Respiration rate: 10 μmol O_2 mg^{-1} chl h^{-1}

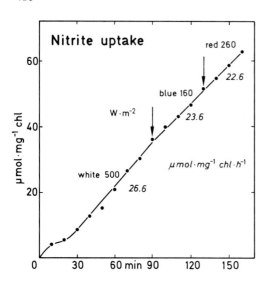

Fig. 5. Nitrite consumption of *Chlorella fusca* in white, red, and blue light at pH 8.0

consumption in red or blue light. This applies also to ammonium uptake whose rate showed only little response to white, red, or blue light of equal energy in the range close to saturation (Table 3).

3.4 Nitrate Reductase Activity

Nitrate reductase was extracted from algae after exposure to white, red, or blue light in parallel experiments in which nitrate uptake had been measured for one hour. The results (Table 4) show only a slight decrease of the activity after exposure to blue or red light, but in spite of possible losses due to the extraction procedure, the activity was in all cases much higher than necessary to sustain the in vivo rates of nitrate reduction. Thus in spite of possible artifacts, it is very unlikely that the activity of nitrate reductase can be responsible for the lower levels of nitrate consumption in monochromatic light.

Table 3. Nitrite and ammonium uptake rates in white, blue, and red light of 200 W m^{-2}. Values in μmol mg^{-1} chl h^{-1}. Air + 2% CO$_2$, 2 mM KNO$_2$ or NH$_4$Cl. For nitrite initial rate was used. (5 experiments)

Illumination	NO$_2$ Uptake	NH$_4^+$ Uptake
White	20	23
Blue	19	19
Red	19	17

Table 4. Nitrate reductase activity after 60 min illumination with white, blue, or red light of *Chlorella fusca* in cell-free extracts. Light intensity 200 W m^{-2} Air + 2% CO$_2$. Rates of NO$_3^-$ uptake and activity of nitrate reductase in μmol mg^{-1} chl h^{-1} 2 mM KNO$_3$

Illumination	NO$_3^-$ reductase activity	Nitrate uptake
White	40 (100%)	22.0 (100%)
Blue	36 (90%)	11.7 (53%)
Red	37 (92%)	11.0 (50%)

Table 5. Effect of weak monochromatic light on glucose-sustained nitrate uptake. 30 mM glucose, 2 mM KNO_3, 30 μM DCMU, air. Cells preilluminated for 30 min with white light prior to addition of glucose, nitrate, and DCMU. Average dark rate 5 μmol mg^{-1} chl h^{-1}. Light intensities for all wavelengths 3.5 W m^{-2}. (3 experiments)

Illumination	Relative nitrate uptake
Dark	100%
Blue 457 nm	213%
Blue 423 nm	169%
Red 666 nm	100%
Red 720 nm	94%
Red 666 + blue 457 nm	218%

Table 6. Effect of weak monochromatic light on glucose-dependent nitrite and ammonia uptake. 30 mM glucose, 2 mM KNO_2 or 1 mM NH_4Cl, 30 μM DCMU, air. Pretreatment for 30 min in the dark. Rates in μmol mg^{-1} chl h^{-1}. (3 experiments)

	Nitrite uptake	Ammonia uptake
Dark	15	9
Blue 457 nm (3.5 W m^{-2})	13.5	9

3.5 Blue Light and Respiratory Nitrate Reduction

In order to separate the effect of monochromatic light on nitrate uptake from that exerted on the photosynthetic apparatus, experiments with red or blue light were carried out also in the presence of DCMU, conditions in which the energy is mainly supplied by mitochondrial metabolism. Respiratory electron transport in algae, *Ankistrodesmus*, or *Chlorella*, can be greatly enhanced by addition of glucose to the medium. With 20 mM glucose, the nitrate uptake rates in the dark are saturated and their usual level is fivefold greater than without glucose. In the presence of glucose and DCMU, red light and far red light turned out to be completely ineffective, while blue light, even at a low intensity, exerted a distinct positive effect (Table 5), which was stronger at 457 nm than at 423 nm, though the relative absorption of chlorophyll a is higher at the latter wavelength. Addition of weak red light of 666 nm to this weak blue light did not produce any further enhancement, in contrast to the photosynthetic conditions (cf. Fig. 2). Also under respiratory conditions, consumption of nitrite or ammonia was unaffected by blue light in the presence or absence of glucose (Table 6), although the uptake rates of these two substrates are considerably higher than those of nitrate uptake.

4 Discussion

4.1 Limiting Steps of Nitrate Utilization

Nitrate utilization in algae may be controlled at various sites. The mechanism of nitrate uptake is not yet adequately known. At the plasmalemma, nitrate must be transported against an electrical gradient. Nitrate uptake induces transient depolarization of the cell, probably due to a cotransport with protons (Novacky et al. 1978). At the same time, nitrate utilization causes stoichiometric alkalinization in the medium due to the production of excess hydroxyl ions during nitrite reduction of at least 1 OH/NO_3^- (Eisele and Ullrich 1975, 1977; Smith and Raven 1979). The depolarization of the membrane of *Lemna* cells suggests a cotransport of more than one H^+ with nitrate. One of them is consumed during protein synthesis, the excess must be released to the medium by a proton pump. Thus nitrate uptake presumably requires energy for the proton pump at the plasmalemma. This energy is most probably supplied by ATP and, therefore, the ATP level or the rate of its distribution within the cell may become rate-limiting. In addition, the nitrate uptake system itself, i.e., the assumed carrier protein, could be directly involved in the regulation of the entrance of nitrate into the cell by changes in affinity or by various kinds of inhibition (Eisele and Ullrich 1977; Ullrich and Eisele 1977). In microalgae as *Chlorella* or *Ankistrodesmus,* nitrate uptake is also controlled by the rate of its reduction, since the cells do not contain large vacuoles for solutes and, therefore, nitrate cannot be stored as in higher plants.

Nitrate reductase catalyzing the first step of nitrate reduction is now mostly regarded as located in the cytoplasmic compartment of the cells, as shown indirectly for algae (Kessler and Zumft 1973) and more directly for higher plants (Dalling et al. 1974), whereas nitrite reduction is closely related to electron transport in plastids. This means that for nitrate reduction, reducing power (NADH) has to be made available at sufficient rates from either the chloroplasts or the mitochondria (Heber 1975). The shuttles for reducing power through chloroplast or mitochondrial envelopes may be severely limited in the absence of carbon sources, due to low levels of carbon intermediates. In addition, nitrate reductase in green algae can exist in two forms, one oxidized and active, converted by reduction to an inactive form. Ferricyanide or blue light are effective reactivating agents. Apparently, this interconversion mechanism is also operating in vivo according with the different nitrogen sources, ammonia being the remote inactivating agent; but this indirect in vivo ammonia effect on the activity of nitrate reductase will be seen usually only after more than 1 h. It was suggested that the reducing power and the energy charge are the actual factors involved in this regulatory mechanism (Losada et al. 1973; Pistorius et al. 1976; see also reviews of Losada and Guerrero 1979; Vennesland and Guerrero 1979). Hence for nitrate reduction either the activity of nitrate reductase or the transport of reducing equivalents may be rate-limiting.

Photosynthetic oxygen evolution is under most conditions higher than the respiratory oxygen consumption, which suggests that the observed higher nitrate uptake rates obtained under phototrophic conditions may correlate with the reducing power generated by either photosynthetic or respiratory metabolism. Under phototrophic conditions there seems to be no competition between reduction of nitrate and CO_2 fixation, since nitrite reduction, a chloroplast reaction, is in most cases faster than nitrate consumption. However, as nitrate reduction is considered as a cytoplasmic reaction, it might depend on the transport rates of reducing power out of the chloroplast.

4.2 Effects of Monochromatic Light on Photosynthetic Nitrate Utilization

At light saturation, the nitrate utilization rate in blue or red light was one half of the rate in white light. Since the maximum rate of O_2 evolution is the same in saturated white, blue, and red light, a limitation of nitrate uptake by noncyclic photosynthetic electron flow seems unlikely in the presence of CO_2. This suggests that also the transport shuttles through the chloroplast envelope to the cytoplasmic compartment can transfer enough reducing power and ATP to be not rate-limiting under monochromatic light. The fact that nitrite assimilation was independent of the light quality at saturation speaks against limitations induced by monochromatic light for the generation of reducing power by the photosynthetic apparatus. At nonsaturating light intensities blue light proved to be more efficient than red light, somewhat less for photosynthetic O_2 evolution than for nitrate uptake. The changes in nitrate uptake rates upon changes in illumination occur almost immediately (within less than 2 min), while the activity of nitrate reductase remains almost unchanged for periods of an hour, and is normally much higher than necessary for the nitrate reduction rates in vivo. This suggests that nitrate uptake is the limiting step. As discussed before, nitrate assimilation does not seem to be limited by the supply of reducing power, so it may become limited by low supply of ATP to the ATP-dependent proton pump associated with the uptake system. Photosynthetic carbon reduction may sometimes require all ATP available from noncyclic electron transport (Heber and Kirk 1975). Therefore, at the plasmalemma, ATP may be scarce, which is sometimes visible by changes of the total ATP content of the cells, e.g., upon light-dark changes in algae (Urbach and Kaiser 1972; Larsson and Tillberg 1978) or less upon addition of nitrate (Ullrich-Eberius 1973). Blue light might contribute to ATP generation by inducing mitochondrial respiration (Strotmann and Murakami 1976) even under conditions of saturated photosynthesis. But the similar efficiency of blue or red light, when added to saturated complementary light, does not fit to this interpretation, since red light is completely ineffective in stimulating respiration and respiratory nitrate utilization. It is suggested that in white light, and in combined red and blue light at saturating intensities, the additional ATP for nitrate uptake is partly supplied from cyclic photophosphorylation. This interpretation implies that monochromatic light is not able to sustain top levels of photosynthetic ATP production. Phosphate uptake and phosphorylation also exhibit wavelength-dependent light effects, blue light being more effective (Steup and Pirson 1974).

4.3 Blue Light Stimulation of Glucose-Dependent Respiratory Nitrate Utilization

Blue light has often been reported to stimulate respiration in algae (Kowallik 1967; Voskresenskaya 1972; Andersag and Pirson 1976; Eichhorn and Augsten 1977) and also nitrite utilization in *Chlorella* (Strotmann 1967). Respiration is also considerably enhanced by glucose in some unicellular algae (Ries and Gauss 1977) which can readily take up glucose as a carbon source (see Eisele and Ullrich 1977).

   In order to avoid direct interference of noncyclic photosynthetic electron flow and of the substrate transport shuttles through the chloroplast envelope, experiments were also carried out in which a glucose-supported respiration was secured by the presence of 30 μM DCMU. Even in the presence of DCMU, some of the light effects observed could be explained by the supply of additional ATP from cyclic photophosphorylation, but this becomes unlikely, since red light, at least at the low light intensities applied, was completely ineffective. Since neither assimilation of nitrite, which requires mainly reducing power, nor assimilation of ammonia, for which the energy requirement is lower, showed the same stimulation by blue light, the blue light stimulation of nitrate assimilation is due to enhancement of the ATP production by respiration and not to the supplementary production of mitochondrial NADH. Thus, also for the blue light effect on the respiratory nitrate utilization, the availability of ATP at the plasmalemma and direct changes of the nitrate uptake system are supposed to be responsible. The optimum wavelength of 457 nm suggests the participation of a flavin or carotenoid, as known from the stimulation of respiration, but does not allow a final decision as yet, whether nitrate uptake is directly enhanced or rather indirectly by increased supply of ATP.

Acknowledgments. This work was supported by grants from the Consejo Superior de Investigaciones Científicas (Madrid) To F.C. and W.R.U., and from the Cooperative Research Programs Spain-USA (III P 7730394/7) to P.J.A. and Richard Malkin, Department of Cell Physiology, University of California, Berkeley, and by grants from the Deutsche Forschungsgemeinschaft to W.R.U. We wish to express our gratitude to Prof. M. Losada and to Dr. M.G. Guerrero for stimulating discussions and for kind and constant support during our common work in Sevilla.

References

Andersag R, Pirson A (1976) Verwertung von Glucose in *Chlorella*-Kulturen bei Blau- und Rotlichtbestrahlung. Biochem Physiol Pflanz 169: 71–85

Aparicio PJ, Roldan JM, Calero F (1976) Blue light reactivation of nitrate reductase from green algae and higher plants. Biochem Biophys Res Commun 70: 1071–1077

Calero F (1978) Estudios a nivel celular de la asimilacion del nitrato en *Chlorella fusca*. Papel regulador del amoniaco y de la luz azul. Thesis, Sevilla

Cawse PA (1967) The determination of nitrate in soil solutions by ultraviolet spectrophotometry. Analyst 92: 311–315

Dalling MJ, Tolbert NE, Hageman RH (1972) Intracellular location of nitrate reductase and nitrite reductase. Biochim Biophys Acta 283: 505–512

Eichhorn M, Augsten H (1977) Die Wirkung von Blau- und Rotlicht auf die Aktivität der Glucose-6-phosphat-Dehydrogenase und das Adenylatsystem bei *Wolffia arrhiza* unter steady state Bedingungen. Z Pflanzenphysiol 85: 147–152

Eisele R, Ullrich WR (1975) Stoichiometry between photosynthetic nitrate reduction and alkalinisation by *Ankistrodesmus braunii* in vivo. Planta 123: 117–123

Eisele R, Ullrich WR (1977) Effect of glucose and CO_2 on nitrate uptake and coupled OH^- flux in *Ankistrodesmus braunii*. Plant Physiol 59: 18–21

Heber U (1975) Energy transfer within leaf cells. In: Avron M (ed) Proc 3rd Int Congr Photosynth, pp 1335–1347. Elsevier, Amsterdam New York Oxford

Heber U, Kirk M (1975) Flexibility of coupling and stoichiometry of ATP formation in intact chloroplasts. Biochim Biophys Acta 376: 136–150

Kessler E, Zumft WG (1973) Effect of nitrite and nitrate on chlorophyll fluorescence in green algae. Planta 111: 41–46

Kowallik W (1967) Action spectrum for an enhancement of endogenous respiration by light in *Chlorella*. Plant Physiol 42: 672–676

Larsson CM, Tillberg JE (1978) Effects of phosphate readdition on ATP levels and O_2 exchange in phosphorus-starved *Scenedesmus*. Z Pflanzenphysiol 90: 21–31

Losada M, Guerrero MG (1979) The photosynthetic reduction of nitrate and its regulation. In: Barber J (ed) Photosynthesis in relation to model systems. Elsevier, Amsterdam New York Oxford

Losada M, Herrera J, Maldonado JM, Paneque A (1973) Mechanism of nitrate reductase reversible inactivation by ammonia in *Chamydomonas*. Plant Sci Lett 1: 31–37

Miyachi S, Miyachi S, Kamiya A (1978) Wavelength effects on photosynthetic carbon metabolism in *Chlorella*. Plant Cell Physiol 19: 277–288

Novacky A, Fischer E, Ullrich-Eberius CI, Lüttge U, Ullrich WR (1978) Membrane potential changes during transport of glycine as a neutral amino acid and nitrate in *Lemna gibba*. G1. FEBS Lett 88: 264–267

Pirson A, Ruppel HG (1962) Über die Induktion einer Teilungshemmung in synchronen Kulturen von *Chlorella*. Arch Mikrobiol 42: 299–309

Pistorius EK, Gewitz HS, Voss H, Vennesland B (1976) Reversible inactivation of nitrate reductase in *Chlorella vulgaris* in vivo. Planta 128: 73–80

Ries E, Gauss V (1977) D-Glucose as an exogenous substrate of the blue light enhanced respiration in Chlorella. Z Pflanzenphysiol 82: 261–273

Roldán JM, Calero F, Aparicio PJ (1978) Photoreactivation of spinach nitrate reductase: role of flavins. Z Pflanzenphysiol 90: 467–474

Smith FA, Raven JA (1979) Intracellular pH and its regulation. Annu Rev Plant Physiol 30: 289–311

Snell FD, Snell CT (1948) Colorimetric methods of analysis. Van Nostrand, New York Toronto London

Steup M, Pirson A (1974) Über den Einfluß des blauen und roten Spektralbereiches auf Phosphatfraktionen, besonders Polyphosphate, bei Grünalgen. Biochem Physiol Pflanz 166: 447–459

Strotmann H (1967) Blaulichteffekt auf die Nitritreduktion von *Chlorella*. Planta 73: 376–380

Strotmann H, Murakami S (1976) Energy transfer between cell compartments. In: Pirson A, Zimmermann MH (eds) Encyclopedia of plant physiology, vol III, pp 398–416. Springer, Berlin Heidelberg New York

Ullrich WR (1974) Die nitrat- und nitritabhängige photosynthetische O_2-Entwicklung in N_2 bei *Ankistrodesmus braunii*. Planta 116: 143–152

Ullrich WR, Eisele R (1977) Relations between nitrate uptake and nitrate reduction in *Ankistrodesmus braunii*. Echanges ioniques transmembranaires chez les végétaux. Colloq Int CNRS Rouen, pp 307–314

Ullrich-Eberius CI (1973) Beziehungen der Aufnahme von Nitrat, Nitrit und Phosphat zur photosynthetischen Reduktion von Nitrat und Nitrit und zum ATP-Spiegel bei *Ankistrodesmus braunii*. Planta 115: 25–36

Urbach W, Kaiser W (1972) Changes of ATP levels in green algae and intact chloroplasts by different photosynthetic reactions. Proc 2nd Int Congr Photosynth Res Stresa, pp 1401–1411. Junk, The Hague

Vennesland B, Guerrero MG (1979) Reduction of nitrate and nitrite. In: Pirson A, Zimmermann MH (eds) Encyclopedia of plant physiology, vol VI, pp 425–444. Springer, Berlin Heidelberg New York

Voskresenskaya NA (1972) Blue light and carbon metabolism. Annu Rev Plant Physiol 23: 219–234

Voskresenskaya NA, Grishina GS (1962) Significance of light in nitrite reduction in leaves. Fiziol Rast 9: 7–15

Flavin-Mediated Photoreduction of Nitrate by Nitrate Reductase of Higher Plants and Microorganisms

W.G. ZUMFT[1], F. CASTILLO[1], and K.M. HARTMANN[1]

1 Introduction

Nitrate reductase (EC 1.6.6.1–1.6.6.3) is the key enzyme in nitrate assimilation by plants and many microorganisms. It is a high molecular weight complex of yet-to-be defined structure which contains as electron transfer groups FAD, cytochrome b-557, and a molybdenum cofactor [1]. The enzyme is usually assayed with one of the following electron donors: (1) reduced pyridine nucleotides, (2) dithionite-reduced flavins, or (3) dithionite-reduced bipyridylium salts. Both dithionite [2] and reduced pyridine nucleotides [3] interfere, however, with the synthesis of the azo dye which is used almost exclusively for analysis of the reaction product, nitrite [4]. To circumvent these difficulties reduction of methyl viologen by hydrogen via hydrogenase may be used, but requires an auxiliary enzyme and does not simplify the assay procedure [5].

A further difficulty in assaying nitrate reductase, which under certain conditions might even prevent the correct determination of the total amount of enzyme in a tissue or organism, arises from the interconvertible nature of the enzyme depending on the redox state. In *Chlorella vulgaris* nitrate reductase is largely found in an inactive form, which can be activated by oxidation by ferricyanide [6]. One physiological factor that poises the balance between the inactive and active enzyme form is ammonia. Metabolic interconversion was also demonstrated for nitrate reductase from other green algae, fungi, higher plants, and bacteria [7]. Inactive spinach nitrate reductase can be reactivated by blue light via a presumed photooxidation, an effect which is still enhanced in the presence of flavins [8]. It thus appears that a photochemical assay procedure for nitrate reductase would have several advantages over the conventional methods. We have further developed and characterized a photochemical assay, introduced by Stoy [9] many years ago with nitrate reductase from wheat leaves, which eliminates the interferences inherent to the chemical assay procedures, and allows facile photoactivation of the enzyme in the absence of an electron donor prior to photoreduction.

Riboflavin [10] and flavoproteins [11] can be reduced in the light in the presence of EDTA. Recently, renewed interest in the controlled photoreduction of flavoproteins, stems from the introduction of deazaflavins, which as powerful reductants reduce a wide range of flavoproteins, heme proteins, and iron-sulfur proteins [12]. We will show here that the photoreduction of riboflavin or FMN constitutes an efficient electron donor system for a wide range of nitrate reductases.

1 Institut für Botanik, Universität Erlangen-Nürnberg, Schloßgarten 4, 8520 Erlangen, FRG

Table 1. Characterization of the photochemical assay with spinach nitrate reductase. The complete reaction mixture contained besides the enzyme in 1 ml final volume: 80 μmol phosphate buffer (pH 7.0); 10 μmol EDTA $\cdot$ Na$_2$; 20 μmol KNO$_3$; and 0.1 μmol FMN. The samples were incubated at 30°C for 5 min in white light of 2.8 kW m^{-2}

System	Rate of nitrite formation (nmol min^{-1} mg protein^{-1})
Complete system	
Anaerobic	215
Aerobic without shaking	205
Aerobic with shaking	12
Anaerobic dark	0
Complete system anaerobic	
Without enzyme	0
Without FMN	0
Without EDTA	15
With riboflavin[a]	240
With FAD[a]	110

[a] Added instead of FMN

2 The Photochemical System

Table 1 shows the requirement of the system for light, enzyme, EDTA, and a flavin component. Riboflavin and FMN had comparable activities, but FAD was considerably less effective. Because oxygen is a very efficient competitor for reduced flavins, the system required anaerobic conditions for best results. However, for most practical purposes it was possible to run the assay in open test tubes, providing that the tubes were not shaken during the incubation period. Continuous irradiation ensured an essentially anaerobic system since diffusion of oxygen into the samples was impeded at the surface due to immediate reaction with the reduced flavin. The reaction rate was constant with time for all nitrate reductases tested (data not shown). Photochemically reduced flavin could be re-oxidized in a dark reaction by adding nitrate to a mixture, from which the electron acceptor had been initially omitted. The re-oxidation reaction proceeded stoichiometrically with respect to the amount of nitrate added and the amount of flavin oxidized, and clearly demonstrated the catalytical role of nitrate reductase.

3 Application to Different Nitrate-Reducing Systems

Nitrate reductases from fungi, algae, and higher plants all possess the same prosthetic groups, reflecting a far-reaching similarity of this enzyme among different phyla [13]. The photochemical assay was found active with representatives of three out of four groups of organisms we have tested so far (Table 2). Their nitrate reductase is specific for NADH or NADPH and has a flavoprotein moiety. The latter might not be the case with the ferredoxin-dependent nitrate reductase from *Anacystis nidulans,* which is the smallest known molybdenum enzyme [14]; the photochemical assay was inactive with

Table 2. Photochemical and chemical assays for assimilatory and dissimilatory nitrate reductases from different organisms. The conditions for the photochemical assay with FMN or riboflavin (Rfl) were as described for the aerobic standard assay in Table 1. The chemical assay was done in open test tubes and contained in 1 ml: 100 μmol Tris-HCl, pH 7.5; 20 μmol KNO_3; 0.1 μmol FMN or 0.2 μmol methyl viologen (MV); 5 μmol sodium dithionite; and 10 μmol $NaHCO_3$. The reaction was run at $30°C$ and stopped after 5 min by oxidizing the excess dithionite

Organism	Specific activity of nitrate reductase (nmol NO_2^- min^{-1} mg $protein^{-1}$)			
	Assay systems			
	FMN + hν	Rfl + hν	FMN + $S_2O_2^{2-}$	MV + $S_2O_4^{2-}$
Assimilatory nitrate reductase[a]				
Anacystis nidulans	0	0	0	30
Neurospora crassa	25	29	30	40
Chlorella fusca	20	30	21	96
Chlamydomonas reinhardii	15	17	15	30
Spinacia oleracea[b]	157	162	178	179
Dissimilatory nitrate reductase				
Pseudomonas perfectomarinus[c]	505	760	480	700

[a] Crude extracts after removal of cell debris by centrifugation were used as enzyme sources
[b] An enzyme partially purified by adsorption and desorption on Ca-phosphate gel and after ammonium sulfate precipitation was used
[c] The enzyme source was the cytoplasmic membrane fraction

this enzyme. Table 2 further compares photochemical and chemical assay procedures which yielded comparable rates. The artificial electron donor, dithionite-reduced methyl viologen, gave the highest activity in most nitrate-reducing systems.

Table 2 shows also one example of a dissimilatory nitrate-reducing system from the marine denitrifying bacterium *Pseudomonas perfectomarinus*. Despite the rather different nature of dissimilatory nitrate reductases, which are molybdenum-containing iron-sulfur proteins [15], the photochemically reduced flavins could also couple to this system. However, the site of electron donation might be remote from the enzyme, since the dissimilatory type nitrate reductase was membrane-bound and pertained its physiological electron transport chain.

4 Kinetic Characterization

For the kinetic characterization of the photochemical system, nitrite formation was studied according to the principles of action spectroscopy [16]. A fluence effect curve (Fig. 1) was obtained with a monochromatic beam of blue light (λ_m = 448 nm, bandwidth 20 nm) which coincided with the long-wavelength absorption band of flavins. For this spectral band, nitrite formation started at an irradiance of about 1 W m^{-2}, was half saturated at approximately 5 W m^{-2} and fully saturated above 30 W m^{-2} (Fig. 1). If nitrite formation from nitrate were due to photocatalysis via photo-reduced flavin, the

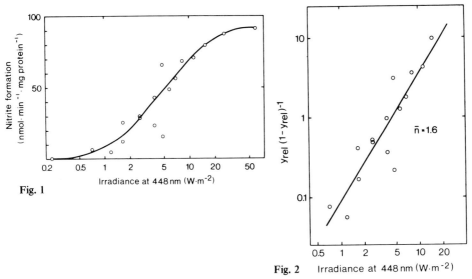

Fig. 1

Fig. 2

Fig. 1. Flavin-mediated nitrite formation as a function of irradiance. The reaction was carried out at 31°C in magnetically stirred and argon-filled Thunberg cuvettes with 10 mm light path. The reaction mixture contained in a total volume of 1.9 ml: 12 μg partially purified spinach nitrate reductase; 0.15 mM riboflavin; 20 mM KNO_3; 80 mM phosphate buffer (pH 7.0); and 10 mM EDTA · Na_2. The cuvettes were kept dark till the onset of the reaction by irradiation from a Leitz slide projector (24 V/250 W), running at controlled and stabilized current. Light was filtered through 4 mm blue glass BG 23 and an interference band pass filter, type AL, with maximal transmission at 448 nm and a bandwidth of 20 nm (both filters Schott & Gen., Mainz). Irradiances in W m^{-2} were measured within the cuvettes by means of a YSI-Radiometer model 65 and corrected for reflection loss by 1.04 for the phase transition glass-air versus glass-water. The reaction was stopped after 5 min by flash-heating to 95°C. Before measuring nitrite by diazotization [4], spectra were recorded to determine the absorbance of the samples

Fig. 2. Linearized fluence effect curve (Hill plot) of the data from Fig. 1. The Hill coefficient for the regression line is $\bar{n} = 1.6$

fluence effect curve should follow a hyperbolic function. Since two reduction equivalents are required to change the formal oxidation state of nitrate to that of nitrite, one would expect a reaction of second order.

A Hill plot, giving a linearized hyperbolic fluence effect curve (Fig. 2), showed that the slope of the regression line was $\bar{n} = 1.6$, thus deviating substantially from the expected value of two. This deviation could be attributed to the decrease of absorbance of the reaction mixture with increasing irradiances (Fig. 3), as a result of the progressing reduction of the flavin component. This decrease of photon absorption, α_{448}, with increasing photon flux, E_{p448}, changed the number of efficiently absorbed photons, the photon dose, D_{p448}, according to the equation:

$$D_{p448} = f \cdot E_{p448} \cdot \alpha(E)_{448}.$$

Herein the term f is the face area of the irradiated cuvette. The photon flux-dependent absorptance $\alpha(E)_{448}$ could be determined from the known absorbance $A(E)_{448}$ of the differently irradiated samples according to

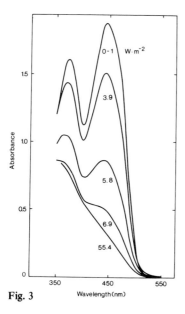

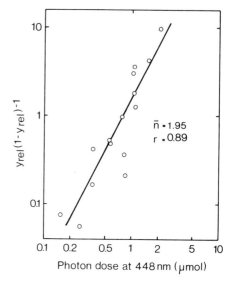

Fig. 3

Fig. 4

Fig. 3. Absorbance changes of the reaction samples for several irradiances. The *curves* show difference spectra of heat-stopped assay mixtures minus identically treated assays without riboflavin (cf. Fig. 1). The corresponding irradiance in $W \cdot m^{-2}$ is shown for each spectrum

Fig. 4. Linearized dose effect correlation for the data from Fig. 1. The Hill coefficient was $\bar{n} = 1.95$ with a correlation coefficient $r = 0.89$

$$\alpha(E)_{448} = 1 - 10^{-A(E)_{448}}.$$

Hence considering the data as a function of the photon dose — instead of the incident fluence — the Hill plot gave a dose effect correlation with the slope $\bar{n} = 1.95$ (Fig. 4) which is close to the expected value and confirms the flavin-mediated photocatalysis as being second-order.

The overall quantum yield of the flavin-mediated photoreduction of nitrate at 448 nm could be determined also from the data. Since both the number of nitrite molecules formed, N, and the number of absorbed photons, $D_{p_{448}}$, were known for each assay mixture, the quantum yield Φ_{448} was calculated as the quotient $\Phi_{448} = N/D_{p_{448}}$. The value of Φ_{448} scattered around 0.03 ± 0.01 (S.D.) for the 15 samples whose rate was $< 90\%$ saturation. The low average quantum yield is not surprising for a flavin-mediated photoctalysis since excited flavins can be efficiently quenched by easily polarizable anions [17].

In the light of the high diversity of oxidation-reduction proteins which can be photo-reduced by 5-deazaflavin, it was of considerable interest to use this donor for nitrate reductase. The powerful reducing capability of the deazaflavin was attributed to the rather low redox potential of the reactive radical dimer [12]. Table 3 compared the efficiency of natural and artificial flavins and their corresponding 5-deaza derivatives in three nitrate-reducing systems. Deazaflavins were nearly inactive as electron donors for nitrate reductase from different organisms, singularizing this system out from many that have been shown to be reducible by deazaflavin. Steric effects appear to be of minor in-

Table 3. Photoreduction of nitrate by various flavins. The concentration of flavins and deazaflavins was 0.1 mM, except for 3-methyl-5-deazaluminflavin which was 0.03 mM. The nitrate reductase from *C. fusca* was assayed in 50 mM Tris-HCl, pH 7.5. Enzyme sources were described in Table 2; irradiance of white light was 4 kW m^{-2}; other conditions were identical to that of the anaerobic standard assay of Table 1

Compound	Relative rate of nitrite formation (% of control with riboflavin)		
	C. fusca	*S. oleracea*	*P. perfectomarinus*
FMN	75	103	85
Riboflavin	100	100	100
5-Deazariboflavin	0	13	5
Lumiflavin-3-acetate	100	96	71
3-Methyl-5-deazalumiflavin	0	13	5

fluence, since lumiflavin was fully active, whereas a 5-deazalumiflavin again was inactive. Mechanistically the 5-deazaflavins are a "flavin-shaped" nicotinamide model with limited analogy to flavin catalysis [18]; within this context it is conceivable that deazaflavins photoreduce an electron transfer component of nitrate reductase but are unable to sustain a catalytical cycle. For lack of a sufficient quantity of a purified nitrate reductase it was at present not possible to check spectrophotometrically for the photoreduction of the enzyme.

Acknowledgments. We wish to thank Prof. P. Hemmerich for his kind gift of flavins. The fellowship of the Krupp Foundation to F.C. is gratefully acknowledged. W.G.Z. would like to express thanks to the Deutsche Forschungsgemeinschaft for financial support.

References

1. Hewitt EJ (1975) Assimilatory nitrate-nitrite reduction. Annu Rev Plant Physiol 26: 73–100
2. Senn DR, Carr PW, Klatt LN (1976) Minimization of a sodium dithionite-derived interference in nitrate reductase-methyl viologen reactions. Anal Biochem 75: 464–471
3. Medina A, Nicholas DJD (1957) Interference by reduced pyridine nucleotides in the diazotization of nitrite. Biochim Biophys Acta 23: 440–442
4. Snell FD, Snell CT (1949) Colorimetric methods of analysis, vol II. Van Nostrand, Toronto New York London
5. Pichinoty F, Piéchaud M (1968) Recherche des nitrateréductases bactériennes A et B: Méthodes. Ann Inst Pasteur 114: 77–98
6. Jetschmann K, Solomonson LP, Vennesland B (1972) Activation of nitrate reductase by oxidation. Biochim Biophys Acta 275: 276–278
7. Losada M (1975/76) Metalloenzymes of the nitrate-reducing system. J Mol Catal 1: 245–264
8. Aparicio PJ, Roldán JM, Calero F (1976) Blue light photoreactivation of nitrate reductase from green algae and higher plants. Biochem Biophys Res Commun 70: 1071–1077
9. Stoy V (1956) Riboflavin-catalyzed enzymic photoreduction of nitrate. Biochim Biophys Acta 21: 395–396
10. Merkel J, Nickerson WJ (1954) Riboflavin as a photocatalyst and hydrogen carrier in photochemical reduction. Biochim Biophys Acta 14: 303–311
11. Massey V, Palmer G (1966) On the existence of spectrally distinct classes of flavoprotein semiquinones. A new method for the quantitative production of flavoprotein semiquinones. Biochemistry 5: 3181–3189

12. Massey V, Hemmerich P (1977) A photochemical procedure for reduction of oxidation-reduction proteins employing deazariboflavin as catalyst. J Biol Chem 252: 5612–5614
13. Zumft WG (1976) Anorganische Biochemie des Stickstoffs. Die Mechanismen der Stickstoffassimilation. Naturwissenschaften 63: 457–464
14. Manzano C, Candau P, Guerrero M (1978) Affinity chromatography of *Anacystis nidulans* ferredoxin-nitrate reductase and NADP reductase on reduced ferredoxin-Sepharose. Anal Biochem 89: 408–412
15. Zumft WG, Cardenas J (1979) The inorganic biochemistry of nitrogen. Bioenergetic processes. Naturwissenschaften 66: 81–88
16. Hartmann KM (1978) Aktionspektrometrie. In: Hoppe W, Lohmann W, Markl H, Ziegler H (eds) Biophysik, pp 197–222. Springer, Berlin Heidelberg New York
17. Calvert JG, Pitts JN (1967) Photochemistry. J Wiley & Sons, New York
18. Duchstein H-J, Fenner H, Hemmerich P, Knappe W-R (1979) (Photo)chemistry of 5-deazaflavin. A clue to the mechanism of flavin-dependent (de)hydrogenation. Eur J Biochem 95: 167–181

Effects of Ammonia on Carbon Metabolism in Photosynthesizing *Chlorella vulgaris* 11 h: the Replacement of Blue Light by Ammonium Ion

SHIZUKO MIYACHI[1] and SHIGETOH MIYACHI[1,2]

1 Introduction

With colorless mutant [2, 6] as well as wild-type cells [7] of *Chlorella vulgaris,* it was shown that dark CO_2 fixation and respiration were suppressed during starvation in phosphate medium in the dark. The main initial dark $^{14}CO_2$ fixation product was aspartate, indicating that $^{14}CO_2$ was fixed via a C_1-C_3 carboxylation reaction to form oxalacetic acid. Illumination with blue light to the starved cells immediately enhanced the C_1-C_3 carboxylation reaction. In both types of algal cell, respiration was also enhanced after a lag period which lasted for several minutes. With colorless mutant cells, it was also shown that prolonged illumination with blue light-induced de novo synthesis of phosphoenolpyruvate (PEP) carboxylase [3]. The blue light effect on C_1-C_3 carboxylation reaction observed immediately after blue light illumination was also concluded to be caused by an enhancement of PEP carboxylase activity [6]. It was, therefore, assumed that blue light exerts a dual effect on PEP carboxylase in *Chlorella* cells; an immediate enhancement of the enzyme reaction and induction of the synthesis of the enzyme [6].

With yellow mutant cells of *Chlorella vulgaris* (Mutant 211), Kowallik and Ruyters [5] reported that blue light induced the synthesis of pyruvate kinase.

Studies on the photosynthetic products with *Chlorella vulagris* 11h (wild-type cells) revealed that the rates of $^{14}CO_2$ fixation into sucrose and starch were greater under red light than under blue light, while blue light specifically enhanced $^{14}CO_2$ incorporation into alanine, other amino and carboxylic acids and lipid fraction [6, 8].

On the other hand, it was found that dark $^{14}CO_2$ fixation in the starved colorless mutant cells of *Chlorella vulgaris* was immediately enhanced by adding ammonium chloride [6]. The main initial $^{14}CO_2$ fixation was aspartate, indicating that PEP carboxylation reaction was enhanced by the addition of ammonia. Several minutes after the addition of ammonia, respiration was also enhanced after a lag period. Likewise, with *Chlorella pyrenoidosa* cells, Hiller [1] and Kanazawa et al. [4] showed evidence suggesting that the reaction catalyzed by pyruvate kinase is enhanced by NH_4^+. These results indicate that various reactions enhanced by blue light are also enhanced by the addition of ammonia. The present experiments were carried out to see whether the bias in photosynthetic $^{14}CO_2$ fixation products which were observed under blue light could also be in-

1 Radioisotope Centre, University of Tokyo, Bunkyo-ku, Tokyo 113, Japan
2 Institute of Applied Microbiology, University of Tokyo, Bunkyo-ku, Tokyo 113, Japan

duced by adding NH_4Cl under the illumination with red light. The results, which also included the effects of nitrate and nitrite, are described in this paper.

2 Materials and Methods

Material used was *Chlorella vulgaris* 11h. The method of the algal culture, apparatus for the determination of photosynthetic $^{14}CO_2$ fixation and of respiratory activity and the analytical method of ^{14}C-products were the same as those described elsewhere [7, 8].

3 Results and Discussion

Figure 1 shows that the rates of $^{14}CO_2$ incorporation into alanine, glycine-plus-serine-plus-glutamate, malate, and aspartate under red light in 3-day-starved *Chlorella vulgaris* 11h cells were enhanced by the addition of NH_4Cl. The effects of NH_4Cl shown in this figure are similar to those reported by Hiller [1] and Kanazawa et al. [4] and essentially the same as those induced by blue light illumination [8]. On the other hand, Fig. 2 shows that in sharp contrast to the effect of blue light, NH_4Cl caused the enhancement in $^{14}CO_2$ incorporation into sucrose. Also NH_4Cl greatly decreased ^{14}C-incorporation into insolubles, while those into phosphate esters and lipid fraction were practically not affected. As shown in Table 1, ^{14}C-incorporation into citrate and fumarate-plus-succinate was also enhanced by the addition of NH_4Cl.

Table 2 shows that essentially the same effects of ammonia were observed in non-starved cells of *Chlorella vulgaris* 11h, indicating that ammonium ion regulates photosynthetic carbon metabolism irrespective of the dark pretreatment of the algal cells.

Table 3 shows that the effects of KNO_3 and $NaNO_2$ are similar to those of NH_4Cl: The percent incorporation of ^{14}C into amino and organic acids as well as sucrose were increased, while that into insolubles was decreased by adding these nitrogenous compounds. (In these particular results shown in Table 3, KNO_3 did not enhance ^{14}C-

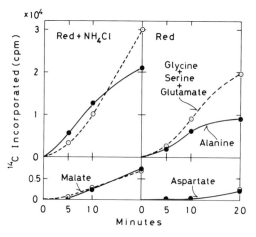

Fig. 1. Effects of ammonia on $^{14}CO_2$ incorporation under monochromatic red light into alanine, glycine-plus-serine-plus-glutamate, aspartate and alanine in 3-day-starved *Chlorella vulgaris* 11h. Cells were suspended in M/400 phosphate buffer (pH 7.6) at a density of 10 packed cell volume/l. Intensity of red light (660 nm, half band width, 10 nm), 10,800 erg cm^{-2} s^{-1}. Concentration of $NaH^{14}CO_3$ and NH_4Cl, 1 mM and 5 mM, respectively. Experimental temperature $25°C$. The data shown in Figs. 1 and 2 were obtained from the same experiment

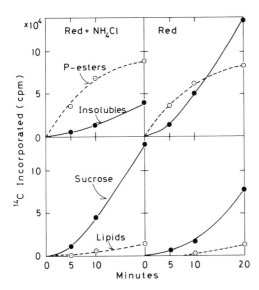

Fig. 2. Effects of ammonia on $^{14}CO_2$ incorporation under monochromatic red light into phosphate esters, insolubles, sucrose and lipid fractions in 3-day-starved cells of *Chlorella vulgaris* 11h. *P-esters* phosphate esters

Table 1. Effects of blue light and ammonia on photosynthetic $^{14}CO_2$ fixation into various products in 3-day-starved *Chlorella vulgaris* 11h

Compound	Radioactivity (cpm) incorporated under		
	Blue light	Red light	Red + NH$_4$Cl
Insolubles	29,300 (24.6)	51,300 (34.4)	13,000 (8.1)
Phosphate esters	60,800 (51.0)	62,300 (41.8)	67,100 (42.0)
PGA	26,700 (22.4)	27,000 (18.1)	33,400 (20.9)
PEP	2,200 (1.8)	3,600 (2.4)	5,300 (3.3)
SBP	7,400 (6.2)	7,900 (5.3)	1,400 (0.9)
Sucrose	2,300 (1.9)	16,100 (10.8)	44,400 (27.8)
Aspartate	1,000 (0.8)	300 (0.2)	2,100 (1.3)
Alanine	9,600 (8.1)	6,000 (4.0)	11,900 (7.4)
Gly-plus-Ser-plus-Glut	10,200 (8.6)	9,000 (6,0)	9,800 (6.1)
Malate	800 (0.7)	— (–)	2,700 (1.7)
Citrate	1,100 (0.9)	— (–)	1,100 (0.7)
Fum-plus-Suc	500 (0.4)	— (–)	800 (0.5)
Glycolate	700 (0.6)	1,000 (0.7)	— (–)
Lipid fraction	2,100 (1.8)	3,000 (2.0)	6,000 (3.8)
Others	700 (0.6)		1,100 (0.7)
Total	119,100 (100.0)	149,000 (99.9)	160,000 (100.1)

Time of $^{14}CO_2$ fixation was 10 min. The figures in parentheses show the percent incorporation of ^{14}C

Abbreviations: *PGA* 3-phosphoglycerate; *PEP* phosphoenolpyruvate; *SBP* Sugar bisphosphates; *Gly* glycine; *Ser* serine; *Glut* glutamate; *Fum* fumarate; *Suc* succinate

incorporation into alanine. However, such enhancement was repeatedly observed by other experiments.) Table 4 shows that nitrate and nitrite exerted similar effects to the nonstarved cells. Figure 3 shows that just as with ammonia [6], dark respiration of

Table 2. Effects of blue light and ammonia on photosynthetic $^{14}CO_2$ fixation into various products in nonstarved *Chlorella vulgaris* 11h

Compound	Radioactivity (cpm) incorporated under		
	Blue light	Red light	Red + NH$_4$Cl
Insolubles	85,500 (29.0)	93,600 (30.7)	34,700 (8.0)
Phosphate esters	134,000 (45.5)	153,600 (50.3)	86,500 (20.1)
Sucrose	3,000 (1.0)	4,000 (1.3)	17,400 (4.0)
Aspartate	22,600 (7.7)	16,800 (5.5)	128,000 (29.7)
Alanine	15,800 (5.3)	5,700 (1.9)	28,100 (6.5)
Glutamate	3,000 (1.0)	3,500 (1.1)	24,000 (5.6)
Ser(-plus-Gly)	5,400 (1.8)	4,500 (1.5)	6,600 (1.5)
Glutamine			17,200 (4.0)
Malate	5,200 (1.8)	2,900 (1.0)	27,100 (6.3)
Citrate	1,200 (0.4)	1,200 (0.4)	13,700 (3.2)
Fumarate	1,000 (0.3)	600 (0.2)	5,900 (1.4)
Lipid fraction	5,900 (2.0)	15,300 (5.0)	5,200 (1.2)
Others	12,500 (4.2)	3,500 (1.1)	37,000 (8.6)
Total	295,100 (100.0)	305,200 (100.0)	431,400 (100.1)

Time of $^{14}CO_2$ fixation was 10 min. The figures in parentheses show the percent incorporation of ^{14}C

Table 3. Effects of nitrogenous compounds on ^{14}C-incorporation during photosynthetic $^{14}CO_2$ fixation under red light in 3-day-starved *Chlorella vulgaris* 11h

Compound	Radioactivity (cpm) incorporated under		
	Red light	Red + KNO$_3$	Red + NaNO$_2$
Insolubles	51,300 (34.4)	19,500 (12.7)	17,000 (11.3)
Phosphate esters	62,300 (41.8)	45,500 (29.6)	76,200 (50.4)
PGA	27,000 (18.1)	23,500 (15.3)	39,700 (26.3)
PEP	3,600 (2.4)	3,900 (2.6)	3,600 (2.4)
SDP	7,900 (5.3)	3,100 (2.0)	3,000 (2.0)
Sucrose	16,100 (10.8)	59,400 (38.6)	29,500 (19.5)
Aspartate	300 (0.2)	1,700 (1.1)	1,200 (0.8)
Alanine	6,000 (4.0)	4,900 (3.2)	8,000 (5.3)
Gly-plus-Ser-plus-Glut	9,000 (6.0)	10,600 (6.9)	10,000 (6.6)
Malate	− (−)	2,100 (1.4)	1,500 (1.0)
Citrate	− (−)	1,000 (0.6)	500 (0.3)
Fum-plus-Suc	− (−)	800 (0.5)	500 (0.3)
Glycolate	1,000 (0.7)	− (−)	− (−)
Lipid fraction	3,000 (2.0)	8,100 (5.3)	5,200 (3.4)
Others	− (−)	300 (0.2)	1,500 (1.0)
Total	149,000 (99.9)	153,900 (100.1)	151,100 (99.9)

Time of $^{14}CO_2$ fixation was 10 min. The figures in parentheses show the percent incorporation of ^{14}C

Table 4. Effects of nitrogenous compounds on ^{14}C-incorporation during photosynthetic ^{14}CO$_2$ fixation under red light in nonstarved *Chlorella vulgaris* 11h

Compounds	Radiactivity (cpm) incorporated under		
	Red light	Red light + KNO$_3$	Red light + NaNO$_2$
Insolubles	126,000 (42.1)	101,000 (35.2)	122,000 (36.8)
Phosphate esters	126,000 (42.1)	131,000 (45.6)	129,000 (38.9)
Sucrose	3,000 (1.0)	6,100 (2.1)	9,200 (2.8)
Aspartate	13,000 (4.3)	16,800 (5.8)	23,000 (7.0)
Alanine	4,700 (1.6)	6,600 (2.3)	9,900 (3.0)
Gly-plus-Ser-plus-Glut	9,100 (3.0)	11,000 (3.8)	12,800 (3.9)
Malate	2,400 (0.8)	2,100 (0.7)	4,500 (1.4)
Citrate	700 (0.2)	700 (0.2)	1,100 (0.3)
Fum-plus-Suc	700 (0.2)	700 (0.2)	1,100 (0.3)
Lipid fraction	9,500 (3.2)	7,100 (2.5)	11,800 (3.6)
Others	4,100 (1.4)	4,200 (1.5)	7,100 (2.1)
Total	299,200 (99.9)	287,300 (99.9)	331,400 (100.1)

Time of ^{14}CO$_2$ fixation was 10 min. The figures in parentheses show the percent incorporation of ^{14}C

Chlorella cells was also enhanced by adding nitrate as well as nitrite. Note that there was a lag period for the enhancement in respiration by both compounds.

We have observed that dark ^{14}CO$_2$ fixation in the colorless *Chlorella* cells is greatly enhanced immediately after the addition of ammonium chloride [6]. Our preliminary experiment showed that dark ^{14}CO$_2$ fixation in one-day-starved *Chlorella vulgaris* 11h cells was not enhanced immediately after the addition of KNO$_3$ or NaNO$_2$. However, the rate started to increase gradually after an induction period which lasted for several minutes (data not shown). These results indicate that both KNO$_3$ and NaNO$_2$ do not affect carbon metabolism directly. Possibly, the regulatory effect on carbon metabolism will be exerted after these compounds are transformed to ammonia. The lag period observed will be the time required for the conversion of these compounds to ammonia.

To study how ammonia affects ^{14}C-incorporation into sucrose, *Chlorella vulgaris* 11h cells were given NaH^{14}CO$_3$ (3 mM) at 25°C under the illumination by metal halide lamp (Yoko Lamp, Toshiba Electric Co. Ltd., Tokyo) from both sides. The light inten-

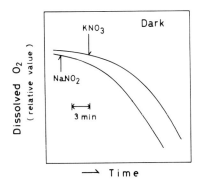

Fig. 3. Effects of KNO$_3$ and NaNO$_2$ on dark respiration in 3-day-starved cells of *Chlorella vulgaris* 11h

sity at the surface of lollipop was $2 \times 24,000$ lx. After 20 min, the cell suspension was washed twice by centrifugation and suspended in M/400 phosphate buffer. These procedures were done in the cold room at $5°C$ and in the dark. Portions (6 ml) of the algal suspension were placed in the reaction vessel and kept in the dark or illuminated with monocromatic red light from one side (660 nm; 10,800 erg cm^{-2} s^{-1} at $25°C$) in the presence or absence of NH$_4$Cl (5 mM). Nonlabeled NaHCO$_3$ (1 mM) was added at the beginning of each experiment. At intervals, aliquots (0.5 ml each) were transferred into the test tube containing 2 ml of methanol. The methanol suspension was kept overnight and centrifuged. The determination of radioactivity in the precipitate (which mostly consists of starch) revealed that, irrespective of illuminating conditions, ^{14}C-starch was greatly decreased by adding ammonia (data not shown). Thus we concluded that the degradation of starch is induced by adding ammonia. Therefore, increase in ^{14}C-sucrose observed by adding ammonia, nitrate, and nitrite would be due to the enhancement in the hydrolysis of ^{14}C-starch. Enhancement of starch degradation by blue light has been reported by various workers (e.g., [6]). We therefore conclude that, in principle, the effects of ammonia and blue light are very similar to each other.

In contrast to the present results, Kanazawa et al. [4] reported that the addition of ammonia (ca. 0.7 mM) to *Chlorella pyrenoidosa* photosynthesizing under white light stopped ^{14}C-incorporation into sucrose. The reason for such discrepancy is not clear at this moment.

Acknowledgments. We thank Dr. A. Kamiya of Teikyo University who cooperated with us during the initial period of this research. This research was aided by grants from Japanese Ministry of Education, Science and Culture as well as from Ministry of Agriculture, Forestry and Fisheries (GEP-54-II-I-10).

References

1. Hiller RG (1970) Transients in photosynthetic carbon reduction cycle produced by iodoacetic acid and ammonium chloride. J Exp Bot 21: 628–638
2. Kamiya A, Miyachi S (1974) Effects of blue light on respiration and carbon dioxide fixation in colorless Chlorella mutant cells. Plant Cell Physiol 15: 927–937
3. Kamiya A, Miyachi S (1975) Blue light-induced formation of phosphoenolpyruvate carboxylase in colorless *Chlorella* mutant cells. Plant Cell Physiol 16: 729–736
4. Kanazawa T, Kirk MR, Bassham JA (1970) Regulatory effects of ammonia on carbon metabolism in photosynthesizing *Chlorella pyrenoidosa*. Biochim Biophys Acta 205: 401–408
5. Kowallik W, Ruyters G (1976) Über die Aktivitätssteigerungen der Pyruvatkinase durch Blaulicht oder Glucose bei einer chlorophyllfreien *Chlorella*-Mutante. Planta 128: 11–14
6. Miyachi S, Kamiya A, Miyachi S (1977) Wavelength effects of incident light on carbon metabolism in *Chlorella* cells. In: Mitsui A et al. (eds) Biological solar energy conversion, pp 167–182. Academic Press, London New York
7. Miyachi S, Kamiya A, Miyachi S (1980) Effects of blue light on respiration and non-photosynthetic CO$_2$ fixation in *Chlorella vulgaris* 11h cells, pp 321–328. This volume
8. Miyachi S, Miyachi S, Kamiya A (1978) Wavelength effects on photosynthetic carbon metabolism in *Chlorella*. Plant Cell Physiol 19: 277–288

Comparative Studies on the Effect of Ammonia and Blue Light on the Regulation of Photosynthetic Carbon Metabolism in Higher Plants

A. GNANAM[1], A. HABIB MOHAMED[1], and R. SEETHA[1]

1 Introduction

Ammonium salts have been shown to divert the photosynthetic carbon more towards the synthesis of amino acids in *Chlorella* [9, 10] and in the leaf cells and discs of higher plants [14, 15, 16]. Blue light has also been reported to have a similar effect on the synthesis of amino acids. It has been suggested that NH_4 brings about the increased synthesis of amino acids by the stimulation of pyruvate kinase [9, 10], while blue light through the activation of PEP carboxylase. Warburg et al. [19] first reported that catalytic amounts of blue light might be necessary to obtain high quantum yields of photosynthesis. The photoreceptor involved in the regulatory role of blue light on the carbon metabolism is presumed to be flavin or carotenoids from the action spectra [11, 18]. The present communication deals with the comparative study on the mechanism of action of NII_4^+ and blue light in bringing about the regulation of the photosynthetic carbon flow, leading to the increased synthesis of amino acids in the leaves of higher plants.

2 Materials and Methods

Plant materials used in this study were *Dolichos lablab* var. *Lignosus* Prain and *Vigna sinensis* L. Plants were grown under field conditions. Fully expanded young leaves were collected for studies. For light activation studies leaves were collected from plants previously darkened for 20 h.

Leaf discs were cut out with 5 mm diameter cork borer. Mesophyll cells were prepared by mechanical grinding following the procedure of Gnanam and Kulandaivelu [8].

2.1 Photosynthetic $^{14}CO_2$ Fixation

The ^{14}C-bicarbonate fixation of the cell suspension or leaf discs was carried out according to the method of Bassham and Calvin [3].

1 School of Biological Sciences, Madurai Kamaraj University, Madurai 625021, India

2.2 Analysis of ^{14}C Labelled Products

Each sample was macerated and successively extracted with 80%, 40% and 20% ethanol (w/v) and then with water. The extracts of each sample were combined and evaporated to dryness. The residue was redissolved in distilled water and was separated into anionic, cationic and neutral fractions following the procedure of Canvin and Beevers [5].

2.3 Enzyme Preparations

Control and NH_4^+-treated leaf cells were homogenized in 20 mM Tris-HCl buffer pH 7.5, 1 mM 2-mercaptothanol with acid washed sand in a prechilled pestle and mortar. The homogenate was strained through eight layers of cheese cloth and centrifuged at 23,000 g for 30 min. All steps were carried out at $-4°$C.

For the assay of PEP carboxylase the leaves were crushed in an ice cold medium containing 100 mM Tris-HCl, pH 7.8, 10 mM $MgCl_2$, 1 mM EDTA, 15 mM 2-mercaptoethanol and 2% PVP. The homogenate was centrifuged at 17,000 g for 30 min. The supernatant was used as the enzyme source. For glycolate oxidase, the enzyme was extracted in an ice cold medium containing phosphate buffer pH 8.0, 100 mM and PVP 2% (w/v).

2.4 Enzyme Assays

Pyruvate kinase (EC 2.71.40) was assayed according to the method of Negelein as described by Bücher and Pfleiderer [4]. The reaction mixture contained in 3 ml 80 μmol Tris-HCl buffer, pH 7.5, 220 μmol KCl, 2.4 μmol $MgSO_4$, 1.5 μmoles ADP, 0.56 μmol NADH, purified LDH 18 units, 4.5 μmol PEP and 100–150 μg of all proteins.

NAD(P)-glyceraldehyde-3-PO_4 dehydrogenase (EC 1.2.1.12; EC 1.2.1.13) were assayed in the direction of reduction of PGA. The reaction was assayed spectrophotometrically at 25°C. The reaction mixture contained in a volume of 1.0 ml 100 μmol Tris-HCl buffer, pH 7.8, 1 μmol dithiothreitol, 15 μmol PGA, 3.4 μmol ATP, 0.6 μmol NAD(P)H and 100–150 μg of protein.

Fructose-1,6-bisphosphatase (EC 3.1.3.11) was assayed in a reaction mixture of 1.0 ml containing 100 μmol Tris-HCl buffer pH 8.4, 10 μmol $MgCl_2$, 10 μmol FBP, 1 μmol sodium EDTA and cell extract 100–150 μg protein. The reaction was stopped by adding 0.5 ml of cold 5% TCA and left at 2°C for 2 h. After centrifugation the released inorganic phosphate in the supernatant was assayed by the method of Fiske and Subbarow [7].

PEP carboxylase (EC 4.1.1.31) was assayed by the method of Maruyama et al. [13] by radiometric method in a reaction medium containing 50 μmol Tris-HCl pH 7.8, 3 μmol $MgCl_2$, 5 μmol $H^{14}CO_3$, 10.5 μci/10 μmol, 0.5 μmol PEP, 0.2 μmol NADH and 60–80 μg enzyme protein. The reaction was initiated with the addition of PEP and stopped at the appropriate time interval by the addition of 10% acetic acid and the acid stable radioactivity was counted in a G.M. counter.

Glycolate oxidase was assayed in a medium containing 50 mM phosphate buffer, pH 8,0, 1 mM FMN, 1 mM glycolate and 80–100 μg enzyme protein; The glycolate-de-

pendent oxygen uptake was measured in an oxygen electrode (Clark type). Total chlorophyll was estimated according to the method of Arnon [2] and protein was estimated
by Lowry's method [12].

2.5 Light Activation of NADP G-3-PO$_4$ Dehydrogenase (EC 1.2.1.13)

Chloroplasts were isolated from dark-adapted leaves of *Vigna sinensis* by the method
of Cockburn et al. [6] as modified by Anderson and Avron [1]. The chloroplast suspension (100 µg chl/ml) was exposed to white light of 20,000 lx at 25°C. Aliquots were
removed and assayed immediately in a medium which is hypotonic to the chloroplasts.
The effect of NH$_4$$^+$ and other inhibitors was studied by adding these at indicated concentrations to the chloroplast suspension just prior to illumination. The effect of dark
activation of this enzyme by DTT, NADPH and ATP was studied by preincubation of
the chloroplasts with these in a hypotonic medium at 25°C in the dark for 20 min.

2.6 Light Activation of Fructose-1,6-Bisphosphatase (EC 3.1.3.11)

The chloroplasts were suspended in a Warburg flask in a medium containing 10 mM
HEPES-KOH pH 7.6, 2 mM EDTA, 5 mM MgCl$_2$, 1 mM MnCl$_2$, 400 mM sucrose, 10 mM
sodium ascorbate, 0.1 mM DCPIP and 0.2 mM benzyl viologen. The central well in the
flask was filled with 0.2 ml of 7N KOH. The air was replaced by flushing with nitrogen
gas and closed airtight before it was exposed to white light of 20,000 lx at 25°C. Aliquots of chloroplast suspension were transferred directly to the assay medium for measuring the enzyme activity. Whenever the effect of NH$_4$$^+$ or other inhibitors was studied
selected concentrations of NH$_4$OH or other inhibitors were added to the activation medium prior to illumination. Appropriate dark controls were routinely maintained simultaneously for all light activation studies. In these studies, NH$_4$$^+$ was added as NH$_4$OH
and the amounts used had no effect whatsoever on the pH of the medium.

3 Results

The effect of ammonium salts on the $^{14}CO_2$ fixation and the relative flow of photosynthetic carbon into neutral, anionic and cationic fractions in the isolated mesophyll
cells of *Dolichos lab lab* is shown in Table 1. Addition of up to 1 mM NH$_4^+$ to the cells
did not change either the kinetics or the net $^{14}CO_2$ fixation, but caused an increased
flow of photosynthetic carbon into amino acid (cationic) fractions with a concomitant
decreased flow into sugar (neutral) fractions. Very little change in the flow of carbon
into the organic acid fractions was observed.

   Ammonium ions caused the same effect in the relative flow of photosynthetic carbon
without affecting the overall photosynthetic rate in the leaf discs of *Vigna* as well. However, the pattern of photosynthetically fixed carbon flow in the isolated chloroplasts
of the same plant species in the presence of ammonium ions is quite distinct (Table 1).
There has been a greater accumulation of labelled carbon in the organic acid fractions

Table 1. Effect of ammonium ions on the distribution of labelled carbon from ^{14}C-bicarbonate into various fractions of the primary photosynthetic products of leaf discs and chloroplasts of *Vigna sinensis* and isolated leaf cells of *Dolichos lab lab*. The leaf discs and isolated leaf cells were incubated with ammonium salts for 1 h and the chloroplasts for 5 min at $20°C$. ^{14}C-bicarbonate fixation was done at saturating light intensity for 30 min

	Isolated leaf cells		Leaf discs		Chloroplasts	
	Control (%)	+ 1 mM NH_4^+ (%)	Control (%)	+ 1 mM NH_4^+ (%)	Control (%)	+ 0.8 mM NH_4^+ (%)
Neutral fraction (sugar)	49	35	48	34	35	25
Cationic fraction (amino acids)	14	24	11	24	10	10
Anionic fraction (organic acids)	37	41	41	42	55	65
	100	100	100	100	100	100
	(74,739 cpm)	(73,469 cpm)	(82,100 cpm)	(83,100 cpm)	(32,900 cpm)	(33,500 cpm)

Table 2. Effect of ammonium ions on the catalytic activities of certain key enzymes in whole cell extracts. Average of at least three determinations

Enzyme	Enzyme source		
	Control[a]	Control + 1 mM NH_4Cl[b]	NH_4Cl treated cell[c]
	nmol NAD(P)H oxidised mg chl^{-1} min^{-1}		
NAD-glutamate dehydrogenase	106	104	108
NADP-glutamate dehydrogenase	689	685	698
	μmol NAD(P)H oxidised mg protein^{-1} h^{-1}		
NAD-glyceraldehyde-3-phosphate dehydrogenase	32.6	32.5	33.0
NADP-glyceraldehyde-3-phosphate dehydrogenase	9.3	9.2	9.4
	Units mg protein^{-1}		
Glutamine synthetase	1.79	1.78	1.68
	nmol NADH oxidised mg protein^{-1} min^{-1}		
Pyruvate kinase	43.8	43.5	41.9
	nmol Pi liberated mg protein^{-1} min^{-1}		
Fructose-1,6-bisphosphatase	45.0	45.6	47.8

[a] An equal quantity of cells or leaf discs were taken for control and ammonium treatment there was no difference between their total enzyme activity

[b] The enzyme from the control cells was incubated with 1 mM NH_4Cl for 10 min at $15°C$ before the assay

[c] The enzyme from the NH_4^+-treated cells was assayed

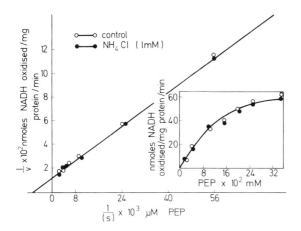

Fig. 1. Saturation curve of pyruvate kinase for PEP in the dialysed crude extract of bean leaf cells

in the NH_4^+-treated chloroplasts as compared to the controls, with the concomitant decrease in sugars.

In an effort to localize the site of action of NH_4^+ in mediating the observed regulation of photosynthetic carbon flow, several of the key enzymes in the photosynthetic carbon flow pathways were assayed. None of the enzyme preparations obtained from the control and NH_4^+-treated bean leaf cells both in the presence and absence of NH_4^+ during the assay showed any significant difference either in their specific activities or in their total activities that could account for the observed diversion of photosynthetic carbon into cationic fractions (Table 2).

Pyruvate kinase has been implicated as a possible enzyme to be activated by NH_4^+ because of its regulatory nature and its pivotal position in the intermediary pathway, through indirect evidence. Careful analysis of the kinetic properties of this enzyme in the bean leaf cell extract did not show any change either in the K_m or in the V_{max} with the addition of ammonium salts either to the cells or to the enzyme extracts (Fig. 1).

In the absence of any significant change in the catalytic actions of most of the key enzymes associated with the photosynthetic carbon flow pathways, the possibility of NH_4^+ interfering with the light activation of some of the photosynthetic enzymes known to be activated during photosynthesis was explored. The kinetics of light activation of the chloroplastic NADP-glyceraldehyde-3-P-dehydrogenase is shown in Fig. 2. The activity of the enzyme in dark-adapted chloroplasts did not show any decay in storage at 25°C, for at least 30 min. On illumination, the activity nearly doubled within 10 min and declined slowly after 20 min (Fig. 2). The effect of different concentrations of NH_4^+ on the light activation of this enzyme showed a biphasic pattern (Fig. 3). When the dark-adapted chloroplasts were incubated with NH_4^+ and exposed to light for 10 min, the extent of light activation declined at low NH_4^+ concentration up to 1 mM but increased steadily with the increasing levels of NH_4^+ and reached the no NH_4^+ level at 5.0 mM.

Since the relative carbon flow into the sugar fractions was considerably lowered in the presence of NH_4^+, the effect of ammonium ions on the light activation of the fructose-1,6-bisphosphatase the chloroplastic enzyme in the biosynthetic pathway of hexose sugars was determined. Interestingly, it was observed that NH_4^+ inhibited the light-dependent activation of this enzyme also, showing a similar biphasic trend (Figs. 2 and 3).

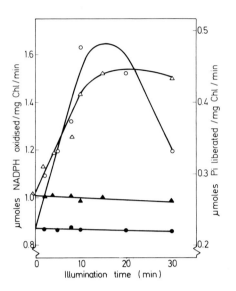

Fig. 2. Time course of light activation of NADP-glyceraldehyde-3-phosphate dehydrogenase (●——○) and alkaline fructose-1,6-bisphosphatase (▲——△) from the chloroplasts of *Vigna sinensis*. *Solid symbols* (●——●, ▲——▲) indicate the enzyme activity in dark; *open symbols* (○——○, △——△) represent the activity in light

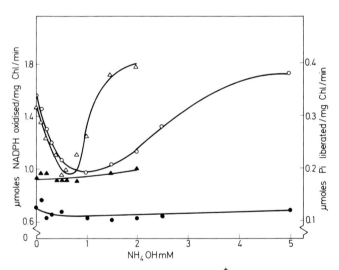

Fig. 3. Effect of different concentrations of NH_4^+ on the light activation of NADP-glyceraldehyde-3-phosphate dehydrogenase (●——○) and the alkaline fructose-1,6-bisphosphatase (▲——△) from the chloroplasts of *Vigna sinensis*. *Solid symbols* (●——●, ▲——▲) indicate the enzyme activity in dark; *open symbols* (○——○, △——△) represent the activity in light

Unlike NH_4^+, blue light (452 nm), when irradiated to cells from bean leaves previously darkened for 10 h, significantly enhanced the rate of dark $^{14}CO_2$ fixation, even though the overall rate was nearly fivefold less than that observed under saturating levels of white light irradiance (Table 3). However, as in the case of NH_4^+, blue light diverted the assimilated carbon more towards amino acids at the expense of sugars (Table 4).

Table 3. Blue light effect on $^{14}CO_2$ fixation in isolated leaf cells of *Dolichos lab lab*

	Expt. 1	Expt. 2	Expt. 3
	μmol CO_2 mg chl^{-1} h^{-1}		
Dark	4.12	2.01	2.06
Blue 452 nm	7.95	5.34	5.31
White	22.10	18.24	18.12

Table 4. Assimilated carbon distribution into various fractions under different light conditions

	Dark	Blue	White
	(% of total)		
Neutral (sugars)	49.5	43.8	56.1
Cationic (amino acids)	30.7	39.9	28.0
Anionic (organic acids)	19.8	18.2	15.9

Table 5. Blue light effect on the PEP-carboxylase activity in *Dolichos lab lab* leaves

	Expt. 1	Expt. 2	Expt. 3	Expt. 4
	(μmol CO_2 mg protein^{-1} h^{-1})			
Dark	0.120	0.232	0.389	0.143
Blue	0.411	0.434	0.797	0.298
White	0.295	0.383	0.473	0.220

Table 6. Effect of translational site specific antibiotics on the light activation of PEP carboxylase

	μmol CO_2 mg protein^{-1} h^{-1}
Dark	0.96
Blue (-CHI)	1.92
Blue (+CHI)	1.80
White (-CHI)	1.80
White (+CHI)	1.62

Table 7. Glycolate oxidase

	μmol O_2 mg protein^{-1} h^{-1}	Control (%)
Control	26.03	100.0
Blue irradiated	12.70	48.8

Since the blue light has been shown to activate the PEP-carboxylase, a primary, dark carboxylating enzyme in algal cultures, an attempt was made to see whether or not a similar activation occurs in the higher plant leaf cells as well, and indeed it was found activated (Table 5). Under the conditions described, the enhanced activity of this enzyme was not due to de novo synthesis but only due to activation through some unknown mechanism. This is corroborated by the lack of any effect of the site-specific protein synthetic inhibitors during activation (Table 6).

Glycolate oxidase, a key enzyme in the photorespiratory pathway, was also inhibited by blue light (Table 7).

4 Discussion

Based on the observations, a possible mechanism of action of NH_4^+ in regulating the photosynthetic carbon flow is explained in Fig. 4. During steady-state photosynthesis, some of the chloroplastic enzymes are activated significantly to promote the relative flow of carbon towards the synthesis of hexose and other sugars. NH_4^+ ions at the critical concentration used, abolish completely the light activation of at least two of the en-

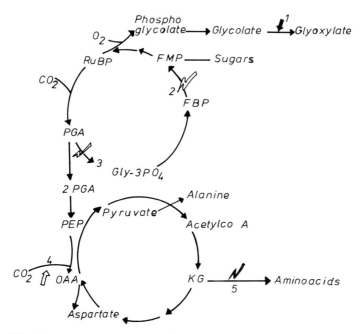

Fig. 4. The photosynthetic carbon flow system.

site of inhibition by ammonium ions;

site of regulation by ammonium ions;

site of inhibition by blue light;

site of activation by blue light

zymes viz. glyceraldehyde-3-phosphate dehydrogenase and fructose-1,6-bisphosphatase, resulting in the accumulation of photosynthetic PGA formed which then moves out of the chloroplasts and further metabolized via pyruvate and Krebs cycle, facilitating the increased supply of carbon skeleton for the biosynthesis of amino acids. Thus NH_4^+ appears to act both as a regulatory and a substrate in promoting the production of amino acids at the expense of sugars.

Even though the ultimate effect of blue light on the photosynthetic carbon metabolism is very similar to that of NH_4^+, obviously the mechanism of its action in bringing out the greater production of amino acids appears to be different. Activation of PEP carboxylase facilitates the additional $^{14}CO_2$ fixation leading to increased synthesis of C_4 acids. It is known that the carbon fixation mediated by C_4 acid metabolism either in dark or in light will lead to the increased synthesis of amino acids and organic acids rather than sugars. Inhibition of glycolate oxidase (Table 7) and the overall enhancement of respiratory metabolism by blue light as reported [17] might contribute to the increased supply of the substrate, PEP.

Apparently, blue light at low intensities supplements the photosynthetic $^{14}CO_2$ fixation via PEP carboxylation and produces C_4 acids directly which would favour the amino acid synthesis rather than sugars, while the NH_4^+ ions seem to regulate the photosynthetic carbon flow by abolishing the light activation of the enzymes which would normally favour the flow of carbon towards sugar biosynthesis, thereby facilitating the increased synthesis of amino acids.

References

1. Anderson LE, Avron M (1976) Light modulation of enzyme activity in chloroplasts. Generation of membrane bound vicinal-dithiol groups by photosynthetic electron transport. Plant Physiol 57: 209–213
2. Arnon DI (1949) Copper enzymes in isolated chloroplasts. Polyphenol oxidases in *Beta vulgaris*. Plant Physiol 24: 1–15
3. Bassham JA, Calvin M (1957) The path of carbon in photosynthesis. Prentice-Hall Inc, Englewood Cliffs New Jersey
4. Bücher T, Pfleiderer G (1955) In: Colowick SP, Kaplan NO (eds) Methods in enzymology, vol I, pp 435–440. Academic Press, London New York
5. Canvin DT, Beevers H (1961) Sucrose synthesis from acetate in the germinating castor bean: Kinetic and pathway. J Biol Chem 236: 988–995
6. Cockburn W, Walker DA, Baldry CW (1968) The isolation of spinach chloroplasts in pyrophosphate media. Plant Physiol 43: 1415–1418
7. Fiske CH, Subbarow Y (1925) The colorimetric determination of phosphorus. J Biol Chem 66: 375–400
8. Gnanam A, Kulandaivelu G (1969) Photosynthetic studies with leaf cell suspensions from higher plants. Plant Physiol 44: 1451–1456
9. Kanazawa T, Kirk MR, Bassham JA (1970) Regulatory effects of ammonia on carbon metabolism in photosynthesizing *Chlorella pyrenoidosa*. Biochim Biophys Acta 205: 401–408
10. Kanazawa T, Kirk MR, Bassham JA (1972) Regulatory effects of ammonia on carbon metabolism in *Chlorella pyrenoidosa* during photosynthesis and respiration. Biochim Biophys Acta 256: 656–669
11. Kowallik W, Gaffron H (1967) Nature (London) 215: 1038–1040
12. Lowry OH, Rosenbrough Farr AL, Randall RJ (1951) Protein measurement with Folin-Phenol reagent. J Biol Chem 193: 265–275
13. Maruyama A, Easterday RL, Chang HC, Lane MD (1966) J Biol Chem 241: 2405–2412
14. Mohamed AH, Gnanam A (1977) Regulation of photosynthetic carbon flow by ammonium ions in isolated bean leaf cells. Plant Biochem J 4: 1–9
15. Platt SG, Plaut JA, Bassham JA (1977) Ammonia regulation of carbon metabolism in photosynthesising leaf discs. Plant Physiol 60: 739–742
16. Rehfeld DW, Jensen RG (1973) Metabolism of separated leaf cells. III. Effects of calcium and ammonium on product distribution during photosynthesis with cotton leaf cells. Plant Physiol 52: 17–22
17. Schmid GH (1969) Effect of blue light on glycolate oxidase. Z Physiol Chem 350: 1513–1520
18. Voskrezsenskaya NP (1973) Dokl Acad Nauk USSR 93: 911–914
19. Warburg O, Krippahl G, Schroder W (1954) Z Naturforsch 9B: 667–675

The Effect of Blue and Red Light on the Content of Chlorophyll, Cytochrome f, Soluble Reducing Sugars, Soluble Proteins and the Nitrate Reductase Activity During Growth of the Primary Leaves of *Sinapis alba*

A. WILD[1] and A. HOLZAPFEL[1]

1 Introduction

The photosynthetic characteristics of many species of plants are influenced by the light intensity under which the plant is grown. This is shown by the light saturation curves for CO_2 uptake. Both the light intensity required for saturation and the light-saturated rate of CO_2 uptake increase with the light intensity under which the plant is grown. Photosynthetic adaptation to different light levels involves balanced changes of many leaf factors. Low-light and high-light plants differ in a number of component steps of photosynthesis as well as in the structure and composition of the photosynthetic apparatus and in the leaf anatomy (see reviews by Boardman 1977; Wild 1979).

In the leaf shade of a plant or plant society not only light intensity, but also the qualitative composition of light differs markedly from sun-exposed habitats. Sunlight contains rather more red than far red light. When daylight passes through a vegetation canopy, much of the red and blue wavelengths are absorbed by the pigments in the leaves. As a consequence the spectral distribution of the canopy light is very rich in far red light compared to unfiltered daylight. The vegetation acts as a far red filter which produces a high far red/red ratio (Seybold 1936; Hanke et al. 1969; Kasperbauer 1971; Björkman and Ludlow 1972). This may well be of ecological significance. One can assume that light adaptation of leaves is not only a matter of energy, but especially of spectral quality.

2 Materials and Methods

Plants of *Sinapis alba* were cultivated under defined conditions ($20°C$; 65%–80% rel. humidity) in Lecaton R and subirrigated with Hoagland nutrient solution. The plants were kept under white, red, and blue light conditions. The light was obtained from Osram tubes of the type L 40 W/25 (white light), Philips tubes of the type 40 W TL 18 (red light, λ max = 655 nm) and Philips tubes of the type 40 W TL 15 (blue light, λ max = 440 nm). The experiments were performed either with a light intensity of 6 W m^{-2} or a quantum flux of 33 μmol m^{-2} s^{-1}. A 16:8 light–dark schedule was chosen.

Abbreviations. *Chl* chlorophyll; *Cyt f* cytochrome f; *Pr* red absorbing phytochrome; *Pfr* far red absorbing phytochrome; *wt* weight

1 Institut für Allgemeine Botanik der Universität Mainz, Saarstr. 21, 6500 Mainz, FRG

Two hours after switching on the light the plant material was harvested because of the fluctuations of nitrate reductase activity and content of reducing sugars during the day (Shibata et al. 1969; Fedtke 1973; Schlesier 1977). The experiments were carried out during the development of primary leaves. In order to avoid shading, secondary leaves were removed together with the shoot apex.

Details about the assay of nitrate reductase activity and content of reducing sugars have already been reported by Wild and Zerbe (1977). The content of soluble proteins was determined by the method of Lowry et al. (1951). The cytochrome f content was assayed by difference spectra according to the method of Bendall et al. (1971) with a millimolar extinction coefficient of 17.7 μmol^{-1} cm^{-2}. The difference spectrum was recorded with an Aminco DW-2 UV-VIS-spectrophotometer. The content of chlorophyll was determined in 80% acetone according to the method of Ziegler and Egle (1965).

All experiments were performed both at equal light intensity and equal quantum flux.

3 Results

Figure 1 shows the dry weight during the ontogeny of the primary leaves. It is obvious that dry weight is higher under blue light than under white light conditions. The lowest values were produced under red light treatment; soon after the beginning of the experiments the values of dry weight decreased.

In Fig. 2 the chlorophyll content is shown. Plants grown under blue light conditions had the highest content of chlorophyll. This was a marked tendency and it made no

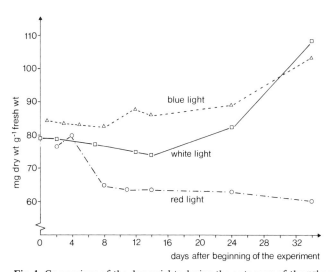

Fig. 1. Comparison of the dry weights during the ontogeny of the primary leaves of *Sinapis alba* plants grown under white (□—□), red (○–·–○; λ max = 655 nm) and blue light (△– –△; λ max = 440 nm) conditions adjusted to equal light quanta density (33 μmol m^{-2} s^{-1}). To avoid shading, the shoot apex of the plants has been removed together with the young secondary leaves

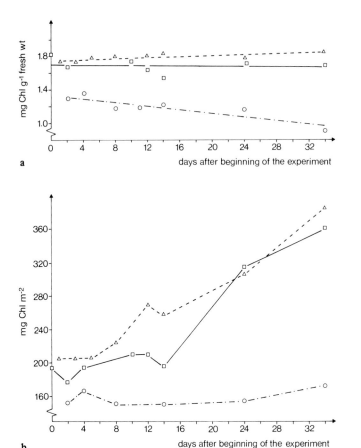

Fig. 2a, b. Chlorophyll content (**a** per fresh weight, **b** per leaf area) of the primary leaves grown under white (□—□), red (○−·−○), and blue light (△− −△) conditions. (The shoot apex has been removed)

difference whether the chlorophyll content was related to leaf area or to fresh weight. Regarding the chlorophyll content per fresh weight under blue and white light conditions a constant level of chlorophyll during the development of the leaves is to be seen. The chlorophyll content related to fresh weight of red light plants deminished with growing age of the test plants. Because of the greater thickness of the leaves the chlorophyll content per leaf area increased under white and blue light conditions. Under red light conditions the increase of leaf thickness is compensated by the decrease of chlorophyll per fresh weight.

If blue light causes an adaptation as normally evoked by high light intensities, the chlorophyll content per fresh weight of blue light plants should be lower than that of white light plants, whereas the chlorophyll content per leaf area should be higher (see review by Wild 1979).

Cytochrome f is a component of the photosynthetic electron transport chain between the two photosystems. Regarding the ratio of chlorophyll/cytochrome f, shown in Fig.

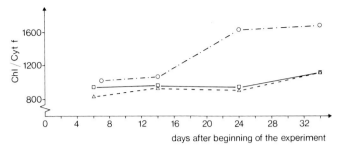

Fig. 3. Molar ratio of chlorophyll per cytochrome f during the ontogeny of the primary leaves under the conditions of different light qualities

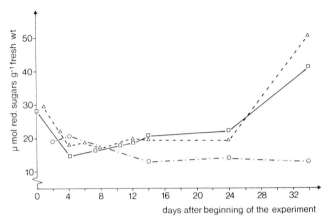

Fig. 4. Comparison of the content of soluble reducing sugars after growing under the different light conditions

3, one observes a significant difference between red and blue light treatment. Especially in older red light plants the ratio Chl/Cyt f was much higher.

The content of soluble reducing sugars (Fig. 4) decreased under each light treatment in the first days of the experimental period. This could have been the effect of growing young cells, which contain only little water. It was not before the last days of our series of tests that the content of reducing sugars increased in white and blue light plants. The content of soluble proteins, however, had the opposite tendency: it was higher at the beginning than at the end of our assays (Fig. 5).

The nitrate reductase activity (Fig. 6) of plants grown under white light showed an increase of activity to a maximum in the first days of the series of experiments. These data were already obtained by Wild and Zerbe (1977). Blue light caused a marked enhancement of nitrate reductase activity, although there were fluctuations at the beginning. Under red light conditions the nitrate reductase activity was rather low, even at the beginning of the experiments. The same level was reached under all light qualities at the end of the experiments. A comparison of the protein content with nitrate reductase

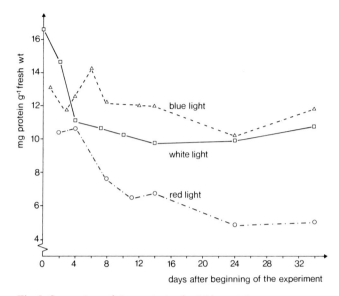

Fig. 5. Comparison of the contents of soluble proteins

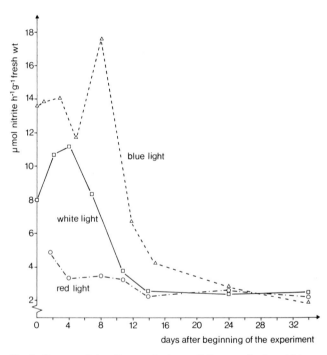

Fig. 6. Changes of the nitrate reductase activity per *g* fresh weight

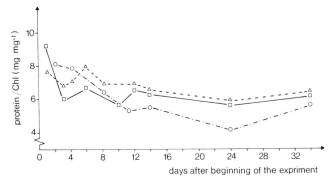

Fig. 7. Ratio of soluble protein per chlorophyll during the ontogeny of the primary leaves

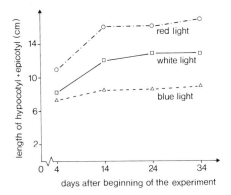

Fig. 8. Comparison of the elongation of the stems of *Sinapis alba* plants grown under white, red, and blue light conditions of equal quantum flux density. (The shoot apex has been removed)

activity makes obvious that the inital high values of protein content corresponded with high values of nitrate reductase activity.

Figure 7 shows the ratio of protein per chlorophyll of primary leaves grown under the different light qualities. The ratio was highest under blue light and lowest under red light conditions, as had been expected.

In Fig. 8 one can see the results of a morphological study. The habit of blue light plants was strong and compact like that of sun plants, while red light plants were elongated and weak.

The experiments, the results of which were shown in Figs. 1–8, were performed under conditions of equal quantum flux (33 μmol m^{-2} s^{-1}). The same kind of experiment has also been carried out under conditions of equal light intensity (6 W m^{-2}). The results of both studies showed the same tendency, which, however, was more pronounced with plants grown under equal light quanta density.

4 Discussion

The effect of light quality on the photosynthetic apparatus and nitrate metabolism was investigated in several studies with higher plants. Blue light has an enhancing effect on protein biosynthesis, nitrate reductase activity, photosynthetic reactions, and on various enzymes of higher plants (Bergfeld 1964; Ziegler et al. 1965; Voskresenskaya and Oshmarova 1969; Poyarkova et al. 1973; Harnischfeger et al. 1974). In addition blue light stimulates the photorespiration and the production and activity of enzymes of the microbodies (Poskuta 1968; Feierabend 1972; Voskresenskaya 1972).

Buschmann et al. (1978) as well as Lichtenthaler and Buschmann (1978) investigated the development of chloroplasts from etioplasts in seedlings of *Hordeum vulgare* and *Raphanus sativus* under weak blue and red light conditions (5.5 μmol m^{-2} s^{-1}). They found that under blue light conditions the content of prenylquinones and the ratio of Chl a/Chl b was higher and the ratio of xanthophyll/carotene was lower than under red light conditions (see Seybold and Egle 1937). The assay of Hill activity showed that "blue light" chloroplasts had a higher activity than "red light" chloroplasts. The differences in the content of prenylquinones and in Hill activity correlated with a difference in the ultrastructure of chloroplasts. Chloroplasts of plants grown under blue light conditions had less thylakoids per granum than "red light" chloroplasts.

These data are in keeping with our results that the effect of blue light on most of the components tested is similar to the effect of high light intensity, whereas the effect of red light resembles the effect of low light intensity. The effect of red light ($\lambda = 653$ nm) is probably caused by the photoreceptor phytochrome. Exposure to red light probably leads to a high photoconversion of Pr to Pfr and thus to a higher ratio of Pfr/Pr.

When daylight passes through a vegetation layer, much of the red and blue wavelengths is absorbed by the pigments in the leaves. As a consequence the spectral distribution of the canopy light is very rich in far red light compared to unfiltered daylight. The vegetation acts as a far red filter which produces a high far red/red ratio and therefore, a low Pfr/Pr-ratio. Consequently an extremely low and an extremely high ratio Pfr/Pr (see Mohr 1969; Kendrick and Frankland 1976) are responsible for the same phenomenon: an adaptation such as is observed under low light conditions.

Under blue light conditions this adaptation may be caused by phytochrome as well as by an additional blue light receptor (cryptochrome); the latter substance probably is a flavoprotein. The increase of nitrate reductase activity and protein synthesis seems to be a major factor in the high light adaptation of plants grown under blue light conditions.

Acknowledgment. This work was supported by the Deutsche Forschungsgemeinschaft.

References

Bendall DS, Davenport HE, Hill R (1971) Cytochrome components in chloroplasts of higher plants. In: San-Pietro A (ed) Methods in enzymology, vol XXIII, part A, pp 327–344. Academic Press, London New York
Bergfeld R (1964) Der Einfluß roter und blauer Strahlung auf die Ausbildung der Chloroplasten bei gehemmter Photosynthese. Z Naturforsch 19b: 1076–1078

Björkman O, Ludlow MM (1972) Characterization of the floor of a Queensland rainforest. Carnegie Inst Yearb 71: 85–94

Boardman NK (1977) Comparative photosynthesis of sun and shade plants. Annu Rev Plant Physiol 28: 355–377

Buschmann C, Meier D, Kleudgen HK, Lichtenthaler HK (1978) Regulation of chloroplast development by red and blue light. Photochem Photobiol 27: 195–198

Fedtke C (1973) Effects of the herbicide methabenzthiazuron on the physiology of wheat plants. Pestic Sci 4: 653–664

Feierabend J (1977) Regulation of organelle development in higher plant cells. Ber Dtsch Bot Ges 85: 601–613

Hanke J, Hartmann KM, Mohr H (1969) The effect of night breaks on flowering of *Sinapis alba*. Planta 86: 235–249

Harnischfeger G, Treharne K, Feierabend J (1974) Studies on the primary photosynthetic processes of plastids from wheat grown under light of different spectral quality. Plant Sci Lett 3: 61–66

Kasperbauer MJ (1971) Spectral distribution of light in a tobacco canopy and effects of end-of-day light quality on growth and development. Plant Physiol 47: 775–778

Kendrick RE, Frankland B (1976) Phytochrome and plant growth. Edward Arnold, London

Lichtenthaler H, Buschmann C (1978) Control of chloroplast development by red light, blue light and by phytohormones. In: Akoyunoglou G (ed) Chloroplast development, pp 801–816. Elsevier/North Holland, Amsterdam

Lowry OH, Rosebrough NJ, Farr AL, Randall RJ (1951) Protein measurements with the folin phenol reagent. J Biol Chem 193: 265–275

Mohr H (1969) Lehrbuch der Pflanzenphysiologie. Springer, Berlin Heidelberg New York

Poskuta J (1968) Photosynthesis and respiration. I. Effect of light quality on the photorespiration in attached shoots of spruce. Experientia 24: 796–797

Poyarkova NM, Drozdova IS, Voskresenskaya NP (1973) Effect of blue light on the activity of carboxylating enzymes and $NADP^+$-dependent glyceraldehyde-3-phosphate dehydrogenase in bean and maize plants. Photosynthetica 7: 58–66

Schlesier G (1977) Nitrate reductase activity in leaves and fruits of various legumes. Biochem Physiol Pflanz 171: 511–523

Seybold A (1936) Über den Lichtfaktor photophysiologischer Prozesse. Jahrb Wiss Bot 82: 741–795

Seybold A, Egle K (1937) Lichtfeld und Blattfarbstoffe. I. Planta 26: 491–515

Shibata M, Kobayashi M, Takahashi E (1969) The possibility of photoinduced induction of nitrate reductase in rice seedlings. Plant Cell Physiol 10: 337–348

Voskresenskaya NP (1972) Blue light and carbon metabolism. Annu Rev Plant Physiol 23: 219–234

Voskresenskaya NP, Oshmarova IS (1969) Photosynthesis and some photosynthetic electron transport chain reactions in pea seedlings grown under red or blue light. In: Metzner H (ed) Prog Photosynth Res, vol III, pp 1669–1674. Metzner, Tübingen

Wild A (1979) Physiology of photosynthesis in higher plants. The adaptation to light intensity and light quality. Ber Dtsch Bot Ges 92: in press

Wild A, Zerbe R (1977) The effect of different light intensities on the nitrate reductase activity, the concentration of soluble proteins, and soluble reducing sugars of *Sinapis alba* during growth from germination to the flowering of the plants. Biochem Physiol Pflanz 171: 201–209

Ziegler H, Ziegler I, Schmidt-Clausen HJ (1965) The influence of light intensity and light quality on the increase in activity of the $NADP^+$-dependent glyceraldehyde-3-phosphate dehydrogenase. Planta 67: 344–356

Ziegler R, Egle K (1965) Zur quantitativen Analyse der Chloroplastenpigmente. I. Kritische Überprüfung der spektralphotometrischen Chlorophyll-Bestimmung. Beitr Biol Planz 41: 11–37

Chloroplast Development

Blue Light Effects on Plastid Development in Higher Plants

L.O. BJÖRN[1]

1 Survey of Light Effects

Light affects the development of plastids in many different ways: Via phototransformation of protochlorophyllide into chlorophyllide a, via the phytochrome system, and via the two photosystems of photosynthesis. All these pigment systems respond to both blue and red light. It is, however, quite clear that in many cases blue light has special effects which cannot be duplicated by red light. This holds both for various algae (as will be shown by several other contributions in this book) and for higher plants. The present treatise will deal with effects induced specifically by blue light (in this term I include violet and near-ultraviolet radiation) in higher plants (among which I include ferns). The topic can be divided into five groups of phenomena:

a) Absolute Blue Light Requirement for Chloroplast Formation. Blue light may induce chloroplast differentiation from proplastids in cells which in the absence of blue light do not form any chloroplasts at all.

b) Major Blue Light Effects on Fern Gametophyte Chloroplasts. Blue light increases the size, chlorophyll content, and protein content of fern gametophyte chloroplasts. Under red light fern gametophytes form only small chloroplasts with a low content of chlorophyll and protein. Because these effects are so drastic, it is motivated to treat them separately from Sect. 3.

c) Blue Light Effects on Lipid Composition, Structure, Enzyme Activity and Electron Transport Capacity of Angiosperm Chloroplasts. This "sun leaf effect" of blue light has been described in different ways by several investigators, e.g., Appleman and Pyfrom (1955), Voskresenskaya and Oshmarova (1969), Harnischfeger et al. (1974), Poyarkova et al. (1973), and Tevini et al. (1978), and is treated in this volume by Lichtenthaler.

d) Blue Light Control of the N/C Assimilation Ratio. Blue light directs assimilation towards more nitrogenous compounds at the expense of carbohydrates. This effect was probably first described by Tottingham and coworkers (Tottingham and Lowsma 1928; Tottingham et al. 1934; Lease and Tottingham 1935) and later literature has been re-

1 Department of Plant Physiology, University of Lund, Box 7007, S-220 07 Lund, Sweden

viewed by Voskresenskaya (1972). This topic be treated by other participants in this symposium.

e) Blue Light Effects on Plastid Replication. The investigation by Possingham (1973) does not provide evidence for an effect of blue light on chloroplast replication which is different from the effect of red light.

2 Discussion

The present review will focus on topics (1) and (2) listed above. An *Absolute Blue Light Requirement* for chloroplast formation (measured as chlorophyll accumulation) was first described by Björn et al. (1963) for excised wheat roots cultured aseptically in an organic medium. It has later been found for cucumber roots (Björn and Odhelius 1966), pea roots (Björn and Odhelius 1966; Richter 1969), cell cultures of tobacco (Bergmann and Berger 1966) and of *Crepis capillaris* (Hüsemann 1970), and dark-adapted pith tissue from potato tuber (Berger and Bergmann 1967; on the contrary, red light alone is sufficient for development of chloroplasts in peripheral cells of potato tuber, as shown by Yamaguchi et al. 1960). The most detailed work has been done on roots.

Plastids in dark-grown roots differ from those of dark-grown leaves in that they lack prolamellar bodies (Newcomb 1967). Dark-grown roots contain protochlorophyll(ide) (Hejnovicz 1958), but its state seems to differ considerably from that of protochloro-phyll(ide) in etioplasts of dark-grown leaves. The protochlorophyll(ide) of the primary seedling roots of maize is mostly in the esterified form (only at the extreme tip is there some unesterified protochlorophyllide), and the in vivo absorption peak is at 634 nm. In red or blue light the protochlorophyll(ide) is very slowly converted to chlorophyll(ide) with an absorption maximum at 675 nm, but unless the light is very weak this chloro-phyll(ide) is destroyed again (Björn 1963, 1976; Figs. 1 and 2). In seedling roots of maize (contrary to seedling roots of wheat and adventitious roots of maize) light trig-gers no increased protochlorophyll(ide) synthesis, and no development of chloroplasts.

When wheat or pea roots are exposed to blue or white light there is a lag phase of three days before chlorophyll starts to accumulate. If wheat roots are kept for three days in blue light, chloroplasts can develop in subsequent red light (see Fig. 3D, E). Sim-ilarly red light preceding blue increases chlorophyll accumulation during the blue-light period (Fig. 4 D–G). On the basis of these and other experiments in which treatments with red and blue light were combined, I proposed the following scheme for chloroplast differentiation, in which A stands for proplastid and D for chloroplast, while B, B', and C are plastids in intermediate stages of development (Björn 1965, 1967d):

$$A \xrightarrow[\text{red or blue light}]{\text{``1}^{\text{st}}\text{ red reaction''}} B \xrightarrow{\text{``dark r.''}} B' \xrightarrow[\text{blue light}]{\text{``blue reaction''}} C \xrightarrow{\text{``2}^{\text{nd}}\text{ red re.''}} D.$$

$$\text{``dark r.''}$$
$$t_{1/2} \approx 2 \text{ days}$$

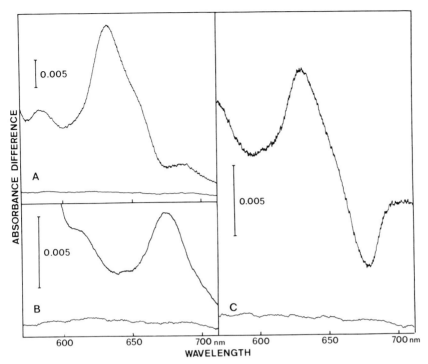

Fig. 1A–C. Absorption difference spectra for maize roots given different light treatments. Vertical absorbance scales as indicated by the *bars.* **A** Apical 5 mm, unilluminated minus bleached (17 h red + 3 h white light). **B** All parts of the roots, red light-treated (30 h dark + 20 h red light) minus bleached (47 h dark + 3 h white). **C** Mature parts of the roots (apical 10 mm removed), unilluminated minus red light-treated (50 h). The *lines below the spectra* are base lines

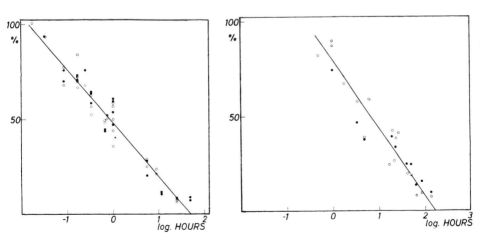

Fig. 2. Protochlorophyll(ide) remaining in maize roots (in percent of initial amount) after various times of irradiation with red light of 7.0–7.4 W m^{-2} *(left)* or 0.13 W m^{-2} *(right). Different symbols* indicate different parts of the roots. (From Björn 1963)

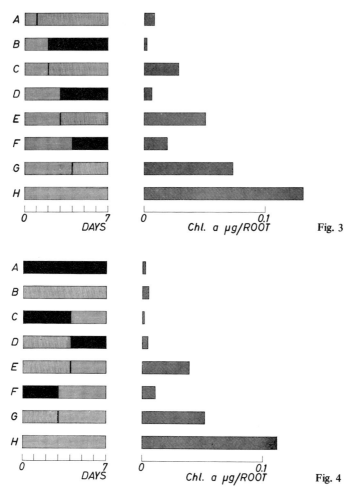

Figs. 3 and 4. Chlorophyll a content in excised wheat roots grown for 7 days under the light conditions indicated (*black* darkness, *vertical hatching* red light 21 W m^{-2}, *horizontal hatching* blue light 8.8 W m^{-2}). (From Björn 1965)

The vertical arrows are degenerative changes from the viewpoint of chloroplast formation, the left one possibly connected with the change from protochlorophyllide to protochlorophyll.

By combining treatments with red light and treatments with shortwave monochromatic light, Björn (1967b) could determine an action spectrum for the specific blue light effect. In Fig. 5 this spectrum is compared with action spectra for two other blue light effects in plants. It shows good agreement with the much more detailed action spectrum that has been determined for blue light-requiring chlorophyll formation in glucose-bleached *Chlorella* (Oh-hama and Senger 1978, Fig. 6). It should be remembered that all the cultures of higher plant cells (with the exception of the carbohydrate-rich potato cells) that show an absolute blue light requirement for chloroplast formation,

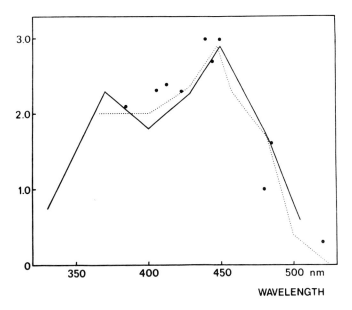

Fig. 5. Action spectra for induction of chloroplast formation in wheat roots (Björn 1967b, *dotted line*), chloroplast rearrangement in the alga *Vaucheria* (Fischer-Arnold 1963, *solid line*), and leaf dropping in *Mimosa* (Fondeville et al. 1967, *dots*)

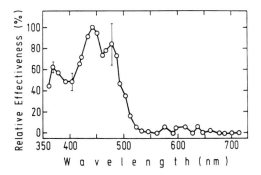

Fig. 6. Action spectrum for chlorophyll formation in glucose-bleached *Chlorella protothecoides*. (From Oh-hama and Senger 1978)

like the *Chlorella* cells, have glucose or sucrose in the culture medium (the sucrose being rapidly hydrolyzed by surface invertase).

In excised wheat and pea roots blue light is necessary not only for the synthesis of chlorophyll (although this is a very convenient indicator of chloroplast development), but for some other chloroplast constituents as well. Thus, a typical chloroplast enzyme, D-glyceraldehyde 3-phosphate:NADP oxydoreductase (phosphorylating, EC 1.2.1.13) is formed in these roots only after irradiation with blue light. The synthesis sets in after the same lag phase as for chlorophyll (Björn 1967c).

An underlying cause for the blue light effects on root plastids seems to be a blue light stimulation of plastid ribosomal RNA synthesis (Dirks and Richter 1975; Richter and Dirks 1978). Figure 7 shows that the patterns obtained with gel chromatography of plastid

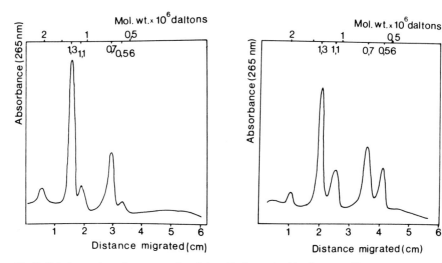

Fig. 7. Gel electrophotetic pattern of nucleic acids from plastids of pea seedling roots cultured for three days in darkness (*left*) or for six days in blue light (*right*). (From Richter and Dirks 1978)

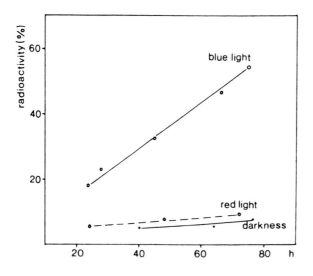

Fig. 8. Synthesis of plastid rRNA under various light conditions. Excised roots were incubated with tritiated uridine and simultaneously irradiated with blue or red light, or kept in darkness. (From Richter and Dirks 1978)

nucleic acids from dark-grown roots differ from that obtained with roots grown in blue light. The synthesis of plastid rRNA can be conveniently followed by incubating the roots with tritiated uridine, and it can then be seen (Fig. 8) that the blue light effect on plastid rRNA is virtually without any lag phase and thus is expressed before the effects on chlorophyll and enzyme synthesis. In leaf etioplasts the plastid rRNA seems to be so abundant before illumination as not to become limiting. Still, there occurs in leaf plastids, as with all probability in root plastids, a red light stimulation of certain plastid mRNA species (Bedbrook et al. 1978), i.e., light-induced transcription (which can, of course, also be induced by blue light) (cf. Gressel, this vol.).

Fern prothallia exhibit a dramatic change in growth pattern (change from one-dimensional to two-dimensional) when exposed to blue light, but this seems to be independent of the effects on plastid development also induced specifically by blue light (Bergfeld 1968). Although chloroplasts develop also under red light, they become much larger under blue (Mohr 1956). This blue light effect on chloroplast size has been investigated in great detail by Bergfeld (1963, 1964, 1968). It is reversible: The time for half-way adaptation to new light conditions is about a week at 20°C, but the effect is perceptible already after 5 h (Cran and Dyer 1975). Blue light-induced growth of chloroplasts is dependent upon blue light-induced RNA and protein synthesis (Ohlenroth and Mohr 1963; Bergfeld 1964, 1968; Raghavan 1968; Payer 1969) by the chloroplasts. Although the effect of blue light (as compared to red) on chloroplast protein content is more rapid and larger than the effect on protein content of the nuclei-rich fraction, the reverse is true for changes in RNA content (Raghavan 1968; Fig. 9). Therefore it is

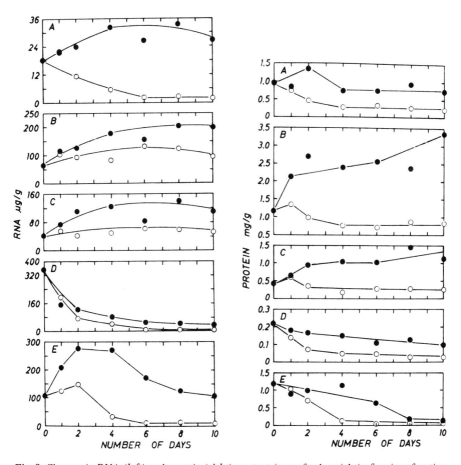

Fig. 9. Changes in RNA (*left*) and protein (*right*) content (per g fresh weight) of various fractions from *Pteridium aquilinum* gametophytes grown in red (*empty circles*) or blue (*filled dots*) light. *A* Nuclei-rich, *B* chloroplast-rich, *C* mitochondria-rich, *D* ribosome-rich, *E* supernatant. (From Raghavan 1968)

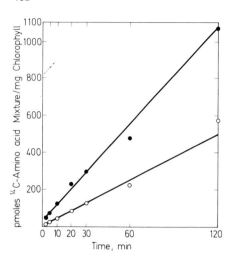

Fig. 10. Time course of incorporation of ^{14}C-amino acid mixture into chloroplasts isolated from *Pteridium aquilinum* gametophytes grown in red (*empty circles*) or blue (*filled dots*) light. (From Raghavan and De Maggio 1971)

likely that part of the blue light effect involves RNA-synthesis in the nucleus. However, chloroplasts from blue light-grown prothallia tested in vitro have a higher capacity for protein synthesis than those of red light-grown prothallia (Raghavan and DeMaggio 1971; Fig. 10). Thus at least part of the blue light effect involves plastid ribosomes, in agreement with the results obtained with roots. RNA synthesis in fern prothallia is further discussed in this volume by Gressel.

Blue light also triggers or greatly favors the development of peroxisomes, and thus of enzymes typical for these bodies, such as glycolate: O_2 oxidoreductase, EC 1.1.3.1. In wheat roots the blue light requirement for this process, as for chloroplast development, is absolute (Björn 1967a), while in other cases (Voskresenskaya et al. 1970; Feierabend 1972) it is quantitative. Although peroxisomes are organelles closely cooperating with chloroplasts, the light effect on peroxisome development can be demonstrated under conditions when chloroplast development does not take place (Feierabend and Beevers 1972). The blue light effect on peroxisome development thus seems to be independent of that on chloroplast development. When interpreting blue light effects on chloroplast development, it should be borne in mind that although synthesis of glycolate: O_2 oxidoreductase as well as that of catalase is favored by blue light, the activity of the former enzyme is reversibly and that of the latter irreversibly inhibited (Björn 1969; Schmid 1969).

Acknowledgments. Dr. A.-S. Sandelius was most helpful in sending me some literature not available in Lund. Mrs. L. Strandh and Mrs. L. Eriksson were both kind enough to interrupt their vacations to get the manuscript typed in time.

References

Appleman D, Pyfrom HT (1955) Changes in catalase activity and other responses induced in plants by red and blue light. Plant Physiol 30: 553–549

Bedbrook JR, Link G, Coen DM, Bogorad L, Rich A (1978) Maize plastid gene expressed during photoregulated development. Proc Natl Acad Sci USA 75: 3060–3064

Berger Ch, Bergmann L (1967) Farblicht und Plastidendifferenzierung in Speichergewebe von *Solanum tuberosum* L. Z Pflanzenphysiol 56: 439–445

Bergfeld R (1963) Die Wirkungen von hellroter und blauer Strahlung auf die Chloroplastenausbildung. Z Naturfrosch 18b: 328–331

Bergfeld R (1964) Der Einfluß roter und blauer Strahlung auf die Ausbildung der Chloroplasten bei gehemmter Proteinsynthese. Z Naturforsch 19b: 1076–1078

Bergfeld R (1968) Chloroplastenausbildung und Morphogenese der Gametophyten von Dryopteris filix-mas (L.) Schott nach Applikation von Chloramphenicol und Actidion (Cycloheximid). Planta 81: 274–279

Bergmann L, Berger Ch (1960) Farblicht und Plastidendifferenzierung in Zellkulturen von *Nicotiana tabacum* var. „Samsun". Planta 69: 58–69

Björn LO (1963) Conversion of protochlorophyll in roots. Physiol Plant 16: 142–150

Björn LO (1965) Chlorophyll formation in excised wheat roots. Physiol Plant 18: 1130–1142

Björn LO (1967a) Some effects of light on excised wheat roots with special reference to peroxide metabolism. Physiol Plant 20: 149–170

Björn LO (1967b) The light requirement for different steps in the development of chloroplasts in excised wheat roots. Physiol Plant 20: 483–499

Björn LO (1967c) The effect of blue and red light on NADP-linked glyceraldehydephosphate dehydrogenases in excised roots. Physiol Plant 20: 519–527

Björn LO (1967d) The effect of light on the development of root plastids. Ph D Thesis, Lund

Björn LO (1969) Photoinactivation of catalases from mammal liver, plant leaves, and bacteria. Comparison of inactivation cross sections and quantum yields at 406 nm. Photochem Photobiol 10: 125–129

Björn LO (1976) The state of protochlorophyll and chlorophyll in corn roots. Physiol Plant 37: 183–184

Björn LO, Odhelius I (1966) Chlorophyll formation in excised roots of cucumber and pea. Physiol Plant 19: 60–62

Björn LO, Suzuki Y, Nilsson J (1963) Influence of wavelength on the light response of excised wheat roots. Physiol Plant 16: 132–141

Buschmann C, Lichtenthaler HK (1979) The influence of phytohormones on prenyllipid composition and photosynthetic activity of thylakoids. In: Appelqvist L-A, Liljenberg C (eds) Advances in the biochemistry and physiology of plant lipids, pp 145–150. Elsevier/North Holland, Amsterdam New York Oxford

Cran D, Dyer AF (1975) The effect of a change in light quality on plastids of protonemata of *Dryopteris borreri*. Plant Sci Lett 5: 57–65

Dirks W, Richter G (1975) Bedeutung von Blaulicht für die Synthese von RNA-Komponenten bei der Chloroplastendifferenzierung in isolierten Wurzeln (Pisum sativum). Biochem Physiol Pflanz 168: 157–166

Feierabend J (1972) Regulation der Entwicklung von Zellorganellen bei höheren Pflanzen. Ber Dtsch Bot Ges 85: 601–613

Feierabend J, Beevers H (1972) Developmental studies on microbodies in wheat leaves. I. Conditions influencing enzyme development. Plant Physiol 49: 28–32

Fischer-Arnold G (1963) Untersuchungen über die Chloroplastenbewegung bei *Vaucheria sessilis*.

Fondeville JC, Schneider MJ, Borthwick HA, Hendricks SB (1967) Photocontrol of *Mimosa pudica* L. leaf movement. Planta 75: 228–238

Harnischfeger G, Treharne K, Feierabend J (1974) Studies on the primary photosynthetic processes of plastids from wheat grown under light of different spectral quality. Plant Sci Lett 3: 61–66

Hejnowicz Z (1958) Protochlorophyll in root tips. Physiol Plant 11: 878–888

Hüsemann W (1970) Der Einfluß verschiedener Lichtqualitäten auf Chlorophyllgehalt und Wachstum von Gewebekulturen aus *Crepis capillaris* (L.) Wallr. Plant Cell Physiol 11: 315–322

Kasemir H, Mohr H (1965) Die Regulation von Chlorophyll- und Proteingehalt in Farnvorkeimen durch sichtbare Strahlung. Planta 67: 33–43

Lease EJ, Tottingham WE (1935) Photochemical responses of the wheat plant to spectral regions. J Am Chem Soc 57: 2613–2616

Mohr H (1956) Die Abhängigkeit des Protonemawachstums und der Protonemapolarität bei Farnen vom Licht. Planta 47: 127–158

Newcomb EH (1967) Fine structure of protein-storing plastids in bean root tips. J Cell Biol 33: 143–163

Oh-hama T, Senger H (1978) Spectral effectiveness in chlorophyll and 5-aminolevulinic acid formation during regreening of glucosebleached cells of *Chlorella protot</hecoides*. Plant Cell Physiol 19: 1295–1299

Ohlenroth K, Mohr H (1963) Die Steuerung der Proteinsynthese und der Morphogenese bei Farnvorkeimen durch licht. Planta 59: 427–441

Payer HD (1969) Untersuchungen zur Kompartimentierung der freien Aminosäure Alanin in den Farnvorkeimen von *Dryopteris filix-mas* (L.) Schott im Rotlicht und im Blaulicht. Planta 86: 103–115

Possingham JV (1973) Effect of light quality on chloroplast replication. J Exp Bot 24: 1247–1260

Poyarkova NM, Drozdova IS, Voskresenskaya NP (1973) Effect of blue light on the activity of carboxylating enzymes and $NADP^+$-*dependent glyceraldehyde-3-phosphate dehydrogenase in* bean and maize plants. Photosynthetica 7: 58–66

Raghavan V (1968) Ribonucleic acid and protein changes in the subcellular components of the gametophytes of Pteridium aquilium during growth in red and blue light. Physiol Plant 21: 1020–1028

Raghavan V, DeMaggio AE (1971) Enhancement of protein synthesis in isolated chloroplasts by irradiation with blue light. Plant Physiol 48: 82–85

Richter G (1969) Chloroplastendifferenzierung in isolierten Wurzeln. Planta 86: 299–300

Richter G, Dirks W (1978) Blue-light induced development of chloroplasts in isolated seedling roots. Preferential synthesis of chloroplast ribosomal RNA species. Photochem Photobiol 27: 155–160

Schmid GH (1969) The effect of blue light on glycolate oxidase of tobacco. Hoppe Seylers Z Physiol Chem 350: 1035–1046

Tevini M, Herm K, Uhrig H, Iwanzik W (1978) Acyllipids and plastid development. In: Akoyunoglou G et al. (eds) Chloroplast development, pp 827–835. Elsevier/North Holland, Amsterdam New York Oxford

Tottingham WE, Lowsma H (1928) Effects of light upon nitrate assimilation in wheat. J Am Chem Soc 50: 2436–2445

Tottingham WE, Stephens HL, Lease EJ (1934) Influence of shorter light rays upon absorption of nitrate by the young wheat plant. Plant Physiol 9: 127–142

Vlasova MP, Drozdova IS, Viskresenskaya NP (1971) Changes in fine structure of the chloroplasts in pea plants greening under blue and red light. Sov Plant Physiol 18: 1–7. Fiziol Rast 18: 5–11

Voskresenskaya NP (1972) Blue light and carbon metabolism. Annu Rev Plant Physiol 23: 219–234

Voskresenskaya NP, Oshmarova IS (1969) Photosynthesis and some photosynthetic electron transport chain reactions in pea seedlings grown in red or blue light. In: Metzner H (ed) Prog Photosynthes Res, vol III, pp 1669–1674. Laupp, Tübingen

Voskresenskaya NP, Grishina GS, Chmora SN, Poyarkova NM (1970) The influence of red and blue light on the rate of photosynthesis and the CO_2 compensation point at various oxygen concentrations. Can J Bot 48: 1251–1257

Yamaguchi M, Hughes DL, Howard FD (1960) Effect of color and intensity of fluorescent lights and application of chemicals and waxes on chlorophyll development of white rose potatoes. Am Pot J 37: 229–236

Blue Light-Induced Development of Thylakoid Membranes in Isolated Seedling Roots and Cultured Plant Cells

G. RICHTER[1], W. REIHL[1], B. WIETOSKA[1], and J. BECKMANN[1]

1 Introduction

Irradiation with blue light (350–550 nm) induces the development of functional chloroplasts from leucoplasts in excised roots of pea seedlings (*Pisum sativum* var. *"Alaska"*; Björn 1965; Dirks and Richter 1973) and in freely suspended callus cells of *Nicotiana tabacum* var. *"Samsun"* (Bergmann and Berger 1966) when cultured in synthetic liquid media under sterile conditions. Exposure to red light (600–700 nm) does not stimulate the synthesis of chloroplast pigments; the formation of membrane elements, however, may occur to a limited extent. An attempt has been made to explore the formation of polypeptides and their insertion into the developing chloroplast membranes triggered by blue light.

2 Material and Methods

2.1 Plant Material

Seeds of *Pisum sativum* var. *"Alaska 7"* treated with Captan were purchased from Asgrow Seed Comp., New Haven. The aseptic technique of germination, the isolation of root tips from seedlings and their axenic culture have been described elsewhere (Dirks and Richter 1973). Freely suspended callus cells from *Nicotiana tabacum* var. *"Samsun"* (Bergmann 1960) were grown at 27°C under sterile conditions in a synthetic medium (Seitz and Richter 1970) with either continuous dark or continuous white, blue, or red light (see below). They were agitated on a horizontal rotary shaker (100 rpm). The cell suspensions were subcultured biweekly. Under these conditions the log phase of growth lasted for about 10 days.

2.2 Illumination

Root fragments were transferred to Fernbach flasks and irradiated from below with blue light (Philips fluorescent tubes TL 40 W/18; 2200–3600 erg/cm^2 s^{-1} at 436 nm) or red light (Philips fluorescent tubes TL 40 W/15; 1450–1800 erg/cm^2 s^{-1} at 665 nm).

1 Institut für Botanik, Universität Hannover, Herrenhäuser Str. 2, 3000 Hannover 21, FRG

Suspended callus cells were illuminated from above with incandescent light (3000 lx) as well as with blue or red light of about the same quantum flux density in a controlled environment incubator shaker.

2.3 Isolation of Plastid Membranes

Membrane fractions were prepared by a modification of the procedure described by Anderson and Levine (1974): Root fragments were harvested, chilled, and homogenized with an equal volume of 50 mM Tris-HCl, pH 7.9, in a mortar. The homogenate was passed through three layers of cheesecloth and centrifuged at 10,000 rpm for 1 min (Christ "Zeta 20").

The pellet containing cell fragments, nuclei, etc. was discarded, and the supernatant subjected to another centrifugation for 15 min at 14,000 rpm. The pellet was resuspended in 3 ml of 50 mM Tris-HCl (pH 7.9) and again spun down at 14,000 rpm. This step was repeated in order to remove all nonplastidic proteins. The final sediment was resuspended in water, lyophilized, and stored at $-20°C$.

2.4 Electrophoresis of Membrane Polypeptides

One mg of lyophilized membrane material was either solubilized in 140 μl of 50 mM Tris-HCl containing 0.5% sodium dodecylsulfate (Kung and Thornber 1971) or dissociated by heating for 1 min in a water bath according to Neville (1971). In another set of experiments the Tris-HCl buffer contained 0.4% sodium dodecylsulfate. Polyacrylamide gel electrophoresis was performed in a 7.5%–15% continuous gradient of acrylamide as described by Neville (1971); the dimension of the gel was 11.5 × 14.5 × 0.2 cm. Densitometric measurements of Coomassie brillant blue stained gels (Fairbanks et al. 1971) were performed at 520 nm (Moltronic) or 550 nm (Zeiß spectrophotometer). Bands associated with pigments from unstained gels electrophorized under nondenaturing conditions were excised and analyzed for pigments and peptide composition. Absorption spectra were measured with a Pye Unicam SP 1800 spectrophotometer.

3 Results and Discussion

3.1 Biosynthesis of Plastid Pigments in Blue Light-Irradiated Seedling Roots

Synthesis of chlorophyll a and chlorophyll b in isolated seedling roots is most prominent during the first nine days of illumination with continuous blue light.

Afterwards the increase in the amounts of both pigments is no longer linear (Fig. 1). At the same time the carotenoids were formed linearly. The chlorophyll a:b ratio being about 1.0 at the beginning of the greening process, changes to a constant value of about 2.3 during the first four days, thus reflecting a preferred synthesis of chlorophyll a (Fig. 2).

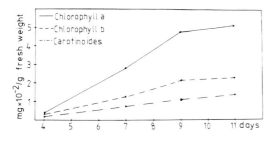

Fig. 1. Changes in the content of chlorophylls and carotenoids during blue light irradiation of isolated seedling roots from *Pisum sativum*

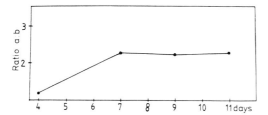

Fig. 2. Changes in the chlorophyll a/b ratio during the greening process in isolated seedling root of *Pisum sativum* in continuous blue light

3.2 Blue Light-Induced Changes in Membrane-Bound Proteins of Seedling Roots

3.2.1 Identification of Chlorophyll-Protein Complexes

Three major bands containing pigments could be electrophoretically resolved from chloroplast membranes of seedling root which had been kept in blue light for 11 days (Fig. 3). The rapidly migrating band (FP) comprising about 64.2% chlorophyll which has been termed "free pigment zone" (Thornber 1975) consisted of detergent-complexed chlorophylls and carotinoids.

This pattern is in good agreement with previously published results from several higher plant species (Thornber 1975). About 14% chlorophyll were found in the band with the lowest mobility designated CP I or "P-700–chlorophyll a–protein complex". The middle band – CP II (= "light-harvesting chlorophyll a/b-protein complex"; Thornber and Highkin 1974) – contained about 22% chlorophyll. From the absorption spectra it becomes obvious that the red peak of CP I compares well with the published value of 674 nm; no chlorophyll b is evident. Spectra of CP II reveal that this band contained both chlorophyll a and b; red absorption peaks were registered at 654 nm and 670 nm. For the FP band the red absorption maximum was found at 668 nm, indicating the presence of both chlorophylls in addition to carotenoids.

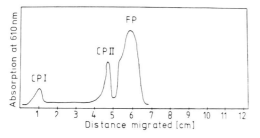

Fig. 3. Densitometric tracing at 610 nm of chlorophyll-containing bands in a gel immediately after electrophoresis of solubilized thylakoid membranes from seedling roots irradiated for 11 d with blue light

The migration of the two chlorophyll—protein complexes was compared to that of denatured proteins of known molecular weights. The corresponding equivalency of CP I was about M_r = 110,000, that of CP II about M_r = 27,000. However, one has to be aware of the fact that the accuracy of molecular size determinations of partially denatured complexes is uncertain. Therefore, the values given here are provided only as a preliminary estimate. The same holds true for the results of the other experiments described below.

With an improved electrophoretic system (see Material and Methods), however, the amount of unassociated pigments was greatly reduced, and two to three additional chlorophyll—protein complexes were resolved with electrophoretic mobilities intermediate between those of CP I and CP II. Moreover, the amount of chlorophyll was increased by two-or threefold over that present in the equivalent complex fractionated by the previous procedure. Similar results were obtained with the suspended callus cells of *Nicotiana* when irradiated with white and blue light, respectively (see Fig. 8). Though the absorption spectra and polypeptide composition of these additional chlorophyll—protein complexes are still unknown, the findings support the notion that all chlorophyll exists as chorophyll—protein complexes in the thylakoid membranes, and that the large amounts of the free pigments observed in earlier experiments resulted from denaturation of the complexes during extraction and/or electrophoretic separation (Henriques and Park 1977: Markwell et al. 1978; Markwell et al. 1979).

3.2.2 Time-Dependent Changes in the Protein Pattern of Greening Plastid Membranes

Significant changes in the protein composition of the developing chloroplast membranes occurred in isolated seedling roots about four days after the onset of blue irradiation (Fig. 4). From the 15—18 membrane-bound proteins detectable in leucoplasts two proteins with apparent molecular weights of M_r = 34,000 and 36,000 disappeared during greening. Among those proteins appearing exclusively during the latter process the apoproteins of CP I and CP II were the main species. From the membrane polypeptides present in both types of plastid and comprising the largest number of proteins, one component with an apparent M_r = 54,000 was most prominent.

In the pattern of membrane proteins from seedling roots grown in red light (4—11 days) only one significant change was detected as compared with that of dark-grown seedlings: only traces of the M_r = 36,000 component were present (Fig. 5).

In heat-dissociated membrane samples CP I was no longer detectable but an additional protein band appeared in the region of M_r = 82,000 (Fig. 6). A simple interpretation of this finding is that the latter represents a chlorophyll-free subunit of CP I. This conclusion, however, must remain conjectural until the existence of another subunit making up the native complex is shown.

On the other hand, the dissociation of CP II did not affect the electrophoretic migration of the corresponding polypeptid of M_r = 27,000 apparently representing the apoprotein of the light-harvesting chlorophyll a/b-complex.

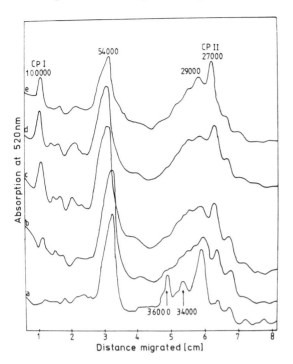

Fig. 4. Blue light-dependent changes in the protein composition of plastid membranes from isolated seedling roots. Densitometric tracings at 520 nm of the gels after treatment with Coomassie brillant blue. Time of illumination with blue light: *b* 4 d, *c* 7 d, *d* 9 d, *e* 11 d; *a* dark control

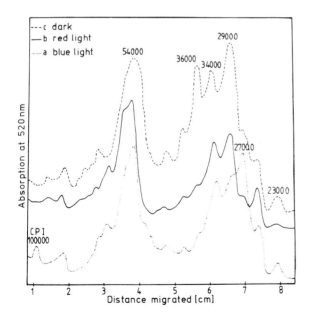

Fig. 5. Effect of continuous red light (11 d) upon the pattern of membrane-bound proteins from isolated seedling roots in comparison with changes induced by blue light and darkness, respectively

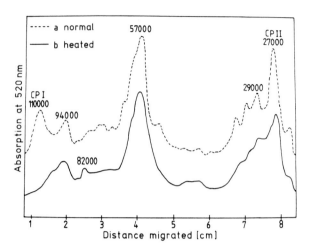

Fig. 6. Electrophoretic separation of proteins from heat-dissociated membrane samples of blue light-irradiated (11 d) seedling roots as compared with untreated SDS-extracted material

3.3 Greening of Freely Suspended Callus Cells (*Nicotiana*) in Either White or Blue Light

3.3.1 Formation of Chlorophyll–Protein Complexes

Solubilization of plastids from callus cells after exposure to white light for 5–10 days followed by electrophoretic separation under improved conditions lead to the visualization of at least four chlorophyll–protein complexes to which the following apparent molecular weights were assigned: M_r = 135,000, 68,000, 60,000, and 25,000 (Fig. 7). When the gels were treated with Coomassie brillant blue, stained bands were located in the positions where the four green bands had concentrated, corroborating that these were indeed chlorophyll–protein complexes. The band with M_r = 135,000 corresponds to CP I, the one with M_r = 25,000 to CP II. The two other complexes seem to agree reasonably with the corresponding ones isolated from plastids of young tobacco leaves (Markwell et al. 1978). This comparison, however, is solely based upon the similarity in molecular size of the polypeptides concerned. Without a thorough knowledge of their pigment and peptide composition further conclusions seem premature at this point.

These results from callus cells after greening in white light are comparable with those obtained from blue light-irradiated cells as well as with isolated seedling roots as desribed above.

3.3.2 Light-Induced Changes in the Protein Composition of Plastid Membranes

Illumination with white light or blue light gave rise to some significant changes in the pattern of plastid membrane proteins characteristic for dark-grown callus cells (Fig. 8).

During the greening process the most prominent feature consisted in the appearance of additional protein bands in the regions of M_r = 135,000 (= CP I), 60,000 (= chlorophyll–protein complex), 30,000 and 25,000 (= CP II) as shown in Fig. 7. Most of the 12–15 membrane-bound proteins resolved were found in fully developed chloroplasts as well as in leucoplasts.

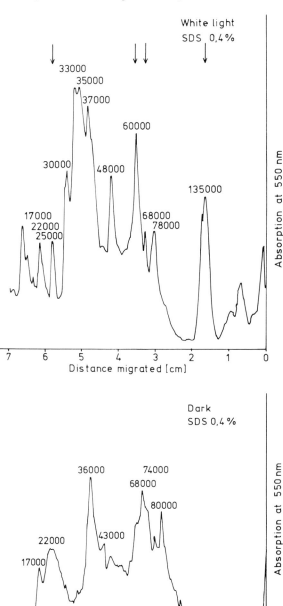

Fig. 7. Densitometric tracing at 550 nm of proteins from solubilized chloroplast membranes of callus cells (*Nicotiana*) cultured in continuous white light for seven days. The gel was stained to reveal proteins; bands originally associated with chlorophylls are indicated by *arrows*

Fig. 8. Densitometric tracing at 550 nm of plastid proteins solubilized from dark-grown callus cells (*Nicotiana*) after treatment with Coomassie brillant blue

The results presented here demonstrate that the differentiation of chloroplast membranes taking place specifically upon irradiation with blue light is a very promising system for studying the mode of synthesis of the proteins involved and their insertion into functional pigment—protein complex of the thylakoid membrane.

Acknowledgments. This investigation was supported by Deutsche Forschungsgemeinschaft and Stiftung Volkswagenwerk. The authors are grateful to Ms. Ch. Balke and Ms. E. Scharfenorth for technical assistance. They thank Dr. K. Kloppstech for valuable suggestions and helpful discussions.

References

Anderson J, Levine RP (1974) The relationship between chlorophyll-protein complexes and chloroplast membrane polypeptides. Biochim Biophys Acta 357: 118–126

Bergmann L (1960) Growth and division of single cells of higher plants in vitro. J Gen Physiol 43: 841–851

Bergmann L, Berger Ch (1966) Farblicht und Plastidendifferenzierung von Zellkulturen von *Nicotiana tabacum* var. *"Samsun"*. Planta 69: 58–69

Björn LO (1965) Chlorophyll formation in excised wheat roots. Physiol Plant 18: 1130–1142

Dirks W, Richter G (1973) Die Wirkung des Cytostaticum „Proresid" auf das Wachstum und die Blaulicht-induzierte Chloroplastendifferenzierung isolierter und normaler Keimlingswurzeln von *Pisum sativum.* Planta 112: 101–120

Fairbanks G, Steck TL, Wallach DFH (1971) Electrophoretic analysis of the major polypeptides of the human erythrocyte membrane. Biochemistry 10: 26-6–2617

Henriques F, Park RB (1977) Polypeptide composition of chlorophyll-protein complexes from Romaine lettuce. Plant Physiol 60: 64–68

Kung SD, Thornber JP (1971) Photosystem I und II chlorophyll-protein complexes of higher plant chloroplasts. Biochim Biophys Acta 253: 285–289

Neville DM Jr (1971) Molecular weight determination of protein-dodecylsulfate complexes by gel electrophoresis in a discontinuous buffer system. J Biol Chem 246: 6328–6334

Markwell JP, Reinman S, Thornber JP (1975) Chlorophyll-protein complexes from higher plants: A procedure for improved stability and fractionation. Arch Biochem Biophys 190: 136–141

Markwell JP, Thornber JP, Boggs RT (1979) Higher plant chloroplasts: Evidence that all the chlorophyll exists as chlorophyll-protein complexes. Proc Natl Acad Sci USA 76: 1233–1235

Seitz U, Richter G (1970) Isolierung und Charakterisierung schnell markierter, hochmolekularer RNS aus frei suspendierten Calluszellen der Petersilie *(Petroselinum sativum).* Planta 92: 309–326

Thornber JP (1975) Chlorophyll-proteins: Light-harvesting and reaction center components of plants. Annu Rev Plant Physiol 26: 127–158

Thornber JP, Highkin HR (1974) Composition of the photosynthetic apparatus of normal barley leaves and a mutant lacking chlorophyll b. Eur J Biochem 41: 109–116

The Effect of Light Quality and the Mode of Illumination on Chloroplast Development in Etiolated Bean Leaves

G. AKOYUNOGLOU[1], H. ANNI[1], and K. KALOSAKAS[1]

1 Introduction

Blue light has been shown to control a number of physiological responses ranging from the conidiation of certain fungi and the phototropism of *Phycomyces* to chloroplast movement and segregation, as well as to cell growth and division. Most of the existing evidence implicates a flavoprotein as the blue light photoreceptor [20, 22, 23]; arguments have also been advanced, however, for the involvement of carotenoids in blue light reception [13, 15].

Blue light has also been shown to control chloroplast development in photosynthetic microorganisms and higher plants. Chlorophyll accumulation and the differentiation of the thylakoid membrane during chloroplast development in *Senedesmus obliquus* C-2A′ mutant is mostly blue light-dependent [10, 16]. Similarly, the development of chloroplasts in *Euglena* seems to be controlled by a cytoplasmic blue light receptor [21].

Etiolated leaves of higher plants exposed to continuous blue or red light form photosynthetically active chloroplasts which differ, however, in structure, composition and activity [11, 17]. "Sun-type" chloroplasts are formed in blue light, with fewer grana and less lamellar material, higher Chla/Chlb ratio, and higher Hill reaction activity; in red light "shade-type" chloroplasts are formed, with more grana, lower Chla/Chlb ratio and lower Hill reaction activity.

It is known that chloroplast development is a stepwise process. The intermediate stages in chloroplast development, however, cannot be readily distinguished when etiolated leaves are exposed to continuous illumination. On the contrary, previous experiments in our laboratory [1, 8, 9] have shown that exposure of etiolated leaves to intermittent illumination prolongs the extent of the greening process and chloroplasts in intermediate stage of development can be isolated. We thus applied the periodic light

---

Abbreviations. *Chl* chlorophyll; *PS I* photosystem I; *PS II* photosystem II; *DCIP* dichlorophenol indophenol; *DPC* 1,5-diphenyl carbazide; *DCMU* 3-(3,4-dichlorophenyl)-1,1-dimethylurea; *LDC* 2-min light—98-min dark cycles; *CL* continuous illumination; *MV* methyl viologen; *SDS* sodium dodecyl sulfate; *P* phytochrome; P_{fr} the far-red absorbing form of phytochrome; F_o the initial level of fluorescence after the dark adaptation of chloroplasts, when all the reaction centers are open; F_m the maximal level of fluorescence when all the reaction centers are closed; $F_v - F_m$ F_o the variable fluorescence.

1 Biology Department, Nuclear Research Center "Demokritos", Athens, Greece

treatment to etiolated bean leaves using blue, red, or white light in an effort to see which steps, if any, are controlled by blue light, and whether, in addition to phytochrome, another blue photoreceptor is acting on chloroplast development.

2 Materials and Methods

Etiolated bean leaves, *Phaseolus vulgaris* (Red kidney var) were used for the study. The growth and handling of the bean seedlings were done as previously described [7]. Five- to six-day-old leaves were exposed to periodic blue, red, or white illumination (2 min light—98 min dark). The blue light source consisted of Sylvania lamps (F 20T12-B) with a 3-mm blue Plexiglas filter (Röhm Darmstadt Germany, No 627); band width 380—550 nm, emission maximum at 450 nm, intensity 12.5 μmol m^{-2} s^{-1}. The red light source consisted of Philips lamps (TL 40W/15) with a 3-mm red Plexiglas (Röhm Darmstadt, No 555); band width 600—700 nm, emission maximum at 660 nm, intensity 20.5 μmol m^{-2} s^{-1}. The white light source consisted of an Osram incandescent lamp (25 W), intensity 7.4 μmol m^{-2} s^{-1}.

The pigments were extracted with 80% acetone as previously described [7], and the Chl concentration was estimated according to Mackinney [18].

For photochemical activity measurements the plastids were isolated as described before [1, 6]; the PS II activity was determined as the rate of the $H_2O \rightarrow$ DCIP or DPC $\rightarrow$ DCIP reduction [24]; the PS I activity was monitored by the rate of the MV-mediated O_2 uptake in a DCMU poisoned reaction mixture containing DCIP and Na ascorbate [14]. For the Chl-protein complexes the plastids were isolated as described before [8], and the complexes resolved by SDS-gel electrophoresis [8]. Fluorescence measurements were done as previously described [1].

For electron microscopy the leaf tissue was fixed for 3 h in 4% glutaraldehyde and for 1.5 h in 1% OsO_4 in Na cacodylate buffer, pH 7.2. Dehydration was done with increasing concentrations of ethanol and propylene oxide. Durcupan resins were used as the embedding medium. The examination was done in a Siemens ELMISCOPE 101 electron microscope.

3 Results and Discussion

3.1 Leaf Growth

Blue, red, or white light—dark cycles stimulate leaf growth in dark-grown plants (Fig. 1). The stimulation, however, is much smaller in blue LDC than in white or red LDC. The plateau in leaf growth is reached earlier in blue than in red LDC; the value at the plateau is 45%, 100%, or 145% higher than that of the dark control, in blue, white, or red LDC respectively. Transfer of the leaves to continuous light of the same quality, after a short preexposure to LDC, further promotes leaf growth; finally all leaves reach the same size. Prolonged preexposure to LDC, however, prevents further leaf growth after their transfer to continuous light.

Since changes in leaf weight are closely correlated with changes in cell number [12], it seems that blue light prevents cell division, whereas red light stimulates it. It is gen-

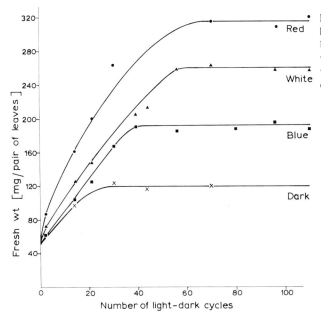

Fig. 1. Time-course analyses of leaf growth in 5-day-old etiolated bean leaves upon illumination with intermittent blue, red, or white light, or remaining in the dark

erally accepted that leaf growth is controlled by phytochrome, and that P_{fr} is its "active" form. It is also known that illumination of dark-grown leaves with blue, white, or red light result in variable P_{fr}/P_{total} ratio, with smaller values in blue than in red light [19]. Red LDC, therefore, are expected to have more pronounced stimulation on leaf growth than blue LDC, as was actually found to be the case.

3.2 Thylakoid Components

3.2.1 Chlorophyll Biosynthesis

Figure 2 shows the effect of light quality on Chl biosynthesis expressed on a g dry wt. basis. It is interesting to note that exposure to blue LDC inhibits the absolute amount of Chl synthesized as well as the rate of its synthesis. Red LDC on the contrary stimulate Chl synthesis much more than white LDC do. The difference becomes more pronounced when the results are expressed on a leaf basis (Fig. 3). Previous experiments in our laboratory have shown that when 5–6-day-old etiolated bean leaves are exposed to red LDC or to red-far red LDC, the same amount of Chl is synthesized (on a g dry wt. basis) [2]. The ratio P_{fr}/P_{total} in the red-far red treatment is very small, smaller than that reached in blue light. Still, the red-far red LDC have no effect on Chl synthesis (on a g dry wt. basis) even though they prevent leaf growth. The phytochrome hypothesis cannot explain, therefore, the blue light effect in reducing Chl synthesis.

The results of Figs. 2 and 3 also indicate that the pattern of Chl accumulation is similar to that described earlier for white LDC (2000 lx) [7]: early in the greening process only Chl a is synthesized; Chl b starts to be formed later in LDC but the Chl a/Chl b ratio remains high even after prolonged LDC treatment. Increased amount of Chl b

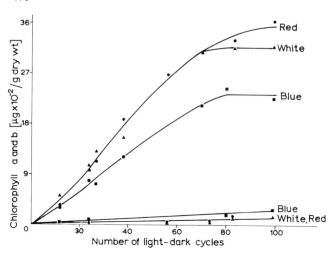

Fig. 2. Time-course analyses of Chl a and Chl b accumulation in 5-day-old etiolated bean leaves upon illumination with intermittent blue, red, or white light. *Upper lines* Chl a, *lower lines* Chl b

is formed only after transfer of these plants to continuous light, with concomitant drop of the Chl a/Chl b ratio.

Figure 4 shows the rate of Chl (a+b) accumulation after transfer of the periodic light plants, which received 25 or 84 LDC, to continuous light of the same quality. Chl a and Chl b accumulate under these conditions and reach a plateau. In accordance with our recent studies [4] the rate of Chl synthesis and the value at the plateau is higher in leaves preexposed to 25 LDC than to 84 LDC. Furthermore, the value at the plateau is higher in leaves preexposed to 84 red LDC than 84 blue or white LDC. It should be mentioned here that blue, red, or white periodic light leaves transferred to continuous illumination with white light of 2000 lx, form 50% more Chl than they do when they are transferred to continuous light of the same quality.

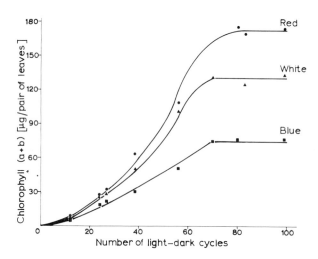

Fig. 3. Time-course analyses of Chl a and Chl b accumulation in 5-day-old etiolated bean leaves upon illumination with intermittent blue, red, or white light

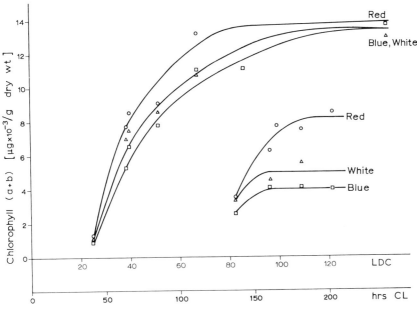

Fig. 4. Time-course analyses of Chl (a + b) accumulation in 5-day-old etiolated bean leaves preexposed to intermittent blue, red, or white light, and then transferred to continuous illumination with light of the same quality

3.2.2 Chlorophyll-Protein Complexes

SDS-PAGE patterns of SDS-solubilized thylakoids prepared from plastids of leaves treated with blue or red LDC, and then transferred to continuous blue or red light are shown in Fig. 5. A mild electrophoretic procedure was used for resolution of the pigment complexes [8], which allows most of the thylakoid Chl of mature green chloroplasts to be complexed with proteins (only 10% runs as free pigment), and up to six Chl-protein complexes to be resolved: CPIa (an oligomeric form of CPI), CPI, LHCP[1], and LHCP[2] (oligomeric forms of LHCP[3]), CPa (considered to originate from the reaction center of PS II) and LHCP[3] (believed to be identical with CP II). Up to 60% of the thylakoid Chl could be resolved bound on the three light-harvesting complexes, LHCP[1], LHCP[2] and LHCP[3] [8].

Recent studies in our laboratory have shown [8] that two Chl—protein complexes, the CPI and CPa, are only resolved from primary thylakoids of leaves exposed to periodic white light for a short time. But, after prolonged exposure to periodic light (vis 80 LDC) the light-harvesting complexes are also easily detected. The results of Fig. 5 clearly show that the same is true for the thylakoids of leaves exposed to 82 blue or red LDC, i.e., they contain mainly the CPI and CPa, and minor amounts of the LHCP[1] and LHCP[3]. However, there is no substantial difference between the SDS-PAGE pattern of the blue and the red LDC thylakoids. It should be mentioned here that the SDS-electrophoretic pattern of the delipilized thylakoids of blue and red LDC protochloroplasts show some

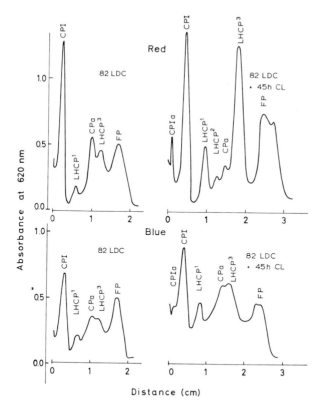

Fig. 5. Distribution of Chl between the Chl−protein complexes resolved by SDS-gel electrophoreis according to [8] from SDS-solubilized thylakoids of etiolated bean leaves exposed to 82 blue or red light−dark cycles before or after transfer to continuous illumination with light of the same quality for 45 h

differences in the relative concentration of their polypeptides. Moreover, the blue LDC thylakoids are deficient in some peptides of high molecular weight.

After transfer of the red LDC leaves to continuous illumination with red light all light-harvesting complexes appear (Fig. 5). The relative concentration of the complexes is similar to the one observed with the white light periodic leaves, which were preexposed to 80 LDC and then transferred to continuous light [8]. The picture, however, is different with the blue LDC leaves, which show only a small increase in the concentration of the LHCP[1] and LHCP[3] during the continuous illumination. Blue light, therefore, seems to have an effect on the synthesis of the light-harvesting complexes, and in addition, on the synthesis of other thylakoid polypeptides.

3.3 Chloroplast Morphology

Figure 6 (Plates 1−6) shows electron micrographs of etiolated bean leaves exposed to blue LDC ± blue continuous illumination (1−3), or to red LDC ± red continuous illumination (4−6). Both blue and red light plastids formed early in periodic light have only primary thylakoids, and they appear similar. They are also similar to the white light LDC protochloroplasts previously described [9]. As the time in periodic light is prolonged, however, a different type of plastid is induced in each case. Prolonged red LDC induce the formation of plastids with the parallel arrays of numerous primary thylakoids usually

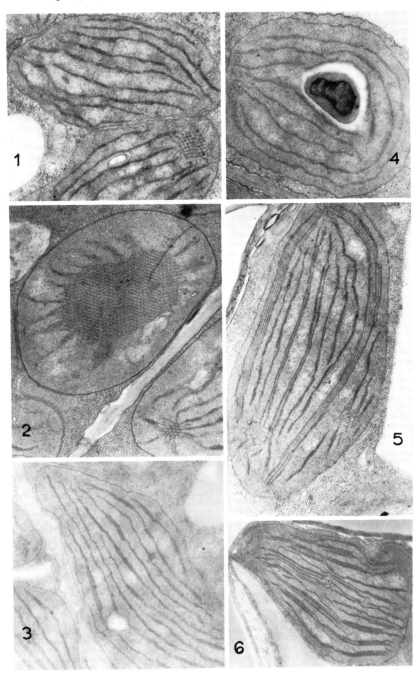

Fig. 6. Electron micrographs of 5-day-old etiolated bean leaves exposed to illumination with intermittent blue or red light (*1, 2:* 24 and 80 blue-LDC; *4, 5:* 24 and 80 red-LDC), or to 83 LDC and then to continuous illumination with light of the same quality (*3:* 83 LDC + 24 h CL, blue; *6:* 83 LDC + 24 h CL, red). *1–5:* $\times$ 25,000; *6:* $\times$ 20,000 magnification

found in white light LDC [5], which fuse to form grana stacks after transfer to continuous light. On the contrary, prolonged exposure to blue LDC induces the formation of
numerous plastids containing huge prolamellar bodies, which upon transfer to continuous
light form normal chloroplasts, which have only a few grana stacks with few appressed
lamellae.

The results, therefore, show that in the periodic blue light the developmental process is drastically controlled so that the synthesis and assembly of some of the thylakoid
components are hindered; this has as a consequence the formation of the huge prolamellar bodies, in addition to the primary thylakoids. Moreover, the results show that continuous illumination with blue light affects the amount of the light-harvesting complexes
formed, and consequently the number and size of grana stacks. Chloroplasts containing
single lamellae and a few grana stacks were also formed when etiolated barley seedlings
were illuminated with blue light [11, 17].

3.4 Photosynthetic Activity

3.4.1 Photosystem I and Photosystem II

As it is known [1], the Chl formed during the intermittent illumination is used for the
formation of small, but very active, photosynthetic units, which increase in size after
transfer to continuous illumination. The organization of Chl into the developing units
is a gradual process. So, our next attempt was to see how blue and red light affect the
development of PS I and PS II units. The results are shown in Table 1 and 2. In accordance with our previous work [5, 6] the PS I and PS II activity is much higher than that
of mature chloroplasts. However, there is no difference in the activity of the plastids
formed in blue, red, or white LDC, indicating that units of the same size are formed,
and that the primary thylakoids in all cases have the same relative concentration of PS I
and PS II units. After transfer to continuous illumination with light of the same quality,
the activity per mg Chl decreases. However, the activity of the blue light plastids remains higher than that of red or white light plastids.

3.4.2 Development of the Photosystem II Unit

The growth, structure, and interaction of the PS II units during greening in blue or red
light has been further studied by fluorescence emission spectroscopy. The results showed

Table 1. Onset of photosystem I activity in developing plastids from 5-day-old etiolated bean leaves
exposed to intermittent illumination (LDC) with blue, red, or white light, and then transferred to
continuous light (CL)

Sample	Blue-LDC	Red-LDC	White-LDC
	μmol O_2 consumed	μmol O_2 consumed	μmol O_2 consumed
	mg Chl · h	mg Chl · h	mg Chl · h
5d + 30 LDC	400	380	380
5d + 82 LDC	395	410	400
5d + 30 LDC + 43 h CL	290	220	240
5d + 82 LDC + 48 h CL	280	220	230

Table 2. Onset of photosystem II activity in developing plastids from 5-day-old etiolated bean leaves exposed to intermittent illumination (LDC) with blue, red, or white light and then transferred to continuous light (CL)

Sample	Blue-LDC		Red-LDC		White-LDC	
	μmol DCIP reduced/mg Chl · h					
	$H_2O \rightarrow$ DCIP	DPC$\rightarrow$ DCIP	$H_2O \rightarrow$ DCIP	DPC $\rightarrow$ DCIP	$H_2O \rightarrow$ DCIP	DPC $\rightarrow$ DCIP
5d + 30 LDC	650	800	800	950	700	850
5d + 82 LDC	800	900	850	900	750	850
5d + 30 LDC + 48 h CL	350	420	300	310	300	320
5d + 82 LDC + 48 h CL	650	700	450	500	400	450

that the blue and red LDC protochloroplasts have fluorescence characteristics similar to those of white LDC protochloroplasts, i.e., high F_m/F_o ratio (twice as high as that of mature chloroplasts), high F_v/F_m ratio (related to the yield of primary photochemistry), and larger half-rise time (at least sevenfold) in the presence or absence of DCMU than the corresponding of mature chloroplasts [1, 3]. The slow rise kinetics in the presence of DCMU is exponential in the beginning of greening in LDC but later it becomes sigmoidal. However, it takes a longer time in LDC for the blue light protochloroplasts to show signoidal kinetics than the red or white light protochloroplasts. The results indicate that in all cases small PS II units are formed, which are several times smaller than those of chloroplasts, and of similar size. The change in the shape of the fluorescence induction from exponential to sigmoidal indicates the onset of the excitation energy transfer from one PS II unit to another, and monitors the gradual connection of the PS II units. The contact of the units is controlled by the structural development and organization of the membrane. In blue LDC, therefore, where more time in LDC is required for the connection of the PS II units to occur, either the organization of the membrane is delayed or a component necessary for the connection of the units is formed with a slow rate, slower than that of the other components of the thylakoid.

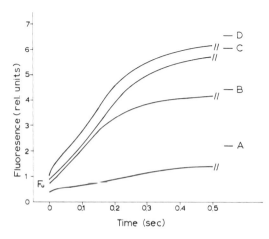

Fig. 7. Fluorescence induction of developing chloroplasts from 5-day-old etiolated bean leaves exposed first to 84 blue, red or white LDC (*A*), and then transferred to continuous illumination with red (*B*), white (*C*) or blue (*D*) light. The curves were taken from actual oscilloscope traces obtained at equal light intensities. Each sample contained the same total Chl concentration [5 μg Chl (a + b)/ml]. Exciting light intensity, $I_0 = 450 \ \mu W/cm^2$

Figure 7 shows the normalized curve for the same amount of Chl fluorescence rise kinetics of plastids of leaves exposed to 84 blue, red, or white LDC, before or after transfer to continuous illumination with blue, red or white light for 48 h. Similar to our previous work [3], transfer of the periodic-light leaves, which received a high number of LDC, to continuous light induces an increase in the F_m level, and a relatively smaller increase in the F_o level, so that the F_m/F_o increases. It is interesting to note, however, that the increase in F_m is higher in the blue and white light than in the red light chloroplasts. Moreover, the half-rise time of the fluorescence induction in the presence of DCMU decreases, but the final value is still three times greater than that of the green control. Since the fluorescence yield of the Chl a of the PS II is related to the structural arrangement of the Chl in the unit, the results indicate a different organization of the PS II units formed in red than in blue or white light.

4 Concluding Remarks

The effect of blue light on chloroplast development can be easily noticed with the system used, where the stepwise formation of chloroplasts can be distinguished, and chloroplasts in an intermediate stage of development can be isolated. The effect of blue light on some of the parameters measured, such as leaf growth and the formation of the light-harvesting complexes, can be easily explained by the "phytochrome hypothesis"; the effect on the fluorescence yield, however, suggests the presence of a different blue light photoreceptor.

The decrease in the rate of Chl biosynthesis and the relatively lower amount of Chl accumulated in blue intermittent light, as well as the huge prolamellar body formed is not due to incomplete phototransformation of the protochlorophyllide during the 2-min irradiation, but rather to the deficiency in some chloroplast components. It seems, therefore, that blue intermittent light inhibits the formation of some component(s) having as consequence the accumulation of the rest of the components formed in the prolamellar body. This component(s), however, is formed in red light as well as in continuous illumination with blue light; moreover, this component does not seem to affect the size of the PS I and PS II units, since the primary thylakoids formed in all cases (blue, red, or white LDC) contain PS I and PS II units of the same structure and size, but it may affect the contact between the individual PS II units, as judged from the delay in the time required for the onset of the sigmoidal fluorescence rise kinetics in blue protochloroplasts. The later effect, however, can also be due to the delay in the general structural development and organization of the thylakoid induced by blue intermittent light.

After transfer of the intermittent light plants to continuous light, and during continuous illumination, the PS II unit increases in size as shown by the decrease in the half-rise time of the fluorescence induction and by the results of the DCIP reduction. This is due to the addition of the light-harvesting complexes to the preexisting small core of the units. On the other hand, the increase in the fluorescence yield in the plastids transferred to continuous light is probably due to the Chl of the light-harvesting complexes, which has a higher probability to emit the excitation energy as fluorescence than the Chl of the core of the unit. Another possibility is that in protochloroplasts excitation energy transfer from PS II to PS I (i.e., spillover) takes place, while after transfer of the leaves to continuous light its control by cations starts to develop.

The observation, however, that the red light chloroplasts have lower fluorescence yield (Fm) than the blue or white light chloroplasts, indicates that some component(s) is missing from these red light plastids and that blue light is necessary for its formation.

This component cannot be the light-harvesting complex, since both white and blue chloroplasts have the same fluorescence yield but different amount of the light-harvesting complexes; on the other hand both red and white light chloroplasts have the same amount of the light-harvesting complexes but different fluorescence yield. This blue light-induced component may either change the organization of the Chl around the reaction center of the PS II unit or it may be necessary for the control of the excitation energy distribution between PS II and PS I. Work is under way to clarify this point.

The results clearly indicate, however, that there is a distinct blue light effect on chloroplast development, which cannot be replaced by red light.

References

1. Akoyunoglou G (1977) Development of the photosystem II unit in plastids of bean leaves greened in periodic light. Arch Biochem Biophys 183: 571–580
2. Akoyunoglou G (1977) Effect of red and far-red light on chloroplast development. Ann Eur Symp Photomorphogen, p 4. Bet Dagan, Israel. Abstr Book
3. Akoyunoglou G (1978) Growth of the PS II unit as monitored by fluorescence measurements, the photoinduced absorbance change at 518 nm, and photochemical activity. In: Akoyunoglou G et al. (eds) Chloroplast development, pp 355–366. Elsevier, Amsterdam
4. Akoyunoglou G, Argyroudi-Akoyunoglou JH (1978) Control of thylakoid growth in *Phaseolus vulgaris*. Plant Physiol 61: 834–837
5. Akoyunoglou G, Argyroudi-Akoyunoglou JH, Christias C, Tsakiris S, Tsimilli-Michael M (1978) Thylakoid growth and differentiation in continuous light as controlled by the duration of preexposure to periodic light. In: Akoyunoglou G et al. (eds) Chloroplast development, pp 843–856. Elsevier, Amsterdam
6. Akoyunoglou G, Michelinaki-Maneta M (1974) Development of photosynthetic activity in flashed bean leaves. In: Avron M (ed) Proc 3rd Int Congr Photosynth, pp 1885–1896. Elsevier, Amsterdam
7. Argyroudi-Akoyunoglou JH, Akoyunoglou G (1970) Photoinduced changes in the chlorophyll *a* to chlorophyll *b* ratio in young bean leaves. Plant Physiol 46: 247–249
8. Argyroudi-Akoyunoglou JH, Akoyunoglou G (1979) The chlorophyll-protein complexes of the thylakoid of greening plastids of *Phaseolus vulgaris*. FEBS Lett in press
9. Argyroudi-Akoyunoglou JH, Feleki Z, Akoyunoglou G (1971) Photoinduced formation of two different chlorophyll-protein complexes at the early stages of greening. Biochim Biophys Res Commun 45: 606–614
10. Brinkmann G, Senger H (1978) Light-dependent formation of thylakoid membranes during the development of the photosynthetic apparatus in pigment mutant C-2A′ of *Scenedesmus obliquus*. In: Akoyunoglou G et al. (eds) Chloroplast development, pp 201–206. Elsevier, Amsterdam
11. Buschmann G, Meier D, Kleudgen HK, Lichtenthaler HK (1978) Regulation of chloroplast development by red and blue light. Photochem Photobiol 27: 195–198
12. Dale JE, Murray D (1969) Light and cell division in primary leaves of *Phaseolus*. Proc R Soc London Ser B 173: 541–555
13. DeFabo E, Harding RW, Shropshire W Jr (1976) Action spectrum between 260 and 800 nm for the photoinduction of carotenoid biosynthesis in *Neurospora crassa*. Plant Physiol 57: 440–445
14. Hall DO, Reeves SG, Baltscheffsky H (1971) Photosynthetic control in isolated spinach chloroplasts with endogenous and artificial electron acceptors. Biochem Biophys Res Commun 43: 359–366

15. Harding RW (1974) The effect of temperature on photoinduced carotenoid biosynthesis in *Neurospora crassa.* Plant Physiol 54: 142–147
16. Klein O, Senger H (1978) Biosynthetic pathways to δ-ALA induced by blue light in the pigment mutant C-2A' of *Scenedesmus obliquus.* Photochem Photobiol 27: 203–208
17. Lichtenthaler H, Buschmann G (1978) Control of chloroplast development by red light, blue light and phytohormones. In: Akoyunoglou G et al. (eds) Chloroplast development, pp 801–816. Elsevier, Amsterdam
18. Mackinney G (1941) Absorption of light by chlorophyll solutions. J Biol Chem 140: 315–322
19. Mancinelli AL (1978) The "high irradiance responses" of plant photomorphogenesis. Bot Rev 44: 129–180
20. Presti D, Hsu WJ, Delbrück M (1977) Phototropism in *Phycomyces* mutants lacking β-carotene. Photochem Photobiol 26: 403–405
21. Schiff JA (1978) Photocontrol of chloroplast development in *Euglena.* In: Akoyunoglou G et al. (eds) Chloroplast development, pp 747–767. Elsevier, Amsterdam
22. Schmidt W, Butler WL (1976) Flavin-mediated photoreactions in artificial systems: A possible model for the blue-light photoreceptor pigment in living systems. Photochem Photobiol 24: 71–75
23. Song P-S, Moore TA (1974) On the photoreceptor pigment for phototropism and phototaxis: Is a carotenoid the most likely candidate? Photochem Photobiol 19: 435–441
24. Vernon LP, Shaw ER (1969) Photoreduction of 2,6-dichlorophenolindophenol by diphenyl-carbazide: A photosystem 2 reaction catalyzed by Tris-washed chloroplasts and subchloroplast fragments. Plant Physiol 44: 1645–1649

The Importance of Blue Light
for the Development of Sun-Type Chloroplasts

H.K. LICHTENTHALER[1], C. BUSCHMANN[1], and U. RAHMSDORF[1]

1 Introduction

Depending on environmental factors, the development of chloroplasts from either pro-plastids or etioplasts will lead to two distinctive types of chloroplast, which are different in composition, ultrastructure, and photosynthetic activity [1, 8]. At low light intensities and in shade leaves the shade-type chloroplast with high grana stacks, lower Hill activity rates and a higher level of chlorophyll b is formed, indicating more light-harvest-ing complex CP II [6, 10]. Sun leaves and plants grown at high light intensities, in turn, develop sun-type chloroplasts with less lamellar material and only few and low grana stacks. Their higher Hill activity is correlated with a higher level of prenylquinones, which function as potential photosynthetic electron carriers, and can also be seen in a changed chlorophyll (higher a/b ratios) and carotenoid composition (lower x/c ratios) [6].

It has been shown before that the sun-type chloroplast growth response can be sim-ulated in barley and radish seedlings with low-intensity blue light [2, 8]. In red light, in turn, the shade-type chloroplast is obtained (Fig. 1). In this report we give details on the accumulation kinetics of pigments during greening in continuous blue and red light. Differences in Hill activity and fluorescence emission are also described.

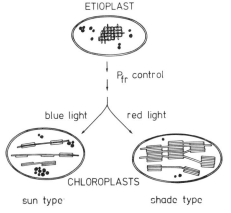

Fig. 1. Light-induced formation of two different types of chloroplast which differ in ultrastructure, prenyllipid composition, and photosynthetic activity. The sun-type chloroplast is formed at high light intensities and in blue light, the shade-type chloroplast at low light intensities and in red light

1 Botanisches Institut (Pflanzenphysiologie) der Universität Karlsruhe, Kaiserstraße 12, 7500 Karlsruhe, FRG

2 Material and Methods

Barley seedlings (*Hordeum vulgare* L., Wilferdingen) were grown on earth in darkness for six days ($25 \pm 1°C$, 60% rel. humidity) and then illuminated with continuous blue (band width 400–500 nm, emission maximum 450 nm, fluence rate 1.5 J/m^{-2} s^{-1}) or red red light (band width 600–700, emission maximum 660 nm, fluence rate 1.0 J/m^{-2} s^{-1}). The light sources consisted of Philips fluorescent lamps (blue: TL 40 W/18, red: TL 40/15) with filter systems as described before [2]. The light quanta density was 5.5 $\mu mol/m^{-2}$ s^{-1} for both light qualities.

The barley shoots were separated into primary and secondary leaves. The photosynthetic pigments were extracted from the upper fully green leaf parts with acetone and transferred to petrolether. The chlorophylls were determined after Ziegler and Egle [12]. The carotenoids were separated by thin layer chromography and measured as described [7].

After isolation [2] the Hill activity of chloroplasts was determined photometrically (dichlorophenol-indophenol 33.4 μmol; chlorophyll 3 μmol). In the light saturation curves the light intensity (incandescent light 250 W) was adjusted with neutral filters.

The in vivo fluorescence emission spectra were taken using an excitation light of 440 nm (light source: xenon lamp and monochromator). The variable and maximal fluorescence was determined using the apparatus already described [9].

The data given in the tables and figures represent mean values from four separate plant cultivations. The values of Figs. 5 and 6 are based on two preliminary experiments.

3 Results

After a growth of six days in darkness the primary leaf of barley seedlings is fully developed. It consists of a lower, white petiol-like part and an upper, broader leaf segment. Upon illumination the latter develops chloroplasts, while the lower part remains predominantly white. The secondary leaf, which after the dark period is still very small within the primary leaf, is developing during the illumination time.

During the greening process in blue or red light, length and leaf area of the primary leaf remain almost the same (Table 1). Blue light plants exhibit a slightly higher dry weight of the primary leaf as compared to red light. In any case it is clear that the differential pigment formation in blue and red light described here proceeds independently of leaf growth and is due to a differential chloroplast development.

3.1 Chlorophyll Formation

Within the first 12 to 24 h of illumination the development of photosynthetically active chloroplasts in the primary leaf appears to proceed faster in blue than in red light. Thereafter the red light plants accumulate significantly more chlorophyll a and b than the blue light plants (Fig. 2). The level of chlorophylls goes through a maximum between the 2nd and 4th day and then decreases to steady-state values. In secondary leaves, which are still growing during the illumination period, the maximum level of chlorophylls is

Table 1. Length, dry weight, and area of green primary leaves of barley seedlings (with standard deviation). The plants were grown six days in darkness and then illuminated with blue or red light for two and seven days

Primary leaf	Blue light	Red light
Length (cm)		
Upper green part 2d	11.0 ± 1.1	11.3 ± 0.8
7d	11.8 ± 1.0	11.6 ± 0.7
Lower "white" part 2d	8.4 ± 1.1	8.6 ± 1.6
7d	8.3 ± 1.6	9.2 ± 1.0
Dry weight (mg/100 leaves)		
Upper green part 2d	625 ±12	550 ±25
7d	635 ±15	624 ±13
Leaf area (cm^2)		
Upper green part 2d	4.92 ± 0.06	4.86 ± 0.07
7d	5.11 ± 0.12	5.40 ± 0.15

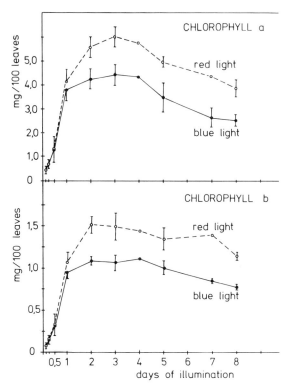

Fig. 2. Kinetic of chlorophyll accumulation in the upper green part of primary leaves of barley seedlings greened in continuous red or blue light

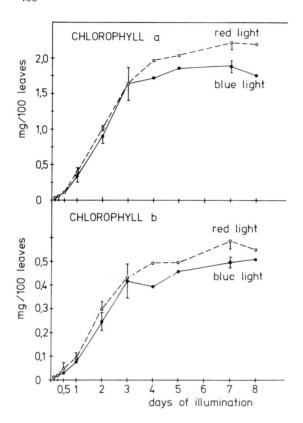

Fig. 3. Accumulation kinetics of chlorophylls in the secondary leaf of barley seedlings greened in continuous red or blue light

reached at a later stage, the higher chlorophyll content is, however, clearly visible after the 3rd day of illumination (Fig. 3). It had been shown before [8] that the chlorophyll a/b ratios are higher in the blue light plants (3.8–4.0) than in red light plants (3.2–3.5). This has also been found in this investigation for the first five to six days of illumination.

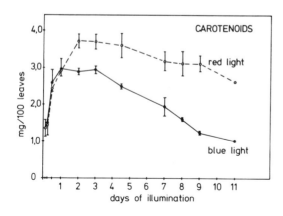

Fig. 4. Kinetics of carotenoid accumulation in primary leaves of barley seedlings greened in blue or red light

Table 2. Percent composition (wt. %) of carotenoids and x/c ratio in blue light and in red light chloroplasts of barley seedlings after two and ten days of illumination

Carotenoid		Blue light		Red light	
		2 d	10 d	2 d	10 d
β-Carotene (c)		35.4 ⎫ 50.8	41.4 ⎫ 55.8	28.8 ⎫ 46.5	28.9 ⎫ 49.9
Violaxanthin	⎫	15.4 ⎭	14.4 ⎭	17.7 ⎭	21.0 ⎭
Lutein	⎬ x	37.6	38.7	41.5	40.4
Neoxanthin	⎭	11.5	7.0	11.9	9.8
x/c		1.82	1.42	2.47	2.46

3.2 Carotenoid Formation

Red light plants accumulate more carotenoids than blue light plants (Fig. 4). In blue light, the maximum level is reached earlier than in red light, and then drops to low values. The decrease of the carotenoid content in red light is much less. It is of interest that the blue light chloroplasts possess a higher percentage of β-carotene (35%–41%) than red light chloroplasts (28%–29%), a difference which remains constant, though the total amount of carotenoids decreases from the 2nd to the 10th day. In red light, in turn, the proportion of lutein and violaxanthin is increased (Table 2). This results in higher values for the xanthophyll to carotene (x/c) ratios in red than in blue light. The percentage of the β-ionone carotenoids (β-carotene and violaxanthin) in blue light amounts to more than 50% of the total carotenoids (Table 2).

3.3 Exchange of Light Quality

When plants are transferred from one light quality to the other, they adapt themselves to the new light conditions. The photoresponse, visualized by changes in the pigment levels, is found in primary and secondary leaves. As compared to the red light controls, blue light leads to a decrease of the chlorophyll a, chlorophyll b, and carotenoid levels which becomes detectable after 24 h. In blue light plants, when transferred to red light, the concentration of chlorophylls and carotenoids, in turn, is increased (Figs. 5 and 6).

3.4 Hill Activity and Fluorescence

The Hill activity (dichlorophenol-indophenol reduction), measured with rising intensity of white light, reaches higher values for blue light chloroplasts than for red light chloroplasts (Fig. 7). For both types of chloroplast the light saturation point is situated around 12,000 lx (12 klx). The data obtained here with different light intensities are in agreement with earlier findings in radish and barley seedlings, where the differences in Hill activity were measured at light saturation [2, 8].

The in vivo fluorescence emission spectrum for primary leaves of barley seedlings illuminated for one day shows a maximum at 688 nm and a shoulder around 735 nm. No significant difference is visible between red and blue light plants (Fig. 8). After a

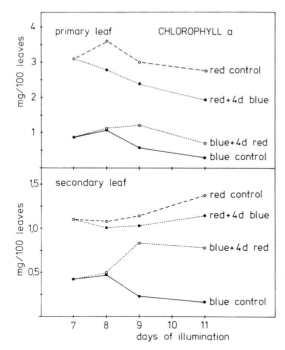

Fig. 5. Change in the accumulation kinetics of chlorophyll a in primary and secondary leaves of barley seedlings after transferring the plants from blue to red or red to blue light

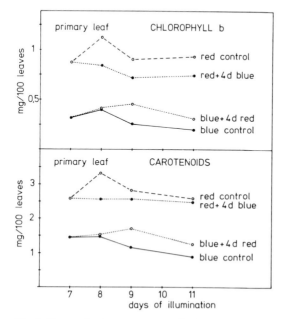

Fig. 6. Changes in pigment content of primary barley leaves after transferring the plants from blue to red or red to blue light

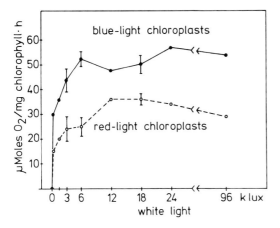

Fig. 7. Hill activity (dichlorophenol-indophenol reduction) of blue light and red light chloroplasts from 8-day-old barley seedlings (2-d illumination) at different intensities of incandescent white light

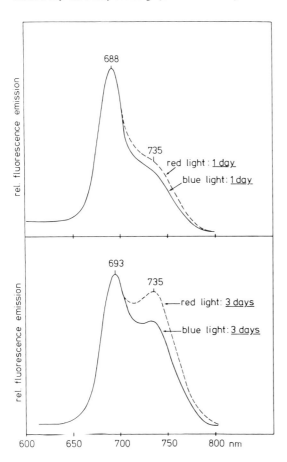

Fig. 8. Fluorescence emission spectra (in vivo fluorescence at room temperature) taken from primary leaves of barley seedlings greened for 1 and 3 days in blue and red light

Table 3. Maximal and variable fluorescence of primary leaves of barley seedlings which greened for three days in red or blue light

	Fluorescence	
	Maximal	Variable
Blue light plants	7.0 ± 1.3	2.2 ± 0.2
Red light plants	7.5 ± 1.2	2.1 ± 0.3

light period of three days the fluorescence spectrum exhibits two peaks, one at 693 nm and the other at 735 nm. The long-wavelength maximum at 735 nm is always higher in leaves of red light plants than of blue light plants.

The appearance of the second fluorescence peak at 735 nm proceeds parallel to the increased formation of grana thylakoids as shown before [2]. The higher 735 nm peak after three days of illumination in red light plants, as compared to blue light plants, coincides with more and higher grana stacks of the red light chloroplasts. From this we conclude that the intensity of the 735 nm fluorescence peak is dependent on the amounts and height of grana stacks present in the chloroplast.

As an additional parameter we measured the variable fluorescence of the intact leaves, which is a good indicator for leaf photosynthesis [9]. Though there exist differences in the Hill activity between the isolated blue light and red light chloroplasts, we did not find significant differences in the maximal or variable fluorescence of both chloroplast types (Table 3). This finding may be explained by a limited CO_2-availability within the whole leaf so that the blue light chloroplasts cannot fully express their apparently higher photosynthetic capacity.

4 Discussion

The results of this investigation show that the Hill activity and the accumulation of photosynthetic pigments upon illumination of etiolated seedlings depends on the light quality applied. Blue light induces the formation of chloroplasts with a chlorophyll (a/b ratios) and carotenoid composition (x/c ratios) different from that of red light chloroplasts [8], which is confirmed in this investigation.

The primary leaf is fully expanded and the cell number does not change during the investigation period. From preliminary studies it appears that also the number of chloroplasts of corresponding cells is the same under both light conditions. The lower pigment content of the blue light plants thus indicates that their chloroplasts possess less chlorophylls and carotenoids than red light chloroplasts. however, a higher proportion of chlorophyll a and β-carotene. In addition they exhibit a higher level of prenylquinones (potential electron carriers) [8], which is correlated with the higher Hill activity. Blue light chloroplasts of barley seedlings possess only few and low grana stacks and less lamellar material than red light chloroplasts [2, 8] and thus behave like sun-type chloroplasts.

Red light chloroplasts, in turn, contain higher pigment levels, which are associated with higher grana stacks [2, 8]. The increased fluorescence emission at 735 nm can be taken as another indicator for an increased amount of grana thylakoids in red light chloroplasts. The higher proportion of chlorophyll b and of xanthophylls, which appear to be predominantly associated with the light-harvesting chlorophyll–protein complex CP II [6, 10], points to a higher CP II content of red light chloroplasts. As in shade-type chloroplasts the antenna system of red light chloroplasts is larger with less photosynthetic units on a chlorophyll basis than that of blue light chloroplasts.

From the results of several laboratories it is clear that active phytochrome Pfr initiates and controls chloroplast formation and the synthesis of many thylakoid components [5, 8]. The formation of shade-type chloroplasts appears to be the normal program at low light intensities and in continuous red light. Blue light modifies this endogenous program, resulting in the formation of sun-type chloroplasts. Since sun leaves receive more blue light than shade leaves, blue light seems to be responsible for the formation of sun-type chloroplasts in nature. The way in which blue light triggers this specific chloroplast differentiation is not known. Sun-type chloroplast formation can also be induced by the application of cytokinins [3, 8]. As compared to red light, blue light increases in pumpkin [4] and in bean plants [11] the endogenous cytokinin levels. On the basis of these observations one may assume that the blue light-induced chloroplast modification may be due to an increased accumulation of cytokinins.

Acknowledgments. This work was sponsored by a grant from the Deutsche Forschungsgemeinschaft, which is gratefully acknowledged. We wish to thank Mrs. U. Prenzel and Miss M. Bächle for excellent technical assistance.

References

1. Boardman NK, Björkman O, Anderson JM, Goodchild DJ, Thorne SW (1974) Photosynthetic adaptation of higher plants to light intensity: Relationship between chloroplast structure, composition of the photosystems and photosynthetic rates. In: Avron M (ed) Proc 3rd Int Congr Photosynth, pp 1809–1827. Elsevier, Amsterdam
2. Buschmann C, Meier D, Kleudgen HK, Lichtenthaler HK (1978) Regulation of chloroplast development by red and blue light. Photochem Photobiol 27: 195–198
3. Buschmann C, Lichtenthaler HK (1979) The influence of phytohormones on prenyllipid composition and photosynthetic activity of thylakoids. In: Appelqvist L-A, Liljenberg C (eds) Advances in the biochemistry and physiology of plant lipids, pp 145–150. Elsevier/North-Holland Biomedical Press, Amsterdam
4. Dörfler M, Göring H (1978) The influence of different light quality (blue and red light) on the cytokinin content of pumpkin seedlings. Biol Rundsch 16: 186–188
5. Kasemir H (1979) Control of chloroplast formation by light. Cell Biol Int Rep 3: 197–214
6. Lichtenthaler HK (1979) Occurrence and function of prenyllipids in the photosynthetic membrane. In: Appelqvist L-A, Liljenberg C (eds) Advances in the biochemistry and physiology of plant lipids, pp 57–78. Elsevier/North-Holland Biomedical Press, Amsterdam
7. Lichtenthaler HK (1979) Effect of biocides on the development of the photosynthetic apparatus of radish seedlings grown under strong and weak light conditions. Z Naturforsch 34c: 936–940
8. Lichtenthaler HK, Buschmann C (1978) Control of chloroplast development by red light, blue light and phytohormones. In: Akoyunoglou G et al (eds) Chloroplast development, pp 801–816. Elsevier/North-Holland Biomedical Press, Amsterdam
9. Lichtenthaler HK, Karunen P, Grumbach KH (1977) Determination of prenylquinones in green photosynthetically active moss and liver moss tissues. Physiol Plant 40: 105–110
10. Thornber JP (1975) Chlorphyll-proteins: light-harvesting and reaction center components of plants. Annu Rev Plant Physiol 26: 127–158

11. Zeinalova SS, Butenko RG, Nichiporovich AA (1967) Kinin activity of extracts from Phaseolus leaves grown in light of different spectral composition. Dokl Akad Nauk SSSR 176: 955–958
12. Ziegler R, Egel K (1965) Zur quantitativen Analyse der Chloroplastenpigmente. I. Kritische Überprüfung der spektralphotometrischen Chlorophyllbestimmung. Beitr Biol Pflanz 41: 11–37

Blue Light and the Photocontrol
of Chloroplast Development in *Euglena*

J.A. SCHIFF[1]

Evolution, through adaptation and selection, has resulted in the complex cellular process of photosynthesis which traps light energy and makes it available to living systems. Evolution has also selected for systems in which light serves as a means of controlling the development and replication of the chloroplast. In order to be adaptive, such control systems must respond to the same wavelengths of light that are effective for photosynthesis since there would be little evolutionary advantage in a system which brought about the formation of a chloroplast in response to wavelengths of light which are ineffective for photosynthesis; for this reason, blue and red light are the most effective regions of the spectrum in controlling plastid development.

The control system shows many resemblances to those already familiar to us from substrate induction in bacterial systems. In *E. coli*, for example, if organisms grown on glucose are presented with, say, lactose, several enzyme activities are induced which enable the cells to utilize lactose. In the case of photosynthesis, however, the substrate to be utilized is light and in order to do this, the induction of many enzymes, structural organization in the form of membranes, pigment complexes, etc. must be formed, all properly coordinated in time. It is this series of substrate-induced enzyme inductions all properly coordinated in time that we call chloroplast development and the end result is the formation of an organelle, rather than just a few enzymes, to enable the organism to utilize light for energy storage through photosynthesis.

In *Euglena* light brings about the development of a comparatively simple proplastid to form a chloroplast (Fig. 1) [1]. This process requires the formation of nucleic acids, proteins, membranes, etc. all properly coordinated in time. Figure 2 summarizes what we know of this process [2, 3]. The developmental needs of the developing plastid for small molecules, energy, and reducing power need not be met from photosynthesis and inhibitors of photosynthesis do not block normal plastid development [4]. Inhibitors of mitochondrial electron transport and phosphorylation, however, are excellent inhibitors and mitochondria and other cellular organelles are found in close juxtaposition to the developing plastids [5]. Out of experiments like this and the fact that many organisms make chloroplasts and chlorophyll in darkness comes the realization that photosynthesis is not necessary for plastid development and that the rest of the cell must provide the necessary energy and intermediates from the breakdown of stored reserves such as paramylum [6].

1 Photobiology Institute, Brandeis University, Waltham, MA 02154, USA

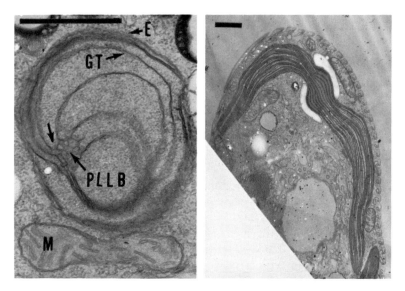

Fig. 1. *GT* girdle thylakoid; *M* mitochondria; *PLLB* prolamellar-like body; *E* envelope; *left* Section through dark-grown resting cell of *Euglena gracilis* var. *bacillaris* showing proplastid. *Marker* indicates 1 μm. *Right* Section through light-grown cell showing chloroplast [1]

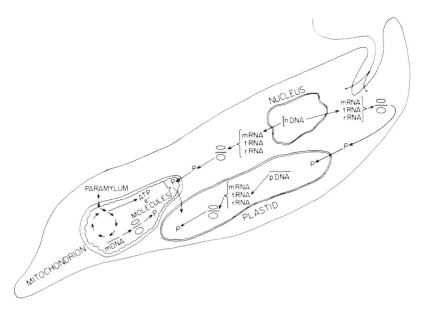

Fig. 2. Summary of nutritional and informational interactions among organelles during chloroplast development in *Euglena*. The DNA's of either the mitochondrion or the chloroplast code for that organelle's tRNA's and rRNA's and for certain proteins which are translated on the ribosomes of that organelle and remain within the organelle. Nuclear DNA codes for rRNA and tRNA's of the cytoplasm and for proteins which are translated on cytoplasmic ribosomes. Some of these remain in the cytoplasm, others enter the developing organelles. Light-induced paramylum breakdown feeding mitochondrial respiration supplies small molecules, energy, and reducing power to the developing chloroplast [26]

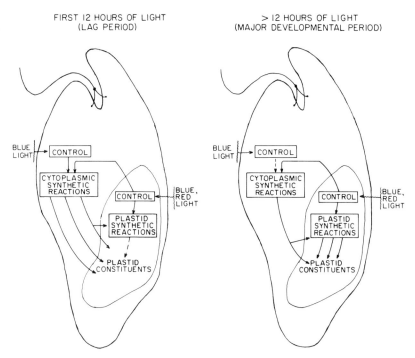

Fig. 3. Model for events of lag period and period of active development [3]

Each cellular compartment also contains the necessary protein-synthesizing machinery for development [2, 3]. The nucleus, plastid, and mitochondrion each contain a DNA genome [7]. The plastid genome codes for proteins of the chloroplast and for the protein-synthesizing machinery of the plastid including the ribosomal RNA's [8] and the tRNA's [9]. Similarly the mitochondrion has its own RNA's which serve as the machinery for its own protein synthesis. The nuclear genome codes for the ribosomal and tRNA's of the cytoplasm and for cytoplasmic proteins [8, 9, 10], but it also codes for proteins which are synthesized in the cytoplasm and which enter the developing plastid or mitochondrion, after some processing en route [10]. Thus, as far as we know, the information for making a chloroplast is contained in both plastid and nuclear DNA.

The events of plastid development are not only organized spatially, they are also arranged temporally (Fig. 3). The lag period which begins chloroplast development (Fig. 4) under normal conditions is a time in which the rest of the cell, under control by blue light, is supplying energy, small molecules, reducing power and proteins to the developing proplastid; activities within the developing plastid are rather low at this time. After the plastid has received sufficient materials from the rest of the cell, the lag period ends and the plastid, under control of the red-blue photoreceptor system, becomes highly active and the grand period of plastid development ensues; the nonplastid systems become correspondingly attenuated during this period.

What are the natures of these two photoreceptor systems which control the plastid and nonplastid portions of plastid development? Preillumination offers a means of sep-

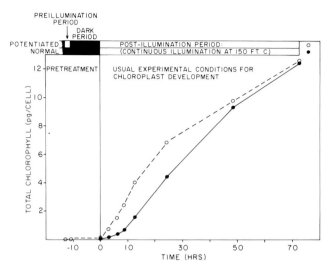

Fig. 4. Time course of chlorophyll accumulation in control (*filled circles*) and potentiated (*open circles*) cells. Zero time is taken as the beginning of the postillumination period. Cells from a 3-day dark-grown resting culture were exposed to 90 min of preillumination (white light, 150 ft-c) starting at −12 h, and to a dark period (potentiated cells) before exposure to continuous illumination with white light (postillumination). Control cells experienced an uninterrupted dark period until they were exposed to continuous illumination at 0 h [11]

arating the early and late events photochemically (Fig. 4) [11]. In normal development when dark-grown resting cells are exposed to continuous light to induce plastid development, a lag of 12 h is evident in chlorophyll formation. If the cells are given a brief exposure to light 12 h before, and are then left in darkness for 12 h when they are then exposed to continuous light, the lag in chlorophyll formation is eliminated. In other words, a brief preillumination enables the cells to go through certain processes in the dark period that they ordinarily would accomplish during the first 12 h of normal development. Development is pushed ahead 12 h by preillumination allowing a separation of early and late events. By giving monochromatic light during the preillumination period followed by white light during the continuous or post-illumination, it is possible to measure the action spectrum for preillumination or the early events of plastid development. Similarly, saturating white light during preillumination and monochromatic light during postillumination permits us to measure the action spectrum of the induction of the later events (Fig. 5) [12].

As may be seen, the action spectrum for postillumination (or later events of plastid development) resembles the absorption spectrum of protochlorophyll(ide) holochrome quite closely with a ratio of about 5 to 1 of the blue to the red peak. Thus the red-blue photoreceptor controlling later development appears to have the absorption of protochlorophyll(ide). The preillumination action spectrum shows a ratio of blue to red approaching 40:1 [a small amount of protochlorophyll(ide) absorption would be expected since 2 h worth of chlorophyll are made during the preillumination period]. The photoreceptor system for the early events, then, is an as yet unidentified blue light-absorbing pigment system [2, 3, 11, 12, 13].

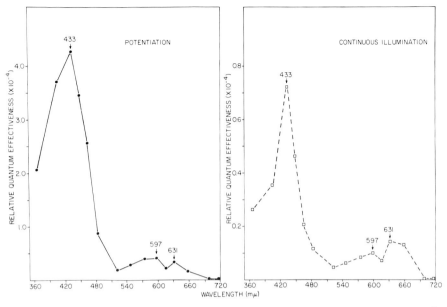

Fig. 5. Action spectra for preillumination in potentiation and for greening in continuous light [12]

We have been able to observe the widely demonstrated [29] flavo-protein-cytochrome photoredox system in *Euglena* [30, 34]. However, since our very early work with systems of this type [31] indicated that added flavins will actively carry out the same reactions, we have been cautious in identifying these systems with those mediating blue light effects in vivo. Further information is required including the comparison of highly detailed action spectra for the flavoprotein-cytochrome systems in vitro with the action spectra for the effects they control in vivo as well as information from mutants blocked in the photoreceptor systems themselves. It seems advisable to keep an open mind concerning the nature of the blue light photoreceptor(s) until such information is available. A blue light-induced photoisomerization of cis zeta carotene to the trans form has also been found in *Euglena* mutants [32, 33] but there is no reason as yet to implicate this system in plastid development; this system may be involved in carotenogenesis or other processes.

Figure 6 shows that paramylum breakdown is light-induced in *Euglena* and is another example of the early events taking place in the cytoplasm [6]. W₃BUL, a mutant in which plastid DNA and protochlorophyll(ide) are undetectable, also shows this breakdown, verifying that it is nonplastidic and under control of the nonplastid blue light system. Since in darkness cycloheximide or levulinic acid will induce paramylum breakdown, we are forced to consider a model for negative control at the transcriptive level by the nonchloroplast blue light receptor system. In Fig. 7 we postulate the existence of a regulatory protein which turns over and blocks paramylum breakdown [6]. If the transcription of this protein is blocked by either blue light or by levulinic acid (acting as an analog of aminolevulinic acid, the normal control compound?), this would lead

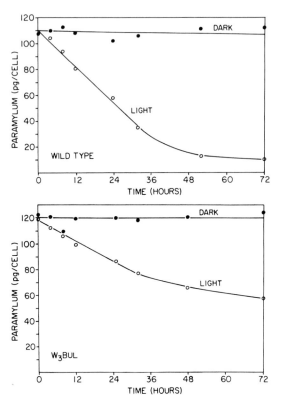

Fig. 6. Kinetics of light-induced paramylum degradation in wild-type cells
of *Euglena* and in the mutant, W₃BUL,
which lacks detectable plastid DNA.
At 0 time dark-grown resting cells were
exposed to continuous illumination
or incubated in the dark and at appropriate times samples were withdrawn for the determination of paramylum [6]

to the lack of the inhibitory protein and the induction of paramylum breakdown. Cycloheximide would induce paramylum breakdown in the dark by preventing the translation
of the inhibitory protein.

The transcription of rRNA's is also under photocontrol in *Euglena*. Figure 8 shows
action spectra for rRNA transcription in wild type and W₃BUL [13]. Consistent with

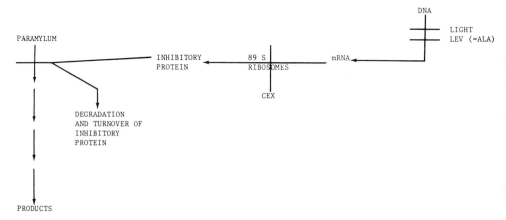

Fig. 7. Negative control model to explain regulation of paramylum degradation in *Euglena. CEX*
cycloheximide; *LEV* levulinic acid; *ALA* aminolevulinic acid [26]

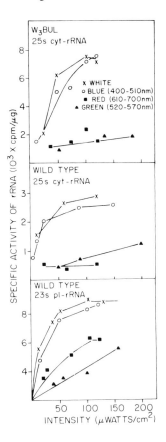

Fig. 8. Effectiveness of the various spectral regions in promoting the labeling of rRNA's of W_3BUL and wild type. Dark-grown resting cells were exposed to increasing intensities of the three spectral regions for 3 h in the presence of $^{32}P_i$; RNA was then extracted [13]

all that has been said, blue light is as effective as white in inducing transcription of the 25S cytoplasmic rRNA from nuclear DNA in both wild type and W_3BUL; green and red are ineffective. Transcription of chloroplast 23S rRNA from plastid DNA in wild type shows a ratio of effectiveness of blue to green to red expected for the absorption cross sections of the protochlorophyll(ide) holochrome. Thus cytoplasmic rRNA transcription is under control of the nonplastid blue receptor system while plastid rRNA transcription is regulated by the plastid blue-red or protochlorophyll(ide) system.

W_3BUL is a mutant of *Euglena* which is widely used for studies of plastid coding and deserves closer examination here. In the analytical ultracentrifuge, plastid DNA is undetectable in this and other "bleached" strains [7]. The absence of plastid DNA is also confirmed by Stutz (personal communication) using columns to isolate the DNA. This strain has been used as a control in RNA-DNA hybridization experiments where in some cases as little as a fraction of one copy of the homologous DNA would have been detected, but was not [9] (R. Hallick, personal communication). Other relevant properties of this strain are shown in Table 1 [10]. Protochlorophyll(ide) [14], chlorophyll and photosynthesis are undetectable, but small amounts of carotenoid are present. Enzymes containing plastid-coded subunits such as RuBP carboxylase are undetectable, while nuclear-coded plastid and cytoplasmic enzymes are present, the former at the re-

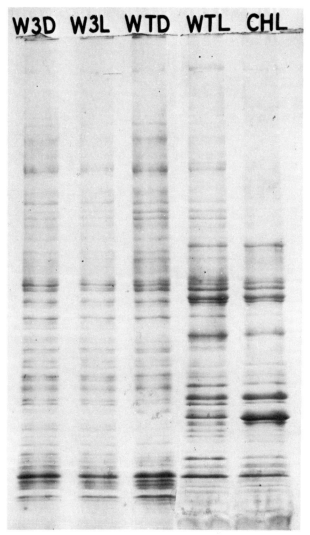

Fig. 9. Polypeptide patterns from plastid thylakoid (*middle*) fractions of dark-grown W₃, light-grown W₃, dark-grown wild type, light-grown wild type and from thylakoid membranes of purified chloroplasts separated on SDS acrylamide gels. Equal amounts of proteins were placed on the gels for each cell type [15, 27, 28]

pressed levels characteristic of dark-grown proplastid-containing cells [10]. The enzymes forming the plastid thylakoid sulfolipid are also present and the sulfolipid can be used as a tag to find the thylakoid membranes on gradients in extracts of W_3 and dark and light-grown wild-type cells; these fractions are relatively uncontaminated by other cell membranes using specific enzymes as markers [15, 27, 28]. Figure 9 shows that W_3 contains all of the plastid thylakoid polypeptides found in the proplastid of wild-type dark-grown cells indicating that all of the proplastid thylakoid polypeptides are nuclear-coded. These nuclear-coded constituents are localized in unusual proplastids in W_3 [16]. Figure 10 shows that dark-grown W_3 contains proplastids with a prominent body associated, light-grown W_3 shows a crystalline prolamellar-like structure near a large vacuole. It is quite possible that this body contains nuclear-coded constituents which cannot organize into normal thylakoids in the absence of the plastid-coded constituents.

Table 1. Selected properties of W_3BUL compared with dark- and light-grown wild-type *Euglena* cells

Parameter	W_3BUL	Wild-type	
		Dark-grown	Light-grown
Plastid DNA	Not detectable	Present	Present
Protochlorophyll(ide)	Not detectable	Present	–
Chlorophyll	Not detectable	Not detectable	Present
Carotenoids	Low	Low	High
Photosynthesis	Not detectable	Not detectable	Present
Various plastid enzymes thought to be nuclear-coded[a]	Low	Low	High
Various plastid enzymes thought to be plastid-coded, at least in part[b]	Not detectable	Not detectable or low	High
Nonplastid enzymes[c]	High	High	High
Plastid rRNA's, ribosomes and tRNA's	Not detectable	Low	High
Cytoplasmic rRNA's, ribosomes and tRNA's	High	High	High

[a] e.g., NADP triose phosphate dehydrogenase, alkaline DNAase, leucyl tRNA synthetase, enzymes forming sulfolipid, proplastid thylakoid polypeptides, etc.
[b] e.g., RuDP carboxylase, cytochrome 552, etc.
[c] e.g., NAD triose phosphate dehydrogenase, etc.

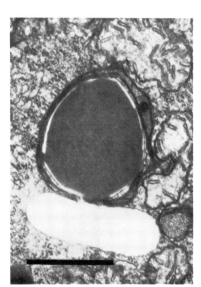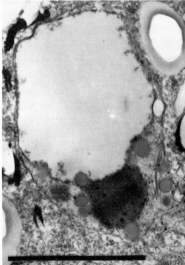

Fig. 10. *Left* section through dark-grown cell of W_3BUL showing proplastid remnant with dense body and associated mitochondrion. *Right* section through dark-grown cell of W_3BUL exposed to light for 48–72 h showing proplastid remnant with prolamellar body and large space [16]

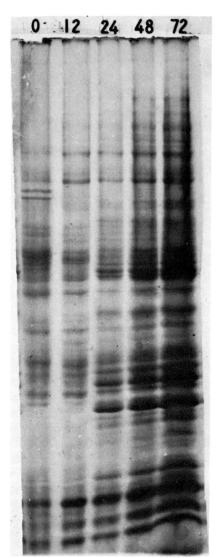

Fig. 11. Polypeptide patterns from plastid thylakoid (*middle*) fractions prepared from equal numbers of dark-grown resting cells exposed to light for various times, separated on SDS acrylamide gels [15, 28]

Perhaps prolamellar bodies in higher plants form when plastid DNA gene products are unavailable during growth in the dark. These situations are similar to the limitations brought about by antibiotics which prevent plastid ribosome formation and, hence, the formation of constituents coded in plastid DNA [17].

The behavior of plastid thylakoid polypeptides during light-induced greening is shown in Fig. 11 [15, 27, 28]. Some bands increase more than others; these include those bands which are thought to carry the photosynthetic pigments. Studies with antibiotic inhibitors of translation show that the formation of all bands is inhibited by 70S inhibitors like streptomycin and chloramphenicol and by 87S inhibitors such as cycloheximide. This is probably a reflection of the tight regulation of each membrane constituent on

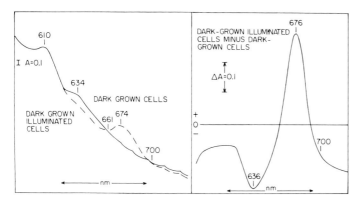

Fig. 12. Absorption spectra in vivo of dark-grown *Euglena* cells before and after illumination (red light, 125 W m^{-2} for 2 min) and the corresponding difference spectrum. The cells were scanned 16 times in each case to produce the absorption spectra shown [14]

the formation of the others. For example, inhibition of carotenoid synthesis blocks the formation of other membrane constituents in *Euglena* [18, 28]. Thus the inhibition by 70S and 87S inhibitors under these conditions does not give a true picture of where individual polypeptides are translated. Other workers have apparently found conditions where these controls may be circumvented to some extent and where 70S and 87S antibiotics will selectively block the synthesis of certain polypeptides [19].

   Another membrane constituent is the protochlorophyll(ide) system in *Euglena*. *Euglena* cells show only a protochlorophyll(ide) 635 which is transformable to chlorophyll(ide) 676; protochlorophyllide 650 appears to be absent (Fig. 12) [14, 21]. Protochlorophyll(ide) regenerates to the same level after each transformation, indicating that it is under tight control, perhaps through feedback inhibition (Fig. 13). The characteristics of the *Euglena* protochlorophyll(ide) system are quite different from those of the older angiosperm leaves which are usually studied, but are very similar to young leaves (Table 2) [20]. It appears that the younger material and *Euglena* exhibit the more normal mode of plastid development, while the older leaves have many characteristics which result from the superposed problems of prolonged etiolation. These lead to extensive

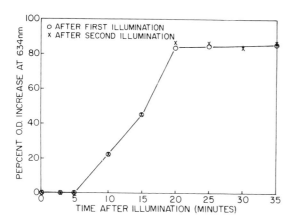

Fig. 13. The kinetics of the regeneration of protochlorophyll(ide) in dark-grown illuminated cells. The ratio: $O.D._{634}$ (t)/$O.D._{634}$ (t_0) $\times$ 100 (where t is the time indicated and t_0 is prior to illumination) is plotted vs time for first and second round of regeneration following each complete transformation of protochlorophyll(ide) [14]

DARK GROWN WILD TYPE ······
DARK GROWN WILD TYPE IMMEDIATELY AFTER ILLUMINATION —·—
DARK GROWN WILD TYPE 30 SECONDS AFTER ILLUMINATION ——
DARK GROWN WILD TYPE 120 SECONDS AFTER ILLUMINATION ———

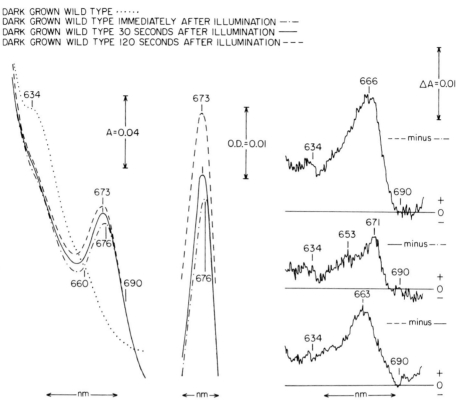

Fig. 14. Absorption spectra and difference spectra of dark-grown wild type cells of *Euglena* before and after illumination. Dark-grown wild-type cells were illuminated for 10 s with red light (1650 W m^{-2}). Each absorption spectrum is the sum of two repetitive scans. The central portion of the figure shows the chlorophyll(ide) maxima magnified for comparison. The spectra were adjusted to coincide as the isosbestic points for phototransformation (660 nm) and for spectra taken subsequently in darkness (690 nm) determined in separate experiments. The pertinent difference spectra computed from these absorption spectra are shown on the right [14]

growth of the etioplast and storage of precursors to enable rapid plastid development when development is induced by light in seedlings which are near the limits of their stored reserves.

In *Euglena* there is a small shift to shorter wavelengths of the chlorophyll(ide) band in the dark after phototransformation (Fig. 14) [14]. Putting together what is known of the in vivo forms of protochlorophyll(ide) and chlorophyll(ide) in old leaves, young leaves, and *Euglena*, the scheme of Fig. 15 results which shows their interrelationships. It might be noted that studies with all of these systems taken together indicate that there is no correlation of the 635 and 650 forms of protochlorophyll(ide) with protochlorophyll and protochlorophyllide contents. Thus we must seek explanations for these forms in their physical interrelationships when bound to sites in the plastid membranes.

The action spectra for the induction of nuclear-coded, cytoplasmically synthesized plastid proteins in *Euglena* indicates that they are under control by the plastid red-blue

Table 2. Comparison of properties of dark-grown *Euglena* 2−3-day-old and 7−9-day-old bean leaves [20]

Property	Dark-grown *Euglena*	2−3 day etiolated bean	7−9 day etiolated bean
Structure	Proplastid	Proplastid	Etioplast
Size (um)	1−2	2−3	4
Prolamellar body	Small, noncrystalline	Absent, or small crystalline	Large, crystalline
Predominant absorption in vivo	Pchl(ide) 635	Pchl(ide) 635	Pchl(ide) 650
Total protopigment (pg/plastid)	1×10^{-4}	0.5×10^{-4}	$7–10 \times 10^{-4}$
Protopigments present	Pchlide and pchl	Pchlide and pchl	Predominantly pchlide
Ratio:mol-ide/mol-yll	3	1	6
First stable photoproduct in vivo	Chl(ide) 676 shifting to chl(ide) 673	Chl(ide) 675	Chlide 685 Shibata shifting to 675 nm
Pigments produced directly on illumination	Chlide and chl	Chlide and chl	Chlide (predominantly)
% conversion pchl(ide) to chl(ide)	10%−50%	40%	80%−90%
Rate of chlide esterification	Fast (no lag)	Fast (no lag)	Slow (lag)
Rate of protopigment regeneration	Fast (short lag)	Fast (no lag)	Slow (lag)

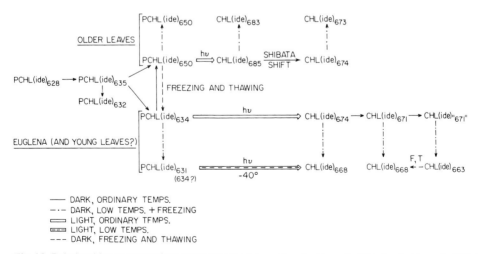

Fig. 15. Relationships among various spectroscopic forms of protochlorophyll(ide) and chlorophyll(ide) in various systems [14]

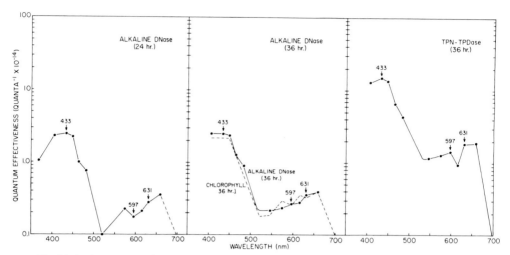

Fig. 16. Action spectrum for alkaline DNase formation at 24 h of continuous illumination; action spectrum for alkaline DNase formation at 36 h of continuous illumination, as compared to the action spectrum for Chl synthesis at 36 h of continuous illumination; and action spectrum for NADP-triose phosphate dehydrogenase formation at 36 h of continuous illumination [12]

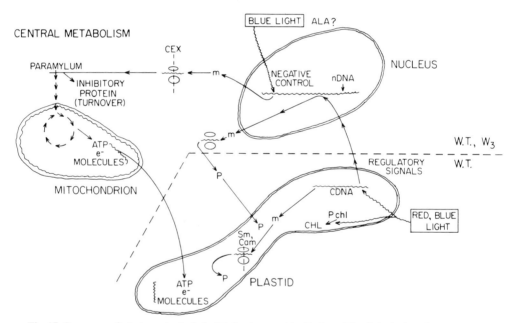

Fig. 17. Summary of photocontrol of plastid development in *Euglena*. Blue light is thought to control paramylum breakdown and nuclear transcirption, possibly through negative control. Blue and red light [a protochlorophyll(ide)-type of photoreceptor] is thought to control plastid transcription and chlorophyll(ide) formation. Since those plastid enzymes which are nuclear-coded show proto-chlorophyll(ide)-type action spectra for induction, it is suggested that illumination of the plastid receptor results in a regulatory signal which passes to the nucleus to derepress the appropriate loci for transcription of these proteins. For further details see text [26]

protochlorophyll(ide) receptor (Fig. 16) [12]. Since it is known that greening of *Euglena* takes place normally in red light alone, a regulatory signal must pass from the chloroplast to the rest of the cell on red light induction which overrides the nonplastid blue receptor to permit normal formation of the plastid constituents formed outside the developing plastid.

This and other features of light control of plastid development already mentioned are summarized in Fig. 17. The blue receptor controls the breakdown of paramylum and the transcription of the nuclear genomes in the nonplastid compartments of wild type and W_3BUL as already discussed. The protochlorophyll(ide) receptor acts to form chlorophyll and to derepress the plastid genome. On induction of the plastid by this system, a signal is sent to the rest of the cell to induce the formation of constituents which must flow to the developing plastid. Thus either blue light or a signal back from the plastid in red light serves to turn on the rest of the cell to provide the nonplastid contributions to chloroplast development.

Thus plastid development in *Euglena* and other organisms appears to involve at least two photoreceptor systems. One controls the nonplastid contributions to plastid development. In *Euglena* this is a blue light receptor system. In some higher plants this function has been assumed by the red-far red phytochrome system and perhaps in some cyanobacteria or blue green algae by other phytochrome-like pigments responding reversibly to red and green light [22, 23]. The second photoreceptor system controls the plastid functions during plastid development. In most organisms showing light control, this appears to be a red-blue system based on protochlorophyll(ide). In certain green algae (such as *Scenedesmus*) which make chlorophyll and chloroplasts in darkness, light-requiring mutants can be isolated. In this case, the mutation appears to have removed a dark route around the blue receptor but has left the dark conversion of protochlorophyll(ide) intact since these mutants show blue light action spectra for induction of plastid development [24].

By analogy to prokaryotic control systems we might expect control to be exerted at all levels from transcription through translation to substrate level control. In *Euglena* we have evidence for control at the genome level by the blue light receptor controlling the nonplastid functions [6, 11, 12, 13]. The red-blue receptor controlling plastid functions acts both at the genome level to derepress RNA synthesis [13, 14, 21] and at the substrate level in chlorophyll formation. It is quite possible that over the course of evolution repressor molecules evolved which recognize specific wavelengths of light rather than organic substrates [13]. These would be pigmented control molecules which repress transcription of various operons in DNA; the repression would be removed on illumination of these molecules with light to remove them from the inhibitory sites on the DNA much like lactose, for example, combines with a specific repressor to unblock the lac operon in *E. coli*. To be adaptive over the course of evolution, these pigmented control molecules would come to have absorption spectra which resemble the processes they control. For plastid development in green organisms, this means that since photosynthesis is most effective in red and blue light, the control systems absorb in blue and/ or red light, as observed. A control system absorbing in green light, for example, would not be adaptive since green light is not very effective for photosynthesis. The search for such pigmented control molecules seems to be an important next step in the investigation of chloroplast development.

References

1. Klein S, Schiff JA, Holowinsky A (1972) Events surrounding the early development of *Euglena* chloroplasts. 2. Normal development of fine structure and the consequences of preillumination. Dev Biol 28: 253–273
2. Schiff JA (1975) The control of chloroplast differentiation in *Euglena*. In: Avron M (ed) Proc 3rd Int Congr Photosynth Intracell Cycl, pp 1691–1717. Elsevier, Amsterdam
3. Schiff JA (1973) The development, inheritance and origin of the plastid in *Euglena*. In: Abercrombie M et al (eds) Advances in morphogenesis, vol X, pp 265–312. Academic Press, London New York
4. Schiff JA, Zeldin M, Rubman J (1967) Chlorophyll formation and photosynthetic competence in *Euglena* during light-induced chloroplast development in the presence of 3, (3, 4-dichlorophenyl) 1, 1-dimethyl urea (DCMU). Plant Physiol 42: 1716–1725
5. Fong F, Schiff JA (1977) Mitochondrial respiration and chloroplast development in *Euglena gracilis* var. *bacillaris*. Plant Physiol 59S: 92
6. Schwartzbach S, Schiff JA, Goldstein N (1975) Events surrounding the early development of *Euglena* chloroplasts. 5. Photocontrol of paramylum degradation. Plant Physiol 56: 313–317
7. Edelman M, Schiff JA, Epstein H (1965) Studies of chloroplast development in *Euglena*. 12. Two types of satellite DNA. J Mol Biol 11: 769–774
8. Rawson J, Haselkorn R (1973) Chloroplast ribosomal RNA genes in the chloroplast DNA of *Euglena gracilis*. J Mol Biol 77: 125–132
9. Schwartzbach S, Hecker L, Barnett W (1976) Transcriptional origin of *Euglena* chloroplast tRNAs. Proc Natl Acad Sci USA 73: 1984–1988
10. Bovarnick J, Schiff JA, Freedman Z, Egan J Jr (1974) Events surrounding the early development of *Euglena* chloroplasts. 4. Cellular origins of chloroplast enzymes in *Euglena*. J Gen Microbiol 83: 63–71
11. Holowinsky A, Schiff JA (1970) Events surrounding the early development of *Euglena* chloroplasts. 1. Induction by preillumination. Plant Physiol 45: 339–347
12. Egan J Jr, Dorsky D, Schiff JA (1975) Events surrounding the early development of Euglena chloroplasts. 6. Action spectra for the formation of chlorophyll, lag elimination in chlorophyll synthesis, and apperance of TPN-dependent triose phosphate dehydrogenase and alkaline DNase activities. Plant Physiol 56: 318–323
13. Cohen D, Schiff JA (1976) Events surrounding the early development of *Euglena* chloroplasts. 10. Photoregulation of the transcription of chloroplastic and cytoplasmic ribosomal RNAs. Arch Biochem Biophys 177: 201–216
14. Kindman L, Cohen C, Zeldin M, Ben-Shaul Y, Schiff JA (1978) Events surrounding the early development of *Euglena* chloroplasts. 12. Spectroscopic examination of the protochlorophyll(ide) phototransformation in intact cells. Photochem Photobiol 27: 787–794
15. Bingham S, Schiff JA (1976) Cellular origins of plastid membrane polypeptides in *Euglena*. In: Bücher Th et al (eds) Proc Conf Genet Biogen Chloroplasts Mitochondira, pp 79–86. Elsevier/North Holland, Amsterdam
16. Osafune T, Schiff JA (1979) Light-induced changes in a proplastid remnant in dark grown resting *Euglena gracilis* var. *bacillaris* W_3BUL. Plant Physiol 63S: 27
17. Galling G (1975) RNA synthesis in synchronous cultures of unicellular algae. In: Les cycles cellulaires et leur blocage chez plusieurs protistes, Gif-sur Yvette, pp 225–231. CNRS, Paris
18. Vaisberg A, Schiff JA (1976) Events surrounding the early development of *Euglena* chloroplasts. 7. Inhibition of carotenoid biosynthesis by the herbicide SAN 9789 and its developmental consequences. Plant Physiol 57: 260–269
19. Gurevitz M, Kratz H, Ohad I (1977) Polypeptides of chloroplastic and cytoplastic origin required for development of photosystem II activity, and chlorophyll-protein complexes, in *Euglena gracilis* Z chloroplast membranes. Biochim Biophys Acta 461: 475–488
20. Lancer H, Cohen C, Schiff JA (1976) Changing ratios of phototransformable protochlorophyll and protochlorophyllide of bean seedlings developing in the dark. Plant Physiol 57: 430–436
21. Cohen C, Schiff JA (1976) Events surrounding the early development of *Euglena* chloroplasts. 11. Protochlorophyll(ide) and its photoconversion. Photochem Photobiol 24: 555–567

22. Lazaroff N, Schiff JA (1962) Action spectrum for developmental photo-induction of the blue-green alga *Nostoc muscorum.* Science 137: 603–604
23. Haury J, Bogorad L (1977) Action spectra for phycobiliprotein synthesis in a chromatically adapting cyanophyte *Fremyella diplosiphon.* Plant Physiol 60: 835–839
24. Brinkman G, Senger H (1978) The development of structure and function in chloroplasts of green-ing mutants of Scenedesmus IV. Blue light-dependent carbohydrate and protein metabolism. Plant Cell Physiol 19: 1427–1437
25. Schmidt G, Lyman H (1974) Photocontrol of chloroplast enzyme synthesis in mutant and wild-type *Euglena gracilis.* In: Avron M (ed) Proc 3rd Int Cong Photosynth, pp 1755–1764. Elsevier, Amsterdam
   Lyman H, Srinivas U (1978) Regulation of chloroplast DNA synthesis: Possible role of chloro-plast nucleases in *Euglena.* In: Akoyunoglou G et al. (eds) Chloroplast development, p 593. Elsevier, Spetsai
26. Schiff JA (1978) Photocontrol of chloroplast development in *Euglena.* In: Akoyunoglou G et al (eds) Chloroplast development, pp 747–767. Elsevier/North Holland, Amsterdam
27. Bingham S, Schiff JA (1979) Events surrounding the early development of *Euglena* chloroplasts. 15. Origin of plastid thylakoid polypeptides in wild type and mutant cells. Biochim Biophys Acta 547: 512–530
28. Bingham S, Schiff JA (1979) Events surrounding the early development of *Euglena* chloroplasts. 16. Plastid thylakoid polypeptides during greening. Biochim Biophys Acta 547: 531–543
29. See other papers (this vol)
30. Zeldin M, Schiff JA (1975) Absorption changes in extracts from *Euglena gracilis* var. *bacillaris* mutant W₃BUL on blue illumination. Plant Physiol 56, Suppl: 33
31. Lewis SC, Epstein HT, Schiff JA (1961) Photooxidation of cytochromes by a flavoprotein from *Euglena.* Biochem Biophys Res Commun 5: 221–225
32. Fong F, Schiff JA (1979) Blue-light absorbance changes associated with isomerization of ζ carotene in *Euglena.* Planta 146: 119–127
33. Steinitz Y, Schiff JA, Osafune T, Green M (1980) Cis to trans photoisomerization of ζ carotene in *Euglena gracilis* var. *bacillaris* W₃BUL: Further purification and characterization of the photoactiv-ity, this volume
34. Zeldin M, Schiff JA (1976) Blue light induced absorption changes in intact cells of *Euglena gra-cilis* var. *bacillaris* mutant W₃BUL. Plant Physiol 57, Suppl: 22

Effects of Blue Light on Greening in Microalgae

E. HASE[1]

1 Introduction

Chloroplast development in microalgae has been studied mainly by using *Euglena*, mutants of *Chlamydomonas, Scenedesmus,* and *Chlorella,* and other algae which require light for greening: the majority of algae, however, form chlorophyll and chloroplasts in the dark as well as in the light. Recent studies have shown that many aspects of chloroplast development in these algae are apparently dependent on light, but their wavelength dependence has been studied only in a limited number of cases, making it difficult to obtain a comprehensive picture of the possible blue light control of greening in algae. In this article attention will be focused on the formation of 5-aminolevulinic acid (ALA) in relation to chlorophyll biosynthesis. An attempt will be made to interpret the sometimes perplexing observations so far reported by proposing a tentative model for greening which emphasizes the importance of transport phenomena associated with the passage of substrates and energy through the plastid envelope.

2 Contribution of Photosynthesis to Greening

In an early phase of greening, in which the plastid is not yet photosynthetically competent, greening is solely dependent on nonphotosynthetic light reactions which are probably less significant after the appearance of photosynthetic activity. To exclude the photosynthetic contribution, use has been made of 3(3,4-dichlorophenyl)-1,1-dimethyl-urea (DCMU) or 3(4-chlorophenyl)-1,1-dimethyl-urea (CMU) (Schiff et al. 1967; Sokawa and Hase 1967; Richard and Nigon 1973) which are inhibitors of photosynthesis; it must be considered, however, that cyclic phosphorylation can take place in the presence of these inhibitors. Studies of the changes in the rate of ALA formation and chlorophyll synthesis in the presence (Richard and Nigon 1973) and absence (Richard and Nigon 1973; Porra and Grimme 1974) of DCMU during the greening of dark-grown *Euglena gracilis* or nitrogen-starved *Chlorella fusca* clearly demonstrated that ALA formation is the rate-limiting step of chlorophyll biosynthesis throughout the greening process. In addition, Richard and Nigon (1973) showed that during the early phase of greening both ALA and chlorophyll synthesis are entirely dependent on DCMU-insensitive light reactions, while in later phases the contribution of the DCMU-sensitive light process

1 The Institute of Applied Microbiology, University of Tokyo, Bunkyo-ku, Tokyo 113, Japan
Present address: Chemistry Laboratory, Faculty of Medicine, Teikyo University, Otsuka, Hachioji, Tokyo 192-03, Japan

(photosynthesis) is dominant. The extent to which greening depends on photosynthesis is different depending on various factors, especially nutritional and light conditions. As pointed out by Nigon et al. (1978), chlorophyll synthesis at low light intensity may depend on the DCMU-insensitive light reactions. Further, when glucose-bleached cells of *Chlorella prototothecoides* are greened in a medium containing an ammonium salt as nitrogen source, a later phase of greening is severely suppressed by CMU (Oh-hama and Hase 1978), but such suppression is significantly reduced when greened in a medium containing glycine (Oh-hama and Hase 1976) or urea (Sokawa and Hase 1967). That photosynthesis plays a dominent role in later phases of greening is suggested also by the finding (Porra and Grimme 1974) that greening in the light of nitrogen-starved *Chlorella fusca* is approximately threefold higher in air supplemented with 1.5% CO_2 than in N_2, or O_2 (20% in N_2). Obviously such nutritional effects must be considered when studying the effects of light quality on greening.

3 Blue Light-Enhanced Respiration and Carbon Metabolism in Relation to Greening

It is known that when dark-grown algal cells are illuminated to induce greening, marked enhancement of respiration concomitant with degradation of polysaccharide occurs supplying raw materials and energy for greening (Senger and Bishop 1972; Schiff 1974; Brinkmann and Senger 1978). The relationship of the wavelength dependence of light-enhanced respiration and of chlorophyll biosynthesis in *Scenedesmus obliquus* mutant C-2A' was studied by Senger and Bishop (1972) and Brinkmann and Senger (1978).They showed that the peaks for light-enhanced respiration were around 380–390, 450, and 460–475 nm with minima at 480 and 475 nm and a similar wavelength dependence has been reported for carbohydrate utilization (Brinkmann and Senger 1978). Studying the effect of cycloheximide on light-enhanced respiration and greening in *Scenedesmus* mutant cells, Oh-hama and Senger (1975) found that light-enhanced respiration is composed of two stages, an initial cycloheximide-insensitive stage, where low concentrations of cycloheximide could actually cause slight stimulation, and a subsequent sensitive stage where cycloheximide inhibited. The effects of cycloheximide at different concentrations seemed to be closely correlated with chlorophyll and ALA formation, thus providing further evidence for the dependency of greening on light-enhanced respiration. Recent studies by Watanabe, Oh-hama and Miyachi (this vol.) showed that the later but not initial stage of light-enhanced respiration in the *Scenedesmus* mutant is also inhibited by 6-methylpurine, an analog of adenine known to block RNA synthesis (cf. McCullough and John 1972). These results suggest that the blue light enhancement of respiration in early greening may be controlled at the transcriptional level; however, further experiments using other transcription inhibitors are needed. They also observed that light enhances CO_2 fixation in the presence of CMU, giving such early products as aspartate and glutamate, which are later incorporated into nucleic acid and protein. These results are consistent with the blue light enhancement of phosphoenolpyruvate carboxylase, which has been extensively studied by Miyachi and colleagues (see present proceedings). When dark-grown cells of *Chlamydomonas reinhardi* y-1 mutant are illuminated, starch degradation is linear with time for up to 2 h (Ohad, pers. comm.; Paglin 1972). Blue light between 400 and 540 nm is two- to threefold more effective than red light above 600 nm and the degradation of starch is not blocked by chloramphenicol or cycloheximide during the first 2 h. Schiff (1974) measured the action spectrum for the preillumination of dark-grown cells of *Euglena gracilis* which seems to "potentiate" or induce early events of plastid development including light-enhanced respiration concomitant with paramylum degradation. The action spectrum shows a major peak at 433 nm with minor peaks at 597 and 631 nm which probably represent the phototransformation of pre-existing protochlorophyllide. Since the ratio of blue to red peaks approaches 40:1, he concludes that the photoreceptor system for the early events of plastid development is a blue light-absorbing pigment system which is, as yet, unidentified.

4 Effects of Light on ALA Formation in Relation to Chlorophyll Synthesis

Since Beale (1970, 1971) demonstrated the formation of ALA in *Chlorella* by using levulinic acid (LA), a competitive inhibitor of ALA dehydratase (Nandi and Shemin 1968), extensive studies have been done on ALA formation in various plant materials using this inhibitor. It has been shown that ALA formation is dependent on light in algae which require light for chlorophyll synthesis (Beale 1971; Richard and Nigon 1973; Oh-hama and Hase 1975, 1978; Oh-hama and Senger 1975, 1978; Nigon et al. 1978). However, information about the wavelength dependence of ALA formation is very limited. Figures 1 and 2 show the wavelength dependence of ALA formation and chlorophyll synthesis during the greening of glucose-bleached cells of *Chlorella protothecoides* (Oh-hama and Senger 1978) and show that both syntheses are entirely blue and near-UV light-dependent, with peaks at 369, 443, and 478 nm. The wavelength dependence of chlorophyll synthesis in this alga was further confirmed by Oh-hama and Hase (1978) who obtained an action spectrum showing the most effective wavelength to be at 444 nm, at which 0.2 W m^{-2} was sufficient for half saturation. These results indicate that

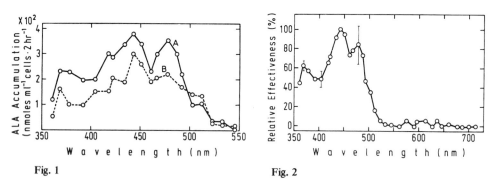

Fig. 1   Fig. 2

Fig. 1. Wavelength dependence of ALA formation in *Chlorella protohecoides* cells partially greened under monochromatic light (Oh-hama and Senger 1978). Cells were irradiated for 20 (*A*) or 28 (*B*) h, then incubated with LA for 2 h in the dark

Fig. 2. Wavelength dependence of chlorophyll synthesis in regreening *Chlorella protothecoides* cells (Oh-hama and Senger 1978). Algal suspension, previously incubated in the dark for about 20 h with greening medium, was irradiated for 30 h with monochromatic light (5 $\times$ 10^{-11} einstein·cm^{-2} s^{-1}) in the presence of DCMU. The effectiveness at each wavelength was expressed as the percent of the maximum value at 443 nm

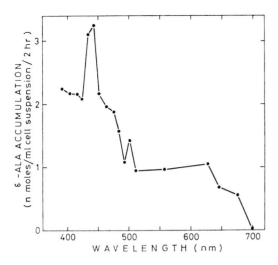

Fig. 3. Wavelength dependence of ALA formation in *Scenedesmus obliquus* mutant C-2A' (Oh-hama and Senger 1975). Aliquots of cells previously greened for 4 h were illuminated for 2 h with monochromatic light (2.5×10^{-10} einstein $\cdot$ cm^{-2} s^{-1}) in the presence of LA

blue light-dependent ALA formation regulates chlorophyll synthesis in this greening alga. It is noteworthy that *Chlorella protothecoides* is different from most other algae in that it can form ALA and chlorophyll in darkness to some extent without any detectable accumulation of protochlorophyll(ide), though the formation is greatly stimulated by light. Porra and Grimme (1974) also found that considerable chlorophyll synthesis occurs in the regreening of nitrogen-starved *Chlorella fusca* in the dark but only in the presence of O_2 (20% in N_2); chlorophyll synthesis will occur also in the light in the presence or absence of oxygen but synthesis in the light with O_2 is greatly stimulated by the presence of CO_2. Oh-hama and Hase (1978) also observed that ALA formation was stimulated in algal cells which had been greened for 20 h under strong white light (10 W m^{-2}) when incubated with LA for 1 h under blue, red, and white light in the presence or absence of CMU: red light (600–680 nm, 6 W m^{-2}) as well as blue light (444 nm, 0.2 and 5.5 W m^{-2}) was effective even in the presence of CMU (see Discussion).

The wavelength dependence of ALA formation in *Scenedesmus obliquus* mutant C-2A' was observed by Oh-hama and Senger (1975) using cells which had been greened for some time in strong white light and then illuminated with monochromatic light in the presence of LA but in the absence of a photosynthesis inhibitor. Their results (Fig. 3) show that the light dependence of ALA formation has a main peak between 434 and 444 nm and a distinct shoulder at about 470 nm. However, all wavelengths tested except 700 nm were effective. This, however, contrasts with the blue light dependence of chlorophyll synthesis in the mutant cells observed by Senger and Bishop (1972) with peaks at 390, 455, and 465 nm and a shoulder at 480 nm as shown in Fig. 4 which indicates that the wavelength dependence of chlorophyll synthesis is similar to that of light-enhanced respiration. As pointed out by Oh-hama and Senger (1975), the wavelength dependence of chlorophyll synthesis (Fig. 4) represents that of the synthesis in cells illuminated with monochromatic light from the beginning of greening, while light dependence for ALA formation (Fig. 3) was obtained with cells which had been preilluminated with strong white light. In fact, a recent measurement of wavelength dependence of chlorophyll biosynthesis by Brinkmann and Senger (1978) showed that in the *Scenedes-*

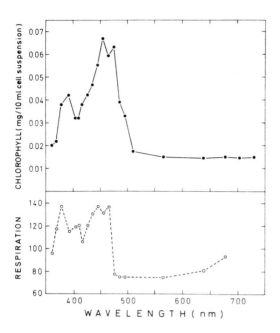

Fig. 4. Wavelength dependence of chlorophyll formation in *Scenedesmus obliquus* mutant C-2A' (Senger and Bishop 1972). Aliquots of cells were irradiated for 8 h with monochromatic light (2.5×10^{-10} einstein $\cdot$ cm^{-2} s^{-1}). See text for explanation

mus mutant cells, preilluminated with strong white light, peaks of effectiveness appeared not only in the blue region (383, 442, and 471 nm) but also in the red region (621 nm) as shown in Fig. 5. This result is in agreement with the action spectrum for chlorophyll synthesis obtained by Ohad and Drews (1974) with preilluminated cells of *Chlamydomonas reinhardi* y-1 mutant, showing peaks at 452 and 632 nm. The action spectrum for greening in dark-grown and preilluminated *Euglena gracilis* obtained by Schiff (1974) showed blue and red peaks with a ratio of 5:1 respectively, which is different from that (40:1) in the action spectrum for preillumination or "potentiation" which was mentioned earlier (Sect. 3). He states that the blue-red photoreceptor system controlling later development in the greening of *Euglena* appears to have the absorption of protochlorophyll(ide).

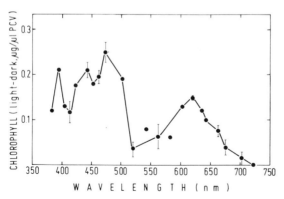

Fig. 5. Wavelength dependence of chlorophyll formation in *Scenedesmus* mutant cells after preillumination (Brinkmann and Senger 1978b). Preillumination was carried out under white light of 20,000 lx for 2 h

To understand the effects of light on ALA and chlorophyll synthesis, one must consider, in addition to the contribution of photosynthesis (see Sect. 2), the following four possible processes: (1) production of necessary substrates, energy, and reducing power, (2) transport of these materials to the site of biosynthesis, (3) development of the necessary biosynthetic enzyme system, and (4) an immediate effect of light on the synthetic reaction(s).

As mentioned earlier in Sect. 3, blue light-enhanced respiration and carbon metabolism occurring during an early phase of greening provide substrates and energy for greening. Therefore, when the wavelength dependence of ALA or chlorophyll synthesis is measured under conditions in which blue light-enhanced respiration and carbon metabolism limit the over-all rate of ALA or chlorophyll synthesis, the results will show complete dependence on blue light. When, on the other hand, the experimental conditions are selected so as to make other processes limit the rate of ALA or chlorophyll synthesis, the result will represent the wavelength dependence of the limiting process selected. It is understandable that the wavelength dependence of chlorophyll formation in *Scenedesmus* mutant cells (Fig. 4) is similar to that of light-enhanced respiration, while the wavelength dependence of ALA and chlorophyll synthesis in preilluminated cells (Figs. 3 and 5) is different from the above result and includes red light, thus suggesting that protochlorophyll(ide) may be a photoreceptor.

It has been demonstrated in greening oat seedlings that marked changes in the permeability of the plastid envelope and mitochondrial membrane occur during greening (Hampp and Schmidt 1976; Hampp and Wellburn 1976; Wellburn and Hampp 1976). They suggested that adaptation of the plastid-envelope permeability to the changing requirements of externally synthesized precursors and intermediates may occur during development. It must be said that the mode of change of permeability of the plastid envelope is different, depending on individual precursors or intermediates. It seems highly likely that the transport of some precursor(s) and/or of energy through the plastid envelope could be the rate-limiting process of ALA and chlorophyll formation in algae as well as in higher plants. According to Heber and Santarius (1965), it is unlikely that reducing power in the form of NAD(P)H is transported through the plastid envelope. Consequently, there must be a mechanism for producing NAD(P)H within the plastid before as well as after it acquires photosynthetic competence. This mechanism may involve glucose-6-phosphate (G-6-P) dehydrogenase which has been shown to exist not only in the cytosol but in the plastid (Anderson et al. 1974; Griffiths 1975) where it is associated with the etioplast oxidative pentose phosphate pathway which may provide the NADPH required for chlorophyll biosynthesis during the early stages of greening (Griffiths 1975). NADPH is required for the photoreduction of protochlorophyllide (Griffiths and Mapleston 1978) and a requirement for NADPH has also been reported for the synthesis of ALA via the C-5 pathway (Harel 1978; Kannangara et al. 1978). However, for G-6-P dehydrogenase to provide NADPH within the plastid, the substrate, G-6-P, must be supplied from the cytoplasm where polysaccharide is degraded in response to blue light. Consequently, the transport of G-6-P across the envelope of the developing plastid may become the rate limiting process for ALA and chlorophyll formation. Anderson et al. (1974) and Anderson and Avron (1976) demonstrated that G-6-P dehydrogenase undergoes very rapid and reversible inactivation and reactivation by illumination and darkness, respectively. Interestingly, this enzyme is active in light when photosynthetic electron transport is blocked by DCMU. This would imply that in the greening of algal cells incubated with DCMU, the plastid enzyme can provide NADPH if the substrate, G-6-P, can reach it. When polysaccharide is contained in the plastids of dark-grown cells, as in *Scenedesmus obliquus* C-2A' mutant (Senger et al. 1974), *Chlamydomonas reinhardi* y-1 mutant (Ohad et al. 1967), and other algae, G-6-P will be supplied within the plastid through degradation of the polysaccharide. In these cases, however, the degradation product(s) will be transported from the plastid to the cytoplasm and mitochondria, where it is metabolized, and the metabolites and energy produced will come back into the plastid to be utilized for its development. It should be cited here that Kowallik and Kirst (1975) inferred from their studies of glucose respiration and uptake of 3-0-methylglucose by a yellow mutant of *Chlorella vulgaris* that blue light may cause a change in permeability of the chloroplast envelope and facilitate the transport of glucose. He also obtained an action spectrum for protein synthesis from added glucose by *Chlorella pyrenoidosa* in the presence of DCMU (Kowallik 1966) which showed a broad peak around 450–500 nm. Not only small molecular precursors and intermediates from the cytosol, but also those proteins made on the cytoplasmic ribosomes which are destined to function in the plastid must enter the plastide through its envelope. This problem of protein transport will not be discussed here other than to refer to the hypothesis of Highfield and Ellis (1978) which suggests that a class of proteins with a "porter" function exists in the chloroplast envelope which can recognize all those proteins of cytoplasmic origin destined to function in the chloroplasts.

Recent studies have suggested that the C-5 pathway may dominate ALA formation in most algae, although the succinyl-CoA glycine pathway contributes to a considerable extent in ALA formation in various algae (Porra and Grimme 1974; Harel 1978; Klein et al. 1978). Also, Harel (1978) suggested that the two pathways are located in different compartments and that their relative contribution is dependent on growth conditions. Klein et al. (1978) showed that light-dependent chlorophyll synthesis in *Scenedesmus* mutant cells occurs via the two pathways, and that the time course of the development of ALA-synthesizing activity was similar to that of the activity of L-alanine-4,5-dioxovalerate aminotransferase. Meller and Harel (1978) studied the contribution of various precursors to the labeling of ALA in *Chlorella vulgaris* (strain 211/βk, Cambridge collection) cells during regreening after bleaching by nitrogen starvation, and found that the C-5 pathway dominates. Salvador (1978) found that chloroplasts isolated from *Euglena gracilis* are able to produce ^{14}C-labeled ALA from ^{14}C-labeled 2-oxo-glutarate or glutamate and that the enzyme catalyzing the conversion of 4,5-dioxovalerate to ALA was present in extracts of whole cells and of isolated chloroplasts. He also demonstrated that the activity of the transaminase in cell-free extracts prepared at different times during the greening of dark-grown *Euglena* cells correlated closely with their rate of chlorophyll synthesis: no detectable activity was observed in dark-grown cells. This suggests that the development of the ALA-synthesizing system can be the rate-limiting process in light-induced greening. He concluded that *Euglena* chloroplasts contain an ALA-synthesizing system using the C-5 pathway. It still remains to be seen whether or not the enzyme system is formed outside the plastid and transported into it through its envelope as is ALA synthetase into mammalian liver mitochondria (cf. Granick and Beale 1978). Information about the light requirement and wavelength dependence of the enzymes of the ALA-forming system is needed and, hopefully, will soon be available.

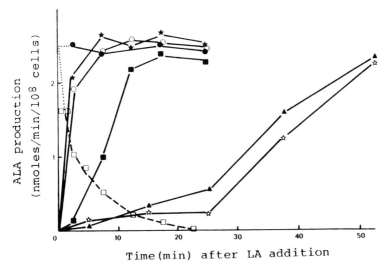

Time(min) after LA addition

Fig. 6. Effects of darkness on the ALA formation in greening *Euglena gracilis* cells in the presence of DCMU (Richard and Nigon 1973). After 8 h greening in the light, dark-grown cells were given DCMU, and 30 min later, the cell suspension was divided into seven aliquots. LA was immediately added to two aliquots, and one of them was immediately replaced in the light (●) and the other transferred to the dark (□). The other five aliquots were incubated in darkness for different periods, after which they were re-illuminated in the presence of LA. The period of dark incubation was 5 min (★), 10 min (○), 30 min (■), 1 h (▲) and 4 h (☆). The *abscissa* represents the time (min) after the addition of LA, and the ordinate the ALA production in nmol/10^8 cells/min

5 Effects of Darkness and Protein Synthesis Inhibitors on ALA Formation in Greening Algal Cells

5.1 "Short-Term", Reversible Effects and "Longer-Term", Irreversible Effects of Darkness

It has been shown that when algal cells accumulating ALA in the light in the presence of LA are transferred to darkness, ALA production ceases very rapidly (Beale 1971; Richard and Nigon 1973; Oh-hama and Senger 1975): this effect of darkness will be called the short-term effect. On the other hand, when greening cells in the light were transferred to darkness and their ALA-forming activity was measured at intervals under re-illumination in the presence of LA, the ALA-forming activity declined more slowly (Richard and Nigon 1973; Oh-hama and Hase 1975, 1978; Oh-hama and Senger 1975), as shown in Fig. 6, 7, and 8: this effect will be called the longer-term effect. As seen from Fig. 6, when *Euglena* cells were re-illuminated after a relatively short period of darkness, their ALA-forming activity was rapidly restored to the pre-darkness level, indicating that the short-term effect of darkness is reversible. In contrast, when the period of darkness became longer the rate of the recovery of the ALA-forming activity went down and it took longer before restoration to the initial level occurred. It must be said that all these experiments were done in the presence of DCMU: as shown later (Fig. 10), the recovery of the ALA-forming activity in the light after long periods in the dark requires protein synthesis which is inhibited by lincomycin. Thus the effect of long

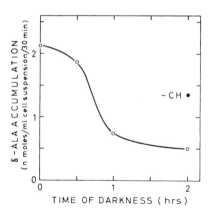

Fig. 7. Effect of darkness on ALA formation in greening cells of *Scenedesmus obliquus* mutant C-2A' (Oh-hama and Senger 1975). After 5.5 h of greening, cells were transferred to darkness. At intervals cells were re-exposed to light for 30 min with addition of LA and cycloheximide. *−CH* shows the control without addition of cycloheximide

dark periods, the longer-term effect, is irreversible in this sense. Salvador (1978) found that the enzyme catalyzing the conversion of 4,5-dioxovalerate to ALA in extracts of greening *Euglena* cells or chloroplasts was stimulated 30% by light. But this is insufficient to account for the much larger effect of darkness. It is noteworthy in this case that the addition of NADPH and $MgCl_2$ to the extract resulted in twofold higher production of ALA. Figure 8 (right) clearly shows that the addition of glycine greatly diminishes the decay of ALA-forming activity during the incubation of greening cells of *Chlorella protothecoides* in the dark. There have been indications that the supply of precursors, energy and/or reducing power limits the rate of ALA or chlorophyll synthesis in early stages of greening (Oh-hama and Hase 1975, 1978).

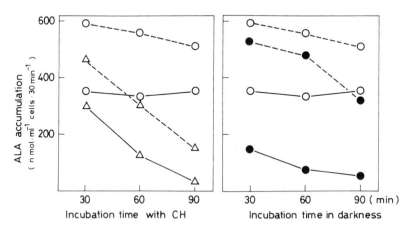

Fig. 8. Effects of feeding glycine on the levels of ALA-synthesizing activity in greening *Chlorella protothecoides* cells after administration of cycloheximide (CH) or exposure to darkness (Oh-hama and Hase 1978). Algal cells, previously greened for 20 h, were fed with glycine (10 mM). After 1 h, CH was added to the cell suspension (see *left figure*) or the suspension was transferred to darkness (see *right figure*). At intervals portions of the suspensions were removed to measure ALA-synthesizing activity. Controls without glycine (——); experiments with glycine (– – –); ○ light; △ light with CH; ● in darkness

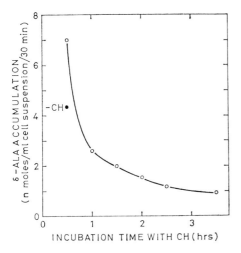

Fig. 9. Time course of decay of ALA-synthesizing activity in greening cells of *Scenedesmus obliquus* mutant C-2A' after addition of cycloheximide (Oh-hama and Senger 1975). After 5 h of greening, cycloheximide (0.5 μg/ml) was administered to the cells. At intervals, LA was added to an aliquot of cell suspension to measure the rate of ALA synthesis. The *closed circle* shows the control without addition of antibiotic

5.2 Effects of Protein Synthesis Inhibitors

It has been demonstrated that when greening algae are given cycloheximide or chloramphenicol (or lincomycin) in the light, the ALA-forming activity declines as in the "longer-term" effect of darkness, the half-life being 30 to 90 min (Beale 1971; Richard and Nigon 1973; Oh-hama and Hase 1975, 1978; Oh-hama and Senger 1975). Figures 8 (left), 9, and 10 summarize some of these observations. From the effects of cycloheximide and chloramphenicol on light-induced chlorophyll synthesis in etiolated barley, Nadler and Granick (1970) concluded that chlorophyll synthesis appears to be limited by a protein(s) related to the synthesis of ALA, with a half life of about 90 min. They inferred that cycloheximide may inhibit the synthesis of the enzyme in the cytoplasm and that chloramphenicol may inhibit the synthesis of some plastid protein(s) required for transport or localization of the enzyme in the plastid. As seen in Fig. 10, the ALA-forming activity in greening *Euglena* cells in the presence and absence of DCMU declined rapidly in darkness, and when illuminated after 1 h in the dark the activity was immediately restored to its original level in the absence of DCMU, but, in the presence of DCMU, more than

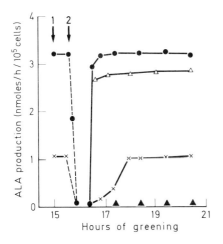

Fig. 10. ALA production rate after transient period of darkness in greening cells of *Euglena gracilis* (Nigon et al. 1978). After 15 h greening in the light, the cell suspension was divided into four aliquots which received (at time indicated by *arrow 1*) different additions. After 30 min (*arrow 2*), these aliquots were transferred to darkness for 1 h and then reilluminated. *Dotted lines* represent the decay of ALA production rate in darkness. ●—● no addition; △—△ lincomycin 5 mg/ml; x—x DCMU 10^{-5} M; ▲—▲ lincomycin + DCMU. *Abscissa* time of greening in h. *Ordinate* ALA production rate in nmol/h/10^5 cells

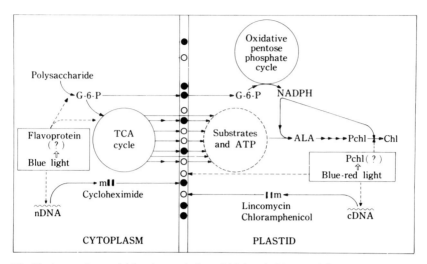

Fig. 11. A tentative model for the regulation of ALA and chlorophyll formation in an early phase
of greening with emphasis on the transfer of precursors through the plastid envelope. *m* messenger
RNA. The *small open* and *closed circles* represent the "protein factors" responsible for the transport
of precursors, the *open ones* being synthesized in the plastid and the *closed ones* in the cytoplasm.
The *broken line* drawn from the square encompassing "blue-red light (Pchl)" to the *open circle* represents the rapid photocontrol of the activity of the envelope-protein factor. See text for further
explanation

1 h was required to regain the original level of activity. The light-induced restoration of
activity in the presence of DCMU was completely inhibited by lincomycin which had
only a small inhibitory effect in the absence of DCMU. The immediate restoration of
ALA-forming activity in the absence of DCMU indicates that this enzyme system was
active and not decaying during the 1 h in the dark, and, upon re-illumination, photosynthetic production of substrate, energy and/or NAD(P)H within the plastid allowed
it to synthesize ALA immediately. In the presence of DCMU, on the other hand, non-
photosynthetic light reactions were required for ALA synthesis. Nigon et al. (1978)
proposed that light is required for the transport into the plastid of substrates and energy
produced in the cytoplasm. They reported that the DCMU-insensitive production of
ALA occurs at light intensities lower than 100 lx. It is further inferred from the inhibitory effect of lincomycin that a protein(s), which is (are) synthesized in the plastid in
response to light (and turning over with a short life time) is involved in the transport
of substrates and energy. The effect of cycloheximide (Figs. 8 and 9) implies that some
other protein(s) made in the cytosol is (are) also involved in the transport system. Figure 9 shows that reillumination of *Scenedesmus* cells incubated in the dark in the presence of LA possessed more ALA-forming activity in the absence than in the presence
of cycloheximide. This indicates that the light-induced development of ALA-forming
activity is inhibited by cycloheximide.
     Based on the above discussion, a tentative model is proposed in Fig. 11 for ALA and
chlorophyll formation in the early phase of greening, with emphasis on transfer of precursors and intermediates through the plastid envelope. This extension of the idea proposed by Nigon et al. (1978) is only one of many possible hypotheses which might be

proposed to explain the experimental observations. Recent unpublished studies by Oh-hama and Hase showed that when cells of *Chlorella protothecoides,* greened for 19 h in light, were transferred to darkness and immediately given glucose together with LA, ALA formation continued as actively for 1 h as in the light in the absence of glucose. When, on the other hand, a similar experiment was done with algal cells which had been incubated in the dark for 2 h after 16 h greening in the light, the addition of glucose did not stimulate the low rate of ALA production observed in the dark in the absence of glucose. In the terms of the model (Fig. 11), these results may be interpreted to mean that glucose added in the dark, but immediately after greening in the light, could more easily enter the plastid than glucose added after 2 h of darkness because of the dark decay of relatively shortlived protein(s) responsible for the transport of glucose (or G-6-P) through the plastid envelope.

To interpret the "short-term", reversible effect of darkness (Sect. 5.1) in terms of the model, it may be assumed that the protein factor synthesized in the plastid in response to the blue-red light is rapidly inactivated in the dark but reactivated on reillumination, as in NADP-linked glyceraldehyde-3-phosphate dehydrogenase involved in photosynthetic carbon metabolism (Anderson et al. 1974).

The similarity of blue-red light dependence of both ALA and chlorophyll formation (Sect. 4) suggests that a common photoreceptor, probably protochlorophyllide, which absorbs blue-red light, may be responsible not only for the transformation of protochlorophyllide to chlorophyllide but also for the rapid photocontrol of the activity of the envelope-protein factor. Control of chlorophyll formation by the feedback inhibition of ALA synthetase by protochlorophyll(ide) has been suggested by the experiments of Ramaswamy and Nair (1973) but, in addition, Granick and Beale (1978) have suggested that protochlorophyll(ide) may also be responsible for a feedback repression of ALA synthetase; in this latter respect, it is interesting to note that in heme biosynthesis in liver mitochondria, heme not only regulates the de novo synthesis of ALA synthetase in the cytosol but also controls its transport through the mitochondrial membrane into the mitochondrial matrix (Hayashi et al. 1972). However, such a feedback control of ALA synthesizing enzyme system or envelope transport protein factor by protochlorophyll(ide) or similar precursor(s) seems unlikely in the short-term blue-red light effects, since the rapid cessation of ALA production in darkness was observed in the presence of LA which blocks further metabolism of ALA.

Since ALA and chlorophyll formation in the dark by greening cells of *Chlorella protothecoides* is stimulated by relatively strong, nonphotosynthetic blue-red light (Sect. 4), it may be assumed that this alga contains two kinds of relevant protein factor: one independent of light for its activity and the other dependent on blue-red light. *Scenedesmus obliquus* (wild-type) is also able to form ALA and chlorophyll in the dark. Oh-hama, in unpublished work, found that ALA formation by this alga is significantly enhanced by light in the presence of CMU: blue light is more effective than red light as is also the case in *Chlorella protothecoides.* This suggests that *Scenedesmus* may also contain the two different kinds of the protein factor, while in the mutant cells (C-2A') the light-independent protein factor is defective. According to Griffiths and Mapleston (1978), both light-dependent and -independent enzymes must be present for the transformation of protochlorophyllide to chlorophyllide in the wild strain of *Chlamydomonas reinhardi,* while in its y-1 mutant only the light-dependent enzyme exists. These observations provide further evidence of the tight coupling between ALA formation and the transformation of protochlorophyllide to chlorophyllide.

Acknowledgments. The author is very grateful to Dr. T. Oh-hama for her valuable suggestions and to Dr. R.J. Porra for assisting with the preparation of the English text.

References

Anderson LE, Avron M (1976) Plant Physiol 57: 209−213
Anderson LE, Ng TCL, Park KEY (1974) Plant Physiol 53: 835−839
Beale SI (1970) Plant Physiol 45: 504−506
Beale SI (1971) Plant Physiol 48: 316−319
Brinkmann G, Senger H (1978a) Plant Cell Physiol 19: 1427−1437
Brinkmann G, Senger H (1978b) In: Akoyunoglou et al (eds) Chloroplast development, pp 201−206. Elsevier/North Holland, Amsterdam New York Oxford
Granick S, Beale SI (1978) In: Meister A (ed) Advances in enzymology, vol 46, pp 33−203. John Wiley & Sons, New York London Sydney Toronto
Griffiths WT (1975) Biochem J 152: 623−635
Griffiths WT, Mapleston RE (1978) In: Akoyunoglou et al (eds) Chloroplast development, pp 99−104. Elsevier/North-Holland, Amsterdam New York Oxford
Hampp R, Schmidt HW (1976) Planta 129: 69−73
Hampp R, Wellburn AR (1976) Planta 131: 21−26
Harel E (1978) In: Akoyunoglou et al (eds) Chloroplast development, pp 33−44. Elsevier/North-Holland, Amsterdam New York Oxford
Hayashi N, Kurashima Y, Kikuchi G (1972) Arch Biochem Biophys 148: 10−21
Heber UW, Santarius KA (1965) Biochim Biophys Acta 109: 390−408
Highfield PE, Ellis RJ (1978) Nature (London) 271: 420−424
Kannangara CG, Gough SP, von Wettstein D (1978) In: Akoyunoglou et al (eds) Chloroplast development, pp 147−160. Elsevier/North-Holland, Amsterdam New York Oxford
Klein O., Dörnemann D, Senger H (1978) In: Akoyunoglou et al (eds) Chloroplast development, pp 45−50. Elsevier/North-Holland, Amsterdam New York Oxford
Kowallik W (1966) Planta 69: 292−295
Kowallik W, Kirst R (1973) Planta 124: 261−266
McCullough W, John PCL (1972) New Phytol 71: 829−837
Meller E, Harel E (1978) In: Akoyunoglou et al (eds) Chloroplast development, pp 51−57. Elsevier/North-Holland, Amsterdam New York Oxford
Nadler K, Granick S (1970) Plant Physiol 46: 240−246
Nandi DL, Shemin D (1968) J Biol Chem 243: 1236−1242
Nigon V, Verdier G, Salvador G, Heizmann P, Ravel-Chapuis P, Freyssinet G (1978) In: Akoyunoglou et al (eds) Chloroplast development, pp 629−640. Elsevier/North-Holland, Amsterdam New York Oxford
Ohad I, Drews G (1974) In: Avron M (ed) Proc 3rd Int Congr Photosynth, pp 1907−1912. Elsevier Sci Publ Co, Amsterdam
Ohad I, Siekevitz P, Palade GE (1967) J Cell Biol 35: 553−584
Oh-hama T, Hase E (1975) Plant Cell Physiol 16: 297−303
Oh-hama T, Hase E (1976) Plant Cell Physiol 17: 45−53
Oh-hama T, Hase E (1978) Photochem Photobiol 27: 199−202
Oh-hama T, Senger H (1975) Plant Cell Physiol 16: 395−405
Oh-hama T, Senger H (1978) Plant Cell Physiol 19: 1295−1299
Paglin S (1972) M Sci Thesis, Hebrew Univ
Porra RJ, Grimme LH (1974) Arch Biochem Biophys 164: 312−321
Ramaswamy NK, Nair PM (1973) Biochem Biophys Acta 293: 269−277

Richard F, Nigon V (1973) Biochim Biophys Acta 313: 130–149

Salvador G (1978) In: Akoyunoglou et al (eds) Chloroplast development, pp 161–165. Elsevier/North-Holland, Amsterdam New York Oxford

Schiff JA (1974) In: Avron M (ed) Proc 3rd Int Congr Photosynth, pp 1691–1717. Elsevier, Amsterdam

Schiff JA, Zeldin MH, Rubman J (1967) Plant Physiol 42: 1716–1725

Senger H, Bishop NI (1972) Plant Cell Physiol 13: 633–649

Senger H, Bishop NI, Wehrmeyer W, Kulandaivelu G (1974) In: Avron M (ed) Proc 3rd Int Congr Photosynth, pp 1913–1923. Elsevier, Amsterdam

Sokawa Y, Hase E (1967) Plant Cell Physiol 8: 495–508

Wellburn AR, Hampp R (1976) Planta 131: 17–20

Blue Light Regulation of Chloroplast Development in *Scenedesmus* Mutant C-2A'

G. BRINKMANN[1] and H. SENGER[1]

1 Introduction

Green algae normally synthesize thylakoid membranes and associated chlorophyll and proteins equally well in both darkness and light. However, the pigment mutant C-2A' of *Scenedesmus obliquus* grown heterotrophically in the dark forms only traces of chlorophyll (Senger and Bishop 1971, 1972). The etioplast of the dark-grown cells shows prethylakoid structures and a high content of starch grains (Senger et al. 1974). Illumination causes a transformation of these structures to a fully active chloroplast (Bishop and Senger 1972; Brinkmann and Senger 1978a). This paper describes the behavior of heterotrophic, dark-grown *Scenedesmus* mutant C-2A' cells during such illumination and discusses the regulatory effects of light quality and quantity on the development of thylakoid membranes of a photosynthetically active chloroplast. The previously presented scheme (Brinkmann and Senger 1978a, b) will be modified in the light of these new results.

2 Materials and Methods

Pigment mutant C-2A' of *Scenedesmus obliquus* obtained by X-ray irradiation (Bishop 1971) was employed in this study.

Conditions of heterotrophic growth, light-induced greening, as well as determinations of photosynthesis and respiration, packed-cell volume (PCV), dry weight, and cell number have been described earlier (Senger and Bishop 1972). Chlorophyll was determined after extraction with hot methanol using the formulas of Holden (1965) based on the specific absorption coefficients obtained by McKinney (1941). Proteins were determined according to Lowry et al. (1951). Carbohydrates were determined by the anthrone method of Roe (1955) and starch by the modified glucose dehydrogenase method of Brinkmann and Senger (1978a). Low-temperature absorption spectra were recorded using liquid nitrogen (77° K) with a Shimadzu MPS-50L spectrophotometer and a 1 mm

---

Abbreviations. *ALA:* 5-aminolevulinic acid, *CAP:* chloramphenicol, *CH:* cycloheximide, *CPI/II:* chlorophyllprotein complex I/II, *DCMU:* 3(3,4 dichlorphenyl)-1,1-dimethylurea, *DNP:* 2,4-dinitrophenol, *KD:* kilodalton, *PCV:* packed-cell volume

1 Fachbereich Biologie – Botanik, Universität Marburg, 3550 Marburg, FRG

Plexiglas cuvette. Thylakoid membranes and soluble protein fraction were prepared as described earlier (Brinkmann and Senger 1980a).

Algae were broken by shaking with glass beads (Senger and Mell 1977) and crude preparations of thylakoid membranes made in 10 mM KCl, 100 mM Tris HCl (pH 7.5). The glass beads were separated from the algal debris by filtering through a coarse fritted-glass funnel and washed with 25 mM Tris HCl pH 7.5, 2.5 mM KCl. To separate starch and unbroken cells, the suspension was centrifuged for 10 min at 600 g and the pellet was discarded. The supernatant was centrifuged at 11,400 g for 30 min. The resulting pellet was used as a crude-thylakoid-membrane fraction. For further purification a flota-tion step-gradient of sucrose (1.8 M sucrose, 5 mM HEPES-KOH buffer pH 7.5; 10 mM EDTA; 1.3 M sucrose, 5 mM HEPES-KOH pH 7.5, 10 mM EDTA; 0.5 M sucrose, 5 mM HEPES-KOH pH 7.5) according to Chua and Bennoun (1975) was used. After centrifuga-tion at 28,000 rpm (= 137,500 g) for 2 h, the band was collected with a syringe, diluted, and pelleted. Analysis of membrane polypeptides by SDS-polyacrylamide gel electro-phoresis was carried out using the system of Chua and Bennoun (1975). Electrophoresis was performed on a slab gel using a 7.5%–15% acrylamide separating gel and a 6% stack-ing gel. Electrophoresis was carried out at a constant current of 5 mA. Gels were stained for 2 h in a solution containing 0.25% Coomassie blue, 50% (v/v) methanol and 7% (v/v) glacial acetic acid and destained in 10% (v/v) glacial acetic acid.

Molecular weights were determined by electrophoretic mobilities (Weber and Osborn 1969) using the following molecular weight standards: bovine serum albumin (68 KD), ovalbumin (45 KD) aldolase subunit (41 KD), horse myoglobin (17 KD) and cytochrome c (12.4 KD).

3 Results and Discussion

3.1 Characterization of the Mutant C-2A'

The mutant C-2A' of *Scenedesmus* has proved absolutely stable: over a period of thirteen years no reversion has been detected. It can be grown indefinitely in the dark on a hetero-trophic medium. The growth curve after inoculation of 100 μl PCV in 250 ml culture medium is shown in Fig. 1. After a lag period of 1 day, the log phase continues for 2 days, then the stationary phase is reached. For all experiments, cells after at least 3 days of growth were used to eliminate further growth during investigations of the differentia-tion process. Since cells use more reserve starch during the stationary phase, age has a considerable effect on their light-dependent reactions during the light-dependent forma-tion of the photosynthetically competent chloroplast. During 12 h of greening, no in-crease in packed-cell volume (PCV), dry weight, or cell number occurred when cells of the stationary phase were employed. During the subsequent 12 h, some increase occurred in cell number, but without a corresponding increase in PCV or dry weight (Fig. 2). This indicates a division of existing large cells rather than new growth.

The dark-grown culture of *Scenedesmus obliquus* mutant C-2A' shows only traces of chlorophyll a and b (Fig. 3). Protochlorophyll accumulation could not be detected by spectrophotometric or chromatographic analysis in the dark-grown mutant (Senger and Bishop 1972). During 12 h illumination, chlorophyll is synthesized and reaches nearly the concentration observed in the *Scenedesmus* wild type (Fig. 3). The kinetics of green-

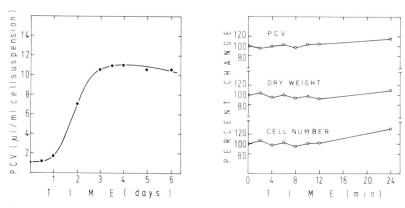

Fig. 1   Fig. 2

Fig. 1. Packed-cell volume of the mutant C-2A' of *Scenedesmus obliquus* grown heterotrophically in the dark. The culture was kept in 500-ml Erlenmeyer flasks in a shaker (Type G. 25 New Brunswick, Scientific Inc.) at 28°C. The inoculum of 100 μl PCV was 4–5 days old

Fig. 2. Packed-cell volume, dry weight, and cell number of *Scenedesmus obliquus* mutant C-2A' during greening in white fluorescent light of 10,000 lx. The culture was 8 days old with a starting density of 10.5 μl PCV/ml. Data from Senger and Bishop (1972)

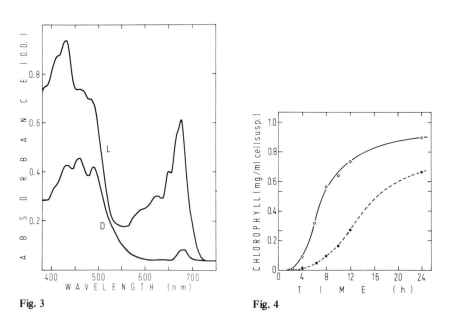

Fig. 3   Fig. 4

Fig. 3. Low-temperature (77° K) absorption spectra of whole cells of a dark-grown and a 24-h-greened culture of *Scenedesmus obliquus* mutant C-2A'. The starting culture was 3 days old. Absorption spectra of aliquots were measured in 1 mm cuvettes in a Shimadzu MPS-50L spectrophotometer

Fig. 4. Chlorophyll formation during greening of *Scenedesmus obliquus* mutant C-2A' in white fluorescent light of 10 W m^{-2}. o——o 6-day-old culture, ●——● 8-day-old culture, the cultures had a density around 10 μl PCV/ml. Data from Senger and Bishop (1972)

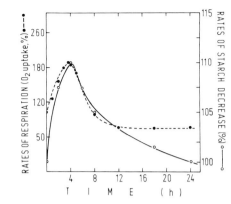

Fig. 5. Rates of respiration ●– – –● and starch degradation (○——○) during greeing of *Scenedesmus obliquus* mutant C-2A'. Greening was induced with monochromatic light of 396 nm and 0.29 W m^{-2}. Oxygen uptake of samples containing 10 μl PCV in 1 ml phosphate buffer (0.05 M, pH 7) was measured polarographically in a Gilson respirometer. For starch decrease equal amounts of PCV were determined. Data from Brinkmann and Senger (1978b)

ing are largely dependent on the age of the culture used: in Fig. 4, the greening of a 6- and 8-day old culture are shown. In both cases, the lag phase is followed by yet another phase of slow chlorophyll formation. Continued growth under suboptimal nutritional conditions causes a prolonged lag phase and a slower but rapid rate of chlorophyll formation during subsequent light-dependent greening.

3.2 Respiration Enhancement and Degradation of Starch

Respiration enhancement and starch degradation are the first measurable changes to occur during greening of the mutant C-2A' of *Scenedesmus*. They appear immediately upon illumination but only with blue light (Fig. 5). Both processes reach maximal rates after 3 to 4 h of greening before falling away: respiration finally decreases to a level of 70% of the initial value which is then comparable to that of the wild type of *Scenedesmus*. Enhanced respiration with the consequent provision of energy, reducing equivalents and precursor molecules is an absolute requirement for greening.

Chlorophyll synthesis does not occur anaerobically under nitrogen or nitrogen + 3% CO_2 during the first 12 h of greening: only at a later stage are small amounts of chlorophyll synthesized (Table 1). On the other hand, during the first 12 h of greening the omission of CO_2 or the inhibition of photosynthesis by DCMU have little effect on chlorophyll synthesis, whereas they affect strongly the later stage of greening. This suggests that following the onset of photosynthesis, substrates, and energy for greening at a later stage may be provided by photosynthetic means rather than by respiration which is required at the beginning.

To discover which photoreceptor is involved in the enhancement of respiration and starch degradation the action spectra of both processes were obtained. The spectrum for respiration enhancement (Fig. 6) shows peaks at 393, 450, and 461 nm: no effect was observed using green or red light. The wavelength dependence for total carbohydrate degradation (Fig. 7) shows identical peaks at 393, 450, and 461 nm; again, no effect was observed using green or red light. The similarity of these spectra with those compiled and interpreted by Munoz and Butler (1975) suggest that flavoprotein is the most probable photoreceptor. Nevertheless, the final proof can only be obtained when the action spectra are further extended to UV, or the photoreceptor is isolated.

Table 1. Chlorophyll formation in *Scenedesmus obliquus* mutant C-2A' under aerobic, anaerobic, photosynthetic and nonphotosynthetic conditions

Conditions during greening	Chlorophyll increase	
	after 12 h	after 24 h
Air + 3% CO_2 ("normal")	45	100
N_2 + 3% CO_2	0	2
N_2	0	5
Air – CO_2	40	55
Air + 3% CO_2 + DCMU	38	50
Air + 3% CO_2 + glucose	40	73

The amounts of chlorophyll are given in percent of the final concentration under "normal" conditions, i.e., after 24 h of illumination with white fluorescent light of 10,000 lx. The starting heterotrophic cultures were 8 days old. (Values from Senger and Bishop 1972)

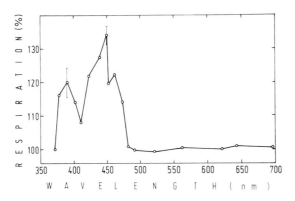

Fig. 6. Wavelength-dependent increase in rates of respiration in *Scenedesmus obliquus* mutant C-2A'. Cells were illuminated for 3 h with monochromatic light of various wavelengths of 1.75×10^{-10} Einstein $cm^{-2} s^{-1}$. Respiration rates were determined as in Fig. 5. Data are expressed as percent of the dark control. Data from Brinkmann and Senger (1978a)

Respiration enhancement and carbohydrate degradation are saturated by blue light at low intensities of 0.5 W m^{-2} (Fig. 8). Respiration enhancement can be induced by a short input (5 min) of blue light (Fig. 9). During subsequent incubation in the dark, respiration rates increase in parallel with the control which is continuously illuminated under blue light. Prolonged preillumination with blue light prior to transfer into darkness slows down the later decrease of the respiration rate which, after 90 min of preillumination, finally parallels that of the light control. Respiration enhancement is not sensitive to inhibitors of plastid protein translation CAP (Pestka 1974) or lincomycin (Fernandez-Muñoz et al. 1971) (Table 2).

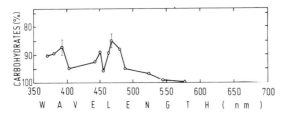

Fig. 7. Wavelength-dependent decrease in total carbohydrates in *Scenedesmus obliquus* mutant C-2A'. Irradiation was described in Fig. 6. Total carbohydrates were determined in equal volumes of cell suspension with the anthrone method. Data from Brinkmann and Senger (1978a)

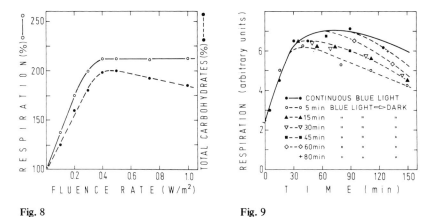

Fig. 8	Fig. 9

Fig. 8. Light-intensity curves for respiration enhancement (○——○) and decrease in total carbohydrates (●– – –●) during greening of *Scenedesmus obliquus* mutant C-2A'. Aliquots were irradiated with monochromatic light of 399 nm at different intensities for 3.5 h. Determinations and data presentations as in Fig. 6. Data from Brinkmann and Senger (1978a)

Fig. 9. Induction of respiration enhancement during greening of *Scenedesmus obliquus* mutant C-2A'. Cells were illuminated for the indicated period with monochromatic light of 399 nm at an intensity of 0.5 W m^{-2} and transferred to the dark. Determinations and data presentation as in Fig. 6

Table 2. Respiration enhancement and chlorophyll formation in *Scenedesmus obliquus* mutant C-2A' in the presence of inhibitors of protein formation

	Control	CH	CAP	Lincomycin
Respiration enhancement	100	130	100	100
Chlorophyll formation	100	0	60–80	60–80

All data are given in percent of the control value. Inhibitors were applied with a final concentration of 20 μg/ml (CH), 200 μg/ml (CAP) and 100 μg/ml (lincomycin)

The kinetics obtained for respiration enhancement with inhibitors are comparable to results obtained in their absence. Oh-hama and Senger (1975) showed that the initial phase of respiration enhancement was not decreased by addition of CH, an inhibitor of cytoplasmic-protein translation (Table 2): indeed, we have shown that the enhancement of respiration in the light was further stimulated by CH. In the dark, also, CH causes an immediate increase in respiration enhancement (Fig. 10). DNP, an uncoupler of phosphorylation (Slater 1967), has the same effect as CH. The effect of CH on dark respiration cannot be increased by the further addition of DNP or vice versa. The similarity of the effects obtained with CH and DNP suggests that CH, in addition to its inhibitory effect on protein formation, may also act as an uncoupler of oxidative phosphorylation. The enhancement of respiration in the dark by CH can be further enhanced by blue light (Fig. 10). This indicates that CH and blue light cause different and independent reactions both enhancing respiration and that the initial blue light-dependent enhancement of respiration is neither dependent on cytoplasmic or noncytoplasmic protein synthesis; rather, it appears to be due to the activation of an already present enzyme.

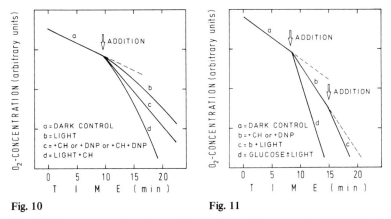

Fig. 10   **Fig. 11**

Fig. 10. Enhancement of respiration in *Scenedesmus obliquus* mutant C-2A' after treatment with saturating white light, CH or DNP, and of light + CH. The O_2 concentration was measured polarographically in a Gilson respirometer and recorded continuously. The final concentration of CH was approx. 2×10^{-5} M and of DNP approx. 5×10^{-4} M

Fig. 11. Enhancement of respiration in *Senedesmus obliquus* mutant C-2A' after treatment with CH or DNP ± light and glucose ± light. Determinations and data presentation as in Fig. 10

That blue light-enhanced respiration is due to the provision of substrate arising from starch degradation was tested by addition of exogenous substrate, glucose, and recording the respiration polarographically. The results (Fig. 11) show that glucose stimulates an immediate enhancement of respiration. With a glucose concentration of 0.5% (w/v) normally used in growth experiments, the respiratory pathway appeared to be saturated because no further stimulation could be obtained by illuminating the cultures or by the addition of uncouplers. However, when glucose was administered in lower ("nonsaturating") concentrations a further stimulation of respiration was evoked by illumination with blue light or by the addition of CH or DNP (Figs. 10 and 11). These results further confirm the notion that blue light enhances respiration by the provision of substrate. Substrate availability may be increased by enhanced enzymic degradation of starch, by increased permeability of the plastid membrane to substrate molecules or by a combination of both processes: this is still to be elucidated.

3.3 Protein Synthesis and Chlorophyll Formation

The formation of soluble proteins starts immediately upon illumination. The synthesis of membrane proteins and/or their integration into the thylakoid membrane reaches its maximal value after 5 h of illumination (Fig. 12); blue light is the most effective. With an intensity of 60–80 W m^{-2} the maximal rate of synthesis is reached after 1 h (Fig. 12).

The membrane polypeptide patterns change considerably during greening. They are shown for a dark-grown culture and after 3, 6, 9, and 12 h of greening (Fig. 13). Between the 3rd to 4th hour the two chlorophyll protein complexes CPI (130 KD) and CPII (32 KD) appear in parallel. During greening three dominant nonpigmented polypeptides

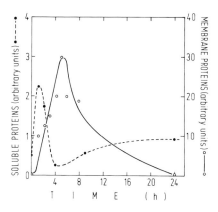

Fig. 12. Rates of soluble protein formation (●— — —●) and of membrane protein formation (○——○) during greening of *Scenedesmus obliquus* mutant C-2A'. Greening was induced with monochromatic light of 396 nm and 50 W m^{-2}. Data are presented in arbitrary units and taken from Brinkmann and Senger (1978b)

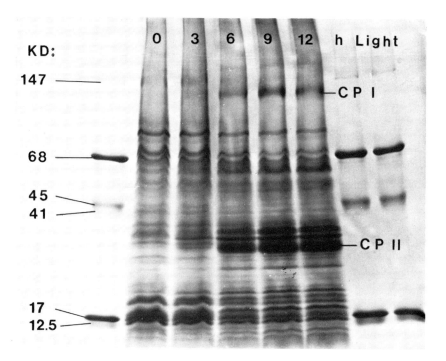

Fig. 13. Electrophoretic pattern on polyacrylamide SDS-gels of membrane polypeptides during greening of *Scenedesmus obliquus* mutant C-2A'. Equal amounts of protein were added on each tracing. For details see methods. The *numbers* given in the figure indicate the hours of greening. Molecular weight markers as indicated were run simultaneously

of 24 KD, 35 KD, and 64 KD molecular weight also appear. The rates of chlorophyll synthesis increase up to the 8th h of greening and then decline for the next 16 h (Fig. 14).

The action spectrum for chlorophyll formation (Fig. 15) shows a distinct peak at 393 nm and a band between 440 and 500 nm. No chlorophyll formation could be ob-

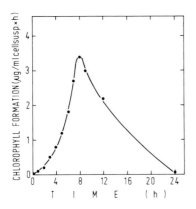

Fig. 14. Rates of chlorophyll formation during greening of *Scenedesmus obliquus* mutant C-2A'. Greening was induced with monochromatic light of 396 nm and 50 W m^{-2}. Data from Brinkmann and Senger (1978b)

tained with red light of even high intensities (10 W m^{-2}) under these experimental conditions.

The wavelength dependence of the synthesis of membrane proteins (Fig. 16) is comparable to that of chlorophyll formation (Fig. 15). It has a peak at 400 nm and a band between 440 and 500 nm. No red or green light was active in membrane protein formation.

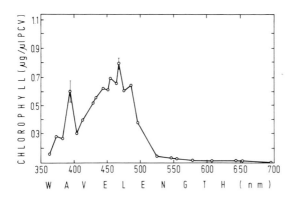

Fig. 15. Wavelength-dependent chlorophyll formation in *Scenedesmus obliquus* mutant C-2A'. Aliquots were irradiated with monochromatic light of various wavelenght of 1.75 × 10^{-10} Einstein cm^{-2} s^{-1} for 15 h. Data from Brinkmann and Senger (1978a)

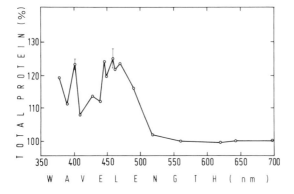

Fig. 16. Wavelength-dependent formation of total protein in *Scenedesmus obliquus* mutant C-2A'. Aliquots were irradiated with monochromatic light of various wavelength of 1.75 × 10^{-10} Einstein cm^{-2} s^{-1} for 13 h. Data from Brinkmann and Senger (1978a)

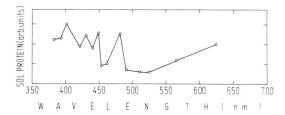

Fig. 17. Wavelength-dependent formation of soluble proteins in *Scenedesmus obliquus* C-2A'. Aliquots were irradiated with monochromatic light of various wavelength of 1.75×10^{-10} Einstein cm^{-2} s^{-1} for 2 h

An action spectrum for the formation of soluble proteins is shown in Fig. 17; because it is a very preliminary result we do not wish to stress it other than to point out that the stimulation occurs best in the blue region but with a very considerable effect obvious in the green and red region.

Both chlorophyll formation and total protein synthesis are saturated by blue light at $60–80\,W\,m^{-2}$ (Fig. 18). Higher intensities of blue light cause both a photooxidation of chlorophyll and a decrease in net protein synthesis.

The action spectra for chlorophyll and protein formation were obtained using three-day old dark-grown cultures, depleted of external carbon sources. Therefore these spectra must include the initial blue light dependence of respiration enhancement and starch degradation. If this forms a prominent part of the action spectra obtained, then the nature of the action spectrum for chlorophyll formation should change if a preillumination is given first to ensure the provision of the products of enhanced respiration.

The light quality dependence for chlorophyll formation after such a preillumination is shown in Fig. 19: peaks appear at 383, 442, 471, and 540 nm, and a broad band between 580 and 670 nm. Thus, in addition to blue light, both green and red light are effective in regulation of thylakoid membrane formation. Chlorophyll formation was also observed under red light after prior addition to glucose (Table 3). After glucose addition under red and blue light chlorophyll synthesis occurred even faster, indicating a complete substitution of the blue light effect by glucose. The similarity of the action spectrum for chlorophyll formation after preillumination with those for soluble protein and ALA formation (Figs. 17, 19, and 20) in *Scenedesmus* mutant C-2A' now becomes obvious.

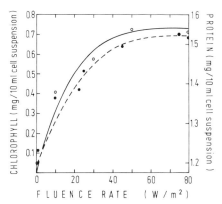

Fig. 18. Light intensity-dependent chlorophyll (O—O) and protein (●— — —●) formation in *Scenedesmus obliquus* mutant C-2A'. Aliquots were irradiated with blue light of 460 nm and various intensities for 9 h

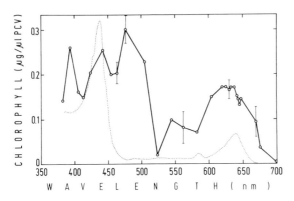

Fig. 19. Wavelength-dependent formation of chlorophyll after preillumination in *Scenedesmus obliquus* mutant C-2A'. Preillumination was carried out under white light of 20 W m^{-2} for 2 h. For methods of illumination with monochromatic light of different wavelengths and of chlorophyll determination see Sect. 2. Data from Brinkmann and Senger (1978b). For comparison, the absorption spectrum of the bean protochlorophyll holochrome is added (Schopfer and Siegelman 1968)

Table 3. Respiration enhancement, formation of soluble and membrane proteins and ALA, and chlorophyll synthesis under blue and red light in the presence and absence of glucose

	Respiration enhancement	Formation of soluble proteins	Formation of ALA	Formation of membrane proteins	Chlorophyll formation
Blue light	+	+	+	+	+
Green or red light	–	+	+	–	–
Blue light transferred to red light	+	+	+	+	+
Glucose (dark)	+	–	–	?	–
Glucose + green or red light	+	+	+	+	+

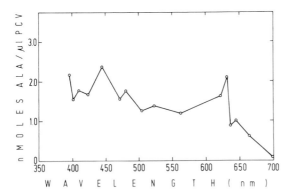

Fig. 20. Wavelength-dependent formation of ALA in *Scenedesmus obliquus* mutant C-2A'. Aliquots were irradiated with monochromatic light of various wavelength of 1.75 × 10^{-10} Einstein cm^{-2} s^{-1} for 15 h. ALA-accumulation was achieved by inhibiting ALA-dehydratase with 10 mM levulinic acid added at the beginning of illumination. ALA was condensed with ethylacetoacetate and measured spectroscopically after addition of Ehrlich's reagent. Data from Senger et al. (1980)

These results suggest that in addition to the blue light effect on respiration which is saturated at low light intensities there is a further effect of blue, green, and red light on the formation of soluble proteins, ALA and chlorophyll, which is saturated only at high light intensities. The accumulation of chlorophyll would in turn be dependent on the formation of membrane proteins into which the chlorophyll is incorporated.

The known absorption spectrum of protochlorophyllholochrome (Fig. 19; Schopfer and Siegelman 1968) does not explain the effectiveness of green, yellow, and red light in the formation of chlorophyll. However, some hemoproteins have absorption spectra with peaks in the green and red regions thus raising the possibility that the photoreceptor concerned may be a heme pigment, as is also suggested in some other papers presented at this symposium.

4 Conclusion

The process of thylakoid membrane differentiation of the mutant C-2A' of *Scenedesmus obliquus* proceeds via a sequence of reactions leading finally to a fully active chloroplast (Bishop and Senger 1972; Senger et al. 1972; Senger and Bishop 1971, 1972; Brinkmann and Senger 1978a, b).

The rates of respiration enhancement, starch decrease, and formation of soluble proteins reach their maximal values during the early stage of chloroplast development (Brinkmann and Senger 1978a, b). They are followed by the formation of ALA (Oh-hama and Senger 1975) and the synthesis and/or incorporation of chlorophyll and proteins into the membrane (Brinkmann and Senger 1978a, b). The parallel appearance of CPI and CPII (Fig. 13) coincides with the attainment of optimal quantum efficiency of the photosynthetic apparatus (Bishop and Senger 1972).

The blue light dependence of both starch degradation and of respiration enhancement are similar. Their requirement for irradiation of low energy for short duration clearly separates them from subsequent processes. A comparison of these action spectra with the absorption spectrum of flavins (Schmidt and Butler 1976) and with the interpretation of other action spectra suggests that the photoreceptor is probably a flavoprotein (Brinkmann and Senger 1978a). The immediate response to blue light and the lower energy requirement suggest that a conformational change is involved in the primary reaction (cf. Schmidt 1969, 1970). The initial stage of respiration enhancement is not dependent on protein formation. Neither CH (inhibitor of cytoplasmic translation) nor CAP (inhibitor of plasmic translation) inhibit the initial respiration enhancement. This finding is supported by the results of Watanabe et al. (1980) showing that 6-methylpurine, an inhibitor of RNA synthesis, does not inhibit the initial phase of respiration enhancement. These data support our hypothesis of an activation rather than a de novo synthesis of an enzyme(s) by blue light (Brinkmann and Senger 1978a, b). Confirmation of activation of enzymes of carbohydrate metabolism by blue light can also be deduced from the effect of external supply of substrates for glycolysis. After addition of glucose in the dark, respiration is immediately enhanced and light does not further increase the respiration rate. This leads to the conclusion that irradiation with blue light leads to the provision of substrates for glycolysis such as glucose or its precursors. Two possible explanations for the action of blue light are either an enhancement of starch breakdown or an enhancement of the transport of the degradation products through the plastid membrane.

The latter is in accordance with the suggestions of Hase (1980) and of Kowallik and Schätzle (1980) who stress the importance of transport phenomena associated with the passage of substrates and energy through the plastid envelope. The action of blue light might be different in *Chlorella vulgaris* (Miyachi et al. 1977; Kowallik and Schätzle 1980; Miyachi et al. 1980) where an activation of the later enzymes of glycolysis was observed.

Application of CH to *Scenedesmus* cells in the dark enhances respiration and thus simulates the light effect. The same effect, obtained for the paramylum breakdown in dark-grown *Euglena* cells, leads Schwartzbach et al. (1975) to the hypothesis that the synthesis of an inhibitory protein for paramylum breakdown is prevented, either by blue light at the transcriptional level or by CH at the translational level. However, in *Scenedesmus,* after addition of CH in the dark, blue light enhances respiration further. This additive effect cannot readily be explained by an inhibition of the synthesis of the same protein. Studies using DNP or CH suggest an uncoupling of oxidative phosphorylation by both.

Whereas the enhancement of respiration by blue light can be replaced by the external application of glucose, the subsequent biosynthesis of chlorophyll and its precursors is absolutely light-dependent. Chlorophyll formation under monochromatic light appears to be strictly blue light-dependent, but after preillumination or addtion of glucose then green and red light also become effective (Table 3). Independent of preillumination or glucose addition soluble protein and ALA formation take place in blue, green, and red light. Since neither the apoproteins nor the chromophore of the chlorophyll holochrome can accumulate separately, it is quite obvious that ALA is only further converted to chlorophyll when the synthesis of membrane proteins is either provided by blue light or added substrate (Table 3). The possibility to form membrane proteins in the dark by addition of glucose cannot be expressed since the chlorophyll chromophore is not present.

Regarding the wavelength-dependent formation of chlorophyll after preillumination (Brinkmann and Senger 1978b) we previously discussed that protochlorophyll might be the photoreceptor. Since ALA formation has a similar wavelength dependence, but precedes protochlorophyll biosynthesis, protochlorophyll is unlikely to be the photoreceptor. This view is supported by comparison of the action spectra for ALA- and chlorophyll formation with the absorption spectrum of the protochlorophyll holochrome (Figs. 19 and 20; Schopfer and Siegelman 1968). These action spectra bear a closer resemblance to the absorption spectra of protoporphyrin IX, bile- and heme-proteins (for further discussion see Senger et al. 1980), but so far no photoreactive heme or bile pigment which could fill the role of a photoreceptor has been detected in green algae.

Our current working hypothesis for the regulation by light of greening in mutant C-2A' of *Scenedesmus obliquus* is summarized in the scheme p. 539.

Acknowledgments. The authors wish to thank Dr. R.J. Porra for reading the manuscript and for valuable discussion, Mrs. I. Koss and Mrs. G. Müller for skillful technical assistance, Mrs. I. Krieger and Mr. H. Becker for valuable help in preparing the manuscript and the Deutsche Forschungsgemeinschaft for financial support.

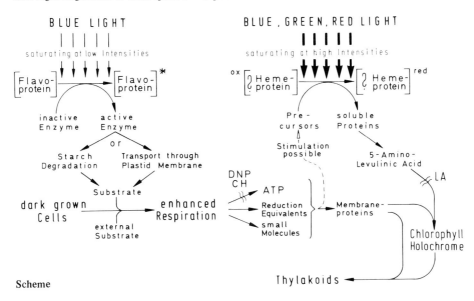

Scheme

References

Bishop NI (1971) Preparation and properties of mutants: *Scenedesmus.* In: Golowick SP, Kaplan NO (eds) Methods of enzymology, vol 23, part A, pp 130–143. Academic Press, London New York

Bishop NI, Senger H (1972) The development of structure and function in chloroplasts of greening mutants of *Scenedesmus.* II. Development of the photosynthetic apparatus. Plant Cell Physiol 13: 937–953

Brinkmann G, Senger H (1978a) The development of structure and function in chloroplasts of greening mutants of *Scenedesmus.* IV. Blue light-dependent carbohydrate and protein metabolism. Plant Cell Physiol 19: 1427–1437

Brinkmann G, Senger H (1978b) Light-dependent formation of thylakoid membranes during the development of the photosynthetic apparatus in pigment mutant C-2A' of *Scenedesmus obliquus.* In: Akoyunoglou G, Argyroudi-Akoyunoglou JH (eds) Chloroplast development. Developments in plant biology, vol II, pp 201–206. Elsevier/North-Holland Biomedical Press, Amsterdam New York Oxford

Chua NH, Bennoun P (1975) Thylakoid membrane polypeptides of *Chlamydomonas reinhardtii:* Wild type and mutant strains deficient in photosystem II reaction center. Proc Natl Acad Sci USA 72: 2175–2179

Fernandez-Munoz R, Mono RE, Vazques D (1971) Ribosomal peptidyltransferase: Binding of inhibitors. In: Moldave K, Grossman L (eds) Methods of enzymology, vol 20, part C, pp 481–490. Academic Press, London New York

Hase E (1980) Effects of blue light on greening in microalgae. This volume

Holden M (1965) Chlorophylls. In: Goodwin TW (ed) Chemistry and biochemistry of plant pigments, pp 461–488. Academic Press, London New York

Kowallik W, Schätzle S (1980) Enhancement of carbohydrate degradation by blue light. This volume

Lowry OH, Rosebrough NJ, Farr AL, Randall RJ (1951) Protein measurement with the folin phenol reagent. J Biol Chem 193: 265–275

MacKinney G (1941) Absorption of light by chlorophyll solutions. J Biol Chem 140: 315–322

Miyachi S, Kamiya A, Miyachi S (1977) Wavelength-effects of incident light on carbon metabolism in *Chlorella* cells. Biological solar energy conversion, pp 167–182. Academic Press, London New York

Miyachi S, Kamiya A, Miyachi S (1980) Effects of blue light on respiration and non photosynthetic CO_2 fixation in *Chlorella vulgaris* 11 h cells. This volume

Munoz V, Butler WL (1975) Photoreceptor pigment for blue light in *Neurospora crassa*. Plant Physiol 55: 421–426

Oh-hama T, Senger H (1975) The development of structure and function in chloroplasts of greening mutants of *Scenedesmus*. III. Biosynthesis of δ-aminolevulinic acid. Plant Cell Physiol 16: 395–405

Pestka S (1974) The use of inhibitors in studies of protein synthesis. In: Moldave K, Grossman L (eds) Methods in enzymology, vol 30, pp 261–282. Academic Press, London New York

Row JH (1955) The determination of sugar in blood and spinal fluid with anthrone reagent. J Biol Chem 212: 335

Schmid GH (1969) The effect of blue light on glycolate oxidase of tobacco. Hoppe Seylers Z Physiol Chem 350: 1035–1046

Schmid GH (1970) The effect of blue light on some flavin enzymes. Hoppe Seylers Z Physiol Chem 351: 575–578

Schmidt W, Butler WL (1976) Flavin-mediated photoreactions in artificial systems. Photochem Photobiol 24: 71–75

Schopfer P, Siegelman HW (1968) Purification of protochlorophyllide holochrome. Plant Physiol 43: 990–996

Schwartzbach SD, Schiff JA, Goldstein NH (1975) Events surrounding the early development of *Euglena* chloroplasts V. Control of paramylum degradation. Plant Physiol 56: 313–317

Senger H, Bishop NI (1971) Light-dependent greening of a yellow *Scenedesmus* mutant. Indian Plant Physiol 14: 164–173

Senger H, Bishop NI (1972) The development of structure and function in chloroplasts of greening mutants of *Scenedesmus*. I. Formation of chlorophyll. Plant Cell Physiol 13: 633–649

Senger H, Mell V (1977) Preparation of photosynthetically active particles from synchronized cultures of unicellular algae. In: Prescott DM (ed) Meth Cell Biol 15: 201–218

Slater EC (1967) Application of inhibitors and uncouplers for a study of oxidative phosphorylation. In: Estabrook RW, Pullman ME (eds) Methods in enzymology, vol X, pp 48–57. Academic Press, London New York

Watanabe M, Oh-hama T, Miyachi S (1980) Light induced carbon metabolism in an early stage of greening in wild type and mutant C-2A'. This volume

Weber K, Osborn M (1969) The reliability of molecular weight determinations by dodecylsulfate-polyacrylamide gel electrophoresis. J Biochem 244: 4406–4412

The Action of Blue Light
on 5-Aminolevulinic Acid Formation

H. SENGER[1], O. KLEIN[1], D. DÖRNEMANN[1], and R.J. PORRA[2]

1 Introduction

In most angiosperms the last step in chlorophyll biosynthesis, the conversion of proto-chlorophyll(ide) to chlorophyll(ide), requires light. In recent years, however, it has been found in several organisms that the first step, the formation of 5-aminolevulinic acid (ALA), is also light-dependent. The formation of ALA is not only the first step in the biosynthesis of tetrapyrroles but is also a rate-limiting step and plays a key role in the regulation of tetrapyrrole biosynthesis. Thus the light dependence of this step may be of considerable importance in the biosynthesis not only of chlorophyll but also of hemes, corrins, and bile pigments including the photomorphogenetically active plant pigment phytochrome. The light dependence of ALA formation has been reported in many higher plants but only in a few cases have investigations been made about the photoreceptor involved and those suggest the participation of phytochrome (Masoner and Kasemir 1975). Among the photosynthetic microorganisms, light-dependent ALA formation has been reported for *Chlorella fusca* (Porra and Grimme 1974), *Chlorella* sp. (Beale 1971), *Chlorella vulgaris* (Meller and Harel 1978), *Chlorella protothecoides* (Oh-hama and Senger 1978), *Scenedesmus obliquus* (Oh-hama and Senger 1975; Klein and Senger 1978a), *Euglena gracilis* (Salvador et al. 1976) and *Rhodopseudomonas spheroides* (Lascelles 1968); however, only in *Chlorella protothecoides* and *Scenedesmus obliquus* has the wavelength dependence of ALA formation been more intensively studied.

In the case of *Chlorella protothecoides* the blue light-dependent ALA formation, observed during the light-dependent regreening (Sokawa and Hase 1967) which follows bleaching by glucose (Hase 1971), seems to be rather uncomplicated and the action spectrum for ALA formation resembles that for chlorophyll formation (Oh-hama and Senger 1978).

Light-dependent ALA formation in the pigment mutant C-2A' of *Scenedesmus obliquus* is known to involve two biosynthetic pathways for ALA formation and at least two different light-dependent steps in chlorophyll synthesis. Current knowledge of the

Abbreviations. *ALA:* 5-aminolevulinic acid, *DOVA:* 4,5-dioxovaleric acid, *LA:* levulinic acid, *PCV:* packed cell volume

1 Fachbereich Biologie/Botanik der Philipps-Universität, Lahnberge, 3550 Marburg, FRG
2 On leave from: C.S.I.R.O., Division of Plant Industry, P.O. Box 1600, Canberra City, A.C.T. 2601, Australia

light dependence of ALA formation in this mutant is summarized here and presented with the results of further experiments which attempt to discover the regulatory role of light of different wavelengths in ALA and chlorophyll formation.

2 Material and Methods

Mutant C-2A' of *Scenedesmus obliquus,* obtained by X-ray irradiation (Bishop 1971), was cultured heterotrophically in the dark (Senger and Bishop 1972) and transferred into light for ALA or chlorophyll formation after at least 72 h when the logarithmic growth phase was terminated. Irradiation with white light was carried out with a combination of five warm white (40 W, Osram 25-1) and four day light (40 W, Osram 19-1) fluorescent tubes. Monochromatic light was obtained with interference filters (Schott and Gen., Mainz). All illumination with monochromatic light was done at a constant quantum flux of 1.75×10^{-10} Einstein cm^{-2} s^{-1}. ALA was measured after the different light treatments by the method of Mauzerall and Granick (1956) as described earlier (Klein and Senger 1978b). Chlorophyll was measured after extracting exhaustively with hot methanol (Holden 1965).

For enzymic assays, cells were harvested and broken as previously described (Klein et al. 1978). ALA-synthetase activity was measured with [2-^{14}C]-glycine as substrate according to Irving and Elliott (1969). The enzymes of the C-5pathway, 4,5-dioxovalerate aminotransferase and 4,5-dioxovalerate dehydrogenase were assayed as described by Klein et al. (1978) with [1-^{14}C]-glutamate as the radioactive substrate. ALA-dehydratase was assayed according to Shemin (1963) and porphobilinogen formed was determined by the method of Mauzerall and Granick (1956).

3 Results and Discussion

3.1 The Light-Dependence and Time Course of ALA Formation

When mutant C-2A' of *Scenedesmus obliquus* was first introduced for greening experiments because of the light dependence of its greening process (Senger and Bishop 1971, 1972), it became quite obvious that the regulatory light-dependent step was not the reduction of protochlorophyll to chlorophyll but occurred much earlier in the biosynthetic sequence; indeed, it became clear that the formation of ALA was the light-dependent step (Oh-hama and Senger 1975). ALA formation under white light starts after a lag period of 2–3 h (Fig. 1) and accumulates in the presence of LA, a competitive inhibitor of ALA dehydratase, during the first 12 h of illumination. Using younger cultures or a preillumination, the lag period can be shortened or even abolished. The light-intensity dependence of ALA formation is shown in Fig. 2. To eliminate the lag period the cells were preilluminated for 4 h, transferred back into the dark for 30 min, LA was added and samples exposed to different intensities of white light for 1 h. Under these experimental conditions saturation with white light was not reached at 20 W m^{-2}, but, by focusing monochromatic light of 450 nm on a small field, saturation of ALA formation could be reached at 25 W m^{-2}; thus, when white light is used, a higher saturating intensity is required, probably near 100 W m^{-2}, as found for light-dependent chlorophyll formation (Senger and Bishop 1972).

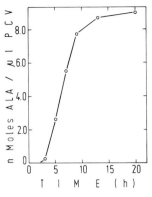

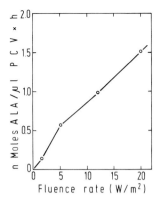

Fig. 1   **Fig. 2**

Fig. 1. ALA accumulation by a suspension (10 μl PCV/ml) of 5-day old, dark-grown cells of mutant C-2A' of *Scenedesmus obliquus* during illumination with white light (20 W m^{-2}) in the presence of LA (10 mM)

Fig. 2. Rate of ALA formation by dark-grown cells of mutant C-2A' of *Scenedesmus obliquus* (10 μl PCV/ml) preilluminated with white light (20 W m^{-2}) for 4 h, transferred back to darkness (30 min). Data were taken from Oh-hama and Senger (1975)

3.2 The Cessation of ALA Formation in the Dark

On the basis of the light intensity required for saturation, two different light-dependent processes occurring during light-dependent greening of *Scenedesmus* mutant C-2A' could be separated (Senger and Bishop 1972; Brinkmann and Senger 1978a, 1980a). The initial enhancement of respiration is saturated at low intensities of blue light and continues in the dark. By contrast, chlorophyll formation is saturated only at high light intensities and ceases immediately the cells are transferred to the dark (Senger and Bishop 1972). It was therefore of interest to transfer light-grown cells to darkness and observe the effect on ALA accumulation. The results in Fig. 3 show the striking and immediate

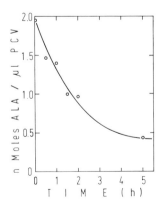

Fig. 3. Decline of ALA concentration in *Scenedesmus obliquus* mutant C-2A' after transfer from light to darkness. Three-day old cells were illuminated with blue light (396 nm, 7.0 W m^{-2}) for 12 h in the presence of LA (10 mM) prior to transferring back to darkness

effect of darkness in causing the cessation of ALA formation which is partly in accord with the known lability and rapid turnover of the enzymes of ALA formation (Beale 1971; Oh-hama and Senger 1975). Nonetheless, these enzymes have a half life of about 30 min (Beale 1971) and one would not, therefore, expect such an immediate cessation of ALA accumulation. One explanation would be the onset, in darkness, of a very rapid degradation of ALA. That this is a likely explanation is confirmed by the results in Fig. 3 which show not only the cessation of ALA accumulation but also an immediate commencement of disappearance of ALA at the onset of darkness. This result is all the more surprising since it occurs in the presence of LA which prevents the further metabolism of ALA to porphobilinogen and tetrapyrroles.

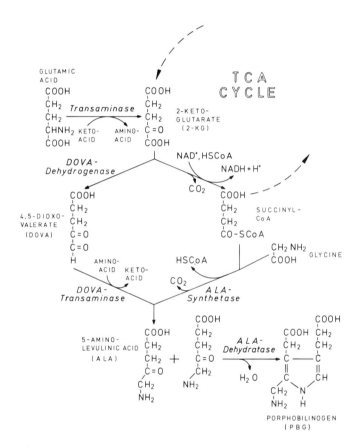

Fig. 4. Schematic representation of the two pathways of ALA formation in the *Scenedesmus* mutant C-2A'

3.3 The Two Alternative Routes for ALA Formation and the Light Dependence of the Synthesis of Their Key Enzymes

The light-dependent formation of ALA was found to be more complicated than first expected. Using labeled precursors it was established that two different pathways of ALA formation were operating in *Scenedesmus obliquus* mutant C-2A' (Klein and Senger 1978a, b); firstly the classical pathway using succinyl CoA and glycine as precursors as in animal (Shemin and Russell 1953) and bacterial cells (Kikuchi et al. 1958) and, secondly the C-5 pathway incorporating the intact C-5 skeleton of glutamate or 2-oxo-glutarate into the ALA as in higher plants (Beale et al. 1975; Harel et al. 1978).

The key enzymes of both pathways (Fig. 4) have been assayed in vitro in crude extracts and their light dependence shown in Fig. 5a, b. That the reactions need continuous illumination (Senger and Bishop 1972), that simultaneous protein formation occurs (Brinkmann and Senger 1978a, b) and that ALA formation is cycloheximide-sensitive

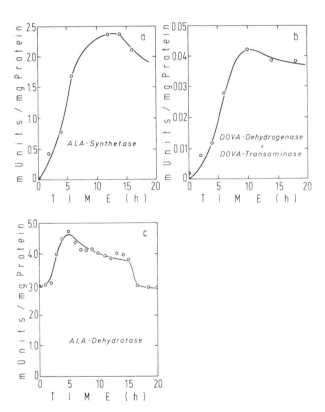

Fig. 5. Development of enzyme activities in dark-grown, 3-day old cells of mutant C-2A' of *Scenedesmus obliquus* (10 μl PCV/ml) after exposure to white light (20 W m^{-2}) for various periods. Enzyme activities were measured in cell-free extracts as described in Section 2. **a** Succinyl-CoA:Glycine-C-succinyl transferase (decarboxylating) (EC 2.3.1.37) = ALA-Synthetase. **b** L-alanine-4,5-dioxovale-rate-aminotransferase (EC 2.6.1.43). 2-Oxo-glutarate-semialdehyde NAD-oxidoreductase (EC NN) = DOVA-Dehydrogenase. **c** 5-aminolevulinate hydro-lyase (EC 4.2.1.24) = ALA-Dehydratase

(Oh-hama and Senger 1975) all suggest that the ALA synthesizing enzymes are not detectable in the dark and are synthesized de novo in the light (Klein et al. 1978).

The following enzyme, ALA-dehydratase, which catalyzes the condensation of two molecules of ALA to form porphobilinogen (Fig. 4), was also assayed for light depen- dence and effect of illumination period (Fig. 5c). This enzyme is already present in the dark-grown culture but its activity increases by 50% in the light, suggesting that it is a constitutive enzyme but, possibly, further induced or activated by the substrate ALA. The enzyme assays clearly indicate that ALA formation by both pathways is strictly light-dependent and that the subsequent step is not directly regulated by light.

3.4 Comparison of the Wavelength Dependence of ALA and Chlorophyll Formation

Results of wavelength dependence measurements of ALA and chlorophyll formation in Scenedesmus mutant C-2A' were more complicated than in Chlorella protothecoides. In the latter alga, as was to be expected, the action spectra for both chlorophyll forma- tion and ALA biosynthesis were similar (Oh-hama and Senger 1978) and showed typical blue light dependence, but these studies have not yet been extended into the red region.

In the Scenedesmus mutant C-2A', however, the action spectrum of ALA formation revealed dependence not only on blue light but also on green and red light (Fig. 6). This spectrum is a mean of three experiments and in all cases the position of the peaks re- mained constant. But further work is required to establish the relative peak heights in the blue and red regions. We also found during these experiments that stimulation of ALA formation by red light is extremely dependent upon the age of the culture; in older cultures the stimulation is less than depicted in Fig. 6. This considerable stimulation of ALA formation by green and red light was puzzling because the action spectrum for chlorophyll formation in this organism showed only the typical blue light dependency

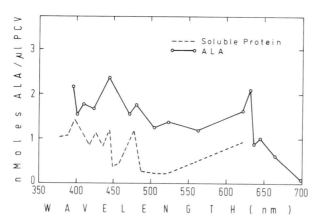

Fig. 6. Wavelength-dependent formation of ALA and soluble protein in Scenedesmus obliquus mutant C-2A'. Three-day old cells were illuminated with monochromatic light (1.75×10^{-10} E cm^{-2} s^{-1}) for 15 h. LA (10 mM) was added at the beginning of illumination. ALA accumulation was de- termined as described in Sect. 2. Protein concentration was measured according to Lowry et al. (1951). The protein action spectrum was taken from Brinkmann and Senger (1978a)

(Senger and Bishop 1972; Brinkmann and Senger 1978a). However, a more recent action spectrum for chlorophyll formation in *Scenedesmus* mutant C-2A' after a short pre-illumination with blue light (Brinkmann and Senger 1978b) showed a considerable stimulation of chlorophyll formation not only by blue light but also by green and red light and the position of the peaks corresponded well with those obtained in the action spectrum for ALA formation.

The difference between the action spectra for chlorophyll formation with and without pre-illumination with blue light can be explained as follows: chlorophyll is formed only when the tetrapyrrole prosthetic group, formed via the ALA synthesizing systems, and the membrane apoprotein can combine to form the holochrome. Thus the formation of membrane-bound protein, which is formed only under blue light, can become a limiting step in chlorophyll production and the resulting action spectrum for chlorophyll formation will be a typical blue light action spectrum identical with that of respiration enhancement (Brinkmann and Senger 1980a). However, if membrane protein formation is ensured by previous illumination with blue light, then the formation of the tetrapyrrolic prosthetic group becomes the limiting factor and thus the action spectrum for chlorophyll formation will now be identical with that for ALA formation, showing dependence not only on blue but also on red and green light.

3.5 Further Consideration of the Wavelength Dependence of ALA Formation

Because more complicated action spectra were obtained for ALA formation in the *Scenedesmus* mutant C-2A' than in *Chlorella protothecoides,* and because it is known that two pathways for ALA formation exist in the *Scenedesmus* mutant it seemed reasonable that the spectrum in Fig. 6 may be composed of two spectra: the blue light stimulating one pathway and green and red light stimulating the other. We therefore selected one wavelength in the blue (393 nm) and in the red (631 nm) regions and fed $[2\text{-}^{14}C]$-glycine for the classical pathway and $[1\text{-}^{14}C]$-glutamate for the C-5 pathway and determined the ratio of incorporation of label into ALA via the two pathways (Table 1) at each wavelength. These experiments with monochromatic light were carried out without any preillumination but in the presence of LA (10 mM). The results indicated no significant change in the ratio of activity of the two pathways at either wavelength, and thus is

Table 1. Ratios of specific radioactivity of in vivo labeled ALA derived from $[2\text{-}^{14}C]$-glycine and $[1\text{-}^{14}C]$-glutamate during illumination with monochromatic light at two wavelengths during the early greening phase. Three-day old cells were illuminated with monochromatic light of 393 or 631 nm with an intensity of 1.75×10^{-10} E cm^{-2} s^{-1}

Period of illumination by monochromatic light (h)	The ratio: at 393 nm	$\dfrac{\text{Spec. act. of ALA from } [2\text{-}^{14}C]\text{-glycine}}{\text{Spec. act. of ALA from } [1\text{-}^{14}C]\text{-glutamate}}$ 631 nm
1	1.75	1.00
2	15.90	13.90
3	0.19	0.31
5	0.16	0.28

consistent with the fact that the development of each enzyme system shows an almost identical light dependence.

3.6 The Relationship of the Action Spectra for the Development of ALA Synthesizing Systems and Protein Synthesis

It was suggested that ALA synthetase, at least that occurring in the mitochondria of rat liver, is formed on cytoplasmic ribosomes but is transferred into the mitochondrion (Hayashi et al. 1969) and is finally located in the soluble matrix compartment between the inner and outer membrane (Jones and Jones 1970). Likewise there is some evidence to suggest that the enzymes of the C-5 pathway of ALA synthesis, irrespective of the mechanism involved, are soluble (Klein et al. 1978) and possibly reside in the soluble stroma of the chloroplast (Gough and Kannangara 1977). Thus one would reasonably predict that the action spectrum for the formation of ALA synthetase should resemble that of the formation of soluble proteins. The wavelength dependence of soluble protein formation is shown in Fig. 6 and has peaks in the blue, green and red regions like the action spectrum for the formation of the ALA-synthesizing system. Nevertheless, we wish to stress that the amounts of protein involved are small and considerable variation was involved, and so this spectrum (Fig. 6) can only be regarded as a preliminary result.

Although the results presented here show that the two different pathways leading to ALA formation appear to have a similar light dependence, a different time sequence for their formation was observed (see Table 1); thus it is possible that these two pathways may occupy different physiological roles and, perhaps, locations within the cell. Such considerations may have to be taken into account in future work.

4 Conclusion

Contrary to some of the blue light effects requiring only brief illumination periods and low energy input for saturation like those discussed by Brinkmann and Senger (1980a), the light-dependent formation of ALA requires a continuous input of light and high intensity for saturation. This is probably due to the fact that a similar continuous supply of high-intensity light is required for the formation of soluble proteins. It is well known that the enzymes of ALA synthesis, to fulfill their regulatory role, are labile (Irving and Elliott 1969) with a half-life of approximately 30 min (Beale 1971; Oh-hama and Senger 1975); thus continuous light must be supplied to maintain the enzyme level. Also the continuous illumination may counter any degradation of free ALA similar to that which has been shown to occur in darkness (see Fig. 3).

The wavelength-dependence studies indicate that ALA formation is dependent not only on blue light but also on green and red light. A cursory examination of the action spectrum suggests the participation of protochlorophyll(ide) as the photoreceptor (Brinkmann and Senger 1978a) but such a possibility is improbable for the following three reasons: firstly, close examination shows that the most effective actinic wavelengths do not exactly correspond with the peaks of protochlorophyll(ide) of higher plants; secondly, protochlorophyll could not be measured in dark-grown cells of *Scenedesmus obliquus* mutant C-2A' (Ellis and Timpson 1979), and, thirdly, it would not be reasonable that a later product of a synthetic pathway should be acting as a photoreceptor to control the first step of the same pathway.

Because of the involvement of green and red light it is interesting to speculate on the possibility that a tetrapyrrole pigment, either protoporphyrin IX or a hemoprotein, may be involved. The action spectrum bears some resemblance to that of hemoproteins, and any discrepancies in the ratios of the peaks in the blue and red regions of the spectra are probably due to the fact that the so-called action spectrum for ALA formation may also have superimposed upon it the blue light dependence of starch degradation for provision of substrates and energy for ALA formation.

From current knowledge of the chemistry of tetrapyrroles the best known models for such photoreactive hemoproteins would be the photolabile carbon-monoxide complexes of hemoglobin and cytochrome oxidase (Chance 1953). The oxygen, nitric oxide, and ethylisocyanide complexes of hemoglobin are also photolabile and it is also known that the ferric form of hemoglobin, leghemoglobin, myoglobin, and catalase can be reduced to the ferrous form by irradiation with UV (cf. Falk 1964). From biological evidence protoporphyrin IX has also been proposed as a photoreceptor in light-dependent induction of carotenogenesis in *Myxococcus xanthus* by Burchard and Hendricks (1969). Further, hemoproteins have been proposed as primary photoreceptors for the biosynthesis of phycobiliproteins in *Cyanidium caldarium* by Bogorad (1965), for the phototaxis of *Dictyostelium* by Poff et al. (1973) and for the conidiation of *Neurospora crassa* by Ninnemann and Klemm-Wolfgramm (1980). Schneider (1980) has obtained an action spectrum for the stimulation of georesponsiveness in maize roots with peaks similar to those observed in our wavelength-dependent response for ALA formation. He discusses a possible cooperation between a hemoprotein and phytochrome as probable photoreceptor, but this cannot apply in our case since it was shown (Brinkmann and Senger 1980b) that phytochrome plays no essential role in the chlorophyll formation of *Scenedesmus* mutant C-2A'.

If tetrapyrrole pigments are to be involved in the photoreceptor process for light-induced ALA formation, then a more intensive study must be made of the photoreactivity of such pigments, more precise action spectra must be obtained and, above, the receptor pigment isolated.

References

Beale SI (1971) Studies on the biosynthesis and metabolism of δ-aminolevulinic acid in *Chlorella*. Plant Physiol 48: 316–319

Beale SI, Gough SP, Granick S (1975) The biosynthesis of δ-aminolevulinic acid from the intact carbon skeleton of glutamic acid in greening barley. Proc Natl Acad Sci USA 72: 2719–2723

Bishop NI (1971) Preparation and properties of mutants: *Scenedesmus*. In: San Pietro A (ed) Methods in enzymology, vol 23, pp 130–143. Academic Press, London New York

Bogorad L (1965) Studies of phycobiliproteins. Rec Chem Prog 26: 1–12

Brinkmann G, Senger H (1978a) The development of structure and function in chloroplasts of greening mutants of *Scenedesmus*. IV. Blue light-dependent carbohydrate and protein metabolism Plant Cell Physiol 19: 1427–1437

Brinkmann G, Senger H (1978b) Light-dependent formation of thylakoid membranes during the development of the photosynthetic apparatus in pigment mutant C-2A' of *Scenedesmus obliquus*. In: Akoyunoglou G et al. (eds) Chloroplast development, pp 201–206. Elsevier/North-Holland Biomedical Press, Amsterdam

Brinkmann G, Senger H (1980a) Blue light regulation of chloroplast development in *Scenedesmus* mutant C-2A'. This volume

Brinkmann G, Senger H (1980b) Is there a regulatory effect of red light during greening of *Scenedesmus* mutant C-2A'? In: De Greef J (ed) Proceedings of photoreceptors and plant development. Antwerpen, in press

Burchard RP, Hendricks SB (1969) Action spectrum for carotenogenesis in *Myxococcus xanthus.* J.Bacteriol 97: 1165–1168

Chance B (1953) The carbon-monoxide compounds of the cytochrome oxidases. I. Difference spectra. J Biol Chem 202: 383–407

Ellis R, Timpson C (1979) The absence of protochlorophyll(ide) accumulation in algal cells with inhibited chlorophyll synthesis. Suppl Plant Physiol 63: 96 Abstr Nr 535

Falk JE (1964) Porphyrins and metalloporphyrins. BBA Library, vol II, p 20. Elsevier Publishing Co, Amsterdam London New York

Gough SP, Kannangara CG (1977) Synthesis of δ-aminolaevulinate by a chloroplast stroma preparation from greening barley leaves. Carlsberg Res Commun 42: 459–464

Harel E, Meller E, Rosenberg M (1978) Synthesis of 5-aminolevulinic acid-^{14}C by cell-free preparations from greening maize leaves. Phytochemistry 17: 1277–1280

Hase E (1971) Studies on the metabolism of nucleic acid and protein associated with the process of de- and regeneration of chloroplasts in *Chlorella protothecoides*. In: Boardman NK et al (eds) Autonomy and biogenesis of mitochondria and chloroplasts, pp 434–446. North-Holland Publishing Co, Amsterdam

Hayashi N, Yoda B, Kikuchi G (1969) Mechanism of allylisopropylacetamide-induced increase of δ-aminolevulinate synthetase in liver mitochondria. IV. Accumulation of the enzyme in the soluble fraction of rat liver. Arch Biochem Biophys 131: 83–91

Holden M (1965) Chlorophylls. In: Goodwin TW (ed) Chemistry and biochemistry of plant pigments, pp 461–488. Academic Press, London New York

Irving EA, Elliott WH (1969) A sensitive radiochemical assay method for δ-aminolevulinic acid synthetase. J Biol Chem 244: 60–67

Jones MS, Jones OTG (1970) Permeability properties of mitochondrial membranes and the regulation of haem biosynthesis. Biochem Biophys Res Commun 41: 1072–1079

Kikuchi G, Kumer A, Talmage D, Shemin D (1958) The enzymatic synthesis of δ-aminolevulinic acid. J Biol Chem 233: 1214–1219

Klein O, Senger H (1978a) Biosynthetic pathways to δ-aminolevulinic acid induced by blue light in the pigment mutant C-2A' of *Scenedesmus obliquus*. Photochem Photobiol 27: 203–208

Klein O, Senger H (1978b) Two biosynthetic pathways to δ-aminolevulinic acid in a pigment mutant of the green alga *Scenedesmus obliquus*. Plant Physiol 62: 10–13

Klein O, Dörnemann D, Senger H (1978) Two pathways for the biosynthesis of δ-aminolevulinic acid in *Scenedesmus obliquus* mutant C-2A'. In: Akoyunoglou G et al (eds) Chloroplast development, pp 45–50. Elsevier/North-Holland Biomedical Press, Amsterdam

Lascelles J (1968) The bacterial photosynthetic apparatus. In: Rose AH, Wilkinson JF (eds) Advances in microbial physiology, vol II, pp 1–42. Academic Press, London New York

Lowry OH, Rosebrough NJ, Farr AL, Randall RJ (1951) Protein measurement with the Folin phenol reagent. J Biol Chem 193: 265–275

Masoner M, Kasemir H (1975) Control of chlorophyll synthesis by phytochrome. I. The effect of phytochrome on the formation of 5-aminolevulinate in mustard seedlings. Planta 126: 111–117

Mauzerall D, Granick S (1956) The occurrence and determination of δ-aminolevulinic acid and porphobilinogen in urine. J Biol Chem 219: 435–446

Meller E, Harel E (1978) The pathway of 5-aminolevulinic acid synthesis in *Chlorella vulgaris* and in *Fremyella diplosiphon*. In: Akoyunoglou G et al (eds) Chloroplast development, pp 51–57. Elsevier/North-Holland Biomedical Press, Amsterdam

Ninnemann H, Klemm-Wolfgramm E (1980) Blue-light-controlled conidiation and absorbance change in *Neurospora* are mediated by nitrate reductase. This volume

Oh-hama T, Senger H (1975) The development of structure and function in chloroplasts of greening mutants of *Scenedesmus*. II. Biosynthesis of δ-aminolevulinic acid. Plant Cell Physiol 16: 395–405

Oh-hama T, Senger H (1978) Spectral effectiveness in chlorophyll and 5-aminolevulinic acid formation during regreening of glucose-bleached cells of *Chlorella protothecoides*. Plant Cell Physiol 19: 1295–1299

Poff KL, Loomis WF, Butler WL (1973) Light-induced absorbance changes associated with phototaxis in *Dictyostelium*. Proc Natl Acad Sci USA 70: 813–816

Porra RJ, Grimme LH (1974) Chlorophyll synthesis and intracellular fluctuations of δ-aminolevulinate formation during the regreening of nitrogen-deficient *Chlorella fusca*. Arch Biochem Biophys 164: 312–321

Salvador GF, Beney G, Nigon V (1976) Control of δ-aminolevulinic acid synthesis during greening of dark-grown *Euglena gracilis*. Plant Sci Lett 6: 197–202

Schneider HAW (1980) Visible and spectrophotometrically detectable blue-light responses of maize roots. This volume

Senger H, Bishop NI (1971) Light dependent greening of a yellow *Scenedesmus* mutant. Indian J Plant Physiol 16: 164–173

Senger H, Bishop NI (1972) The development of structure and function in chloroplasts of greening mutants of *Scenedesmus*. I. Formation of chlorophyll. Plant Cell Physiol 13: 633–649

Shemin D (1963) Δ-aminolevulinic acid dehydrase from *Rhodopseudomonas spheroides*. In: Colowick SP, Kaplan NO (eds) Methods in enzymology, vol V, pp 883–885. Academic Press, London New York

Shemin D, Russell CS (1953) δ-aminolevulinic acid, its role in the biosynthesis of porphyrins and purines. Am Chem Soc 76: 4873

Sokawa Y, Hase E (1967) Effect of light on the chlorophyll formation in the glucose bleached cells of *Chlorella protothecoides*. Plant Cell Physiol 8: 495–508

Physiology of Blue Light Effects

Blue-Light Photomorphogenesis
in Mushrooms *(Basidiomycetes)*

G. EGER-HUMMEL[1]

Morphogenesis in mushrooms is induced and controlled by environmental factors (Fig. 1). This can be microbial stimuli as in *Psilocybe paneoliformis* (Urayama 1960) and *Agaricus bisporus* (Eger 1961, 1962b, 1972; Hayes et al. 1969; Couvy 1974), which are partially replaceable by activated charcoal (Eger 1961; Couvy 1974; Long and Jacobs 1974), or light (see Brefeld 1889; Buller 1909; Schenck 1919; Madelin 1956; Garnett 1958; Alasoadura 1963; Lu 1965; Schwantes and Hagemann 1965; Miller 1967; Kitamoto et al. 1968; Perkins 1969; Tsusué 1969; review of Volz and Beneke 1969; McLaughlin 1970; Eger 1970a, b, Lavallée and Lortie 1971; Chapman and Fergus 1973, Morimoto and Oda 1973; Rogers 1973; Eger et al. 1974; Horikoshi et al. 1974; Uno et al. 1974), mainly in dung- and wood-inhabiting species. In some cases lowering temperature is also required (Aschan-Åberg 1960b; Kinugawa and Furukawa 1965; Eger et al. 1976; Li and Eger 1978). Fruiting is sensitive to excess CO_2 (Plunkett 1956; Tschierpe 1959; Niederpruem 1963; Tschierpe and Sinden 1964; Long 1966; Ingold and Nawaz 1967; Long and Jacobs 1968; Schwantes and Gessner 1974) and other volatile metabolites (Mader 1943; Schisler 1957; Eger 1961, 1962b, 1972; Lockard and Kneebone 1962) such as detected by Eger (1962a), Tschierpe and Sinden (1965), Turner et al. (1975), and Wood and Hammond (1977). The latter can act directly or antagonize the effect of CO_2 (Visscher 1978). Whilst fruiting stages II and III exude water (Eger 1973), later stages depend on transpiration for normal development (Borris 1934; Plunkett 1956). Sensitivity to fruiting stimuli is also influenced by nutrition (Heim 1958; Robbins and Hervey 1960; Eger 1961, 1970a; Schwantes 1968; Volz and Beneke 1969; Uno and Ishikawa 1974) and age of culture (Madelin 1956; Kitamoto et al. 1968; Eger 1970a; Morimoto and Oda 1973).

In regard to light four types of mushroom can be distinguished (Fig. 1). It is, however, impossible to classify all investigated strains accordingly since "darkness" and "fruiting initiation" have diverse meanings in different publications. A "dark" culture wrapped in aluminum foil eventually can fruit at very low light intensities because of multiple light reflection. It can also be "conditioned" for weak light stimuli by exposure of the inoculum to light (Robbins and Hervey 1960). Keeping dark cultures in closed containers or wrapped in black paper can inhibit primordia formation by restricted ventilation. Thus a D-type mushroom can turn out to be actually B-type, if the culture is well aerated (Schwantes and Hagemann 1965; Kinugawa 1977). Sometimes mechanical stimuli can replace the action of light (Borris 1934; Leonard 1973). Also a few primordia

1 Institut für Pharmazeutische Technologie der Universität, Marbacher Weg 6, 3550 Marburg, FRG

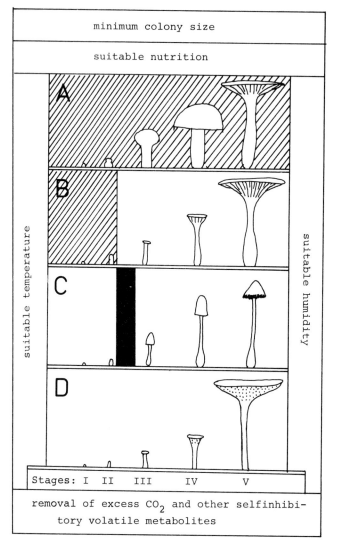

Fig. 1. Requirements for fruiting in mushrooms. *A Agaricus bisporus, B Lentinus tigrinus* (Schwantes and Hagemann 1965), *C Coprinus congregatus* (Manachère 1970), *D Favolus arcularius* (Kitamoto et al. 1968). ▨ = light and darkness irrelevant, ☐ = light required, ▬ = darkness required

can form spontaneously in aged or dry cultures. Spontaneous fruiting initiation is, however, poor as compared with that induced in cultures that received the proper amount of light at the right time (Eger 1970a, b; von Netzer 1978). This striking difference was already observed by Brefeld (1889).

Using bottles with solutions of a copper-ammonium complex and potassium dichromate, respectively, Brefeld stated already in 1889 that blue light is most stimulating on fruiting initiation and sporocarp development in *Coprinus* species and *Sphaerobolus*, whilst yellow is not. With various types of filter, later investigators confirmed the effect of blue light (Schenck 1919; Borris 1934; Schneiderhöhn 1954; Madelin 1956; Aschan-Åberg 1960a; Manachere 1961; Schwantes and Hagemann 1965; Perkins and Gordon 1969; Kitamoto et al. 1972, 1974; Chapman and Fergus 1973; Morimoto and Oda 1973; Rogers 1973; Durand 1977) but also found that the green and in one case even the yellow to orange part of the spectrum reveal some activity (Ingold and Nawaz 1967; Ingold and Peach 1970; Alloway and Ingold 1971, 1972). Red light was always ineffective. Perkins and Gordon (1969) obtained an action spectrum for fruiting initiation in *Schizophyllum commune* Fr. ex. Fr. starting at 320 nm. Activity was found up to 525 nm. The action spectra elaborated by Durand (1977) for primordia formation and sporophore development in *Coprinus congregatus* were equal. Only at high irradiance levels activity exceeded 520 nm. In *Favolus arcularius* (Fr.) Ames, however, a different fine structure of the action spectra for primordia formation (Kitamoto et al. 1972) and pileus expansion (Kitamoto et al. 1974) was observed. For pileus expansion the activity ended at 510 nm already whilst that for primordia formation extended to 560 nm and the maxima around 398 nm and 515 nm were lacking. The authors therefore suggested the existence of two receptor pigments, which Tan (1978) classified as the "characteristic" and the "atypical blue photoreceptor". However, mycelium and primordia of *F. arcularius* are buff in color, whilst young sporophores are rather dark brown and turn more or less ochraceous during later development. The difference in the action spectra may be caused by a shielding effect of yellow pigments which are not related to the photoreceptor. The open question of number and identity of photoreceptors in the process of reproduction in fungi is discussed in detail by Durand (1976).

In order to attack the problem of blue light photomorphogenesis in mushrooms the author chose *Pleurotus ostreatus* (Jacq. ex. Fr.) Kummer, a wood-rotting fungus. The pilei of its fruit bodies are described as being more or less gray to brown or bluish (Eger et al. 1979). Strains with white sporophores which are inherited by a single recessive gene exist (Arita 1974). White sporophores also develop in temperature-tolerant strains at low light intensities and/or at 24°C and above (Eger et al. 1976; Li and Eger 1978). *P. ostreatus* is a D-type mushroom according to Fig. 1. It requires light for both fruiting initiation and sporophore development. When kept for more than 3 days in continuous darkness each fruiting stage will start to convert into mycelium again. Even mature sporophores will do so. Light inhibits mycelial growth (Eger 1970a). To make mycelium fruit, a minimum colony size is necessary. There is a short period during colony development in which the response to light is optimal in regard to the time required for primordia formation. In sensitive strains a few lux given continuously is sufficient for fruiting initiation. However, the number of primordia formed per culture strongly depends on the amount of light given (Eger 1970a, b; Eger et al. 1974; von Netzer 1978). In strain "868x381", up to 4000 primordia (covering the whole surface of the colony) can be obtained in one petri dish (von Netzer 1978). To obtain such high numbers, a colony with poor aerial mycelium has to be used and a suitable N-source must be added at the time of illumination (Eger 1970a, b; Eger et al. 1974). In "868 x 381" glutamine is the most suitable (von Netzer 1978). The formation of fruiting stages II to V can as yet not be controlled so well. However, enough material can regularly be obtained to allow physiological and biochemical investigations. Since *P. ostreatus* cultivation is being commercialized now, any demand for fruit-body material can be satisfied in future.

P. ostreatus is tetrapolar. Fruiting occurs only in the dikaryotic stage (Eger 1974, 1978). In wild-type strains spores germinate within 2 days at high rates (Eger 1978; Li and Eger 1978). After 5 days successful mating can be detected microscopically by the presence of clamp connections. The dikaryon requires 6 to 8 days for mycelial growth, 4 to 6 days for primordia formation after light induction and 4 to 6 days more to develop mature sporophores under controlled conditions. Thus a life cycle can be completed within 4 to 5 weeks. P. ostreatus is therefore suitable for genetic analyses.

Physiological studies revealed that in P. ostreatus as in some other light-dependent mushrooms (Brefeld 1889; Buller 1909, 1922, 1931; Borris 1934; Plunkett 1961; Chapman and Fergus 1973; Schwalb and Shanler 1974), stages II and III can react positive phototropically and stage IV negative geotropically (Fig. 2). However P. ostreatus is unique in having a second phototropic phase. It is characterized in that the pileus margins, which receive more light, grow faster and toward the light source. This gives the fruit bodies their typical excentric shape (Block et al. 1959; Gyurkó 1972). Whilst the first phototropic phase and the geotropic phase exclude each other, the geotropic and second phototropic phase overlap. During phototropic phase 1 a curvature up to 90° can be enforced. The geotropic phase is more spectacular. Curvatures up to 180° were observed. With scattered light from above perfectly circular pilei develop on vertical stipes. The phototropic responses are more pronounced in weak light and/or at reduced ventilation. If air exchange is gradually reduced, pileus expansion, geotropic response, and phototropic response 2 are affected by the accumulating volatiles earlier than stipe elongation and phototropic phase 1. Therefore pileus expansion and phototropic reaction 2 can depend on common internal factors.

Phototropic responses require less light energy and time than fruiting initiation and development of stipes and pilei do. They depend also on blue light (Borris 1934). Therefore phototropism in P. ostreatus could be a good tool to attack the question of existence of more than one photoreceptor system in the morphogenesis of mushrooms. Our first approach is a genetic one. By treatment of dikaryons with 1-methyl-3-nitro-1-nitroso-guanidine so far three types of mutant were obtained (Fig. 3). Type a is completely atropic. The sporophores can grow even into the nutrient medium and the pilei can turn upside down. Type b lacks the geotropic and the second phototropic phase. Its pilei are therefore always circular and directed toward the light. Type c is blocked at stage II, III or IV. Yellow (like flavins) to orange (like carotines) pigments accumulate in bright light. Thus striking colored sporophores result. In weak light or darkness the pigments

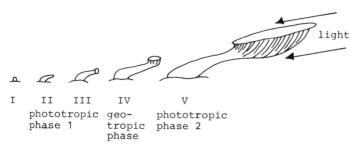

I   II   III   IV   V
       phototropic   geo-   phototropic
       phase 1   tropic   phase 2
                     phase

Fig. 2. Fruiting stages of *Pleurotus ostreatus* and tropic phases

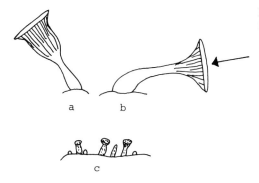

Fig. 3. Type of mutants in *Pleurotus ostreatus* in regard to tropic reactions

bleach, leaving the fruiting structures as white as mycelium. All three types of mutant require much more light for fruiting initiation than wild-type strains. The mutants demonstrate firstly that mushrooms with gray to brown sporophores can be capable of producing yellow to orange pigments. Normally these pigments might be present in invisible amounts which may nevertheless be sufficient to distort action spectra. Secondly phototropic phases 1 and 2 do not necessarily depend on the same factors, since some mutants are affected in phototropic phase 2 but not in phototropic phase 1. A new dedikaryotization method (Eger and Leal-Lara 1978) allows to separate the two nuclei in a mutant dikaryon and use them for complementation tests. Also the neohaplonts obtained this way can be crossed with wild-type monokaryons and the segregation of mutant genes in the progenies can then be studied. In this way we hope to gather information about the number of factors involved in the single tropic reactions and of factors common to them. Later we will try to find out if the single mutations affect photoreception, energy conversion, energy transfer or the response. We hope to get on the way to understand the blue light photomorphogenesis in mushrooms.

References

Alasoadura SO (1965) Fruiting in *Sphaerobolus* with special reference to light. Ann Bot NS 27: 123–145

Alloway JM, Ingold CT (1971) Physiological observations on *Sphaerobolus*. Trans Br Mycol Soc 57: 411–416

Alloway JM, Ingold CT (1972) Interaction of light of different wave-length in the late stage of sporophore development of *Sphaerobolus*. Trans Br Mycol Soc 58: 504–507

Arita J (1974) Genetic study on white fruit-bodies of *Pleurotus ostreatus* (Fr.) Kummer. Rep Tottori Mycol Inst 11: 58–68

Aschan-Åberg K (1960a) The production of fruit bodies in *Collybia* velutipes. III. Influence of the quality of light. Physiol Plant 13: 276–279

Aschan-Åberg K (1960b) Studies on mono- and di-caryotic mycelia of *Collybia velutipes*. Physiol Plant 13: 280–297

Block SS, Tsao G, Han L (1959) Experiments in the cultivation of *Pleurotus ostreatus*. Mushroom Sci 4: 309–325

Borris H (1934) Über den Einfluß äußerer Faktoren auf Wachstum und Entwicklung der Fruchtkörper von *Coprinus lagopus*. Planta 22: 644–684

Brefeld O (1889) Untersuchungen aus dem Gesamtgebiet der Mykologie. H. 8, Basidiomyceten III: 275–290

Buller AHR (1909) Researches on fungi, vol I. Longmans Green and Co, London

Buller AHR (1922) Researches on fungi, vol II. Longmans Green and Co, London

Buller AHR (1931) Researches on fungi, vol III. Longmans Green and Co, London

Chapman ES, Fergus CL (1973) An investigation of the effects of light on basidiocarp formation of *Coprinus domesticus*. Mycopathologia 51: 315–326

Couvy J (1974) La fructification d'*Agaricus bisporus* en milieu aseptique: un modele experimental pour l'etude des substances impliquees dans l'initiation fructifere. Mushroom Sci 9/1: 157–164

Durand R (1976) Influence des radiations lumineuses sur les processus de reproduction des champignons; hypotheses sur l'identite des photorecepteurs. Revue bibliographique. Mycopathologia 60: 3–16

Durand R (1977) Equal quantal spectra for different light sensitive phases during fruiting in a basidiomycete, *Coprinus congregatus*. In: Bigelow HE, Simmons EG (eds) 2nd Int Mycol Congr Abstr, p 150. IMC-2 Inc, Tampa, Florida

Eger G (1961) Untersuchungen über die Funktion der Deckschicht bei der Fruchtkörperbildung des Kulturchampignons, *Psalliota bispora* Lge. Arch Microbiol 39: 313–334

Eger (1962a) Ein flüchtiges Stoffwechselprodukt des Kulturchampignons, *Agaricus (Psalliota) bisporus* (Lge) Sing., mit antibiotischer Wirkung. Naturwissenschaften 49: 261

Eger G (1962b) Untersuchungen zur Fruchtkörperbildung des Kulturchampignons. Mushroom Sci 5: 314–320

Eger G (1970a) Die Wirkung einiger N-Verbindungen auf Mycelwachstum und Primordienbildung des Basidiomyceten *Pleurotus* spec. aus Florida. Arch Microbiol 74: 160–173

Eger G (1970b) Die Wirkung des Lichts auf die Primordienbildung des Basidiomyceten *Pleurotus* spec. aus Florida. Arch Microbiol 74: 174–192

Eger G (1972) Experiments and comments on the action of bacteria on sporophore initiation in *Agaricus bisporus*. Mushroom Sci 8: 719–725

Eger G (1973) *Flammulina velutipes*, Lebenszyklus eines Basidiomyceten. Film C 1083. Inst Wiss Film 3400 Göttingen

Eger G (1974) Rapid method for breeding *Pleurotus ostreatus*. Mushroom Sci 9/1: 567–573

Eger G (1978) Biology and breeding of *Pleurotus*. In: Chang ST, Hayes WA (eds) The biology and cultivation of edible fungi, pp 497–519. Academic Press, London New York

Eger G, Leal-Lara H (1978) Verfahren zur Dedikaryotisierung dikaryotischer Stämme von Basidiomyceten. Deutsches Patentamt München, P 28 13 521.9

Eger G, Gottwald H-D, von Netzer U (1974) The action of light and other factors on sporophore initiation in *Pleurotus ostreatus*. Mushroom Sci 9/1: 575–583

Eger G, Eden G, Wissig E (1976) *Pleurotus ostreatus*-breeding potential of a new cultivated mushroom. Theoret Appl Genet 47: 155–163

Eger G, Li S-F, Leal-Lara H (1979) Contribution to the discussion on the species concept in the *Pleurotus ostreatus* complex. Mycologia 71: 577–588

Garnett E (1958) Studies of factors affecting fruiting body formation in *Cyathus stercoreus* (Schw.) de Toni. PhD Thesis, Indiana University

Gyurkó P (1972) Die Rolle der Belichtung bei dem Anbau des Austernseitlings (*Pleurotus ostreatus*). Mushroom Sci 8: 461–469

Hayes WA, Randle PhE, Last FT (1969) The nature of the microbial stimulus affecting sporophore formation in *Agaricus bisporus* (Lge.) Sing. Ann Appl Biol 64: 177–187

Heim R, Brack A, Kobel H, Hofmann A, Cailleux R (1958) Déterminisme de la formation des carpophores et des sclérotes dans la culture du *Psilocybe mexicana* Heim, agaric hallucinogène du Mexique, et mise en évidence de la psilocybine et de la psilocine. CR Seances Acad Sci 246: 1346–1351

Horikoshi T, Kitamoto Y, Kasai Z (1974) Effect of light on the initiation of pileus formation in a basidiomycete, *Favolus arcularius*. Plant Cell Physiol 15: 903–911

Ingold CT, Nawaz M (1967) Carbone dioxide and fruiting in *Sphaerobolus*. Ann Bot 31: 351–357

Ingold CT, Peach J (1970) Further observations on fruiting in *Sphaerobolus* in relation to light. Trans Br Mycol Soc 54: 211–220

Jablonský I (1975) Einfluß der Belichtungsintensität und anderer Faktoren des Milieus auf die Entwicklung der Fruchtkörper des Austernseitlings *Pleurotus ostreatus* (Jacq. ex. Fr.) Kumm. Česka Mykol 29: 140–152

Kinugawa K (1977) *Collybia velutipes* can fruit under total darkness. Trans Mycol Soc Jan 18: 353–356

Kinugawa K, Furukawa H (1965) The fruit-body formation in *Collybia velutipes* induced by the lower temperature treatment of one short duration. Bot Mag Tokyo 78: 240–244

Kitamoto Y, Takahashi M, Kasai Z (1968) Light-induced formation of fruit-bodies in a basidiomycete, *Favolus arcularius* (Fr.) Ames. Plant Cell Physiol 9: 797–805

Kitamoto Y, Suzuki A, Furukawa S (1972) An action spectrum for light-induced primordium formation in a basidiomycete *Favolus arcularius* (Fr.) Ames. Plant Physiol 49: 338–340

Kitamoto Y, Horikoshi T, Suzuki A (1974) An action spectrum for photoinduction of pileus formation in a basidiomycete, *Favolus arcularius.* Planta 119: 81–84

Lavallée A, Lortie M (1971) Notes sur la production, in vitro, de sporophores par *Pholiota aurivella.* Can J Bot 49: 567–572

Leonard TJ (1973) Induction of haploid fruiting by mechanical injury in *Schizophyllum commune.* Mycologia 65: 809–822

Li S-F, Eger G (1979) Characteristics of some *Pleurotus*-strains from Florida, their practical and taxonomic importance. Mushroom Sci 10: 155–169

Lockard JD, Kneebone LR (1962) Investigation of the metabolic gases produced by *Agaricus bisporus* (Lange) Sing. Mushroom Sci 5: 281–299

Long PE, Jacobs L (1968) Some observations on CO_2 and sporophore initiation in the cultivated mushroom. Mushroom Sci 7: 373–384

Long PE, Jacobs L (1974) Aseptic fruiting of the cultivated mushroom. *Agaricus bisporus.* Trans Br Mycol Soc 63: 99–107

Long TJ (1966) Carbon dioxide effect in the mushroom *Collybia velutipes.* Mycologia 58: 319–321

Lu BC (1965) The role of light in fructification of the basidiomycete *Cyathus stercoreus.* Am J Bot 52: 432–437

Madelin MF (1956) The influence of light and temperature on fruiting of *Coprinus lagopus* Fr. in pure culture. Ann Bot NS 20: 467–480

Mader EO (1943) Some factors inhibiting the fructification and production of the cultivated mushroom, *Agaricus campestris* L. Phytopathology 33: 1134–1145

Manachère G (1961) Influence des conditions d'éclairement sur la fructification de *Coprinus congregatus.* Bull Fr CR Séances Acad Sci 252: 2912–2913

Manachère G (1970) Recherches physiologiques sur la fructification de *Coprinus congregatus.* Action de la lumiere; rythme de production de carpophores. Ann Sci Nat, Bot.11: 1–96

McLaughlin DJ (1970) Environmental control of fruitbody development in *Boletus rubinellus* in axenic culture. Mycologia 62: 307–331

Miller OK (1967) The role of light in the fruiting of *Panus frogilis.* Can J Bot 45: 1939–1943

Morimoto N, Oda Y (1973) Effects of light on fruit-body formation in a basidiomycete, *Coprinus macrorhizus.* Plant Cell Physiol 14: 217–225

Netzer U von (1978) Induktion der Primordienbildung bei dem Basidiomyceten *Pleurotus ostreatus.* Bibliotheca mycologia, vol 62. J Cramer, Vaduz

Niederpruem DJ (1963) Role of carbon dioxide in the control of fruiting of *Schizophyllum commune.* J Bacteriol 85: 1300–1308

Perkins JH (1969) Morphogenesis in *Schizophyllum commune.* I. Effects of white light. Plant Physiol 44: 1706–1711

Perkins JH, Gordon AS (1969) Morphogenesis in *Schizophyllum commune.* II. Effects of monochromatic light. Plant Physiol 44: 1712–1716

Plunkett BE (1956) The influence of factors of the aeration complex and light upon fruit-body form in pure cultures of an agaric and polypore. Ann Bot NS 20: 563–586

Plunkett BE (1961) The change of tropism in *Polyporus brumalis* stipes and the effect of directional stimuli on pileus differentiation. Ann Bot NS 25: 206–233

Robbins WJ, Hervey A (1960) Light and the development of *Poria ambigua.* Mycologia 52: 231–247

Rogers MA (1973) Photoresponses of *Coprinus stercorarius:* Basidiocarp development and matura-
   tion. Mycologia 65: 907–913
Schenck E (1919) Die Fruchtkörperbildung bei einigen *Bolbitius-* und *Coprinus*-Arten. Beih Bot
   Zentralbl 36 I: 355–413
Schisler LC (1957) A physiological investigation of sporophore initiation in the cultivated mush-
   room, *Agaricus campestris* L. ex Fr. Diss Abstr 17, 958. Publ 20973, Pennsylvania University
Schneiderhöhn G (1954) Das Actionsspectrum der Wachstumsbeeinflussung durch Licht bei *Coprinus
   lagopus.* Arch Mikrobiol 21: 230–236
Schwalb MN, Shanler A (1974) Phototropic and geotropic responses during development of normal
   and mutant fruit bodies of the basidiomycete *Schizophyllum commune.* J Gen Microbiol 82:
   209–212
Schwantes H-O (1968) Wirkung unterschiedlicher Stickstoffkonzentrationen und -verbindungen auf
   Wachstum und Fruchtkörperbildung von Pilzen. Mushroom Sci 7: 257–272
Schwantes H-O, Gessner E (1974) Untersuchungen zur Fruchtkörperbildung von *Lentinus tigrinus*
   (Bull. ex Fr.) Fr. und *Polyporus melanopus* (Swartz ex Fr.) Fr. in Abhängigkeit von der Zusam-
   mensetzung des umgebenden Gasraumes. Biol Zentralbl 93: 561–570
Schwantes H-O, Hagemann F (1965) Untersuchungen zur Fruchtkörperbildung bei *Lentinus tigrinus*
   Bull. Ber Dtsch Bot Ges 78: 89–101
Tan KK (1978) Light-induced fungal development. In: Smith JE, Berry DR (eds) The filamentous
   fungi, vol III, pp 334–357. Edward Arnold, London
Tschierpe HJ (1959) Die Bedeutung des Kohlendioxyds für den Kulturchampignon. Gartenbauwis-
   senschaft 24: 18–75
Tschierpe HJ, Sinden JW (1964) Weitere Untersuchungen über die Bedeutung von Kohlendioxyd
   für die Fruktifikation des Kulturchampignons, *Agaricus campestris var. bisporus* (L.) Lge. Arch
   Mikrobiol 49: 405–425
Tschierpe HJ, Sinden JW (1965) Über leicht flüchtige Produkte des aeroben und anaeroben Stoff-
   wechsels des Kulturchampignons, *Agaricus campestris var. bisporus* (L.) Lge. Arch Microbiol 52:
   231–241
Tsusué YM (1969) Experimental control of fruit-body formation in *Coprinus macrorhizus.* Dev Growth
   Differ 11: 164–177
Turner EM, Wright M, Ward T, Osborne DJ, Self R (1975) Production of ethylene and other volatiles
   and changes in cellulase and laccase during the life-cycle of the cultivated mushroom, *Agaricus
   bisporus.* J Gen Microbiol 91: 167–176
Uno I, Ishikawa T (1974) Effect of glucose on the fruiting body formation and adenosine 3',5'-
   cyclic monophosphate levels in *Coprinus macrorhizus.* J Bacteriol 120: 96–100
Uno I, Yamaguchi M, Ishikawa T (1974) The effect of light on fruiting body formation and adenosine 3':5'-cyclic monophosphate metabolism in *Coprinus macrorhizus.* Proc Natl Acad Sci USA
   71: 479–483
Urayama T (1960) Studies on fruit body formation of *Psilocybe paneoliformis* in pure culture. Mem
   Fac Lib Arts Miazaki Univ 9: 393–462
Visscher HR (1979) Fructification of *A. bisporus* (Lge) Imb. in relation to the relevant microflora
   in the casing soil. Mushroom Sci 10: 641–654
Volz PA, Beneke ES (1969) Nutritional regulation of basidiocarp formation and mycelial growth
   of *Agaricales.* Mycopathologia 37: 225–253
Wood DA, Hammond JBW (1977) Ethylene production by axenic fruiting cultures of *Agaricus
   bisporus.* Appl Environ Microbiol 34: 228–229

Blue Light Induced Differentiation in *Phycomyces blakesleeanus*

V.E.A. RUSSO[1], P. GALLAND[1], M. TOSELLI[1], and L. VOLPI[1]

1 Introduction

We have heard at this conference that blue light has several effects in many plants and microorganisms. Blue light has several physiological effects in the *same* organism: *Phycomyces blakesleeanus.* The oldest effects known are phototropism and a light-induced growth response (for a review see Bergman et al. 1969; Russo and Galland 1979). For thirty years it has been known that blue light also induces β-carotene (Garton et al. 1951). A few years ago Bergman (1972) showed that short pulses of blue light on a growing mycelium leads to an ordered spatial distribution of the production of sporangiophores. We have further investigated the action of blue light on the differentiation of *Phycomyces.*

The three major morphogenetical changes in the asexual life of *Phycomyces* are depicted in Fig. 1:

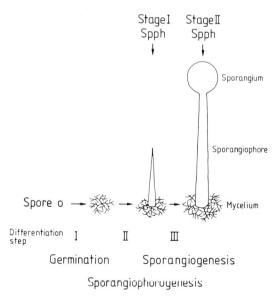

Fig. 1. Schematic representation of the asexual life cycle of *Phycomyces* with the three major differentiation steps: germination, sporangiophorogenesis and sporangiogenesis

Abbreviation: *Spph:* Sporangiophore

1 Max-Planck-Institut für Molekulare Genetik, Abt. Trautner, Ihnestraße 63/73, D-1000 Berlin 33

1. The germination of the spore with the consequent formation of the mycelium (differentiation step I).

2. The formation of the sporangiophore (differentiation step II).

3. The formation of the sporangium (differentiation step III).

Under normal laboratory conditions the three differentiation steps are "automatic", so that the asexual cycle is completed in about four days. There are, however, conditions which permit one to uncouple the growth of mycelium from the formation of spphs. Under these conditions blue light can induce the formation of spphs we name this phenomenon photophorogenesis. There are other conditions in which the growth of spph is uncoupled from the formation of sporangia; blue light can induce the formation of sporangia, we name this phenomenon photosporangiogenesis.

No conditions have been found until now in which blue light has an effect on germination.

2 Photophorogenesis

When mycelia are grown in shell vials and kept in a closed beaker in the dark, it is found that the number of spph per vial is a function of the number of vials per unit volume of the closed beaker (Russo 1977b; Galland and Russo 1979a). Figure 2 shows this crowd effect in the dark. The presence of 10 ml 5 M NaOH in a small container placed in the closed beaker eliminates this crowd effect. Continuous blue light also eliminates this crowd effect.

The experiments with NaOH indicate that the crowd effect is not due to a lack of oxygen but to the presence of an unknown gas X which is made by the mycelium and which is absorbed by NaOH.

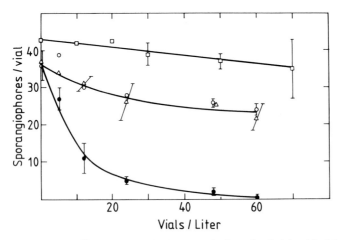

Fig. 2. Crowd effect on sporangiophorogenesis. Inoculated vials with rich medium were closed in a 1-l beaker. Beakers have different amounts of vials as indicated on the *abscissa*. Zero vial/liter indicates an open beaker. ● wt continuous dark, ○ wt continuous light, △ wt continuous dark but with 10 ml 5 N NaOH in a small beaker placed in the 1-l beaker, □ B454 (a mutant isolated in our laboratory) in continuous dark

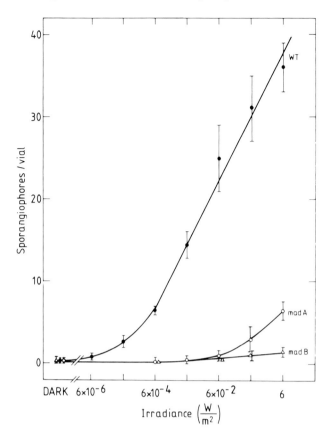

Fig. 3. Threshold of photophorogenesis for *wt*, mutants in *mad*A and *mad*B genes. Seventy inoculated vials were closed in a 1-l beaker; at 50h after inoculation the samples were irradiated for 1 min with blue light at the indicated fluence rate. After 100 h the spphs were counted. (After Galland and Russo 1979a)

We have been able to select mutants which are resistant as well as mutants which are supersensitive to the gas X. It is clear that the chemical nature of the gas X could give some clue as to the action of blue light. The existence of the mutants resistant and super-sensitive to the gas X should help in the identification of this gas. We know that the gas X is not CO_2 or HCN or O_2.

We have shown elsewhere (Galland and Russo 1979a) that 1 min of blue light of quite low intensity, given at 50 h after the spore seeding, is sufficient to induce the phorogenesis.

The thresholds of *wt* and of the phototropic abnormal mutants in the genes *mad*A and *mad*B are shown in Fig. 3. The threshold for *mad*A is shifted by a factor at 10^4, and for *mad*B by a factor of 10^5. The phototropic abnormal mutants in the genes *mad*C, D, E, F, G behave like *wt* (Galland and Russo 1979a).

This implicates the products of the genes *mad*A and *mad*B in the transduction of the blue light information in phototropism and photophorogenesis. Beside the gas X

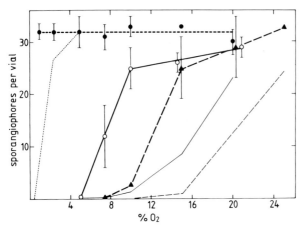

Fig. 4. Effect of blue light irradiation on spph formation under limiting oxygen conditions. At hour 56 the mycelia were irradiated for 1 min with blue light (20 W m^{-2}). A bar indicates the standard error of the mean of at least 4 experiments. ●——● wt, ○——○ C171 (caroteneless mutant), ▲——▲ C173 (caroteneless mutant). The value of the unirradiated samples are indicated by *thin lines*. (After Galland and Russo (1979b)

there are other molecules which can uncouple mycelial growth from phorogenesis. One molecule is oxygen. In Fig. 4 is shown the threshold of oxygen for the photogenesis in *wt* and two mutants lacking β-carotene (C171, C173), in continuous dark and with 1 min blue light at 6 W m^{-2}, given at 50 h after the seeding of the spores. Why the mutants lacking β-carotene have a higher oxygen threshold than *wt* is discussed at length by Galland and Russo (1979b).

Briefly, the hypothesis is that β-carotene is the precursor of vitA and vitA is necessary for phorogenesis. This is based on the observation that vitA added to the culture medium brings the oxygen threshold of the caroteneless mutants to the level of *wt*.

The other molecule is cAMP. In Fig. 5 is shown the effect of cAMP on the number of spph per vial in the dark and in white light. It is evident that cAMP is also able in an open system to inhibit the phorogenesis at a concentration above 10^{-3} M while white light can eliminate this inhibition.

3 Photosporangiogenesis

We have shown previously that it is possible to obtain a syncronous culture of spph stage I in vials of 1 cm diameter (Russo 1977b). In these vials we obtain about thirty spph per vial. If we transfer the vials to a closed beaker when the spph are about 1.5 cm high, then the sporangiogenesis will take about 40 h to occur. If the closed beaker is in continuous blue light the sporangiogenesis is completed in about 12 h. Red light has no effect. The same fast sporangiogenesis (in about 12 h) occurs in an open beaker or in a closed system with a stream of air. This indicates that a self-made gas Y inhibits sporangiogenesis in the dark. This gas is made only by the spphs and not by the mycelium which indicates that the gas Y is different from the gas X which inhibits the spor-

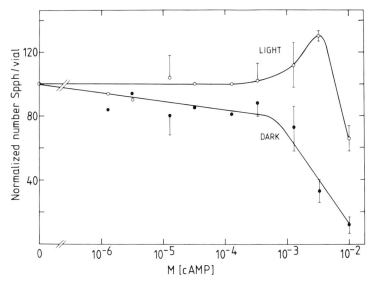

Fig. 5. Effect of cAMP on the sporangiophorogenesis in the dark ●——● and in continuous light ○——○. The vials were in an open system. Average of 4 experiments with the standard error of the mean. The absolute value of spph/vial were: in the dark 22 ± 3; in the light 14 ± 1. The spphs were counted after 150 h

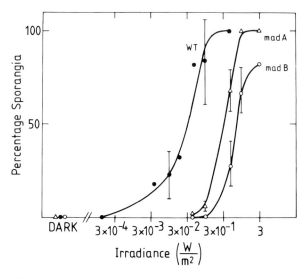

Fig. 6. Threshold of photosporangiogenesis for *wt* and mutants in *mad*A and *mad*B genes. Syncronous culture of stage I spph on vials, were closed in a 1-l beaker and continuously illuminated with blue light at the given intensity. After 24 h the percentage of sporangia was counted. **A** *bar* indicates the standard error of the mean of at least 4 experiments

angiogenesis. If we give continuous light at different intensities to *wt* and to the mutants abnormal in phototropism for 24 h and then measure the number of sporangia made, we obtain the results shown in Fig. 6. The threshold for *wt* is about 10^{-4} W m^{-2} while for mutants in the genes *mad*A and *mad*B it is about 10^{-2} W m^{-2}. All the other mutants in genes *madC to madG* behave like *wt*. This indicates that the action of the gene products of *mad*A and *mad*B is also necessary for photosporangiogenesis. More details will be published elsewhere.

4 Discussion

The genes *mad*A and *mad*B are necessary for the transduction of the blue light information in all the five blue light effects known in *Phycomyces:* phototropism (Bergman et al 1973); light-induced growth response (Lipson 1975), photocarotenogenesis (Jayaram et al. 1979), photophorogenesis (Bergman 1972; Galland and Russo 1979a), photosporangiogenesis Section 3, Fig. 7.

It is therefore conceivable that all these responses are mediated by the same blue light photoreceptor. We do not know the biochemical effect of blue light in photodifferentiation of *Phycomyces* but our physiological results with cAMP help us to speculate.

Other effects of cAMP are known in *Phycomyces.* Cohen (1974) has shown that the level of cAMP in *Phycomyces* spph drops transiently by a factor of two after a step-up of light. Cohen and Atkinson (1978) have further shown that blue light in vitro activates *Phycomyces* cAMP phosphodiesterase.

On the other hand Tu and Malhotra (1977) have found that 40 μM cAMP decreases the glycogen content of mycelium by a factor of two and thickens the cell wall of *Phycomyces* mycelium up to ten times its normal value. They assume that the inhibition of glycogen synthesis is responsible for the thickening of the cell wall. It is reasonable to assume that a thickening of the cell wall of mycelium will inhibit the phorogenesis. In light of these results it is conceivable that blue light either lowers the amount of cAMP in the cell or increases the production of glycogen; in both cases preventing the thickening of the cell wall, therefore allowing the phorogenesis. We are currently testing this hypothesis.

On the biological significance of all these light responses, of geotropism (Bergman et al. 1969) and of the avoidance response (Cohen et al. 1975; Russo 1977a; Russo et al. 1977) can be speculated. One has to keep in mind that *Phycomyces* grows on animal dung and on rotten wood and that the spores can be dispersed only by contact. It would therefore be a selective advantage for the sporangiophore and the sporangium to be made

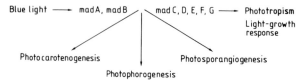

Fig. 7. Diagram showing that of the seven known genes necessary for phototropism, the first two *mad*A and *mad*B are also necessary for the other three blue light responses: photocarotenogenesis, photophorogenesis, photosporangiogenesis

only when light is present. The geotropic response will help the spornagiophore to grow upward. The phototropic and the avoidance response will enable the sporangium to get out of the small cavity where the mycelium is eventually grown, so that the spores can be dispersed by contact with animals.

Acknowledgment. We thank David Musgrave for critically reading the manuscript and making it more readable.

References

Bergman K (1972) Blue light control of sporangiophore initiation in *Phycomyces.* Planta 107: 53–67

Bergman K, Burge PV, Cerdá-Olmedo E, David CN, Delbrück M, Foster KW, Goodell EW, Heisenberg M, Meissner G, Zalokar M, Dennison DS, Shropshire W Jr (1969) *Phycomyces.* Bacteriol Rev 33: 99–157

Bergman K, Eslava AP, Cerdá-Olmedo E (1973) Mutants of *Phycomyces* with abnormal phototropism. Mol Gen Genet 123: 1–16

Cohen RJ (1974) Cyclic AMP levels in *Phycomyces* during a response to light. Nature (London) 251: 144–146

Cohen RJ, Atkinson MM (1978) Activation of *Phycomyces* adenosine 3', 5'-monophosphate phospho-diesterase by blue light. Biochem Biophys Res commun 83: 616–621

Cohen RJ, Jan YN, Matricon J, Delbrück M (1975) Avoidance response, house response, and wind responses of the sporangiophore of *Phycomyces.* J Gen Physiol 66: 67–95

Galland P, Russo VEA (1979a) Photoinitiation of sporangiophores in *Phycomyces* mutants deficient in phototropism and in mutants lacking β-carotene. Photochem Photobiol 29: 1009–1014

Galland P, Russo VEA (1979b) The role of retinol in the differentiation of *Phycomyces blakesleeanus.* Planta (in press)

Garton GA, Goodwin TW, Liginsky W (1951) Studies in carotenogenesis. I. General considerations governing β-carotene synthesis by the fungus *Phycomyces blakesleeanus Burgeff.* Biochem J 48: 154–163

Jayaram M, Presti D, Delbrück M (1979) Light induced carotene synthesis in *Phycomyces.* Exp Mycol (in press)

Lipson ED (1975) White noise analysis of *Phycomyces* light growth response system II. Photomutants. Biophys J 15: 1033–1045

Russo VEA (1977a) Ethylene-induced growth of *Phycomyces* mutants abnormal for autochemotropism. J Bacteriol 130: 548–551

Russo VEA (1977b) The role of blue light in synchronization of growth and inhibition of differentiation of stage I sporangiophore of *Phycomyces blakesleeanus.* Plant Sci Lett 10: 373–380

Russo VEA, Galland P (1979) Sensory physiology of *Phycomyces blakesleeanus.* In: Hemmerich P (ed) Sensory physiology and structures of molecules. Springer, Berlin Heidelberg New York (in press)

Russo VEA, Halloran B, Gallori E (1977) Ethylene is involved in autochemotropism of *Phycomyces.* Planta 134: 61–67

Tu JC, Malhotra SK (1977) The significance of cAMP induced alterations in the cellular structure of *Phycomyces.* Can J Microbiol 23: N4, 378–388

Effect of Blue Light on Metabolic Processes, Development and Movement in True Slime Molds

L. RAKOCZY[1]

1 Introduction

In acellular slime molds (Myxomycetes), as in other organisms, light affects a number of metabolic and physiological processes. Most of the information concerns the plasmodium. Mainly white light has been studied, though there is evidence for blue light effects in many slime mold responses.

The plasmodium, main stage in the development cycle of myxomycetes, possesses a number of features which make it interesting and useful for photobiological studies. Plasmodia of many species of slime molds can be maintained under laboratory conditions and some of them can be grown in pure culture on semi-defined or defined media (Daniel and Rusch 1961; Brever et al. 1964; Daniel and Baldwin 1964; Ross 1964; Ross and Sunshine 1965; Henney and Henney 1968; Henney and Lynch 1969; Mohberg and Rusch 1969; Carlile 1971). Some of the species belonging to *Physarales* produce the plasmodial phase in large amounts which is useful for the study of metabolic events affected by light. Moreover, the great ability of plasmodia to regenerate makes it possible to obtain many individuals in the same physiological state from one organism.

The plasmodium is a coenocyte with a large number of nuclei which divide synchronously. It is therefore one giant cell, which can cover many hundred of cm^2 in some species. Some organelles of the plasmodium, for example nuclei and mitochondria, can be isolated and thus studied separately. As no rigid wall surrounds the plasmodium it can be easily disrupted and its macromolecules isolated readily. The plasmodium can migrate over the substratum for long distances. The migration and its direction are controlled by internal and external factors, including light. Growth of the plasmodium and differentiation proceed separately, which facilitates a separate study of the effect of light on each of these processes.

The action of light on plasmodia is manifested by its effects on migration, growth rate, sporulation, and pigmentation.

2 Effect of Light on Migration

The ability of a plasmodium to migrate over a solid substratum presumably enables it to choose optimal environmental conditions (Rakoczy 1963b; Carlile 1970; Tso and

1 Department of Plant Physiology, Polish Academy of Sciences, Sławkowska str. 17, 31-016 Kraków, Poland

Mansour 1975; Denbo and Miller 1976; Knowles and Carlile 1978a, b). Light is one of the external factors influencing the direction of movement of the plasmodia. This was realized long ago, although early reports were contradictory. Hofmeister (1867), for example, held that slime molds migrated toward light, whereas Baranetzki (1876) found a light-avoiding response by plasmodia of a number of species. This was confirmed by Stahl (1880). Employing a light gradient method it was found for *Physarum nudum* that the response to light depends on the age of the plasmodium (Rakoczy 1963a). In given conditions the young plasmodia migrated to dark places, whereas fully developed organisms in which vegetative growth had terminated showed migration towards illuminated areas. Such behavior is understandable, since it has been demonstrated for this species (Rakoczy 1966) as well as for *Physarum polycephalum* (Daniel and Baldwin 1964) that light inhibits growth of plasmodia. However, after the period of growth has terminated light is required for induction of sporulation (Daniel and Rusch 1962; Rakoczy 1962). The investigations performed with the gradient method were qualitative in nature. The quantitative measurements (Białczyk and Rakoczy 1974, 1975) for young and old plasmodia are presented in Fig. 1. As can be seen in Fig. 1 the positive response

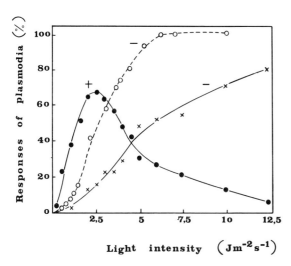

Fig. 1. Behavior of the young and old plasmodia of *Physarum nudum* under light conditions. One half of each specimen was illuminated for 3 h with white light. + = responses of old (12 to 13 days) plasmodia towards light, − = light avoidance responses: ●——● young (4-day old), x——x old plasmodia (Białczyk and Rakoczy 1974, 1975)

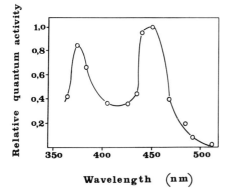

Fig. 2. Action spectrum for the avoidance of light by *Physarum nudum* plasmodia. The curve was drawn by plotting the reciprocal of quantum flux density that induce response in 50% of plasmodia as a function of wavelength (Białczyk 1979)

of old plasmodia (toward illuminated areas) is limited to a narrow range of light intensity. The active ranges of radiation for the light-avoiding responses of young plasmodia
are blue and near UV. The action spectrum (Fig. 2) for this type of response has been
obtained for *Physarum nudum* (Białczyk 1979). It shows two maxima: at 375 nm and
452 nm.

3 Growth, Utilization of Glucose and Respiration of Plasmodia Affected by Light

It was frequently observed (Baranetzki 1876; Gray 1938; Sobels and van der Brugge
1950; Daniel and Baldwin 1964; Rakoczy 1966) that the growth rate of plasmodia decreases markedly under illumination. The quantitative data were obtained for *Physarum
nudum* cultured on oat flakes agar medium in petri dishes under continuous illumination
with white "day" light omitted by fluorescent lamps. Control plasmodia were kept in
darkness. Figure 3 shows that changes in dry weight in the course of growth are different
for light and dark conditions. The inhibitory effect of light on growth rate was also found
by Daniel and Baldwin (1964) for the *Physarum polycephalum* plasmodia grown on
semidefined media and illuminated by fluorescent lamps. Unfortunately, no quantitative
data were presented. The spectral ranges active in delay and decreasing the growth rate
of plasmodia are not known for any species. It appears from the paper by Daniel and
Baldwin (1964) that the short wavelength range of the spectrum up to 500 nm is involved in this process because they stated that they carried out inoculations under light
above 500 nm to avoid harmful light effects on growing plasmodia.
      The inhibitory effect of light on growth corresponds with the light effect on utiliza-
tion of glucose by plasmodia. As was stated by Daniel (1966), glucose uptake was strongly inhibited in *Physarum polycephalum* plasmodia exposed to light. Lynch and Henney
(1974), using specifically labeled glucose, found for another species, *Physarum flavicomum*, that illuminated plasmodia metabolize glucose at a reduced rate as compared to
light-protected cultures in actinic flasks. The growing plasmodia of *Physarum flavicomum*
metabolize glucose by both the EMP and pentose-phosphate pathways (Lynch and Hen-

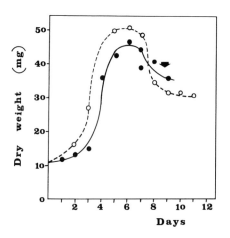

Fig. 3. Growth curves of the *Physarum nudum*
plasmodia. Fresh weight of the inocula = 100 mg.
Continuous line cultures exposed to light, *broken
line* cultures from darkness. *Arrow* shows the
mean time of sporulation for illuminated cultures
(Rakoczy 1966)

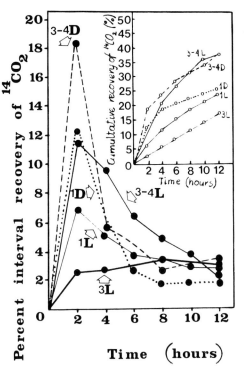

Fig. 4. Radiorespirometric patterns of *Physarum flavicomum* during light exposure and protection, expressed as percent of total administered radioisotope. The plasmodia (55 mg of protein) were incubated in 10 ml of R medium containing 0.20 μCi of ^{14}C-substrate. The respire $^{14}CO_2$ from the shake cultures was collected in hyamine for radioactivity analyses. 1D-$^{14}CO_2$ evolution from glucose 1-^{14}C in the actinic flasks (light-protected), 3-4D - $^{14}CO_2$ evolved from glucose 3-4-^{14}C in actinic flasks, 1L - $^{14}CO_2$ evolved from glucose 1-^{14}C in light, 3-4L - $^{14}CO_2$ evolved from glucose 3-4-^{14}C in light (Lynch and Henney 1974)

ney 1973). These two pathways were equally depressed in plasmodia under light (Fig. 4). The effect of light on the metabolism of glucose in *Physarum flavicomum* was not due to an interference with glucose uptake, since transport was unaffected by light (Lynch and Henney 1974). Although the final (12 h) yields of $^{14}CO_2$ were the same (Fig. 4, upper curves), production of $^{14}CO_2$ by light-exposed plasmodia in the first 2-h interval was only about 57% of that produced by dark-incubated cultures. The data also suggest that a light had no specific effect on either EMP or the pentose-phosphate paths, since both appeared to be equally depressed. The inhibition of glucose uptake (Daniel 1966) and/or of the metabolism of this compound under light (Lynch and Henney 1974) may be the cause of the decreased growth rate of plasmodia.

Light inhibits respiration of plasmodia, as was found for *Physarum polycephalum* (Daniel 1966). Polarographic measurements of oxygen uptake by plasmodial suspension showed that light strongly inhibited respiration of growing plasmodia. The inhibition could also be demonstrated in isolated mitochondria (Daniel 1965, 1966). In both cases inhibition occurred rapidly, within 15 s, and was reversible in the dark. The plasmodia exposed to light for 4 h regained approximately the original respiration rate when returned to the dark and retained the same sensitivity to light. The experiments were performed with white light (tungsten lamps) of various intensities (200–1250 foot candles).

4 Photoinduction of Sporulation

Sporulation is an essential process for myxomycetes, which enables survival of the organisms under unfavorable growth conditions by dispersing them in the form of spores. Meiosis, leading to genetic recombination and variability, also occurs during sporulation.

It has been established for various species that light is indispensable for sporulation. The earlier reports claim that light is necessary for sporulation in species with colored plasmodia, whereas species with nonpigmented plasmodia can sporulate both in light and in darkness. However, it is well known that some slime molds with white plasmodia require light for sporulation, e.g., *Stemonitis fusca* (McManus 1961), *Physarum nicaraguense* (Solis 1962), and *Physarum gyrosum* (Fergus and Schein 1963; Koevenig 1963). Obviously, plasmodia that require light for sporulation must contain light-absorbing pigments, but not necessarily in amounts visible in a living plasmodium.

The plasmodia become capable of sporulation when nutrients are exhausted from the medium, but light is required at this stage for induction of the process. It has been found for *Physarum nudum* that in given culture conditions and under continuous illumination the time of sporulation, e.g., the time elapsing since inoculation, depends on the age of the plasmodium. In old plasmodia in which growth has terminated, sporulation occurs after a shorter time than required for young cultures, which still have nutrients in the medium (Rakoczy 1962). It has also been found that light is necessary to initiate sporulation, whereas further morphological transformations accompanying the formation of sporangia can take place in darkness. With 14-day-old plasmodia of *Physarum nudum* 6 h illumination (4.5 J m^{-2} s^{-1}) is sufficient for inducing sporulation. For *Didymium iridis* plasmodia grown for 7 days on corn meal agar and then incubated for 4 days on plain agar in darkness, 2 min exposure with white light at 30 ft-c was found to induce sporulation in about 80% of cultures. However, comparison of the results obtained for different species is difficult, as only some authors clearly specify the dosage of light and conditions of culturing.

The spectral ranges active in sporulation have been tested for various species: *Physarum polycephalum* (Gray 1938, 1941, 1953; Daniel and Rusch 1962; Nair and Zabka 1965), *Physarum gyrosum* (Nair and Zabka 1965), *Physarum nudum* (Rakoczy 1963b, 1965, 1967), *Didymium nigripes* (Straub 1954; Lieth 1956), and *Didymium iridis* (Nair and Zabka 1965). In all cases blue light was found to be active in the induction of sporulation. Although red light also promoted sporulation, the intensity in this range of the spectrum had to be much (about 10 times) higher. The action spectrum for sporulation was obtained for *Physarum nudum* (Fig. 5). The activity was the highest in the UV and extended to 540 nm, its effectiveness gradually decreasing. Some, however weaker, activity was detected in the red range of the spectrum (Rakoczy 1965). Different spectral ranges were found to manifest distinct differences in their activity for inducing sporulation. The ranges active in inducing sporulation are: UV, blue, and red, whereas green and far-red do not promote the process. The active ranges display some quantitative and qualitative differences. The three curves (Fig. 6) are representative of the effect exercised by the ranges active in inducing sporulation. The differences concern the intensity of radiation and the percentage of sporulation under supraoptimal light intensity. In both the UV and red the highest percent of sporulation was less than 100, and at higher intensities it decreased. In the blue region 100% of sporulation was obtained, and this

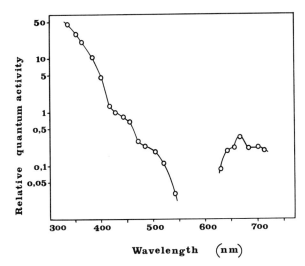

Fig. 5. The action spectrum of sporulation of the myxomycete *Physarum nudum*. The mature (12-day-old) plasmodia with a decreasing of growth rate due to exhaustion of nutrients in the medium were transferred onto filter paper and illuminated 12 h. Following irradiation, plasmodia were kept in darkness and the percentage of sporulated cultures was calculated. Relative quantum activity was calculated in comparison to the activity at the wavelength of 429 nm which was set as 1 (Rakoczy 1965)

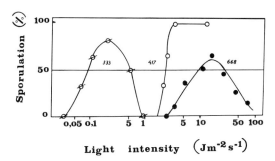

Fig. 6. Dependence of the sporulation of *Physarum nudum* on the intensity of the applied radiation. The *numbers* at the curves indicate the wavelength in nm (Rakoczy 1965)

level was not changed by further increasing of radiation intensity to the highest available level. Thus, in UV and red light the curves of the correlation between sporulation and light intensity (Fig. 6) are of optimal character, while the curve in the blue range is of maximum character.

Daniel and Rusch (1962) obtained a highly synchronous sporulation of illuminated *Physarum polycephalum* plasmodia. They found three conditions necessary for the vegetative plasmodia to sporulate: an optimum age of the culture at the time of harvest, an additional incubation period in the dark on a nonnutrient salts medium with nicotinic acid and nicotinamide, and illumination following this treatment. The plasmodia were grown in a liquid semi-defined media, harvested in the phase of exponential growth, and then incubated 4 days in darkness on the salts medium. Following the period of starvation plasmodia were illuminated for 3–4 h and subsequently they were kept in darkness until sporulation occurred (Daniel and Rusch 1962; Daniel and Baldwin 1964; Rusch 1970). After the period of starvation the plasmodium became sensitive to light, so that a short period of illumination was sufficient to induce sporulation. During the period of starvation the amounts of protein, nucleic acids, and polysaccharides decrease in plasmodium (Daniel 1966; Sauer et al. 1969). In spite of this, mitosis and DNA synthesis still occur, though there is less synchrony and the periods between mitoses become much

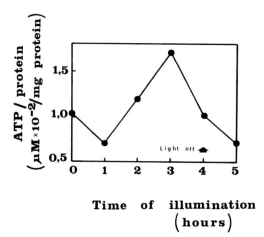

Fig. 7. The effect of light on the plasmodial ATP level. Cultures illuminated at 5 days. Plasmodia extracted in pH 8 Tris buffer at 100°C for 1 min and the ATP content estimated by a fluorometric firefly method (other nucleoside triphosphate contributed less than 10%). Protein estimated by Folin method (Daniel 1966)

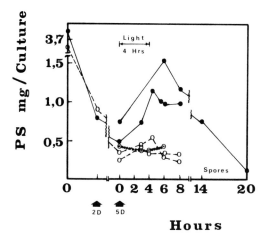

Fig. 8. Changes in plasmodial polysaccharide in light-exposed cultures. Cultures (in the absence of exogenous glucose) illuminated at 5 days. Plasmodia digested in 30% KOH, polysaccharide (*PS*) precipitated with hot ethanol and estimated by the anthrone reaction. ●——● illuminated cultures (two experiments), ○——○ cultures from darkness, ☉——☉ cultures in Corning red "actinic" flasks (absorb approximately 90% of light less than 500 nm) exposed to white light (Daniel 1966)

longer (Guttes et al. 1961; Sauer et al. 1969). At least one mitosis and DNA replication must occur in the plasmodia prior to the period of illumination, since after its inhibition sporulation does not appear (Sauer et al. 1969). In starved plasmodia, as in growing ones, no G_1 period exists (Mohberg and Rusch 1970) and light is very effective in the induction of sporulation during the S phase. Significant differences were found in the phenol-soluble acidic nuclear protein fraction from starved and vegetative plasmodia. The electrophoretic separation of these proteins revealed in the preparations from starved cultures four additional bands, and one band present in growing plasmodia was missing in starved ones (LeStourgeon and Rusch 1971). Starvation by itself is not sufficient to cause sporulation in *Physarum polycephalum*, which is apparent from the refeeding experiments in which cultures would resume growth if placed on fresh nutrient medium (Ward 1959) or would no sporulate on salts medium containing glucose (Daniel and Rusch 1962). After the period of starvation, plasmodia become capable of sporulation and require light for its induction.

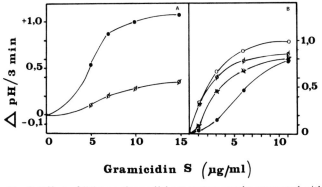

Fig. 9. Effect of light on plasmodial net proton uptake promoted with gramicidin S. *A* plasmodia treated with gramicidin S (GS) at 0 time. ●━━━━● dark, ⌀━━━━⌀ illuminated 15 min before GS-treatment. *B* GS-pretreated plasmodia. ●━━━━● dark, illumination: ○━━━━○ 5 min, ⌀━━━━⌀ 2 min, ⊠━━━━⊠ 0.5 min (Daniel and Eustace 1972)

During the period of illumination a transient increase in ATP level (Fig. 7) and in the amounts of glycogen-like polysaccharide (Fig. 8) was found in *Physarum polycephalum* plasmodia (Daniel 1966) and a sustained, large increase in the pH of culture medium was detected (Daniel 1966). Preliminary reports suggest that light facilitates the plasmodial uptake of exogenous Ca^{2+} (Daniel and Järlfors 1972) and sensitized H^+ plasmodial uptake (Fig. 9) promoted by gramicidin S. The H^+ uptake is accompanied by K^+ release into the culture medium (Daniel and Eustace 1972). Dibutyryl cAMP mimics, in the dark, the light-dependent reactions and can partly replace light to induce sporulation (Daniel 1974, 1975). Light inhibits respiration in starved plasmodia, similarly as in growing ones (Daniel 1965, 1966). Sporulation is dependent on the synthesis of DNA prior to the illumination period, on the continued synthesis of protein throughout the entire period, and on the synthesis of RNA until approximately 3 h following the exposition (Sauer et al. 1969; Sauer 1973). 3 h after illumination also marks the time at which the processes associated with early stages of sporulation attain "a point of no return" – the moment at which the plasmodium is irreversibly committed to sporulation and even though it is returned to a growth medium in darkness the sporangia are formed (Ward 1959; Daniel and Rusch 1962; Sauer et al. 1969).

The sporulation-inducing effects of light can be transferred from an illuminated to an unilluminated plasmodium. It was found by Straub (1954) that slime mold *Didymium nigripes* could sporulate in darkness after feeding with homogenized illuminated plasmodia. Wormington et al. (1975) stated that also *Physarum polycephalum* could sporulate in darkness following injection of plasmodia with ultrafiltrate from starved or growing plasmodia exposed 6 h to light. The compound promoted sporulation isolated from illuminated *Physarum polycephalum* plasmodia is yellow (maximal absorption in visible range of the spectrum at 365–370 nm and elevated absorbance in UV) and of a small (< 500) molecular weight. The compound purified and illuminated in vitro was also capable of inducing sporulation (Wormington and Weaver 1976). It was found for *Physarum nudum* that the light-induced effects in plasmodia were preserved in sclerotia derived from exposed cultures of the slime mold and plasmodia obtained in turn from such sclerotia, even after few weeks of storage, would sporulate in darkness (Rakoczy, unpublished data).

5 Effect of Light on Plasmodial Pigmentation

In Myxomycetes, the color of plasmodia is widely differentiated. There are species with colorless, watery transparent plasmodia (Martin 1962), however, slime molds with pigmented plasmodia are encountered in nature more frequently. The pigmentation can vary from milky white, through yellow, orange, red, violet, to black. It has not yet been definitely settled whether the color of plasmodium is species-specific. Probably, it depends on strain differences within the species (Brandza 1927), and to some extent the color may change with the acidity of the medium (Seifriz and Zetzmann 1935; McManus 1962) or with the acidity in plasmodium (Seifriz and Zetzmann 1935). Besides, light is a factor which, in some degree, influences the change in plasmodial pigmentation. It has been found for *Fuligo septica* that lemon-colored plasmodia of this species turned to cream-yellow under illumination (Baranetzki 1876; Gray 1938; Rakoczy, unpublished data). White plasmodia of *Physarum gyrosum* changed to yellow when exposed to light (Koevenig 1964; Nair and Zabka 1965). The photobleaching of plasmodia and isolated pigments has been found for *Physarum polycephalum* (Daniel 1966). *Physarum nudum* plasmodia, yellow in darkness, turned to brown under light conditions (Rakoczy 1962, 1966). Figure 10 shows the absorption spectra of the *Physarum nudum* plasmodia from light and dark conditions.

Only two reports pertain to the effect of chromatic or monochromatic light on a change of plasmodial color. Nair and Zabka (1965) have found the color of *Physarum gyrosum* to change from white in darkness to yellow under blue light and to brown following exposition to red radiation. Green did not induce any change in color. The activity of monochromatic light in the induction of color change was studied in *Physarum nudum* plasmodia (Rakoczy 1965). It has been found that plasmodia turned brown after prolonged illumination with wavelengths from the blue region of the spectrum, whereas UV, green, red, and far red did not evoke any change in plasmodial pigmentation (Fig. 11). The activity of radiation in the change of color in *Physarum nudum* plas-

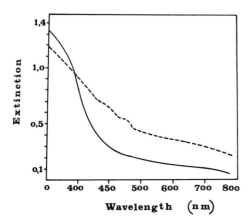

Fig. 10. Absorption spectra of *Physarum nudum* plasmodia. *Full line* plasmodia from darkness, *broken line* plasmodia cultured 7 days under continuous white light ($6 \text{ J m}^{-2} \text{ s}^{-1}$) (Rakoczy 1966)

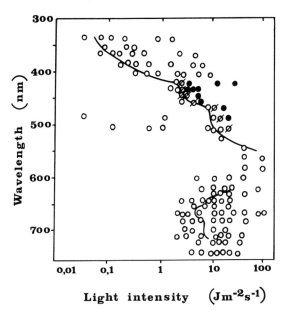

Fig. 11. The change in color of *Physarum nudum* plasmodia dependent on the intensity of applied radiation. *White* and *black circles* plasmodia exposed 12 h to various intensities of different wavelengths, *black circles* plasmodia which changed color from yellow to brown. *Full lines* the action spectrum of sporulation (Rakoczy 1965)

modia did not coincide with the activity of radiation in induction of sporulation (Figs. 5 and 11), and there is no correlation with the changed plasmodial color and the process of sporulation. Plasmodia exposed to UV or red light were able to sporulate, although the change in their pigmentation was not observed until the maturation of sporangia. On the other hand, young plasmodia illuminated since inoculation with white or blue light change color for deep brown, though they did not sporulate, because they were not yet competent in this process (Rokoczy 1962, 1965).

The chemical nature of the pigments present in plasmodia of the myxomycetes is not known. The suggestions in the matter found in reports by various investigators concerning even the same species are contradictory. It has been found only for *Physarum nudum* that the brown color of the plasmodia is due to melanin synthetized during illumination. In exposed cultures carotenoids are also present, whereas in plasmodia from darkness only phytofluen was found (Rakoczy 1973). Yellow pigments, the main pigments in plasmodia cultured in darkness, and present also in illuminated plasmodia, though in small quantity, are not yet identified. Melanin and carotenoids present in illuminated plasmodia of *Physarum nudum,* play probably a protective role against a harmful effect of light on plasmodia. The results of experiments on sporulation induced with different spectral ranges may support this supposition. Contrary to UV and red, in the blue range of the spectrum under which melanin and carotenoids are synthetized, the sporulation in 100% was obtained, and this value was attained even under the highest light intensity (Fig. 6).

6 General Remarks

The above review shows that light (for the most part or exclusively blue) controls many
different physiological processes in myxomycetes. The molecular mechanism of this
control remains unknown. One can suppose that there is one photoreceptor, and probably the primary processes induced by light are common for all these phenomena. The
chain of the reactions leading from the excited photoreceptor molecules to the finally
observed responses is certainly complex and unknown. Nevertheless, some available data,
though scarce and fragmentary, can be combined in a hypothetical chain of events, as
presented in Fig. 12. The photoreceptor molecules, probably a flavin-type pigment,
may be supposed to be localized in the plasma membrane in plasmodium. The excitation
of the photoreceptor leads to some changes in the permeability of the plasma membrane
and consequently, e.g., plasmodial uptake Ca^{2+} and H^+ can proceed (Daniel and Eustace
1972; Daniel and Järlfors 1972; Daniel 1973, 1975, 1977). Calcium is known as a factor
controlling the activity of the actomyosin system in the muscles as well as in the plasmodia of the Myxomycetes (Komnick et al. 1973). There is evidence that the actomyosin system involved in the contractile activity of protoplasmic veins is localized in the
ectoplasmic layer of the plasmodial veins (Wohlfarth-Bottermann 1964, 1965) into
which plasma membrane makes a number of invaginations (Wohlfarth-Bottermann 1965,
1975). Changes in the calcium level may act directly on the motility phenomena in the
plasmodium. the calcium may also alter the cAMP level in the cytoplasm, and vice versa,
this factor may influence calcium concentration in plasmodia, and the mitochondria
can play a role as a regulatory system of free calcium concentration in the cytoplasm
(Daniel and Järlfors 1972; Nicholls 1972; Borle 1973). One can postulate that cAMP
can control the transcription processes in the nucleus and in some way lead to the production of new proteins essential for morphogenetic processes accompanying sporangia
formation. Although the physiological processes such as motility, growth, and sporulation are not involved in the same path, one cannot exclude that the primary processes
induced by light are common for all these phenomena, and the secondary processes may
trigger the metabolic events of the various pathways.

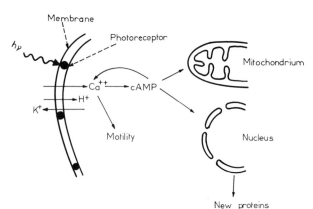

Fig. 12. Scheme illustrated a
hypothetical chain of primary
events effected by blue light
in slime mold plasmodia

References

Baranetzki J (1876) Influence de la lumiere sur les plasmodia des Mycomycetes. Mem Soc Sci Nat Cherbourg 19: 321–360

Białczyk J (1979) An action spectrum for light avoidance by *Physarum nudum* plasmodia. Photochem Photobiol 30: 301–303

Białczyk J, Rkoczy L (1974) Phototaxis in the Myxomycetes: Migration of the young plasmodia in white light. Bull Acad Pol Sci Ser Biol 22: 817–873

Białczyk J, Rakoczy L (1975) Phototaxis of the Myxomycetes. Behaviour of old plasmodia in white light. Bull Acad Pol Sci Ser Biol 23: 571–575

Borle AB (1973) Calcium metabolism at the cellular level. Fed Proc 32: 1944–1950

Brandza M (1927) Sur la fusion ou la separation des plasmodiums prises comme criterium dans les Myxomycetes. CR Acad Sci 185: 10772–10774

Brever EN, Kuraishi S, Garver JC, Strong FM (1964) Mass culture of a slime mold *Physarum polycepahlum*. Appl Microbiol 12: 161–164

Carlile MJ (1970) Nutrition and chemotaxis in the myxomycete *Physarum polycephalum:* the effect of carbohydrates on the plasmodium. J Gen Microbiol 63: 221–226

Carlile MJ (1971) Myxomycetes and other slime moulds. In: Methods in microbiology, vol I, pp 237–265. Academic Press, London New York

Carlile MJ (1974) The myxomycete Physarum nudum: life cycle and pure culture of plasmodia. Trans Br Mycol Soc 62: 213–215

Daniel JW (1965) Control of respiration by light during the sporulation and growth of a myxomycete. J Cell Biol 27: 23 A–24A

Daniel JW (1966) Light-induced synchronous sporulation of a myxomycete – the relation of initial metabolic changes to the establishment of a new cell state. In: Cameron IL, Padilla GM (eds) Cell synchrony, pp 117–152. Academic Press, London New York

Daniel JW (1973) The role of calcium in light-induced ion movements of a myxomycete. Am Soc Photobiol Meet, p 182

Daniel JW (1974) The replacement of light by cAMP for the stimulation of differentiation in inducing photometabolic ion transport. Miami Winter Symposia: Membrane transformations in Neoplasia

Daniel JW (1975) Light controlled ion movements in a myxomycete: The use of cafeine or dibutyryl cAMP to replace light. Am Soc Photobiol Abstr, p 47

Daniel JW (1977) Photometabolic events programming expression of growth and development in *Physarum polycephalum*. 2nd Int Mycol Congr Abstr, vol A–L, p 126

Daniel JW, Baldwin HH (1964) Methods of culture for plasmodial myxomycetes. In: Prescott DM (ed) Methods in cell physiology, vol I, pp 9–41. Academic Press, London New York

Daniel JW, Eustace J (1972) The effect of light on the permeability of a myxomycete. FEBS Lett 26: 327–332

Daniel JW, Järlfors U (1972) Light-induced changes in the ultrastructure of a plasmodial myxomycete. Tissue Cell 4: 405–426

Daniel JW, Rusch HP (1961) The pure culture of *Physarum polycephalum* on a partially defined medium. J Gen Microbiol 25: 47–59

Daniel JW, Rusch HP (1962) Method for inducing sporulation of pure cultures of the myxomycete *Physarum polycephalum*. J Bacteriol 83: 234–240

Denbo JR, Miller DM (1976) Factors affecting the movement of slime mold plasmodia. Comp Biochem Physiol 55 A; 5–12

Fergus CL, Schein RD (1963) Light effects on fruiting of *Physarum gyrosum*. Mycologia 55: 540–548

Gray WD (1938) The effect of light on the fruiting of Myxomycetes. Am J Bot 25: 511–522

Gray WD (1941) Some effects of heterochromatic ultra violet radiation on myxomycete plasmodia. Am J Bot 28: 212–216

Gray WD (1953) Further studies on the fruiting of *Physarum polycephalum*. Mycologia 45: 817–824

Guttes E, Guttes S, Rusch HP (1961) Morphological observations on growth and differentiation of *Physarum polysephalum* grown in pure culture. Dev Biol 3: 588–614

Henney HR, Henney MR (1968) Nutritional requirements for the growth in pure culture of the myxomycete *Physarum rigidum* and related species. J Gen Microbiol 53: 333–339

Henney HR Jr, Lynch T (1969) Growth of *Physarum flavicomum* and Physarum rigidum in chemically defined minimal media. J Bacteriol 99: 351–534

Hofmeister W (1867) Die Lehre von der Pflanzenzelle. Handb Physiol Bot 1, Leipzig

Knowles DJC, Carlile MJ (1978a) The chemotactic response of plasmodia of the myxomycete *Physarum polycephalum* to sugars and related compounds. J Gen Microbiol 108: 17–25

Knowles DJC, Carlile MJ (1978b) Growth and migration of plasmodia of the myxomycete *Physarum polycephalum:* the effect of carbohydrates, including agar. J Gen Microbiol 108: 9–15

Koevenig JL (1963) Effect of environment on fruiting in the myxomycete *Physarum gyrosum* Rost. Am Midl Nat 69: 373–375

Koevenig JL (1964) Studies on the life-cycle of *Physarum gyrosum* and other myxomycetes. Mycologia 56: 170–184

Komnick H, Stockem W, Wohlfarth-Bottermann KE (1973) Cell motility: mechanism in protoplasmic streaming and ameboid movement. Int Rev Cytol 34: 170–249

LeStourgeon WM, Rusch HP (1971) Nuclear acidic protein changes during differentiation in *Physarum polycephalum*. Science 174: 1233–1236

Lieth H (1956) Die Wirkung des Grünlichtes auf die Fruchtkörperbildung bei Didymium eunigripes. Arch Microbiol 24: 91–104

Lynch TI, Henney HR Jr (1973) Carbohydrate metabolism in the plasmodium of the myxomycete *Physarum flavicomum*. Can J Microbiol 19: 803–810

Lynch TI, Henney HR Jr (1974) Effect of light on the carbohydrate metabolism of *Physarum flavicomum* plasmodia. Microbios 10: 39–43

Martin GW (1962) The genus Schenella. Mycologis 53: 25–30

McManus SMA (1961) Culture of *Stemonitis fusca* on glass. Am J Bot 48: 582–588

McManus SMA (1962) Some observations on plasmdia of the Trichiales. Mycologia 54: 78–90

Mohberg J, Rusch HP (1969) Growth of large plasmodia of the myxomycete *Physarum polycephalum*. J Bacteriol 67: 1411–1416

Mohberg J, Rusch HP (1970) Nuclear histone in *Physarum polycephalum* during growth and differentiation. Arch Biochem Biophys 138: 418–432

Nair P, Zabka G (1965) Light quality and sporulation in myxomycetes with special reference to a red, far-red reversible reaction. Mycopathol Mycol Appl 26: 123–128

Nicholls TJ (1972) The effects of starvation and light on intramitochondrial granules in *Physarum polycephalum*. J Cell Sci 10: 1–14

Rakoczy L (1962) The effect of light on the fructification of the slime mold *Physarum nudum* Macbr. as influenced by the age of the culture. Acta Soc Bot Pol 31: 651–665

Rakoczy L (1963a) Application of crossed light and humidity gradients for the investigation of slime molds. Acta Soc Bot Pol 32: 393–403

Rakoczy L (1963b) Influence of monochromatic light on the fructification of Physarum nudum. Bull Acad Pol Sci Ser Biol 11: 559–562

Rakoczy L (1965) Action spectrum in sporulation of slime mold *Physarum nudum* Macbr. Acta Soc Bot Polon 34: 97–112

Rakoczy L (1966) Further studies on the physiology of the sporulation of the myxomycete *Physarum nudum*. Acta Soc Bot Pol 35: 315–324

Rakoczy L (1967) Antagonistic action of light in sporulation of the myxomycete *Physarum nudum*. Acta Soc Bot Pol 36: 153–159

Rakoczy L (1971) Studies on pigments of the myxomycete *Physarum nudum*. I. Absorption spectra of the crude extracts of pigments from plasmodia cultured in continuous light and in darkness. Acta Soc Bot Pol 40: 483–497

Rakoczy L (1973) The myxomycete *Physarum nudum* as a model organism for photobiological studies. Ber Dtsch Bot Ges 86: 141–164

Ross IK (1964) Pure cultures of some myxomycetes. Bull Torrey Bot Club 91: 23–31

Ross IK, Sunshine LD (1965) The effect of quinic acid and similar compounds on the growth and development of *Physarum flavicomum* in pure culture. Mycologia 57: 360–367

Rusch HP (1970) Some biochemical events in the life cycle of *Physarum polycephalum.* In: Prescott DM, Goldstein L (eds) Advances in cell biology, vol I, pp 297–327. Appleton-Century-Crofts, Educational Division Meredith Corporation, New York

Sauer HW (1973) Differentiation in *Physarum polycephalum.* In: Symp Soc Gen Microbiol 33, 375–405

Sauer HW, Babcock KL, Rusch HP (1969) Sporulation in *Physarum polycephalum.* A model system for studies on differentiation. Exp Cell Res 57: 319–327

Seifriz W, Zetzmann M (1935) A slime mold pigment as indicator of acidity. Protoplasma 23: 175–179

Sobels JC, van der Brugge HFH (1950) Influence of day light on the fruiting of two orange-yellow pigmented myxomycete plasmodia. Proc K Ned Acad Wet 53: 1610–1616

Solis BC (1962) Studies on the morphology of *Physarum nicaraguense* Macbr. MS Thesis, Univ Iowa

Stahl E (1880) Über den Einfluß von Richtung und Stärke der Beleuchtung auf einige Bewegungs-erscheinungen in Pflanzenreiche. Bot Ztg 38: 229

Straub J (1954) Das Licht bei der Auslösung der Fruchtkörperbildung von *Didymium nigripes* und die Übertragung der Lichtwirkung durch Plasma. Naturwissenschaften 41: 219–220

Tso WW, Mansour TE (1975) Termotaxis in a slime mold *Physarum polycephalum.* Behav Biol 14: 499–504

Ward JM (1959) Biochemical systems involved in differentiation of the fungi. In: Nickerson WJ (ed) 4-th Int Congr Biochem, Vienna, 1958, Symp VI: Biochemie der Morphogenese, pp 33–58. Pergamon Press, Oxford

Wohlfarth-Bottermann KE (1964) Differentiations of the ground cytoplasm and their significance for the generation of the motive force of ameboid movement. In: Allen RD, Kamiya N (eds) Primitive motile systems in cell biology, pp 79–109. Academic Press, London New York

Wohlfarth-Bottermann KE (1965) Weitreichende fibrilläre Protoplasmadifferenzierungen und ihre Bedeutung für die Protoplasmaströmung. III. Entstehung und experimentell induzierbare Muster-bildungen. Roux' Arch Entwicklungsmech 156: 371–403

Wohlfarth-Bottermann KE (1975) Weitreichende fibrilläre Protoplasmadifferenzierungen und ihre Bedeutung für die Protoplasmaströmung. X. Die Anordnung der Actomyosin-Fibrillen in experi-mentell unbeeinflußten Protoplasmaadern von Physarum in situ. Protistilogica 11: 19–30

Wormington WM, Weaver RF (1976) Photoreceptor pigment that induces differentiation in the slime mold *Physarum polycephalum.* Proc Natl Acad Sci USA 73: 3895–3899

Wormington WM, Cho CG, Weaver RF (1975) Sporulation-inducing factor in slime mould *Physarum polycephalum.* Nature (London) 256: 413–414

Blue-Light Photoreception in the Inhibition and Synchronization of Growth and Transport in the Yeast *Saccharomyces**

L.N. EDMUNDS, Jr.[1]

1 Introduction

Visible light has been shown to inhibit growth [8, 10, 18], respiration [10, 12, 14], protein synthesis [9], and membrane transport [18] in bakers' yeast and to have a deleterious effect on membrane integrity [18]. Several observations suggest that cytochromes participate in these inhibitory light effects. Blue light has been identified as the maximally inhibitory wavelength in investigations on growth, protein synthesis, and respiration [8, 9] and on respiratory adaptation [11] of anaerobically grown yeast cells. Ninnemann et al. [15] have shown that the inhibition of respiration by high intensity blue light was due to the destruction of cytochromes a/a_3 and partial destruction of cytochrome b. Finally, we have recently demonstrated [18] that respiratory-deficient petite (rho^-) yeasts lacking cytochromes b and a/a_3 are resistant to light intensities that photokill wild-type (rho^+) parent cells.

In addition to these direct inhibitory effects, white light (at least at lower intensities) may act also by entraining endogenous, self-sustaining oscillators which, under alternating cycles of light and darkness, would synchronize cell growth and division, membrane transport, and other processes [6, 7]. This paper summarizes further evidence that the inhibitory effects of light on growth and transport in *Saccharomyces* require the presence of cytochromes b and a/a_3 (from a comparison of light effects in wild-type cells and in mutants lacking one or more of these respiratory pigments and an examination of the action of light of different wavelengths) and that light-entrainable, persisting circadian and ultradian oscillations do indeed occur in cell division and amino acid transport activities.

Abbreviations. *LD: 10,* 14, a repetitive, diurnal (diel) light cycle (period $T = 24$ h) consisting of 10 h of light (L) followed by 14 h of darkness (D); *LL,* continuous illumination; *DD,* continuous darkness; τ, the period of a rhythm observed in a culture in either LD (τ_{LD}) or in DD (τ_{DD}); *ss,* stepsize, or factorial increase in cell concentration after a synchronized or phased "burst" of cell division in the population has been completed; *g,* generation (or doubling) time of a culture

* Preliminary portions of these results were presented at the VII International Congress on Photobiology, Rome, Italy (29 August–3 September 1976); at the 77th Annual Meeting of the American Society for Microbiology, New Orleans, Louisiana (8–13 May 1977); at the 5th Annual Meeting of the American Society for Photobiology, San Juan, Puerto Rico (11–15 May 1977); and at the XIII and XIV International Conferences of the International Society for Chronobiology, Pavia, Italy (4–7 September 1977) and Hannover, Federal Republic of Germany (8–12 July 1979)

1 Department of Biology, State University of New York, Stony Brook, NY 11794, USA

2 Materials and Methods

2.1 Yeast Strains and Growth Conditions

Saccharomyces cerevisiae Y185 *rho*⁺ was obtained from Dr. H.O. Halvorson of Brandeis University. The isolation and characterization of all other strains and mutants used has been detailed elsewhere [16]. Cells were maintained on supplemented agar slants and were grown in filter-sterilized synthetic medium containing 5% glucose and proline as a nitrogen source either on a rotary shaker or with stirring by magnetic bars in constant-temperature water baths or environmental chambers [6, 16, 18]. For the photoeffect experiments, 100-ml cultures were inoculated with 2-ml inoculants (previously grown in liquid medium at 28°C for one day in the dark) and then were grown at 12° ± 0.1°C in the light or dark without stirring, except during sampling of cell number once per day. For the longer-term rhythm experiments, 6-l master cultures were grown at 12° ± 0.25°C in serum jugs (inoculated from smaller liquid cultures pregrown in DD) that were aerated and magnetically stirred to insure a homogeneous cell suspension. This relatively low temperature yielded culture generation times in excess of 24 h in growing cultures [18]. The cultures were maintained in small chambers furnished with clock-programmed banks of 15- or 40-Watt cool-white fluorescent bulbs surrounding the vessels. Light intensity (lx incident at the surface of the vessels as measured with a Weston Model 756 illumination meter and quartz filter) was varied by changing the distance of the cultures from the light source.

2.2 Population Growth and Amino Acid Uptake

Cell number was monitored with a miniaturized fraction collector system (in the rhythm experiments) and Coulter Model B Electronic Particle Counter [6]. The 10-ml aliquots withdrawn automatically from the master cultures at 2-h intervals were killed with four drops of 37% formaldehyde and were later diluted with a salt solution. The raw Coulter counts could be used with confidence after correcting for dilution and coincidence. In synchronous cultures, the intervals between successive onsets of cell number increase were taken as phase reference points for determination of the period. Amino acid transport was measured as ^{14}C-histidine or ^{14}C-lysine uptake [18]. Assays were carried out at 12°C over a timespan of 10 to 30 min, and rates were calculated from the slopes of the plots of cpm versus time and expressed as nmol min^{-1} 10^7 cells^{-1}. A similar protocol was followed for the assay of both growth and transport in rhythm experiments: cultures were first synchronized during the exponential growth phase by LD and then were placed into DD and constant temperature to determine if any rhythmicities observed would persist in the absence of obvious Zeitgeber and ultimately, in the absence of cell division.

2.3 Action Spectra and Cytochrome Absorption Spectra

Broad-band, isoenergy action spectra for both growth and amino acid transport were derived as previously described [16] using magnetically stirred 100-ml cultures (pregrown in DD) in flasks that were masked so that light could enter only through sharp cutoff Corning Glass filters taped to the front of the flasks. Each filter allowed all light above a given wavelength to be transmitted but permitted virtually no transmittance below the cutoff value. Transmitted energies were measured with a Yellow Springs Instrument Co. radiometer. The cytochrome absorption spectra of cells harvested from agar plates grown at 12° or 28°C were determined using a cell paste reduced with powdered $Na_2S_2O_4$, pressed into special cuvettes (final paste thickness of 2 mm), and chilled in liquid nitrogen [16]. The spectra were then measured with a Cary 17 recording spectrophotometer and a Lucalox disc as a reference blank; those presented in the figures were reproduced by tracing over actual recordings.

3 Results

3.1 Light Inhibition of Growth and Transport

Growth and transport patterns for rho^+ cells of the wild type Y185 strain of *Sacharomyces* and of the haploid wild-type D225-5A strain, as well as their respective rho^- petite mutants, are shown in Fig. 1. Cells were grown in the dark or were illuminated with cool-white fluorescent light (5400 lx). During growth in the dark, amino acid transport in both strains showed a transient derepression followed by a sharp decrease during the stationary phase of growth [18]. Exposure of rho^+ cells to light resulted in severe inhibition of growth and transport (Fig. 1a, b). The low-temperature absorption spectrum (Fig. 1c) revealed that rho^+ cells had a normal cytochrome complement.

Plating of cells from light-inhibited Y185 or D225-5A rho^+ cultures showed very low viability; surviving cells were almost exclusively light-resistant rho^- petites [16]. The rate of overgrowth of the rho^- variants was faster in the D225-5A strain, however, so that the lag period before net growth observed in the illuminated rho^+ culture was shorter. The effect of light on ethidium bromide-treated, light resistant rho^- cells [18] was also studied. The rho^- cells of both strains grew as well as the corresponding rho^+ cells in the dark (Fig. 1a) and exhibited a normal transport pattern (Fig. 1b). Cultures exposed to the light went through an adaptive phase for two or three days, after which they grew almost as quickly as dark-grown rho^- cells. This period of adaptation to light was reflected also in the transport pattern (Fig. 1b), the peak in transport being delayed by at least one day. The absorption spectra of the rho^- cells (Fig. 1c) suggest that this partial resistance to light may be due to the absence of cytochromes a/a_3 and b. Identical results were obtained with rho^- strains isolated from the light-inhibited rho^+ cells that had not been treated with ethidium bromide (data not shown).

The correlation between the resistance of rho^- cells to light and their lack of cytochromes b and a/a_3 suggested that these respiratory pigments are photoreceptors for light inhibition and ultimate photokilling of yeast. This suggestion was confirmed by the

Fig. 1a–c. Growth, transport, and low-temperature absorption spectra of rho^+ and rho^- cells of *S. cerevisiae* strains Y185 (*upper panels*) and D225 (*lower panels*). Cells were grown at 12°C in proline-supplemented medium in the dark (●) or light (○) (5400 lx white fluorescent light). Aliquots were removed at intervals, and cell number (**a**) was determined after suitable dilution with a Model B Coulter Electronic Particle Counter. Histidine transport (**b**) was assayed as described in the text. Low-temperature absorption spectra of the cellular cytochromes (**c**) was carried out with a Cary Dual-Beam Spectrophotometer by comparing the spectrum of a 0.2 mm-thick paste of reduced cells frozen in liquid nitrogen with reference to a Lucalox disc

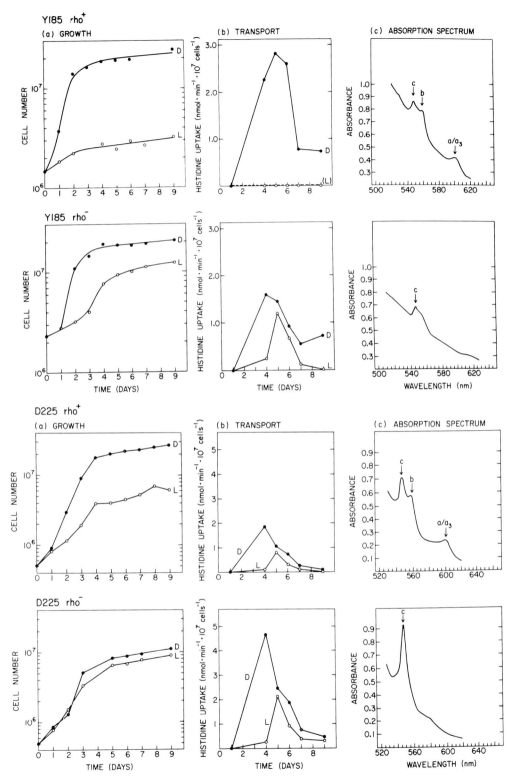

Fig. 1a–c

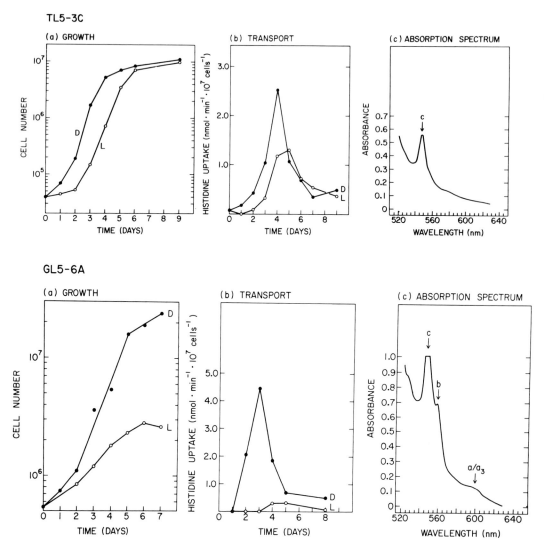

Fig. 2a–c. Effect of light on growth and transport in *S. cerevisiae* mutant strains TL5-3C (*upper panels*), a nuclear petite that lacks cytochromes b and a/a₃; and GL5-6A (*lower panels*), a nuclear petite that has reduced amounts of cytochromes a/a₃, as indicated by their low-temperature absorption spectra (c). Cells were grown at 12°C either in darkness or in light (5000 lx), assayed and characterized as described in the legend to Fig. 1. The TL5-3C was resistant to light, while GL5-6A was only partially resistant

effect of light on selected yeast strains with various defects in their respiratory chains: There was a close correlation between the presence of cytochromes b and a/a₃ and photosensitivity [16]. Thus, strain TL5-3C (Fig. 2), a nuclear petite which lacks cytochromes b and a/a₃, was resistant to light; strain GL5-6A, a nuclear petite with reduced amounts of cytochromes a/a₃, was partially resistant (Fig. 2); and strains MB127-20C

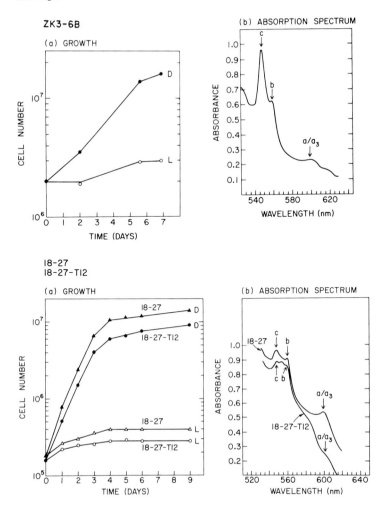

Fig. 3a, b. Effect of light on the growth of *S. cerevisiae* strain ZK3-6B, a nuclear petite that is respiratory-deficient, and on strain 18-27 (a wild-type strain) and 18-27/T12, its oxidative phosphorylation-deficient derivative. All strains have a full cytochrome complement, as indicated by their low-temperature absorption spectra (**b**), and are fully sensitive to light. Cells were grown and assayed as described in Fig. 2

and MB1-6C (not shown), nuclear petites which lack only cytochrome b, were also partially light-resistant. Finally, nuclear petites containing all three cytochromes but having their respiratory chain either nonfunctional (ZK3-6B) or uncoupled (18-27/T12) were fully sensitive to inhibition by light (Fig. 3). Nevertheless, the observation that dark-grown *rho⁻* cultures of strains Y185 and D225-5A (as well as the dark-grown TL5-3C and GL5-6A strains that lack normal cytochrome complements) typically showed peaks in the rate of cell growth (Fig. 1a) and transport activity (Fig. 1b) that occurred earlier than in light-grown *rho⁻* cells, together with the fact that light-grown Y185 (but *not*

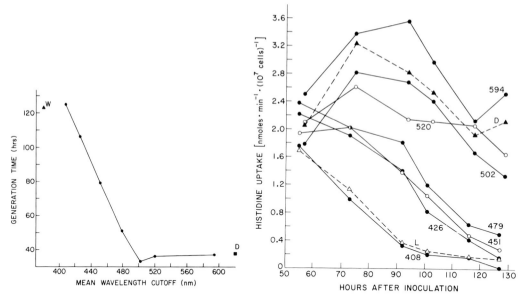

Fig. 4. Effects of light of various wavelengths but equal energies (4.0×10^3 and 5.2×10^3 ergs cm^{-2} s^{-1}, respectively) on the inhibition of growth (*left panel*) and on histidine transport (*right panel*) in 100-ml cultures of *S. cerevisiae* Y185 *rho*$^+$ grown at 12°C. Sharp-cutoff Corning filters were utilized; the values (nm) given correspond to the mean wavelength cutoff (transmittance of at least 37%), with progressively shorter wavelengths being added to the spectrum received by any given culture as one moves left along the abscissa of the action spectrum (*left panel*), or toward lower-wavelength curves in cultures at various stages of population growth (*right panel*). White-light (no filter) and dark controls are indicated, respectively, by (▲) and (■), or by the *dashed-line curves* labeled L and D

D225-5A) *rho*$^-$ cells never achieved quite the same titer (Fig. 1a) or demonstrated the same viability [16] as dark-grown Y185 *rho*$^-$ cultures, suggests that processes mediated by either some undetected residual cytochrome a/a$_3$ or a nonheme photoreceptor may also participate in light inhibition.

3.2 Action Spectra for Light Inhibition of Growth and Transport

The action spectrum for the inhibition of growth by light of various (equal energy) wavelengths is shown in Fig. 4 (left panel). Light in the 400- to 420-nm region of the visible spectrum was equally effective as white light in causing growth inhibition. Little or no inhibition was seen at wavelengths above 500 nm. Comparable results were obtained for the action of different wavelengths of light on histidine transport (Fig. 4, right panel). Above 500 nm, light had little effect on transport at various stages of population growth. As the light utilized, however, included progressively more irradiation from the blue region, inhibition increased; light of 408 nm was as effective as white light. The inhibitory action of white light on growth and transport in yeast, therefore, can be attributed primarily to irradiation in the blue region of the spectrum.

3.3 Light-Entrained Rhythms of Cell Division and Transport

Yeast cultures (6-l) were grown first in DD at several different temperatures, all below the optimum of 28°C. At 12°C, the typical exponential growth curve had an average generation time (g) of about 31 h [6]; this temperature was then utilized in all ensuing experiments inasmuch as it afforded a growth rate sufficiently low to permit synchronization by 24-h LD cycles [5]. Cultures maintained in LL ($\cong$ 3100 lx) also grew exponentially, but more slowly; light in this intensity range inhibits growth and transport.

In contrast to the growth curves obtained in LL or DD, synchronization of cell division occurred in LD: 10,14 (Fig. 5): intervals of 8 to 10 h during which there was virtually no increase in cell number alternated with intervals of population increase (confined largely to the 14-h dark spans, although the onsets of division often would precede the beginning of darkness by 2 or 3 h). For the five successive onsets in the *population* shown in Fig. 5, the period (τ) $\cong$ 24 h. The stepsize (ss) — an indication of the amplitude of the rhythm in a population of cells [some (ss $<$ 2.00) or all (ss $\cong$ 2.00) of which may divide at about the same time] — in this experiment was 1.54; only about 50% of the cells budded each 24 h during a given step (assuming that all cells were progressing through the cell division cycle). The value of g (45 h) for this synchronous culture in LD, obtained from the slope of the line drawn through the "staircase", fell between that found in DD ($\cong$ 31 h) and in LL ($\leqslant$ 80 h) of similar intensity [18]. Similar results were obtained in LD: 12,12 and LD: 14,10 [6]. This overt rhythm of phased cell division in a LD cycle was endogenous: in yeast cultures entrained by LD:14,10 (ss $\cong$ 1.48; τ $\cong$ 24.0 h) and placed into DD at 12°C, the rhythm persisted throughout the ensuing four days with ss and τ_{DD} of the freerunning oscillations approximating 1.47 and 26 to 28 h, respectively (data not shown). In other experiments the rhythm persisted for as long as seven to eight days in DD with no signs of a decrease in amplitude; for four separate runs (a total of 18 cycles in DD), τ_{DD} = 25.6 h [6].

Oscillations in transport capacity occurred also in nondividing, or very slowly dividing, infradian (g > 5-to-10 days), LD-entrained cultures in the "stationary" phase of growth. Results for the uptake of ^{14}C-histidine in nongrowing yeast exposed to LD: 14, 10 (3130 lx) at 12°C are shown in Fig. 5 (lower panel). For the five days in LD during which uptake was measured, five major peaks were found, each occurring shortly after the onset of light, with τ $\cong$ 24 h; the onsets of increase in uptake usually preceded the beginning of light. Such oscillations were absent in nondividing cultures maintained in DD without having been exposed to a prior entraining LD cycle. Secondary "shoulders" were sometimes seen at about the time of the onset of dark. A gradual reduction in the absolute level and the amplitude of the oscillations was observed as the culture aged. The rhythm appeared to freerun in DD for at least three days. Similar results were obtained for ^{14}C-lysine transport in yeast cultures that had been synchronized by LD:14,10 released into DD at 12°C for five days [6]. The persisting rhythm was clearly bimodal: once again, major peaks occurred at subjective "dawn" (when light would have commenced had the LD cycle been continued) and secondary peaks or shoulders (in some cases of equal amplitude) at subjective "dusk", with τ_{DD} for either $\cong$ 22.0 h. Thus, either a "major" or a "minor" peak was seen at intervals of $\cong$ 11 h.

Finally, the *entrainability* of the transport rhythm was examined in nondividing cultures of the Y185 rho^{-} mutant of *S. cerevisiae* (lacking cytochromes a/a$_3$, b and c$_1$)

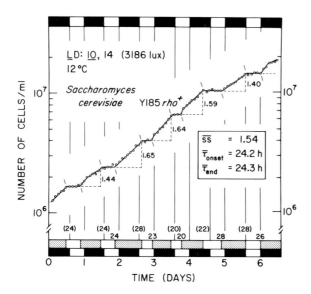

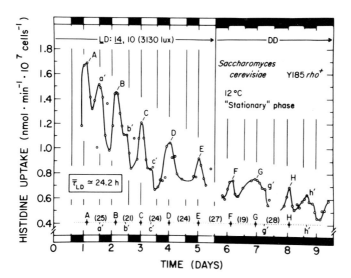

Fig. 5. Synchronization of cell division (*upper panel*) and amino acid transport (*lower panel*) in populations of *S. cerevisiae* Y185 *rho*[+] batch-cultured in 24-h light-dark cycles at $12°C \pm 0.25°C$ on filter-sterilized Wickerham's medium containing 5% glucose and proline as a nitrogen source. *Upper panel* Entrainment of the rhythm of cell division (bud release) by LD: 10,14 (3180 lx, cool-white fluorescent). The 6-l, magnetically stirred culture was automatically sampled at intervals of 2 h; cell concentration was then determined with a Coulter Electronic Particle Counter after appropriate dilution. Dark intervals are indicated by the *heavy black bars. Stippled bars* denote the times during which increase in cell number was actually observed. Beside each fission "burst" on the curve, the stepsize (*ss*), or factorial increase is given. Time intervals (h) between successive onsets of division, or between endpoints (in parentheses), are shown at the *bottom of the figure*; the average stepsize (*ss*), period of onsets (τ_{onset}), and period of endpoints (τ_{end}) are shown in the box. Although not every cell divides during a given 24-h timespan, those that do, do so during intervals restricted pri-

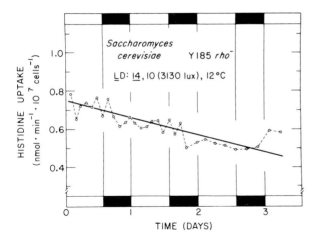

Fig. 6. Failure to synchronize ^{14}C-histidine uptake activity by *LD*: *14*,10 (3130 lx) in a "stationary"-phase culture (12°) of *S. cerevisiae* Y185 *rho⁻*, a petite mutant lacking cytochromes a/a₃, b and c₁. Contrast with Fig. 5 (*lower panel*)

exposed to *LD: 14*,10 (Fig. 6). This LD cycle, imposed on the culture from the time of inoculation and throughout the 80-h timespan during which the uptake of ^{14}C-histidine was monitored at 2- to 3- intervals, failed to entrain the rhythm: the data were best fitted by a straight line [in striking contrast to the *rho⁺* strain (Fig. 5, lower panel)], whose negative slope reflected the aging of the population.

4 Discussion

4.1 Photoinhibition of Growth and Transport

Visible light is known to inhibit cell growth, protein synthesis, respiration, and other functions in the yeast, *Saccharomyces cerevisiae,* as well as in other microorganisms (see Epel [10] for review). Recently, we have added inhibition of membrane transport activity to the extensive list of damaging effects of light: The inhibition of sugar and amino acid transport were found to be due to the development of cell membrane leakiness [18]. Respiratory-deficient (*rho⁻*) cells devoid of cytochromes b and a/a₃, however, were found to be resistant to light damage, suggesting that the site of action of light was in the mitochondrion and involved these cytochromes.

In this paper and elsewhere [16], we have confirmed that the presence of all three cytochromes is required for maximal photoinhibition of cell viability, growth, and amino

---

marily to darkness. *Lower panel* Entrainment and persistence of the rhythm of ^{14}C-histidine uptake in a yeast population that had reached the mid-to-late "stationary" phase of growth, where little, if any net increase in cell number occurred. Cultures were initially batch-cultured in *LD*: *14*,10 at 12°C and were subsequently released into DD. "Subjective" light and dark intervals are indicated to facilitate comparison with LD. Each point of the curve represents a rate determination from a time-course assay of ^{14}C-histidine uptake. Major and minor (shoulders) peaks are labeled *A* to *H* or *a'* to *h'*, respectively. Intervals between the major peaks are given on the *dotted horizontal line* below the curve. The period (τ_{LD}) approximates that of the imposed 24-h LD cycle. (After Edmunds et al. 1979)

acid transport (Fig. 1); a reduction in the cytochrome complement (Fig. 2, upper panel), or the absence of cytochrome b alone (Fig. 2, lower panel) or of cytochromes a/a_3, confers only partial resistance, although a more rigorous quantitative relationship between the amount of chromophore and the degree of photosensitivity is desirable (e.g., by the use of yeast mutants incapable of synthesizing δ-aminolevulinic acid to regulate cytochrome levels). Finally, the conferral of light sensitivity by the cytochromes is independent of their participation of normal mitochondrial function: respiratory-deficient mutants lacking usual complements of cytochromes, and oxidative phosphorylation-deficient mutants are fully sensitive to light (Fig. 3). The results suggest that the heme chromophores of the three cytochromes are the endogenous photosensitizers responsible for the photodynamic damage ultimately observed, although the contribution of other molecules, such as flavins [1], is not excluded [16]. This suggestion is supported by the increase in the inhibitory effects of light of constant energy successively enriched with wavelengths that constitute the Soret region of the cytochrome absorption spectrum (Fig. 4) and is consistent with the conclusions of Ninnemann et al. [14, 15]. The loss of mitochondrial activity alone cannot account for the observed inhibitory effects and development of membrane leakiness because selective photodestruction only of cytochromes b and a/a_3 produced petites that were light-resistant. The events that cause the loss of mitochondrial activity [16] must ultimately cause irreversible damage to the cell membrane — either directly (e.g., by singlet oxygen or the superoxide ion), or indirectly (e.g., by the action of these reactive species on the vacuolar membrane that would cause the release of hydrolytic enzymes).

4.2 Light-Entrained Oscillations in Cell Division and Transport

Notwithstanding the direct inhibitory action of light, we have noted that light may also exert its effects by synchronizing or entraining endogenous oscillatory mechanisms (or biological "clocks") when imposed as a LD cycle [5, 6, 7], and indeed, Ninnemann et al. [15] have pointed out the consequences of the naturally occurring diurnal day–night cycle for the inhibition of respiration in *Saccharomyces*. Our results now implicate light-sensitive, self-sustaining circadian and ultradian clocks in the generation and modulation of rhythmicities in cell division and transport capacity of yeast (Fig. 5). To the best of our knowledge, this represents the first report for circadian organization in *Saccharomyces* (although Wille [17] has found light-entrainable circadian oscillations in the growth rate of the yeast *Candida utilis*).

The phased pattern of cell division seen in slowly dividing batch cultures of *Saccharomyces* in LD cycles (Fig. 5, upper panel) cannot be explained solely by the direct inhibitory effects of light on growth [darkness could either release inhibition, negate it (e.g., by dark repair), or actively promote division (e.g., by production of some critical component)] since it persists in DD with no significant diminution in amplitude and a period only approximating 24 h [6]. Indeed, a variety of unicellular systems, comprising both protozoans and algae, have similar endogenous, light-entrainable circadian clocks underlying cell division (and many other physiological and biochemical variables) that continue to oscillate for long timespans when the LD (or temperature) cycle is removed [4, 5].

The relatively low "permissive" growth temperature utilized (12°C) yielded *culture* generation times of two to ten days, thereby enabling a putative circadian clock not only to be synchronized by the imposed 24-h LD cycle, but also, in turn, to entrain the cell division cycle (in a manner as yet unknown) [6]. Although a step increase in cell number was observed in the population at 24-h intervals, not every cell divided each time (ss < 2.00); by inference, the lengths of individual cell division cycles must be longer (e.g., averaging $\cong 48$ h when ss $= 1.50$), although it is not known whether they are continuously distributed about some mean or discontinuously clustered (quantized) at nodal points [5], as Klevecz [13] has suggested for mammalian cells. Nevertheless, our yeast cultures, although not developmentally synchronous (i.e., a one-to-one correspondence of individual cell developmental cycle stages), are synchronized in the sense of event simultaneity: cell division (and no doubt other markers) in DD, when it does occur, takes place at periodic intervals irrespective of amplitude that can be defined relative to the previous entraining LD cycle [4, 5].

The longer-period oscillations in transport capacity observed in nondividing (or very slowly dividing) cultures (Fig. 5, lower panel) might be anticipated on the various membrane models for circadian rhythms [7] but have not been hitherto reported. One would expect to find changes in transport activity in rhythmically dividing cultures of unicells [7], including *Saccharomyces* [3], if for no other reason than that the cell division cycle constitutes a driving force itself. We found (Fig. 5) that the uptake of histidine or lysine could be entrained by a 24-h LD cycle and that the rhythm persisted in DD with $\tau_{DD} \cong$ 22 h [6]. "Secondary" peaks often were observed at roughly the onset of dark in LD (or subjective "dusk" in cultures freerunning in DD), about 11 to 12 h after the primary peaks.

The occurrence of transport maxima at 11- to 12-h intervals is open to several formal interpretations [6]: (a) a true bimodal circadian rhythm exists; (b) there are two (or more) subpopulations of cells, each having a circadian rhythm in uptake but 180° out-of-phase; (c) two different circadian oscillators govern the primary and secondary peaks; or (d) the rhythm is truly ultradian, with a period of about 12 h. We are unable to distinguish among these alternative hypotheses at present.

The unsuccessful attempt (Fig. 6) to entrain transport activity by LD cycles in the Y185 *rho*⁻ mutant (lacking cytochromes a/a₃, b and c_1) is consistent with — but by no means demands — the hypothesis that cytochrome(s) are photoreceptors for the synchronization of biological oscillators as well as for the direct inhibitory effects discussed previously. More rigorous tests for such a role of the heme chromophore in clock function will require, however, that the mutants be affected in only one (or very few) genes and that they exhibit some assayable biological rhythm. If, indeed, yeast cytochromes a/a₃ and b participate in the observed inhibitory and entraining effects of (blue) light, they would fall into the larger class of blue-light photoreceptors that are being reported with increasing frequency for a number of biological phenomena (reviewed by Briggs [2]).

Acknowledgments. The work reported represents the concerted efforts of my laboratory; especial thanks are due R.I. Apter, V.P. Cirillo, P.J. Rosenthal, W.-K. Shen, S. Ułaszewski and J.R. Woodward for their contributions. I thank S.M. DiRienzo for her expert technical assistance. Supported by National Science Foundation grants GB-43543 and PCM76-10273 to L. Edmunds and V. Cirillo.

References

1. Aggarwal BB, Avi-Dor Y, Tinberg HM, Packer L (1976) Effect of visible light on the mitochondrial inner membrane. Biochem Biophys Res Commun 69: 362–368
2. Briggs WR (1976) The nature of the blue light photoreceptor in higher plants. In: Smith H (ed) Light and plant development, pp 7–18. Butterworths, London
3. Carter BLA, Halvorson HO (1973) Periodic changes in rate of amino acid uptake during yeast cell cycle. J Cell Biol 58: 401–409
4. Edmunds LN Jr (1975) Temporal differentiation in *Euglena:* circadian phenomena in non-dividing populations and in synchronously dividing cells. Colloq Int CNRS 240: 53–67
5. Edmunds LN Jr (1978) Clocked cell cycle clocks: implications toward chronopharmacology and aging. In: Samis HV Jr, Capobianco S (eds) Aging and biological rhythms, pp 125–184. Plenum Press, New York
6. Edmunds LN Jr, Apter RI, Rosenthal PJ, Shen W-K, Woodward JR (1979) Light effects in yeast: persisting oscillations in cell division activity and amino acid transport in cultures of *Saccharomyces cerevisiae* entrained by light-dark cycles. Photochem Photobiol 30: 595–601
7. Edmunds LN Jr, Cirillo WV (1974) On the interplay among cell cycle, biological clock and membrane transport control systems. Int J Chronobiol 2: 233–246
8. Ehrenberg M (1966a) Wirkungen sichtbaren Lichtes auf *Saccharomyces cerevisiae*. 1. Einfluß verschiedener Faktoren auf die Höhe des Lichteffektes bei Wachstum und Stoffwechsel. Arch Mikrobiol 54: 358–373
9. Ehrenberg M (1966b) Wirkungen sichtbaren Lichtes auf *Saccharomyces cerevisiae*. 2. Die Gärung unter dem Einfluß von Pasteur-Effekt und Lichteffekt. Arch Mikrobiol 55: 26–30
10. Epel BL (1973) Inhibition of growth and respiration by visible and near visible light. In: Giese AC (ed) Photophysiology, vol VIII, pp 209–229. Academic Press, New York
11. Guerin B, Jacques R (1968) Photoinhibition de l'adaptation respiratoire chez *Saccharomyces cerevisiae*. 2. Le spectre d'action. Biochim Biophys Acta 153: 138–142
12. Guerin B, Sulkowski E (1966) Photoinhibition de l'adaptation respiratoire chez *Saccharomyces cerevisiae*. 1. Variations de la sensibilité à l'adaptation. Biochim Biophys Acta 129: 193–200
13. Klevecz RR (1976) Quantized generation time in mammalian cells as an expression of the cellular clock. Proc Natl Acad Sci USA 73: 4012–4016
14. Ninnemann H, Butler WL, Epel BL (1970a) Inhibition of respiration in yeast by light. Biochim Biophys Acta 205: 499–506
15. Ninnemann H, Butler WL, Epel BL (1970b) Inhibition of respiration and destruction of cytochrome a_3 by light in mitochondria and cytochrome oxidase from beef heart. Biochim Biophys Acta 205: 506–512
16. Ułaszewski S, Mamouneas T, Shen W-K, Rosenthal PJ, Woodward JR, Cirillo VP, Edmunds LN Jr (1979) Light effects in yeast: evidence for participation of cytochromes in photoinhibition of growth and transport in *Saccharomyces cerevisiae* cultured at low temperatures. J Bacteriol 138: 523–529
17. Wille JJ Jr (1974) Light entrained circadian oscillations of growth rate in the yeast *Candida utilis*. In: Scheving LE, Halberg F, Pauly JE (eds) Chronobiology, pp 72–77. Igaku Shoin, Tokyo
18. Woodward JR, Cirillo VP, Edmunds LN Jr (1978) Light effects in yeast: inhibition by visible light of growth and transport in *Saccharomyces cerevisiae* grown at low temperatures. J Bacteriol 133: 692–698

Role of Light at Shorter Wavelength
in Photobiological Phenomena in Blue-Green Algae

K. OHKI[1], T. ISONO[1,2], and Y. FUJITA[1]

1 Introduction

In the life of blue-green algae, several photobiological phenomena are known, from morphological to metabolic. A morphological development of *Nostoc muscorum* A is one of the examples. Lazaroff and his colleagues reported that this alga has two morphologically different stages in its life cycle. It takes a coccoid form (aseriate stage) when grown in the dark and differentiates into a filamentous form (seriate stage) in the light [11]; the differentiation from the aseriate to the seriate is induced by the red light, and the green light cancels the red effect [10]. Secondly, a photocontrolled heterotrophic growth was found by Diakoff and Scheibe with *Fremyella diplosiphon* [3]; the heterotrophic growth of this alga is suppressed by a short illumination with green light, and the suppression is cancelled by a short red illumination. The third is well known as the complementarily chromatic adaptation of phycobilin composition [1]. In this case, green and red lights again control the formation of two phycobiliproteins, phycoerythrin and phycocyanin [2, 8]; green light induces phycoerythrin but suppresses phycocyanin formation, and red light acts reversely.

In all three, the green (around 540 nm) and the red (around 650 nm) lights act antagonistically, and the photocontrol occurs not only during illumination but also lasts for a long time in the dark after the illumination. Thus one can imagine that a common phototrigger mechanism is operating in these phenomena, and the effective wavelengths suggest an occurrence of the photoreceptor characteristic to blue-green algae. To evaluate this possibility, we tried to re-examine these three phenomena. Our results preliminarily indicate that the photocontrol by green and red lights also occurs in the heterotrophic respiration and that the respiration control provides morphological regulation in *N. muscorum* A and regulation in the heterotrophic growth in *F. diplosiphon.* However, the photocontrol of phycoerythrin formation seems to be independent of the respiration control.

2 Materials and Methods

Fremyella diplosiphon (M-100), *Nostoc muscorum* A (M-14) and *Tolypothrix tenuis* (M-29) were obtained from the Algal Collection at the Institute of Applied Microbiology, University of Tokyo.

1 Ocean Research Institute, University of Tokyo, Nakano, Tokyo 164, Japan
2 Present address: Department of Botany, Faculty of Science, Kyoto University, Kita-shirakawa, Kyoto 606, Japan

2.1 Experiments for Morphological Differentiation in *Nostoc muscorum* A

Four culture media, MDM (MDM, cf. [15]), KNO_3-free MDM (MDM-N), Lazaroff's medium (LM, cf. [11]) and KNO_3-enriched (10 mM) Lazaroff's medium (LM+N) were used. Glucose (1% w/v) was always added. For growth at the seriate stage, cells were grown in the liquid MDM under red light (130 μW cm^{-2}) at 22°C. Under these conditions, all cells were at the seriate stage at least by the late exponential growth phase. Cells at the aseriate stage were prepared from the seriate cells by culturing on agar plates of LM (solidified by addition of 1% agar) in the dark. For convenience to transplant to experimental media, a cellophane sheet was placed on the agar plate, and the seriate cells were inoculated on the cellophanes.

Expeimental cultures for seriate-aseriate transformation were done on agar plates of various media at 30°C under various light conditions or in the dark. After 3 to 4 weeks, seriate and aseriate colonies were microscopically counted, respectively. For aseriate-seriate transformation, the aseriate cells, together with cellophane, were asceptically transferred to the new agar plates of various media. Green (520 μW cm^{-2}) or red (520 μW cm^{-2}) preillumination was given for 30 min, then they were incubated in the dark at 30°C. These light conditions gave a saturating effect on the transformation. After 4 days, colonies of respective forms were counted.

Green and red lights were obtained with use of colored cellophanes; the light source, cool-white fluorescent light. Green cellophane mainly transmitted the light of wavelength range from 450 to 600 nm, and red, longer than 590 nm. Algal growth was measured as chlorophyll a increase.

2.2 Experiments for Heterotrophic Growth and Respiration in *Fremyella diplosiphon*

Algal cells were first grown in the organic MDM (O-MDM, cf. [13]) at 30°C under red illumination (longer than 590 nm, 250 μW cm^{-2}). The cells at the late exponential growth phase were used for all experiments.

For experiments of photocontrolled respiration, cells were grown in the O-MDM or in the medium of Diakoff and Scheibe (DSM, cf. [3]) in the dark at 30°C. The incubation was done in Monod-type flasks (200 ml). Cells at various growth phases were used for respiration measurements immediately after sampling or after a short centrifugation (5 s). Oxygen uptake was measured by a Clark-type O_2 electrode (YSI 4004, cf. [4]) at 30°C. Experimental regime for respiration measurements was as follows: cell suspensions (cells after exponential growth phase were resuspended in the fresh medium) were placed in the electrode vessel and incubated in the dark for 15 min. Then green (with a 40% [w/v] $CuCl_2$ solution filter; light path, 5 mm; transmitting the wavelength range between 450 and 600 nm; intensity, 500 μW cm^{-2}; cf. [7]) and red (with a Toshiba VR 62 sharp cut filter; longer than 600 nm; 800 μW cm^{-2}) lights were alternately given with 15-min interval; the illumination time, 5 min. The above dose of green or red light gives a full effect. Respiration rates were measured 5 min after each illumination. Increase in O_2 concentration due to photosynthetic O_2 evolution was avoided by opening the cap of the electrode vessel, and measurements were done almost at the same O_2 concentration.

    For experiments of photocontrolled growth, cells in the above two media were grown
in Monod-type flasks (200 ml) in the dark at 30°C. Red (with a red cellophane, longer
than 590 nm, 7000 μW cm^{-2}) or green (with CuCl$_2$ filter, 220 μW cm^{-2}) light was given
for 7 min every 24 h. Growth was followed as chlorophyll a increase.

2.3 Experiments for Photocontrolled Phycoerythrin Formation in *Tolypothrix tenuis*

Experimental procedures were the same as those reported previously [13]. Cells were
first grown in the O-MDM under red illumination (longer than 590 nm, 250 μW cm^{-2}).
Cells at the late exponential growth phase were inoculated in the fresh medium, and
after incubation for 15 h in the dark, the first green light (with CuCl$_2$ filter, 200 μW
cm^{-2}) was given for 7 min. Then the second shot was made 9 or 23 h after the first shot.
After the green illumination, pigment increase in the dark was followed as described
previously [13].

2.4 Measurements of Pigment Contents

Phycobiliproteins were extracted and measured by the method of Hattori and Fujita
[9], and chlorophyll a was measured with acetone extracts with use of the extinction
coefficient of Mackinney [12].

3 Results and Discussion

3.1 Morphological Differentiation in *Nostoc muscorum* A

According to Lazaroff [10], the differentiation from aseriate to seriate stage occurs only
when cells receive the red light, and the green light cancels the red effect. If this process
is strictly photocontrolled, cells should be always at the aseriate stage in the dark or
under the green light and at the seriate stage under the red light. However, occurrence
of the seriate-aseriate transformation under different nutritional conditions was surprisingly different from the above expectation (Table 1). When combined nitrogen was
present in the medium, most cells grew at the seriate stage even in the dark. Further,
the green light which should stimulate the transformation was not effective in MDM.
Therefore, the cells can grow at the seriate stage without the red light and the green light

Table 1. Effect of growth conditions on the seriate-aseriate transformation in *Nostoc muscorum* A

| | Appearance of aseriate colonies (%) | | | |
	MDM	MDM-N	LM+N	LM
Red	0 (24)	30 (20)	5 (19)	70 (19)
Green	0 (10)	35 (10)	25 (11)	80 (8)
Dark	25 (10)	65 (15)	40 (9)	80 (15)

Red or green light was illuminated throughout the incubation; light intensities, 60 to 100 μW cm^{-2}.
Number of repetition of each experiment is indicated in parenthesis, and values are the mean of all
experiments. Experimental details, see text

does not necessarily keep the cells at the aseriate stage. Though data are not shown here, algal growth measured as cell yield was larger in order of MDM, LM+N, MDM-N and LM in medium, and red, green, and dark in light conditions (T. Isono and Y. Fujita, in prep.). Thus, the occurrence of the transformation is more closely related to algal growth than to the light conditions. Cells probably grow at the seriate stage under the conditions for rapid growth while they have to stay at the aseriate stage under unfavorable conditions. Indeed, 90% of cells stayed at the aseriate stage even in MDM and under the red light when the growth was suppressed at lower temperature (15°C) or pH (6.0) (T. Isono and Y. Fujita, in prep.).

However, the aseriate-seriate transformation in the dark was strongly dependent on the short red illumination, and the green light canceled the red effect (Table 2). These results agreed with those of Lazaroff [10]. However, the transformation to the seriate occurred again more often in MDM than LM indicating a close correlation with algal growth. If the algal growth governs this transformation, the red light must enhance the dark growth and the green light must suppress the enhanced growth. Thus, we examined the effect of short green illumination on the growth at the aseriate stage. Cells at the aseriate stage were grown on the agar plates of LM, and a short illumination (30 min) was given every 24 h. Results (Table 3) indicate that the green light strongly suppresses the algal growth, as was found in the heterotrophic growth of *Fremyella diplosiphon* by Diakoff and Scheibe [3]. Therefore, differentiation between the seriate and the aseriate stage cannot be a strictly photocontrolled process but it depends on the growth conditions as in the case of *Chlorogloea fritschii* [5, 6]; red light causes the transformation to the seriate stage because of its action for enhancement of growth via photosynthesis, and the green light inhibits the process through suppression of the growth. Then, there remains a question why the green light suppresses the algal growth in the dark. To open this question, we studied green light effect on the heterotrophic growth of *Fremyella diplosiphon* in relation to the respiration activity.

3.2 Heterotrophic Growth and Respiration of *Fremyella diplosiphon*

The finding that the green light suppresses the heterotrophic growth of not only *F. diplosiphon* but also *N. muscorum* A suggests that the photocontrol of this kind occurs rather generally in the heterotrophic growth of blue-green algae. In most cases, the hetero-

Table 2. Effect of short red or green illumination on the aseriate-seriate transformation in *Nostoc muscorum* A

| | Appearance of seriate colonies (%) | |
	MDM	LM
Red	80	50
Green	25	0
Red then green	20	15
Dark	0	0

Experimental details, see text

Table 3. Effect of short red or green illumination on the heterotrophic growth of *Nostoc muscorum* A at the aseriate stage

Illumination	Relative growth[a]
Dark	1.00
Green	0.22
Red then green	0.25

[a]Values relative to the dark control are presented. Experimental details, see text

Table 4. Effect of short illumination on the dark respiration in heterotrophically grown cells of *Fremyella diplosiphon*

Respiration rate (μmol O_2/mg chlorophyll $\cdot$ h)

Dark	$\rightarrow$	Red	$\rightarrow$	Green	$\rightarrow$	Red	$\rightarrow$	Green
18		20		18		20		18

Fremyella diplosiphon was grown in DSM in the dark for 13 days. Respiration restes were estimated from the time courses (10 min) in the dark 5 min after each illumination; the value for the first dark was obtained in the time course (10 min) 5 min after placing the cell suspension in electrode vessel. The transfer was made in the dark. Experimental details, see text

Table 5. Control of respiration by red or green light observed in *Fremyella diplosiphon* grown heterotrophically in DSM or O-MDM

Culture days		Respiration rate[a]		Changes due to green illumination	Increase in chlorophyll a
		Before illumination	After illumination		
		(μ mol O_2 / mg chlorophyll $\cdot$ h)		(%)	(μM)
DSM	4	37	38	+ 2.7	0.8
	9	26	26	0	7.6
	13	25	22	−12	10.2
O-MDM	4	29	29	0	1.5
	9	33	32	− 3.1	20.2

[a] Values before and after the first green illumination are presented (cf. Table 4).
Experimental details, see text

trophic growth of blue-green algae is limited by the energy supply from the respiration system [14]. Thus, the site to be photocontrolled possibly presents in the heterotrophic respiration. This possibility was evaluated with heterotrophic cells of *F. diplosiphon*. One of the results is shown in Table 4. Though the first illumination, irrespective of color of light, was somewhat stimulative, antagonistic effects on the respiration rate were observed in the subsequent illumination. Green light caused a partial suppression of the dark respiration and red light gave a recovery from the suppression. In most cases, around 10% of the activity was photoregulated, but in extreme case, it attained 20%.

However, the photoregulation did not always appear. Growth phases or growth conditions are critical. As seen in Table 5, the regulation tends to be discernible in cells at later growth phase when their growth has been nutritionally limited. Thus, such regulation was less observable in the respiration by cells grown in O-MDM which provided more rapid heterotrophic growth (Table 5). The respiratory activity was generally lower when the regulation became observable (cf. Table 5). As the substrate for the respiration, glucose, was present at saturating amount, the low activity is not due to limitation in the substrate, but may result from characteristics of a respiratory system in cells grown under poor conditions, and in such a system, the step to be photocontrolled may determine the overall reaction rate. Mechanism of the photocontrol is not known, and has been under investigation.

Noteworthy is a correlation in occurrence of two photocontrols in heterotrophic respiration and growth. As reported by Diakoff and Scheibe [3], we observed the red enhancement and the green suppression on the heterotrophic growth of this alga when they grew in DSM. However, the light effect was always insignificant on the growth in O-MDM which gave a more rapid proliferation (cf. Table 5). A similar feature was also found in the growth of *N. muscorum* A; the light effect appeared in a slow heterotrophic growth at the aseriate stage. A rapid growth at the seriate stage in MDM did not respond to the light. This correlation strongly suggests that the photocontrol of heterotrophic growth observed in *F. diplosiphon* and *N. muscorum* A results from the photoregulation of the respiration which is an energy-yielding metabolism for the heterotrophic growth. Thus, we consider that the light effect on the dark transformation to the seriate stage in *N. muscorum* A originates from the photoregulation of the respiration.

3.3 Photocontrolled Phycoerythrin Formation in *Tolypothrix tenuis*

As reported previously [13], a short green illumination can induce phycoerythrin formation not only in autotrophic cells but also in heterotrophic cells. Thus, the green light may simultaneously cause induction of phycoerythrin formation and suppression of algal growth under heterotrophic conditions. The action spectra for both phenomena are very similar. The action maximum of green effect is located around 540 nm and of red light, around 650 nm in both cases (cf. [2, 8, 10]), suggesting that the photocontrol mechanism is common in both phenomena. Indeed, the suppressive effect of green light was observed in biosyntheses of chlorophyll a and phycocyanin (Fig. 2) and even in phycoerythrin formation (Fig. 1). The suppression was removed by red light as similar to the suppression of heterotrophic growth of *F. diplosiphon* and *N. muscorum* A. However, the life time of the suppression was rather short. After 2 to 4 h, biosynthetic rates recovered to the levels before illumination (Fig. 2). Therefore, the heterotrophic growth of *T. tenuis* is inhibited by the green light as similar to those of *F. diplosiphon* and *N. muscorum* A, though the effect can survive only for a short time differently from the effect observable with the latter two algae. The phycoerythrin formation induced by the first shot was once suppressed by the second shot, but it again started after 2 to 5 h rest. Further, the second shot always provided more rapid and extensive phycoerythrin formation (Fig. 1 vs. Fig. 2), indicating that the effect of green light on phycoerythrin formation lasts for long period even after the suppressive effect of green light disappeared. Difference in two green light effects can also be seen in conditions for their occurrence. As discussed above, the green effect on the heterotrophic growth or respiration occurs in cells at a limited physiological state. However, the photocontrolled phycoerythrin formation occurs in cells growing in any trophic type. These differences between two green effects suggest that the photocontrol mechanism of phycoerythrin formation is different from that for the heterotrophic growth or respiration.

Our results presented here, therefore, indicate a possibility that despite similarity in the action spectra, two kinds of photocontrol mechanism are acting in the green light regulation occurring in the life of blue-green algae. However, our investigation has not directly analyzed the photocontrol mechanism yet. Thus, there still remains a possibility that a single photoprocess governs all green-controlled phenomena. The final con-

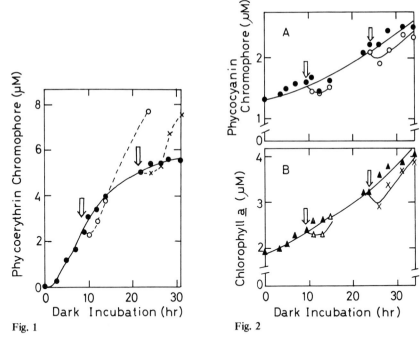

Fig. 1   Fig. 2

Fig. 1. Effect of the second green illumination on the dark phycoerythrin formation in *Tolypothrix tenuis.* The first short illumination was given at 0 h and the second after 9 h or 23 h (indicated by *arrows*). *Closed circles:* phycoerythrin formation without the second illumination; *open circles* and *crosses:* with the second illumination. Experimental details, see text and [14]

Fig. 2A, B. Effect of short green illumination on phycocyanin (**A**) and chlorophyll (**B**) formation in the dark. Short green illumination was given at 9 h or 23 h (indicated by *arrows*). *Closed circles* and *closed triangles:* without illumination; *open circles, open triangles* and *crosses:* with illumination. Experimental details, see text and [14]

clusion must wait for further investigation on the detailed mechanism of photocontrol process.

Acknowledgment. This work was partly supported by Grants-in-Aid for Scientific Research from the Ministry of Education, Sciences, and Culture, Japan.

References

1. Bogorad L (1976) Phycobiliproteins and complementary chromatic adaptation. Annu Rev Plant Physiol 26: 369–401
2. Diakoff S, Scheibe J (1973) Action spectra for chromatic adaptation in *Tolypothrix tenuis.* Plant Physiol 51: 382–385
3. Diakoff S, Schcibe J (1975) Cultivation in the dark of the blue-green alga *Fremyella diplosiphon.* A photoreversible effect of green and red on growth rate. Physiol Plant 34: 125–128
4. Ebata T, Fujita Y (1971) Changes in photosynthetic activity of the diatom *Phaeodactylum tricornum* in a culture of limited volume. Plant Cell Physiol 12: 533–541

5. Evans EH, Foulds I, Carr NG (1976) Environmental conditions and morphological variation in the blue-green alga *Chlorogloea fritshii.* J Gen Microbiol 92: 147–155
6. Fay P, Kumar HD, Fogg GE (1964) Cellular factors affecting nitrogen fixation in blue-green alga *Chlorogloea fritshii.* Ibid 35: 351–360
7. Fujita Y, Hattori A (1960) Effect of chromatic light on phycobilin formation in a blue-green alga, *Tolypothrix tenuis.* Plant Cell Physiol 1: 293–303
8. Fujita Y, Hattori A (1962) Photochemical interconversion between precursors of phycobilin chromoproteins in *Tolypothrix tenuis.* Plant Cell Physiol 3: 209–220
9. Hattori A, Fujita Y (1959) Crystalline phycobilin chromoproteids obtained from a blue-green alga, *Tolypothrix tenuis.* J Biochem 46: 633–644
10. Lazaroff N (1973) Photomorphogenesis and Nostocean development. In: Carr NG, Whitton A (eds) The biology of the blue-green algae, pp 297–319. Blackwell Scientific Publication, Oxford
11. Lazaroff N, Vishniac W (1961) The effect of light on the developmental cycle of *Nostoc muscorum* A, a filamentous blue-green alga. J Gen Microbiol 25: 365–374
12. Mackinney G (1941) Absorption of light by chlorophyll solution. J Biol Chem 140: 315–322
13. Ohki K, Fujita Y (1978) Photocontrol of phycoerythrin formation in the blue-green alga *Tolypothrix tenuis* growing in the dark. Plant Cell Physiol 19: 7–15
14. Stanier RY (1973) Autotrophy and heterotrophy in unicellular blue-green algae. In: Carr NG, Whitton A (eds) The biology of the blue-green algae, pp 501–518. Blackwell Scientific Publication, Oxford
15. Watanabe A (1960) List of algal strains in collection at the Institute of Applied Microbiology, University of Tokyo. J Gen Appl Microbiol 6: 283–292

Blue Light Effects on Some Algae Collected from Subsurface Chlorophyll Layer in the Western Pacific Ocean

A. KAMIYA[1] and S. MIYACHI[2]

1 Introduction

It has been shown that the concentration of chlorophyll in the ocean often shows a rather sharp peak in a narrow layer at the depth between 50 to 100 m from the surface ("deep chlorophyll maximum" or "subsurface chlorophyll layer") [1]. The light intensity penetrating to this subsurface chlorophyll layer is usually near or below 1% of the surface light. The main composition of the light at this layer is blue and green, with a maximum at ca. 480 nm [14].

It is generally assumed that this subsurface chlorophyll layer is due to the phytoplanktons which concentrated at the specific depth with weak short wavelength light. It is therefore highly likely that the algae at this layer are metabolically adapted to the blue light.

We have been studying the effects of blue light on photosynthetic and nonphotosynthetic carbon metabolism with *Chlorella* as material [6–11]. To study the effects of blue light with these algae, we succeeded in collecting one *Ochromonas* sp. and another alga belonging to the Chryptophyceae from the subsurface chlorophyll layer in the western Pacific Ocean.

In this paper it is revealed that low light intensity and blue light irradiation, similar to the light conditions in the subsurface chlorophyll layer, is favored for the formation of chlorophyll as well as of photosynthetic capacity of these algae.

2 Materials and Methods

2.1 Isolation and Culture of Algae from the Subsurface Chlorophyll Layer

Figure 1 shows the vertical distribution of chlorophyll observed during Expedition KT-77-6 by Tansei-Maru of the Ocean Research Institute, University of Tokyo. The peak concentration of chlorophyll was found at a depth of 90 m. The sea water at various depths was collected in 250 ml flasks containing 2 ml of Enrichment for sea water (Provasoli ES Enrichment [13]). The *Ochromonas* sp. was isolated from sea water collected

1 Department of Chemistry, Faculty of Pharmaceutical Sciences, Teikyo University, Sagamiko, Kanagawa 199-01, Japan

2 Institute of Applied Microbiology and Radioisotope Centre, University of Tokyo, Bunkyo-ku, Tokyo 113, Japan

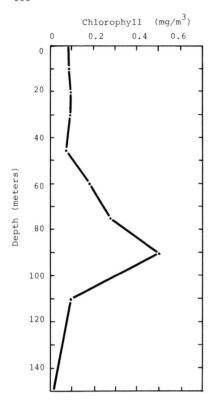

Chlorophyll (mg/m³)

Fig. 1. Vertical distributions of chlorophyll in the western Pacific Ocean (Station 16, lat. 33°20.2′ N. and long. 139°11.5′ E., Expedition KT-77-6, May 20, 1977)

from 75 m depth. The alga belonging to Chryptophyceae was also isolated from sea water at 85 m depth collected during Expedition KT-78-9. These algae were grown in 100-ml Erlenmeyer flasks kept stationary at 18°C, with a light−dark rhythm of 18 and 6 h. Each flask contained 100 ml of filtered sea water containing 2 ml of Provasoli Enrichment.

2.2 Blue, Red, and White Light During the Algal Culture

Light source was provided by a bank of 40 W cool-white fluorescent tubes (Fishlux; Toshiba Co. Ltd., Tokyo), whose distribution of spectral energy was mainly blue (350 $< \lambda < 600$ nm) and red (600 $< \lambda < 700$ nm) (Fig. 2a). The lamps were placed beneath an acrylic plastic plate supporting the culture flasks. These were used as white light. The blue light (370 $< \lambda < 560$ nm, max. 470 nm) was obtained by placing a blue filter (Paraglas 315K, Kyowa Gas Chemical Industry Co. Ltd., Tokyo) on a plastic plate. The transmittance (Fig. 2b) of the blue filter resembles closely that of clear oceanic water at 100 m depth (max. 470 nm). The red light (> 580 nm) was obtained with a red filter (Paraglas 102K, Fig. 2b). Light intensity was determined with a radiometer (Spectra, Model PR-1,000, Photo Research, Burbank, Calif.). When necessary, light intensity was changed by using cotton clothes or changing the distance between the algal culture and the light source. The sides of the flask were wrapped with aluminum foil. Therefore, the suspension received light only through the bottom of the flask.

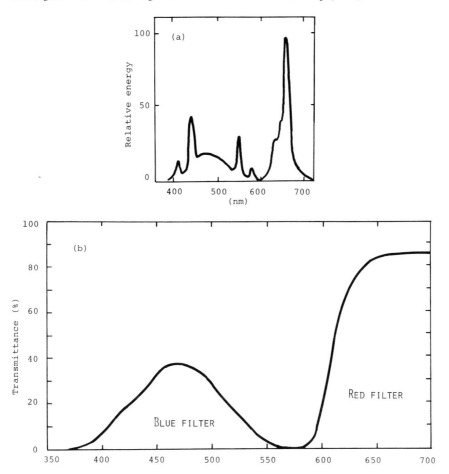

Fig. 2a, b. Spectral energy distribution of white light (**a**) and spectral transmittances of blue and red filters (**b**)

2.3 Determination of Oxygen Gas Exchange

The rate of oxygen gas exchange was determined with a Clark-type oxygen probe (Rank Brothers, London). The light source for the measurement of photosynthesis was a reflex lamp which gave the saturated intensity (100,000 lx).

2.4 Pigment Analysis

Chlorophylls a and c were extracted according to Jeffrey [3] and separated by one-dimensional (descending) paper chromatography with n-propanol-petroleum ether (0.8: 99.2, v/v) as solvent. Immediately after the development, chlorophylls a and c were eluated with diethyl ether and methanol, respectively, to determine respective absorption spectra.

Pigments were extracted from the frozen cells in the dark by adding 5 ml of 90% acetone. Amounts of chlorophyll a and c were determined spectrophotometrically, using an equation developed by Jeffrey and Humphrey [4].

3 Results and Discussion

3.1 Growth and Chlorophyll Formation at Different Intensities of White Light in Ochromonas sp. Collected from Subsurface Chlorophyll Layer

Concentrations of chlorophyll a and c increased in parallel with the rise of light intensities from 100 to 500 lx (Fig. 3a). The chlorophyll concentration decreased when the intensities were raised higher than 500 lx. The concentration of cell mass determined by optical density at 750 nm also attained maximum at intensity around 500 lx. Figure 3b shows that chlorophyll a/c ratio kept increasing as the light intensity was raised to 8000 lx. In plants having chlorophyll a and b, chlorophyll a/b ratio in sun plants is known to be higher than that in shade plants [2]. The present result indicates that the relationship between chlorophyll a/c ratio and the light intensities under which the plants had been grown, would be the same as observed in chlorophyll a/b ratio and light intensities.

Figure 4 shows that the maximum rate of photosynthesis on pcv-basis in the cells grown at low intensities (500 lx) was 2.5 times higher than that in the cells grown at high intensities (8000 lx). On the other hand, the light intensities during growth did not affect the intensity at which the rate of photosynthesis attains one-half its maximum level; the apparent K_m-value was 5600 lx. Since the chlorophyll concentration was greater in the cells grown at lower intensities, the rate of photosynthesis per chlorophyll concentration did not vary depending on the light intensities during growth, $\mu l\ O_2\ \mu g$ chl^{-1} h^{-1} being 8.29 in both types of the algal cells.

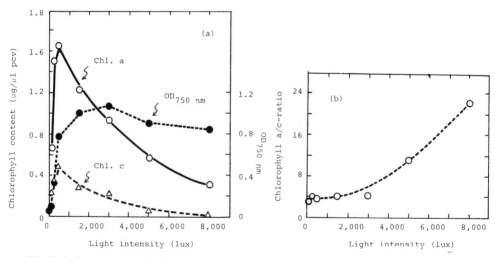

Fig. 3a, b. Concentration of chlorophylls a and c per pcv (**a**) and chlorophyll a/c ratio (**b**) in *Ochromonas* sp. grown under various light intensities of white light for 10 days. Temperature was kept at 18°C

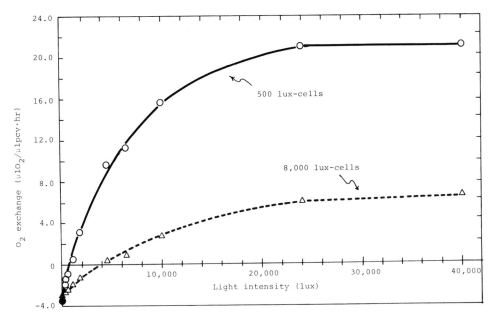

Fig. 4. Oxygen exchange of *Ochromonas* sp. grown under 500 or 8000 lx of white light for 10 days. Oxygen exchange was determined under illumination with a reflex lamp at varied intensities. The experimental temperature was 25°C

With regard to adaptation to different light intensities, two types were described by Jorgensen [5]. The *Chlorella* type adapts to a new light intensity mainly by changing the pigment contents. On the other hand, the *Cyclotella* type adapts by changing the maximum rate of photosynthesis on chlorophyll basis, but without changing the chlorophyll content. The present result indicates that *Ochromonas* sp. collected from subsurface chlorophyll layer belongs to *Chlorella* type in this respect.

3.2 Effects of Blue Light on Chlorophyll Content and Photosynthetic Capacity in Ochromonas sp. Collected from Subsurface Chlorophyll Layer

Table 1 shows that the change in the chlorophyll concentration induced by the different spectral composition of light was not so significant during the exponential growth period (3rd day). However, by the late stationary period (13th day) the chlorophyll concentration in the cells grown under white and red light became only 25% of that in the cells grown under blue light.

Figure 5 shows that the blue light-enhancing effect on chlorophyll formation during growth is saturated at intensity as low as 500 erg cm^{-2} s^{-1}. Figure 5 also shows that with the whole intensity range tested, chlorophyll concentration in the cells grown under blue light was higher than in those grown under red light. If we assume that the light intensities at the surface of the ocean is 100×10^4 erg cm^{-2} s^{-1} and 0.06% of the surface light reaches the subsurface chlorophyll layer, the maximum light intensities in that region will be ca. 600 erg cm^{-2} s^{-1} [12]. Since the light reaching this region mostly

Table 1. The chlorophyll formation and light quality provided during growth of *Ochromonas* sp.

Growth time	Light conditions	Chlorophyll content (μg/μl pcv)		
		Chl a	Chl c	a/c
3rd day (exponential period)	White light[a] (4000 erg cm^{-2} s^{-1})	1.492	0.446	3.35
	Blue light (1400 erg cm^{-2} s^{-1})	1.688	0.498	3.39
	Red light (5000 erg cm^{-2} s^{-1})	1.126	0.356	3.16
13th day (stationary period)	White light[a] (4000 erg cm^{-2} s^{-1})	0.420	0.134	3.13
	Blue light (1400 erg cm^{-2} s^{-1})	1.894	0.710	2.67
	Red light (5000 erg cm^{-2} s^{-1})	0.574	0.204	2.81

[a] 2700 lx

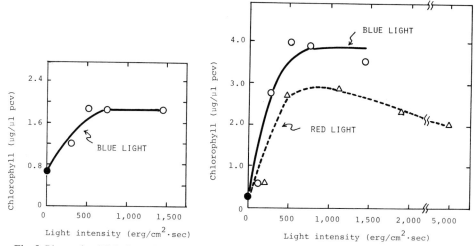

Fig. 5. Blue and red light intensity dependency of chlorophyll formation in *Ochromonas* sp. Cells were grown under blue or red light at indicated intensities for 12 days (18°C)

consists of blue and green, we assume that the alga used in this experiment adapted to utilize the light in the subsurface chlorophyll layer efficiently.

Figure 6 shows the experimental results in which the algal cells which had been cultured for 10 days under blue light were suspended in sea water (nongrowing condition) and then the changes in chlorophyll content and the photosynthetic capacity followed under illumination with red light, and vice versa. Change in the color of illuminating light did not cause appreciable change in chlorophyll content. On the other hand,

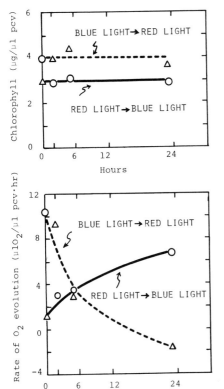

Fig. 6. Effects of light quality on chlorophyll content and photosynthetic capacity in *Ochromonas* sp. The alga which had been grown under blue light was transferred into sea water (nongrowing condition) and then illuminated with red light for 24 h, and vice versa. Photosynthetic capacity (rate of oxygen evolution) was determined under 10^5 lx light from a reflex lamp and at 25°C

transferal from blue to red light greatly decreased the photosynthetic capacity, while the change from red to blue light increased the same capacity.

When the algal cells which had been grown under red light at 4300 erg cm^{-2} s^{-1} for 10 days were suspended in sea water and then kept under various intensities of blue light for one day, the photosynthetic activity increased in parallel with the blue light intensity to attain the maximum level at ca. 500 erg cm^{-2} s^{-1} (Fig. 7). On the other hand, the same capacity in the cells kept under red light for one day was the greater the lower the intensity applied. On the other hand, when the algal cells which had been grown under blue light at 1300 erg cm^{-2} s^{-1} for 10 days were suspended in sea water and kept under red light at various intensities, the photosynthetic capacity in the cells kept under weak red light was higher than that in those kept in the dark (Fig. 8). When the cells were kept under red light at intensities higher than 400 erg cm^{-2} s^{-1}, the photosynthetic capacity decreased in parallel with the intensities of red light applied. At 1500 erg cm^{-2} s^{-1}, photosynthetic capacity was only a little higher than the respiratory activity (-1 μl O_2/μl pcv h). The similar changes in photosynthetic activity were observed in the algal belonging to Chryptophyceae which had been collected from the subsurface chlorophyll layer.

Thus, the above results clearly indicate that the algae in the subsurface chlorophyll layer are photosynthetically adapted to the wavelength and intensity of light which reaches that region.

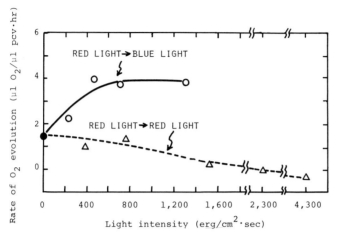

Fig. 7. Effects of light quality on photosynthetic capacity in *Ochromonas* sp. grown under red light. Algal cells grown under red light for 10 days were suspended in sea water (nongrowing condition) and then illuminated with blue or red light of varied intensities. Photosynthetic capacity (rate of oxygen evolution) was determined after one day. Other conditions were the same as those described in Fig. 6

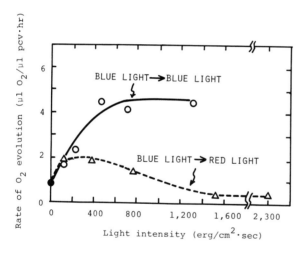

Fig. 8. Effects of light quality on photosynthetic capacity in *Ochromonas* sp. grown under blue light. Algal cells grown under blue light for 10 days were suspended in sea water (nongrowing condition) and then illuminated with blue or red light of varied intensities. Photosynthetic capacity (rate of oxygen evolution) was determined after one day. For other conditions, see the explanation in Fig. 6

4 Summary

1. A kind of alga belonging to *Ochromonas* (Chrysophyceae) and another kind of alga belonging to Chryptophyceae were isolated from the subsurface chlorophyll layer in the western Pacific Ocean.

2. The *Ochromonas* sp. cells grown under weak white light (500 lx) contained much more chlorophyll and exhibited higher photosynthetic capacity (O_2 evolution per pcv) than those grown under strong white light (8000 lx). The chlorophyll a/c-ratio increased significantly in parallel with the increase in the light intensities during growth.

3. Chlorophyll content of the *Ochromonas* sp. grown under blue light was higher than that grown under red light of the same intensities. Chlorophyll formation was saturated at blue light intensities as low as 500 erg cm^{-2} s^{-1}.

4. When the *Ochromonas* sp. grown under red light was illuminated with red or blue light at varied intensities under non-growing condition, blue light illumination caused increase in photosynthetic capacity, while red light did not show such an enhancement. The enhancing effect of blue light was saturated at 500 erg cm^{-2} s^{-1}. The similar enhancing effects of blue light were observed in the alga belonging to Chryptophyceae.

These results indicate that the algae in the subsurface chlorophyll layer are photosynthetically adapted to the short wavelengths and the low intensities of light which arrive this layer.

Acknowledgments. Thanks are due to Dr. A. Hattori and Dr. Y. Fujita of the Ocean Research Institute, University of Tokyo, Chief investigator in the Expedition KT-77-6, and Expedition KT-78-9, respectively. Taxonomical identification of algae was done by Dr. K. Nozawa, Faculty of Fisheries of Kagoshima University. This research was supported by a grant from the Japanese Ministry of Education, Science and Culture. Thanks are also due to the Deutsche Forschungsgemeinschaft for the travel grant to attend the "Blue Light Syndrome".

References

1. Anderson GC (1969) Subsurface chlorophyll maximum in the northeast Pacific Ocean. Limnol Oceanogr 14: 386–391
2. Boardman NK (1977) Comparative photosynthesis of sun and shade plants. Annu Rev Plant Physiol 28: 355–377
3. Jeffrey SW (1968) Quantitative thin-layer chromatography of chlorophylls and carotenoids from marine algae. Biochim Biophys Acta 162: 271–285
4. Jeffrey SW, Humphrey GF (1975) New spectrophotometric equations for determining chlorophyll a, b, c_1 and c_2 in higher plants, algae and natural phytoplankton. Biochem Physiol Pflanz 167: 191–194
5. Jorgensen EG (1969) The adaptation of plankton algae IV. Light adaptation in different algal species. Physiol Plant 22: 1307–1315
6. Kamiya A, Miyachi S (1974) Effects of blue light on respiration and carbon dioxide fixation in colorless *Chlorella* mutant cells. Plant Cell Physiol 15: 927–937
7. Kamiya A, Miyachi S (1975) Blue light-induced formation of phosphoenolpyruvate carboxylase in colorless *Chlorella* mutant cells. Ibid 16: 729–736
8. Miyachi S, Kamiya A, Miyachi Shizuko (1977) Wavelength effects of incident light on carbon metabolism in *Chlorella* cells. In: Mitsui A et al (eds) Biological solar energy conversion, pp 167–182. Academic Press, London New York
9. Miyachi S, Miyachi Shizuko, Kamiya A (1978) Wavelength effects on photosynthetic carbon metabolism in *Chlorella*. Plant Cell Physiol 19: 277–288
10. Ogasawara N, Miyachi S (1969) Effects of wavelength on CO_2-fixation in *Chlorella* cells. In: Metzner H (ed) Progr Photosynth Res, vol III, pp 1653–1661. C Lichtenstern, München
11. Ogasawara N, Miyachi S (1970) Regulation of CO_2-fixation in *Chlorella* by light of varied wavelengths and intensities. Plant Cell Physiol 11: 1–14
12. Okami N (1978) Light environment in the sea (in Japanese). In: Shibata K et al (eds) Biological and chemical utilization of solar energy, pp 75–84. Center for Academic Publications Japan/ Japan Scientific Societies Press, Tokyo
13. Starr RC (1971) The culture collection of algae at Indiana University-Additions to the collection July 1966–July 1971. J Phycol 7: 350–362
14. Waterman TH (1974) Underwater light and the orientation of animals. In: Jerlov NG, Steeman Nielsen E (eds) Optical aspects of oceanography, pp 415–441. Academic Press, London New York

Visible and Spectrophotometrically Detectable Blue Light Responses of Maize Roots

H.A.W. SCHNEIDER[1]

1 Introduction

Blue light responses of plants are ubiquitous and manifold (see the other contributions of this volume), but there is scarcely a single organ which shows such a variety of blue light responses as maize roots. We may distinguish between responses coming into existence after a few seconds of illumination, those becoming manifest after hours of illumination, and those needing days of illumination to produce a visible change. The first group of fast responses is comprised of light-induced, dark reversible absorbance changes in the roots [13] and the stimulation of georesponsiveness [6], which does not seem to be solely mediated by phytochrome. Phototropic responses belong to the second, slower group. In order to obtain measurable curvatures, prolonged illumination is necessary [12]. The curvatures in response to blue light may be negative or positive and both proceed concomitantly with a negative light-stimulated negative growth reaction, which is also found if the roots are illuminated from all sides [12]. Finally, and as an example of the third, very slow responses, we note that maize roots form chlorophyll if they are exposed to light for days [3, 14]. From studies of other roots chlorophyll biosynthesis and also changes in enzyme activites are known to be mainly evoked in blue light [1, 2].

The present communication deals with the different blue light requirements for phototropic reactions, for the stimulation of the georesponsiveness, the negative light growth reaction and the light-induced, dark-reversible absorbance changes, respectively. The question is also raised whether the light-induced, dark-reversible spectral absorbance changes can be correlated with one of the light-induced macroscopic reactions of the root.

2 Experimental Annotations

After soaking for 24 h at room temperature, kernels of maize (*Zea mays*, var. *vulgata*, Badischer Landmais) were germinated in petri dishes at 24° to 26°C. After two days when the tips of the roots emerged, the roots were put into vertical holes of 1.5 mm in diameter drilled into blocks of "Steckmoos" (as used by florists; Compo; BASF, Ludwigshafen) or into a bed of moist sand. When the roots reached a length of 3 to 4

1 Botanisches Institut der Universität zu Köln, 5000 Köln 41, FRG

cm, that is after an additional 24 h, they were used for the experiments on phototropic and geotropic reactions [6, 12].

During geotropic experiments, the seedlings were pinned horizontally to bars in transparent plastic boxes which were kept moist with wet filter paper. The roots were oriented antiparallel to the light beam. Care was taken to avoid any exposure to stimulating illumination before the experiments. Phototropic experiments were performed in cuvettes ($210 \times 30 \times 70$ mm) made of UV transparent glass. The roots of intact seedlings were immersed in Knop solution diluted with water 1:1. Curvatures of the roots were followed and measured by shadow photographs.

Changes in the spectral absorbance of maize roots (*Zea mays,* EFG TMS x BS 7 of FR^c_{ms} x FR 37) were demonstrated in 7-day-old tips of seedlings grown in vermiculite at $25°C$ [13].

Prior to the recording of the absorbance changes in an AMINCO DW 2 spectrophotometer, 3-cm long root tips were cut into 1- to 2-mm pieces and layered into cuvettes of 1-cm lightpath filled with water or with a nutritional medium which had the composition as the grinding medium but lacked sucrose (see below). The spectrophotometer was equipped with a device to illuminate the root sections with monochromatic light independently of the measuring light beam.

Monochromatic light was produced by using either a xenon or hydrogen light source and appropriate interference filters. The light intensities were measured by means of a thermopile (Kipp & Zonen, Holland) and a galvanometer or by means of an ISCO model SR spectroradiometer. Appropriate light intensities were adjusted by using neutral filters, or by changing the electric current to the light source (if possible) or by changing the distance of the light source to the sample. Some experiments were performed in the light of colored glass filters and fluorescent tubes.

Cell-free extracts, used for demonstrating in vitro absorbance changes, were prepared by grinding the roots in twice their volume of precooled grinding medium containing sucrose (15% w/v) in 10^{-2} M NH_4Cl, 10^{-3} M $MgSO_4$, $5 \cdot 10^{-4}$ M $CaCl_2$ and $2 \cdot 10^{-3}$ M KH_2PO_4. The slurry was centrifuged for 5 min at $500 g$.

3 Results and Discussion

3.1 Positive and Negative Phototropism

Maize roots show a response to light which is seldom seen in roots: namely positive phototropism. If maize roots are exposed to sunlight they grow toward the light. It has been shown that positive phototropism is evoked by light in the range of the wavelengths transmitted by a Schott UV filter UG 2(Fig. 1). The magnitude of the curvature is linearly correlated with the light intensity (Fig. 2). Although the roots are strongly curved to the light source, the extreme root tip proceeds to be directed positively in a geotropic sense (Fig. 1). Elimination of geoperception by cutting the first half millimeter of the root tip caused an almost twofold increase in the magnitude of phototropic curvatures. Thus the sites of the perception of the geotropic and phototropic stimuli are different. The finding that the roots of *Helianthus* and *Sinapis* show negative phototropic curvatures under 360 nm light, while a variety of other roots show no reactions at all, argues

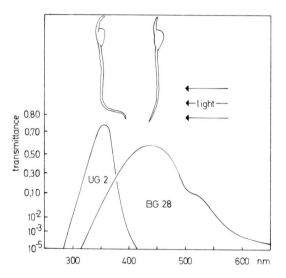

Fig. 1. The spectral transmission ranges of Schott UG 2 and BG 28 glass filters of 3 mm thickness (supplied by the handbook of Schott). The type of phototropic reaction induced by the light of these filters and the typical pattern of associated curvatures is indicated

against the possibility that the curvatures of positive phototropism are caused by irreversible damage in the roots.

Negative phototropic curvatures of maize roots are mainly evoked in the blue and green light transmitted by a Schott filter BG 28 (Fig. 1): this reaction is also remarkable since roots of most other species do not react phototropically at all (see [12]). The magnitude of the negative curvatures is generally smaller than that of the positive ones. Maximum curvatures are about $15°$ at 500 erg cm^{-2} s^{-1} and cannot be increased by higher light intensities. In contrast to the positive phototropic curvature, the negative one has a lag phase of about 6 h [12].

All the available data suggest different photoreceptors and different reaction chains for the two photoreactions. However, spectral response curves will be necessary to make more precise statements because both of the active spectral regions are within the absorption range of the typical blue light receptor.

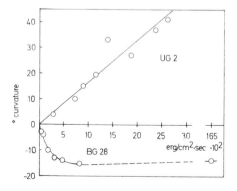

Fig. 2. The magnitude of the negative and positive phototropic curvatures after 20 h of illumination with different intensities of blue (BG 28) and UV (UG 2) light. (1000 erg cm^{-2} s^{-1} correspond with 301 pE cm^{-2} s^{-1} of 360 nm and 377 pE cm^{-2} s^{-1} of 450 nm monochromatic light)

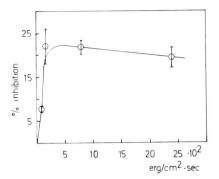

Fig. 3. Negative light-induced growth under blue light (BG 28) as a function of light intensity. The roots were illuminated for 20 h

3.2 Light Growth Reaction

In contrast to the expectation that a positive phototropic reaction is linked to a negative light growth reaction and a negative phototropic reaction with a positive light growth reaction, it was found that blue light which initiates negative phototropism also mediates a negative light growth reaction (Fig. 3). The system is saturated at low light intensities. The light-mediated inhibition of growth is about 20% except for very low light intensities. It will be a task for further investigations to elucidate whether the two different light-mediated reactions, negative phototropism and negative growth reaction, are caused by one or two photoreceptors.

3.3 Stimulation of Georesponsiveness

More precise spectral data are available for the light-induced stimulation of geotropism than is available for phototropic reactions in maize roots. It has been known for half a century or more that horizontally placed maize roots show little or no response to gravity [9]. These respond, however, if exposed to light [12, 19]. The magnitude of geotropism seems to depend on the variety of maize under investigation [15, 16]. After a 6 h period in a horizontal posture the roots of the variety used in the present experiments (Badischer Landmais) show a positive geotropic curvature of about $28°$ to $29°$ in darkness and of about $50°$ to $60°$ if illuminated.

A spectral response curve recently established by Klemmer and Schneider [6] reveals two main regions of response: in the blue and red regions. The active spectrum in the blue region largely resembles that of the photoreceptor involved in many other blue light responses of plants (see other papers in this vol.): peaks at 455 and 484 nm can clearly be distinguished.

Although the rest of the spectral response curve does not closely resemble the typical action spectrum of phytochrome, nonetheless phytochrome appears to play an important role in this phenomenon because the light-induced stimulation of georesponsiveness can be fully reversed by far red light; even stimuli in the region of the blue light photoreceptor can be reversed. Obviously then, the blue light photoreceptor is functionally linked to the phytochrome system.

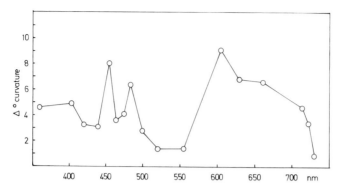

Fig. 4. Spectral response curve for the stimulation of georesponsiveness. The roots were illuminated for 5 s with a quanta flux of 257 pE cm^{-2} s^{-1}. Each *point* represents an average of between 40 to 160 single measurements. The standard error about the mean is 1° or less

In contrast to the light requirements for phototropic curvatures, the stimulation of georesponsiveness needs only small light doses: in the present experiments about 1.3 nE cm^{-2} given during a 5-s illumination period. The same light dose applied during a longer illumination period, as well as higher light doses, resulted in the leveling of the troughs and peaks of the spectral response curve; however, it was found that light in the red region was then relatively more effective than in the blue region [6]. In contrast to the present results, which were confirmed by repetition of the experiments after more than a year, spectral response curves with other varieties of maize did not show the pronounced effectiveness in the absorbance region of the blue light photoreceptor [15, 16].

3.4 Light-Induced, Dark-Reversible Absorbance Changes

In search of possible photoreceptors and the primary reactions involved in light-mediated visible responses of maize roots Schneider and Bogorad [13] found a hemoprotein-like compound which was responsible for light-dependent absorbance changes in maize roots. If the roots are illuminated, their absorbance decreases at 419, 537, and 575 nm and increases at 439 nm (Fig. 5). The absorbance changes are reversed by darkness. A spectral response curve for the light-induced absorbance changes shows peaks at 420 and 575 nm (Fig. 6). The spectral regions at which the changes take place as well as the spectral response curve suggests the involvement of hemoprotein-like compounds, similar to those found in coleoptiles [13, 18]. At present it is not clear what kind of reaction causes the absorbance changes, but it could be shown [13] that a prerequisite for the light-induced changes is the reduced hemoprotein: only if cell-free preparations of maize roots were treated with dithionite, NADH or NADPH, did they show light-induced, dark-reversible changes with maxima very similar to those in intact roots. The absorbance changes may be those associated with changes in either the redox state or ligand binding of the hemoprotein-like substance. The light minus dark difference spectrum is similar

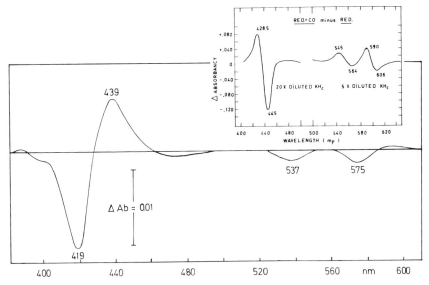

Fig. 5. Light-induced absorbance changes (1 cm light path) in 1 to 2 nm high cylinders of maize root tips: only the root tip and the next 3 cm were used. The absorbance of the sample was about 2.0. *Inset* Difference spectrum of reduced CO-cytochrome c oxidase complex minus reduced cytochrome c oxidase (Vanneste 1966)

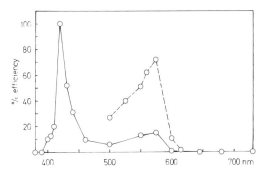

Fig. 6. Spectral response curve for the light-induced absorbance changes in maize roots. The quanta fluxes were 288 pE cm^{-2} s^{-1} (○——○) and 1440 pE cm^{-2} s^{-1} (○– – –○); the illumination time was 15 s

but not identical to the difference spectrum of reduced-CO cytochrome c oxidase complex minus reduced cytochrome c oxidase ([17], inset of Fig. 5); nonetheless this similarity does suggest the involvement of a hemoprotein ligand change.

But be that as it may; at any rate the maxima of absorbance and response are in the blue region of the spectrum and this also raises the possibility that hemoproteins may be photoreceptors for some light-dependent physiological reactions of maize roots such as light growth reaction and negative phototropism. A decisive participation of light absorption of hemoproteins in the stimulation of georesponsiveness is excluded by the nature of the spectral response curve and this also seems to be true for the positive phototropism since the hemoprotein absorption spectra and the spectral region for positive phototropism scarcely overlap and even high doses of blue light containing the 420 nm region do not evoke positive phototropism (Fig. 2).

4 Concluding Remarks

The number of blue light responses in maize roots is exceptionally high and it is astonish-
ing that they have attracted so little attention and have not been studied more intensively. Further we do not know the exact spectral response curves of the photoreceptors
involved in phototropic responses or the exact spectral response curves of the photoreceptor responsible for growth retardation of maize roots in the blue and UV regions.
We do know, however, that the stimulation of georesponsiveness by blue light is mediated by a pigment with an action spectrum typical for the blue light photoreceptor, and
that this photoreceptor system seems to be functionally linked to the phytochrome
system. As to the elucidation of the chain of reactions linking photoreceptors to visible
reactions of maize roots, the detection of what appears to be a photo-reactive hemoprotein complex may be a beginning, however, the reaction time of such componds is
in the range of seconds and minutes and the involvement of componds with much shorter reaction times may have escaped detection. We do not yet know what physiological
function the light-inducible absorbance changes of hemoproteins serves in the plant,
but it is possible that such changes may stimulate many other reactions in the cell, so
causing quite dramatic changes in its physiological state. It seems unlikely that these
reactions of hemoproteins are wholly fortuitous since hemoproteins play a vital role in
cell metabolism. Also the reduction of cytochromes has been shown to be associated
with phototactic and phototropic reactions [5, 10]. Thus the changes in hemoprotein
absorption reported here would be sure to influence the physiological state of the cell
no matter whether the light-induced absorbance changes are concerned with changes
in redox state or in ligand binding. In any case, the changes require the presence of a
reduced hemoprotein and therefore depend on the redox state of the cell. It is also
known, of course, that the photolabile CO-hemoprotein complex is a reduced hemoprotein, rather than an oxidized hemiprotein-complex; however, it is not necessary that
CO is the actual ligand involved.

    If the changes on the hemoprotein-like compound are not directly involved in one
of the phototropic reactions they may be associated in a less direct manner in an enhancing or antagonistic way. At present we can neither exclude phototropic nor light-
induced growth responses from our considerations. Spectra such as found for the light-
induced growth reactions of *Sinapis* roots [7] or for the basis reaction of *Avena* [4] may
possibly in part, at least, be accounted for by light absorption of photoreactive hemoproteins. Negative light-induced growth reactions of light-scattering organs like roots
acting concomitantly with negative phototropism in blue light still await explanation.
It may be mentioned that some action spectra presented in this volume also show action
peaks in the region of hemoprotein light absorption.

    In order to link primary reactions with visible responses of maize roots we may also
use transfer models and hypotheses worked out for organs other than maize roots (see
e.g., [5, 10]), but since we do not have more spectral data, models, reactions of hemoproteins and visible responses can neither be correlated nor compared. Even if flavoproteins and hemoproteins are involved in the primary photoreceptor reactions, the
reaction chain linking the primary reactions to the physiological responses is unknown:
cells unlikely to possess any natural photophysiology have also been found to show
blue light-induced absorbance changes [8]. It may be, however, that studies of maize

roots with their great variety of blue light responses may provide a key to the solution of such problems.

References

1. Björn LO (1967) The light requirement for different steps in the development of chloroplasts in excised wheat roots. Physiol Plant 20: 483–499
2. Björn LO (1967) The effect of blue and red light on NADP-linked glyceraldehydphosphate de-hydrogenases in excised roots. Physiol Plant 20: 519–527
3. Fiedler H (1936) Entwicklungs- und reizphysiologische Untersuchungen an Kulturen isolierter Wurzelspitze. Z Bot 30: 385–436
4. Haig C (1934) The spectral sensibility of *Avena.* Proc Natl Acad Sci USA 20: 476–479
5. Jesaitis AJ, Heners PR, Hertel R, Briggs WR (1977) Characterization of a membrane fraction containing a *b*-type cytochrome. Plant Physiol 59: 941–947
6. Klemmer R, Schneider HjAW (1979) On a blue light effect and phytochrome in the stimulation of georesponsiveness of maize roots. Z Pflanzenphysiol 95: 189–197
7. Kohlbecker R (1957) Die Abhängigkeit des Längenwachstums und der phototropischen Krüm-mung von der Lichtqualität bei Keimwurzeln von *Sinapis alba.* Z Bot 45: 507–524
8. Lipson ED, Presti D (1977) Light-induced absorbance changes in *Phycomyces* photomutants. Photochem Photobiol 25: 203–208
9. Porodko TM (1924) Über den Diageotropismus der Hauptwurzel bei Maiskeimlingen. Ber Dtsch Bot Ges 42: 405–419
10. Poff KL, Butler WL (1975) Spectral characterization of the photoreducible *b*-type cytochrome of *Dictyostelium discoideum.* Plant Physiol 55: 427–429
11. Richter G (1969) Chloroplastendifferenzierung in isolierten Wurzeln. Planta 86: 299–300
12. Schneider HjAW (1965) Kritische Versuche zum Problem des Phototropismus bei Wurzeln. Z Bot 52: 451–499
13. Schneider HjAW, Bogorad L (1978) Light-induced, dark-reversible absorbance changes in roots, other organs, and cellfree preparations. Plant Physiol 62: 577–581
14. Segelitz G (1938) Der Einfluß von Licht und Dunkelheit auf Wurzelbildung und Wurzelwachs-tum. Planta 28: 617–643
15. Shen-Miller J (1978) Spectral response of corn (*Zea mays*) in root geotropism. Plant Cell Physiol 19: 445–452
16. Suzuki T, Fujii T (1978) Spectral dependence of the light-induced geotropic response in *Zea* roots. Planta 142: 275–279
17. Vanneste WH (1966) The stoichiometry and absorption spectra of components a and a_3 in cyto-chrome c oxidase. Biochemistry 5: 838–848
18. Widell S, Björn LO (1976) Light-induced absorption changes in etiolated coleoptiles. Physiol Plant 36: 305–309
19. Wilkins H, Wain RL (1975) The role of the root cap in the response of the primary roots of *Zea mays L.* seedlings to white light and gravity. Planta 123: 217–222

Synergistic Action of Red and Blue Light on Stomatal Opening of *Vicia faba* Leaves

T. OGAWA[1]

1 Introduction

Stomata open in the light. There are two different effects of light on stomata, one which operates in response to reduction of CO_2 concentration inside the leaf and another which is independent of the concentration. Mouravieff (1958), Kuiper (1964), Mansfield and Meidner (1966) and Hsiao et al. (1973), using leaves of *Veronica, Senecio, Xanthium* and *Vicia,* respectively, found greater stomatal opening in blue light than in red light. Mansfield and Meidner (1966) showed that the effect of blue light on stomata is independent of the CO_2 concentration inside the leaf.

It was shown in a previous paper that blue light is more effective than red light for the formation of malate in guard cells of *Vicia faba,* and that the rate of malate formation with blue and red light applied simultaneously was much higher than the sum of the rates measured with red and blue light applied separately (Ogawa et al. 1978). These previous experiments led us to the present study on the effects of red and blue light on transpiration of *V. faba* and on the action spectra for the effects of red and blue light on malate formation.

2 Materials and Methods

2.1 Preparation of Sample Leaves and Sonicated Epidermal Strips

Plants of *Vicia faba,* c.v. Ryosai issun (seeds purchased from Yamato Seeds Co., Tokyo) were grown in a green house at $25°C$ on a vermiculite bed with irrigation with a solution of 0.1% Hyponex (Hyponex Chemical Co., Copley, Ohio, U.S.A.) for a period of 20–30 days. They were placed in darkness for 6 h and used for experiments.

Epidermal strips were peeled from the abaxial (lower) leaf surface and were sonicated in a solution of 0.01 mM $CaCl_2$ according to the procedure described previously (Ogawa et al. 1978). The sonicated strips contained guard cells as the only viable cells.

---

1 Laboratory of Plant Physiology, The Institute of Physical and Chemical Research (Riken), Wako-shi, Saitama 351, Japan

2.2 Measurement of Transpiration

Transpiration of *V. faba* leaves was measured with an open gas-analysis system, which directly records the transpiration rate against time. A pair of leaves on a plant was placed in an acrylic chamber (width = 70 mm, height = 50 mm, volume = 35 ml) with the roots dipped in water through a hole on a bottom of the chamber; the space between the hole and the petiol was filled with vaseline. The air containing 0.04% CO_2 was led to the chamber, and the exchanged gas in the chamber was led to a hygrometer with an aluminum oxide sensor (developed by Dr. Furuichi of our institute). The relative humidity of the air was 50% at 25°C and the flow rate was adjusted with a needle valve to be 1 l/min throughout the experiments.

2.3 Illumination of Leaves and Sonicated Epidermal Strips

When measuring transpiration, the leaves were illuminated with red (> 600 nm, 3.0 mW cm^{-2}) and blue (460 nm, 0.08 mW cm^{-2}) light. The red light was obtained with a glass filter (VR-60, Toshiba Co., Tokyo) placed in front of a slide projector (Procabin.667, Cabin Co., Tokyo) equipped with a 650-W halogen lamp, and the blue light with an interference filter (transmission max. at 460 nm, half-band width = 10 nm, Nihon Shinku Kogaku Co., Tokyo) placed in front of a Canon Slide Star Projector (Canon Co., Tokyo) equipped with a 300-W incandescent lamp. A solution of 0.5% $CuSO_4$ (7 cm in thickness) was placed behind each filter for heat absorption.

When measuring malate formation in guard cells, the sonicated epidermal strips (about 3–4 mg in dry weight) were put into 2.5 ml of 5 mM potassium phosphate buffer, pH 7.0, in a test tube and were illuminated under various light conditions; (1) monochromatic red light (675 nm) with and without background blue light, (2) monochromatic blue light (430 nm) with and without background red light, and (3) monochromatic light of various wavelengths with red or blue light background. Monochromatic light was obtained from the same light sources as described previously (Ogawa et al. 1978). The background red light (> 600 nm) was obtained with a glass filter (VR-60, Toshiba Co.) and the background blue light with a blue glass filter (V-B48, Toshiba Co.), each placed on a heat-absorbing $CuSO_4$ solution in front of the Canon projector.

Action spectra were measured under two different light conditions with background red illumination or with background blue illumination superimposed on the spectrum from a Jasco Spectroirradiator (Nihon Bunko Co., Tokyo) equipped with a 2 kW-Xenon lamp as the light source. Ten different wavelengths were chosen between 356 and 544 nm with red background illumination and eight wavelengths between 612 and 737 nm with blue background. The band width of the monochromatic light was about 12 nm. The light intensity of the monochromatic light was reduced with neutral films (Fuji Photo Co., Tokyo) to 0.05–0.07 nE cm^{-2} s^{-1} in the measurement with red light background and to 1.8–3.0 nE cm^{-2} s^{-1} in the measurement with blue background. Three projectors, each with a glass filter (VR-60 or V-B48) and the $CuSO_4$ solution, as described above, were used to illuminate the sample tubes uniformly. Light intensities were measured with a thermocouple (model E2, Kipp and Zonen, Deft, Netherlands) standardized with a standard lamp (U.S. National Bureau of Standards, Washington D.C.).

2.4 Malate Analysis

Malate was extracted from a sample epidermis according to the procedure described
previously (Ogawa et al. 1978). The content of malate in the extract was determined
by the method of Möllering (1974).

After extraction, the epidermis was dried in an oven at 60°C for 1 h and weighed.
The content of malate in the epidermis was calculated on the basis of this dry weight.
About 4 cm^2 of fresh epidermis gave 1 mg dry weight.

3 Results

3.1 Synergistic Action of Red and Blue Light on Transpiration of *Vicia faba* Leaves

The rate of transpiration is considered as a measure of stomatal opening. When *Vicia*
leaves were illuminated with red light ($>$ 600 nm) at a medium intensity (3.0 mW cm^{-2}),
stomata gradually opened. This was reflected as a gradual increase of transpiration rate,
which was 130 mg H_2O dm^{-2} h^{-2}. This acceleration was enhanced by applying blue
light (460 nm) at a low intensity (0.08 mW cm^{-2}) simultaneously with the red light.
This effect of blue light was much less when applied without background red light.

The acceleration (A_{R+B} in Fig. 1) of transpiration obtained with red and blue light
applied simultaneously and those (A_R and A_B) with red and blue light applied separately,
which were calculated from the data in Fig. 1, are listed in Table 1 as the data of sample

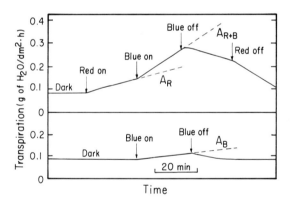

Fig. 1. Effects of red and blue light
on transpiration of *Vicia faba* leaves.
Red light ($>$ 600 nm, 3.0 mW cm^{-2})
and blue light (460 nm, 0.08 mW
cm^{-2}) were applied simultaneously
or separately during incubation of
leaves in normal air containing
0.04% CO_2. A_{R+B}, A_R and A_B
are defined in the text in relation
to Table 1

Table 1. The accelerations of transpiration of *V. faba* leaves under illumination with red (A_R), blue
(A_B) and red *plus* blue (A_{R+B}) light, and the ratios of $A_{R+B}/(A_R + A_B)$. The intensities of the red
and blue light were the same as in Fig. 1

Sample	A_R	A_R	A_{R+B}	$A_{R+B}/(A_R + A_B)$
	in mg H_2O dm^{-2} h^{-2}			
No. 1	130	86	403	1.9
No. 2	144	72	432	2.0
No. 3	187	113	504	1.7
No. 4	173	62	446	1.9

1, together with those obtained for other samples. It is evident from the ratios of $A_{R+B}/(A_R + A_B)$ listed in the last column that the acceleration obtained with simultaneous illumination is 1.7–2.0 times the sum of the values measured with separate illumination. This clearly demonstrates a synergistic action of red and blue light on the rate of stomatal opening.

3.2 Synergistic Action of Red and Blue Light on Malate Formation in *Vicia faba* Guard Cells

The rate of malate formation in sonicated epidermal strips of *V. faba* was measured under illumination with blue light at various intensities with and without background illumination with red light (> 600 nm, 3.0 mW cm^{-2}). Figure 2, showing the results as a function of blue light intensity, indicates that the blue light with red light background was much more effective for malate formation than that without the background. The blue light at ca. 0.2 nE cm^{-2} s^{-1} was strong enough to obtain the saturation rate with red light background while the rate measured with the same blue light without the background light was nearly zero. A much higher intensity of 3 nE cm^{-2} s^{-1} was required for saturation without red light background.

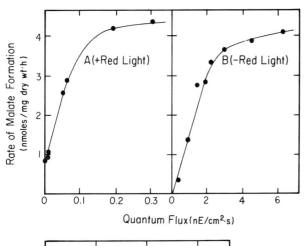

Fig. 2. Rate of malate formation in *V. faba* guard cells as a function of the light intensity of 430-nm monochromatic light with (curve *A*) and without (curve *B*) background illumination with red light (>600 nm, 3.0 mW cm^{-2})

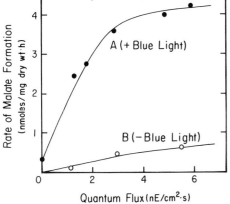

Fig. 3. Rate of malate formation in *V. faba* guard cells as a function of the light intensity of 675-nm monochromatic light with (curve *A*) and without (curve *B*) background illumination with blue light (360–580 nm, 0.1 mW cm^{-2})

Curves A and B in Fig. 3 show the dependency of the rate of malate formation on the intensity of red light (675 nm) with and without background illumination with blue light (360–580 nm, 0.1 mW cm^{-2}). The results clearly show that the red light is much more effective with the background blue light, and that the effect reaches saturation at ca. 4 nE cm^{-2} s^{-1}. For example, the rate with blue (430 nm, 0.2 nE cm^{-2} s^{-1}) and red (> 600 nm, 3.0 mW cm^{-2}) light applied simultaneously was 4.2 mol malate formed mg dry weight^{-1} h^{-1}, which was about four times the sum of the rates, 0.9 and 0.2 mol mg dry wt^{-1} h^{-1}, respectively, measured by separate illumination with red and blue light. It is evident from Figs. 2 and 3 that the rate of malate formation with red and blue light applied simultaneously is much higher than the sum of the rates obtained by separate illumination with red and blue light.

3.3 Action Spectra for Malate Formation

The quantum fluxes of monochromatic light used in the measurement of action spectra for malate formation in guard cells were chosen from the proportionality range of the curves in Figs. 2 and 3; 0.05–0.07 nE cm^{-2} s^{-1} for the blue light region with red light background and 1.8–3.0 nE cm^{-2} s^{-1} for the red light region with blue light background.

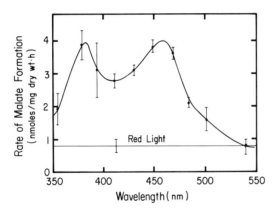

Fig. 4. Action spectrum for malate formation in *V. faba* guard cells determined with background red light (> 600 nm, 3.0 mW cm^{-2}). The rate is expressed in nmol malate formed per mg of dry wt. during 1 h of illumination at blue light quantum flux of 0.07 nE cm^{-2} s^{-1}. The rate at each wavelength, as shown by a *solid circle*, was determined from three separate experiments; the *bars* indicate the standard deviation from the mean of these data

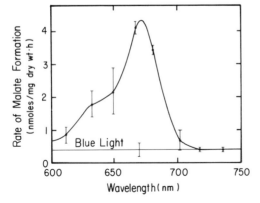

Fig. 5. Action spectrum for malate formation in *V. faba* guard cells determined with background blue light (360–580 nm, 0.1 mW cm^{-2}). The rate is expressed in nmol malate formed per mg dry wt. during 1 h of illumination at red light quantum flux of 2.0 nE cm^{-2} s^{-1}. The rate of each wavelengths, as shown by a *solid circle*, was determined from three separate experiments; the *bars* indicate the standard deviation from the mean of these data

Figure 4 shows the action spectrum obtained with red light background, which indicates two maxima comparable in height at 380 and 460 nm, respectively. These maxima agree with the absorption maxima of flavin. The action spectrum in Fig. 5 determined with background red light shows a high peak at 670–680 nm, which is in approximate agreement with the absorption maximum of chlorophyll a in chloroplasts in guard cells.

4 Discussion

The results of these experiments clearly demonstrate a synergistic action of red and blue light on transpiration of *V. faba* leaves and malate formation in guard cells. It was reported by Brogårdh for *Avena* leaves that blue light-induced transpiration is enhanced either by pre-illumination of leaves with red light or by pre-incubation of leaves with CO_2-free air; he concluded that the red light enhancement is due to lowering of intercellular CO_2 concentration by photosynthesis in mesophylls. The red light effect found in this study using sonicated epidermis, which does not contain mesophyll cells, indicates that photosynthetic reaction occurring in guard cells is involved in the mechanism of stomatal opening. This was confirmed by the action spectrum for the red light effect with blue light background (Fig. 5).

Photosynthesis may be involved in two ways in stomatal opening. First, photosynthesis in guard cells will produce ATP needed for active uptake of potassium ion. Second, photosynthesis in mesophyll cells will lower the CO_2 concentration in substomatal space, and this will enlarge stomatal opening. The red light effect on transpiration may result from both of these reactions. In the case of the red light effect on malate formation in isolated guard cells, however, the second mechanism of lowering CO_2 concentration may not play an important role. The activity of CO_2 fixation seems to be too low to reduce the CO_2 concentration to a level sufficient for stomatal opening.

The maximal wavelengths of 380 and 460 nm for the blue light effect with red light background, observed in terms of malate formation, agree with the absorption maxima of flavin. Zeiger and Hepler (1977) found that onion guard cells contain a blue light-absorbing pigment which fluoresces green light (maximum at 525–535 nm) when illuminated with blue light. This supports the above view that the pigment responsible for the blue light effect is a flavin.

Willmer et al. (1973) demonstrated that phosphoenolpyruvate (PEP)-carboxylase is located in guard cells and mediates the formation of malate. Blue light enhanced PEP-carboxylase activity in *Chlorella* cells (Kamiya and Miyachi 1975). These observations indicate the possibility that PEP-carboxylase activity in guard cells is enhanced by blue light.

The following mechanism is inferred from these observations to be involved in the synergistic action of red and blue light. The red light effect could be due to production of ATP needed for K^+ influx into guard cells, while the blue light effect could be due to activation of PEP-carboxylase. Neither of these effects alone enhances stomatal opening, and only balanced influx of K^+ caused by red light and activation of PEP-carboxylase by blue light accelerate stomatal opening.

Acknowledgment. The author expresses his gratitude to Prof. K. Shibata for his valuable discussion and criticism throughout this study and in preparing this manuscript. The present study was supported by a research grant for "Solar Energy Conversion-Photosynthesis" given by Japan Science and Technology Agency.

References

Brogårdh T (1975) Regulation of transpiration in *Avena*. Responses to red and blue light steps. Physiol Plant 35: 303–309

Hsiao TC, Allaway WG, Evans LT (1973) Action spectra for guard cell Rb[+] uptake and stomatal opening in *Vicia faba*. Plant Physiol 51: 82–88

Kamiya A, Miyachi S (1975) Blue light-induced formation of phosphoenolpyruvate carboxylase in colorless *Chlorella* mutant cells. Plant Cell Physiol 16: 729–736

Kuiper PJC (1964) Dependence upon wavelength of stomatal movement in epidermal tissue of *Senecio odoris*. Plant Physiol 39: 925–955

Mansfield TA, Meidner H (1966) Stomatal opening in light of different wavelengths: effects of blue light independent of carbon dioxide concentration. J Exp Bot 17: 510–521

Möllering H (1974) Determination with malate dehydrogenase and glutamate-oxaloacetate transaminase. In: Bergmeyer HU (ed) Methods of enzymatic analysis, vol III, pp 1589–1593. Chemie Verlag, Weinheim

Mouravieff I (1958) Action de la lumière sur la cellule végétale. Bull Soc Bot Fr 105: 467–475

Ogawa T, Ishikawa H, Shimada K, Shibata K (1978) Synergistic action of red and blue light and action spectra for malate formation in guard cells of *Vicia faba* L. Planta 142: 61–65

Willmer CM, Kanai R, Pallas JE Jr, Black CC Jr (1973) Detection of high levels of phosphoenolpyruvate carboxylase in leaf epidermal tissue and its significance in stomatal movements. Life Sci 12: 151–155

Zeiger E, Hepler PK (1977) Light and stomatal function: blue light stimulates swelling of guard cell protoplasts. Science 196: 887–889

The Blue Light Response of Stomata and the Green Vacuolar Fluorescence of Guard Cells

E. ZEIGER[1]

1 Blue Light and Stomatal Function

Guard cells modulate gas exchange between the leaf and the environment by regulating the aperture of the stomatal pore. The changes in aperture are mediated by fluctuations in the turgor of the guard cells, which require active ion transport (Hsiao 1976). Light is indisputably a primary factor affecting stomatal opening and closing but its mode of action remains controversial. Two current points of view postulate a purely incidental role for light on the one hand (Raschke 1977) and a primary role in driving ion transport on the other (Zeiger et al. 1977, 1978). The wavelength dependence of the stomatal responses to light provides us with a useful tool to examine the mechanism whereby light affects the opening and closing of stomata. Published data on the subject can be grouped in three classes:

1. The responses show an action spectrum which resembles the absorption spectrum of chlorophyll.
2. The responses show maxima in the blue and the red but the effect of blue light is markedly enhanced.
3. The responses can be generated by blue light only, red light being ineffective.

Kuiper's classical action spectrum of stomatal closing in *Senecio* is the best known example of class 1 (Kuiper 1964). Most of the data, however, cluster in class 2 (Meidner 1968; Hsiao et al. 1973; Nelson and Mayo 1975; Voskrensenskaya and Polyakov 1976; Lurie 1978; Ogawa et al. 1978). They encompass a significant number of species in different families of monocots and dicots and point to the presence of a specific non-chlorophyllic blue light response in stomata. The presence of a specific blue light response is also indicated by the findings in class 3 which include the swelling of *Allium* guard cell protoplasts (Zeiger and Hepler 1977) and the rapid increase in transpiration in *Avena* and other grasses (Johnsson et al. 1976). It is noteworthy that there is an overlap between classes 2 and 3, as evident in experiments in which both an enhanced blue and a red light response were observed at high light intensities, whereas at low fluxes only a blue light response could be detected (Hsiao et al. 1973; Ogawa et al. 1978).

Abbreviations. *CCCP* Carbonyl cyanide m-chlorophenyl hydrazone; *HEPES* N-2-Hydroxyethyl-piperazine-N'-2-ethanesulfonic acid; *MES* 2(N-morpholino) ethane sulfonic acid

1 Department of Biological Sciences, Stanford University, Stanford, CA 94305, USA

The best interpretation of these results points to two different photoreceptors in the guard cells, one being the chlorophyll in the chloroplasts and the other a chromophore which absorbs blue light only (Hsiao et al. 1973; Brogårdh 1975; Zeiger and Hepler 1977). Within this framework, the three classes of observation can be interpreted as differential expressions of the two photoreceptors as affected by experimental conditions.

A better understanding of the nature of the chromophores in the guard cells and their specific roles in stomatal function should prove valuable to both stomatal physiologists and photobiologists. The blue light response has important implications for a broad spectrum of biological processes (Presti and Delbruck 1978), yet our understanding of the subject remains limited mainly because of the lack of an effective physiological probe with which to analyze it. The guard cells might provide such a probe because of the localization of the response in a single cell type and the possible direct connection between the reception of blue light and the active transport of ions (Hsiao et al. 1973).

2 Blue Light-Induced, Vacuolar Fluorescence of Guard Cells

2.1 The Fluorescence in *Allium*

The intrinsic, vacuolar fluorescence of the guard cells might have a bearing on the cellular nature of the blue light response of stomata. The fluorescence was originally discovered in *Allium* guard cells (Zeiger and Hepler 1977, 1979). It is specific for that cell type and it is absent from neighboring epidermal cells. Guard mother cells exhibit the fluorescence at a very early stage of differentiation, indicating a close correlation between the cell's ability to fluoresce and the process of stomatal development. The fluorescing compound is clearly localized in the vacuole and indirect evidence suggests that it might be associated with the tonoplast. The emission spectrum peaks at 520 nm and initial studies on the excitation spectrum showed a peak at 450 nm. The fluorescence was found in eleven species of the genus *Allium* but not in several other monocots or dicots tested (Zeiger and Hepler 1979).

One interpretation of these findings is that the fluorescing compound might be a tonoplast-bound flavin or flavoprotein that could be associated with the reception of blue light and the driving of active ion transport in the guard cells (Zeiger and Hepler 1979). The restriction of the fluorescence to the *Allium* genus was, however, intriguing. The blue light response has been observed in many different species; hence, if the fluorescing compound is associated with that phenomenon, one would expect a more widespread distribution of the fluorescence. We reasoned that our failure to detect it outside of the *Allium* genus could be due to variations in the intracellular properties of the guard cells that would preclude the fluorescence under our experimental conditions (Zeiger and Hepler 1979).

2.2 The Fluorescence in *Vicia*

Recent findings reported here show that this is indeed the case in *Vicia faba*. Guard cells of *Vicia* lack the fluorescence when bathed in H_2O but exhibit it in the presence of ammonia, some buffers, and metabolic inhibitors. These findings have provided a new approach to the problem and demonstrate that the vacuolar fluorescence is not a peculiarity of *Allium* guard cells.

2.2.1 Methods

Epidermal peels were obtained from 10–25-day-old plants of *Vicia faba* (W. Atlee Burpee Co., Riverside, CA, USA) grown in a greenhouse. Young, fully expanded leaves gave the most consistent results. Epidermal peels were mounted in microchambers (Zeiger and Hepler 1979) in demineralized H_2O. During the observations, the chambers were continuously perfused at 1.6 cm min^{-1} with a Razel (Razel Scientific Instruments, Stamford, CT, USA) perfusion pump connected to two Butterfly infusion sets (Abbott, South Pasadena, CA, USA) inserted through the o-ring of the chamber. An arrangement including two pumps connected in parallel to a Hamilton (Hamilton, Reno, NV, USA) valve allowed the changing of solutions without interrupting the perfusion. Besides facilitating the continuous perfusion of the preparations, the use of microchambers eliminated the spurious fluorescence occasionally observed when using conventional microscope slides and coverslips. Single cells were observed under fluorescence microscopy in an AO optical system with incident fluorescence (Zeiger and Hepler 1979) and filters and objectives as specified in the figure legends. Fluorescence intensity and emission spectra were measured with a Nanospec/10 microfluorospectrophotometer equipped with a gallium-arsenide photomultiplier and a motor-driven diffraction grating monochrometer (Nanometrics, Sunnyvale, CA, USA) and recorded on a chart recorder (Zeiger and Hepler 1979). The area of the cell to be measured was delimited by a variable slit located between the specimen and the photomultiplier.

2.2.2 Results

In contrast with *Allium, Vicia* guard cells lack the green vacuolar fluorescence when bathed in H_2O. Under these conditions, they exhibit only a bright, red fluorescence in their chloroplasts and a faint yellow-green fluorescence in the ridge surrounding the stomatal pore. Exposure to 1 mM NH_4OH, however, induces the green vacuolar fluorescence, its intensity increasing with NH_4OH concentration (Fig. 1). At concentrations of 50 mM or higher, the fluorescing vacuoles appear distinctly swollen and there is considerable disruption of the cell protoplast, indicating an osmotic effect. At 1 mM the NH_3-induced fluorescence readily disappears upon washing with H_2O (Fig. 2). Vacuoles from many epidermal cells also fluoresce in the presence of ammonia.

The effects of some buffers and metabolic inhibitors on the fluorescence of the guard cell vacuoles are being studied and initial results are shown in Table 1. MES-HEPES buffers at 50 mM induced the fluorescence at pH 6.6 and 7.6 but not at pH 5.6. The intensity of the fluorescence increased with alkalinization. The electron uncoupler CCCP (Sigma) and the H^+/K^+ ionophore nigericine (a gift from Eli Lilly Co.) induced the fluorescence at 10^{-5} M in aqueous solutions. Buffering with 50 mM MES-HEPES at pH 5.6 for nigericine and 6.6 for CCCP increased the fluorescence intensity. The fluorescence induced by either CCCP or nigericine in buffer was more intense than that seen

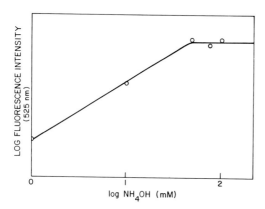

Fig. 1. Vacuolar fluorescence intensity at 525 nm of *Vicia* guard cells as a function of NH_4OH concentration. Epidermal peels in microchambers continuously perfused with NH_4OH at the indicated concentrations. The slit aperture was adjusted to expose an area within the vacuole. BG12 exciting filter, 450 nm cut off dichroic beamsplitter and OG515 barrier filter with a 40X objective. Each data point is the average of 30 guard cells

Table 1. Effect of MES-HEPES buffers, CCCP, and nigericine on the blue light-induced vacuolar fluorescence of *Vicia faba* guard cells

Treatment	Fluorescence[a]
H_2O	None
NH_4OH, 1 mM	+++++
MES-HEPES buffer, 50 mM	
pH 5.6	None
pH 6.6	+
pH 7.6	++
CCCP, 10^{-5} M, in H_2O	+
buffered at pH 6.6	++
pH 7.6	+++
Nigericin, 10^{-6} M, in H_2O	None
10^{-5} M, in H_2O	+
buffered at pH 5.6	++
pH 6.6	+++

[a] Visual estimation of relative fluorescence on an arbitrary scale of (none) (H_2O) to (+++++) (1 mM NH_4OH)

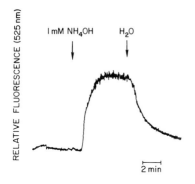

Fig. 2. Reversal by H_2O of the vacuolar fluorescence of *Vicia* guard cells induced by NH_4OH. Record of a single cell bathed in H_2O and perfused with 1 mM NH_4OH and then with H_2O (*arrows*). Conditions as in Fig. 1

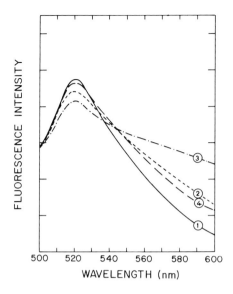

Fig. 3. Emission spectra of the guard cell vacuoles of *Allium* bathed in H_2O (*1*) and in 3 mM NH_4OH (*2*), those of *Vicia* in 3 mM NH_4OH (*3*), and the emission spectrum of 10^{-6} M riboflavin (*4*). Blue, 436 nm exciting filter, 450 nm dicroic beamsplitter and 100X oil immersion objective. Scanning speed: 100 nm/min. The initial intensity of all four spectra were equalized by adjusting the gain of the photomultiplier

with the buffer alone. The vacuole from epidermal cells did not fluoresce in the presence of CCCP.

The emission spectra of fluorescing vacuoles from guard cells of *Allium* and *Vicia* and that from a 10^{-6} M solution of riboflavin (Sigma) are shown in Fig. 3. All spectra peak at around 520 nm.

2.3 Conclusions

These findings show that like those of *Allium,* the guard cells of *Vicia* can exhibit the green vacuolar fluorescence when excited with blue light. Unlike those of *Allium,* however, they require special conditions in order to fluoresce. The identical cellular localization of the fluorescing compounds in both *Allium* and *Vicia* and the similarity of their emission spectra indicate that the fluorescing chromophores might be identical. If that is the case, the different conditions required by the guard cells of each species to fluoresce probably reflect intrinsic differences in their cellular physiology. In fact, the nature of the agents capable of inducing the fluorescence in *Vicia* indicates that the intracellular pH may be a prime factor in determining the manifestation of the fluorescence and suggest that the guard cells of *Allium* bathed in H_2O have a more alkaline pH than those of *Vicia.*

The ability of ammonia to induce the vacuolar fluorescence in *Vicia* is probably a result of the alkalinization of the intracellular pH (Jacobs 1940; Boron and De Weer 1976). This is consistent with the higher levels of fluorescence seen with increasing concentrations of NH_4OH and with the induction of the fluorescence in the presence of nigericine at pH 6.6 or higher. Nigericine should be expected to increase the membrane permeability to protons (Ramos and Kaback 1977) and therefore allow a rise of the intracellular pH in response to the buffers. On the other hand, exposure of *Vicia* guard cells to the buffer alone, to which the plasmalemma is supposedly impermeable, would not be expected to alter the intracellular pH. The fact that the fluorescence is enhanced by exposure to buffer alone may be due to the disruption of the pH gradient across the guard cell membranes (Zeiger et al. 1978).

The induction of the fluorescence by CCCP is of special interest because of the reported ability of the uncoupler to inhibit light-induced stomatal opening (Lurie 1978). The induction of the fluorescence could therefore be a result of the inhibition of photochemical activity in the guard cells. On the other hand, CCCP is also an effective proton translocator (Felle and Bentrup 1977), hence its effect could also be related to intracellular pH changes.

The observation that in *Vicia,* the vacuoles of epidermal cells as well as those of guard cells fluoresce in the presence of ammonia and buffers is inconsistent with the hypothesis that the fluorescing compound is a specific feature of the guard cells, as suggested by the observations in *Allium.* It is quite possible, on the other hand, that identical chromophores have distinct functions in each cellular type, arising from overall differences in the physiology of the cells.

The chemical nature of the vacuolar fluorescent compound(s) remains uncertain. The similarity in the emission spectra of riboflavin and the vacuolar fluorescing compound (Fig. 3) and the kinetics of the decay of the fluorescence in *Allium* (Zeiger and Hepler 1979), point to a flavin or flavoprotein. Recently determined action spectra for malate synthesis in *Vicia* guard cells (Ogawa et al. 1978) and for the rapid increase in transpiration in *Avena* (Skaar and Johnsson 1978) are also indicative of a flavin. Thus the vacuolar fluorescing compound may be the photoreceptor responsible for those blue light-induced phenomena.

In conclusion, this study shows that the blue light-induced, green, vacuolar fluorescence is a sensitive probe of the physiological status of the guard cells. This and other recently developed techniques, such as single cell chemical microanalysis (Outlaw and Kennedy 1978), experiments with guard cell protoplasts (Zeiger and Hepler 1977; Schnabl 1978) and electrophysiological studies (Moody and Zeiger 1978) should provide effective approaches to the investigation of the blue light response of stomatal guard cells.

3 Prospective

The understanding of the role of blue light in stomatal movement demands the consideration of the physiological processes involved in the functioning of the guard cells. Active ion transport is clearly a central one, and the mode of action of blue light is likely to be coupled to ion fluxes in a specific way. We have postulated a chemiosmotic mechanism driving ion transport in the guard cells (Zeiger et al. 1978). This hypothesis predicts specific proton fluxes generating an electrochemical gradient across the guard cell membranes and also affecting the prevailing pH in the different cell compartments. Within this hypothesis, a blue light-driven, tonoplast-bound electron transport chain could extrude protons into the vacuolar sap, causing cytoplasmic alkalinization which would substantially increase malate synthesis (Outlaw and Kennedy 1978), as observed by Ogawa et al. (1978). This mechanism is distinctly different from the less specific blue light-enhanced respiration which is often invoked to explain the blue light response (Kowallik and Gaffron 1966) and deserves closer investigation.

The blue light-driven transport mechanism could also be capable of sustaining osmotically active levels of ion uptake (Hsiao et al. 1973). In this capacity, the blue light-dependent energy source would complement what appears to be the major light-dependent energy-producing pathway driving stomatal opening: the generation of ATP and/or reducing agents by the guard cell chloroplasts with the wavelength dependence of chlorophyll (Zeiger et al. 1978). The cellular regulation of these different sources might constitute one of the most interesting properties of the exquisitely complex biology of the guard cells.

Acknowledgments. I thank Eleanor Crump for editing the manuscript. Supported by NSF grant PCM 77-17642.

References

Boron WF, De Weer P (1976) Intracellular pH transients in squid giant axons caused by CO_2, NH_3, and metabolic inhibitors. J Gen Physiol 67: 91–112

Brogårdh T (1975) Regulation of transpiration in *Avena*. Responses to red and blue light steps. Physiol Plant 35: 303–309

Felle H, Bentrup FW (1977) Evidence for a CCCP-induced proton permeability of the plasmalemma of *Riccia fluitans*. In: Thellier M, Monnier A, Demarty M, Dainty J (eds) Transmembrane ionic exchanges in plants. Colloq CNRS 258: 193–198

Hsiao TC (1976) Stomatal ion transport. In: Lüttge U, Pitman MG (eds) Encyclopedia of plant physiology, New Ser, vol II, part B, pp 195–221. Springer, Berlin Heidelberg New York

Hsiao TC, Allaway WG, Evans LT (1973) Action spectra for guard cell Rb^+ uptake and stomatal opening in *Vicia faba*. Plant Physiol 51: 82–88

Jacobs MH (1940) Some aspects of cell permeability to weak electrolytes. Cold Spring Harbor Symp Quant Biol 8: 30–39

Johnsson M, Issaias S, Brogårdh T, Johnsson A (1976) Rapid, blue-light-induced transpiration response restricted to plants with grasslike stomata. Physiol Plant 36: 229–232

Kowallik W, Gaffron H (1966) Respiration induced by blue light. Planta 69: 92–95

Kuiper PJC (1964) Dependence upon wavelength of stomatal movement in epidermal tissue of *Senecio odoris*. Plant Physiol 39: 952–955

Lurie S (1978) The effect of wavelength of light on stomatal opening. Planta 140: 245–249

Meidner H (1968) The comparative effects of blue and red light on the stomata of *Allium cepa* L. and *Xanthium pennsylvanicum*. J Exp Bot 19: 146–151

Moody W, Zeiger E (1978) Electrophysiological properties of onion guard cells. Planta 139: 159–165

Nelson SD, Mayo JM (1975) The occurrence of functional non-chlorophyllous guard cells in *Paphiopedilum* spp. Can J Bot 53: 1–7

Ogawa T, Ishikawa H, Shimada K, Shibata K (1978) Synergistic action of red and blue light and action spectra for malate formation in guard cells of *Vicia faba* L. Planta 142: 61–64

Outlaw WH, Kennedy J (1978) Enzymic and substrate basis for the anaplerotic step in guard cells. Plant Physiol 62: 648–652

Presti D, Delbruck M (1978) Photoreceptors for biosynthesis, energy storage and vision. Plant Cell Environ 1: 81–100

Ramos S, Kaback HR (1977) The electrochemical proton gradient in *Escherichia coli* membrane vesicles. Biochemistry 16: 848–854

Raschke K (1977) The stomatal turgor mechanism and its response to CO_2 and abscisic acid. In: Marre E (ed) Regulation of cell membrane activities in plants, pp 173–183. Elsevier/North Holland, Amsterdam

Schnabl H (1978) The effect of Cl⁻ upon the sensitivity of starch-containing and starch-deficient
    stomata and guard cell protoplasts towards potassium ions, fusicoccin and abscisic acid. Planta
    144: 95–100
Skaar H, Johnsson A (1978) Rapid, blue-light induced transpiration in *Avena*. Physiol Plant 43:
    390–396
Voskresenskaya NP, Polyakov AM (1976) Regulatory action of blue light on photosynthetic gas ex-
    change: action spectrum of light saturation of CO_2 exchange in lily of the valley leaves. Sov
    Plant Physiol 23: 6–11
Zeiger E, Hepler PK (1977) Light and stomatal function: blue light stimulates swelling of guard
    cell protoplasts. Science 196: 887–889
Zeiger E, Hepler PK (1979) Blue light-induced, intrinsic vacuolar fluorescence in onion guard cells.
    J Cell Sci 37: 1–10
Zeiger E, Moody W, Hepler P, Varela F (1977) Light sensitive membrane potentials in onion guard
    cells. Nature (London) 270: 270–271
Zeiger E, Bloom A, Hepler PK (1978) Ion transport in stomatal guard cells: a chemiosmotic hypo-
    thesis. What's New in Plant Physiol 9: 29–32

Light Induced Changes
in the Centrifugability of Chloroplasts Mediated
by an Irradiance Dependent Interaction
of Respiratory and Photosynthetic Processes

K. SEITZ[1]

1 Introduction

Light-induced intracellular motion responses of chloroplasts in *Vallisneria* and in other
plants are controlled by an irradiance-dependent variation of the mobility of chloroplasts in cytoplasmic streaming. The mobility of chloroplasts depends upon their ad-
hesional contact to the ectoplasm at the interface between the streaming endoplasm
and the stationary ectoplasm. In this interface the motive mechanism of cytoplasmic
streaming is located. In the motive mechanism an ATP-dependent actomyosin-like molecular system causes the motive force of cytoplasmic streaming and controls the mobility of chloroplasts (Seitz 1979a). Experimentally the effect of light upon the mobility
of chloroplasts can be investigated by taking their centrifugability as a measure.

The dose response curve for the effect of blue light on the centrifugability of chloroplasts shows three ranges: at very low irradiance centrifugability is increased (relative
to the dark level), at moderately low irradiance ("low intensity" range between about
0.3 and 50 pmol cm^{-2} s^{-1}) centrifugability is decreased and at high irradiance it is increased again (Virgin 1951; Zurzycki 1962; Seitz 1971). The form of this dose response
curve for blue light-induced changes in the centrifugability of chloroplasts with ranges
of positive and negative response is similar to the dose response curve for many other
blue light-dependent responses. Such dose response curves can be caused by an interaction of different photoreaction systems in the action mechanism of light. In the experiments on the light-induced changes in the centrifugability of chloroplasts the effect
of light has been investigated in *Vallisneria* in comparison for the low and high intensity
range of the dose response curve (Seitz 1974, 1975, 1979b).

2 Results and Discussion

Action spectra for the effect of light on the centrifugability of chloroplasts in the low
and high intensity range were obtained by an irradiation with equal quantum flux density in the blue range of the spectrum. For the irradiation a nonsaturating quantum flux
density of 750 pmol cm^{-2} s^{-1} was chosen in the high intensity range and of 0.8 pmol
cm^{-2} s^{-1} in the low intensity range. The action spectrum for the induction of increased
centrifugability (high intensity response) has a peak at 454 nm, a minimum at 400 nm

---

1 Institut für Botanik der Universität Erlangen-Nürnberg, 8520 Erlangen, FRG

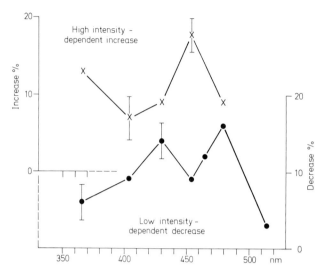

Fig. 1. Action spectra of the light-induced change in centrifugability of chloroplasts in the low and high intensity range. *Ordinate* Increase or decrease of percent centrifugability as difference between irradiated and dark sample. *Irradiation* 20 min with equal quantum flux density. Low intensity: 0.8; high intensity: 750 pmol cm^{-2} s^{-1}

and shows an increase in the response at 370 nm, which indicates a flavin as photoreceptor (Fig. 1, upper curve). The action spectrum for the induction of decreased centrifugability (low intensity response) in contrast has peaks at 430 and 480 nm with a small minimum at 454 nm. The action spectrum does not show an increase of the response in the near UV at 370 nm (Fig. 1, lower curve). The form of this action spectrum suggests neither a flavin nor carotins, but rather chlorophylls a and b as photoreceptors (Seitz 1979b). The different form of the action spectra in the low and high intensity range indicates that different photoreaction systems are involved.

A different effect in the low and high intensity range has also been found for the influence of metabolic inhibitors. In the experiments with inhibitors a higher irradiance was chosen in both intensity ranges in order to obtain a somewhat higher level in the response and the same wavelength was taken to ensure a good comparability (Seitz 1979b). The results of the experiments with metabolic inhibitors are given in Table 1. PCMB inhibits the light-induced change in the centrifugability of chloroplasts in both intensity ranges. This indicates that free sulfhydryl groups are essential for the response to light. The inhibition by PCMB may be caused either by an effect on the actomyosin system of the motive mechanism of cytoplasmic streaming or by an effect on respiratory or photosynthetic processes. Some evidence for such difference in the action of PCMB may be seen in the different extent of the inhibition in the low and high intensity range. DCMU and other photosynthetic inhibitors have only an effect on the light-induced decrease in centrifugability in the low intensity range, so this response depends upon the photosynthetic electron flow. Uncouplers of the oxydative phosphorylation in contrast inhibit only the light-induced increase in centrifugability in the high intensity range. The same effect has carboxyatractyloside, which specifically blocks the transfer of ATP

Table 1. Influence of different inhibitors on the light-induced changes in the centrifugability of chloroplasts

Irradiation (454 nm, 20 min)		Dark control	Low intensity (6 pmol cm^{-2} s^{-1})	High intensity (2200 pmol cm^{-2} s^{-1})
		% centrifugability	Light-induced change of centrifugability (difference to dark control)	
H$_2$O		41 ± 2	− 13	+ 25
PCMB	5·10^{-4} M	45 ± 3	− 8	+ 3
DCMU	3·10^{-6} M	40 ± 2	− 5	·: 28
o-Phenanthroline	10^{-3} M	35 ± 5	− 2	+ 24
Atrazine	3·10^{-4} M	38 ± 3	− 7	+ 18
Na-azide	5·10^{-3} M	43 ± 3	− 15	+ 8
DNP	2·10^{-4} M	58 ± 4	− 22	+ 6
FCCP	2·10^{-5} M	44 ± 4	− 16	+ 4
Carboxyatractyloside	10^{-3} M	42 ± 2	− 13	− 1

Table 2. Influence of different redox compounds on the light-induced changes in the centrifugability of chloroplasts

Irradiation (454 nm, 20 min)		Dark control	Low intensity (6 pmol cm^{-2} s^{-1})	High intensity (2200 pmol cm^{-2} s^{-1})
		% centrifugability	Light-induced change of centrifugability (difference to dark-control)	
H$_2$O		62 ± 3	− 21	+ 20
Methyl viologen	3·10^{-3} M	66 ± 2	− 9	+ 17
Benzyl viologen	10^{-2} M	65 ± 4	− 4	+ 22
Thionine	3·10^{-3} M	78 ± 5	− 8	+ 9
Ascorbate	10^{-3} M	75 ± 4	− 11	+ 1
DCIP	10^{-4} M	79 ± 3	− 14	+ 7
p-amino-dimethyl-aniline	10^{-3} M	65 ± 3	− 20	+ 9
Cysteine	3·10^{-3} M	63 ± 2	− 20	+ 9

through the mitochondrial membrane (Klingenberg 1976). The increase in centrifugability thus depends upon ATP from the oxydative phosphorylation.

The different influence of inhibitors in the low and high intensity range indicates that the two photoreaction systems known from the action spectra depend upon different metabolic processes. A different influence has also been found with the ATP-analog adenylyl-imidodiphosphate (AMP-PNP), which is known as a competitive inhibitor of ATP-dependent responses (Yount et al. 1971). AMP-PNP inhibits only the light-induced increase in centrifugability, but has no effect on the decrease (Seitz 1974).

The participation of different reaction systems in the low and high intensity range is further confirmed by new experiments with redox compounds. The results in Table 2 show that redox compounds with a highly negative redox potential, i.e., viologenes, inhibit only the decrease in centrifugability and have no effect on the increase. Viologenes affect the response to light in a way similar to the photosynthetic inhibitors. Viologenes are known as electron acceptors of photosystem I in the photosynthetic electron flow and could thus interfere with photosynthetic CO_2 fixation. Due to their high redox potential viologenes have no direct effect on the respiratory electron transport chain and oxydative phosphorylation (Nultsch 1968). This probably explains why viologenes do not inhibit increase in centrifugability. Substances with a lower redox potential like ascorbate, dichlorphenole-indophenole, and thionine, known as electron acceptors of photosystem II, inhibit the light-induced increase in centrifugability as well as the decrease. But these redox compounds also have a rather great increasing effect on the centrifugability of chloroplasts in the unirradiated controls and thus simulate the effect of high intensity light. Two other substances, p-amino-dimethyl-aniline, which is also an electron acceptor of photosystem II, and cysteine inhibit only the increase in centrifugability, and have no effect on the decrease. The inhibitory effect of all these latter redox compounds, in contrast to the viologenes, can be caused by different mechanisms. These substances could function as electron acceptors of photosystem II of photosynthesis. But they could also interfere with the respiratory electron transport chain and could furthermore affect the primary redox reaction mediated by the flavin photoreceptor. Thus a clear conclusion cannot be drawn from these results. But the different influence of the redox compounds in the low and high intensity range also suggests, like the other experimental results, that different photoreaction systems are involved in the effect of light on the centrifugability of chloroplasts.

3 Model of the Action Mechanism of Light

On the basis of the experimental results described above the following action mechanism of light can be considered. ATP is a limiting factor for the induction of increased centrifugability, as the influence of the inhibitors has shown (Table 1). An increase in centrifugability can in fact be induced by adding ATP to the medium (Seitz 1971). So light could cause the changes in the centrifugability of chloroplasts by an irradiance-dependent effect on the intracellular availability of ATP. Such a light-induced variation in the supply of ATP could be caused via the two photoreaction systems in dependence upon an interaction of respiratory and photosynthetic processes (Seitz 1971, 1979b).

In one system blue light induces an increase in the rate of respiration, which in turn causes an increased availability of ATP, as is known from investigations in *Chlorella* (Kowallik and Scheil 1976). This response is induced by a flavin as photoreceptor in the range of very low and presumably also of higher intensities. Such light effects upon the rate of respiration are known to occur in *Vallisneria* (Prins and Wolff 1974) and have been analyzed in relation to the light-induced motion response of chloroplasts in *Lemna* (Zurzycki 1970).

The other system begins to operate at moderately low irradiance, in the "low intensity" range. Here light is effective with chlorophylls as photoreceptors via photosynthetic processes. In this system two different reactions can modify the availability of ATP depending upon the irradiance. These are the different types of photophosphorylation and the noncyclic electron flow leading to CO_2 fixation. In CO_2 fixation more ATP is consumed than is synthesized by the concurrent noncyclic photophosphorylation. Under limiting light conditions, in the low intensity range, the electrons are channeled preferentially into the reduction of NADP and not into cyclic or pseudocyclic photophosphorylation. This will cause a reduced availability of ATP. At high intensity, in the range of saturation of CO_2-fixation, the level of $NADPH_2$ rises and more electrons are diverted into cyclic or pseudocyclic photophosphorylation and a surplus of ATP will be available (Heber 1973).

The blue light-induced increase in the rate of respiration, together with the photosynthetic reactions, can thus control the intracellular availability of ATP in dependence upon the irradiance. Besides ATP, probably changes in the redox state or in ionic relations or other parameters could also be involved. An irradiance-dependent change in the intracellular availability of ATP could be effective in the motive mechanism of cytoplasmic streaming and could thus control the centrifugability of chloroplasts (Seitz 1979a, b). A light-induced change in the supply of ATP could also be of importance for other blue light-dependent responses (Seitz 1964, 1975).

According to this model of the action mechanism of light the change from the low to the high intensity range in the dose response curve should occur at that irradiance which causes saturation of photosynthetic CO_2-fixation. Such a correlation between the change in the direction of the response and the saturation of photosynthesis is known from the orientation movement of chloroplasts in *Lemna* (Zurzycki 1970) and has been reported for the phototactic responses of *Euglena* (Diehn 1969). Due to the interaction of the different reaction systems with different kinetics, sudden changes in the irradiance can furthermore cause oscillatory changes in the intracellular ATP, as have been observed in *Chlorella* (Lewenstein and Bachofen 1972).

Acknowledgment. The experiments of this investigation have been supported by the Deutsche Forschungsgemeinschaft.

References

Diehn B (1969) Phototactic response of *Euglena* to single and repetitive pulses of actinic light. Exp Cell Res 56: 375–381

Heber U (1973) Stoichometry of reduction and phosphorylation during illumination of intact chloroplasts. Bioch Biophys Acta 305: 140–152

Klingenberg M (1976) The ADP-ATP carrier in mitochondrial membranes. In: Martonosi A (ed) The enzymes of biological membranes, vol III, pp 383:438. Wiley, New York

Kowallik W, Scheil J (1976) Lichtbedingte Veränderungen des ATP Spiegels einer chlorophyllfreien *Chlorella* Mutante. Planta 131: 105–108

Lewenstein A, Bachofen R (1972) Transient induced oscillations in the level of ATP in *Chlorella fusca.* Biochem Biophys Acta 267: 80–85

Nultsch W (1968) Einfluß von Redox-Systemen auf die Bewegungsaktivität und das phototaktische Reaktionsverhalten von *Phormidium uncinatum.* Arch Mikrobiol 63: 295–320

Prins HBA, Wolff W (1974) Photorespiration in leaves of *Vallisneria spiralis.* Proc K Ned Akad Wet Ser C 77: 239–265

Seitz K (1971) Die Ursache der Phototaxis der Chloroplasten: ein ATP-Gradient? Z Pflanzenphysiol 64: 241–256

Seitz K (1974) Lichtabhängige Orientierungsbewegungen und ihre Regelung. Ber Dtsch Bot Ge 87: 195–206

Seitz K (1975) Orientation in space: plants. In: Kinne O (ed) Marine ecology, vol II, part 2, pp 451–497. Wiley, New York

Seitz K (1979a) Cytoplasmic streaming and cyclosis of chloroplasts. In: Haupt W, Feinleib ME (eds) Encycl plant physiol, New Ser, vol VII, pp 150–169. Springer, Berlin Heidelberg New York

Seitz K (1979b) Light induced changes in the centrifugability of chloroplasts: Different action spectra and different influence of inhibitors in the low and high intensity range. Z Pflanzenphysiol 95: 1–12

Virgin HJ (1951) The effect of light on the protoplasmic viscosity. Physiol Plant 4: 255–357

Yount RG, Ojala D, Babcock (1971) Interaction of P-N-P and P-C-P analogs of ATP with heavy meromyosin, myosin, and acto myosin. Biochem 10: 2490–2498

Zurzycki J (1962) The mechanism of the movement of plastids. In: Ruhland W (ed) Encycl plant physiol, vol 17, part 2, pp 940–978. Springer, Berlin Heidelberg New York

Zurzycki J (1970) Light respiration in *Lemna trisulca* Acta Soc Bot Pol 39: 485–495

Growth Rate Patterns
Which Produce Curvature and Implications
for the Physiology of the Blue Light Response

W. KUHN SILK[1]

1 Introduction

Since the observations of Darwin (1897), it has been well known that plants bend in response to oriented light. The purpose of this paper is to show the growth rate patterns which produce observed curvatures. An equation (Silk and Erickson 1978) will be derived and tested to relate elemental curvature ot elemental growth rate distributions.

   An important theme in this article is that curving plant structures are often, like the wake of a boat, steady structures composed of changing elements. Hans Mohr (1962) recognized this in his study of hook opening in lettuce when he stated, "Der geschlossene Plumulahaken. . .behält seinen morphologischen Zustand weitgehend bei, obgleich die ihn aufbauenden Zellen sich beständig ändern. Die Zellen, so kann man sagen, 'fließen' durch den Haken hindurch." Professor Mohr's statement, that the plumular seedling hook is a structure which retains its form although the cells which comprise it continually change, is illustrated in Fig. 1. In this time lapse set of photographs an epidermal hair which is, of course, firmly attached to the hypocotyl surface, appears to flow through the hook. This set of photographs tells us that each element of hypocotyl tissue first curves and then straightens as it becomes located farther from the growing tip; and, at any one time, the stable hook is composed of tissue elements which are curving (on the apical side of the hook summit) and other elements which are straightening (on the basal side of the hook summit). This same sort of picture has been obtained for the plumular hook of bean (Rubinstein 1971). The curvature could be said to move up the stem, from the hypocotyl into the epicotyl and stem if we take the ground or the nongrowing region as our reference origin. A more convenient choice of origin is the stem apex. Although it may be intuitively difficult to think of cells being displaced from a stationary apex, the choice of apex as reference origin leads to the recognition of the steady, or time-independent, pattern of curvature. But recognition that the steady pattern results from changing elements leads to the central question in finding the relationship between growth rates and curvature: if local curvature is not changing in time, how can we express the kinematics of the changing tissue elements? The answer can be found in the methods of fluid flow characterization, as described below.

1 Land, Air and Water Resources, University of California, Davis, CA 95616, USA

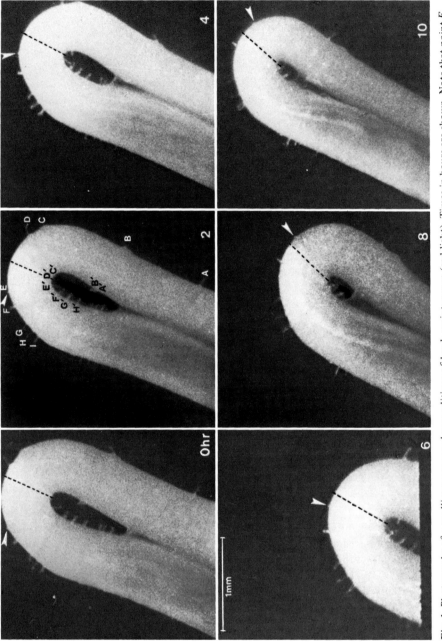

Fig. 1. Photographs of a seedling grown under conditions of hook maintenance (red light). Times in hours are shown. Note that point *E*, one of a number of epidermal hairs, has passed the vertical line, the hook bisector, between 6 and 8 h (Silk and Erickson 1978)

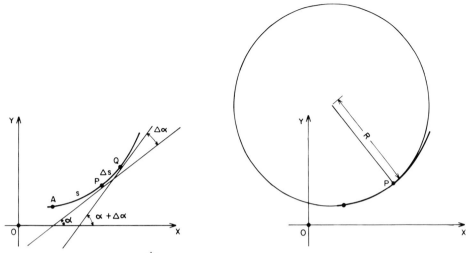

Fig. 2. Local curvature, $K = \dfrac{d\alpha}{ds}$ (*left*) and R, the radius of curvature (*right*)

2 Definitions of Variables in the Equation Relating Growth Rate to Curvature

The quantitative relationship between growth rate and curvature requires two definitions from calculus (curvature and velocity gradient) and an idea from fluid flow theory (the material derivative as the sum of the local derivative and the convective change).

2.1 Curvature

Curvature, K, is the rate of change with respect to arc length, s, of α, the angle of inclination of the tangent to a curve. As illustrated in Fig. 2 (left),

$$K = \frac{\partial \alpha}{\partial s},$$ (1)

where α is measured in radians and K is in mm^{-1}. A more intuitively obvious measure of curvature is given by the length, R, of the radius of the circle which just fits the curve at a given point as shown in Fig. 2 (right). The smaller the radius of the fitting circle, the more curved is the line; in fact it can be shown that R is the reciprocal of K, the curvature defined above:

$$R = \frac{1}{K}.$$ (2)

From these definitions it is clear that only a few shapes have constant curvature along their profiles. A circle has constant curvature along its circumference; a straight line has everywhere zero curvature (and an infinite radius of curvature); while a parabola has a curvature maximum at its vertex. Curvature is plotted against arc length for two circles and a parabola in Fig. 3.

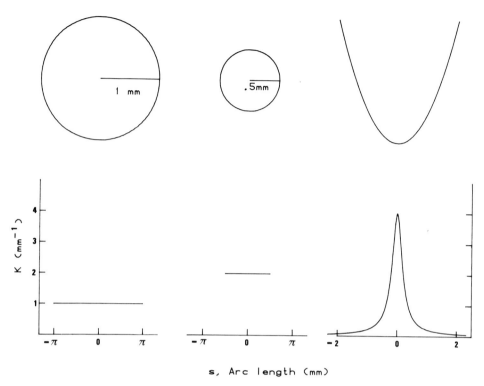

Fig. 3. Curvature plotted against arc length for a circle of radius 1 (*left*), a circle of radius 1/2 (*middle*) and the parabola $y = 2x^2$ (*right*)

The mathematical definition of curvature was used in early studies of phototropism of *Avena* coleoptiles (Dolk 1929) and *Phycomyces* sporangiophores (Castle 1962). In recent years, curvature has been measured as an angle in degrees. The "angle of curvature" has the advantage of ease of measurement, while estimates of K as a function of distance from a plant apex must be made by numerical methods (Castle 1962; Silk and Erickson 1978) or by fitting circles to photographs (Dolk 1929). The relationship between the "angle of curvature" and K is analogous to the relationship between total elongation rate and the relative elemental growth rate, defined below.

2.2 Material Derivative of Curvature

From the definition of curvature, it is clear that only a straight stem or a stem bent in a perfectly circular arc will have constant curvature along its length. And if curvature varies with position, then the cells, or material elements which comprise the stem, will experience curvature change as they are displaced from the tip during growth. The rate of curvature change associated with a real or *material* element of stem tissue is termed the material derivative of curvature and is symbolized $\frac{DK}{Dt}$. The curvature change experienced by a stem element is the sum of the local curvature change at the stem position instantaneously occupied by the element and the convective curvature change acquired by movement to a new position. The convective change is the product of displacement rate and spatial curvature gradient. In symbols,

$$\frac{DK}{Dt} = \frac{\partial K}{\partial t} + u(s)\frac{\partial K}{\partial s}, \qquad (3)$$

material curvature change = local curvature change + convective
(cell specific)　　　　　　　(site specific)　　　　　　curvature change

where t is time, s is distance from the apex, and u is velocity of displacement of the element from the apex.

More generally, the material derivative of any growth field variable (i.e., any function of position in the plant and of time) can be expressed as the sum of local change plus convective change (Silk and Erickson 1979).

2.3 Relative Elemental Growth Rate

To find the relationship between curvature and growth rate we need a quantitative description of growth at a point in an expanding tissue. Such a description is given by the relative elemental growth rate (also called the stretch rate). We begin with a classical growth curve which is a plot of the length of a stem, branch, or root against time. The slope of the growth curve, known as g, the "growth velocity" of the apex, gives the rate at which the tip of the organ is being pushed away from the nongrowing region. For instance, the primary root of corn has a nearly constant growth velocity for a long period under controlled conditions; the root tip moves away from the seed linearly with time. The constant growth velocity of the tip is produced by the expansions of the tissue elements between the tip and the nongrowing region. To measure the expansion rate of a small tissue segment in the continuum of expanding cells which make up the root, the growth velocity of the segment base must be subtracted from the growth velocity of the segment tip. If the tip is growing faster than the base the segment is expanding, and the segment expansion rate might be measured in mm h^{-1}. This measure of local expansion has the disadvantage that it depends on the amount of tissue initially in the segment; the disadvantage is circumvented if we divide the segment expansion rate by the initial segment length to obtain a relative expansion rate which gives fractional change per unit time. As the segment is made shorter and shorter, the relative expansion rate begins to refer to a point, or spatial element, on the root and is called the relative elemental growth rate at point x. In calculus notation

$$\lim_{\Delta x \to o}\left[\frac{1}{\Delta x}\left(\begin{array}{c}\text{growth}\\ \text{velocity of}\\ \text{segment tip}\end{array} - \begin{array}{c}\text{growth}\\ \text{velocity of}\\ \text{segment base}\end{array}\right)\right] = \frac{\partial g}{\partial x} = L(x), \qquad (4)$$

where $L(x)$ is the relative elemental growth rate at x and has the units of reciprocal time. In a tissue which is expanding in three dimensions, L is the divergence of the vector g which gives the growth velocities of points in the expanding volume

$$L = \nabla \cdot g = \frac{\partial g_i}{\partial x_i}. \qquad (4a)$$

The relative elemental growth rate was defined and experimentally evaluated in primary corn roots by Erickson and colleagues (Erickson and Goddard 1951; Erickson and Sax 1956a; Erickson 1976b). By plotting the growth velocity of closely spaced marks against distance from the apex, then differentiating this curve with respect to distance, Erickson was able to estimate the relative elemental growth rate as a function of position on the root. Interested readers should study the original papers and the review by Green (1976). The variable L has also been termed the local stretch rate in studies of *Phycomyces* sporagniophore expansion (Cohen and Delbrück 1958).

The graph of $L(x)$ versus x in the primary corn root proved to be asymmetric, with a sharp rise and gradual decline. An important observation made by Erickson (1976a) in his early work is that, if distances are measured from the root tip, the values of the relative elemental growth rate at given positions do not change appreciably for many hours. For more than 48 h the fastest growing region is located about 4 mm behind the tip, where the relative elemental growth rate is almost 40% h^{-1}, and the nonexpanding region begins 10 mm behind the tip. This confirms the intuitions of the classical morphologists that the developmental history of the root is evident from its longisection, so that (if the growth pattern is not changing) the history of a mature cell can be inferred from the characteristics of cells closer to the root tip, while the fate of a young cell can be predicted by examination of cells farther from the tip.

Another aspect of growth theory should be mentioned before curvature relations can be discussed. The relative elemental growth rate at a position is often constant in time, but the cells which are found in a particular spot are continually displaced. Thus, even in a tissue with a constant growth pattern, the relative elemental growth rate of the cell in the growing regions changes continually as the cell becomes farther from the tip. Cohen and Delbrück (1958) and Silk and Erickson (1978, 1979) have discussed the importance of distinguishing the growth rate, L, at a position from the growth rate, M, associated with a material or real element such as a cell. The material (cell-specific) growth rate can be calculated from spatial (site-specific) growth data (Silk and Erickson 1979; Erickson and Silk 1980).

3 Model for Bending in Response to Gradients in Growth Rate

The simplest conceivable model is that any difference in growth rate across the stem results in bending. Growth is assumed to be symplastic; quantitatively this means points initially opposite each other continue to have coinciding centers of curvature as they are displaced during growth. Further assumptions are absence of both torsion and shear growth. Then the arc lengths of outer, δs_O, and inner, δs_i, edges of an element of stem projection will be in the ratio of the radii of curvature to the edges. As illustrated in Fig. 4:

$$\frac{\delta s_O}{\delta s_i} = \frac{R + w}{R} = 1 + Kw. \tag{5}$$

This simple geometrical relationship is the basis for the equation derived in the next section.

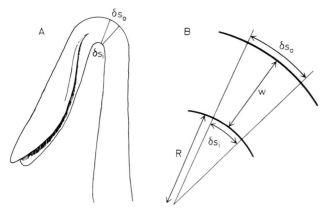

Fig. 4A, B. Diagrams of the hypocotyl hook. **A** Portion of seedling showing cotyledons at the left, curved hypocotyl at the right. δs_o and δs_i represent elements of arc length, s, measured along the outer and inner edges of the hypocotyl. **B** Enlargement of the element of longisection to show δs_o, δs_i, R (the radius of curvature to the inner edge of the hook) and w (the width of the hypocotyl) (Silk and Erickson 1978)

4 Derivation of the Equation Relating Growth Rate to Curvature

From the definition of the material relative elemental growth rate we can relate the length of an element to its history of growth rates:

$$\int M \, dt = \ln \delta s. \tag{6}$$

Now, from Eqs. (5) and (6), it can be seen that

$$\int M(s,o) \, dt - \int M(s,i) \, dt = \ln [1 + Kw]. \tag{7}$$

The essential trick in the problem is to find the material derivative of $[\ln (1 + Kw)]$. By analogy with Eq. (3)

$$\frac{D}{Dt}[\ln (1 + Kw)] = \frac{\partial}{\partial t} [\ln (1 + Kw)] + u(s) \cdot \frac{\partial}{\partial s}[\ln (1 + Kw)]. \tag{8}$$

Substituting (7) in (8) and differentiating we find

$$M(s,o) - M(s,i) = \frac{\partial}{\partial t}[\ln (1 + Kw)] + u(s) \cdot \frac{\partial}{\partial s}[\ln (1 + Kw)]. \tag{9}$$

Equation (9) gives the desired relationship between growth rate and curvature and can be generalized to

$$M (s,r) - M (s,i) = \frac{\partial}{\partial t} [\ln (1 + Kr)] + u(s) \cdot \frac{\partial}{\partial s} [\ln (1 + Kr)], \tag{10}$$

where r is the distance toward the outer edge along the radius of curvature from the point (s,i) on the inner edge of the projection.

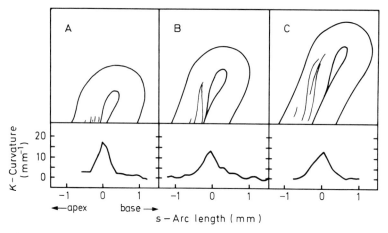

Fig. 5A–C. Local curvatures in the lettuce hypocotyl of Fig. 1 plotted against distance from the hook bisector under conditions of hook maintenance (red light). **A** Beginning of the observation period; **B** after 2 h when the seedling has grown 1 mm; **C** after 4 h when the seedling has grown 2 mm

5 Local Curvatures in Hypocotyls and Coleoptiles

Curvature of the inner edge of the projection of the lettuce hypocotyl hook of Fig. 1 is shown in Fig. 5A–C. Curvature increases from 0 to a maximum of about 16 mm^{-1} one millimeter from the apex and declines again to 0 about 2 mm below the apex. In red light, the hook is maintained while the plant grows. The plot of curvature vs. distance from the apex can be seen to be steady,i.e., it does not change in time. Quantita-

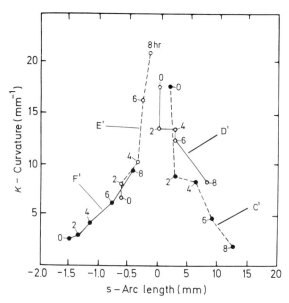

Fig. 6. Material curvatures at the location of real points, i.e., hairs, which arc designated by letters in Fig. 1. With time, each hair is displaced from the apex (Silk and Erickson 1978)

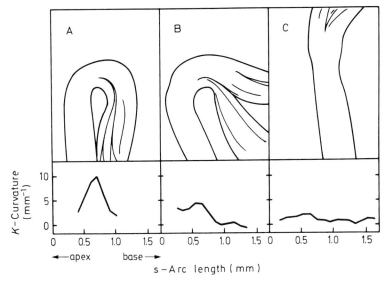

Fig. 7A–C. Local curvatures in the lettuce hypocotyl plotted against distance from the seedling apex under conditions of hook opening (blue light). **A** Beginning of the observation period; **B** after 14 h; **C** after 28 h

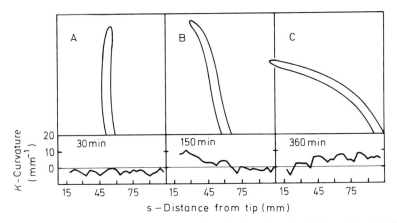

Fig. 8A–C. Local curvatures in the oat coleoptile plotted against distance from the tip under conditions of unilateral illumination with white light. Times after light was turned on are indicated. Between 150 and 360 min the bend migrates from the tip toward the base of the coleoptile

tively, we can say that the local derivative is zero. Yet, as discussed in the introduction, curvature of the material elements changes continually. Figure 6 shows curvature of four different marked tissue elements during an 8-h observation period. Those bits of stem on the apical side of the hook bisector increase in curvature with time, while the elements initially on the basal side of the summit decrease in curvature as they are displaced toward the nongrowing zone.

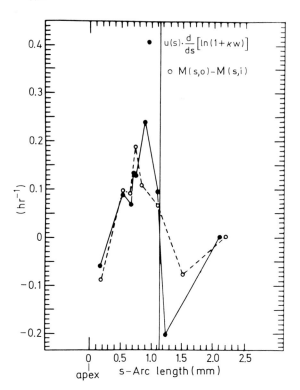

Fig. 9. Empirical test of Eq. (9) during hook maintenance. Cross-sectional growth rate difference, $M(s,o) - M(s,i)$, *(solid line)* and curvature term, $u(s) \cdot (d/ds)\left[\ln (1 + Kw)\right]$ *(dashed line)* have been evaluated at different values of arc length. Both functions show the oscillation expected from Fig. 1 (Silk and Erickson 1978)

Figure 7A–C shows curvature vs. arc length during hook opening. The obvious feature of this graph is that the local curvatures change in time. Curvature is no longer steady, and the local derivative has become important.

Figure 8 shows the well-known progression of curvature from tip to base during phototropism of the *Avena* coleoptile. As in Fig. 6, local curvature change is important; in this case the apical side of the bend decreases in curvature, while the basal side increases in curvature.

6 Tests of Equation Relation Growth Rate to Curvature

Figure 9 is an empirical test of Eq. (9) during hook maintenance in the hypocotyl of lettuce (*Lactuca sativa* cv. "Grand Rapids"). Left side of equation $M(s,o) - M(s,i)$, and right side $u(s) \cdot (\partial/\partial s)[\ln (1 + Kw)]$ have been evaluated independently and can be seen to be in satisfactory agreement. On the apical side of the hook the spatial gradient in curvature is positive. Here the outer elements grow faster than the corresponding elements on the inside of the hook, and the elements increase in curvature as they are displaced. On the basal side of the hook bisector the spatial curvature gradient is negative; the elements decrease in curvature as they are displaced; and the inner growth rate exceeds the outer growth rate.

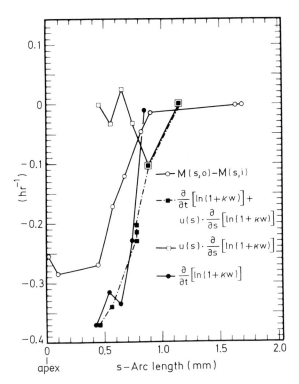

Fig. 10. Empirical test of Eq. (9) during hook opening. Terms have been evaluated during the interval of most rapid straightening which occurs 14–16h after exposure to blue-rich light (Silk and Erickson 1978)

Legend within figure:

$-\!\!\circ\!\!-$ $M(s,o)-M(s,i)$

$-\cdot\blacksquare\cdot-$ $\dfrac{\partial}{\partial t}\left[\ln(1+\kappa w)\right]+ u(s)\cdot\dfrac{\partial}{\partial s}\left[\ln(1+\kappa w)\right]$

$-\!\!\square\!\!-$ $u(s)\cdot\dfrac{\partial}{\partial s}\left[\ln(1+\kappa w)\right]$

$-\!\!\bullet\!\!-$ $\dfrac{\partial}{\partial t}\left[\ln(1+\kappa w)\right]$

y-axis: (hr^{-1}); values 0.1, 0, −0.1, −0.2, −0.3, −0.4

x-axis: s−Arc length (mm); values apex 0, 0.5, 1.0, 1.5, 2.0

Figure 10 is an empirical test of Eq. (9) during opening of the lettuce hypocotyl hook. The dominant curvature term in this example is the local derivative $\dfrac{\partial}{\partial t}[\ln(1+Kw)]$ which is strongly negative and which is produced by $M(s,i)$, the large relative elemental growth rate on the inside of the hook. Again, the left hand side of the equation is in fairly good agreement with the right hand side.

7 Physiological Implications

Any causal agent of light-induced curvature will be consistent with Eqs. (9) and (10). For instance, the mechanism for plumular hook maintenance must involve a reversal in sign of the spatial gradient in growth rate as demonstrated in Fig. 9 and shown schematically in Fig. 11A. Growth rate must decrease between A and a and increase between B and b. Thus, the hook could be maintained by growth promotors at A and b and/or by growth inhibitors at a and B. Any postulated curvature-regulating factor, such as wall extensibility or hormone sensitivity must be shown to have the same reversal in sign of gradient between the apical and basal sides of the hook center.

An understanding of the growth rate pattern which maintains the hook comes from recognition of the importance of convective change. And even when local curvature changes dominate, during the period of most rapid opening of the hook of lettuce or bean, convective change may be important. There is a lag period of 8–12 h between

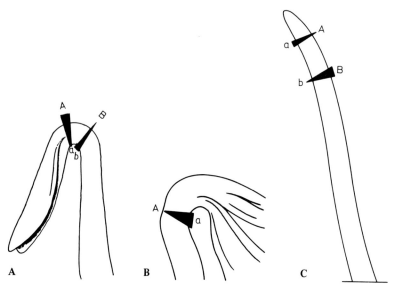

Fig. 11A–C. Growth rate gradients which cause curvatures. **A** During hook maintenance the relative elemental growth rate must decrease between *A* and *a* and between *b* and *B*. **B** During hook opening, growth rate on the inside of the hook increases relative to the growth rate on the outside. **C** During basipetal movement of the bend in the *Avena* coleoptile, growth rate must decrease between *a* and *A* and increase between *B* and *b*. Width of the dark wedges indicates magnitude of the local stretch rate

applications of active light wavelengths [550 mμ for lettuce (Mohr and Noble 1960) and 630–700 mμ for bean (Klein et al. 1956)], and the time of most rapid hook opening (Rubinstein 1971; Gee and Vince-Prue 1976). This is also the length of time required for a cell near the apex to reach the hook center (Silk and Erickson 1978). It is tempting to speculate that there may be a young developmental stage competent to receive the light-induced stimulus and that the stimulus is expressed in the more adult stage corresponding to location at (convection to) the hook summit. Studies of the temperature dependence of both growth velocity and lag period for hook opening could test this possibility. Whether or not hook opening depends on convection of stimulated cells to the hook center, the presumed causal agent for hook opening must also be shown to obey Eq. (9).

Figure 11B, C shows the growth rate distributions responsible for lettuce hoop opening and for oat coleoptile phototropism in constant blue light. It is worth noting that the structures of both 11A and 11C could have constant "degrees curvature" in time, yet their material curvature changes are great and are produced by growth rate gradients of considerable magnitude. It can be concluded that the material derivative of curvature and the possible importance of convective curvature change should be investigated in studies of phototropism and bending.

Acknowledgments. Curvature studies were performed in collaboration with Ralph O. Erickson and had support from National Science Foundation grants GB37632 and PCM-7823710. I thank D.G. Stavenga and W. Shropshire for helpful criticism.

References

Castle ES (1962) Phototropic curvature in *Phycomyces*. J Gen Physiol 45: 743–756

Cohen R, Delbrück M (1958) Distribution of stretch and twist along the growing zone of the sporangiophore of *Phycomyces* and the distribution of response to a periodic illumination program. J Cell Comp Physiol 51: 361–388

Darwin C (1897) The power of movement in plants. D Appleton & Co Inc, New York

Dolk HE (1929) Über die Wirkung der Schwerkraft auf Koleoptilen von *Avena sativa*. II. K Akad van Wet Proceed XXXII: 1127–1140

Erickson RO (1976a) Modeling of plant growth. Annu Rev Plant Physiol 27: 407–434

Erickson RO (1976b) Growth in two dimensions, descriptive and theoretical studies. In: Lindenmayer A, Rozenberg G (eds) Automata, languages and development. North-Holland Publ Co, Amsterdam New York

Erickson RO, Goddard DR (1951) An analysis of root growth in cellular and biochemical terms. Growth Symp 10: 89–116

Erickson RO, Sax KB (1956a) Elemental growth rate of the primary root of *Zea mays*. Proc Am Philos Soc 100: 487–498

Erickson RO, Silk WK (1980) The kinematics of plant growth. Sci Am 242(5): 134–151

Gee H, Vince-Prue D (1976) Control of the hypocotyl hook angle in *Phaseolus mungo* L.: the role of parts of the seedling. J Exp Bot 27: 314–323

Green PB (1976) Growth and cell pattern formation on an axis: critique of concepts, terminology, and modes of study. Bot Gax 137: 187–202

Klein WH, Withrow RB, Elstad V (1956) Response of the hypocotyl hook of bean seedlings to radiant energy and other factors. Plant Physiol 31: 289–294

Mohr H (1962) Primary effects of light on growth. Ann Rev Plant Physiol 13: 454–488

Mohr H, Haug A (1962) Die histologischen Vorgänge während der lichtabhängigen Schließung und Öffnung des Plumulahakens bei den Keimlingen von *Lactuca sativa* L. Planta 59: 151–164

Mohr H, Noble A (1960) Die Steuerung der Schließung und Öffnung des Plumulahakens bei Keimlingen von *Lactuca sativa* durch sichtbare Strahlung. Planta 55: 327–342

Rubinstein B (1971) The role of various regions of the been hypocotyl on red light-induced hook opening. Plant Physiol 48: 183–186

Silk WK, Erickson RO (1978) Kinematics of hypocotyl curvature. Am J Bot 65: 310–319

Silk WK, Erickson RO (1979) Kinematics of plant growth. J Theoret Biol 76: 481–501

Organism Index

Subject Index

absorbance changes 31, 178, 618
accessory pigments 177
acellular slime molds 570ff.
acetylcholine 291
acid production 353
acid-base-equilibria 158
actin 66, 266
actinomycin 112, 182, 357, 359
action dichroism 57, 75, 76, 79, 81, 84, 89–91
– potential 233
– spectrum 38, 54, 70, 71, 84, 85, 105, 115, 175, 187ff., 234, 278, 286, 424, 548, 557, 585, 623, 637
actomyosin 65, 66, 93, 637
afterpotential 19
ageing 373ff.
ALA dehydratase 545ff.
ALA formation 514ff., 536, 538, 541ff.
ALA synthetase 545ff.
alanine 430
amino acid uptake 585
amitrole 222
ammonium 415, 429, 435ff., 630
amphiflavins 212
anthocyanin synthesis 97, 309
apical growth 123, 131
– swelling 125, 131
artificial flavin systems 205ff.
– membranes 212
– photoreceptors 289
aspartate 321, 338, 340, 430
atebrin 80f.
ATP 80ff., 93, 352, 637, 639
ATP-ase 255
auxin 162

bacteriorhodopsin 7, 28, 30, 177
bending 643ff.
betalain synthesis 100
biological clock 3, 594
– rhythms 189, 405
biopotential changes 221ff., 230ff.
biplanar growth 141

Bunsen-Roscoe law of reciprocity 190, 224 234

calcium ions 46, 93
callus cells 465, 468
canopy light 444
carbohydrate degradation 344, 347, 348
carbon metabolism 368, 381ff., 429, 435ff.
2-carboxy-3-keto ribitol-1,5 bisphosphate 381ff.
carboxyatractyloside 638
C_1-C_3 carboxylation reaction 328, 429
carotenogenesis 135, 136, 175, 181, 183, 191, 278, 283, 285, 300, 301, 305, 309ff., 489, 568
carotenoids 14, 41, 42, 111, 115, 135, 170, 172, 174, 187, 190, 212, 222, 224, 229, 230, 234, 269, 273, 277ff., 283, 286, 303, 316, 466, 467, 473, 488, 503, 638
carotenoprotein 40, 42, 44, 157, 164, 169, 175, 176, 178, 182, 187, 191, 194
CCCP 631, 633, 634
cell cycle 119, 126, 129
– division 119, 126, 131, 144, 584
centrifugation experiments 58, 81, 92, 93, 637
chemiosmotic mechanism 634
chemotaxis 3
chitin synthetase 116
chloramphenicol 504, 521
chlorophyll 165, 466, 473, 485, 488, 503, 608, 638
– biosynthesis 102, 338, 466, 475, 486, 517, 523, 532, 546ff., 608, 609
– fluorescence 89f.
chlorophyll-protein complexes 467, 468, 470, 477, 532
chloroplast 455ff., 465, 470
– aggregation 265
– development 368, 456, 473ff., 485, 495
– displacement 50, 54, 60
– membranes 465, 468, 470
– morphology 478
– movements 69–96, 261, 264
chromatic adaptation 597
circadian rhythms 3, 238, 239, 595

Volume 5

Photosynthesis I

Photosynthetic Electron Transport and Photophosphorylation

Editors: A. Trebst, M. Avron

1977. 128 figs. XXIV, 730 pages.
ISBN 3-540-07962-9

"This excellent reference and source book contains
42 articles... written by active research workers...
An impressive amount of new knowledge has been
acquired about these subjects in the past fifteen
years, and this is well summarized through 1975..."
Amer. Scientist

Volume 6

Photosynthesis II

Photosynthetic Carbon Metabolism and Related
Processes

Editors: M. Gibbs, E. Latzko

1979. 75 figs., 27 tab. XX, 578 pages.
ISBN 3-540-09288-9

This book examines the mechanism and regulation
of photosynthetic CO_2 fixation and related carbon
metabolism, as well as those parts of hydrogen,
nitrogen, and sulfur metabolism that are closely
connected to photosynthesis. Information on the
Calvin-cycle of CO_2 fixation, and on C_4 metabolism
in C_4 plants and plants with crassulacean acid
metabolism is presented from studies using whole
tissues, isolated cells and isolated chloroplasts.
Interactions between photosynthesis, respiration,
and photorespiration are described, and the metabolism of starch and sucrose is examined.
A section on the enzymes of the Calvin-cycle and C_4
metabolism includes an introduction to enzyme
regulation.

Volume 7

Physiology of Movements

Editors: W. Haupt, M. E. Feinleib

1979. 185 figs., 19 tab. XVII, 731 pages
ISBN 3-540-08776-1

Plant movements cover a wide field of plant physiology. This volume emphasizes those fields where
substantial progress in understanding has been
made, or where major new aspects are evolving. It is
devoted particularly to the mechanisms of perception, transduction and response common to
completely diverse types of movement. On the
other hand, these steps of the reaction chain may be
different even within a given type of movement for
different examples. This volume clarifies and analyzes the signal chains that lead to movements in
plants.

Volume 8

Secondary Plant Products

Editors: E. A. Bell, B. V. Charlwood

1980. 176 figs., 44 tab. and numerous schemes
and formulas. XVI, 674 pages.
ISBN 3-540-09461-X

The first comprehensive exposition of this important and timely field, illuminates recent research
results on a variety of secondary plant by-products.
Using numerous illustrations and tables, Drs. Bell
and Charlwood discuss the biochemical and
physiological phenomena involved in the synthesis
and accumulation of compounds such as alkaloids,
isoprenoids, plant phenolics, non-protein amino
acids, amines, cyanogenic glycosides, gluco-
sinolates, and betalains. **Secondary Plant Products**
is sure to become a standard reference to all
botanists, biochemists, pharmacologists, and
pharmaceutical chemists.

Springer-Verlag
Berlin
Heidelberg
New York